Elements of Environmental Pollution Control

Om Prakash Gupta
Ex DGM, HRD
Bokaro Steel Plant
Steel Authority of India Limited (SAIL)

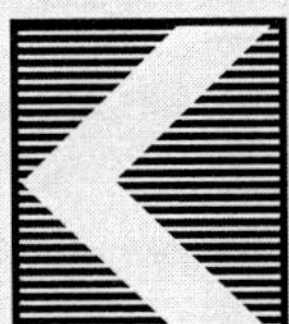

KHANNA BOOK PUBLISHING CO. (P) LTD.

Publisher of Engineering and Computer Books

4C/4344, Ansari Road, Darya Ganj, New Delhi-110002
Phone : 011-23244447-48 **Mobile:** +91-9910909320
E-mail : contact@khannabooks.com
Website : www.khannabooks.com

Price: ₹ 575.00

ISBN: 978-93-82609-66-7

Elements of Environmental Pollution Control
by O.P. Gupta

First Edition: 2019
Reprint: 2023, **2025**

To view complete list of books, please scan the QR Code:

Published by:
Khanna Book Publishing Co. (P) Ltd.
CIN: U22110DL1998PTC095547

Visit us at: www.khannabooks.com
Write us at: contact@khannabooks.com

Printed in Bharat.

Preface

It gives me immense pleasure to introduce this book titled Elements of environmental pollution control. I have taken the opportunity to compile the informations pertaining to environmental pollution control with special emphasis on air pollution control from various sources. This book is meant for the students, teachers, practicing engineers and policy makers on environmental/ air pollution control.

Extensive use of internet resources have been made in gathering the relavant informations and the subject matter has been presented in a simple lucid manner for easy understanding of the subject.

This book is targeted to benefit the following:

- Diploma in engineering students
- Degree in engineering students B.Tech (chemical engg.), B.Tech (Civil Engg.), B.Tech (Environmental Engg.), B.Tech (Metallurgical Engg), B.Tech (Minning Engg.), B.Tech (Aeronautical Engg.) etc.
- AMIE, AMIIM, AMIICHE, AIC Students.
- M.Tech Students of various disciplines pursuing courses on environmental Engg.
- Faculty members/ Teaching staff of Engineering colleges /IIT's/NIT's etc.
- Practising environmental engineers in various private sector/public sector undertakings, state/ central government departments, NGOs etc.
- Students of foreign universities of developing countries pursuing diploma/degree/postgraduate courses in vaious engg disciplines having a paper in environmental pollution control .
- Environmental Auditors.
- Policy makers on environmental pollution control in various institutions.

The scope and coverage of the subject matter presented can be gauged from glancing the contents portion of the book to judge its suitability and relevance for a particular target group.

Lot of literatures were consulted while writing this book for making this compilation of informations for various target groups, hence I gratefully acknowledge their assistance & contribution. Latest and updated informations/data/statistics pertaining to the subject matter has been included in this edition for the benefit of the readers.

Constructive criticisms and suggestion for further improvement of the book are solicited.

Om Prakash Gupta

Abberviation

A

ALARA – As Low as Reasonably Achievable

AMD – Acid Mine Drainage

ASHRAE – American Society of Heating, Refrigerating and Air Conditioning Engineers

ASME – American Society of Mechanical Engineers

ASTM – American Society for Testing and Materials

ATSDR – Agency for Toxic Substances and Disease Registry (U.S.)

B

BACT – Best Available control Technologies

BAP – Benzo-(a)-Pyrene

BHC – Benzene Hexachloride

BOD – Biochemical Oxygen Demand

Bq – Becquerel

BRI – Building Related Illness

BS – British Standard

BTU – British Thermal Unit

BTX – Benzene-Toluene-Xylene

C

CDC – Centers for Disease Control and Prevention (U.S.)

CDD – Chlorinated Dibenzo-p-Dioxin

CDF – Chlorinated Dibenzofuran

CDM – Clean Development Mechanism

CFB – Circulating Fluidized Bed

CFBC – Circulating Fluidized Bed Combustion

CFC – Chlorofluorocarbons

CFC – 1,1,2-Trichloro-1,2,2-Trifluoroethane

CI – Curie

CNG – Compressed Natural Gas

COD – Chemical Oxygen Demand

COH – Coefficient of Haze

COHb – Carboxyhemoglobin

COP – Conference of the Parties (UNFCCC)

COPD – Chronic Obstructive Pulmonary Disease

CO_{2eq} – Carbon Dioxide Equivalent

CPCB – Central Pollution Control Board (India)

CSIR – Council for Scientific and Industrial Research (India)

D

dB – Decibel

dB (A) – Decibel A- Weighted

DDT – Dichlorodiphenyltrichloroethane

D/F – Dioxin/Furan

DME – dimethyl ether

DMT – Dimethyl Tetraphthalate

DNA – Deoxyribonucleic Acid

DO – Dissolved Oxygen

E

EA – Environmental Audit

EDTA – Ethylenediaminetetraacetate

EIA – Environmental impact Assessment

EMF – Electromagnetic Field

EMP – Emergency Management Plan

EMS – Environmental Management Systems

ENSO – El Nino-Southern Oscillation
EPA – Environmental Protection Agency (U.S.)
ERA – Environmental Risk Assessment
ETBE – Ethyl Tert-Butyl Ether
ETS – Environmental Tobacco Smoke
EU – European Union
EV – Electron Volt

F

F – Fahrenheit
FBC – Fluidized Bed Combustion
F/CC – Fiber per Cubic Centimeter
FEV – Forced Expired Volume
FGD – Flue Gas Desulfurization
FGR Flue Gas Recirculation
FID – Flame Ionization Detector
FPD – Flame Photometric Detector
FRP – Fiberglass Reinforced Plastics
FVC – Forced Vital Capacity

G

GATT – General Agreement on Trade and Tariffs
Gb – Giga Barrels
GC – Gas Chromatograph
GCCI – Global Climate Change Initiative
GC-MS – Gas Chromatograph and Mass Spectrometry
GDI – Gasoline Direct Injection
GDP – Gross Domestic Product
GEF – Global Environment Facility
GHG – Greenhouse Gas
GJ – Giga Joules
g.l.c. – Ground Level Concentration
GNP – Gross National product
GPD – Gallons per Day
GPH – Gallons per Hour
GPM – Gallons per Minute
GtC – Gigatons Carbons
GWP – Global Warming Potential
Gy – Gray

H

Ha – Hectare
HAP – Hazardous Air Pollutant
HAZOP – Hazard and Operability Study
HCFC – Hydrochlorofluorocarbon
HDPE – High Density Polyethylene
HEV – Hybrid Electric Vehicle
HF – Hydrogen Fluoride
HFC – Hydrofluorocarbon
HIV – Human Immuno- Deficiency Virus
HLW – High Level Radioactive Waste
HM – Heavy Metal
HVAC – Heating, Ventilating and Air Conditioning

I

IAEA – International Atomic Energy Agency
IAP – Indoor Air Pollution
IAQ – Indoor Air Quality
IARC – International Agency for Research on Cancer
IC – Internal Combustion
IDLH – Immediately Dangerous to Life and Health
IGCC – Integrated Gasification Combined Cycle
ILO – International labor Organization
IP – Inhalable Particles
IPCC – Intergovernmental Panel on Climate Change
IPM – Integrated Pest Management
IR – Infrared
ISI – Indian Standards Institution

ISO – International Standards Organizations

IWM – integrated Waste Management

J

J – Joule

K

K – Kelvin

KV – Kilovolt

KW – Kilowatt

Kwh – Kilowatt Hour

L

LASER – Light Amplification by Stimulated Emissions of Radiation

LCA – Life-Cycle Assessment

LCC – Life Cycle Costing

LC_{50} – Mean Lethal Concentration

LD – Lethal Dose

LD_{50} – Mean Lethal Dose

LDPE – Low Density Polyethylene

LEA – Low Excess Air Operation

LED – Light Emitting Diode

LEL – Lower Explosive Limit

LFG – Landfill Gas

LFL – Lower Flammability Limit

LLDPE – Linear Low Density Polyethylene

LNB – Low Nox Burner

LNG – Liquefied Natural Gas

LPG – Liquefied Petroleum Gas

M

MAC – Maximum Acceptable (Allowable) Concentration

MAC – (EEC) – Maximum Acceptable Concentration (European Economic Community)

MCL – Maximum Contaminant Level

MEA – Monoethanolamine

MEK – Methyl Ethyl Ketone

MGD – Million Gallons per Day

Mg/L – Milligrams per Liter (ppm)

MHD – Magnetohydrodynamics

MLD – Mean Lethal Dose

MLD – Million Liters per Day

MLSS – Mixed Liquor Suspended Solids

MMA – Methyl methacrylate

MOEF – Ministry of Environment and Forests (India)

MPD – Maximum Permissible Dose

MPE – Maximum Permissible Exposure

MPN – Most Probable Number

Mppcf – Millions of Particles per Cubic Foot

MSAT – Mobile Source Air Toxics

MSDS – Material Safety Data Sheets (U.S.)

MSL – Mean Sea Level

MSW – Municipal Solid Waste

MTBE – Methyl Tert-Butyl Ether

Mtoe – Million Tons Oil Equivalent

MW – Megawatt

N

NAAQS – National Ambient Air Quality Standards

NDIR – Nondispersive Infrared

NEERI – National Environmental Engineering Research Institute (India)

NIOSH – National Institute for Occupational Safety and Health (U.S.)

NMHC – Non- Methane Hydrocarbon

NOx – Nitrogen Oxides

NSPS – New Source Performance Standards (U.S.)

NTU – Nephelometric Turbidity Unit

O

ODP – Ozone Depletion Potential

ODS – Ozone Depleting Substance
OEM – Original Equipment Manufacturer
OHS – Occupational Health and Safety
O & M – Operation and Maintenance
OSHA – Occupational Safety and Health Administration (U.S.)

P

Pa – Pascal
PAH – Polynuclear Aromatic Hydrocarbons
PC – Pulverized Coal
PCB – Polychlorinated Biphenyl
PCDD – Polychlorodibenzo-p-Dioxin
PCDF – Polychlorinated Dibenzofuran
PCI/L – Picocuries per Liter
PEL – Permissible Exposure Limit
PF – Pulverized Fuel
PFBC – Pressurized Fluidized Bed Combustion
PM – Particulate Matter
$PM_{0.1}$ – Ultrafine Particle
$PM_{2.5}$ – Particles Less than 2.5 Micrometers
PM_{10} – Particles Less than 10 Micrometers
PMMA – Polymethylmethacrylate
POM – Polycyclic Organic Matter
POP – Persistent Organic Pollutant
Ppb – Parts per Billion
Ppbv – Parts per Billion by Volume
PPE – Personal Protective Equipment
ppm – Parts per Million (Milligrams per Liter) Grams per Cubic Meter)
ppmv – Parts per Million by Volume
ppmw – Parts per Million by Weight
ppt – Parts per Trillion
PSI – Pounds per Square Inch
PTFE – Polytetrafluoroethylene
PVA – Polyvinyl Acetate
PVC – Polyvinyl Chloride

Q

QC – Quality Control
QRA – Quantitative Risk Assessment

R

r – Roentgen
Rad – Roentgen – Absorption-Dose
RBC – Rotating Biological Contactor
R & D – Research and Development
RDF – Refuse-Derived Fuel
REL – Recommended Exposure Limit
Rem – Roentgen-Equivalent-Man
Rep – Roentgen-Equivalent-Physical
RF – Radio Frequency
RFG – Reformulated Gasoline
RH – Relative Humidity
RNA – Ribonucleic Acid
RO – Reverse Osmosis
RVP – Reid Vapor Pressure

S

SBS – Sick Building Syndrome
Sc – Schmidt Number
SCF – Standard Cubic Feet
SCR – Selective Catalytic Reduction
SD – Standard Deviation
SI – International System of Units (Systems International d'Units
SNCR – Selective Noncatalytic Reduction
SO_x – Sulfur Oxides
SS – Suspended Solids
STEL – Short Term Exposure Limit
STP – Standard Temperature and Pressure
Sv – Sievert
SVI – Sludge Volume Index
SVOC – Semivolatile Organic Compound

T

TB – Tracheobronchial

TC – Total Coliforms

TCA – Trichloroethane

TCDD – 2,37,8- Tetrachlorodibenzo-p-Dioxin

TCE – Trichloroethylene

TCER – Target Carbon Emission Rate

TEL – Tetraethyl Lead

THC – Total Hydrocarbon

TLV – Threshold Limit Value

toe – Tonnes Oil Equivalent

TSS – Total Suspended Solids

TWA – Time Weighted Average

U

UEL – Upper Explosive Limit

UF – Ultrafiltration

UFL – Upper Flammability Limit

UNEP – UN Environment Program

UNFCCC – United Nations Framework Convention on Climate Change

UOD – Ultraviolet

UVR – Ultraviolet Radiation

V

VOC – Volatile Organic Compounds / Chemicals

VTOC – Volatile Toxic Organic Compound

v/v – Volume/Volume

W

W – Walt

Wh – Watt Hour

WHO – World Health Organization

X

XLPE – Crosslinked Polyethylene

XRD – X-Ray Diffraction

XRF – X-Ray Fluorescence

Z

ZEV – Zero Emission Vehicle

Contents

CHAPTER 1

Introduction, Classification, Sources and Effects of Environmental Pollution

Introduction

Natural events and human activities are adding number of substances, both toxic and non-toxic to our environment these are undesirable consequences of modern lifestyle, use of pesticides in agriculture, increasing population, increased use of natural resources, increasing number of vehicle and excessive industrial growth. These substances continued being discharged to air, water and soil bring about undesirable changes in physical, chemical and biological characteristics of our environment and adversely affect the life of human being, animal, plants/vegetables buildings, crops, materials etc. This contamination of any part of environment is termed as pollution and the substances, which cause pollution are called pollutants which may cause air pollution, water pollution or land pollution.

With the advancement in technology and industrialisation, undesirable material being added into the environment, has disturbed the ecological balance (suitable co-existence of living organisms that is animals and plants with the non-living components like soil, water and air.) The air we breath is polluted by the exhaust gases of vehicles/industries, fruits and vegetables are polluted due to use of pesticides added in the soil and water is polluted due to toxic effluents (wastes) being thrown into lakes) rivers/seas by the industry.

Environmental pollution brings illness to human beings, slows down the growth of the plants and also adversely affects the animals, buildings and materials. Pollution cause habitat destruction in all habitats-land, sea, fresh water and in the atmosphere.

Definition of Environmental Pollution

"Pollution is the introduction of contaminants into the environment that cause harm or discomfort to humans or other living organisms, or that damage the environment which can come in the form of chemical substances or energy such as noise, heat or light". It is the addition of any substances or form of energy to the environment at a rate faster than environment can accommodate it by dispersion, breakdown, recycling or storage in some harmless form. Pollutants can also be naturally occurring substances or energies when present in excess of natural levels. A few environmental pollution episodes/accidents that have occured in the past has been tabulated below:

S.No.	Year of Occurrence	Place of Occurrence	Mortality	Remarks/causes and effects
1.	1930 (December) Persisted for 3 days	Belgium (Meuse valley)	63	All age group people Suffered chest pain, cough, eyes and nasal irritation, caused by inversion, smoke, SO_2 (9.6-38.4ppm), H_2 So_4 mist
2.	1948 (October) Persisted for 4 days	Pennsylvania (Donora)	20 (6000 ill)	Older people mainly affected with cough, chest pain and nasal & eye irritation, caused by inversion and fog, SO_2 (0.5-2ppm), smoke, H_2SO_4 mist, Zn particles Complete darkness was observed due to gases and soot.
3.	1952 (November)	Mexico (Poza Rice)	22	320 people hospitalized, all age group people affected, irritation of respiratory tract, caused by shallow inversion, fog and calm condition H_2S released due to burner failure.
4.	1952 (December) Persisted for 5 days	London	4000	Died due to respiratory diseases caused by London smog Thousands hospitalized for respiratory and heart disease, chronic bronchitis, Bronchopneumonia, caused by low temperature inversion, thick fog stagnant air, smoke (4500 microgram per m^3) and SO_2 (1.4ppm) accumulation.
5.	1953 (November)	New York	220	-
6.	1956 (January)	London	1000	Extended fog conditions similar to 1952 episodes, particulate (3250 microgram per m^3), ,SO_2(0.57 ppm)
7.	1957 (December)	London	750	-
8.	1953-1960	Japan	-	Hundreds of Japanese died due to eating of fish from mercury polluted-Minamata bay.
9.	1962 (December)	London	700	Caused by shallow inversion, thick fog smoke. (2000 microgram per m^3), SO_2 (1-26ppm)
10.	1963 (January)	New York	300	-
11.	1966 (November)	New York	168	Caused by SO_2, particulates and Smoke
12.	1972	Iraq	500	Died after Consuming polluted wheat.

S.No.	Year of Occurrence	Place of Occurrence	Mortality	Remarks/causes and effects
13.	1984 December	India (Bhopal)	2500 (100000 people affected)	Methyl isocynate vapor (30 tons), also affected about 100000 people who are still suffering from respiratory, lung, throat and eyes diseases, chemical conjunctivitis, cardiac failure, suffocation.
14.	1986 (April)	Chernobyl (USSR)	2500	People died in the world's worst nuclear disaster
15.	-	Greece and Italy	-	Historic marble statues have been damaged by acid rain.
16.	-	Agra (India)	-	Exhaust gases (SO_2) from Mathura oil refinery is affecting the whiteness of the marble of Tajmahal
17.	-	Hiroshima and Nagasaki (Japan)	>1000	Nuclear bomb drop out during world war has subsequently affected the people by emission of nuclear radiation.
18.	1976	Sevasco, Italy	No death	187 cases of skin chloraene, 32 abortions, 15 deformed births, 6 premature delivery, dizziness, diarrhea, caused by reactor explosion releasing dioxine plume and cloud settled over an area of 1430 hectares

The effect of pollutants does not remain localized but in many cases, these pollutants are carried away by air to remote places and thereby affecting the people, plants and materials there also. For example, the radioactive fall out of nuclear test carried out in any part of the world can spread to other parts of the globe very rapidly. Similarly ill effects of use of any part of the world can spread to other parts of the globe very rapidly. Similarly, ill effects of use of pesticides (DDT) on animals can be felt living about a thousand kilometer away. The gradual increase in pollution has now becoming alarming. Both the developed and the developing countries have now initiated steps to minimize the pollution level. One of the steps taken by Indian government is to set up Central Ganga Authority to take the necessary step to purify the Ganga water to make it pollution free. Government has further initiated steps to use CNG (compressed natural gas) in place of petrol and diesel in metropolitan cities to minimize vehicle pollution.

Classification of Environmental Pollution

Environment pollution is not only due to the addition of obnoxious substances into the environment, but also due to factors like noise, electromagnetic waves (like ultraviolet rays) etc.

Environmental pollution is classified in various ways as follows.

a. **Depending on the part of environment polluted, it is classified as:** (i) Air pollution (ii) Water pollution (iii) Soil/land pollution (It is also called third pollution)

b. **Depending on the nature of the pollutants added into the environment ,it is classified as:** (i) Noise pollution (ii) Thermal pollution (iii) Radioactive pollution (iv) Oil pollution (v) Smog pollution (vi) Metal toxicity pollution (vii) Soap/detergent pollution (viii) Effluent pollution (ix) Acid rain pollution (x) chemical pollution (xi) Solid waste pollution (xii) Plastic pollution (xiii) Automatic exhaust pollution (xiv) Light pollution (xvi) Invasive species pollution

c. **By economic sources of pollution, it is classified as:** (i) Agricultural pollution (ii) Industrial pollution (iii)Transport pollution (a) car/vehicle pollution (b)ship pollution (c) Air plane pollution (iv) Commercial and domestic sector pollution

d. **Natural sources of pollution are:** (i) Volcanic eruptions (ii) Dust storms (iii) Smoke from forest and grass fires

However, the first classification method is the most widely used that is, Air, Water and soil Pollution.

List of Environment Pollutants

Most common air, water and soil pollutants are listed below:

Air pollutants:

a. Sulfur dioxide
b. Nitrogen oxides
c. Ammonia
d. Carbon monoxide
e. Ozone
f. Volatile Organic compounds (VOCs)
g. Airborne particulates
h. Toxic metals
i. Radioactive pollutants
j. Persistent organic pollutants (POPs)

Water pollutants -- Organic	Water pollutants -- Inorganic
• Insecticides and herbicides • Bacteria from livestock operations • Food processing wastes including pathogens. • Tree debris from logging operation • Voc's (Volatile organic compounds) including solvents and hydrocarbons • Dense non-aqueous phase liquids such as chlorinated solvents • Detergents and chemical compounds found in cosmetics products	• Chemical wastes • Heavy metals • Fertilizers from agricultural use • Silts from construction sites, logging, slash and burn operation • Acidity caused by industrial discharges such as sulfur dioxide from power plants.

Soil Pollutants

a. pesticides
b. solvents
c. hydrocarbons
d. Lead and other heavy metals

Fossil Fuels Sources of Environmental Pollution

Combination of fossil fuels (oil, gas and cool) produces extremely high levels of air pollution and hence is the most important target area for reduction and control of environment pollution. Fossil fuels also contribute to soil contamination and water pollution example, when oil is transparent from the point of its production it further destinations by pipelines, and oil leak form pipeline may occur and pollute soil and subsequently groundwater. When oil is transported by tankers in ocean, an oil spill may occurs and pollute ocean water.

Power generating plants and transport are the biggest sources of fossil fuel pollution.

Common sources of fossil fuel pollution are:

Industrial Sources	Transport Sources
• Power generating plants • Petroleum Refineries • Petrochemical plants • Production and distribution of fossil fuels • Other manufacturing facilities	• Road transport (motor vehicle) • Shipping industry • Aircrafts

Fossils fuel combustion is a major sources of carbon dioxide (CO_2) emissions which is the most important cause of global warming.

Other (Non –fossil fuel) sources of Environmental Pollution:

Agriculture sources	Trading activities sources	Residential sector sources
• Agriculture (livestock farming) is the largest generator of ammonia emissions resulting in air pollution. • Pesticides and fertilizers used in agriculture leads to water pollution and soil contamination.	• Packaging materials for the products sold in markets generates large quantities of solid wastes that ends up either in land file or municipal incinerators leading to soil contamination and air pollution	• Residential sectors generates solid municipal wastes that ends up in landfills or incinerators leading to soil pollution and air pollution.

Fossils fuels among the biggest sources of pollution, hence we have to find and develop alternative renewable sources of energy which can replace fossil fuels in the future such as solar electricity.

Excessive levels of environmental pollution are causing a lot of damage to human and animal health, plants and trees as well as the wider environment. All types of environmental pollution-air, water and soil pollution have an impact on the living environment.

Environment pollution can also be defined as “the unfavorable alteration of our surroundings”.

Types of Pollutants

- Bio degradable pollutants-decompose rapidly by natural processes.
- Non-degradable pollutants –do not decompose or slowly decompose in the environment.

Classification of Environmental Pollution

- Air pollution
- Water pollution
- Soil pollution

- Marine pollution
- Noise pollution
- Thermal pollution
- Nuclear pollution

Air Pollution

The presence of one or more contaminants like dust, smoke, mist and odor in the atmosphere which are injurious to human beings, plants, animals and building/materials etc.

Sources of Air pollution

- Natural pollution-volcanic eruptions, forest, fires, biological decay.
- Man-made –Thermal power plants, agricultural activities, fossil fuels combustion, industries etc.

General Classification of Air Pollution

- Primary pollutant-these are those emitted directly in the atmosphere in harmful form like CO, NO, SO_2, etc.
- Secondary pollutants-these may react with one another or with the basic components of air to form new pollutants eg, O_3, H_2S, smog etc.

Control Measures

- Sources Control
- Use only unleaded petrol in vehicles.
- Use petroleum products and other fuels that have low sulphur and ash content.
- Plant trees roads streets because they remove particulars and carbon monoxide and absorb noice.
- Industries and waste disposal sites should be situated outside the city.
- Use catalytic converters to help control the emissions of carbon monoxide and hydrocarbons from vehicle exhaust.

Air pollution control measures in Industrial sectors:

- Emission rates should be restricted to permissible levels for all the pollutants.
- Incorporation of air pollution control equipments in the design of the plants.

Water Pollution

It is defined as " the alteration in physical, chemical and biological characteristics of water that may cause harmful effects on human beings and aquatic life.

Types, effects and sources of water pollution

Infectious: Bacteria, viruses, protozoa and parasitic worms.

Sources: Human and animal wastes.

Effects: Variety of diseases like diarrhoeas, cholera etc.

Oxygen demanding wastes: Animal manure and plant debris that can be decomposed by aerobic bacteria

Sources: Sewage, paper mills, and food processing facilities.

Effects: wastes can degrade water quality by reducing dissolved oxygen in water.

Oxygen demanding wastes: Animal manure and plant debris that can be decomposed by aerobic bacteria.

Sources: Sewage, paper mills and food processing facilities.

Effects: Wastes can degrade water quality by reducing dissolved oxygen in water.

Inorganic Chemicals: Water soluble inorganic chemicals. Compounds of toxic metals such as lead, arsenic and selenium. Salts such as in ocean water.

Effects: Genetic mutations, birth defects and certain cancers.

Thermal Pollution (heat) of water

Example: Excessive heat of cooling water in power plant when discharged into river, lakes, cooling pond etc.

Human source

Water cooling of thermal power plants and some types of industrial units. Almost all water withdrawn for cooling in thermal power plants.

Effects

Lower dissolved oxygen levels in water and makes acquatic organisms more vulnerable to disease and toxic chemicals.

When a power plant first opens or shuts down for repair, fish and other organisms adapted to a particular temperature range can be killed by the abrupt change in water temperature known as thermal shock in rivers, lakes, cooling ponds, etc.

Control measures of water pollution

The administration of water pollution should be in hands of state or central government, besides the polluting industrial units.

Industrial plants should be based on recycling operations, because it will not only stop the discharge of industrial wastes into natural water sources but by products can be extracted from the wastes.

Plants, trees and forests control pollution and they acts as natural air pollution conditioners.

Land/Soil Pollution

It is defined as "the contamination of soil by human and natural activities which may cause harmful effects on living beings."

Types of Soil Pollution

Industrial Wastes Leading to Soil Pollution

Sources and effects: Pulp and paper mills, chemical industries, oil refineries, sugar factories. These pollutants affects and alter the chemical and biological properties of soil.

As a result, hazardous chemicals can enter into human food chain from the soil; disturb the bio chemical process and finally lead to serious effects/diseases.

Urban waste Leading to Soil Pollution

Sources and effects: Plastic, Glasses, metallic cans, fibers, papers, rubbers, street, sweeping, and other discarded manufactured products. These are also dangerous for human beings.

Agriculture Practices Leading to Soil Pollution

Sources and effects: Huge quantities of fertilizers, pesticides, herbicides, and weedicides are added to increase the crop yield. Apart form these farm waters, manure, slurry etc also.

Radioactive/Nuclear Pollutants Leading to Soil Pollution

Sources and effects: These are resulting form explosive of nuclear bombs and radio active wastes penetrate the soil and accumulate there by creating land pollution.

Biological Agents Leading to Soil Pollution

Sources and effects: Soil gets large quantities of human, animal and birds excreta which constitute the major source of land pollution by biological agents.

Control Measure of Soil Pollution

- Population growth should be restricted.
- Decrease of the available farm land due to urbanization should not be allowed.
- Forestry and farm practices should be improved.
- Proper dumping of unwanted materials should be ensured.
- Production and use of natural fertilizers should be ensured.
- Reducing, Recycling and Reuse of wastes should be done.
- Ban on Toxic chemicals should be ensured.

Solid Waste Pollution

Solid waste means any garbage, trash, waste, tire, sludge from a waste water treatment plant and other discarded material, including solid, liquid, semisolid or contained gaseous material.

In other, words, solid wastes may be defined as the organic and inorganic waste produced by various activities of the society which have lost their value to the first user.

Types of Solid Waste

Municipal solid waste consists of households waste, construction and demolition debris, sanitation residue etc.

Hazardous waste : Industrial and hospital waste is considered as hazardous waste as they contain toxic substances.

Infectious waste: Biomedical waste or hospital waste, generated during diagnosis treatment etc. Include needles, chemicals wastes, discarded medicines and human excreta etc.

Solid wastes are generally stored at the generation site and then transported either to an intermediate place or treatment prior to final disposal or to the final disposal site directly.

Sources of solid wastes are categorized as:

- Residential wastes
- Industrial wastes
- Commercial wastes
- Institutional wastes
- Construction and demolition wastes
- Municipal wastes

- Process (manufacturing, wastes)
- Agriculture wastes

Solid waste management is required for avoiding:

- Ground water pollution and environmental degradation.
- Serious health problems.
- Clogging of drains.
- Decomposition of organic materials.
- Air water and soil pollution.

Solid waste management (SWM) process consists of following four steps

- Collection
- Transportation
- Treatments
- Ultimate disposal

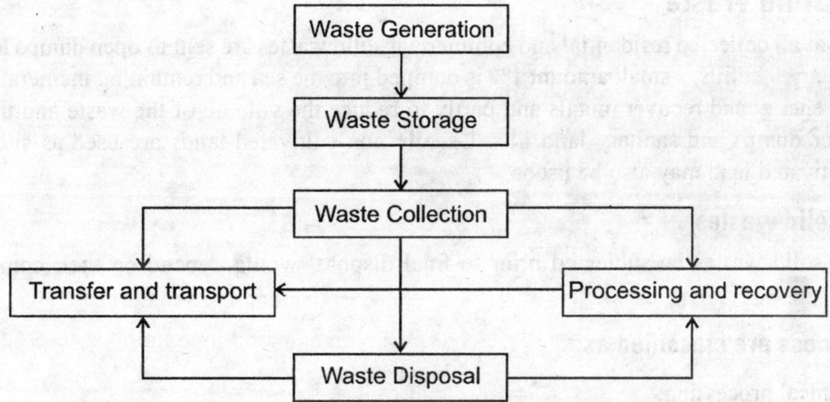

Flowchart of the Functional Elements in SWM

Main ways to manage solid waste pollution include:

- Source reduction
- Recycling
- Composting
- Incineration
- Land filling

Storage of solid wastes

Solid wastes are stored till a sizable amount accumulate. While storing, special containers are to be used for poisonous materials and hazardous wastes. The storage should be convenient for the user and it should facilitate safe and efficient collection. Storage devices should prevent access to odours, vectors, and emissions of excessive odours.

Collection and handling of solid wastes

When the quantities of solid wastes to be managed are relatively small then collection, handling and short distance transport may be done manually. However, when the quantities is large, mechanical devices be used.

Transportation of solid wastes

To transport relatively small quantity of solid wastes over, a short distance horizontally or along a slight incline a wheel barrow may be used However, when the quantity to be transported is large or the distance is long or both, suitable vehicles should be used.

Recovery and Recycling of solid wastes

Solid wastes particularly mixed refuse may contain non-renewable resources like metal pieces etc, as well as renewable resource-based refuse like food waste and agricultural wastes. In some instances the waste from a particular process may find useful application else-where wherever possible reusable components of wastes should be recovered and recycled. Such practices would reduce the waste disposal cost by reducing the quantity of wastes to be handled.

Disposal of Solid Waste

Nearly 75-80% of all collected residential and commercial solid wastes are sent to open dumps less than 10% is buried in sanitary landfills, a small amount 1% is dumped into the sea and remaining incinerated converted either to obtain energy and recover metals and partly to reduce the volume of the waste and the amount of land required for dumps and sanitary land fills. Usually non-cultivated lands are used as sites for wastes disposal but cultivated land may also be used.

Treatment of solid wastes

The treatments solid wastes be subjected prior to final disposal would depend on their composition and quantities.

Treatment process are classified as:

- Mechanical processing.
- Biological processing.
- Thermal processing.

Mechanical processing is used to "recover salvable materials as well as to reduce the size of refuse constituents and to compact the refuses size. Reduction and compaction operations makes it easier to handle and transport solid wastes".

Biological processing is used to convert the organic constituents of solid refuse into some by products."

Thermal processing aims at energy recovery from organic matter bearing refuse prior to its final disposal.

Marine Pollution

It is defined as "the discharge of waste substances into the sea resulting in harm to living resource hazards to human health, hindrance to fishery and impairment of quality for use of sea water."
The coastal zones contain rich heritage, coral reefs, wetlands, and sea grass beds.

Effects of marine pollutants

- The presence of heavy metals and organic pollutants cause more damage in birds as thinning of eggshells and tissue damage of eggs.
- Oil spilling causes abnormally low body temperature in birds resulting in hypothermia.

- Oil films are able to retard significantly the rate of oxygen uptake by water.

Control measures of marine pollution

- Plants for conserving marine biodiversity must be taken into accounts of human needs.
- People should be educated about marine ecosystems and the benefits offered by them.
- Local communities must be involved in protecting and managing their coastal resource. Social and economic incentives must be offered for conserving and sustainable use of marine resource.

Noise Pollution

It is defined as "the unwanted, unpleasant or loud sound that causes annoyance and hearing problems.

Type of noise sources

- Industrial noise sources.
- Transport noise sources.
- Neighborhood noise sources.

Effects of Noise pollution

- This affects human health, comfort and efficiency.
- It cause muscles to contract leading to nervous breakdown and tension.
- It affects health, efficiency and behavior.
- In addition to serious loss of hearing due to excessive noise, impulsive noise also causes psychological and pathological disorders.
- Brain is also adversely affected by loud and sudden noise as that of jet and aero plane noise.

Control and preventing measure for noise pollution:

- Sources control-acoustic treatment of machine surface, design changes, limiting the operational timings.
- Transmission path intervention-the source inside a sound insulting enclosure constructing of noise barrier or provision of sound absorbing materials.
- Lubrication and proper oiling reduce the noise from the machines.

Thermal Pollution Of Water

It is defined as the " addition of excess of undesirable heat to water the makes it harmful to man, animal and aquatic life or otherwise causes significant departure from the normal activities of aquatic communities in water".

Sources of thermal pollution

- Nuclear power plants discharging hot water into river, lakes, cooling ponds etc.
- Thermal power plants discharging hot water into river, lakes, cooling ponds etc.
- Industrial effluents discharging hot water into river, lakes, cooling ponds etc.
- Domestic effluents discharging hot water into river, lakes, cooling ponds etc.

Effects of thermal pollution

- Increase in Toxicity of water in the river bodies.
- Interference with biological activities of aquatic creatures.
- Interference with reproduction of aquatic creature.
- Direct mortality of aquatic creatures: example fish etc.

Control measures of thermal pollution

- Cooling towers-This is used to cool water in cooling tower,
- Cooling ponds and spray ponds are also used for cooling hot water.
- Artificial lakes-The heated effluents can be discharged into the lake at one end and water for cooling purpose can be drawn from the other end.

Nuclear/ Radioactive Pollution

The radiation hazard in the environment comes from ultraviolet, cosmic, rays and micro wave radiation which produces genetic mutation in human beings.

Sources of Nuclear / Radioactive pollution

Natural sources: This is space which emits cosmic rays.

Man made sources: (Anthropogenic sources) these are nuclear power plants, X-rays, nuclear accidents, nuclear bombs, diagnostic kits for medical purposes using radioisotopes.

Effects of Nuclear / Radioactive Pollution:

- Exposure of the brain and central nervous system to high dose of radiation cause delirium, convulsions and death within hours or days.
- The use of eye is vulnerable to radiation. As its cell die, they become opaque forming cataracts that impair eye sight.
- Acute radiation sickness is marked by vomiting; bleeding of gums and in severe cases mouth ulcer.
- Nausea and vomiting often begin a few hours after the gastrointestinal tract is exposed. Infection of the intestinal wall can kill weeks afterwards.
- Unborn children are vulnerable to brain damage or mental retardation especially if irradiation occurs during formation of the central nervous system during early pregnancy.

Control measure for nuclear / radioactive pollution

- Nuclear devices should never be exploded in air.
- In nuclear reactors, closed cycle coolant system with gaseous coolant may be used to prevent extraneous activation products.
- Contaminants may be employed to decrease the radio active emissions.
- Extreme care should be exercised in the disposal of industrial wastes contaminated with radio nuclides.
- Use of high chimneys and ventilations at the working place where radioactive contamination is high for dispersing pollutants.

Solid Waste Management

Management of solid waste is very important in order to minimize the adverse effects of solid wastes.

Types of solid wastes

Urban wastes sources

- **Domestic wastes:** Food waste, Cloth, Waste paper.
- **Commercial wastes:** Packaging materials, cans, bottles, polythenes.
- **Construction wastes:** wood, concrete, debris.
- **Bio medical wastes:** Medical wastes, infectious wastes.

Industrial wastes sources

- **Nuclear power plants:** generates radioactive wastes.
- **Thermal power plants:** produces fly ash in large quantities.

Chemical industries sources

Produces large quantities of hazardous and toxic materials.

Steps involved in solid waste management

- Reduce, Reuse and Recycle of materials should be done.
- Land fill: Solid wastes are placed in sanitary landfill system in alternate layers of refuse, covered with selected earth fill.
- Incineration: It is hygienic way of disposing the solid waste. It is a thermal process and is very effective for detoxification of all combustible pathogens.
- Composting: It is another popular method practiced in many cities in our country. In this method, bulk organic waste is converted into fertilizing manure by biological action.

Role of an individual in prevention of pollution

- Plant more trees.
- Help more in pollution prevention than pollution control.
- Use water, energy and other resource efficiently and optimally.
- Purchase recyclable, recycled and environmentally safe products.
- Reduce deforestation.
- Remove NO_x form motor vehicular exhausts.
- Use eco friendly products.

Effects of Environmental Pollution

Pollution Effects are categorized in following three classes:

- Effects of environmental pollution(air, water and soil) on human beings.
- Effects of environmental pollution on animals.
- Effects of environmental pollution on plant and trees.

a. **Effects of air, water and soil pollution on human beings are summarized below in the table:**

Effects of air pollution on human beings	Effects of water pollution on human beings	Effects of soil pollution on human beings
• Irritation of eyes, nose, throat and mouth • Asthma attack. • Reduced lung functioning • Respiratory problems such as coughing and wheezing. • Increased respiratory diseases such as bronchitis. • Cardiovascular problems. • Headaches and dizziness. • Reduced lung energy levels. • Neurobehavioral disorders • Cancer • Disruption of endocrine, reproductive and immune systems. • Premature death.	• Following water borne disease are caused by polluted drinking water: – Hookworm – Typhoid – Amoebiasis – Giardiasis – Ascariasis • Following water borne diseases are caused by polluted sea beach water: – Respiratory infections – Pink eye, ear ache, rashes – Diarrhoea, vomiting, hepatitis encephalitis, gastroenteritis, stomach aches. • Diseases caused by water polluted by chemicals (such as pesticides, heavy metals, hydrocarbons, persistent organic pollutants etc): – Damage to nervous system. – Liver and kidney damage. – Cancer including prostrate cancer – Harmonal problems that can disrupt reproductive and development process – Exposure to mercury(heavy metals)causes Parkison's disease, Alzeimer's disease, heart disease, multiple sclerosis and even death in adults. In the womb of mother, it may cause neurological problems including slower reflexes, learning deficit, delayed or incomplete Mental development, austism and brain damage. Damage to people may be caused due to fish food coming from polluted water and vegetables crops grown/ washed with polluted water.	• Causes cancers including leukemia • Lead in soil causes developmental damage to the brain of young children. • Mercury increases the risks of kidney damage while cyclodiene lead to liver toxicity • Causes headache, nausea, fatigue, eye irritation and skin rash. • Causes neuromuscular damage and depression of the central nervous system • Contamination of crops grown in polluted soil brings up problem with food security. • Water pollution may result from interaction between water and the contaminated soil. • Contact with contaminated soil may be direct (from using schools, parks etc.)or indirect (by in holding soil contamination which have vaporised). • Soil contamination results from secondary contamination of water supplies and from depositing air contaminants (eg, acid rain)

b. Environmental pollution effects on animals are summarised below in the table:

Effects of air pollution on animals	Effects of water pollution on animals	Effects of soil pollution on animals
• Acid rain (formed in air destroys fish life in the lakes and rivers/sea streams). • Ozone present in the lower atmosphere(tropospehre) damages lung tissues of animals. • Excessive ultraviolet radiation coming from sun through the ozone layer in the upper atmosphere(stratosphere)which is eroded by some air pollutants may cause skin cancer in some animals. • Presence of fluorine, arsenic and ,lead in air is responsible for most live stock damage. Exposure to ionizing radiation fallout from nuclear bomb testing in the atmosphere causes cancer (including leukemia) shortening of life span and genetic or mutation effect in animals.	• Mercury in water causes abnormal behavior slower growth and development, reduced reproduction and death of aquatic creature. • Persistent organic pollutants (POPs) may cause declines, deformities and death of fish life. • Chemical contamination of water causes declines in frog biodiversity and tadpole mass. • **Nutrient pollution:**(nitrogen, phosphate etc.) causes overgrowth of toxic algae eaten by other aquatic animals. It also cause outbreak of fish disease and death. • **Oil pollution:** negates the development of marine organisms, increase susceptibility to disease and affect reproductive process. It also causes gastrointestinal irritation, liver and kidney damage and damage to the nervous system. • Too much of sodium chloride (ordinary salt) in water may kill animals. • Higher forms of non-aquatic animals may have similar effects from water pollution as those experienced by human beings.	• Soil contamination alters the metabolism of microorganisms and arthropods in a given soil environment, which may destroy some layers of the primary food chain and thus have a negative effect on predator animal species. • Small life forms may consume harmful chemicals which may then be passed up the food chain to larger animals leading to increased mortality rates and even animals extinction.

c. Environmental pollution effects on trees & plants are summarised below:

Effects of air pollution on trees & plants	Effects of water pollution on trees & plants	Effects of soil pollution on trees and plants
• Acid rain(formed in air) kills trees, destroys the leaves of the plants and can infiltrate soil making it unsuitable for purposes of nutrition and habitation. • Ozone holes in the upper atmosphere (stratosphere) can allow excessive UV radiation from the sun to enter the earth causing damage to trees and plants. • Ozone present in the lower atmosphere (troposphere) prevents the plant respiration by blocking stomata (openings in leaves) and negatively effects plant's photosynthesis rates which will still stunt plant growth. Ozone can also decay plant cells directly by entering stomata.	• It disrupts photosynthesis in aquatic plants and thus effecting ecosystems that depend on these plants. • Terrestrial and aquatic plants may absorb pollutants from water (as their main nutrient source) and pass them up the food chain to consumer animals and human beings. • Plants may be killed by too much of sodium chloride present in water. • Plants may be killed by herbicides present in water. Herbicides are chemicals which are most harmful to plants. • Plants may be killed by mud from construction sites as well as bits of wood and leaves, clay and other similar materials.	• Soil contamination alters plant metabolism and reduces crop yields. • Tress and plants may absorb soil contaminants and pass them up the food chain.

Apart from destroying the aquatic life in lakes and streams, acid rain can be also corrode metals, damage surface of buildings and monuments and cause soil acidification. Pollution of water may cause oxygen depletion in marine environments and severely affect the health of whole ecosystems.

Appendices

Contents — **Page No.**

Contents			Page No.

Contents			Page No.

		Contents	Page No.

Appendix 1.1: Pollutants Generated by Various Artificial Sources

S.No	Artificial sources	Main pollutants generated
1.	Agriculture: • Field ploughing • Crop spraying and dusting • Field burning	Dust and odours. Pesticides and insecticides, chlorinated hydrocarbons, lead, arsenic, phosphates. Smoke, flyash, soot.
2.	**Domestic combustion of fuels** like coal, wood, gas.	CO, CO_2, SO_2, smoke, metallic, oxides.
3.	Cigarette smoking	NO_x, 4% CO, 0.5% of Pb per cigarette
4.	**Incinerators**: • Open burning of wastes • Municipal, commercial and industrial incinerator	Smoke CO, NO_x, smoke and flyash
5.	**Food processing:** • Drying, preserving and packaging	Vapors, odours, dusts
6.	**Transportation**: • Aircrafts, trains, motor vehicles, bullock carts	Dust, noise, lead, NO_x, hydrocarbons, carbons monoxide (95% of total CO and HC emissions) are from transportation sector only).
7.	**Power generation**: • Thermal power plant. • Diesel generators. • Nuclear Power Plants. • Hydroelectric power plants	CO, CO_2, SO_x, NO_x, dusts HC, CO, NO_x, noise Radioactive compounds eg, I-131, Ar-41, Sr-90, Cs-137, C-14 etc. Methane as marshy gas from water logged areas.
8.	Pulp and paper plants	Mercaptons(RSH), SO_2, H_2S, odour, noise, SPM
9.	**Petroleum oil refineries**: • Process heaters, boilers, regenerators. • Storage tanks, compressors, reactors, generators.	SO_x, HC, noise CO, NOx, odour, SPM
10.	**Organic chemicals manufacturing plants:** • Soaps and detergents, polymers, paints varnishes, insecticides, alcohols, phenols, etc.	CO, SO_x, gases and vapors, noise, SPM, Odour
11.	**Inorganic chemicals and Fertiliser plants:**	NH_3, HF, H_3PO_4, odour, noise, sulphuric acid mists, SPM, fertilizers dust.
12.	Ferrous metallurgical process/ Irons steel plants	CO, H_2S, SOx, NOx, smoke, fumes, odours, SPM, vapours, flourides
13.	Non-ferrous metallurgical process	Dust, fumes of Cu, Zn etc.

S.No	Artificial sources	Main pollutants generated
14.	**Roasting and heating process of nonmetallic minerals**: • Gravel; sand stone; crushed stone, sand processing; cement, glass, ceramic, and refactory-manufacturing, coal cleaning	SOx, NOx, dusts, minerals, organics, SPM
15.	Tanneries and leather industry	Meracaptons and sulphides
16.	Coal tar industry	Polynuclear hydrocarbon, carbon soot and hydrogen sulphide.
17.	Cement industry	Cement and lime dust
18.	Carbon black manufacture	Polynuclear, hydrocarbon, carbon soot and hydrogen sulphide
19.	Fertiliser industry and aluminium manufacturing plants	Hydrogen fluoride, ammonia, fluorides, fertilizer dust, sulphuric acid mist
20.	Lead casting and melting, Pigment etc.	Tin, lead etc, fumes and oxides, solvents and thinners
21.	Electrolytic manufacture of chloride	Chlorine
22.	Paints, pigments and dye industry	Nitrobenzene and aniline, thinners, solvent and base materials

Appendix 1.2: Concentration of Common Air Pollutants in Clean Atmospheric Air and Polluted Air over a Densely Populated Area

S.No.	Constituents of air	Clean atmospheric air, ppm	Polluted air over densely populated area,ppm
1.	Carbon dioxide (CO_2)	310-330	350-370
2.	Carbon monoxide (CO)	<1	5-200
3.	Sulfur dioxide (SO_2)	10^{-3} to 10^{-2}	0.02-2
4.	Oxides of nitrogen (NO_x)	10^{-3} to 10^{-2}	10^{-2} to 10^{-1}
5.	Hydrocarbons (HC)	<1	1-20
6.	Particulate matter (PM)	0.01-0.02 mg/m³	0.07-0.7 mg/m³

Appendix 1.3: Natural Activities and Pollutants Released in the Environment

S.No.	Natural activities	Environmental Pollutants released
1.	Forest fire	Smoke, ash, CO, CO_2, NO, NO_2 unbrunt hydrocarbons
2.	Volcano eruption	Particulate matter, H_2S, SO_2 CH_4
3.	Plants and trees	Pollen grains and hydrocarbons
4.	Oceans	Salt particles which show corrosive properties to metals and paints

NOTES

Following sources of natural air pollution exist:

- Volcanoes: Volcanic activity produce smoke
- Geysers: The air pollutants emitted by geysers include hydrogen sulfide, arsenic and other heavy metals.
- Digestive gases: Methane and other gases generated by the digestion of food and emitted by animals such as cattle.

Oceans, Rivers and Estuaries: These are sources of methane emissions thought to be caused by the digestive systems of marine life, methanogenesis in sediments and drainage areas along **coastal regions**, and possibly seepage from methane hydrates on the ocean floors.

Dust: Windblown dust from areas with little or no **vegetation** such as **desert areas.**

Sea salt: Wind-blown sea water which evaporates in the atmosphere and releases sodium chloride and other particulates into the atmosphere.

Radioactive decay: Radon gas is released into the atmosphere by radioactive decay occurring in the Earth's crust.

Forest fires: Forest fires created by lightning, or other natural causes, result in the formation and release of smoke, ash, dust, carbon dioxide, nitrogen oxides and other air pollutants.

Plants and trees: Biogenic sources such as pine trees and certain other plants and trees which release volatile organic compounds (VOC). About 80% of the overall emissions of VOC are from biogenic sources.

Wetlands: Microbial action in wetlands result in significant amounts of methane being formed and released to the atmosphere. In fact, wetlands are the largest natural source of methane emissions.

Termites: Termites are the second largest natural source of methane emissions. The methane is produced by their normal digestive process.

Lightning: Lightning converts atmospheric nitrogen to nitrogen oxides.

***Soil outgassing*:** Another biogenic source wherein microbial action in soils result in the formation and release of significant amounts of nitrogen oxides.

However two significant natural causes of air pollution are **volcanic eruptions** and **forest fires.**

Air pollution from Volcanoes

One of the most important natural causes of air pollution is volcanic eruption.

The most problematic gases emitted in a volcanic eruption include sulfur dioxide, carbon dioxide and hydrogen fluoride.

Sulfur dioxide gas can cause acid rain in nearby places and air pollution in downwind areas from the volcanic site.

Besides the volcanic gases, there is also volcanic ash. Volcanic ash can move hundreds to thousands of miles downwind from a volcano. Fresh volcanic ash is gritty, rough, and at times corrosive. The ash can cause respiratory problems for young children, the elderly or those already with respiratory ailments.

Volcanic eruptions can generate so much polluting gases and ash into the air that the sun's rays could be blocked, and land temperature in the affected area lowered

Air pollution from Forest Fires

All forest fires emit carbon monoxide, sulfur dioxide, nitrogen dioxide, ozone, and particulate matter.

Carbon monoxide, which is a poisonous gas, can be emitted in large amounts during forest fires.

Particulates, which are mixtures of soot, tars, and volatile organic substances, either solid or liquid, are emitted in large quantities from forest fires. The particulates, which can be smaller than 2.5 micrometers in diameter, if deeply inhaled into the lungs, can damage lung tissues and cause respiratory and cardiovascular problems.

Nitrogen oxides (NO_x) are released at temperatures greater than 1, 500 °C.

Therefore nitrogen oxides are released in significant quantities only during the most severe fires.

Sulfur dioxide emissions are significantly less serious with forest fires, as concentration most forest fuels is usually less than 0.2 percent. However, forest fire sites with "peat" and "muck soil" may be exceptions.

There are many other air pollution causes like wind erosion, pollen dispersal, evaporation of organic compounds, and natural radioactivity.

However, these causes are usually not significant.

Although there is no doubt that the impact of natural causes for air pollution can sometimes be drastic, for example volcanic eruptions, but they are usually not frequent. Rather, the more significant air pollution causes are in fact caused by human activities, such as through the combustion of fuel by vehicles and power stations, waste incineration, etc.

Whilst man-made pollution and poor air quality is major environmental concern, there are many natural sources of pollution which are often much greater than their man-made counterparts.

Natural sources of sulphur dioxide include release from volcanoes, biological decay and forest fires. Actual amounts released from natural sources in the world are difficult to quantify. In 1983 the United Nations Environment Programme estimated a figure of between 80 million and 288 million tonnes of sulphur oxides per year (compared to around 69 million tonnes from human sources world-wide).

Natural sources of **nitrogen oxides** include volcanoes, oceans, biological decay and lightning strikes. Estimates range between 20 million and 90 million tonnes per year nitrogen oxides released from natural sources (compared to around 24 million tonnes from human sources worldwide).

Ozone is a secondary photochemical pollutant formed near ground level as a result of chemical reactions taking place in sunlight. About 10 to 15% of low level ozone, however, is transported from the upper atmosphere (called the stratosphere), where it is formed by the action of ultraviolet (UV) radiation on oxygen (the ozone layer).

Natural sources of **particulate matter** are less important than man-made sources. These include volcanoes and dust storms. However, such sources do account for intense high particulate pollution episodes, occurring over relatively short times scales. It is not unknown for Saharan dust to be deposited in the UK after being blown thousands of kms.

Volatile organic compounds (VOCs) are naturally produced by plants and trees. **Isoprene** is a common VOC emitted by vegetation, and some believe it to be a more significant trigger for **asthma** and other allergic reactions than man-made irritants. **Plant, grass and trees** are also a source of **pollen**, which can act as triggers in some **asthmatics**. Pollen is in the air year-round, but the concentration is highest during the growing season, from March to the first frosts in autumn.

Natural pollutants found indoors include the dust mite, mould spores and radon gas.

Forests also produce natural hydrocarbon products, which can undergo photochemical oxidation to produce pollutant haze. This formation of wastes is part of the cycling and recycling of matter in natural ecological systems. Wastes released into the atmosphere are diluted and dispersed in the air. and are processed and recycled through a variety of natural physical, chemical and biological mechanisms. For example, many particles of waste are removed by settling or are washed out by rain. Many gaseous wastes are oxidised to form particles, which then settle out or are washed out while others are converted to other gases in the atmosphere or absorbed by plants or other soil micro-organisms. In this way the atmosphere is continually cleaned of these substances.

Atmospheric problems are made worse when weather conditions such as a lack of rain or wind cannot disperse pollutants. This not only holds them in the air, increasing human exposure, but also allows them to undergo chemical reactions to produce secondary pollutants that are far more dangerous than those substances from which they were made.

The residence time of waste products in the atmosphere before they are broken down or brought back to the ground varies between a few hours and many years. Air pollution problems occur whenever any of these normal functions of the atmosphere are disturbed or overloaded.

Global Emissions of air pollutants form Natural and Man Made Sources are shown below in figure.

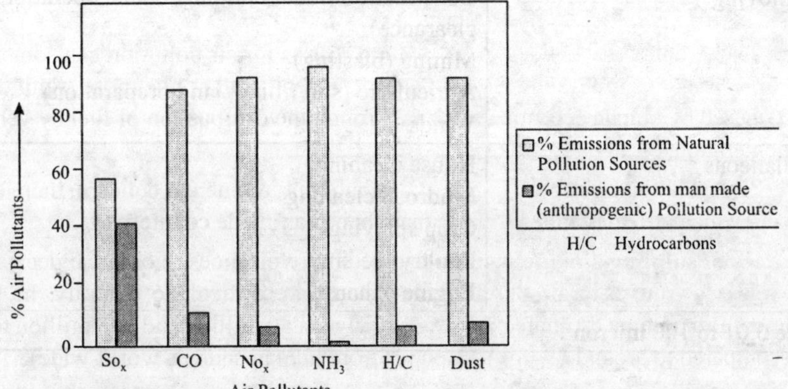

Figure above shows that compared to natural sources, man's activities produce a much smaller amount of global pollution. Indeed natural pollutants may pose a serious threat to air quality when they are generated in large amounts near human settlements.

The principal natural phenomena that contribute acid-producing gases to the atmosphere are emissions from **volcanoes.** for example, fumaroles from Laguna Caliente crater of Poas Volcano create extremely high amounts of acid rain and fog with acidity of pH <2 clearing an area of any vegetation and frequently causing irritation to the eyes and lungs of inhabitants in nearby settlements, Acid-producing gasses are created also by **biological processes** that occur on the land, in wetlands, and in the oceans. The major biological source of sulphur containing compounds is dimethyl sulphide.

Nitric acid in rainwater is an important source of fixed nitrogen for plant life, and is also produced by electrical activity in the atmosphere such as lightning.

Acidic deposits have been detected in **glacial ice** thousands of years old in remote parts of the globe.

Soils of **coniferous forests** are naturally very acidic due to the shedding of needles (this phenomenon should not be confused with acid rain)

Appendix 1.4: Sources of Atmospheric Dust Pollution

S.No.	Dust sources	Examples
1.	Combustion	Fuel burning (coal, wood, fuel oil) Incineration (house and municipal garbage) Others (Open fires, forest fires, tobacco smoke)
2.	Materials handling and processing	Loading and unloading (sand, gravel, coal, ores, lime, cement) Missing and packaging (chemical, fertilisers) Food processing (flour, corn, starch, grains) Cutting and forming (saw mills, wall board, plastics Metallurgical (foundries, smelters) Industrial (Paper, textiles manufacture)
3.	Earth moving	Construction (road, building, dams, operations site clearance) Mining (blasting) Agriculture (soil filling, land preparation) Winds
4.	Miscellaneous	House cleaning Mudroad cleaning Crop spraying Poultry feeding Engine exhaust
Aerosols – Size 0.01 to 100 micron		

NOTES

The fractions of the airborne particles inhaled and deposited in the ' various regions depend on many factors. However, for sampling purposes, terminology used is aerodynamic diameter

Inhalable particulate fraction is that fraction of a dust cloud that can be breathed into the nose or mouth. Examples of dusts for which any inhalable particle is of concern include certain hardwood dusts (which may cause nasal cancer), and dusts from grinding lead- containing alloys (which can be absorbed and cause systemic poisoning).

Thoracic particulate fraction is that fraction that can penetrate the head airways and enter the airways of the lung. Examples of dusts for which this fraction is of particular concern include cotton and other dusts causing airway disease.

Respirable particulate fraction is that fraction of inhaled airborne particles that can penetrate beyond the terminal bronchioles into the gas-exchange region of the lungs.

Examples of dusts for which the respirable fraction offers greatest hazard include quartz and other dusts containing free crystalline silica; cobalt-containing and other hard metal dust produced by grinding masonry drill bits; and many others.

Other dust characteristics besides composition and particle aerodynamic diameter can be important in dust control, for example, adhesion, light scattering, absorption capacity, solubility and hygroscopicity. Various dust deposition sites in human respiratory system has been shown below in figure

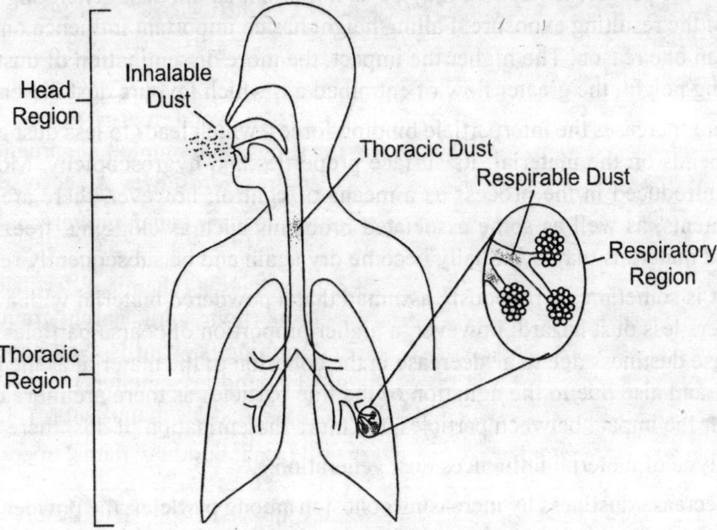

Mechanism of Dust Generation & Release

Mechanical Breakdown

Dusts usually originate from larger masses of the same material, through a mechanical breakdown process such as grinding, cutting, drilling, crushing, explosion, or strong friction between certain materials (e.g., rocks). Dust thus generated is called "primary airborne dust." The composition of mineral dusts is not necessarily the same as that of the parent rock since different minerals may break down or be removed at different rates.

Vegetable dusts can originate in the same manner from a work process, for example wood dusts produced in sawing and sanding, cotton dust in ginning, carding spinning operations, and wool dust in shearing sheep.

The rate of dust generation increases with the energy associated with the process in question. For example, a grinding wheel will produce more dust when it operates at higher speeds. Although friability, that is ability to be broken down, is another important characteristic, more friable does not necessarily mean more hazardous; for example, very hard quartz, once submitted to strong enough forces that break it down to microscopic sizes, is a much more serious health hazard than the more friable marble.

Dust Dispersal

Instead of resulting directly from the breakage of a bulk material, airborne dust may arise from dispersal of materials in powder or granular form. Dust is released whenever processes involve free falling or handling of such materials, e.g., transferring, dumping, filling (bagging) or emptying bags or other containers, dropping material from a hopper to a weighing station, weighing, mixing, conveying and so on. Moreover, air currents over powdered materials may be important.

These mechanisms not only release dust, they also generate it, because smaller particles may be formed from larger ones by impaction and friction. The particle size distribution of a dust cloud may be different from that of the powder it originated from; this should be investigated for each situation, as it depends on the type of material and on the forces it underwent during its handling or processing.

In order to decrease dust emissions from such operations, it is important to understand the mechanisms of its generation and release. Studies on dust generation by free falling powders have demonstrated that the manner in which the powder is handled may be as important as the dust generating capacity of the bulk material, in terms of the resulting exposure. Falling height has an important influence on dust generation and release for more than one reason. The higher the impact, the more dissemination of dust there is. Moreover, the greater the falling height, the greater flow of entrained air, which favours dust dissemination.

Moisture content increases the interparticle binding forces, which leads to less dust generation; however, how much less depends on the material, its surface properties and hygroscopicity. Moisture - in the form of water - can be introduced in the process as a means of'control; however, there are limitations in view of process requirements, as well as some associated problems such as clogging, freezing, or evaporation. Furthermore, wetted materials may eventually become dry again and be subsequently redispersed.

For example, it is sometimes erroneously assumed that a powdered material with a larger proportion of coarse particles offers less dust hazard; however, a higher proportion of coarse particles in the bulk material may actually increase dustiness due to a "decrease in the cohesion of the material as the proportion of coarse particles increases" and also due to the agitation of the fine particles as there are more collisions with large particles. The higher the impact between particles, the more dissemination of dust there is.

Moreover, the type of material influences dust generation.

If one tries to decrease/dustiness by increasing cohesion among particles, the powder handling equipment might get clogged; in certain situations, exposure could even be increased because workers would have to shake the equipment. The implications of process changes, in terms of maintenance requirements, must also be considered.

The nature of dust dispersion during materials handling has not provided much basic insight into the physical processes involved.

Dusty Occupations

Dust exposure is linked to occupations and workplaces, both in the industrial and agricultural settings, for example:

- mining, quarrying and tunnelling;
- stone-working and construction;
- foundries and other types of metallurgical activity;
- shipbuilding (abrasive blasting);
- manufacture of glass, ceramics (pottery, porcelain and enamel) and stone objects;
- etching glass;
- manufacture of cleansing agents and abrasives;
- chemical and pharmaceutical industry (handling of powdered chemicals);
- rubber manufacturing industry;
- manufacture of lead storage batteries (bulk lead oxide);
- removing paint and rust from buildings, bridges, tanks and other surfaces;

- formulation of pesticides;
- agricultural work (ploughing, harvesting, grain storage);
- food industry (bakeries, animal products);
- forestry and woodworking.

Dusty Processes

Dust releases in the workplace may result from any form of mechanical breakdown, such as occurs in mining and quarrying, machining and other process operations, or from the movement of dusty materials.

Specific dust-producing operations include sandblasting, rock drilling, jack hammering, stone cutting, sawing, chipping, grinding, polishing, breaking of sand moulds, "shake-out", cleaning foundry castings, use of abrasives, plus all the powder and granule handling operations such as weighing and mixing and transferring dusty raw materials and products.

Appendix 1.5: Common Properties of Unpolluted Air at Different Temperatures

Temperature -t- (°C)	Density -- (kg/m³)	Specific heat capacity - Cp - (kJ/kg.K)	Thermal conductivity -t- (W/m.K)	Kinematic viscosity -v- x 10^6 (m²/s)	Expansion coefficient -b- x 10^3 (1/K)	Prandtl number -Pr-
-150	2.793	1.026	0.0116	3.08	8.21	0.76
-100	1.980	1.009	0.0160	5.95	5.82	0.74
-50	1.534	1.005	0.0204	9.55	4.51	0.725
0	1.293	1.005	0.0243	13.30	3.67	0.715
20	1.205	1.005	0.0257	15.11	3.43	0.713
40	1.127	1.005	0.0271	16.97	3.20	0.711
60	1.067	1.009	0.0285	18.90	3.00	0.709
80	1.000	1.009	0.0299	20.94	2.83	0.708
100	0.946	1.009	0.0314	.23.06	2.68	0.703
120	0.898	1.013	0.0328	25.23	2.55	0.70
140	0.854	1.013	0.0343	27.55	2.43	0.695
160	0.815	1.017	0.0358	29.85	2.32	0.69
180	0.779	1.022	0.0372	32 29	2.21	0.69
200	0.746	1.026	0.0386	34.63	2.11	0.685
250	0.675	1.034	0.0421	41.17	1.91	0.68
300	0.616	1.047	0.0454	47.85	1.75	0.68
350	0.566	1.055	0.0485	55.05	1.61	0.68
400	0.524	1.068	0.0515	62.53	1.49	0.68

Appendix 1.6: Environmental Fate of Pollutants

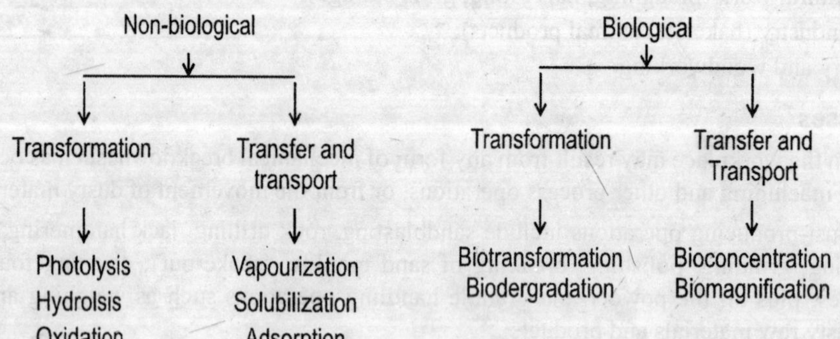

Appendix 1.7: Depiction of Relationship Between Energy and Environment

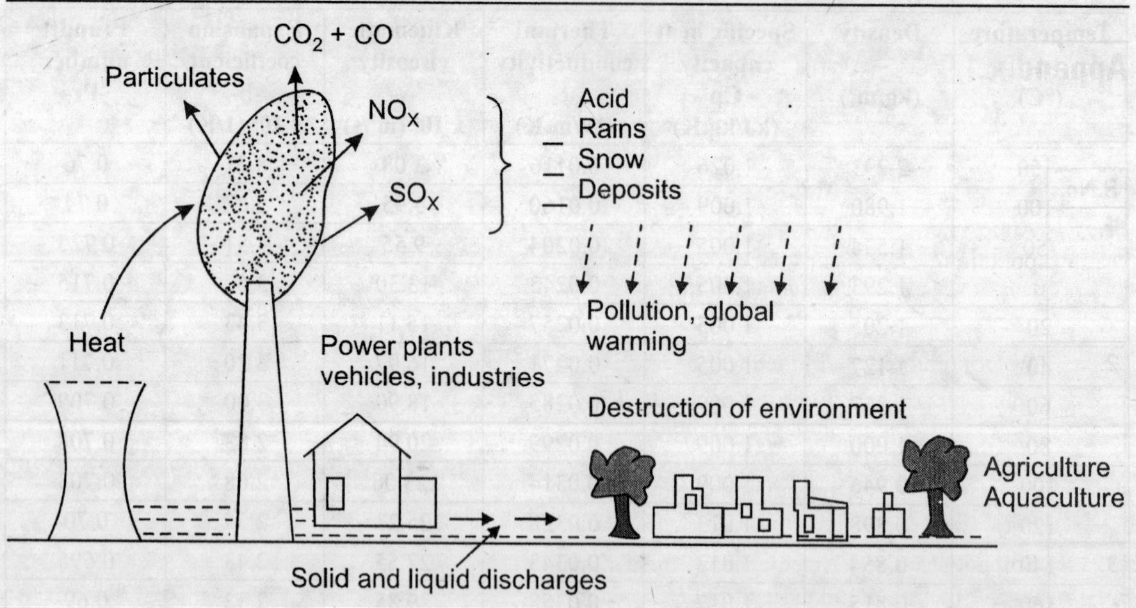

Appendix 1.8: List of Pollutants released from Energy Conversion Plants

S.No.	Plants	Pollutants released
1.	Fossil fuel plants	Sulphur oxides (SO_x)
		Nitrogen oxides (NO_x)
		Carbon dioxide (CO_2)
		Carbon monoxide (CO)
		Particulate matter (PM)
		Thermal pollution
2.	Nuclear power plants	Radioactive emission
		Radioactive wastes
		Thermal pollution
3.	Motor Vehicles	Similar to fossil fuel plants

Appendix 1.9: Seven Categories of Environmental Pollution: Their Causes, Symptoms and Examples

S.No.	Category	Major causes of pollution	Major symptoms	Examples
1.	Atmospheric pollution	Smoke, dust, exhaust fumes, toxic substances (such as sulfur dioxide and nitrogen dioxide)	Asthma, bronchitis	Photochemical smog, "Yokkaichi Asthma"
2.	Water pollution	Polluted waste water, waste fluids (such as petroleum), sludge, household sewage, sewage discharge, general waste, agricultural chemicals	Noxious odors, poisoning	Minamata Disease, "Itai-Itai" Disease (cadmium poisoning), PCB poisoning
3.	Soil pollution	Arsenic, heavy metals (especially in agricultural chemicals)	---	Poisoning
4.	Noise Pollution	Factories, construction work, road traffic, trains and aircraft, late-night commercial operations, advertising	Headaches, insomnia, depression, hearing loss, impaired development	Osaka Airport noise
5.	Vibration	Factories, construction work, road traffic, trains and aircraft	Dizziness, discomfort, structural damage to homes	Shinkansen (bullet train) vibration

6.	Ground subsidence	Upswelling of groundwater, gravel quarrying, coal mining	Structural damage to buildings	Koto Ward, Tokyo
7.	Noxious odors	Exhaust fumes, river contamination, sanitation facilities, accumulated sewage, livestock farms, etc.	Headaches, discomfort	Sewage in the Sumida River

Appendix 1.10: Sources (Natural and Anthropogenic), Physical Properties and Global Background Concentration Levels of Important Air Pollutants

S.No.	Air pollutants	Physical properties	Concentration levels[a]	Anthropogenic sources	Natural sources
1.	SO_2 (Sulfur dioxide)	Colorless gas with irritating, pungent odor; detectable by taste at levels of 0.3 to 1 ppm; highly soluble In water (10.5 g/100 cm^3 at 293 K)	Global background concentration levels in the range 0.04 to 6 ppb; hourly averaged maximum concentrations in urban areas have occasionally exceeded 1 ppm	Fuel combustion in stationary sources; industrial process emissions; metal and petroleum refining	Atmospheric oxidation of organic sulfides
2.	H_2S (Hydrogen Sulphide)	Colorless, flammable gas; highly toxic; characteristic rotten egg odor	Global background about 3 μg m^{-3}; urban levels have been observed as large as 390 μg m^{-3}	Kraft pulp mills; natural gas and petroleum refining; rayon and nylon manufacture; coke ovens	Biological decay processes; volcanoes and geothermal activities
3.	NO (Nitric oxide)	Colorless, odorless gas; nonflammable and slightly soluble in water; toxic	Global background level from 10 to 100 ppt; urban levels have been observed as large as 500 ppb	Combustion	Bacterial action; natural combustion processes; lightning
4.	NO_2 (Nitrogen dioxide)	Reddish-orange-brown gas with sharp, pungent odor; toxic and highly corrosive; absorbs light over much of the visible spectrum	Global background level from 10 to 500 ppt; urban concentrations have reached values exceeding 500 ppb	Combustion	

S.No.	Air pollutants	Physical properties	Concentration levels[a]	Anthropogenic sources	Natural sources
5.	NH_3 (Ammonia)	Colorless gas with pungent odor; detectable at concentrations exceeding 500 ppm; highly soluble in water	Global background level of 1 ppb; urban concentrations in range of 5 ppb	Combustion	Bacterial decomposition of amino acids in organic waste
6.	CO_2 (Carbon dioxide)	Colorless, odorless, nontoxic gas moderately soluble in water	Global background concentration has increased from 290 ppm in 1900 to about 345 ppm in 1985	Combustion of fossil fuels	-
7.	CO (Carbon monoxide)	Colorless, odorless, flammable, toxic gas, slightly soluble in water	Global average concentration of 0.09 ppm; concentrations in northern hemisphere are about twice those in southern hemisphere; urban levels in the vicinity of heavily traveled roadways can exceed 100 ppm	Combustion of fossil fuels	Atmospheric oxidation of methane and other biogenic hydrocarbons
8.	O_3 (Ozone)	Colorless, toxic gas, slightly soluble in water	Global background concentrations range from 20 to 60 ppb; polluted urban levels range from 100 to 500 ppb	No primary sources; formed as secondary pollutant from atmospheric reactions involving hydrocarbons and oxides of nitrogen	Natural tropospheric chemistry; transport from stratosphere to troposphere
9.	Nonmethane hydrocarbons (NMHC)	-	Global background concentrations range from 10 to 20 ppb; polluted urban levels range from 500 to 1200 ppb	Incomplete combustion; industrial sources	Vegetation

Appendix 1.11: Global Environmental Pollutants Circuit

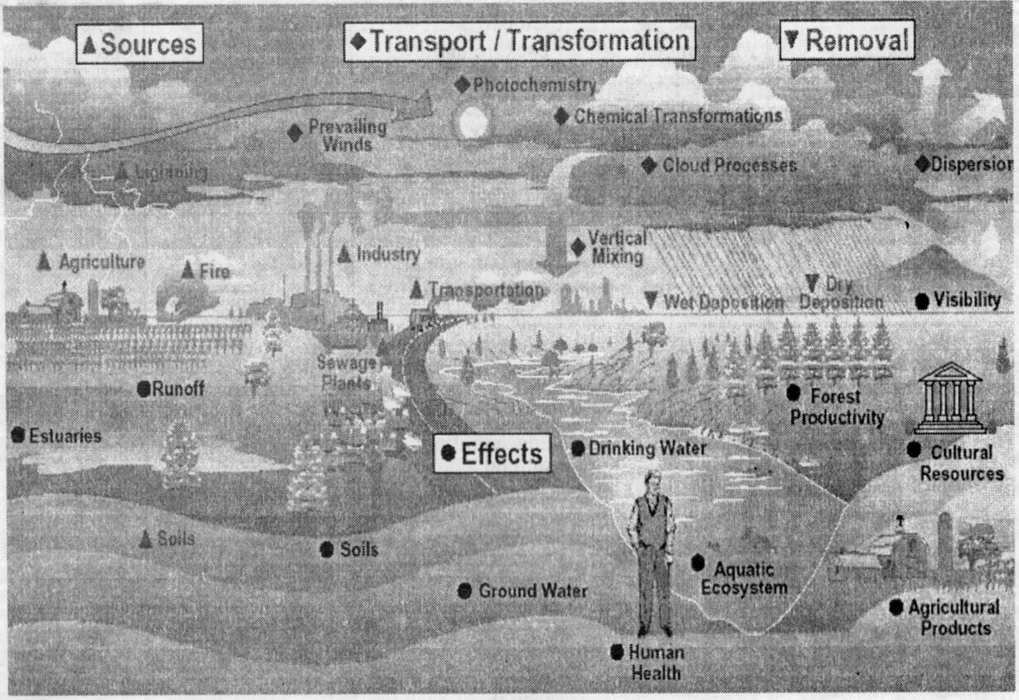

NOTES

The above circuit incorporates production of pollutants by different sources, their transport and transformation in atmosphere, their removal and finally left out pollutants reaching the earth having different effects on different biocoenosis and biotopes (air, soil, water) components of ecosystems.

Appendix 1.12: Type of Environmental Pollutants, Origin and their Effects on Global Level

Name of pollutants	Origin of pollutants	Effects of pollutants
Natural sources: Sulphur, chlorine, and ash particulates, smoke and carbon monoxide methane volatile organic compounds (VOC_S) Aerosol from deforestation and burning: CO, CO_2, NO, NO_2, N_2O, NH_3	Volcanoes, wildfires, cattle and other animals, pine trees	• acid rain, • somg, • respiratory irritant • increased respiratory • diseases • dmage cell membranes of plants The effects are high only for volcanoes.

Anthropic/anthropogenic sources:	**Industry:**	
Carbon monoxide, carbon dioxide, sulphur dioxide, nitrogen oxide, fluorides and substances with fluorine, chlorine (Cl_2), bromine (Br_2) and iodine (I_2), small dust particles, VOC, methane, ammonia and radioactive radiation.	The mining industry, oil and natural gas energy industry based on fossil fuels – coal, oil, natural gas, the production of brick, tile, enamel frit, ceramics, and glass; the manufacture of aluminium and steel; and the production of hydrofluoric acid, phosphate chemicals and fertilizers.	• respiratory irritant, • acid rain, • smog, • increased respiratory • formation of secondary pollutants (PAN, O_3) • effect on soil fertilizer • Respiratory diseases • Toxic effects on living cells • Greenhouse gas effect
CO, CO_2, NO, NO_2, NH_3, CH_4, SO_2, oxides of heavy metals, H_2SO_4, SPM, HC, VOC, background aerosols: seasalt oxidation of sulphur containing gases, same organics, nitrous oxide (N_2O) pesticides	Central heating, chemical and metallurgical industry, engineering internal combustion machinery industry, industrial waste, noises Agriculture: The vegetation fire, the denitrification process, in soils excessively fertilized and excessive use the pesticides, paddy field, intensive husbandry, deforestation	• Toxic effects • Carcinogenic proprieties • Accumulation in tissues • Blocking of different processes • Stratospheric ozone depletion • Formation of secondary pollutants (PAN, O_3) • Effect on soil fertilizer • Respiratory diseases • Greenhouse gas effect
Aerosols from transport and constructions NO_X, CO, HCl, Lead and other heavy metals, SPM	The motor vehicle pollution, noises	• Toxic effects • Acid rain, • Stratospheric ozone depletion • Smog • Increased respiratory diseases
Domestic aerosols CFC, HC, FC H_2S, CH_4, CO_2	Sewage plans, authorized landfill site	• Damage cell membranes of plants • Carcinogenic proprieties • Accumulation in tissues • Blocking of different processes • Stratospheric ozone depletion • Carcinogenic proprieties • Accumulation in tissues • Blocking of different processes • Stratospheric ozone depletion

Appendix 1.13: Human Exposure Pathways of Environmental Pollutants

We come in contact with environmental contaminants by

- Ingestion (water, food, pharmaceuticals, hand-to-mouth)
- Inhalation
- Dermal (skin) uptake
- Ocular (eyes)
- Hand to body

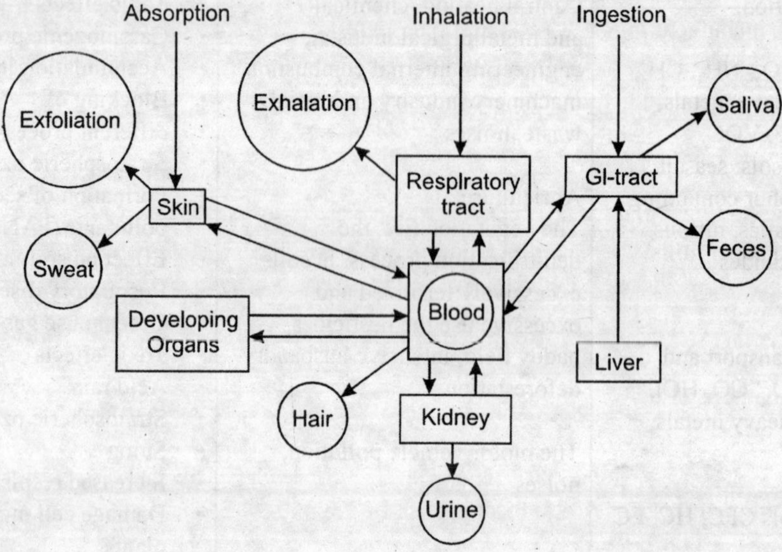

Appendix 1.14: Various Industrial Sectors Contribution to Air Pollution, Water Pollution and to Workplace Hazards

S.No.	Industrial sectors	Air pollution caused by	Water pollution caused by	Work Place hazards
1.	Base metal and iron ore mining	PM	Toxic metal sludge	Silica
2.	Cement manufacturing	PM	Sludge	Silica
3.	Coal mining and production	PM, coal dust	Sludge Coal dust	Silica
4.	Copper smelting	Arsenic	Arsenic	Arsenic, cadmium
5.	Electricity generation	PM, SO_2	Hot water	SO_2
6.	Foundries	PM	Solvents	Silica, solvents

S.No.	Industrial sectors	Air pollution caused by	Water pollution caused by	Work Place hazards
7.	Iron and steel smelting	PM	Sludge	Carbon monoxide, nickel
8.	Lead and zinc smelting	PM, SO_2, lead, cadmium, Arsenic	Lead, cadmium, arsenic	PM, SO_2, lead cadmium, arsenic
9.	Meat processing and rendering	Odour	High biological oxygen demand	Infections
10.	Oil and gas development	SO_2, carcinogens	Oil	Hydrocarbons
11.	Pesticide manufacturing	Pesticides and toxic Intermediates	Pesticides and toxic intermediates	Pesticides and toxic intermediates
12.	Petrochemicals manufacturing	SO_2	Oil	Hydrocarbons
13.	Petroleum refining	SO_2	Sludge, hydrocarbons	Hydrocarbons
14.	Phosphate fertilizer plants	PM	Nutrients	
15.	Pulp and paper mills	Odour	High biological oxygen demand, mercury	Chlorine
16.	Tanning and leather finishing	Odour	Chromium, acids	Chromium acids
17.	Textile manufacturing		Toxic dyes	

NOTES

In all the cases, the workplaces are subject to risk of injury, noise, dust, and excessively hot or cold temperatures.

Appendix 1.15: Recycling as a tool for Environmental Pollution Control

Recycling is the collection, processing, and reuse of materials that would otherwise be thrown away. Materials ranging from precious metals to broken glass, from old newspapers to plastic spoons, can be recycled. The recycling process reclaims the original material and uses it in producing new products.

Recycling can be done internally (within a company) or externally (after a product is sold and used).

Internal recycling occurs in the paper industry, for example, when leftover stock and trimmings are salvaged to help make more new product. Since the recovered material never left the manufacturing plant, the final product is said to contain preconsumer waste.

External recycling occurs when materials used by the customer are returned for processing into new products. Materials ready to be recycled in this manner, such as empty beverage containers, are called post consumer waste.

Types of Materials Recycled

Just about any material can be recycled. On an industrial scale, the most commonly recycled materials are those that are used in large quantities–metals such as steel and aluminum, plastics, paper, glass, and certain chemicals.

Steel

There are two methods of making steel using recycled material: the *basic oxygen furnace* (BOF) method and the *electric arc furnace* (EAF) method. The BOF method involves mixing molten scrap steel in a furnace with new steel. Steel made by the BOF method typically is used to make sheet-steel products like cans, automobiles, and appliances. The EAF method normally uses 100 percent recycled steel. Scrap steel is placed in a furnace and melted by electricity that arcs between two carbon electrodes. Limestone and other materials are added to the molten steel to remove impurities. Steel produced by the EAF method usually is formed into beams, reinforcing bars, and thick plate. Approximately 65 percent of all steel is recycled; making it one of the world's most recycled materials.

Aluminum

The overall recycling rate of all forms of aluminum is about 35 percent. Cans brought to collection centers are crushed, baled, and shipped to regional mills or reclamation plants. The cans are then shredded to reduce volume and heated to remove coatings and moisture. Next, they are put into a furnace, melted, and formed into ingots, or bars. The ingots go to another mill to be rolled into sheets. The sheets are sent to a container plant and cut into disks from which new cans are formed. The cans are printed with the beverage makers' logos and are shipped (with tops separate) to the filling plant.

Plastics

Plastics are more difficult to recycle than metal, paper, or glass. One of the problems is that any of the seven categories of plastics can be used for containers alone. For effective recycling, the different types cannot be mixed. Plastic containers should have identification codes so that they can be more easily identified and separated. The code assigns a particular number to each of the seven plastics used in packaging. Number 1 refers to polyethylene teraphthalate (PET) and number 2 refers to high-density polyethylene (HDPE). PET can be made into carpet, or fiberfill for ski jackets and clothing. HDPE can be recycled into construction fencing, landfill liners, and a variety of other products. Plastics coded with the number 6 are polystyrene (PS), which can be recycled into cafeteria trays, combs, and other items.

The recycling process for plastic normally involves cleaning it, shredding it into flakes, then melting the flakes into pellets. The pellets are then melted into a final product. Some products work best with only a small percentage of recycled content. Other products, such as HDPE plastic milk cases, can be made successfully with 100 percent recycled content. The plastic container industry has concentrated on weight reduction and source reduction.

Paper and Paper Products

Paper products that can be recycled include cardboard containers, wrapping paper, and office paper. The most commonly recycled paper product is newsprint.

In newspaper recycling, old newspapers are collected and searched for contaminants such as plastic bags and aluminum foil. The paper goes to a processing plant where it is mixed with hot water and turned into pulp in a machine that works much like a big kitchen blender. The pulp is screened and filtered to remove smaller contaminants. The pulp then goes to a large vat where the ink separates from the paper fibers and floats to the surface. The ink is skimmed off, dried and reused as ink or burned as boiler fuel. The cleaned pulp is mixed with new wood fibers to be made into paper again.

Glass

Scrap glass taken from the glass manufacturing process, called cullet, has been internally recycled for years. The scrap glass is economical to use as a raw material because it melts at lower temperatures than other raw materials, thus saving fuel and operating costs.

Glass that is to be recycled must be relatively free from impurities and sorted by color. Glass containers are the most commonly recycled form of glass, and their colors are flint (clear), amber (brown), and green. Other glass, such as window glass, pottery, and cooking utensils, are considered contaminants because they have different compositions from glass used in containers. The recycled glass is melted in a furnace and formed into new products.

Chemicals and Hazardous Waste

Household hazardous wastes include drain cleaners, oven cleaners, window cleaners, disinfectants, motor oil, paints, paint thinners, and pesticides. Most municipalities ban hazardous waste from the regular trash. Periodically, citizens are alerted that they can take their hazardous waste to a collection point where trained workers sort it, recycle what they can, and package the remainder in special leak-proof containers for safe disposal. Typical materials recycled from the collection drives are motor oil, paint, antifreeze, and tires.

Although large quantities of chemical solvents are used in cleaning processes, technology has been developed to clean and reuse solvents that used to be discarded. Even the vapors evaporated from the process are recovered and put back into the recycled solvent.

Nuclear Waste

Certain types of nuclear waste can be recycled, while other types are considered too dangerous to recycle. Low-level wastes include radioactive material from research activities, medical wastes, and contaminated machinery from nuclear reactors. Nickel is the major metal of construction in the nuclear power field and much of it is recycled after surface contamination has been removed.

High-level wastes come from the reprocessing of spent fuel (partially depleted reactor fuel) and from the processing of nuclear weapons. These wastes emit gamma radiation, which can cause birth defects, disease, and death. High-level nuclear waste is so toxic it is not normally recycled. Instead, it is fused into inert glass tubes encased in stainless steel cylinders, which are then stored underground.

Reasons for Recycling

Rare materials, such as gold and silver, are recycled because acquiring new supplies is expensive. Other materials may not be as expensive to replace, but they are recycled to conserve energy, reduce pollution, conserve land, and to save money.

Resource Conservation

Recycling conserves natural resources by reducing the need for new material. Some natural resources are renewable, meaning they can be replaced, and some are not Paper, corrugated board, and other paper products come from renewable timber sources. Trees harvested to make those products can be replaced by growing more trees. Iron and aluminum come from nonrenewable ore deposits. Once a deposit is mined, it cannot be replaced.

Energy Conservation

Recycling saves energy by reducing the need to process new material, which usually requires more energy than the recycling process. To make aluminum can from recycled metal takes only 5 percent of the total energy needed to produce the same aluminum can from unrecycled materials, a 95 percent energy savings. Recycled paper and paperboard require 75 percent less energy to produce than new products. Significant energy savings result in the recycling of steel and glass, as well.

Pollution Reduction

Recycling reduces pollution because recycling a product creates less pollution than producing a new one. For every ton of newspaper recycled, 7 kg of air pollutants are pumped into the atmosphere. Recycling can also reduce pollution by recycling safer products to replace those that pollute. Some countries still use chlorofluorocarbons (CFCs) (to manufacture foam products such as cups and plates) harm the atmosphere's protective layer of ozone. Using recycled plastic instead for those products eliminates the creation of harmful CFCs.

Land Conservation

Recycling saves valuable landfill space, land that must be set aside for dumping trash, construction debris, and yard waste. Landfills fill up quickly and acceptable sites for new ones are difficult to find because of objections by neighbours to noise and smells, and the hazard of leaks into underground water supplies. The two major ways to reduce the need for new landfills are to generate less initial waste and to recycle products that would normally be considered waste. Solid waste can also be burned instead of buried in the ground.

Economic Savings

Recycling in the short term is not always economically profitable or a break-even financial operation. The economic consequences of recycling are positive in the long term. Recycling will save money, if potential landfill sites are used for more productive purposes and by reducing the number of pollution-related illnesses.

In general, using recycled materials to make new products costs less and requires less energy than using new materials and recycling can also reduce pollution, either by reducing the demand for high-pollution alternatives or by minimizing the amount of pollution produced during the manufacturing process. Recycling decreases the amount of land needed for trash dumps by reducing the volume of discarded waste.

In nutshell

- Recycling is the collection, processing, and reuse of materials that would otherwise be thrown away.
- The most commonly recycled materials are those that are used in large quantities.
- Recycling conserves natural resources by reducing the need for new material.
- Recycling saves energy by reducing the need to process new material, which usually requires more energy than the recycling process.
- Recycling reduces pollution because recycling a product creates less pollution than producing a new one.
- Recycling saves valuable landfill space, land that must be set aside for dumping trash, construction debris, and yard waste.

Recycling Fact Sheet

Energy savings are a very important environmental benefit of recycling, since using energy requires the consumption of scarce fossil fuels and involves emissions of numerous air *and* water pollutants. The steps in supplying recycled materials to industry (including collection, processing and transportation) typically use less energy than the steps in supplying virgin materials to industry (including extraction, refining, transportation and processing).

Additional energy savings associated with recycling accrue in the manufacturing process itself, since the materials have already undergone processing.

Recycling paper cuts energy usage to half. Recycling 1 ton of paper saves 380 gallons of oil. Recycling one aluminum can saves the energy equivalent of one cup of gasoline. Recycling used aluminum cans requires only about five percent of the energy needed to produce aluminum from bauxite. Enough energy is saved by recycling steel. Recycling one ton of glass saves the equivalent of nine gallons of fuel oil.

Benefits of Recycling

Good For Our Economy

Companies should rely on recycling programs to provide the raw materials they need to make new products.

Creates Jobs

Recycling is an industry because recycling and reuse enterprises employ workers nationwide.

Reduces Waste

People discards lot of garbage every day. Most of this garbage goes into to landfills, where it's compacted and buried.

Good for The Environment

Recycling requires far less energy, uses fewer natural resources, and keeps waste from piling up in landfills.

Saves Energy

Recycling offers significant energy savings over manufacturing with virgin materials. (Manufacturing with recycled aluminum cans uses 90% less energy.)

Preserves Landfill Space

No one wants to live next door to a landfill. Recycling preserves existing landfill space.

Prevents Global Warming

In 2015, recycling of solid waste globally prevented the release of 45 million metric tons of carbon equivalent (MMTCE, the unit of measure for greenhouse gases) into the air.

Reduces Water Pollution

Making goods from recycled materials generates far less water pollution than manufacturing from virgin materials.

Protects Wildlife

Using recycled materials reduces the need to damage forests, wetlands, rivers and other places essential to wildlife.

Creates New Demand

Recycling and buying recycled products creates demand for more recycled products, decreasing waste and helping our economy.

Common Recycling Materials

Aluminum Cans

- Recycling aluminum reduces energy use by 90%.
- Energy saved from recycling one aluminum can will run a TV for 3 hours.
- Recycling one aluminum can saves the energy equivalent of one cup of gasoline.
- Recycling aluminum reduces air pollution by 95%.

Paper

- Recycling 1 ton of paper saves 17 trees, 7, 000 gallons of water, 380 gallons of oil, three cubic yards of landfill space, and 4000 kilowatts of energy. This represents a 64% energy savings, a 58% water savings, and 25 kg. less of air pollution.
- Recycling paper reduces air pollution by 75% and water pollution by 35%.
- Paper products use up at least 35% of the worlds commercial wood harvest.
- More than 1/3 of all fiber used to make paper comes from recycled paper.
- The 17 trees saved (above) can absorb a total of 250 pounds of carbon dioxide from the air each year. Burning that same ton of paper would *create* 1500 pounds of carbon dioxide.
- The construction costs of a paper mill designed to use waste paper is 50 to 80% less than the cost of a mill using new pulp.

Steel

- Annually, enough electrical energy is saved by recycling steel.
- Recycling one kg of steel saves enough energy to light a 100 watt light bulb for 30 hours.
- Recycling one ton of steel saves 1400kg of iron ore, 600kg of coal and 20kg of limestone.
- Recycling steel saves 40% of water used to make steel from ores.
- Recycling steel reduces air pollution by and water pollution by 75%.

Glass

- The energy saved from recycling one glass bottle will operate a 100-watt light bulb for four hours.
- Recycling one ton of glass saves the equivalent of nine gallons of fuel oil.
- Recycling glass reduces air pollution by 15-20%.
- Recycling glass saves 25-30% of the energy used to make glass from virgin materials.

Plastic

- It takes 5 recycled 2-liter plastic(PET) bottles to make one square foot of carpet.
- PET bottles and containers are actually a form of polyester, which is why it is so easy to recycle bottles into T-shirts, sweaters, and socks.
- It takes 35 2-liter recycled PET bottles to make the soft filling inside a sleeping bag, called "fiberfill"
- About 1, 200 soda plastic bottles could carpet the average living room.
- Plastic makes up tesser percent of trash by weight, but is greater % by the volume. volume. volvolume. *
- Recycling plastic saves twice as much energy as burning it in an incinerator.

Trash and Landfills

- The U.S. is the no. 1 trash- producing country in the world at 650 kg per person per year. This means that 5% of the world's people generate 40% of the world's waste.
- Recycling one kg of steel saves enough energy to light a 100 watt light bulb for 30 hours.
- Recycling one ton of steel saves 1400kg of iron ore, 600kg of coal and 20kg of limestone.
- Recycling steel saves 40% of water used to make steel from ores.
- Recycling steel reduces air pollution by and water pollution by 75%.

Glass

The energy saved from recycling one glass bottle will operate a 100-watt light bulb for four hours.

Recycling one ton of glass saves the equivalent of nine gallons of fuel oil. Recycling glass reduces air pollution by 15-20%.

Recycling glass saves 25-30% of the energy used to make glass from virgin materials.

Plastic

It takes 5 recycled 2-liter plastic(PET) bottles to make one square foot of carpet. PET bottles and containers are actually a form of polyester, which is why it is so easy to recycle bottles into T-shirts, sweaters, and socks.

- It takes 35 2-liter recycled PET bottles to make the soft filling inside a sleeping bag, called "fiberfill"
- About 1, 200 soda plastic bottles could carpet the average living room.
- Plastic makes up tesser percent of trash by weight, but is greater % by the volume. volume. volvolume. *

Recycling plastic saves twice as much energy as burning it in an incinerator.

Trash and Landfills

- The U.S. is the no. 1 trash- producing country in the world at 650 kg per person per year. This means that 5% of the world's people generate 40% of the world's waste.
- Between 5 and 15% of what we throw away contains hazardous substances.
- Packaging represents about 60% of household trash.
- On average, costs per ton to recycle trash is less compares to sending it to the landfill.

Appendix 1.16: Sources of Common Environmental Contaminants and their Health Effects

S.No.	Common Sources	Contaminants	Potential Health Effects
1.	Household Items, such as Batteries, Thermometers, and Paints	mercury	Toxic to kidneys. Can cause eye and skin irritation; chest pain; tremor; fatigue; weakness.
2.	Car Radiators and De-icing Agents	ethylene glycol	Can cause abdominal pain; vomiting; weakness; dizziness; central nervous system depression.
3.	Photocopy Machines	chromium	Toxic to kidneys; potential human carcinogen.
4.	Dry Cleaning Agents and Degreasers	trichloroethane and trichloroethylene	Central nervous system depression: decreased alertness, headaches, sleepiness, loss of consciousness. Kidney changes: decreased urine flow, swelling (especially around eyes), anemia. Liver changes: fatigue, malaise, dark urine, liver enlargement.
5.	Herbicides for Vegetation Control	chlorophenoxy compounds; 2;4-dichlorophenoxyacetic acid	Chloracne, weakness or numbness of arms and legs, long-term nerve damage.
		dioxin	Dioxin causes chloracne and may aggravate pre-existing liver and kidney disease.
6.	Pesticides	chlorinated ethanes; DDT; lindane	Acute symptoms of apprehension, irritability, dizziness, disturbed equilibrium, tremor, and convulsions.
		Cyclodienes (aldrin; chlordane; dieldrin; endrin); chlorocyclohexanes	
		Organophosphate: diazanon; dichlorovos; dimethoate; trichlorfon; malathion; methyl parathion; parathion **carbamate:** aldicarb; baygon; zectran	Acute symptoms of apprehension, irritability, dizziness, disturbed equilibrium, tremor, and convulsions. Liver toxicity and permanent kidney damage. Chlorocyclohexanes can cause anemia.

S.No.	Common Sources	Contaminants	Potential Health Effects
7.	Electrical Transformers and Other Industrial Uses	polychlorinated biphenyls (PCBs)	All cause a chain of internal reactions leading to neuromuscular blockage. Acute symptoms include headaches, fatigue, dizziness, increased salivation and crying, profuse sweating, nausea, vomiting, cramps, diarrhea, tightness in the chest, and muscle twitching.
8.	Commercial Solvents	benzene; ethyl benzene; toluene; xylene	Various skin ailments, including chloracene. May cause liver toxicity. Carcinogenic to animals.
		carbon tetrachloride; chloroform; ethyl bromide; ethyl chloride; ethylene dibromide; ethylene dichloride; methyl chloride; methyl chloroform; methylene chloride; tetrachloroethane; tetrachloroethylene; trichloroethylene; vinyl chloride	Benzene suppresses bone marrow function, causing blood changes; chronic exposure can cause leukemia. Central nervous system depression: decreased alertness, headaches, sleepiness, loss of consciousness. Defatting dermatitis.
9.	Various Commercial and Industrial Manufacturing Processes	arsenic; beryllium; cadmium; chromium; lead; mercury	Central nervous system depression: decreased alertness, headaches, sleepiness, loss of consciousness. Kidney changes: decreased urine flow, swelling (especially around eyes), anemia. Liver changes: fatigue, malaise, dark urine, liver enlargement, jaundice.
			brain damage. Cadmium can cause kidney and lung disease. Chromium, beryllium, arsenic, and cadmium have been implicated as human carcinogens.
		PCBs (Polychlorobiphenyls)	Various skin ailments, including chlor acne; may cause liver toxicity; carcinogenic to animals.

S.No.	Common Sources	Contaminants	Potential Health Effects
10.	Chemical Manufacturing	benzene; ethyl benzene; toluene; xylene	Benzene suppresses bone marrow function causing blood changes; chronic exposure can cause leukemia. Central nervous system depression: decreased alertness, headaches, sleepiness, loss of consciousness. Defatting dermatitis.
11.	Steel and Glass Manufacturing	chromium; lead; mercury	All are toxic to kidneys. Lead causes decreased mental ability, weakness, headache, abdominal cramps, diarrhea, and anemia. Also affects blood-forming mechanisms and the peripheral nervous system. Long-term exposure to lead can cause permanent kidney and brain damage. Chromium has been implicated as a human carcinogen.
12.	Chrome Plating Operations	chromium	Toxic to kidneys; potential human carcinogen.

Appendix 1.17: Environmental Benefits of a Tree

Benefits from a tree of 50 tons over a period of 50 years	
Other benefits of trees	**World's forest statistics**
Oxygen production	Myanmer = 68%
Conservation of mineral proteins	Japan = 62 %
Control of soil erosion	Sweden = 55.5 %
Recycling of water and control of humidity	USA = 33 %
Shelter to birds, insects etc.	India = 20 %
Control of air pollution	For ecological balance – minimum required forest is 30%

Trees "breathe", or respirate, and exchange gases similar to the way humans do, yet the methods and results are quite different. Trees intake gases through stomatos, or holes, on their leaves; these gaseous air pollutants. Once inside the leaf, gases diffuse into the spaces between the cells of the leaf to be absorbed by water films or chemically altered by plant tissues. Trees also reduce air pollution by intercepting airbone particles and retaining them on the leaf surface itself, although most remain on the plant surface.

Leaf surfaces are most efficient at removing pollutants that are watersoluble including sulfur dioxide, nitrogen and ozone. Pollutant removal rates are highest when vegetative surfaces are wet or damp; these conditions can increase removal rates ten-fold because the entire trees surface (leaves, twigs, trunk, and branches) is available for pollutant uptake. Because of the drier climate, trees of the front Range are not a efficient at removing airborne chemicals as trees in a humid climate.

Pollutants travel through plants by translocation via the xylem and phloem. Xylem is responsible for bringing minerals and water from the roots to the foliage, while phloem transport sugars and other dissolved food from the foliage to all non- photosynthesis plant cells. Chemical pollutants absorbed by the leaves are translocated to the roots areas where they can be broken down and translocated to the leaves where they are released into the atmosphere.

Soils are also active in removing gaseous pollutants from the air; they have a significant ability to assimilate and convert these gases in or on the soil through microbial, physical and chemical processes. Healthy soil is most efficient at the process. During the process of photosynthesis, trees leaves absorb airborne pollutants and translocate them along with carbohydrates to the root zone, where microbes are responsible for breaking them down.

Trees are also responsible for reducing summer time high temperatures, and therefore slowing the rates at which many air pollutants are formed and volatized. Reduction of air temperatures in summer time help to reduce pollution in three ways:

1 By reducing chemical reaction rates in the atmosphere that result in Ozone formation. Lower temperatures reduce ozone-precursor emission rates, thus influencing ozone formation.
2. By decreasing temperature-dependant emissions of hydrocarbons from both natural and manmade sources.
3. By decreasing emissions of pollutants form electric is also extremely efficient at removing the toxic chemicals benzene and formaldehyde from the air conditioning demands.

In addition to all of the above, vegetation is also extremely efficient at removing the toxic chemicals benzene and fomaldehyde from the air.

Interception of particles by vegetation has been shown to be much greater for street due to their proximity to high intensities of road traffic.

Role of trees in improving air quality by reducing temperature is shown below dramatically:

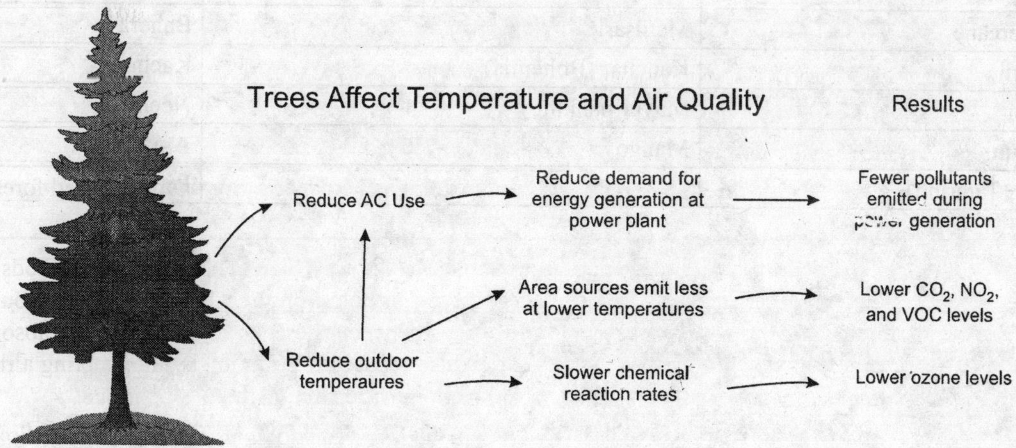

NOTES

- Well-maintained trees and shrubs can increase property value.
- Trees properly placed around buildings can reduce air conditioning needs by 30%
- A mature tree removes almost 70 times more pollution than a newly planted tree.
- The shade and wind buffering provided by trees reduces annual heating and cooling Costs.
- Each average-sized tree provides savings in annual environmental benefits, including energy conservation and reduced pollution.
- A single tree produces approximately 120 kgs of oxygen per year. That means two mature trees can supply enough oxygen annually to support a family of four
- Water originating in our forests provide drinking water for communities, and individuals.
- One tree can absorb as much carbon in a year as a car produces while driving 42000 Kms
- Over the course its life, a single tree can absorb one ton of carbon dioxide.
- An average American uses about 340 kgs of paper every year, and 95% of homes are built using wood. That means each person uses the equivalent of one 30 metres tall, 40 cms diameter, tree every year for their paper and wood product needs.
- About one third of the United States of America is covered by forests.
- Threr are about 247 billion trees over 40 cm in diameter in the U.S.A
- The average tree in an urban/city area has a life expectancy of only 8 years.

Trees are the longest living organisms on the planet and one of the earth's greatest natural resources. They keep our air supply clean, reduce noise pollution, improve water quality, help prevent erosion, provide food & building materials, create shade, and help make our landscapes look beautiful.

Appendix 1.18: Environmental Friendly Trees

Trees releasing relatively higher amount of oxygen	Sulphur dioxide absorbing plants	Dust collection trees
Peepal	Bamboo	Teak
Banyon	Imli (Tamrid)	Saal
Sugarcane	Maulsari	Bahera
Neem	Kanchar (Bohemia)	Kachnar
Tulsi	Mayur pankhi	Neem
Jamun	Mango	Amaltas
Imli (Tamrind)	-	Peepal

Appendix 1.19: Categories of Risk identified by the Environmental Protection Agency (EPA) of U.S.A

Cancer risks	Non-cancer health risks		Ecological risks	Welfare risks
	Type of risk	Examples		
Lung	Cardiovascular	Increased heart attacks	**At ecosystem level:-** Direct physical destruction or major alterations	Material damages and soiling
Colon	Developmental	Birth defects	Alterations	-
Breast	Hematopoietic	Impaired heme synthesis	Changes in species richness and diversity	Reduced recreational opportunities
Pancreas	Immunological	Increased infections	Changes in species richness and diversity	Damage to natural resources (eg, crop, timber)
Prostate	Kidney dysfunction	Threats to/loss of rare or endangered species	---	---

Appendix 1.20: Example of People at Work Encountering Various Environmental Stresses

S.No.	Basic classes of environmental stresses	Examples
1.	Physical	Noise, heat, light, vibration, ionizing radiation, pressure, ultra violet light.
2.	Chemical	Exposure of workers to dust, vapor, fumes, gases mists, etc.
3 .	Biological	Insects, mites, yeasts, hormones, bacteria, viruses, proteolytic enzymes
4.	Ergonomic	Man-machine interaction eg, body position in relation to task on machine

Appendix 1.21: Symptoms, Signs and Diagonoses Associated with Exposure to Air Pollutants

Upper respiratory Tract Effects	
Rhinitis Sinusitis Pharyngitis	
Lower respiratory Tract effects	
Cough (wet or dry) Chest tightness/pain Shortness of breath Wheezing	Worsening of the asthama or COPD Bronchitis Pneumonia Lung cancer
Non-respiratory effects	
Eye irritation Fatigue Malaise Dizziness Headache Nausea Skin irritation Fever	Decrease exercise performance Worsening of angina pectoris(cardiac chest pains) Malignancy

Appendix 1.22: List Of Environmental Health Hazards

S.No.	Environmental Health Hazards
1.	Allergens
2.	Anthrax
3.	Antibiotic agents in animals destined for human consumption
4.	Antibiotic resistance
5.	Arbovirus
6.	Arsenic – a contaminant of fresh water sources (water wells)
7.	Asbestos – carcinogenic
8.	Avian influenza
9.	Bovine spongiform encephalopathy (BSE)
10.	Carcinogens
11.	Cholera
12.	Cosmic rays
13.	DDT

S.No.	Environmental Health Hazards
14.	Dioxins
15.	Drought
16.	Dysentery
17.	Electromagnetic fields
18.	Endocrine disruptors
19.	Epidemics
20.	e-waste
21.	Explosive material
22.	Floods
23.	Food poisoning
24.	Fungicides
25.	Furans
26.	Haloalkanes
27.	Heavy metals
28.	Herbicides
29.	Hormones in animals destined for human consumption
30.	Lead in paint
31.	Light pollution
32.	Lighting
33.	Lightning
34.	Malaria
35.	Marine debris
36.	Mercury
37.	Molds
38.	Mutagens
39.	Noise pollution
40.	Onchocerciasis (river blindness)
41.	Pandemics
42.	Pathogens
43.	Pesticides
44.	Pollen for allergic people
45.	Polychlorinated biphenyls
46.	Quicksand
47.	Rabies
48.	Radon and other natural sources of radioactivity

S.No.	Environmental Health Hazards
49.	Severe acute respiratory syndrome (SARS)
50.	Sick building syndrome
51.	Soil pollution
52.	Tobacco smoking
53.	Toxic waste
54.	Ultraviolet light
55.	Vibration
56.	Wildfire
57.	X-rays

Appendix 1.23: Industrial Pollutants and their Related Health Hazards

S.No.	Industrial Pollutants	Heath effects of the pollutants
1.	Nitrogen Oxides(NO_x)	Irritation of respiratory tract
2.	Ozone	Irritation of eye, nose, throat: risk asthmatics, children and those involved in heavy exercise
3.	Hydrocarbons	Drowsiness, Eye irritation, cough
4.	Mercury	Nervous diseases viz. Control power of muscle damage, Mental disability, Paralysis of hands, legs, tongue and lip, deafness
5.	Lead	Obstructs haemoglobin creation and activates Anemia, Damage of brain cells, Abortion, growth of child obstruction
6.	Cadmium	Pain in joint and bones, blood pressure increase, kidney and lung diseases
7.	Arsenic	Anemia, skin disease, cancer of lung, skin and liver
8.	Nitrate and Nitrite	Reduction of haemoglobin, Cancer of liver and intestine damage of foetus of womb
9.	Chromium	Cancer and skin diseases
10.	Fluoride	Fluorosis of bone and teeth
11.	Iron	Constipation, Piles, Gastric

Appendix 1.24: Environmental Effects of Construction Activities

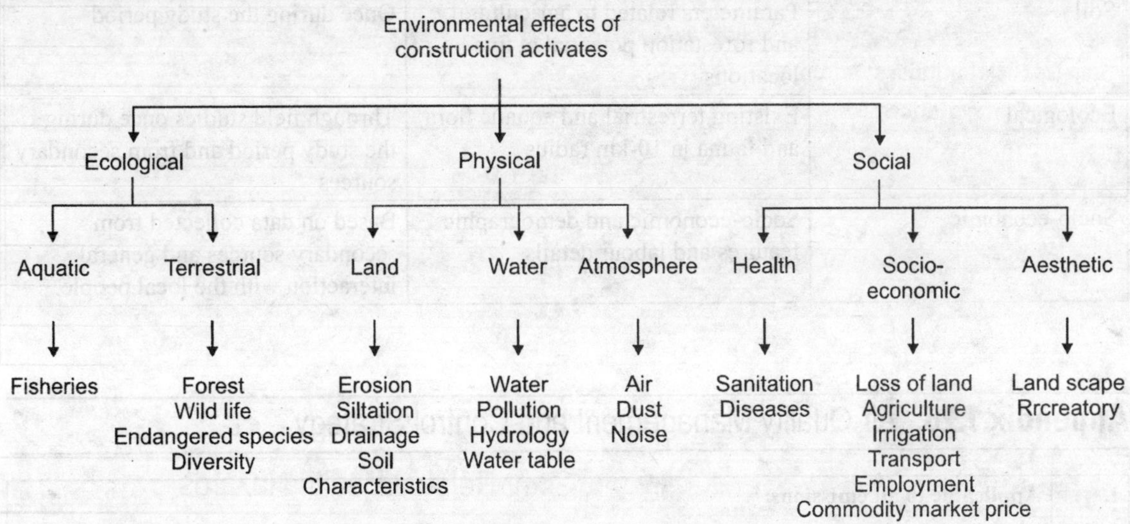

Appendix 1.25: Environmental Attributes and Frequency of Monitoring

Environmental Attributes	Environmental parameters	Monitoring frequency
Meteorology	Wind speed and direction, rainfall, temperature, relative humidity, atmospheric pressure, solar radiation	Continuous hourly recording using automatic weather station. Secondary data collection from IMD station
Ambient Air Quality	SPM, RSPM, SO_2, NO_X	24 hourly samples for SPM, RSPM, SO_2 and NO_2 twice a week at six locations for three months
Water quality	Physical, chemical and bacterio-logical parameter and heavy metals and 8 locations (surface water and ground water)	Grab samples are collected from surface water and ground water locations once during study period
Noise and Vibration levels	Noise and vibration level at dB (A) at 8 different locations	Data monitored continuously for 24 hours using Integrated Sound Level Meter (with automatic data analysis and logging facilities). Hourly Leq, Lmax, Lmin parameter are monitored twice during study, during day and night

Environmental Attributes	Environmental parameters	Monitoring frequency
Soil	Parameters related to agricultural and forestation potential at five locations	Once during the study period
Ecological	Existing terrestrial and aquatic flora and fauna in 10-km radius	Through field studies once during the study period and from secondary sources
Socio-economic	Socio-economic and demographic features and labour details	Based on data collected from secondary sources and general interaction with the local people.

Appendix 1.26: Air Quality Management and Control Strategy

I.	**Applicable to all emissions:** A. Decrease or eliminate production of emission 1. Change specification/design of product 2. Change process temperature, pressure, or cycle 3. Change specification of materials 4. Change the product B. Confine the emissions: 1. Enclose the source of emissions 2. Capture the emissions in an industrial exhaust system C. Separate the contaminant from effluent gas stream: 1. Scrub with liquid
II.	**Applicable specifically to particulate matter emissions:** A. Decrease or eliminate particulate matter production 1. Change to process that does not require blasting, blending, calcining, chipping, crushing, drilling, drying, grinding, milling, polishing, pulverizing, sanding, sawing, spraying, tumbling, etc. 2. Change from solid to liquid or gaseous material 3. Change from dry to wet solid material 4. Change particle size of solid material 5. Change to process that does not require particulate material B. Separate the contaminant from effluent gas stream 1. Gravity separator 2. Centrifugal separator 3. Bag filter 4. Electrostatic precipitator

III.	**Applicable specifically to gaseous emissions:**
	A. Decrease or eliminate gas or vapor production 1. Change to process that does not require annealing, baking, boiling, burning, casting, coating, cooking, dehydrating, dipping, distilling, expelling, galvanizing, melting, pickling, plating, quenching, reducing, rendering, roasting, smelting, etc. 2. Change from liquid or gaseous to solid material 3. Change to process that does not require gaseous or liquid material
	B. Burn the contaminant to CO_2 and H_2O 1. Incinerator 2. Catalytic burner /converter
	C. Adsorb the contaminant by Activated carbon

Appendix 1.27: Human Health Effects Of Some Major Air Pollutants

S.No.	Air pollutant	Body system affected	Health effects
1.	Particulate matters ($PM_{2.5}$ and PM_{10})	Lower respiratory system. Particulate matter less than 10 μm in diameter gets into the lower respiratory system, particularly during exercise	Asthma Bronchitis Reduced lung function Cancer Heavy metal poisoning
2.	Lead (Pb)	Organs and soft tissue. Lead concentrations and accumulates in bones and soft tissue. A high concentration of lead in the body poisons the blood, nervous and renal systems	Anaemia High blood pressure Cancer Neurological disorders Intellectual function
3.	Carbon Monoxide (CO)	Circulatory system. Carbon monoxide interferes with the blood ability to provide an adequate supply of oxygen to body tissue	CO poisoning Angina pectoris Neurological dysfunction Brain damage Fatel abnormalities Asphyxiation
4.	Nitrogen Dioxide (NO_2)	Respiratory system. (NO_2) is very toxic at high levels	(NO_2)poisoning Asthma Lowered resistance to infection
5.	Sulphur Dioxide (SO_2)	Respiratory system. (SO_2) poisoning directly affects the respiratory system and increases airway resistance adding to heart lung load	Asthma Bronchial constriction (SO_2) poisoning Heart attack

S.No.	Air pollutant	Body system affected	Health effects
6.	Ozone (O_3)	Respiratory system. Tropospheric ozone results in irritation of the respiratory system	Lung inflammation Reduced lung elasticity Transient cough Chest pain Throat irritation Nausea

Note:

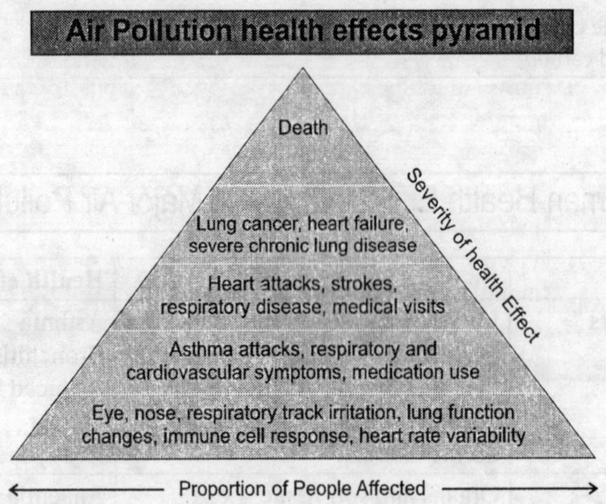

Health effects of air pollution are portrayed as a pyramid, with the mildest and most common effects at the bottom of the pyramid, and the more severe but less frequent effects at the top of the pyramid. The pyramid shows that as severity decreases the number of people affected increases. Exposure to air pollution can affect both the respiratory and the cardiac systems.

Appendix 1.28: Sources and Effects of Some Important Environmental Pollutants on Human Beings & Animals

S.No.	Environmental pollutants	Major Sources of Pollutants	Typical Effects of Pollutants
1.	Carbon monoxide (CO)	Incomplete combustion of fuels, automobile exhausts, jet engine emissions, blast furnaces, mines and tobacco smoking	Toxicity, blood poisoning, increased proneness to accidents, CNS impairments, CO combines with haemoglobin, forming carboxyl haemoglobin, which causes asphyxiation and hence leads to death.

S.No.	Environmental pollutants	Major Sources of Pollutants	Typical Effects of Pollutants
2.	Sulphur dioxide (SO_2)	Combustion of coal, combustion of petroleum products, burning of refuse, petroleum industry, oil refining, power houses, sulphuric acid plants, metallurgical operations and from domestic burning of fuels.	Increased breathing rate and feeling of air-starvation, suffocation, aggravation of asthma and chronic bronchitis, impairment of pulmonary functions respiratory irritation, sensory irritation, irritation of throat and eyes.
3.	Oxides of Nitrogen (NO_x)	Automobile exhausts, coal-fired and gas-fired furnaces, boilers, power stations, explosives industry, manufacture of HNO_3, combustion of wood and refuse.	Respiratory irritation, headache, bronchitis, pulmonary emphysema, impairment of lung defences, edema of lungs, lachrymatory effect, loss of appetite, corrosion of teeth.
4.	Hydrogen sulphide (H_2S)	Coke ovens, kraft paper mills, petroleum industry, oil refining, viscose rayon manufacturing plants, manufacture of dyes, tanning industry and sewage treatment plants.	Headaches, conjunctivitis, sleeplessness, pain in the eyes, irritation of respiratory tract, respiratory paralysis, asphyxiation, malodorous. In high concentrations, it may lead to blockage of oxygen transfer, poisoning cell enzymes and damaging nerve tissues
5.	Chorine (Cl_2)	Accidental breakage of chlorine cylinders, electrolysis of brine, bleaching of cotton pulp and other process industries using chlorine	Irritation to eyes, nose and throat, toxicity, respiratory irritation, lachrymatory effects. In large doses it may cause edema, pneumonitis, emphysema and bronchitis.
6.	Hydrogen fluoride (HF)	Glass fibre manufacture, chemical industry, fertilizer industry, aluminium industry, ceramic industry, phosphate rock processing.	Irritation, respiratory diseases, fluorosis of bones, mottling of teeth.
7.	Carbon dioxide (CO_2)	Combustion of fuels, automobile exhausts, jet engine emissions.	Toxic in large quantities, Hypoxia.
8.	Hydrocarbons (HC)	Organic chemical industries, petroleum refineries, automobile exhausts, rubber manufacture.	Some hydrocarbons have carcinogenic effects, lachrymatory effect.
9.	Oxidants (e.g., O_3)	Photochemical reactions in atmosphere involving organic materials and NO_2 etc. Reactions induced by silent electrical discharge and intense u.v. radiations in the atmosphere.	Irritation of lungs, eyes and respiratory tract. Accumulation of fluids in lungs and damage to lung capillaries. These biochemical effects of O_3 mostly arise from generation of free radicals which attack the -SH groups present in the enzymes.

S.No.	Environmental pollutants	Major Sources of Pollutants	Typical Effects of Pollutants
10.	Dusts	Asbestos factories, mining activities, power stations, metallurgical industries, ceramic industry, factory stacks, glass industry, cement industry, foundries.	Respiratory diseases, toxicity from metallic dusts, silicosis and asbestosis from the specific dusts. Asbestos dust causes pulmonary fibrosis, pleural calcification and lung cancer.
11.	Ammonia (NH_3)	Chemical industries, coke oven, refineries, stocks yards, fuel incineration.	Damage to respiratory tracts and eyes, corrosive to mucous membranes.
12.	Formaldehyde (HCHO)	Waste incineration, automobile exhausts, combustion of fuels, photochemical reactions.	Irritation to eyes, skin and respiratory tract.
13.	Hydrochloric acid (HCl)	By-product from chlorination of organic compounds, combustion of gasoline in which ethylene dichloride is present, burning of coal, paper and chlorinated plastics	Inflammation and ulceration of upper respiratory tract, clouding of cornea, coughing and choking while inhalation.
14.	Radioactive gases and dusts	Radioactive gases and suspended dusts from natural and artificial radioisotope sources.	Somatic effects such as leukaemia and other types of cancer, cataracts and reduction in life expectancy. Genetic effects such as mutations in human.
15.	Arsenic (As)	Arsenic containing fungicides, pesticides and herbicides, metal smelters, by-product of mining activities, chemical wastes.	Inhalation, ingestion or absorption through skin can cause mild bronchitis, nasal irritation or dermatitis. Carcinogenic activity also is suspected. Attack -SH groups of enzymes, coagulate proteins.
16.	Beryllium (Be)	Coal, nuclear power and space industries, production of fluorescent lamps, motor fuels and other industrial use.	Damage to skin and mucous membranes, pulmonary damage, perhaps carcinogenic.
17.	Boron (B)	Boron producing units, production and use of petroleum fuel and additives, burning coal and industrial wastes, detergent formulations.	Ingestion or inhalation as dust causes irritation and inflammation. Boron hydrides can damage CNS and may result in death.

S.No.	Environmental pollutants	Major Sources of Pollutants	Typical Effects of Pollutants
18.	Cadmium (Cd)	Cadmium producing industries, electroplating, welding. By product from refining of Pb, Zn and Cu, fertilizer industry, pesticide manufacture, cadmium nickel batteries, nuclear fission plants, production of TEL used as additive in petrol.	Inhalation of fumes and vapours causes kidney damage, bronchitis, gastric and intestinal disorders, cancer, disorder of heart, liver and brain. Chronic and acute poisoning may result in renal dysfunction, anaemia, hypertension, bone-marrow disorder and cancer.
19.	Chromium (Cr)	Metallurgical and chemical industries, processes using chromate compounds, cement and asbestos units.	Toxic to body tissues, can cause irritation, dermatitis, ulceration of skin, perforation of nasal septum. Carcinogenic action suspected.
20.	Lead (Pb)	Automobile emissions, lead smelters, burning of coal or oil, lead arsenate pesticides, smoking mining and plumbing.	Absorption through gastrointestinal and respiratory tract and deposition in mucous membranes, cause liver and kidney damage, gastro-intestinal damage, mental retardation in children, abnormalities in fertility and pregnancy.
21.	Zinc (Zn)	Zinc refineries, galvanizing processes, brass manufacture, metal plating, plumbing.	Zinc fumes have corrosive effect on skin and can cause irritation and damage mucous membranes.
22.	Manganese (Mn)	Ferromanganese production, organo-manganese fuel additives, welding rods, incineration of manganese containing substances.	Poisoning of CNS, absorption, ingestion, inhalation, or skin contact may cause manganic pneumonia.
23.	Nickel (Ni)	Metallurgical industries using nickel, combustion of fuels containing nickel additives, burning of coal and oil, electroplating units using nickel salts, incineration of nickel containing substances, vanaspati manufacture.	Respiratory disorders, dermatitis, cancer of lungs and sinus.
24.	Vanadium (V)	Vanadium refining, production of vanadium containing alloys, power plants, burning of oil rich in vanadium.	Gastro-intestinal and respiratory disorders, inhibition of synthesis of cholesterol, heart disease and cancer in case of chronic exposures.
25.	Selenium (Se)	Burning of fuels and residual oils, fumes and gases from refinery wastes, incineration of paper and other wastes, natural sources.	Irritation of gastro-intestinal and respiratory tracts, irritation of eyes, nose and throat, damage to lungs, liver and kidneys.

S.No.	Environmental pollutants	Major Sources of Pollutants	Typical Effects of Pollutants
26.	Mercury (Hg)	Mining and refining of mercury, organic mercurials used in pesticides, laboratories using mercury.	Inhalation of mercury vapours may cause toxic effects and protoplasmic poisoning. Organo-mercurials are highly toxic and may cause irreversible damage to nervous system and brain.

Appendix 1.29: Group of Population Having Higher Probability Of Developing Illness Than General Population

S. No.	Group of population	Air pollutants
1.	Healthy young adults who exercise or work outdoors	Ozone
2.	Children and adolescents	Ozone, lead
3.	Pregnant women and foetuses	Carbon monoxide
4.	Patients with asthma	SO_2, O_3, NO_2, H_2SO_4, particulates, sulphates
5.	Patient with chronic obstructive pulmonary disease(COPD)	CO, SO_2, particulates
6.	Patient with ischemic heart disease	CO, SO_2, particulates

Appendix 1.30: Observed Relation Between Pollutant Levels and Human Health Effects

S.No.	Pollutants	Concentration level producing adverse health effect	Adverse health effects observed
1.	Carbon monoxide	a. 50 ppm (58 mg/m^3 for 1.5 hours [similar effects observed upon exposure to 0-15 ppm (10-17 mg/m^3) for 8 or more hours. b. 30 ppm (35 mg/m^3) for 8 or more hours.	a. Impaired time interval discrimination. b. Impaired performance in psycho-motor tes increase in visual threshold.

2.	Photo chemical oxidant (O_3 and PAN)	a. In excess of 0.07 ppm (130 µg/m^3) b. 0.01 ppm (200 µg/m^3) maximum daily value. c. 0.25 ppm (490 µg/m^3) maximum daily value.	a. Impairment of performance by student athlete. b. Eye irritation c. Aggravation of asthma check.
3.	Particulate matter and sulphur oxides	a. 0.028-0.035 ppm (80-100 µg/m^3) particulates (annual geometric mean). b. 0.046 ppm (130 µg/m^3) of SO_2 (annual mean) accompanied by particulate concentration of 130 µg/m^3. c. 0.068 ppm (190 µg/m^3) of SO_2 (annual mean) accompanied by particulate concentration of 177 µg/m^3. d. 0.037-0.092 ppm (105-265 µg/m^3) of SO_2 (annual mean) accompanied by particulate concentration of 177 µg/m^3. e. 0.05-0.09 (140-260 µg/m^3) of SO_2 (24 hrs average). f. 0.11 – 0.19 ppm (300-500 µg/m^3) of SO_2 (24hrs. mean) with a low particulate levels. g. 300 µg/m^3 particulate for 24 hours accompanied by SO_2 concentration of 0.22 ppm (630 µg/m^3)	a. Increased death rate for persons above 50 years of age. b. Increased severity and frequency of respiratory disease in school children. c. Increased severity and frequency of respiratory disease in school children. d. Increased frequency of respiratory symptoms and lung disease. e. Increased illness rate of older persons with severe bronchitis. f. Increased hospital admissions for respiratory disease and absentism from work of older persons. g. Chronic bronchitis patient suffering acute worsening of symptoms.

NOTES

Human health and environmental pollution

Pollution is any undesirable change in the physical, chemical or biological characteristics of air, water or land. Pollution can harm the health and threaten the survival or activities of human beings and other living organisms. The environment is made up of air, water and land, technically known as atmosphere, hydrosphere and lithosphere respectively which together constitute the biosphere. In the biosphere, apart from human beings, plants, animals, birds, fishes, insects and microorganisms (algae, bacteria and virus) also exist. The atmosphere provides oxygen, while the hydrosphere and lithosphere provide food, water and space. Whenever a change, physical or chemical, occurs in the atmosphere, hydrosphere or lithosphere, all living beings are affected.

Degradable pollutants are those that can be decomposed, removed or consumed or reduced to acceptable levels either by natural or artificial means. However, pollutants such as human sewage and animal and crop wastes can decompose only if the system is not overloaded. Certain chemicals decompose slowly, and can persist at harmful levels for decades like detergents and pesticides.

Nondegradable pollutants include many radioactive materials, heavy metals and some plastics which cannot be degraded by nature or artificial means. They must be controlled or prevented from reaching the environment.

Air Pollution

Air pollution can be due to natural or man-made causes. The former is beyond our control as natural disasters like dust storms, earthquakes and volcanic eruptions throw up large quantities of dust and gases into the atmosphere. Man –made causes, however, should be prevented or controlled as they pose a greater danger by way of toxic emissions from factories, power-plants, vehicular traffic, etc. Industries such as mining, thermal plants, brick kilns, etc. also pollute the air. These emissions are particularly intense in urban conglomerations where the density of human habitation is very high.

Primary pollutants are those which are emitted directly into the atmosphere, like sulphur dioxide, nitric oxides and carbon monoxide.

Secondary pollutants are pollutants formed by the photochemical reaction of primary pollutants. For example, "smog", is a combination of smoke and fog. Smoke consists of carbon particles and fog is an emulsion of water vapour in air. Smog has become very common in large cities, especially during winter. Similarly acid rain is formed by the combination of sulphur dioxide and water vapour present in the air.

Pollutants in the air can be dispersed by wind movement, temperature and topography.

Major air pollutants and their effect on human health		
Pollutant	**Origin of Pollutant**	**Effects**
Sulphur dioxide	Industries, especially where coal or oil are used as fuel	Irritation of eyes, and respiratory system, increased mucus production, cough and shortness of breath
Carbon monoxide	Automobile exhaust and industries	Reduction in oxygen-carrying capacity of blood
Oxides of nitrogen	Automobile exhaust	Irritation of pulmonary tract affecting functions of lungs
Hydrocarbons	Automobile exhaust	Lung cancer
Chlorine	Chloralkali industry	Irritation of mucous membrane
Ammonia	Fertiliser, industry, agriculture and in poultry farming	Irritation of mucous membrane
Hydrogen sulphide	Manufacture of coke, viscose rayon, distillation of tar and petroleum	Excessive inhalation leads to death
Acids and aldehydes	Chemical industries	Eyes, nose and throat irritation
Suspended particulate matter (SPM)	Industries, automobile exhaust	Respiratory disease
Dust	Industries and automobile exhaust	Silicosis
Asbestos	roofings, brake linings	Asbestosis
Lead	Automobile exhaust	Cumulative poison, impairment of central nervous system

Major air pollutants and their effect on human health		
Pollutant	**Origin of Pollutant**	**Effects**
Beryllium	Aerospace industry, manufacture of house-hold appliances	Fatal to heart and lungs
Manganese	Mining operations	Damages nerves and reproductive system
Benzene	Automobile exhaust and manufacture of chemicals	Leukemia, chromosomal damage
Pesticides	Manufacture and application of pesticides	Depression, leads to death if inhaled in excess
Arsenic	Thermal power plants	Toxic

Other Effects of Air Pollution

- Ozone causes reduced pollination and yellow spot formation on leaves, thus affecting the rate of photosynthesis
- Sulphur dioxide is converted to sulphuric acid in the presence of moisture and iron, which are present in dust, nails, etc. This results in yellowing, weakening and corrosion of materials.

Water Pollution

The limited availability of fresh water and its unequal distribution make water pollution a matter of great concern. Water pollution is generally localised and confined, making it more severe. In underdeveloped countries, sewage is a major source of water pollution. Human excreta contains 400 different species of bacteria and viruses. Even well-treated sewage contains pathogenic bacteria and virus, unless properly chlorinated before being discharged into any water course. Sewage is a major contributor to water-borne diseases and affects the health of people and other organisms in the environment in many ways.

Industrial effluents from sugar factories, distilleries, tanneries and paper industries are accompanied by very high organic loads by products of paper and pulp industry cause depletion of fish upto as far as 40 km downstream. The wastes from oil refineries and steel industries contain phenol which imparts a strong odour, apart from poisoning the water body. Fertiliser industry wastes contain ammonia, urea, phosphate and sulphate which, in water, causes algal blood and are toxic to aquatic fauna and flora. Alkaline industry wastes contain mercury which can kill human beings who consume mercurised fishes. Lead generated from battery, printing, petrol and paste-processing industries, trace and toxic elements such as zinc, copper etc., and effluents from mining industries are injurious to aquatic organisms.

Water in which maximum permissible concentration of any single or more constituents is in excess is unfit for drinking and human health.

There are definite tolerance levels for water used for different activities such as drinking, bathing, irrigation and industrial purposes. Depending on its use, there are different physio-chemical and bacteriological standards for water.

Major water pollutants and their effect on human health		
Pollutant	**Origin**	**Effects**
Arsenic	pesticide industries and agriculture	toxic, eruptions on skin, carcinogenic
Cadmium	industrial waste discharge	many substitute calcium in bone structure, causes a diseases – Itai-Itai in Japan
Chlorine	industries using chlorine bleach	Physiological
Chromium	industrial waste, especially from tanneries	toxic and carcinogenic
Cyanide	pesticides	toxic and fatal
Lead	industrial, mine and smelter discharges, water pipes	causes a disease called plumbism
Nitrates and Nitrites	fertilizer industries	causes metaemo globinaemia in infants
Sulphates	leaching of gypsum or industrial waste discharge	have iaxative effect in combination with magnesium or sodium in drinking water may from hydrogen sulphide gas which is fatal for workers in sewers
Selenium	pesticides, industrial waste	carcinogenic, causes caries
Mercury	pesticides brain damage	toxic, causes foetal
Bacteria	sewage	cholera, typhoid, hepatitis, dysentery

Soil Pollution

Land is a very valuable but limited resource, as the population increases rapidly.

India loses 20 tons of top soil per hectare in a year due to floods, rainfall and deforestation. 20% to 50% of lands under irrigation can go out of cultivation at this rate because of water logging and salinity.

This scenario of desertification is compounded by pollution which includes

- Indiscriminate discharge of industrial effluents on land and into water bodies
- An increase in the use of fertilisers for agriculture
- Open defecation by animals and human beings
- Accumulation of solid waste; this is a major problem in developed countries like India where the garbage and refuse products are not degraded
- Radioactive substances from nuclear plants which are released into the soil

Major soil pollutants and their effect on human health		
Metal	**Source**	**Effects**
Arsenic	Occurs naturally	Chronic poisoning leads to a loss of appetite and weight, diarrhoea, alternating with constipation, gastro intestinal disturbances, peripheral neuritis, conjunctivitis and sometimes skin cancer

Major soil pollutants and their effect on human health		
Metal	**Source**	**Effects**
Cadmium	Mining, metallurgy chemical industry and electroplating	Leads to chronic poisoning and affects the proximal tubules of the kidney, causing formation of kidney stones
Lead	Lead smelters storage battery	Lead poisoning can lead to severe mental retardation or death
Mercury	Industrial wastes	Methyl mercury compounds are much more toxic than other forms of mercury, causes neurological problems and damages renal glomeruli and tubules
Cyanides	Wastes from heat treatment of metals, dismantling of electroplating shops, etc.	Rapid death may follow due to exposure to cyanide as a result of inhibition of cellular respiration

Food Pollution

Removal of nutrients or the addition of adulterants into the food chain also harms our health.

Foreign substances are deliberately mixed with food materials for economic gain. Undesirable substances enter the food chain through agencies such as contaminated fish and food, microbes, chemical pollutants in water including fluoride, lead, tin, cadmium, etc. Pollution of food comes from the pollution of water and soil. If that is prevented, our food will be nourishing and healthy and fit for eating.

Major food pollutants and their effect on human health		
Pollutant	**Source**	**Effects of health**
Organochlorine DDT, BHC (lindane), aldrin, dieldrin	Compounds used to control mosquitoes	Affects nervous system
Organophosphorous compounds (parathion, malathion, toxaphene)	Used in agriculture to kill pests	Affects nervous system, through vegetable breast milk, fish and meat
Bacteria	Handling and water used in food-preparation	Typhoid, dysentery, amoebiasis, jaundice, etc.

Noise Pollution

Noise is unwanted sound and has become a part of urban life and industrial activity. Noise pollution may come from loudspeakers, factories, aeroplanes, moving trains, construction activity etc.

Noise level of 80 decibels or more for more than 8 hours a day increases tension and changes in breathing pattern. Continued exposure to high levels of noise results in fatigue, hearing loss or even total loss of hearing, changes in blood circulation, changes in breathing, etc. Noise pollution above 120 decibels can cause many adverse biochemical changes. Cholesterol levels in the blood and white cell counts increase, besides causing hypertension.

Psychological and physical effects of noise at different decibel levels (db)		
Noise level in decibels	**Source**	**Effect**
135	Pneumatic drill	Painful
110	Rock band	---
88	Industry/ city traffic	Hearing impairment on prolonged exposure
80	Alarm clock	Annoying
65	Average city traffic	Intrusive

Appendix 1.31: Anthropogenic & natural sources, background concentration, estimated life time and removal mechanisms of trace gases present in atmosphere

S.No.	Atmospheric trace gases	Anthropogenic sources of emission	Natural sources of emission	Background concentration, ppm	Estimated lifetime	Removal mechanisms from atmosphere
1	SO_2	Combustion of coal and oil	Volcanoes	0.002-0.01	3days	Oxidation to sulphate photochemically or catalytically.
2	H_2S	Chemical Processes	Biological	0.002-0.02	1days	Oxidation to SO_2
3	CO	Combustion	Oxidation of CH_4, oceans	0.12-0.15	0.1yr	Reaction with OH in troposphere and stratosphere .soil removal.
4	NO-NO_2	Combustion	Bacterial action in soil	NO:0.2-2ppb NO_2:0.5-4ppb	5days	Oxidation to nitrate by photochemical reactions or on aerosol particles.
5	NH_3	Waste treatment	Biological decay	6-20ppb	2weeks	Reaction with SO_2 to from $(NH_4)_2SO_4$. Oxidation to nitrate.
6	N_2O	None	Bacterial action in soil	0.25	4yr	Photodissociation in stratosphere.

S.No.	Atmospheric trace gases	Anthropogenic sources of emission	Natural sources of emission	Background concentration, ppm	Estimated lifetime	Removal mechanisms from atmosphere
7	CH_4	Combustion, Chemical Processes	Swamps, paddy	1.5	1.5yr	Reaction with OH.

*.Trace gaseous Components of atmosphere are classified into two groups :-

Group I: Relatively inert gases like CO_2, CO, N_2O, CH_4 with lifetimes of the order of month to years .life cycles of these species are heavily influenced by the rate of emission from natural sources.

Group II: Gases which are rapidly converted to particles like SO_2, H_2S, NH_3, HC, NO with the lifetime of the order of days to weeks.

NOTES

Air pollution may be caused by various natural and anthropogenic processes. Below are listed some of the main air pollution causes.

Natural Causes:

- **Volcanic activities**: volcanic eruptions emit a serious of toxic gases (including sulfur and chlorine) as well as particulate matter (ash particles) but are usually restricted to localized areas;
- **Winds and air currents:** can mobilize pollutants from the ground and transport them over large areas;
- **Wildfires:** add to the atmosphere smoke, carbon monoxide as well as particulate matter (containing organic contaminants such as PAHs) and could affect significant areas although in general are restricted and may be contained;
- **Microbial decaying processes:** microorganisms which are present in any environment have a major role in natural decaying processes of living organisms as well as environmental contaminants; this activity results in the natural release of gases especially methane gas;
- **Radioactive decay processes:** for example radon gas is emitted due to natural decay processes of Earth's crust which as potential to accumulate in enclosed spaces such as basements;
- **Increasing temperatures** - contribute to an increase in the amounts of contaminants volatilizing from pollution soil and water into air.

Anthropogenic Sources

- **Mining and smelting:** emit into the air a variety of metals adsorbed on particulate matter that is suspended in the air due to crushing and processing of mineralogical deposits;
- **Mine tailing disposal:** due to their fine particulate nature (resulting after crushing and processing mineral ores) constitute a source of metals to ambient air which could be spread by wind over large areas;

- **Foundry activities:** emit into the air a variety of metals adsorbed on particulate matter that is suspended in the air due to processing of metallic raw materials (including the use of furnaces)
- **Various industrial processes** may emit both organic and inorganic contaminants through the following mechanisms:
  - Accidental spills and leaks of stored chemicals
  - Handling and storage of chemicals - especially of volatile inorganic chemicals
- **Transportation**: emits a series of air pollutants (gases - including carbon monoxide, sulfur oxides and nitrogen odides - and particulate matter) through the tailpipe gases due to internal combustion of various fuels (usually gases such as oxides of carbons, of sulfur, of nitrogen, as well as organic chemical such as PAHs) /
- **Construction and Demolition activities**: pollute the air with various construction materials. of special threat is the demolition of old buildings which may contain a series of banned chemicals such as PCBs, PBDEs, asbestos.
- **Coal Power Plants**: by burning coal they may emit a series of gases as well as particulate matter with metals (such as As, Pb, Hg) and organic compounds (especially PAHs);
- **Heating of buildings:** emits a series of gases and particulate matters due to burning of fossil fuels;
- **Waste Incineration**: depending on waste composition, various toxic gases and particulate matter is emitted into the atmosphere.
- **Landfill disposal practices:** usually generate methane due to intensification of natural microbial decaying activity in the disposal area;
- **Agriculture:** pollute the air usually through emissions of ammonia gas and the application of pesticides/herbicdes/insecticides which contain toxic volatile organic compound (VOC_s)
- **Control burning in forest and agriculture management:** includes controlled burning that will emit gases and particulate matter (similar to wildfires described above)
- **Military activities**: may introduce toxic gases through practices and training;
- **Smoking** - emits a series of toxic chemicals including a series of organic and inorganic chemicals, some of which are carcinogenic;
- **Storage and use of household products:** such as paint, sprays, varnish, etc that contains organic solvents which volatilize in the air (hence the smell we all feel while using them);
- **Dry cleaned clothes:** may retained and emit in the atmosphere small amounts of chlorinated solvents (such as PCE) or petroleum solvents that have been used by the dry cleaners; this could eventually create a health risk if the clothes returned from the dry cleaners are stored in enclosed indoor spaces.

Appendix 1.32: Sources, Effects & Control Methods for 10 Important Air Pollutants

S.No.	Air Pollutants	Sources of Pollutants	Effects of Pollutants	Prevention and Control method of Pollutants
1.	Ozone (O_3)	Formed when reactive organic gases and nitrogen oxides-react in the presence of sunlight. ROG sources include any source that burns fuels, (e.q., gasoline, natural gas, wood, oil) solvents, petroleum processing and storage and pesticides	Breathing Difficulties, Lung Tissue Damage to Rubber and Some Plastics	Reduce motor vehicle reactive organic gas (ROG) and nitrogen oxide emissions reformulated fuels, reduced vehicle use. Limit ROG emissions from commercial operations and consumer products. Limit ROG and NOx emissions from industrial sources such as power plants and refineries, Conserve energy.
2.	Respirable Particulate Matter (PM_{10})	Road Dust, Windblown Dust (Agriculture) and Construction (Fireplaces) Also formed from other pollutants (acid rain, SOx, NOx, organics), Incomplete combustion of any fuel	Increased Respiratory Disease, Lung Damage, Cancer, Premature Death, Reduced visibility, Surface Soiling	Control Dust Sources, Industrial Particulate Emissions, Wood Burning Stoves and Fireplaces Reduce secondary pollutants which react to form PM_{10}, conserve energy.
3.	Fine Particulate Matter ($PM_{2.5}$)	Fuel Combustion in Motor Vehicles, Equipment and Industrial Sources, Residential and Agricultural Burning. Also formed from reaction of other pollutants (acid rain, NOx, SOx, organics).	Increases Respiratory Disease, Lung Damage, Cancer, Premature Death, Reduced Visibility, Surface Soiling	Reduce Combustion Emissions from Motor Vehicles, Equipment, Industries and Agriculture and Residential Burning. Precursor controls, like those for ozone, reduce fine particle formation : in the atmosphere.

S.No.	Air Pollutants	Sources of Pollutants	Effects of Pollutants	Prevention and Control method of Pollutants
4.	Carbon Monoxide (CO)	Any source that burns fuel such as automobiles, trucks, heavy construction equipment, farming equipment and residential heating Industrial sources.	Chest Pain in Heart Patients, Headaches, Reduced Mental Alertness	Control motor vehicle and industrial emissions. Use oxygenated gasoline during winter months, Conserve energy.
5.	Nitrogen Dioxide (NO_2)	Some as Carbon Monoxide	Lung Irritation and Damage. Reacts in the atmosphere to form ozone and acid rain	Control/motor vehicle and industrial combustion emissions, Conserve energy.
6.	Lead(Pb)	Metal Smelters, Resource Recovery, Leaded Gasoline, Deterioration of Lead Paint		Control metal smelters, no lead in gasoline. Replace leaded paint with non-lead substitutes
7.	Sulfur Dioxide ($S0_2$)	Coal or Oil Burning Power Plants and Industries, Refineries, Diesel Engines.	Increases lung disease and breathing problems for asthmatics. Reacts in the atmosphere to form acid rain.	Reduce/ the use of high sulfflr fuels (e.g., use low sulfer reformulated diesel or natural gas) Conserve energy.
8.	Visibility Reducing Particles	Same as $PM_{2.5}$	Reduces visibility (e.g., obscures mountains and other sceneries), reduced airport safety, lower real estate value, discourages tourism.	'Same as $PM_{2.5}$
9.	Sulfates	Produced by the reaction in the air of $S0_2$, a component of acid rain.	Breathing Difficulties, Aggravates Asthma, Reduced Visibility	Same as SO_2
10.	Hydrogen Sulfide	Geothermal Power Plants, Petroleum Production and Refining, Sewer Gas	Nuisance Odour (Rotten Egg Smell), Headache and Breathing Difficulties (Higher Concentrations)	Control emissions from geothermal power plants, petroleum production and refining, sewers, sewage treatment plants

Appendix 1.33: Human Health Effects of Air, Water and Soil Pollution

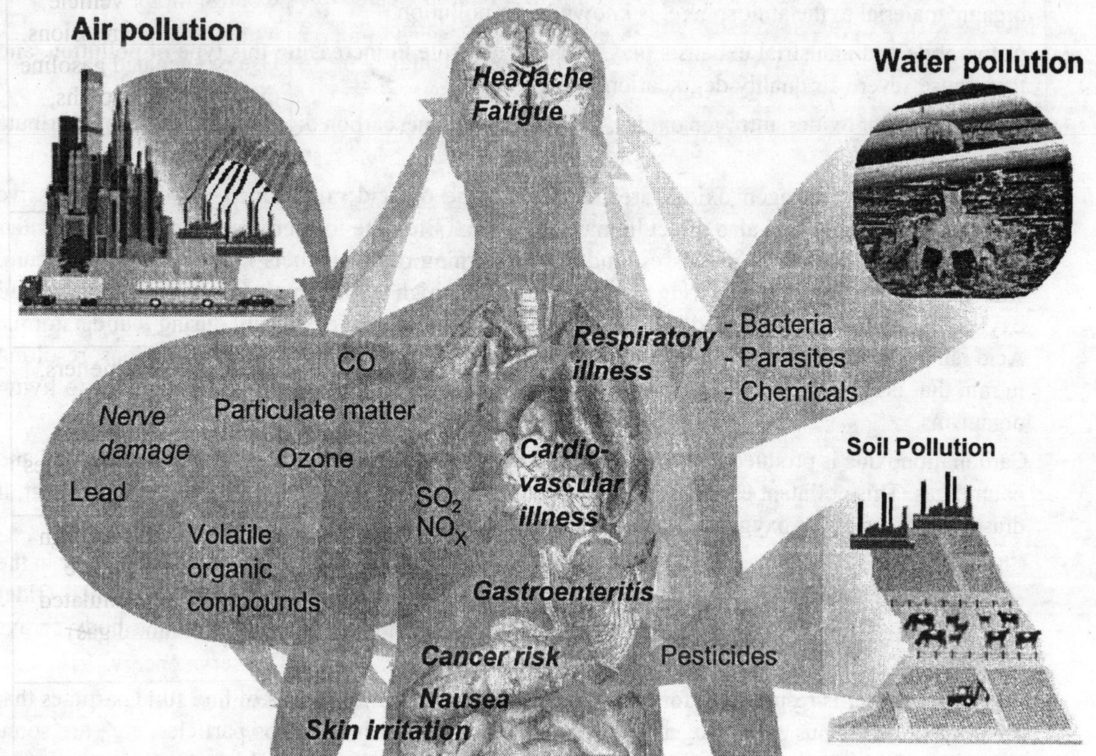

NOTES

Earth's environment consists of four basic components:

Atmosphere: It is a layer of gases that envelope the Earth. The air we breathe is a part of the atmosphere. ***Hydrosphere-*** It includes all the elements that are a part of the water bodies and their sources.

Lithosphere: It includes everything land-based; every minor and major constituent that exists on the Earth's crust is included.

Biosphere: It includes all the living organisms that inhabit the planet.

The introduction of contaminants to the first three components affects the fourth component - the biosphere, in many different ways. The contamination of the atmosphere pollutes air, that of the hydrosphere pollutes water, and that of the lithosphere pollutes land. And these are the three major types of pollution.

- **Air pollution**
- **Water Pollution**
- **Land Pollution**

The quality of the environment deteriorates in many ways, due to the various harmful substances that pollute it.

Air Pollution Fact sheet

- The addition of harmful chemicals, particulate matter, aerosols, toxic and hazardous gases, and organic material to the atmosphere, is known as air pollution.
- Automobile and industrial exhausts play an important role in increasing this type of pollution, and they cause severe air quality degradation.
- Gases like sulfur oxides, nitrogen oxides, ammonia, methane, carbon dioxide, etc., mainly contribute to air pollution.
- Sulfur oxides and nitrogen oxides are the main cause of acid rain, which can lead to massive destruction of. crops, and also affect human and animal life. The former is released from vehicular exhausts, during industrial processes, and due to burning of fossil fuels like coal and petroleum, whereas, the latter is released during processes involving high temperatures (reaction of nitrogen and oxygen during combustion), and also when there is a discharge of electricity during thunderstorms. Acid rain is formed due to the interaction of these gases with the natural cloud formations, resulting in rain that is mixed with dilute amounts of sulfuric and nitric acid, which are hazardous to living organisms.
- Carbon monoxide is produced when there is incomplete combustion of fossil fuels like coal, oil and natural gas. This pollutant causes severe respiratory problems, and being a strong reducing agent, it drastically reduces the oxygen percentage in the air.
- Chlorofluorocarbons (CFCs) are harmful chemical compounds which damage the ozone layer in the stratosphere, thereby causing the sun's hazardous ultraviolet rays to reach the Earth's surface. Older refrigerators and air conditioners released CFCs. Aerosol sprays like perfumes and body sprays release them.
- Particulate matter is responsible for many respiratory diseases. It consists of fine solid particles that are a result of various anthropogenic processes. Dust, soot, ash, asbestos particles, etc., are some examples of this category of pollutants. Smoke consists of particulates, which on mixing with fog, form smog. Formation of smog is a serious concern in the urban areas.
- Apart from these, other pollutants like radioactive elements, ammonia, methane, and secondary pollutants (PAN and ozone) also harm the atmosphere.

Water Pollution Fact sheet

- Harmful chemicals, gases, and waste products enter water bodies, and lead to contamination of water causing water pollution.
- Industrial wastes, untreated sewage, oil spills, toxic chemicals and compounds that are discharged from factories and refineries, and radioactive wastes are some other causes of water quality degradation.
- ntreated sewage drained into water bodies can affect the aquatic flora and fauna. It also leads to waterborne diseases like typhoid, cholera, and diarrhea.
- Oil spills caused due to shipwrecks, oil rig disasters, pipeline leaks, etc., damage the aquatic fauna. This mainly affects fish and other organisms living in water. Also, the process of cleanup is difficult and the costs involved are very high.
- Eutrophication causes water pollution wherein there is an increase in the nutrients needed by aquatic plants and animals. This causes a drastic drop in the oxygen level of water (anoxia), thus degrading

the water quality and posing a risk to the aquatic life. An example of this is the rampant growth of water hyacinth in rivers and streams causing depletion of the oxygen content of water.

- Ground water pollution is caused due to the seepage of toxic chemicals and compounds like insecticides and pesticides, below its surface into the natural water-storing formations known as aquifers. This significantly degrades the water quality and can make it unfit for consumption. Groundwater pollutants also include disease-carrying pathogens, radioactive wastes, and leachates.
- Apart from this, toxic heavy elements like arsenic, zinc, cadmium, and even mercury, when introduced in water bodies, can severely affect the water quality.

Soil/Land Pollution Fact sheet:

- A majority of the human population lives on land or the lithosphere. If a pollutant gets introduced in the lithosphere, it directly affects the surrounding environment, causing soil or land pollution.
- It impacts the entire food chain and thus turns out to be the most hazardous of all types of pollution.
- Untreated solid wastes like sewage, insecticides, pesticides, herbicides, litter contamination, slash and burn- techniques, radioactive wastes, and mining activities lead to soil pollution.
- When inorganic and organic sewage is left untreated, it degrades the soil quality. Synthetic materials like plastic, nylon, thermocol, and other complex polymer compounds do not decompose for millions of years. They keep accumulating and there is no possible solution to this problem, except for treatment and proper disposal.
- Organic wastes, if left untreated, can form leachates and other harmful chemical compounds, causing extensive damage to the soil and its fertility. In such cases, the soil cannot be used for cultivation, unless the sewage is properly treated and disposed off in landfill areas.
- Improper use of pesticides, insecticides, herbicides, and fertilizers, can cause permanent or long-term damage to the agricultural land. Soil salinity drastically increases due to the use of chemical compounds on a large scale, and may cause it to become infertile.
- Nuclear detonations release a large amount of nuclear wastes that pollute the land and water bodies. The Chernobyl disaster, Hiroshima and Nagasaki nuclear explosions, and the recent Fukushima disaster, are well -known examples of such incidents.
- Improperly conducted mining activities can destroy the landscape of the area where they are undertaken. A barren wasteland results mostly due to mining activities, and is also considered as a form of land pollution.
- Apart from the above-mentioned sources, electronic wastes, acid rain, deforestation, rampant urbanization, and open defecation lead to land pollution.

Earth is the only planet capable of sustaining life. And that underlines the importance of maintaining it in a state conducive to living. Though we may not be able to eradicate pollution in totality, we can surely take steps towards reducing it.

Appendix 1.34: Key Characteristics of Various Pollution Control Devices and/or Systems

S.No.	Factor considered	Characteristic of concern
1.	General factors	Collection efficiency Legal limitations such as best available technology Initial cost Lifetime and salvage value Operation and maintenance costs Power requirement Space requirements and weight Materials of construction Reliability Reputation of manufacturer and guarantees Ultimate disposal/use of pollutants
2.	Carrier gas characteristics	Temperature Pressure Humidity Density Viscosity Dewpoint of all condensibles Corrosiveness Inflammability Toxicity
3.	Process characteristics	Gas flow rate and velocity Pollutant concentration Variability of gas and pollutant flow rates, temperature, etc. Allowable pressure drop
4.	Pollutant (if gaseous)	Corrosiveness In flammability Toxicity Reactivity
5.	Pollutant (if particulate)	Size range and distribution Particle shape Agglomeration tendencies of particles Corrosiveness Abrasiveness Hygroscopic tendencies Stickiness In flammability Toxicity Electrical resistivity Reactivity

Appendix 1.35: Properties of the Atmosphere

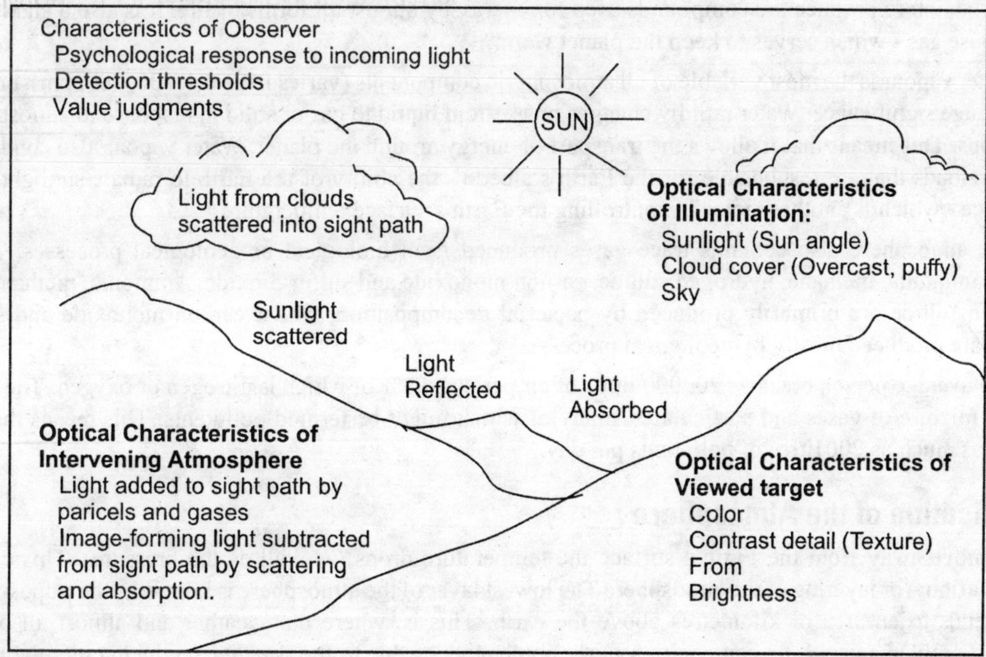

NOTES

The Earth's atmosphere is approximately 160 kilometers deep, but 95% of its air mass lies within 20 kilometers of the surface. Moving up from the earth's surface the density of the atmosphere decreases rapidly and the air "thins". Hence the atmosphere is neither uniform, nor static in nature. Its characteristics vary widely with altitude, season, location and solar flare activity.

The concentrations of nitrogen, oxygen, argon, neon, helium, krypton, hydrogen and xenon atmosphere remain essentially constant. These gases are fairly inert and play little or no role in atmospheric chemistry.

Even though it is the most abundant gas, nitrogen has little effect on major atmospheric processes or to the sustenance of life forms. It does however, serve as a precursor for other species such as NO3", as well as amino acids and nucleic acids (amongst others) which are essential for life.

Nitrogen reacts with oxygen - the second most abundant gas in the atmosphere to form oxides of nitrogen (NO_x). These include NO, NO_2, N_2O_4 and N_2O. The NO_x compounds in the atmosphere are found in trace levels and their concentration varies with place and time of day. Nitrous oxide (N_2O) was thought to remain constant in atmospheric concentration, but synthetic fertiliser compounds containing nitrogen have been found to increase the amount of nitrification by soil bacteria, which has in turn increase the amount of N_2O being generated by natural sources.

As the great majority of species on Earth use an oxidative metabolism, oxygen is the most important atmospheric gas for the nurturing of life. Its significance as an atmospheric gas goes far beyond sustaining life however. Oxygen is also present in the atmosphere as ozone (O_3), which exists in high concentrations in the upper atmosphere and acts as a heat and radiation shield for the earth maintaining fairly constant temperatures that allow life to exist.

At 0.035%, the concentration of carbon dioxide in the atmosphere is very low compared to molecular oxygen or nitrogen. It is still of enormous significance however as it is the raw material used by plants for carbon fixation to produce the compounds used for energy by almost all forms of life. It is also a significant greenhouse gas - which serves to keep the planet warm.

Water vapour is the most variable of all atmospheric components (varies from 0.1 – 30, 000ppm), but it is also of huge significance. Water rapidly changes phase from liquid to gas or solid in response to atmospheric conditions. This means that it allows the transport of energy around the planet. Water vapour also condenses to form clouds that are responsible for the Earth's albedo - the ability of the Earth to radiate sunlight back into space - which is another factor in controlling the Earth's surface temperature.

The atmosphere also contains trace gases produced from biological or geological processes. These include ammonia, methane, hydrogen sulfide, carbon monoxide and sulfur dioxide. Ammonia, methane and hydrogen sulfide are primarily produced by bacterial decomposition, whilst carbon monoxide and sulfur dioxide are produced mostly by geological processes.

The average person breathes 20, 000 litres of air per day, 99% of which is nitrogen or oxygen. The other 1% is a mixture of gases and particulates, many of which might be termed pollutants. This means that we breathe as much as 2001itres of pollutants per day.

Stratification of the Atmosphere

As we move away from the Earth's surface the temperature drops (as well as the pressure). This causes **stratification** – or layering of the atmosphere The lowest layer of the atmosphere is known as the troposphere, and extends to an area of kilometres above the earth. This is. where our weather and almost all of our pollution problems occur. Its composition fairly homogeneous due to the weather producing circulating air masses that allow constant mixing. 95% of the atmosphere's air mass is found in the troposphere. The upper troposphere has a temperature of – 56°C.

At the top of the troposphere is the **tropopause layer.** It serves as a barrier to prevent water vapour rising much higher as it causes ice formation. Water vapour cannot pass through it. If it did, it would photodissociate in contact with ultraviolet radiation, and the hydrogen would be lost to space. The tropopause separates the stratosphere from the troposphere.

Above the tropopause layer is the **stratosphere,** between 11 and 50 km above the earth.. Moving up in this layer we find the temperature rises, and may reach up to –2°C., The ozone layer is within the stratosphere, and reaches levels of up to l0ppm in the middle of the stratosphere. The stratosphere can reach temperatures of up to –2°C due to the absorption of ultraviolet radiation by the ozone.

The **mesosphere** is above the stratosphere, covering an area of between 50 and 85km above the earth. At 85km the temperature is –92°C. Above the mesosphere is an area called the **exosphere** where molecules and ions can be lost to space.

An area extending far into space is called the thermosphere, an area of high temperatures (1200°C) due to the absorption of radiation from space.

Appendix 1.36: Mutual Relationship of Air Pollutants

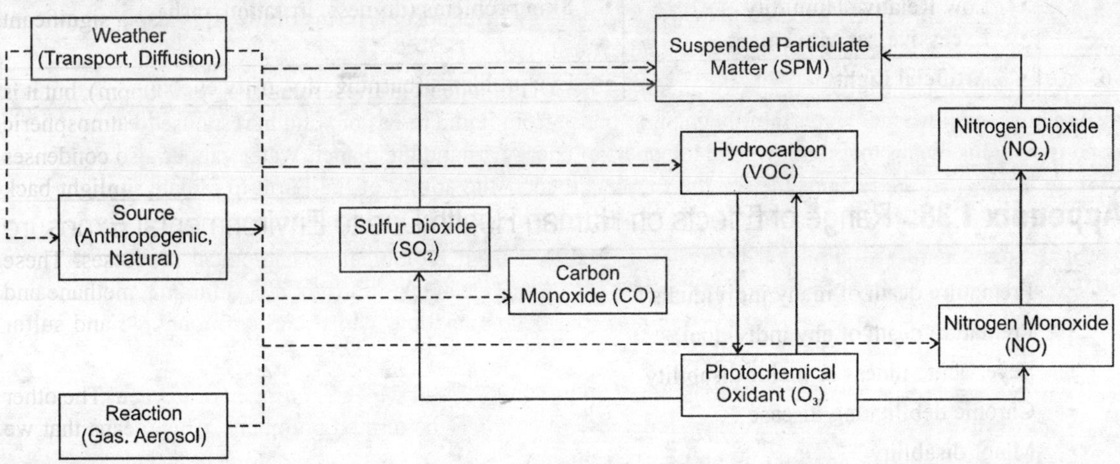

NOTES

Dashed line indicates the flow of primary air pollution and solid line indicates that of secondarily generated air pollution.

Appendix 1.37: Health Symptoms Associated with Various Environmental Conditions

S.No.	Environmental	Health Symptoms
1.	• Ergonomic Conditions	• Headache
2.	• Noise and Vibration	• Fatigue • Poor Concentration • Dizziness • Tiredness • Headache with nausea • Ringing in ears • Pounding heart
3.	• Relative Humidity	• Dry throat • Shortness of breath or bronchial asthma • Irritation and infection of respiratory tract • Nasal problems (stuffiness, irritation)
4.	• Relative humidity and High Temperatures	Nasal problems (Stuffness, irritation)

5.	• Warm Air	
	• Low Relative Humidity • Excessive Air Movement	• Skin problems (dryness, irritation, rashes)
6.	• Artificial Light	• Eye problems (burning, dry gritty eye)

Appendix 1.38: Range of Effects on Human Health Due to Environmental Exposure

- Premature death of many individuals
- Premature death of any individual
- Sever acute illness or major disability
- Chronic debilitating disease
- Minor disability
- Temporary minor illness
- Discomfort
- Behavioral changes
- Temporary emotional effects
- Minor physiological change

Appendix 1.39: Examples of Useful biological Markers of Environmental Exposure

Substance	Biological Marker
Carbon Monoxide	COHb in blood
Lead	Pb in blood
Pentachlorophenol (PCP)	PCP in urine
Alcoholic Beverages	Ethanol in exhaled breath
Volatile Organics (VOCs)	VOCs in exhaled breath

Appendix 1.40: Recommended Standard Values for Daily Intake of Air, Water and Soil by Human Beings

Age (years)	Air Inhalation (m^3/day)	Water Ingestion (L/day)	Soil Ingestion (mg/day)	Total soil Adhered (mg/day)
0-<0.5 0.5-<5 5-<12 12-<20 20^+	2 5 12 21 23	F:0/0 NBF:0.2/0.8 0.2/0.8 0.3/0.9 0.5/1.3 0.4/1.5	35 50 35 20 20	2200 3500 5800 9100 8700

Appendix 1.41: Selected Industrial Sectors and theirs Contribution to Air & Water Pollution and to Workplace Hazards

S.No	Industrial sectors	Air pollutants	Water pollutants	workplace hazards
1.	Base metal and iron ore mining	Particulate Matter (PM)	Toxic metal sludge	Silica
2.	Cement manufacturing	Particulate Matter (PM)	Sludge	Silica
3.	Coal mining and production	Particulate Matter (PM), coal dust	Coal dust and sludge	Silica
4.	Copper smelting	Arsenic	Arsenic	Arsenic, Cadmium
5.	Electricity Generation	Particulate Matter (PM) sulpher dioxide	Hot water	Sulphur dioxide
6.	Foundries	Particulate Matter (PM)	Solvents	Silica and solvents
7.	Iron and steel Smelting	Particulate Matter (PM)	Slude	Carbon monoxide and nickel
8.	Lead and zinc Smelting	Particulate Matter (PM) sulpher dioxide, cadmium, Arsenic	Lead, Cadmium and Arsenic	Particulate Matter (PM), Sulpher dioxide, Lead cadmium, Arsenic
9.	Meat processing and rendering	Odour	High biological oxygen demand	Infections
10.	Oil and gas Developing	Sulphur dioxide, carcinogens	Oil	Hydrocarbons

S.No	Industrial sectors	Air pollutants	Water pollutants	workplace hazards
11.	Pesticide Manufacturing	Pesticide and toxic intermediates	Pesticide and toxic intermediates	Pesticide and toxic Intermediates
12	Pesticide manufacturing	Sulphur dioxide	Oil	Hydrocarbons
13	Phosphate refining	Sulphur dioxide	Sludge and hydrocarbons	Hydrocarbons
14.	Phosphate fertilizer Plants	Particular Matter (PM)	Nutrients	-
15.	Pulp and paper mills	Odour	High biological oxygen demand, mercury	Chlorine
16	Tanning and Leather Industry	Odour	Chromium and acids	Chromium and acids
17	Textile manufacture	-	Toxic dyes	-

Appendix 1.42: Health Effects of Use or Exposure of Biological, Chemical and Physical Agents during Hospital Investigations

Biological Agents-Blood and Body Fluids

S.No.	Use or exposure	Health effects (Acute and chronic effects)
1.	Contact with blood and body fluids may occur as a result of medical and surgical procedures, such as labor and delivery, blood or body fluid collection and analysis, the handling of contaminated waste (e.g., gloves, linens, bandages, protective clothing, etc.) or the suctioning of airways. Exposure usually occurs because inadequate infection control procedures are in use.	**Acute:** The severity of infection depends on: • The number of pathogens encountered; • The worker's resistance, which is affected by such things as state of health, predisposing diseases, age, sex, and hereditary factor. • Portal of entry (via inhalation, ingestion, mucous membrane or skin contact, or direct inocultaion); and • Virulence of the organism. **Chronic:** Reproductive consequences ranging from congenital anomalies to death of the fetus and other chronic diseases, such as cirrhosis of the liver and primary liver cancer, may result from some viruses including hepatitis B, rubella, cytomegalovirus herpes and human immunodeficiency virus (HIV)

Chemical Agents

The following are specifically mentioned because of the severity of their health effects. This list is by no means all-inclusive

2.	**Ethylene oxide:** A disinfectant and sterilant, it is usually used in the central supply area Exposure usually occurs from improper aeration of the ethylene oxide chamber after the sterilizing process. It can also occur in outpatient surgery clinics and in the cardiac catheterization laboratory.	**Acute:** Respiratory and eye irritation, vomiting, and diarrhea. **Chronic**: Altered behavior, anemia, secondary respiratory infections skin sensitization, miscarriages, and reproductive problems. Carcinogenesis.
3.	**Waste anesthetic gases such as nitrous oxide, Halothane, Enfluorane:** Waste gases result from poor work practices during the anesthetization of patients, improper or inadequate maintenance of the machine, and/ or patient exhalation after the surgical procedure (recovery).	**Acute**: Drowsiness, irritability, depression, headaches, nausea, and problems with coordination and judgement. **Chronic**: Embryotoxicity, liver and kidney disease, and cancer.
4.	**Antineoplastic (cancer) drugs**: Antineoplastic drug, used in the treatment of cancer and other tumors, are usually given as intravenous fluids. Mixing usually occurs in the pharmacy area of the hospital in a biological safety cabinet. Exposure may occur during preparation, administration or disposal of the drug and equipment.	**Acute:** Severe soft-tissue damage, fetotoxicity, headaches, lightheadedness, dizziness, and nausea. **Chronic:** Chromosomal damage, teratogenesis, and carcinogenesis
5.	**Methyl methacrylate:** An acrylic cement-like substance used to secure prostheses to bone during orthopedic surgery. Exposure usually occurs during mixing, preparation, and in the operating room.	**Acute:** Eye, skin and mucous membrane irritant. Acute effects have varied from a decrease in blood pressure to (rarely) cardiac arrest. **Chronic:** Degeneration of the liver, mutagenesis, and teratogenesis.
6.	**Ribavirin:** An antiviral drug used to treat some infants and young children with lower respiratory syncytial virus infections. This drug is aerosolized to a respirable size of approximately 1.3 microns and is usually administered to the patient in an oxygen tent or face mask. This is when exposure can occur.	**Acute:** Headaches, coughing, dry upper respiratory tract and dry-burning eyes. **Chronic:** Carcinogenesis, fertility impairment and fetotoxicity

7.	**Formaldehyde:** Used as a fixative and is commonly found in most laboratories and the morgue.	**Acute:** Eye and respiratory irritation from the liquid and vapor forms Severe abdominal pains, nausea, vomiting and possible loss of consciousness could occur, if ingested in large amounts. **Chronic:** High concentration of vapor inhaled for long periods can f cause laryngitis, bronchitis or bronchial pneumonia. Prolonged exposure may cause conjunctivitis. Nasal tumors have been reported in animals suspected carcinogen.
8.	**Toluene or Xylene :** Solvents used to fix tissue specimens and rinse stains. They are primarily found in the histology, hematology, microbiology, and cytology laboratories.	**Acute:** Eye and mucous membrane irritation from vapor and liquid forms. Dizziness, headache, and mental confusion from inhalation of vapor. Ingestion or absorption through the skin can cause poisoning There is a potential for thermal burns as it is extremely flammable. **Chronic:** If the xylene or toluene contains benzene as an impurity, repeated breathing of the vapor over long periods may cause leukemia Prolonged skin contact may cause dermatitis. Tolune has been implicated in reproductive disorders.
9.	**Acrylamide :** The resin, usually found in research labs, is used to make gels for biochemical separations.	**Acute: Eye and skin irritation.** **Chronic:** Central nervous system disorders, i.e., polyneuropathy Suspected carcinogen. Mutagen.

Physical Agents

10.	**Laser :** Used in the operating rooms for excision and cauterization of tissue. Exposure usually occurs from unintentional operation and/or when proper controls are not in effect. The high electrical energy used to generate the beam is a potential shock hazard. The smoke plume during a surgical procedure and the laser's reaction to certain explosive or flammable agents also present hazards in the operating room.	**Acute:** From direct beam exposure, bums to skin and eyes possibly resulting in blindness. Chemical by-products in the smoke plume ma cause irritation to the eyes, nose and throat, and nausea. Biological and insert particulates can be found in the smoke plume but these have not been well studied for their effects. **Chronic:** Unknown.

11.	**Ionizing radiation :** Portable and fixed X-ray machines are used for diagnostic procedures. Exposure occurs when unprotected employees are near a machine in operation. The degree of exposure depends on the amount of radiation, the duration of exposure, the distance from the source and the type of shielding in place. Kits containing radioactive isotopes or specimens and excreta of humans and animals who have received radio-nucleotides may pose a hazard. Exposure may also result from handling of radioactive spills	Effects of radiation exposure are somatic (body) and/or genetic (offspring) in nature. **Acute:** Erythema and dermatitis. Large whole-body exposures cause nausea, vomiting, diarrhea, weakness, and death. **Chronic:** Skin cancer and bone marrow suppression. Genetic effects may lead to congenital defects in the employee's offspring.
12.	**Magnetic radiation :** Magnetic resonance instrumentation.	No conclusive effects.
13.	**Electrical hazards :** Exposure may occur when there is lack of maintenance to any electrical equipment, abuse, and lack of understanding of the equipment and/ or its controls. Oxygen-enriched atmospheres and water may contribute to hazardous conditions	**Acute:** Painful shocks, respiratory inhibition, deep burns (electric am thermal) heart rate irregularities, death. **Chronic:** No conclusive effects
14.	**Ultraviolet radiation :** Ultraviolet lamps are sometimes used in biological safety cabinets.	**Acute:** Skin bums, damage to the eye. **Chronic:** No conclusive effects other than cataracts
15.	**Compressed gases :** Compressed gases are used in many clinical laboratories. They are found in varying sizes and in pure or mixed states. Examples: ammonia, carbon dioxide, and nitrogen.	Compressed gases can be toxic, radioactive, flammable, and explosive These effects arise from the compression of the gas and the health effects of the chemical itself.
16.	**Glass :** Glassware is used as bottles, beakers, flasks, test tubes, pipettes, and tubing. Chipped, cracked, badly etched glassware and sharp edges present hazards.	Cuts, scratches and abrasion are potential locations for infection.

NOTES

Airborne (or Respiratory) Hazards may result from either an oxygen deficient atmosphere or breathing air contaminated with toxic particles, vapors, gases, fumes or mists.

Airborne hazards generally fall into the following basic categories:

- **Dusts.** Particles that are formed or generated from solid organic or inorganic materials by reducing their size through mechanical processes such as crushing, grinding, drilling, abrading, or blasting.
- **Fumes.** Particles formed when a volatilized solid, such as a metal, condenses in cool air. This physical change is often accompanied by a chemical reaction, such as oxidation. Examples are lead oxide fumes from smelting, and iron oxide fumes from arc-welding. A fume can also be formed when a material such as magnesium metal is burned or when welding or gas cutting is done on galvanized metal.
- **Mists.** A mist is formed when a finely divided liquid is suspended in the air. These suspended liquid droplets can be generated by condensation from the gaseous to the liquid state or by breaking up a liquid into a dispersed state, such as by splashing, foaming, or atomizing. Examples are the oil mist produced during cutting and grinding operations, acid mists from electroplating, acid or alkali mists from pickling operations, paint spray mist from spraying operations, and the condensation of water vapor to form a fog or rain.
- **Gases.** Gases are formless fluids that occupy the space or enclosure and which can be changed to the liquid or solid state only by the combined effect of increased pressure and decreased temperature. Examples are welding gases such as acetylene, nitrogen, helium and argon; and carbon monoxide generated from the operation of internal combustion engines. Another example is hydrogen sulfide, which is formed wherever there is decomposition of materials containing sulfur under reducing conditions.
- **Vapors.** Vapors are the gaseous form of substances that are normally in the solid or liquid state at room temperature and pressure. They are formed by evaporation from a liquid or solid, and can be found where parts cleaning and painting takes place and where solvents are used.
- **Smoke.** Smoke consists of carbon or soot particles resulting from the incomplete combustion of carbonaceous materials such as coal or oil. Smoke generally contains droplets as well as dry particles.
- **Oxygen deficiency.** An oxygen deficient atmosphere has an oxygen content below 19.5% by volume. Oxygen deficiency may occur in confined spaces, which include, but are not limited to, storage tanks, process vessels, towers, drums, tank cars, bins, sewers, septic tanks, underground utility tunnels, manholes, and pits.

Appendix 1.43: Methadology of Study of Health Effects of Air Pollutants

An average person breathes 22, 000 times a day and takes in 16 kg of air each day, which implies that individuals cannot live without air for more than five minutes. Contaminated air is a major culprit in most of the health effects associated with humans, animals, plants, microbes or other living matter. Damage to materials is also important while studying the effects.

The effect of an air pollutant is a measurable or perceivable detrimental change. These effects could be synergistic, additive, potentate or antagonistic, which are explained below:

Synergistic	Combined effect of pollutants is worse than the simple sum of effects of individual pollutants
Additive	Combined effect of pollutants is the same as the sum of effects of individual pollutants
Potentate	Toxic substance made worse by presence of non toxic substance
Antagonistic	Effect of toxic substance lessened by presence of nontoxic substance

The precise effect of pollutants on health is virtually difficult to obtain, because individual differences in humans are significant. The effects depend on the genetic make up, health, history of exposure, preconditioning and pollutant reaction time. The health studies are generally conducted to develop cause and effect relationships.

The studies of health effects are classified into three groups as discussed below:

Human Exposure Studies

These are based on the occupational or accidental exposures resulting in adverse effects. The studies also include subjects exposed to atmospheric pollutants in uncontrolled conditions. For example, truck drivers, traffic policemen, and individuals working in toll booths are used for determining cause and effect relationships.

Data collected from major air pollution episodes are used to establish a relationship between air pollution and reported diseases.

Laboratory Research/Animal Studies

This is conducted on animals or cells or biochemical systems. Actual exposure experiments are carried out. Different types of animals are critically studied under controlled conditions of concentrations and dosages. This provides immense information regarding the mode of various pollutants and their effects. Statistical analysis is rigorously used. The problem, however is that the extrapolation of these results to the human population is not precise. Moreover, one can only test one cause and effect relationship at a time.

Epidemiological Studies

In epidemiological studies, the relationship between the distribution of specific diseases in a human population and possible causes are determined. The studies focus on communities. A typical study uses hospital records, morbidity records and mortality records. To avoid misinterpretation, the population under study must be carefully observed for smoking habits, occupational exposures and any other factor that might prejudice the results of the study. The studies are not repeatable and often have no control over timing and other variables.

Appendix 1.44: Environmental Pollution Control in Underground Mines

Air is required in mines for four important reasons namely:

- to supply oxygen for breathing
- to remove heat from underground environment
- to dilute the presence of dust and various unwanted gases
- to remove dust and unwanted gases

Classification of underground mines environment impurities is shown below in the figure.

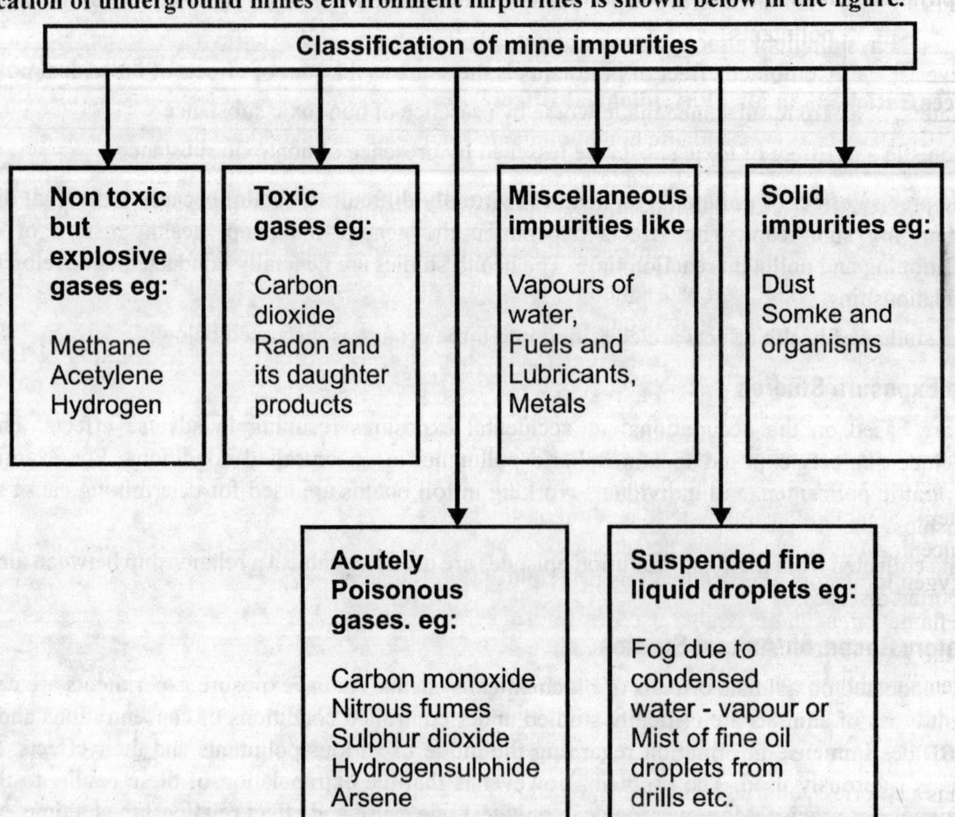

Classification of mine impurities

When air flows through the mine environment, its composition changes due to addition of different kinds of impurities along its path. These impurities and requirements in mines environment is discussed below:

Oxygen (O_2)

Oxygen is a colorless, odorless, and tasteless gas with a specific gravity of 1.1047(heavier than air = 1), a typical miner at rest consumes about 0.005 litres of oxygen every second due to respiration. Similarly, a miner exposed to moderate and extreme workloads consumes 0.03 litres and 0.05 litres of oxygen every second respectively due to his breathing.

Oxygen can cause a dangerous condition in a mine only by its absence. Air which has been stagnant for a considerable period may have much of its oxygen removed by the oxidation of metals and minerals and the decaying of timber, and some or all of it may be replaced by carbon dioxide. Such a mixture of gas which is deficient of oxygen, but is neither poisonous nor explosive, is called black-damp in underground mine environment and ventilation. Normal air contains 21% of oxygen by volume (23% oxygen by weight).

Mine safety laws in India require a TLV greater than 19% for oxygen. This means that any part of a given mine must have an oxygen concentration of atleast 19% in the mine air/atmosphere.

The physiological effects of staying/working in an atmosphere deficient in oxygen are tabulated below:

Physiological effects of staying/working in an atmosphere deficient in oxygen

O_2 Concentration, % in air	Physiological Effect
19	Flame height on flame safety lamp gets reduced by 50%.
17	Noticeable increase in rate of breathing. Increase in the concentration of carbon dioxide in the atmosphere.
15	Spinning sensation in head and increased heartbeat.
13-9	Disorientation, fainting, vomiting sensation, headache, blue lips, and coma may occur.
7	Coma, convulsions (violent uncontrollable contractions of muscles) or death may occur.
<6	The conditions may eventually lead to death.

The percentage of oxygen present in an atmosphere can be estimated by using the principle of electrochemical method, paramagnetic method or the flame safety lamp. In the electrochemical method, very small concentrations of the gas are detected by its influence on the output from an electrochemical cell. The MSA Oxygen Indicators are based on this principle and use fuel cells as oxygen sensors.

The flame safety lamp is also used to find out the percentage of oxygen in the atmosphere. In this method, the length of the flame produced by the lamp because of the burning of the gas is used. Accordingly, the oxygen percentage can be found out by noticing the flame height.

Nitrogen (N_2)

Nitrogen is a colorless, odorless and tasteless gas. It is slightly lighter than air with a specific gravity of 0.967. It is inert and does not support life or ignition. It has low solubility in water. There are three major sources of nitrogen in mines:

- Production by the decomposition of organic substances
- Production from blasting using explosives (1 kg of nitroglycerine releases 0.135 m3 of nitrogen)
- Production from the strata through cracks

Nitrogen has no known harmful effects on the human system but a higher concentration of nitrogen leads to deficiency of oxygen in the mine air. Thus, increase in nitrogen concentration indirectly leads to the physiological effects caused by a lack of oxygen on humans.

Carbon Dioxide (CO_2)

Carbon dioxide is colorless, odorless and has a slightly acidic taste. The specific gravity of carbon dioxide is 1.519 which is almost one-and-a-half times that of the specific gravity of air. That is why carbon dioxide is found in low-lying areas in the mines. It is fairly soluble in water and forms carbonic acid when dissolved in water. Its solubility in water increases with decrease in temperature {An increase in inhaled CO_2 and subsequent reaction with water in the blood forms carbonic acid (H_2CO_3), which then dissociates into hydrogen ions [H^+] and bicarbonate [HCO_3]. The excess CO_2 shifts the equilibrium towards the creation of more hydrogen ions, thus creating an acidic environment. That is why the pH of the blood becomes less than 7.35 in such conditions.

Carbon dioxide occurs both in coal and metal mines. It is produced from a variety of sources including strata emissions, oxidation of carbonaceous materials, internal combustion engines, blasting, fires, explosions and respiration. A mixture of carbon dioxide and nitrogen in which the concentration of CO_2 may vary from almost negligible to 20%, is known as black damp. Blackdamp is usually heavier than air, but becomes lighter when the percentage of CO_2 in it falls below 5.25%. Carbon dioxide liquefies at -5°C under a pressure of 31.4 bar. If it is further cooled, it solidifies into dry ice which derives its name from the fact that it evaporates in air without melting (undergoes sublimation process).

Carbon dioxide dilutes oxygen in air and acts as a stimulant to the respiratory and central nervous system. Diffusion of gas in bloodstream is rapid and it affects the rate and depth of breathing. The physiological effects of carbon dioxide are tabulated below:

Physiological effects of carbon dioxide

CO_2 Concentration, % in air	Physiological Effect
0.03	No effect- it is the normal concentration of $C0_2$ in air
0.50	Lung ventilation gets increased by 5%
2.00	Lung ventilation gets increased by 50%
3.00	Lung ventilation gets doubled, heavy breathing or panting on exertion
5 to 10	Violent panting leading to fatigue from exhaustion(extreme fatigue), headache
10 to 15	Intolerable panting, severe headache, rapid exhaustion and collapse

As per Indian Standards (DGMS), the CO_2 concentration should not be allowed to exceed 0.5 % (5000 ppm) in air

All active workings in coal mines must be ventilated by an air current containing not less than 19.5 % of O_2 and not more than 0.5% of CO_2. If we need to calculate the air quantity required only to satisfy the respiratory requirement, we need to consider both the cases i.e., the oxygen consumed by each worker and the CO_2 produced by him. Since, CO_2 is a contaminant, accurate measurement of quantity needed is a must for the safety of the workers. Table given below is useful in calculating quantity of air requirements under different conditions.

Respiratory, Requirements Under Different Situations

Type of activity of human beings	Respiration rate (Breath/min)	Air inhaled/ respiration (10^3 mm^3)	Air inhaled ($10^{-4} m^3/s$)	O_2 consumed ($10^{-5} m^3/s$)
At rest	12-18	377-705	0.82-2.18	0.47
Moderate	30	1476-1968	7.64-9.83	3.3
Vigorous work	40	2460	16.4	4.7

The term **respiratory quotient** is defined as the ratio of carbon dioxide expelled to the oxygen consumed by person in the process of respiration.

The presence of CO_2 is usually detected from oxygen depletion indicated by the extinguishing of oil lamps at 17-17.5%O_2.CO_2 also turns lime water milky and this property can be used in detection of CO_2 in mines. Optical methods like non-dispersive infrared gas analyzers can also be used in finding out the concentration of CO_2. Portable instruments are also used to estimate CO_2 concentration. Hand held detectors based on the principle of colorimetric indication are also used for finding our CO_2 percentage. These detectors have detector tubes which are filled with an indicating chemical substance. When a sample of air is drawn through a detector tube, this substance changes its color over a length. This length is proportional to the concentration of CO_2. The concentration of gas can be read off on the concentration scale printed on the tube.

Carbon Monoxide (CO)

CO is a colorless tasteless odorless gas with a specific gravity of 0.972.Its specific gravity is almost equal to that of air and therefore it exists at all levels in an underground opening. It burns with a blue flame and is explosive in presence of air at concentrations between 12.5% and 75%. Though CO is an inflammable gas, this fact is of no practical importance in mining because it never occurs in significant concentration to burn or to cause an explosion. The ignition temperature of CO is 873K. CO is produced by the incomplete combustion of carbonaceous materials. It is also produced by internal combustion engines, blasting and spontaneous combustion in coalmines. It can also be generated as a component of water gas (mixture of CO and H_2), when water is applied to coal for controlling the fire. That is why it is advisable not to apply water at the centre of coal fire because it will lead to the formation of hydrogen and CO, and because of the formation of hydrogen which is an explosive gas, this fire will become more violent.

CO is considered as the most dangerous gas in mines because of the following reasons:

- It is highly toxic in nature.
- As it is colorless tasteless and odourless; it is not noticed easily.
- As its specific gravity is almost equal to the specific gravity of air, it exists at all the levels in an underground opening. Hence the chances of its inhalation are very high.

As per Indian Standards (DGMS), the CO concentration should not be allowed to exceed 0.005 % (50 ppm) The TWA for CO is 0.005% and STEL is 0.04% (400 ppm).

Hemoglobin present in human blood has 300 times more affinity towards CO than O_2.The new substance formed by the combination of CO and hemoglobin is known as carboxy-haemoglobin. This is relatively stable and accumulates in the bloodstream. This results in a reduction in the number of red cells for carrying oxygen to vital parts of the body. Thus the physiological effects of CO arise because of the reduction in oxygen supply to vital parts of the body. The symptoms depending upon the saturation of blood by carboxy-haemoglobin are given below in Table.

Physiological effects of CO poisoning

Blood saturation, % CO.Hb	Symptoms
5–10	Slight loss of concentration
10–20	Sensation of tightness across forehead, slight headache
20–30	Throbbing headache, judgment impaired
30–40	Severe headache, dizziness, disorientation, dimmed vision, nausea(vomiting), possible collapse
40–60	Increased probability of collapse, rise in rates of pulse and respiration, convulsions
60–70	Coma, depressed pulse and respiration, possible death
70–80	Fatal- will lead to death

CO.Hb - Carboxyhemoglobin

Carbon monoxide imparts a bright pink color to blood, and the patient poisoned by CO has a typically pink color as a result of this. The best remedy for CO poisoning is to quickly expose the patient to fresh air and provide him pure oxygen. The patient should be covered by blankets so that he can be kept warm. Black coffee is also very useful. The property of blood turning pink by absorbing carbon monoxide has been used for testing the gas. The air with CO contained in it is passed through a light straw colored solution of blood and the coloration produced is compared with a standard color chart calibrated for different colorations of the gas. This method gives a good accuracy within the range of 0.01 to 0.2% CO.

Modern CO detectors, tubes and ammonium-palladium-complex colorimetric detectors use detector tubes containing suitable gels such as alumina, silica gel, iodine pentoxide and fuming sulphuric acid soaked in pumice stone, silica gel impregnated with palladium sulphate and ammonium molybdate etc. The CO is detected using the principle of the charges in the:

- color of the chemical present in the detector tube
- length of the color of the detector tube.

Warm blooded birds like munia/canary or mouse are also used for detecting CO as they are affected sooner than human beings by CO. Only fresh birds are used in 'this method as repetition of same bird may lead to the acclimatization of the bird to the low percentages of CO. There are no immediate signs of distress observed when birds are exposed to 0.1% of CO. But at 0.15% of CO, a bird shows distress (pronounced chirruping and loss of liveliness) in 3 minutes. And at 0.3% of CO in air, the bird shows almost immediate distress and falls off its perch (when a bird perches on something such as a branch, it lands on it and stands there) in 2-3 minutes.

Hydrogen Sulphide (H_2S)

H_2S is a gas with a smell similar to that of a rotten egg and has a sweetish taste with a specific gravity of 1.175. It burns with a light blue flame and is soluble in water. It is also known as Stink Damp. It is combustible and is explosive over a wide range of its concentration from 4.3 % to 45.5 %. It is more poisonous than CO and has a TWA value of 10 ppm and an STEL value of 15 ppm. H_2S is not very common in mines and usually occurs in firedamp and gob fires in sulphurous coal. It is formed naturally by bacterial/chemical decomposition of organic compounds and is often detected near stagnant pools of water in underground mines. It may also occur in natural gasand petroleum reserves and migrate through the strata in a weakly acidic water solution. In metal mines it is produced by the action of acidic water on iron pyrites which can be represented by the equation:

$$[FeS_2 + 2H_2O \rightarrow Fe\,(OH_2) + H_2S + S]$$

Hydrogen sulfide has a very low odor threshold, with its smell being easily sensed by the human nose at concentrations well below 1 part per million (ppm) in air. The odor increases as the gas becomes more concentrated, with the strong rotten egg smell recognizable up to 30 ppm. Above this level, the gas has a sickeningly sweet odor up to around 100 ppm. However, at concentrations above 100 ppm, a person's ability to detect the gas is affected by rapid temporary paralysis of the Olfactory nerves in the nose, leading to a loss of the sense of smell. This means that the gas can be present at dangerously high concentrations, with no perceivable odour. Prolonged exposure to lower concentrations can also result in similar effects of olfactory fatigue. This unusual property of hydrogen sulfide makes it extremely dangerous to rely totally on the sense of smell to detect the presence of the gas. The physiological effects of H_2S poisoning are given below in Table:

Physiological effects of H_2S Poisoning

Concentration of H_2S (ppm) in air	Physiological Effects
50-100	Slight irritation to eyes and respiratory tract, headache
200	Intensified irritation of nose and throat
500	Serious inflammation of eyes, nasal secretions, coughing, palpitations, fainting
600	Chest pains due to corrosion of respiratory system, may be fatal
700	Depression, coma, probable death
1000	Paralysis of respiratory system, very rapid death

A person who recovers from H_2S poisoning may have conjunctivitis and bronchitis for a long period after recovery.

Although hydrogen sulfide is an extremely poisonous gas, miners are rarely affected by it. This is mainly because it seldom occurs in dangerous concentrations and also a very small concentration can be easily detected by virtue of smell.

H_2S is easily detected by its smell at very low concentrations of up to 0.000075% (0.75 ppm).Another detection test for this gas is done by exposing a filter paper soaked in lead acetate solution to an atmosphere containing H_2S gas. The filter paper turns brown and then black if the concentration of the gas is sufficiently high. H_2S.detector is an accurate instrument for detecting H_2S. It consists of a glass tube filled with white granules of activated aluminum oxide coated with silver cyanide. When gas containing H_2S is drawn through the tube, the gas combines with silver cyanide forming black silver sulphide which turns the granules black. The percentage is calculated with the help of a scale placed along the side of the tube which measures the length up to which the change of color has taken place.

Nitrous Fumes (NOx)

Nitrous fumes are rarely found in mines and mainly consist of nitric oxide (NO) nitrogen dioxide (NO_2) and nitrogen tetra-oxide(N_2O_4). The original product is nitric oxide (NO) which quickly combines with oxygen to form nitrogen dioxide (NO_2), which has a pungent smell like that of fuming nitric acid. As NO_2 cools down it is slowly converted to nitrogen tetra-oxide (N_2O_4), which is a colorless gas. The specific gravity of NO is 1.036, specific gravity of NO_2 is 1.519 and specific gravity of N_2O_4 is 1.588. The nitrous fumes are highly soluble in water. These fumes are very poisonous and TWA value for NO is 25 ppm, for NO_2 is 50 ppm and for N_2O_4 is 3 ppm. The nitrous fumes are mainly produced by explosives containing nitroglycerine and in exhaust fumes of diesel locomotives. They are also found in shafts and tunnels where heavy shot firing takes place in a confined place.

The brown fumes of NO_2 get dissolved in water and produce nitrous acid (HNO_2) and nitric acid (HNO_3). These acids cause irritation and corrosive effect on eyes and respiratory system.

Various physiological effects of these fumes at different concentrations are given below in Table.

Physiological Effects of Nitrous Fumes at Different Concentrations in Air

Concentration of NO_2 (ppm) in air	Physiological Effects
60	Minor throat irritation
100	Coughing may commence
150	Severe discomfort, may cause pneumonia later
200	Conditions are likely to be fatal to human beings

The first-aid for NO_2 poisoning includes:

- Administration of oxygen to the affected person,
- The person should rest and should not be allowed to move,
- The person should be kept warm so that body temperature doesn't fall to a lower value.

The common test for nitrous fumes is to soak a filter paper with starch and potassium iodide solution and expose it to the air containing nitrous fumes. The paper turns blue by the liberation of iodine. Tube type detectors are also %available which indicate the percentage of nitrous fumes by the length of change in color in the tube. For example tubes manufactured by Drager use diphenyl benzidine as the reactant.

Sulphur Dioxide (SO_2)

This is a colorless gas with a very suffocating odor and a specific gravity of 2.264. It is highly soluble in water. The TWA value for this gas is 2 ppm and STEL value is 5 ppm. It is found in smaller quantities in afterdamp (Afterdamp is a mechanical mixture of gases found in a mine after an explosion. Afterdamp contains nitrogen and carbon dioxide as the chief constituents along with carbon monoxide, methane, water vapour, hydrogen, oxygen and small quantities of H_2S and SO_2. A typical composition of afterdamp may be: CO-1.5%, CO_2-8%, O_2-5%, CH_4-0.3%, H_2-0.2%, N_2-85%) in some coal mines; but occurs abundantly in sulphide ore mines with fires. Blasting in rich sulphide ores also produces large quantities of SO_2 and H_2S. It is also released from the exhaust of internal combustion engines.

SO_2 is an irritating and toxic gas. At low concentrations it causes intense burning sensation to the eyes and the respiratory tracts. Table below the gives the physiological effects at various concentrations of sulphur dioxide.

Physiological Effects of SO_2 Poisoning

Concentration of SO_2 (ppm) in air	Physiological Effects
20	Irritation of eyes and respiratory system
50	Severe burning sensation in eyes, nose and throats
400	Immediately dangerous to life

SO_2 can easily be detected by its odor even at low concentrations of 0.0005% (5 ppm). However, tube-type detectors using iodine, starch, and potassium iodide as reagent are available for more accurate estimation.

Hydrogen

It is colorless, odorless, tasteless and non-toxic in nature. It is the lightest of all gases. It has specific gravity of 0.07. This makes it rise to the roof.

The sources of hydrogen in underground mines includes:-

- Charging of batteries
- Action of acids on metals
- Action of water on hot coal (as water gas) or even some of the hot minerals
- Rarely as strata gas and in afterdamp

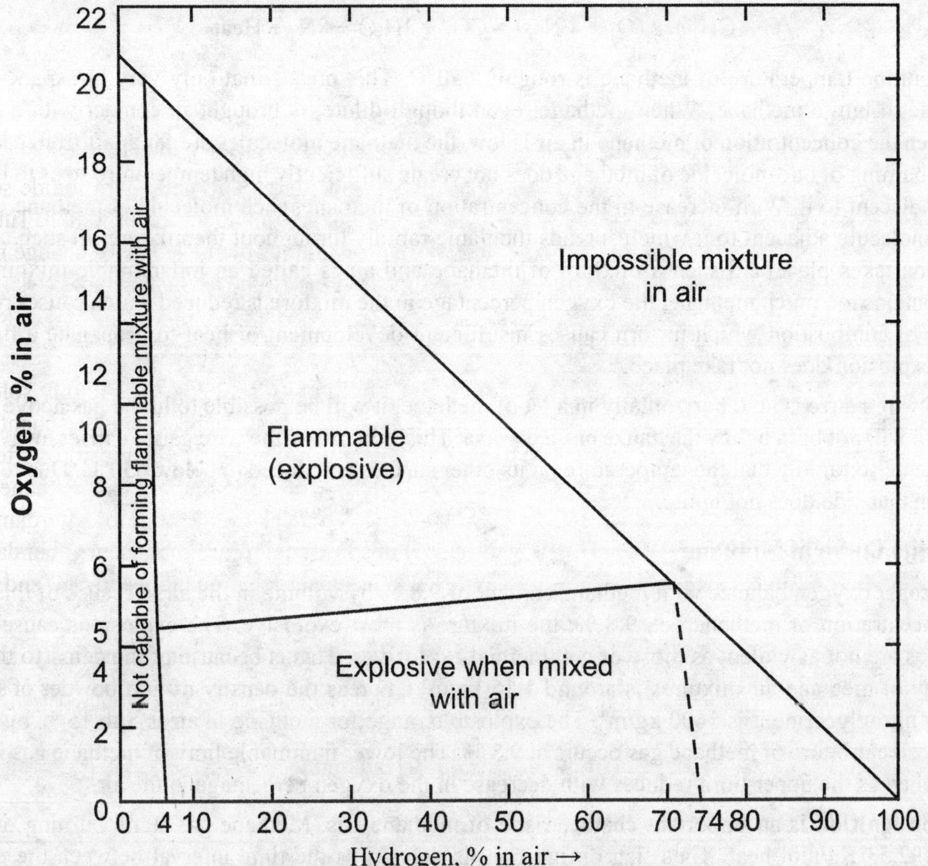

Expolsibility diagram for hydrogen

The explosibility curve of hydrogen is shown above in Fig. As can be seen from the figure, the explosibility range of hydrogen is 4–74% in air. **Also, it burns/ignites at 480° C, lower than that of methane.** It ignites at oxygen concentration in air as low as 5%. Methane requires at least 12% of oxygen in air for ignition. **Further, hydrogen requires almost half of the energy required by methane to ignite.** This makes hydrogen a very explosive gas.

Methane (CH_4)

Methane is a colorless, odorless and tasteless gas. This gas is lighter than air as its specific gravity is equal to 0.559. That is the reason why it tends to rise to the roof of a mine working. Methane becomes liquid below 112 K and solidifies below 90.5 K. Methane gas is poorly soluble in water, but is soluble in organic solvents like alcohol and ethers. This property of methane is utilized during drainage of coal bed methane. It burns with a blue flame and produces carbon dioxide and water as products.

$$CH_4 + 2O_2 \rightarrow CO_2 + 2H_2O + \text{Heat}$$

This exothermic reaction is a chain reaction and under suitable conditions gets self accelerated and leads to an explosion. The combustion of methane in air can be represented as follow:

$$CH_4 + 2\,(O_2 + 4\,N_2) \rightarrow CO_2 + 2H_2O + 8N_2 + \text{Heat}$$

The ignition temperature of methane is roughly 650°C. This means that only very hot spark or flame will be able to ignite methane. When methane, even though dilute, is brought in contact with a flame, it burns. When the concentration of methane in air is low, the methane molecules are far apart from each other. Therefore burning of one molecule of methane does not create sufficiently high temperature so as to ignite the molecule adjacent to it. With increase in the concentration of methane, each molecule of methane is able to ignite the molecule adjacent to it which spreads the flame rapidly throughout the mixture. In such condition an explosion takes place and such a mixture of methane and air is called an inflammable mixture. If the mixture contains too much methane, the oxygen percentage in the mixture is reduced which causes reduction in the rate of combustion which in turn causes insufficient development of heat to propagate a flame and therefore explosion does not take place.

When wire gauze is held horizontally in a jet of methane, it will be possible to lit the gas above the wire gauze and it will not burn below the gauze or vice-versa. This is because, the wire gauze carries away the heat from the flame so rapidly that the temperature on its other surface is reduced to below 650°C. Due to this, the methane on that side does not ignite.

Explosibility Curve of Methane

There is proper oxygen balance with methane content of 9.8 % by volume in the air. Because of this reason, at this concentration of methane i.e., 9.8 %, the mixture is most explosive. The explosions caused due to methane gas are not as violent as those of commercial explosives. This is because of the density difference. The density of methane-air mixtures is around 1.15 kg/m^3 whereas the density of gun powder is 1000 kg/m^3 and for nitroglycerine it is 1600 kg/m^3. The explosible range for methane in air is 5 to 15 % by volume. Most explosive mixture of methane gas occurs at 9.8 %. The lower flammable limit of methane gas is almost constant whereas the upper limit reduces with decrease in the oxygen percentage in the air.

Lag on ignition is an important characteristic of methane gas. Methane gas starts burning only after absorbing 92.53 KJ/mol heat. Thus 'lag on ignition' is defined as the time interval between the exposure of "CH_4 - Air" to an igniting source to the appearance of flame. This lag on ignition is dependent on the temperature of the igniting source. As for instance, at 650 °C, the delay is around 10 seconds, at 1000 °C, it is 1 second and 1200 °C it is 1/15th of a second. This property of methane gas is utilized in designing of permitted explosives which are prescribed for use in underground coal mines. These permitted explosives produce a flame of a very short duration. In this time period/duration, methane doesn't get enough heat required for its ignition. Thus the permitted explosives can be used safely in underground coal mines. The presence of hydrogen or other gases reduces the ignition lag.

In India, coal mines are classified into three categories based on methane emission. The details of this classification is given below in Table:

Classification of Coal Mines Based on Methane Gas Emission

Degree of gassiness	Percentage of inflammable gas in general body of air	Rate of emission of gas in m3/ton of coal raised
I	< 0.1 and	<1
II	> 0.1 and/or	1–10
III	–	>10

As can be seen from above Table for underground coal mines to come under Degree - II gassy mine category, both the "percentage of inflammable gas in general body of air" as well as "rate of emission of methane gas in m³/ton of coal raised" must be satisfied. For underground coal mines to come under Degree - II gassy mine category, any one of the conditions i.e., "percentage of inflammable gas in general body of air" or "rate of emission of methane gas in m³/ton of coal raised" is sufficient.

Methane is usually found in coal mines and sometimes in the tunnels passing through carbonaceous shale. Methane is the major constituent of firedamp [a mixture of air which mainly consists of methane (80–96%) with other minor gases such as nitrogen, ethane, carbon dioxide etc], which is a common gas mixture in mines.

The vegetable matter composed mainly of cellulose and lignin is decomposed by bacterial action which produces methane and carbon dioxide. That is why methane is sometimes called as marsh gas. The process continues and in the early stage of coalification, much of methane escapes because of poor confinement but largely retained in higher rank coal. Coal has large number of small pores and correspondingly high internal surface area. Porosities of coal may vary from 1 % to more than 20 %.

Methane exists in coal in two different forms which are referred as free gas and adsorbed gas. In free gas, the methane molecules are free to move within the pores. In adsorbed state, the methane molecules are adsorbed on to the coal surface. Around 95 % of the total methane content in coal is in adsorbed state which is due to very high gas pressure.

Methane is not the only gas that is trapped in the coal seam or strata. The other gases which are also found in coal seams are carbon dioxide, carbon monoxide, nitrogen, etc. Of these, carbon dioxide is the most important one. It is the carbon dioxide which is a major constituent in coals of lower grade, while methane is the major constituents in the higher rank coal. For example, in lignite, bituminous and sub-bituminous coal, carbon dioxide is the most significant one. However, in anthracite (which is the highest rank coal) methane has major part compared to carbon dioxide. One tone of anthracite may have approximately 765 m³ of methane and 565 m³ of carbon dioxide.

During coalification, a fraction of these gases escape out. The amount of gas retained per ton of coal is called seam gas content (includes all gases). It is generally expressed in m³/ton. Another important term to know is, the specific methane emission, defined as the volume of methane emitted per ton of coal. It is used as an index for/of seam gas content.

Methane is emitted mainly from coal seams and associated strata in an underground mine. Methane in the coal seam has considerable pressure. But the mine workings have pressure equal to that of atmospheric pressure. Hence the adsorbed gas in the coal gets desorbed because of the lowering of pressure when the rock containing the gas is drilled/cut. Methane tends to migrate to low pressure zones. Speedy migration of methane takes place through macro cracks and fissures and slow migration takes place by permeation through micropores and micro cleavages. Major channels for gas migration are faults and joints. Dykes and sills also provide a path for methane migration.

When coal is cut, transported etc, desorption of methane occurs. The size of coal produced has a great influence on the amount of gas emitted at the face. Large lumps of coal release only 2 % of their total gas content in the first 10 minutes whereas coals of 0.25–1.0 mm size release 40 % and coal of size less than 0.25 mm release 66 % of their total gas content.

Some of the Indian laws regarding the methane concentration are:

- Methane concentration should not exceed 0.75 % in the return of a ventilation district.
- Methane concentration should not exceed 1.25% in any part of the underground mine.
- Electric supply should be cut-off from the district if the methane concentration exceeds 1.25%. Also workers should be withdrawn from the place.
- Charging, stemming or firing of shot holes is not allowed in areas where methane is found /detected.
- In gassy mines, where electricity is used, detection/testing of methane is carried on the intake side of the first working face and on the return side of the last working face in a district as per the following rule:

Concentration of methane in air, %	Testing to be carried out every
<0.6	30 days
0.6% – 0.8	week
>0.8%	day

Methane in mine air can be detected either by using chemical analysis in laboratory or by using flame safety lamp and special instruments called methanometers. Using safety lamp or methanometers, methane can be detected on the spot in underground.

Flame Safety Lamp

It was discovered by SIR HUMPHREY DAVY in 1815. It was discovered with dual purpose of detecting methane as well as mine illumination. Some of the main/major modifications made to the safety lamp since its discovery are:

- Introduction of glass cylinder around the flame
- Addition of internal chimney
- Two gauzes instead of one
- Safety locks, arrangement for relighting, controlled air-feeding arrangement

The flame safety lamp can be divided into three main sections – lower, middle and upper sections. There are various components in all the three sections. Table below lists the different components found in different section along with their functional details.

Components of Flame Safety Lamp in Different Sections along with their Functional Details

S.No.	Section	Name of the components	Functions/details
1.	Lower	Fuel vessel, burner, wick assembly, a screw, locking arrangement, and re-lighting device (if available)	Locking arrangement is generally of magnetic type. It can be opened only in the lamp room. Thus, once it is locked, it cannot be opened in the underground.
2.	Middle	Two sets of rings interconnected by five steel rods, glass (may be one or two set-inner and outer) around the flame, chimney, two asbestos washer	Ring and rod combination serves to protect the glass. Rings have holes to allow air in/out. Chimney is introduced to increase the illumination produced. Asbestos washer is used to make the assembly air tight.
3.	Upper	A bonet, two steel wire gauzes (of 20 or 28 mesh size), handle/hook	It is the wire gauze which is responsible for the working of flame safety lamp and not allowing the ignition of methane, even if present in the general body of air. Handle serves the purpose of holding the lamp while detecting methane.

Working Principle of Flame Safety Lamp

It is very interesting to know that wire gauze does not let methane to ignite even if there is methane in the general body of air. The ignition of methane remains limited only to the interior of the lamp. For a lay man it would be like a magic.

In a flame safety lamp the distance between the wire gauze and the mouth of the Bunsen burner is kept to be around 30 mm. The gas is turned on and then by using match box or suitable device it is ignited below the gauze. It is observed that the flame of the burning gas is restricted below the gauze only and is not able to cross the gauze wire until the gauze wire is red hot. Once the wire is red hot, the flame can be seen above as well as below the gauze wire.

If gauze wire is lowered to the mouth of the Bunsen burner or moved upward, the flame gets extinguished, even if gas supply is turned on. However, this is possible only when it is done before the gauze wire becomes red hot.

Flame safety lamp to pass through a performance test at different concentrations of methane in the air

Performance Test on Flame Safety Lamp

Methane gas, %	Minimum length of blue flame (mm)	Condition of blue flame
0.0	–	Near the top of the standard flame a slight cobalt-blue lined orange-yellow flame is seen
1.0	7.0	Blue flame is not observed because the flame color is light and hence it is difficult to measure the length
1.5	8.0	Lower part of flame turns blue
2.0	9.0	Blue flame is distinctly visible except at the top

Methane gas, %	Minimum length of blue flame (mm)	Condition of blue flame
2.5	10.0	Blue color becomes more visible, but still top is invisible
3.0	11.5	Top of flame is still invisible and blue flame is seen clearer
3.5	14.5	Blue flame is clearly visible
4.0	20.0	Blue flame is extremely clear, highly sensitive to slight change in methane concentration

Illumination in Mines

Mine illumination refers to the lighting up of the mine. Surface mines are sufficiently illuminated by the sun in the day times but needs to be provided with some artificial illumination during the night times. But in case of an underground mine artificial illumination is required right from the starting of the mine till its closure.

This condition of complete darkness in any underground mine is quite similar to what we experience every day.after the sunset. We cannot light up the underground mine with tube lights, CFL or LED lights at frequent locations, the reason being methane gas, which can ignite and cause an underground explosion with devastating results. Consequently any illumination system must be designed to be either "intrinsically safe" or "explosion proof". An intrinsically safe light source is one in which the current feeding the light has very little energy so that any short in the circuit will not produce a spark which can ignite the methane gas. In an explosion proof lamp, any explosion triggered by the lamp's electrical activity will be contained within the device. In addition, the device itself will not become hot enough to cause an explosion.

Then how is mine lighting unique from what is being used in the other industries? It is the low level of luminance, which is usually not encountered in any other industry. In such conditions the eye is highly susceptible to glare.

Before the invention of the flame safety lamp, naked lamps were used in underground coal and metal mines. Now a day's worker in coal mines and a large number of metal mines are using the electric cap lamps.

A flame safety lamp lights up the mine without the danger of igniting inflammable gases, especially in underground coal mines. It is also a very convenient and handy tool for detecting the presence and the percentage of fire damp in the underground coal mines.

General Lighting Arrangements in Mines

We have to ensure the provision of adequate general lighting arrangements during working hours:

a. **On surface**, at places with insufficient natural light, in every engine house, in the vicinity of every working shaft, at every opencast working and at places where workers are employed.

b. **Underground:**

   i. At every shaft inset and shaft bottom or siding which is in regular use.

   ii. In every travelling roadway in which 50 or more miners travel in any shift, and, in every working stope, in metalliferrous mines.

   iii. At every place on a haulage roadway, at which tubs are regularly coupled or uncoupled, or attached to or detached from a haulage rope.

   iv. At every place where mechanical filling of the tubs is being carried out.

v. At the top and bottom of every self-acting incline, which is in regular use.

vi. At every first aid station in the underground.

vii. At every room and place where engine, motor or any other apparatus is present.

viii. In coal mines, at every place where pillar is under extraction.

As per the Indian Electricity Rules, in every mine illuminated by electricity, one or more flame safety lamps, or other lights, approved by the Regional Inspector, should be maintained in a state of continuous illumination at all places where failure of the electric light at any time would be prejudicial to safety.

Recommended Standards of General Lightings in Underground Mines is Tabulated Below:

Place in the	Minimum average illumination level (lux or lumens/m^2)
Pit bottom	10
Main junction	15
Roadways	4.5
Haulage engine room	
i. Floor	10
ii. Drum	25
iii. controller	20

Recommended Standards of General Lightings in Opencast Mines is Tabulated Below:

S.No	Place/area to be illuminated	Manner in which it is to be illuminated	Minimum standard Illumination (Lux)	Plane/Level in which illumination is to be provided
1.	General working areas as determined by the manager in writing	–	0.2	At the level of the surface to be illuminated
2.	Work place of heavy machinery	So as to cover the depth and height through which the machinery of the rig operates.	5.0	Horizontal
3.	Area where drilling rig works	So as to illuminate the full height of the rig.	10.0	Vertical
4.	Area where bull dozer or other tractor mounted machine works.	–	10.0	At the level of the crawler tracks.

S.No	Place/area to be illuminated	Manner in which it is to be illuminated	Minimum standard Illumination (Lux)	Plane/Level in which illumination is to be provided
5.	Places where manual work is done.	To be provided at the level of the surface on which such work is done.	5.0 10.0	Horizontal Vertical
6.	Places where loading, unloading or transfer loading of dumpers, trucks or train is carried on	–	3.0	Horizontal
7.	Operator cabins of machines or mechanisms	To be provided up to a height of 0.8 meters from the surface	30.0	Horizontal
8.	At hand picking points along a conveyor belt.	To be provided up to a distance of not less than 1.5 meters from the picker	50.0	On the surface of the conveyor belt
9.	Truck haulage roads	To be provided at the level of the road	0.3 to 3.0	Horizontal
10.	Rail haulage track in the pit	To be provided at the level of the rail head	0.5	Horizontal
11.	Roadways and footpaths from bench to bench	–	3.0	Horizontal
12.	Permanent paths for use of persons employed, etc.	–	1.0	Horizontal

Control of Heat (Thermal) in Mines

Human beings are capable of working efficiently within a certain range of temperatures which is concerned with degree of hotness. Temperature of a substance or body is liable to change when we have a source of heat. Heat always flow from high temperature to low temperature body. This flow can take place in any one of the three ways of transferring heat- conduction, convection and radiation. These modes of transferring heat mainly cause flow of sensible heat. Through evaporation, transfer of latent heat takes place. In this case, we do not sense the rise in temperature, but heat content of the body is increased. In underground mines, there are many sources of heat which cause rise of temperature of air during its travel in mine airways. The condition may worsen if the temperature of the air increases beyond a certain limit. The situation becomes more critical especially when humidity of air also increases simultaneously. In such situation, we require air conditioning of mine air. We have to set up refrigeration plants, spot coolers, etc. so as to ensure comfort, safety and high working efficiency of workers in mines. On the other hand, in areas with extreme cold climatic conditions, we need to supply heat to the air on the surface before sending it to underground mines.

There are various sources of heat in underground mines. Some are major contributors of heat addition to air and some are minor. Some of them are unavoidable, while some of them require utilization of proper techniques for their reduced effect.

The major sources of heat in underground mines are:

- Strata heat
- Auto-compression
- Machinery and lights
- Underground water

The minor sources of heat in mines include:

- Human metabolism
- Oxidation
- Blasting
- Rock movement
- Pipelines
- Energy losses in airflow

Strata Heat

Strata heat is a major source of heat in underground mines. In some cases especially in deep mines, it becomes Intolerable and we require installation of refrigeration systems so as to provide safe and comfortable working conditions. The term strata heat means the heat emitted from the surrounding rocks and getting added to the mine atmosphere. Subsurface rocks mainly have their heat reservoirs in the core portion of the earth, which emits heat and that heat gets transferred to the upper portion.

Metabolic Heat Balance in Human Body

The food taken by human beings serves as fuel for their body. The oxidation of food is an exothermic reaction. The energy released in the process is called metabolic energy. This metabolic energy produced by oxidation of food is used for working against gravity, growth of body (accumulates in the body) and rest is transferred to the surroundings. The energy released depends upon the oxygen consumed in the oxidation of food. More is the amount of oxygen consumed, more is the energy produced. When we work hard, the breathing rate increases. This increases consumption of oxygen. That's why as the amount of work load increases, the amount of metabolic heat produced by human beings increases. The metabolic heat produced can be transferred to the surrounding by convection, radiation, respiratory exchange (breathing) and evaporation.

In the form of equation we can write the energy/heat flow for a human body as follow:

$$\textit{metabolic energy } (M) = \textit{metabolic heat} + \textit{work against gravity}(W) + \textit{stored energy } (A_c)$$

$$\text{or, } M = W + A_c + (Rad + Con + Evp + Br)$$

Heat flow due to radiation and convection can be reverse also. It happens when the skin temperature of human body is less than the dry bulb temperature (DBT) of the mine air. In this case, human body mainly removes heat by evaporation of sweats produced. Heat removal through breathing is less than 5% of the metabolic energy produced and hence neglected.

Human body can be thought of divided into two-parts (interior to exterior) namely -central core body and the skin. Blood vessels separate these two. Human body has to be maintained at 36.9 °C for proper physiological functions. The heat to maintain this temperature is provided by oxidation of food. This temperature is called temperature of the central core body (T_c). However, T_c may vary ±1°C depending up on the atmospheric conditions and individual response to the atmospheric conditions. Hence, we can say that the normal central core body temperature of human being lies in the range of 35.9–37.9 °C. Blood acts as cooling agent for skin. In general, mean skin temperature (T_{sk}) is lower than T_c. At 34 °C, the skin does not feel any sensation of hotness or cold with the environment. T_{sk} depends on many factors including dry bulb and wet bulb temperature, rate of perspiration, types of clothing, etc.

Hypothermia: A state when the rate of heat production in the body is lower than the heat removed to the surrounding leading to reduction of the central core body temperature i.e., below 36 °C.

Hyperthermia: A state when the rate of metabolic heat production is more than the rate of metabolic heat removal, leading to increase in the central core body temperature i.e., above 37.9 °C.

Effects of Execessive Heat and Humidity on Human Beings

Heat Loss Greater than Heat Generated

When the rate of heat loss to the surroundings (H_{loss}) is more than the rate of metabolic heat generation (H_{gen}), the effects observed can be shown as in Fig. below:

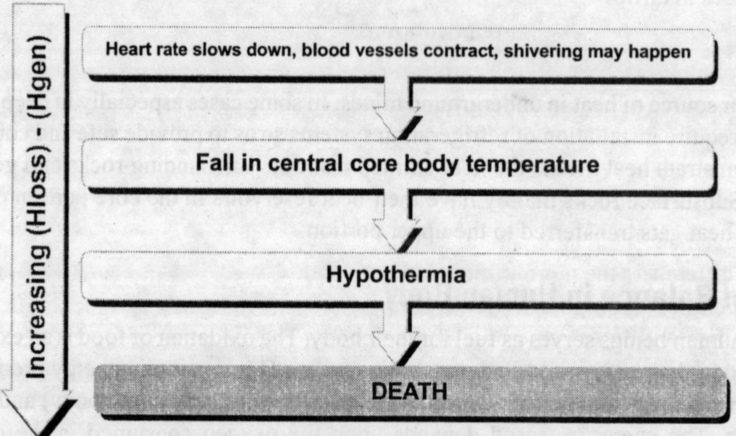

Physiological effects observed when H_{loss} is more than H_{gen}

Heat Loss Lower than Heat Generated

This is more common in mine atmosphere. This condition to a miner/worker may arise due to the combined effect of wet bulb and dry bulb temperatures, psychrometric properties of the mine air and velocity of air-current. Besides these factors, the effects observed on an individual are also dependent on time duration of exposure to such condition. This condition generally arises when the dry bulb temperature of the air increases beyond a certain limit such that heat removal from the body is reduced. If the dry-bulb temperature of air/atmosphere goes on increasing, heat removal from body through radiation and convection may reverse the directions (heat may be added to the human body from the atmosphere through the process of radiation and convection). In such cases, the metabolic heat from the body is removed through evaporation of sweat. Sweat, produced by the sweat glands, adds moisture to the surrounding air. If the humidity of the air is already high, the evaporation of sweat is reduced and therefore the central core body temperature of an individual rises rapidly. Even if air is moderately humid, sweat glands reduce their capacity to produce sweat with time. A time comes when no more sweat is produced due to fatigue of the sweat glands. If exposure to such environmental conditions continues, it may lead to heat stroke ultimately leading to the death of the person. Table below lists the physiological effects on a miner/individual with the rise in temperature of air and human body.

Effect of Increase in The Dry Bulb Temperature of Air

Dry bulb temperature (°C)	Body temperature (°C)	Physiological effects on human beings
≤25	–	Normal blood circulation, no observable effect, heat removal from body mainly through convection and radiation
25–29	–	Heat removal rate increases, slight rise in central core body temperature, vaso-motor control of body increases blood circulation
29–37.5	–	Body starts sweating, heat removal mainly through evaporation
≥36.9	36.9	Body temperature equals dry bulb temperature, heat transfer through convection and radiation reverses the direction, heat removal from body through evaporation only.
–	39	Heart beat rises above 140 beats/min, fatal
–	41	Unconsciousness, coma, may lead to death
–	> 43.3	Sudden death

Effect of High Wet Bulb Temperature of air

The difference between the wet bulb temperature and dry bulb temperature decide the relative humidity of the air. The lesser is the difference, the more is the relative humidity. We know that air at high relative humidity hampers rate of evaporation. At higher wet bulb temperature, the reduction in evaporation of sweat causes body temperature to rise and in extreme case if the rise in temperature exceeds 2°C, heart beat becomes faster, which on continuation may lead to unconsciousness or even death. Table below lists the rise in body temperature with wet-bulb temperature.

Increase in Human Body Temperature with Rise in Wet Bulb Temperature of Air

Wet – bulb temperature in K	Rise in body temperature in K
≤ 302.15	0.11 – 0.66
302.65 – 304.85	0.33 – 0.77
305.35 – 307.65	0.66 – 1.55
≥307.65	1.44 – 1.90

Heat Hazards

Heat Cramps

- It happens when an individual is exposed for a period of longer duration to temperature moderately higher than the normal environmental temperature.
- Heavy perspiration from body leading to salt loss from the body and muscle cramps.

Sign and Symptoms of Heat Cramps

- Severe muscle cramps
- Exhaustion
- Dizziness

Remedial Measurers

- Patient should be taken to cool place.
- Salted water should be given to the patient.
- Massage should be done to get relief from cramps.
- Moist towels/ice packs should be applied on forehead.

Heat Exhaustion

- It is caused because of exposure to excessive heat while individual is at work. It is like a form of shock triggered by loss of fluid and salt from the body. In worst condition, it may develop to heat stroke.

Sign and Symptoms of Heat Exhaustion

- Rapid and shallow breathing
- Cold and clammy skin
- Dizziness o Heavy perspiration
- Weak pulse

Remedial Measures

- Patient should be moved to cool place and allowed to rest
- Clothes should be removed
- Fan patient's skin to reduce the temperature
- Salted water should be given
- If necessary, oxygen should be supplied

Heat Stroke

- It happens when the body is unable to remove heat because of failure of individual's temperature-regulating mechanisms like, sweat glands, vasomotor control, etc. It is mainly due to exposure to high temperature and humidity, especially when body temperature exceeds 41°C.
- Heat stroke is fatal in nature and immediate cooling of body should be done and doctor should be called.

Signs and Symptoms of Heat Stroke

- Deep breaths followed by shallow breathing
- Rapid strong pulse followed by rapid weak pulse
- Dry, hot skin

- Dilated pupils, skin turns blue
- Loss of consciousness

Remedial Measures

- Patient should be immediately cooled by any possible means
- Clothes should be removed and person should be kept in dark room
- Treatment for shock should be done by administering oxygen
- Ice packs, if available, should be applied near pressure points
- Person should be immediately taken to doctors

Effect of Air Velocity on Human Body Temperature

- Higher air velocity aids in evaporative cooling of the body
- Generally air velocity upto 1 m/s provides good comfort conditions
- Air velocity greater than 2 m/s produces dust and causes discomfort to the miners.
- Higher air velocity of the order of 3 m/s + High dry bulb temperature (322 K) + Low relative humidity = Causes burning sensation + rise of body temperature = This may lead t:o heat stroke

With the rise in temperature and relative humidity of the mine atmosphere, a decrement in miners' efficiencies is observed. Initially efficiency decreases gently and becomes steeper when the wet bulb temperature goes on increasing.

Other Sources of Heat Addition to Underground Mine Environment

.Machinery and Light

Most of the machine used in mines add sensible heat to the mine air either through friction or electric losses. All machines add heat to the mine air by non mechanical work produced by them. Different types of machinery used in mines can be classified on the basis of their source of power supplied. They are.

- Electrical equipments
- Diesel engine
- Compressed autorun equipments/machinery

Electrical Machinery

The main electrical equipments utilized in the mines is fans, haulage, road headers, conveyors, hoists, pumps transformers, shearer loaders, etc. Most of the inputs provided to them are converted into heat except for the part which is utilized in doing the work against gravity. Conveyors transporting ore in an incline or shaft, hoists, and pumps are the examples of machines which work against gravity.

Air Flows In Mines

Whenever the word "ventilation" comes to our mind, we immediately get a picture wherein a person with a cardiac problem is put on a ventilator, which involves supplying of oxygen to him through hoses and gas cylinders. In no way, the Ventilation in mines is different from it. Compared to the circulatory system in human beings, the ventilation in mines is no less complex. Analogous to the circulatory system, the intakes (intake roadways) are the arteries and the returns (return roadways) are the veins. The upcast and downcast shafts can be considered as pulmonary artery and the pulmonary veins respectively. The bottom of the downcast shaft can be taken as the left chamber and that of upcast shaft as right chamber of the heart. The atmosphere at the surface can be considered as that of the lungs. The various machinery, persons and workings in the mine can be considered as different parts of the body.

Ventilation System in Mines

The vast network of ventilation system that runs throughout the underground mines can be divided into three broad categories on the basis of the area covered by them. They are:

- Primary ventilation system or main ventilation system
- District system
- Auxiliary system

The first one runs throughout the mines, the second one, in districts or workings or panels and third one in the areas with blind openings. By blind openings, we mean an opening in underground that has only one open end and the other end is closed.

These elements or components are the characteristics of any ventilation system and are must for their proper functioning. These are the components which along with the efficiency of ventilation are also responsible for quality management of air supply.

Downcast and Upcast

Downcast is the entry gate for fresh air from the atmosphere i.e. surface to the underground. Downcast can be in the form of a shaft, an incline or an adit. On the other hand, upcast can also be anyone of these, but it serves just the opposite function i.e. it is an exit gate for polluted air from the mine to the surface. Please note that we have mentioned fresh air in case of downcast and polluted air in case of upcast. In the course of fresh air being supplied to different workings, it is added up with methane, coal dust, fumes, dust and other noxious gases. Hence, ventilation engineer also as to look into the quality aspect of the air.

Intakes and Returns

The network of airways running from the bottom of downcast to different workings in underground are termed as intakes and the network of airways that run from different workings to the bottom of the upcast are termed as returns. The former carry the fresh air or air to be used up, while the latter carry the polluted air after being used.

Fans

It is the pressure difference, which is a must condition for air to flow in mines. The necessary pressure for air to flow in mines is built up by providing a surface fan. Now the types of fan we can make use of depends on the way in which we want to create this pressure difference

The types of fan used for underground ventilation can be

- Forcing type (push system)
- Exhausting type (pull system)
- Push-pull type (a combination of the above two)

The forcing fan is generally located on tine surface near downcast side or at the bottom of the downcast and it forces air from the atmosphere to underground. The second one i.e. the exhausting type of fan is located either on the surface near upcast side or at the bottom of the upcast. It sucks air (polluted air) from underground and releases it to the surface via upcast.

The push-pull system is a combination of both forcing and exhausts types. It is generally used in metallic mines and is very advantageous as it maintains neutral pressure in underground with respect to the surface and minimizes or reduces the degree (amount) of air leakage in the course of travelling.

In most of the cases fan is on the surface and may be a mandatory condition in case of coal mines. This is for the purpose of better performance, safety, maintenance and lower cost. Location on the surface has got advantages in terms of installation testing, maintenance and access as on the surface, we can make use of space. On the other hand, in underground we cannot as the space is restricted and confined. In case of an emergency situation like explosion, etc, the surface location of fan is always advantageous. Further, when the fan is located on the surface it does not have to deal with problems posed by doors, airlocks and leakage routes existing in mines, as opposed to fan located underground.

The choice of a fan system is made after assessing the advantages and disadvantages of each system. However in most of the mines exhausting type of fan is used.

Natural Ventilation

Air will always flow from a region of high pressure to a region of low pressure. The difference in pressure may be either by natural means or by artificial means using fan. The ventilation in a mine due to the former is referred to as natural ventilation and that due to the latter is called mechanical ventilation. In case of mechanical ventilation, natural ventilation also exists. In some situations, natural ventilation will aid the mechanical ventilation whereas in some case, it may oppose mechanical ventilation.

For natural ventilation to take place, there must be difference in density between the air of the two shafts.

The density difference between the air of the two shafts may be created by the following factors:

- Presence of firedamp in one of the shafts which makes the air lighter.
- Steam jet introduced in one of the shafts making the air lighter.
- Introducing furnaces at the bottom of the upcast shaft. The heated air coming out at the bottom of the upcast shaft gets further rarefied by the furnace and is forced upwards. The air in the mines gets heated up due to addition of the from rocks men machinery, lights, spontaneous heating etc.
- Cage movement in the shaft.
- Flow of cold water down one of the shafts, which makes the air cooler and denser.
- Addition of gasses like methane and carbon dioxide in the mine air changes the density of the air. While adding of methane reduces the density of return air, addition of carbon dioxide increases the density of the return air and causes the opposite effect to take place.
- In multilevel mine workings, the leakage of denser downcast air to the upcast shaft results in an increase in the density of upcast air, thereby reducing natural ventilation
- Cool air circulation through the downcast shaft, increases the density of downcast air and thus increasing natural ventilation in the mine.

Appendix 1.45: Causes, Effects and Control of Deforestation

Deforestation is an act of indiscriminate cutting of trees for lumber or pulp, or to clear the land for agriculture, ranching, construction, or other human activities. Although humans have been practicing deforestation since ages, it was in the mid – 1800s that forests began to be destroyed at an unprecedented rate. As a matter of fact, throughout the earlier part of the medieval age, Europeans used to live amongst vast areas of forested land. But later, they began deforestation at such a high rate that they started to run out of wood for cooking and heating. Also, due to the depletion of their natural habitat, wild game too began disappearing, which the Europeans largely depended upon for their nutritional requirements. Today, parallel can clearly be observed in the deforestation that is occurring in most developing countries.

Causes of Deforestation

Encroachment of forest land to agriculture: Rapid growth of population, inequitable land distribution and expansion of export agriculture have reduced the crop land to greater extent. Therefore, the subsistence farmers are formed to grow food by clearing forests.

Over grazing: The overgrazing in the forest lands causes serious damage to both seedling and older trees.

High demand for timber: High demand of developed and developing countries for hard timber and plywood have ruined the world's tropical forests.

Shifting cultivation: Shifting cultivation is adopted mainly by the tribal peoples which involves clearing of forest land for cultivation. It is estimated that nearly 5000 km^2 of forest land is cleared for this purpose.

Clearance for settlement: Sometimes, the people from over-crowded areas are settled by the government by clearing forests.

Natural factors: Natural factors like flood, drought, fire, decline of water level, cyclonic storm, lightening, snowfall, and diseases earth quake etc. cause massive deforestation.

Forest fire: Forest fire (deliberate or natural) causes deforestation to a large extent.

Wrong government policies: The deforestation is also due to wrong policies adopted by the government to collect revenue from products in the name of providing employment to tribals.

Clearance for forest plantation: Now a days, an attempt has been to replace natural forests by new plantations that are high yielding in timber production.

Urban growth and construction activities: The cutting down of trees for lumber that is used for building materials, furniture, and paper products. Forests are also cleared in order accommodate expanding urban areas.

Agricultural practices: Forests are also cut down in order to clear land for growing crops and / or in order to create land for grazing cattle.

Fuel: Trees are cut down in developing countries to be used for cooking and heating purposes.

Other Causes: clearing forests for oil and mining exploitation; to make highways and roads; slash and burn farming techniques; wildfires; and acid rain.

Effects of Deforestation

There are a number of adverse effects of deforestation, such as:

- **Erosion of soil:** When forest areas are cleared, it results in exposing the soil to the sun, making it very dry and eventually, infertile, due to volatile nutrients such as nitrogen being lost. In addition, when there is rainfall, it washes away the rest of the nutrients, which flow with the rainwater into waterways. Because of this, merely replanting trees may not help in solving the problems caused by deforestation, for by the time the trees mature, the soil will be totally devoid of essential nutrients.

Ultimately, cultivation in this land will also become impossible, resulting in the land becoming useless. Large parts of land will be rendered permanently impoverished due to soil erosion.

- **Disruption of the Water Cycle:** Trees contribute in a large way in maintaining the water cycle. They draw up water via their roots before it is released into the atmosphere. A large part of the water that circulates in the ecosystem of rainforests, for instance, remains inside the plants. When these trees are cut down it results in the climate getting drier in that area.
- **Loss of Biodiversity:** The unique biodiversity of various geographical areas is being lost on a scale that is quite unprecedented. Even though tropical rainforests makes up just 6 percent of the surface area of the Earth, about 80-90 percent of the entire species of the world exist here. Due to massive deforestation, about 50 to 100 species of animals are being lost each day. The outcome of which is the extinction of animals and plans on a massive scale.
- **Flooding and Drought:** One of the vital functions of forests is to absorb and store great amount of water quickly when there are heavy rains. When forests are cut down, this regulation of the flow of water is disrupted, which leads to alternating periods of flood and then drought in the affected area.
- **Climate Change :** It is well known that global warming is being caused largely due to emissions of greenhouse gases like carbon dioxide into the atmosphere. However, what is not known quite as well is that deforestation has a direct association with carbon dioxide emissions into the atmosphere. Trees act a major storage depot for carbon, since they absorb carbon dioxide from the atmosphere, which is the used to produce carbohydrates, fats, and proteins that make up trees. When deforestation occurs, many of the trees are burnt or they are allowed to rot, which results in releasing the carbon that is stored in them as carbon dioxide. This, in turn, leads to greater concentrations of carbon dioxide in the atmosphere.

Environmental refugees: Depletion of forests in developing countries is interlinked with population growth, poverty, illiteracy, urban migration, rural ecosystem etc. Being unable to support their growing population due to deforestation, there happens migration rural people towards urban areas which causes big slums, pollution and sanitation problems, break out of epidemics etc.

Global warming: The concentration of carbon dioxide in the environment becomes higher and higher due to the depletion of forests which are well know absorber of CO_2. Increase in CO_2 concentration in environment causes global warming due to green house effect.

Fuel wood scarcity: The availability of fuel wood becomes difficult due to deforestation and there happens a sharp rise in the price of fuel wood. Increase in price of fuel wood exerts high pressure on rural people and induces illegal extraction of wood.

Depletion of large number of species: Depletion of forests has caused extinction of a number of species and their natural habitats.

Control of Deforestation: In order to check the detrimental effects due to deforestation, the following control measures should be undertaken:

- All nations should promote the regeneration of degraded forest land to increase permanent forest cover, reduce pressure on natural forest, conserve biodiversity, protect waster sheds and soils and stabilize climate.
- Effort to prevent encroachment pressures by providing land security to small and landless farmers through land tenure law, land reform or agrarian reform should be done.
- There should be proper modification in the development schemes (i.e., mining, hydroelectric, road building, plantation etc.) to minimize their direct and indirect impacts on natural forests.

- Over grazing in forest areas should be regulated by government.
- There should be efforts to relieve pressure on forests by decreasing waste in wood processing, increasing the efficiency of wood burning stoves and seeking alternative raw materials.
- Ecological stability should receive precedence over the economic stability in the matter of forest growth.
- Efforts should be made to involve local people, NGOs and forest department in developing degraded forests land as well as conserving the existing forests.
- There should be effort to encourage sustainable forests management in order to maintain ecological balance.
- Public awareness should be created against the hazards of deforestation.
- Forests should be honoured not as a source of revenue but as a global asset to be protected for the well being of the globe.

In view of the vital importance of the forest in maintaining ecological balance and preserving biophysical system of the environment, the Government of India has revised the old forest policy of 1952 in 1988.

- To maintain ecological stability by adopting appropriate measures of environment protection, conservation and restoration.
- To prevent the natural traditions by conserving the remaining natural forests as well as to conserve the vast original resources for the benefit of future generation.
- To meet the basic requirements of the rural people and tribal people for the firewood, fodder and small timbers.

Note:

Facts on Deforestation

- Forests cover 30% of the earth's land.
- It is estimated that within 100 years there will be no rainforests.
- Agriculture is the leading cause of deforestation.
- One and a half acres of forest is cut down every second.
- Loss of forests contributes between 12 percent and 17 of annual global greenhouse gas emissions.
- If the current rate of deforestation continues; it will take less than 100 years to destroy all the rainforests on the earth.
- There are more than 121 natural remedies in the rain forest which can be used as medicines.
- United States has less than 5% of the world's population yet consumes more than 30% of the world's paper.
- The over exploitation of forests is making it extremely difficult to replant a new ecology.
- 20% of the world's oxygen is produced in the Amazon forest.
- Up to 28, 000 species are expected to become extinct by the next quarter of the century due to deforestation.
- 25% of cancers fighting organism are found in the amazon.
- 13 million hectare per year in South America and Africa and south East Asia is converted from a forest to an agriculture land.

- Deforestation has considerably stopped in places like Europe, Pacific, North America and some parts of Asia due to lack to agricultural land.
- Half of the world's tropical forests has already been cleared.
- 4500 acres of forests are cleared every hour by forest fires, bull dozers, machetes etc.
- Poverty, over population and unequal land access are the main causes of man- made deforestation.
- The total world forest loss till date is 7.3 million hectares per year.
- 1.6 billion people across the globe depend on forest products for their livelihoods there by adding more to deforestation.
- Almost half of world's timber and up to 70% of paper is consumed by Europe. United States and Japan alone.
- Industrialized countries consume 12 times more wood and its products per person than the non-industrialized countries.
- Fuel wood in sub Saharan African countries is consumed up to 200% times more than the annual growth rates of the trees. This is causing deforestation, lack of timber resources and loss of habitat for the species living in it.
- Trees are important constituents of the ecosystem by absorbing carbon.
- Soil erosion, floods, wildlife extinction, increase in global warming, and climate imbalance are few of the effects of deforestation.
- Tropical forests, where deforestation is most prevalent, hold more than 210 gigatonnes of carbon.
- Tropical rainforests which cover 6-7% of the earth's surface, contain over half of all the plant and animal species in the world.
- Deforestation affects water cycle. Trees absorb groundwater and release the same into the atmosphere during transpiration. When deforestation happens, the climate automatically changes to a drier one and also affects the water table.
- The world's forests store 283 billion tons of carbon present in the biomass.
- A lot of paper and cardboard is used unnecessarily for packing. This means more tree felling.
- Re-use paper and plastic bags to discourage deforestation.
- Pick products which require less packaging.
- Be active and plant trees

Appendix 1.46: Prevention of Environmental Pollution by Modification of Industrial Equipments

S.No.	Industrial Equipments	Potential environment problems	Suggested Possible Approach/modifications: Design Related issues	Suggested Possible Approach/modifications: Operational Related issues
1.	Compressors, blowers, fans	• Shaft seal leaks, piston rod seal leaks, and vent streams	• Seal-less designs (diaphragmatic, hermetic or magnetic) • Design for low emissions (internal balancing, double inlet, gland eductors) • Shaft seal designs (carbon rings, double mechanical seals, buffered seals) • Double seal with barrier fluid vented to control device	• Preventive maintenance program
2.	Concrete pads, floors, sumps	• Leaks to groundwater	• Water stops • Embedded metal plates • Epoxy sealing • Other impervious sealing	• Reduce unnecessary purges, transfers, and sampling • Use drip pans where necessary
3.	Controls	• Shutdowns and start-ups generate waste and releases	• Improve on-line controls • On-line instrumentation • Automatic start-up and shutdown • On –line vibration analysis • Use "consensus" system (e.g., shutdown trip requires 2 out of 3 affirmative responses)	• Continuous versus batch • Optimize on-line run time • Optimize shutdown interlock inspection frequency • Identify safety and environment critical instruments and equipment
4.	Distillation column	• Impurities remain in process streams • Large amounts of contaminated water condensate from stream stripping	• Increase reflux ratio • Add section to column • Column intervals • Change feed tray • Insulate to prevent heat loss • Preheat column feed • Increase vapor line size to lower pressure drop • Use reboilers or inert gas stripping agents	• Change column operating conditions : • Reflux ratio • Feed tray • Temperature • Pressure • Clean column to reduce fouling • Use higher temperature steam

5.	General manufacturing equipment areas	• Contaminated rainwater • Contaminated sprinkler and fire water • Leaks and emissions during cleaning	• Provide roof over process facilities • Segregate process sewer from storm sewer (diking) • Hard-pipe process streams to process sewer • Seal floors • Drain to sump • Route to waste treatment • Design for cleaning • Design for minimum rinsing • Design for minimum sludge • Provide vapor enclosure • Drain to process	• Return samples to process • Monitor stormwater discharge • Use drip pans for maintenance activities • Rinse to sump • Reuse cleaning solutions
6.	Heat exchangers	• Increased waste due to high localized temperatures Contaminated materials due to tubes leaking at tube sheets Furnace emissions	• Use intermediate exchangers to avoid contact with furnace tubes and wails • Use staged heating to minimize product degradation and unwanted side reactions. (Waste heat >> low pressure steam >> high pressure steam) • Use scraped wall exchangers in viscous service • Using falling film reboiler, piped recirculation reboiler or high-flux tubes • Use lowest pressure steam possible • Use welded tubes or double tube sheets with inert purge. Mount vertically • Use superheat of high-pressure steam in place of a furnace	• Select operating temperatures at or near ambient temperature when-ever possible. These are generally most desirable from a pollution prevention standpoint. • Use lower pressure steam to lower temperatures • Monitor exchanger fouling to correlate process conditions which increase fouling, avoid conditions which rapidly foul exchangers • Use on-line tube cleaning techniques to keep tube surfaces clean • Monitor for leaks
7.	Piping	• Leaks to groundwater; fugitive emissions • Releases when cleaning or purging lines	• Design equipment layout so as to minimize pipe run length • Eliminate underground piping or design for cathodic protection if necessary to install piping underground. • Welded fittings • Reduce number of flanges and valves • All welded pipe • Secondary containment • Spiral – wound gaskets • Use plugs and double valves for open end lines • Change metallurgy • Use lined pipe • Use "pigs" for cleaning • Slope to low point drain • Use heat tracing and insulation to prevent freezing • Install equalizer lines	• Monitor for corrosion and erosion • Paint to prevent external corrosion • Flush to product storage tank

8.	Pumps	• Fugitive emissions from shaft seal leaks • Residual "heel" of liquid during pump maintenance • Injection of seal flush fluid into process stream	• Mechanical seal in lieu of packing • Double mechanical seal with inert barrier fluid • Double machined seal with barrier fluid vented to control device • Seal-less pump (canned motor magnetic drive) • Vertical pump • Use pressure transfer to eliminate pump • Low point drain on pump casing • Use double mechanical seal with inert barrier fluid where practical	• Seal installation practices • Monitor for leaks • Flush casing to process sewer for treatment • Increase the mean time between pump failures, by: • Selecting proper seal material; • Good alignment; • Reduce pipe-induced stress • Maintaining seal lubrication
9.	Reactors	• Poor conversion or performance due to inadequate mixing • Waste by-product formation	• Static mixing • Add baffles • Change impellers • Add horsepower • Add distributor • Provide separate reactor for converting recycle streams to usable products	• Add ingredients with optimum sequence • Allow proper head space in reactor to enhance vortex effect • Optimize reaction conditions (temperature, pressure, etc.)
10.	Relief Valve	• Leaks • Fugitive emissions • Discharge to environment from over pressure • Frequent relief	• Provide upstream rupture disc • Vent to control or recovery device • Pump discharges to suction of pump • Thermal relief to tanks • Avoid discharge to roof areas to prevent contamination of rainwater • Use pilot operated relief valve • Increase margin between design and operating pressure	• Monitor for leaks and for control efficiency • Monitor for leaks • Reduce operating pressure • Review system performance
11.	Sampling	• Waste generation due to sampling (disposal, containers, leaks, fugitives, etc.)	• In-line insitu analyzers • System for return to process • Closed loop • Drain to sump	• Reduce number and size of samples required • Sample at the lowest possible temperature • Cool before sampling
12.	Tanks	• Tank breathing and working losses • Leak to groundwater • Large waste heel	• Cool materials before storage • Insulate tanks • Vent to control device (flare, condenser, etc.) • Vapor balancing • Floating roof • Higher design pressure • All aboveground (situated so bottom can routinely be checked for leaks • Secondary containment • Improve corrosion resistance • Design for 100% de-inventory	• Optimize storage conditions to reduce losses • Monitor for leaks and corrosion • Recycle to process if practical

13.	Vacuum	• Waste discharge from jets	• Substitute mechanical vacuum pump • Evaluate using process fluid for powering jet	• Monitor for air leaks • Recycle condensate to process
14.	Valves	• Fugitive emissions from leaks	• Bellow seals • Reduce number where practical • Special packing sets	• Stringent adherence to packing procedures
15.	Vents	• Release to environment	• Route to control or recovery device	• Monitor performance

Appendix 1.47: Major Global Contributors of CO_2 emissions (1996)

S.No.	Countries	CO_2 emissions, % of worlds total	Population, % of world's total
1.	U.S.A	25	4.7
2.	Europe	22.1	10.0
3.	China	13.5	21.5
4.	Former Soviet union	10.0	5
5.	Japan	5.6	2.2
6.	India	3.6	16.3
7.	Republic of Korea	2.2	0.8
8.	Canada	2.1	0.5
9.	Australia	1.3	0.3

Note:

Some Facts about Carbon Dioxide (CO_2)

- Under atmospheric conditions, CO_2 never condenses or freezes into liquid CO_2 or solid CO_2 We never get clouds, rain, or snow made of CO_2.
- CO_2 is chemically very stable. In the atmosphere, other gases (such as methane and carbon monoxide) turn into CO_2. But CO_2 hardly ever turns into other gases.
- CO_2 is a greenhouse gas. It absorbs and then re-emits heat radiation in such a way as to restrict its escape, making the Earth's surface warmer. CO_2 is our atmosphere's most important *non-condensing greenhouse gas.*
- Water vapour is also a greenhouse gas. It too is important. However, because it can condense or freeze, its behaviour in the atmosphere is very different from that of CO_2
- Going back in time from 200 years ago to 800, 000 years ago, atmospheric concentrations of CO_2 varied within a range of values between about 180 and 290 ppmv (parts per million by volume), as the Earth went through about eight ice-age cycles. This is known from very careful measurements of air trapped in Antarctic ice. The measurements tell us that, whatever the complexities of the natural

carbon cycle, the natural variation of atmospheric CO_2 has been, within the range 180 to 290 ppmv for a very long time.

- After the end of the last ice age around 10, 000 years ago, and before industrial times, atmospheric CO_2 concentrations stayed close to 280 ppmv, near the high end of the natural range. However, since then they have reached values **close to 400 ppmv**, In the year 2015 taking them far outside the natural range.
- Most of this rise is due to human activity. It represents a large disturbance to the Earth's carbon cycle, by any natural measure, even though the actual CO_2 concentrations appear small. It therefore represents a large disturbance to the climate system.

Note: Although it is a normal constituent of the atmosphere essential to supply the carbon required for plant photosynthesis, carbon dioxide may turn out to be a harmful air pollutant at higher levels. This is because of its role in re-absorbing the infrared radiation by which Earth radiates back into space the solar energy received from the sun. This greenhouse effect is essential to life on Earth because it keeps the surface warm enough for living organisms to survive. However, if too much carbon dioxide is present in the atmosphere, earth is surface will become too warm due to detrimental effects. As shown in the plot below, atmospheric carbon dioxide levels from the burning of fossils fuels and other sources have been increasing by about I part per million per year, now closer to 2 parts per million (ppm) per year. The resulting doubling of atmospheric carbon dioxide levels within the next century could have substantial, catastrophic, effects upon the global climate.

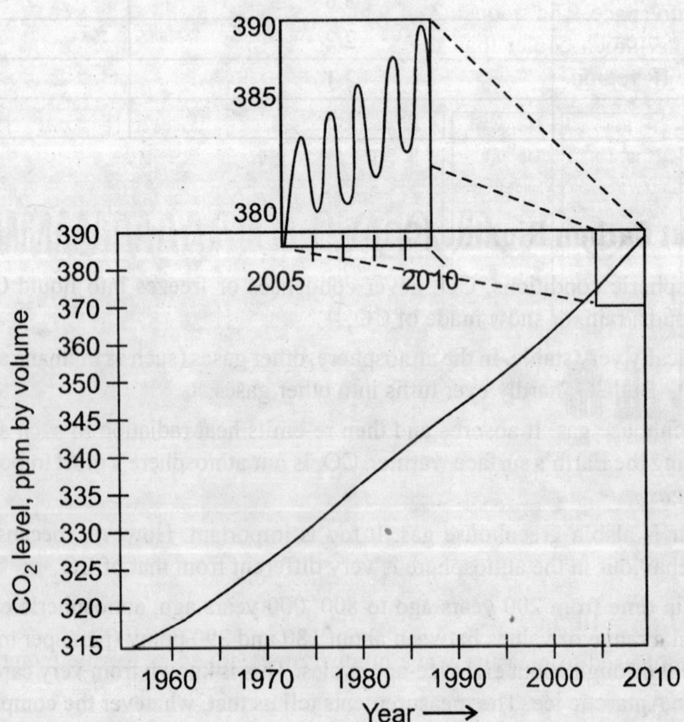

Appendix 1.48: The World in year 2050 (if no action is taken to reduce green house gases)

1. Alps ski resorts will be closed because of lack of snow.
2. Mediterranian beaches will disappear because of the sea-level rise.
3. The Sahara moves across the mediterranian to southern Spain and sicily.
4. Water shortage will occur in the middle east. The Nile delta will disappear.
5. One third of Bangladesh will be submerged.
6. Forests (in Canada, Russia, and Amazon) will be damaged by heat and drought.
7. Arctic ice will melt. Polar bear will disappear.
8. The wheat belt (USA, Russia) will move to the North.
9. Storms threaten Florida and the southern U.S.A.
10. Maldives will disappear.

Note:

The Greenhouse Effect and Global Warming

The sun produces radiation mainly in the ultraviolet (UV), visible (vis) and infrared (IR) regions of the electromagnetic spectrum. When these reach the Earth, part is reflected back into space and part of it is absorbed by the Earth's surface. The part which is absorbed heats up the Earth which in turn then radiates some of its energy out into space. The frequency at which any object emits radiation depends on its temperature. The Earth, being that much cooler than the Sun, emits energy at a lower frequency and therefore longer wavelength - in the IR region.

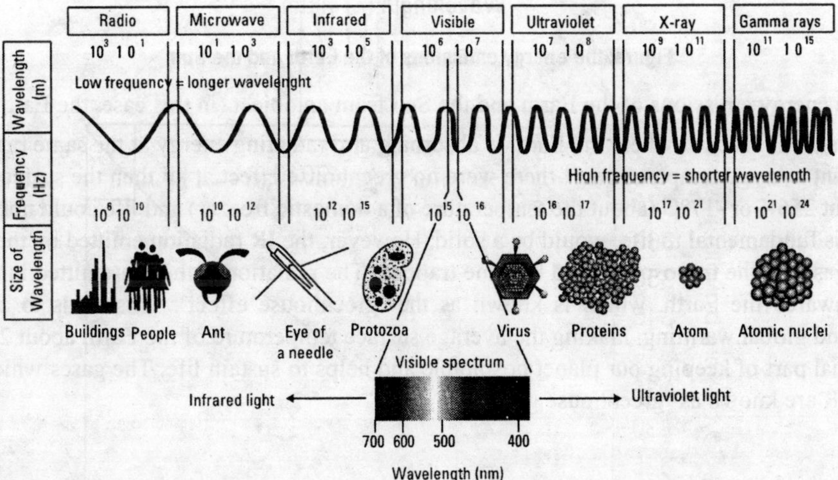

$1\mu m = 10^{-6} m \cdot 1\mu m = 10^{-9}$ m

1 cm^{-1} is the unit used to measure wave number and defined as the the reciprocal of wavelength (1 / wavelength)

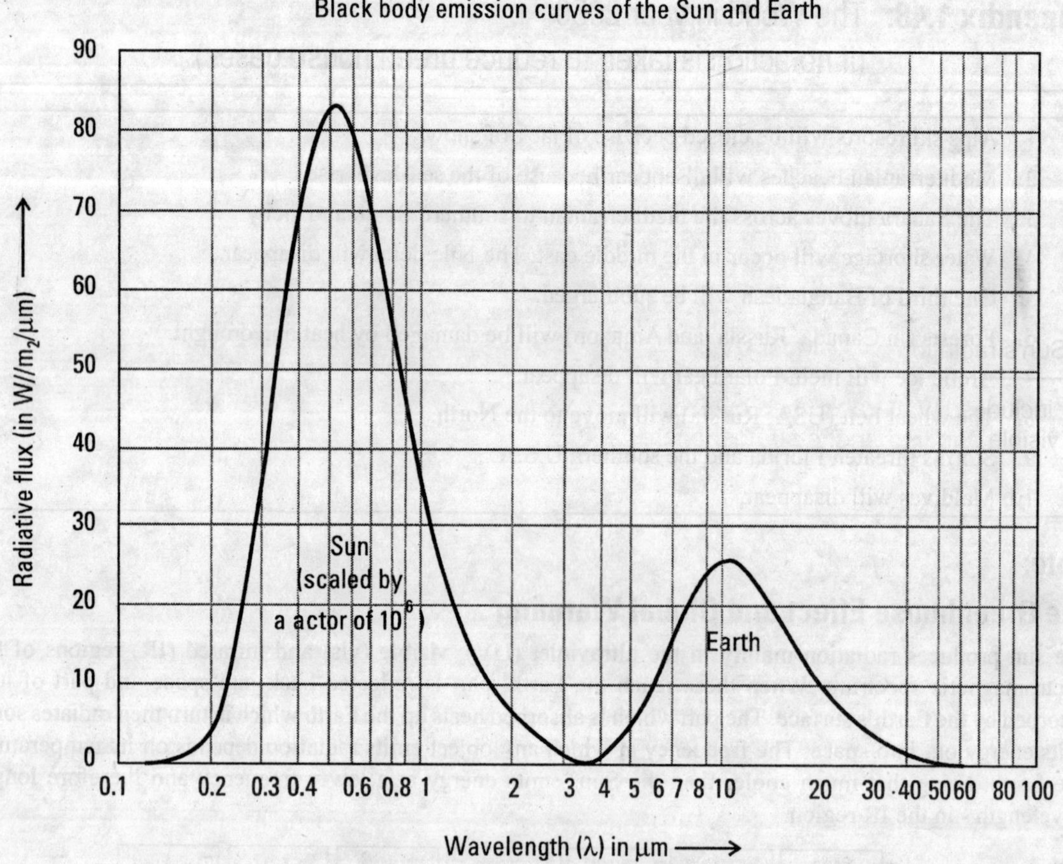

Figure: the energy emissions of the Earth and the Sun.

Note: the energy emissions of the Earth and the Sun.from an object (in this case, the Earth.)

A steady state is reached where the Earth is absorbing and radiating energy at the same rate, resulting in a fairly constant average temperature. If there were no greenhouse effect at all then the surface temperature would be about 256K or -17°C (about the temperature of a domestic freezer) and life could not exist because water, which is fundamental to life, would be a solid. However, the IR radiation emitted by the Earth can be absorbed by gases in the troposphere and become trapped. The radiation is then re-emitted in all directions; some back towards the Earth, which is known as the 'greenhouse effect'. This leads to an increase in temperature and global warming, making the average surface temperature of the Earth about 286K or 13°C. It is an essential part of keeping our planet hospitable and helps to sustain life. The gases which absorb and then re-emit IR are known as 'greenhouse gases.'

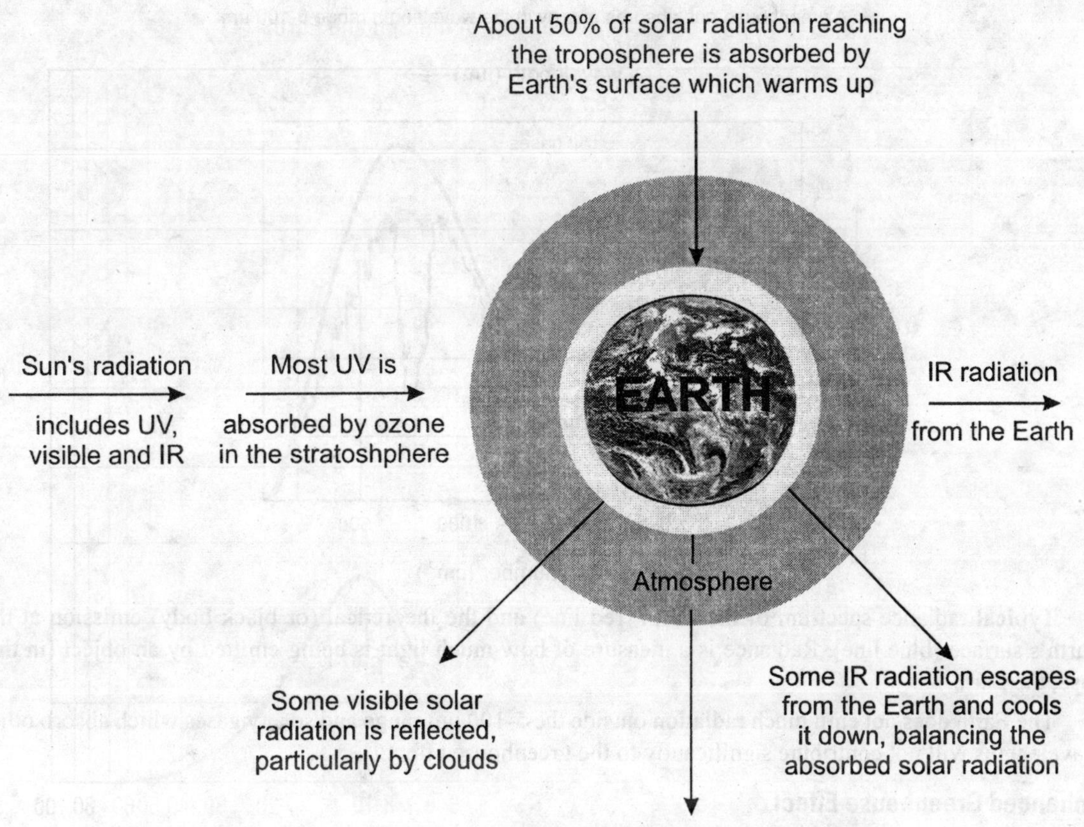

Properties of a Greenhouse Gas

A greenhouse gas absorbs infrared radiation

When molecules interact with UV radiation the energy they absorb can break bonds. IR radiation is less powerful as the photons contains less energy. Rather than breaking bonds it causes them to vibrate more energetically. Advanced theory shows that a molecule will absorb IR radiation if the vibration causes a change in its dipole moment.

Oxygen, O=O, vibrates as a stretching and compression of the bond. Oxygen does not absorb IR radiation as the molecule is symmetrical and does not have a dipole moment - so there can be no change in dipole moment on vibration.

Methane, CH_4 is another greenhouse gas. Althouh it is symmetrical and has no permanent dipole moment, It can vibrate in ways which change its dipole moment and so will absorb IR radiation.

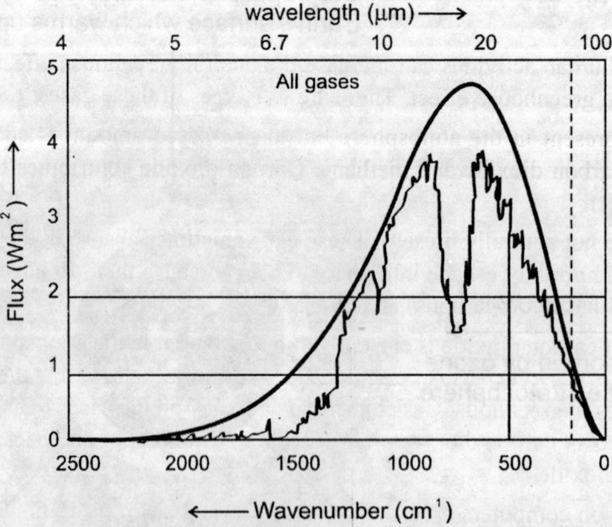

Typical radiance spectrum of the Earth (red line) and the theoretical (or black body) emission at the Earth's surface (blue line). Radiance is a measure of how much light is being emitted by an object (in this case, the Earth.)

The Earth does not emit much radiation outside the 5-100 μm range and so any gases which absorb other wavelengths will not contribute significantly to the greenhouse effect.

Enhanced Greenhouse Effect

The two most significant greenhouse gases in the atmosphere are carbon dioxide and water vapour. Water makes the bigger contribution (about 60%) to the natural greenhouse effect. Between the absorptions caused by carbon dioxide and water there is a 'window' where the majority of the infrared radiation can escape with relatively little absorption (except for a narrow band where ozone absorbs.) About 70% of Earth's radiation escapes into space through this 'window.'

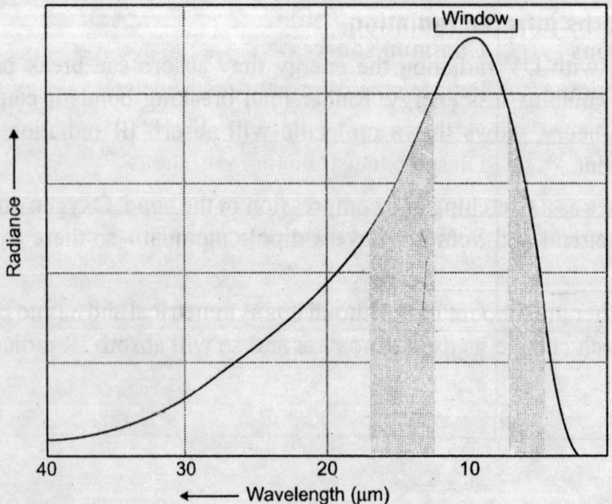

Simplified version showing the Earth's radiation spectrum with the regions where CO_2 and H_2O absorb, (Between these two absorptions is the 'window' where the majority of the IR emitted by the Earth escapes)

Gases produced by human activities can increase the natural greenhouse effect of the atmosphere. This is known as the enhanced greenhouse effect. There are two types of these gases:

- Gases already present in the atmosphere but an increased amount is added by human activities. This includes carbon dioxide and methane. Carbon dioxide contributes the most to the enhanced greenhouse effect.
- Gases which are not naturally present. These can sometimes absorb in the 'window' through which radiation would normally escape into space. This can cause them to have a very large greenhouse effect. CFCs (chlorofluorocarbons) are an example.

The concentration of carbon dioxide is currently about 390 ppm in the atmosphere and is rising by about 0.45% per year. Modellers use a scenario of doubled $C0_2$ concentration to assess the sensitivity of the Earth's climate system to changes in greenhouse gas concentration. They predict rises of about 1.5-4.5°C. This is not just due to the direct effect of carbon dioxide - there are a large number of feedback processes which amplify the effect of the carbon dioxide.

Climate models run on computers are used by climate scientists to try to predict changes to the Earth's climate. The water in the atmosphere is crucial for these models. The amount of water can be modelled well - it just depends on evaporation and condensation. As the temperature increases the amount of water which the atmosphere can hold increases. Predicting the behaviour of the water is much harder and is one of the causes of uncertainty in the predictions Water vapour in the atmosphere can condense to form clouds consisting of water droplets or ice particles. The processes operating in clouds are relatively poorly understood and therefore harder to model and predict, as is the resulting effect on temperature.

Carbon dioxide is about 0.04% of the Earth's atmosphere; water is more variable, but at the surface is usually about 1-4% of the atmosphere.

Appendix 1.49: How Common Poisons in Air Affects Human Health

S.No.	Gaseous poisons	Common sources	Health Effects
1.	Carbon monoxide	Petrol vehicles(mostly 2/3 wheeler)	Reduces ability of blood to carry oxygen.
2.	Sulphur dioxide	Diesel exhaust, factory emissions	Affects upper respiratory tracts.
3.	Oxides of nitrogen	Vehicles and thermal power plants.	Eye irritation, bronchial infections
4.	Benzene	Catalytic converters of petrol vehicles	Lung cancer, leukemia
5.	Lead	Leaded petrol vehicles	Hypertension, renal damage, affectly, central nervous system

S.No.	Gaseous poisons	Common sources	Health Effects
6.	Organochlorines	Treated water effluents from industries using chlorines, eg paper, viscose rayon, Pvc, organochlorine, pesticides etc.	Affects liver, kidney cancer
7.	Suspended particulate matter (SPM)	Factory emissions and vehicles gases	Respiratory diseases

Appendix 1.50: Sources and ill Effects of Various Hazardous Industrial Dusts

S.No.	Specific dust type	Principal sources of dust	Lung disease caused	Remarks
I. Quartz and mixture containing quartz				
1.	Powdered quartz	Plastering (Mixing)	Silicosis	Involves nodular fibrosis, found also in conjunction with tuberculosis and chronic non-specific lung disease (CNSLD), crystalline silica and tridymite act like quartz
2.	Coal, mineral ores, fluorspar, rock sand	Mining and metallurgy, building materials and construction, stone cutting, foundry work (moulders), sand blasting	Silicosis, pneumoconioses	--do--
3.	Kaolin	Ceramic industries (porcelain, pottery, earthware, sanitaryware and electrical ceramics)	Silicosis	--do--
4	Quartzite	Refractories (fireclay, silica bricks)	--do--	--do--
5	Keiselguhr (burnt)	Manufacture of the filtering and insulating materials	--do--	--do--
II. Asbestos and mixtures containing asbestos				
6.	Raw asbestos, chrysolite, amphibole	Asbestos mining, Manufacture, treatment and preparation (insulation textile, friction materials, packing and jointing materials, fire prevention)	Asbestosis, mesothelioma, bronchial carcinoma, upper respiratory tract carcinoma	Involves diffuse fibrosis and carcinoma, Pleural hyalinosis, CNSLD, may cause stomach carcinoma
7.	Asbestos cement	Building materials and construction industry	--do	--

S.No.	Specific dust type	Principal sources of dust	Lung disease caused	Remarks
8.	Talc	Rubber industry, pharmaceuticals, cosmetics, paint, paper and printing textiles	Talcosis, mesothelioma, bronchial carcinoma of upper respiratory tract	Involves diffuse fibrosis and carcinoma (nodular fibrosis in rare cases), Frequently due to admixture of asbestos and quartz
III. Metals and metal compound dust:-				
9.	Aluminium, aluminium oxide	Aluminium powder, manufacture of corundum (aluminium smoke from smelting of bauxite), light metal industries (welding and flame cutting)	Aluminium lung, bauxite smelter's lung, CNLSD	Involves diffuse fibrosis, irritation
10.	Beryllium, beryllium oxide	Manufacture of the luminescent tubes, metallurgy	Tracheobronchitis, pneumonotis, beryllosis	Involves cellular granuloma, diffuse fibrosis, irritation, immune reaction, granuloma may also develop in liver, spleen, skin or muscles, often acute development; latent period up to 25 years possible
11.	Cadmium, cadmium oxide	Metallurgy, paint industry (pigments), electroplating	Tracheobronchitis, bronchopneumonia, emphysema of the lung	Causes irritation, systemic poison, kidney damage, osteoporosis
12.	Chromium, chromic oxide, chromate	Electroplating, welding and flame cutting, metallurgy, pigments, austenitic stainless steel manufacture	Ulceration and perforation of nasal septum, bronchial asthma, carcinoma of nasal cavity, CNSLD	Causes irritation, carcinoma, immune reaction, only hexavalant chromium compounds are carcinogenic (eg. Alkali chromate, chronic oxide)
13.	Hard metals	Sintering	Fibrosis, immune reaction (?)	Involves part diffuse, part fibrosis, pathogenic effects of cobalt not fully ascertained
14.	Iron, iron oxide	Metallurgy, metal working (welding, flame cutting, grinding), paint industry (pigments)	Siderosis	Involves accumulation type reaction, after Termination of exposure, x-ray shadow continues to build up

S.No.	Specific dust type	Principal sources of dust	Lung disease caused	Remarks
15.	Lead, lead oxide	Metallurgy, lead battery manufacture lead shot manufacture, pain industry (pigment) glazing, scrap lead working (flame cutting of lead painted materials)	Systemic poison effects	Causes lead poisoning (anemia, colic, polyneurological symptoms, encephalopathy), possible local effects in the lung
16.	Manganese, manganese oxides	Metallurgy, metal working (welding with electrodes containing manganese), preparation and use of manganese ore	Manganic pneumonia, CNSLD	It is a systemic poison, causes irritation
17.	Nickel, nickel oxide, nickel salt	Electroplating, metallurgy, chemical industry	Bronchial carcinoma, carcinoma of nasal cavity	Causes irritation, immune reaction, systemic poisoning with nickel tetracarbonyl (liver dermatitis)
18.	Platinum compounds (salt)	Metallurgy	Allergic rhinitis, bronchial asthma	Causes irritation, involve immune reaction (type I), dermatitis
19.	Vanadium pentoxide	Manufacture of the vanadium catalysts, power plants (cleaning of residue in oil burning furnaces)	Tracheobronchitis, bronchial asthma, CNSLD	Causes irritation, dermatitis
IV. Plant and animals dusts (Organic)				
20.	Crushed grain and bran`	Grain milling and storage, bakeries	Allergic and chronic rhinitis, bronchial asthma, CNSLD	Causes irritation, involves immune reaction (type I), Mould fungi produce similar symptoms
21.	Exotic type wood dusts	Manufacture of veneers, furniture polishing, wood turning	Allergic rhinitis, CNSLD, bronchial asthma, carcinoma of nose and nasal cavity	Cause irritation, carcinoma, immune reaction, dermatitis
22.	Animals hides and skins, hair, feathers and scales	Agriculture, zoo attendants, laboratory animal keepers, furrier and fur dealers	Allergic rhinitis, bronchial asthma	Causes irritation, dermatitis, involves immune reaction
23.	Enzymes (Proteases)	Pharmaceutical industries, washing powder manufacture, food and drink industries	Allergic rhinitis, bronchial asthma	Causes irritation, dermatitis, involves immune reaction
24.	Excrements from hens, pigeons, parakeets	Poultry keeping and zoos	Allergic alveolitis, diffuse fibrosis	Caused by protein serum in bird excrement, involves immune reaction (type III)

S.No.	Specific dust type	Principal sources of dust	Lung disease caused	Remarks
25.	Mouldy hay, straw, cereals and bagasse dust	Agriculture, grain silos, sugar industry	Former's lung, bagassosis, allergic alveolitis, diffuse fibrosis	Caused by thermophilic actinomycetes, rarely caused by mould fungus, involves immune reaction (type III)
26.	Cotton, flax, hemp, sisal, jute dust	Cotton cording, cotton and flax spinning combing (hackling)	Byssinosis, CNSLD	"Monday" syndrome due to release of histamines and related substances, chronic symptoms due to long lasting irritation of mucous membranes, causes irritation, immune reaction uncertain
V. Other dusts				
27.	Carbon dust, soot, graphite	Manufacture of electrodes, rubber industry, any work involving soot	Graphite pneumoconiosis, CNSLD	Causes irritation, possible Fibrogenic effects and bronchial carcinoma
28.	Arsenic, arsenic trioxide, arsenic salts	Mining, Pb and Zn smelters, metal etching, electroplating, coloring and finishing materials	Ulceration and perforation of nasal septum, tracheobronchitis, carcinoma of nasal cavity, bronchial carcinoma	Causes irritation, systemic toxic effects, polyneurological symptoms, diseased lever, dermatitis
29.	Maleic anhydride, phthalic anhydride dust	Chemical industry, Manufacture of plastics	Conjunctivitis, rhinitis with ulceration, asthmatic bronchitis	Causes irritation, dermatitis, immune reaction possible

Note:

Health Problems caused by dusts

Routes of Exposure

Maximum attention is given to dust exposure by inhalation, However, other routes are also important.

Skin absorption (or percutaneous absorption) can occur, if water-soluble materials dissolve in sweat and pass through the skin into the bloodstream, causing systemic intoxication.

Ingestion is likely when poor hygiene allows eating, drinking or smoking in contaminated or dirty workplaces. Particles do not need to be airborne. For example, many cases of lead poisoning have occurred in poorly kept small potteries, in which ingestion of lead salts has been an important route. Entry by this route can be reduced by good housekeeping, personal hygiene and adequate work practices. Many inhaled particles are swallowed and ingested, but for control and measurement purposes these are usually considered with the inhalation route.

Effects on the Skin. In addition to the risk of absorption through the skin, many dusts may affect the skin directly, causing various types of dermatoses, which are a widespread and often serious problem, or even skin cancer. Cement is an important cause of dermatitis. For such substances, dust of any size has health significance, even if it never becomes airborne. Some allergens act on the skin, including many wood dusts. This is important for the woodworking industry as well as for rural workers, e.g., in agricultural and forestry.

Potential Health Effects by Dust Inhalation

If dust is released into the atmosphere, there is a good chance that someone will be exposed to it and inhale it. If the dust is harmful, there is a chance that someone will suffer from an adverse health effect, which may range from some minor impairment to irreversible disease and even life-threatening conditions.

The **health risk** associated with a dusty job depends on the **type of dust** (physical, chemical and mineralogical characteristics), which will determine its toxicological properties, and hence the resulting health effect; and the **exposure,** which determines the dose. Exposure depends on the **air (usually mass) concentration** and **particle aerodynamic diameter** of the dust in question, and **exposure time** (duration). The dose actually received is further influenced by conditions that affect the uptake, for example, breathing rate and volume. Particle aerodynamic diameters will determine if and for how long dusts remain airborne, their likelihood of being inhaled, and their site of deposition in the respiratory system. Dust concentration in the air and the aerodynamic diameter of the particles will determine the amount of material deposited, hence the dose received at the critical site. Very soluble substances can be absorbed from all parts of the respiratory tract, so for soluble particles the site of deposition (and hence aerodynamic diameter) is of less importance. For insoluble particles, the site of deposition in the respiratory system is of fundamental importance, which means that the aerodynamic properties of the particle, shape (fibres), dimensions of the airways and breathing patterns are relevant.

Health effects resulting from exposure to dust may become obvious only after long-term exposure; this is often the case with pneumoconioses. It may happen that effects appear even after exposure has ceased, thus being more easily overlooked or mistakenly attributed to non-occupational conditions. For example, mesothelioma resulting from exposure to crocidolite has appeared after latency periods of 40 years or more after beginning of exposure

However, many dusts have effects that result from shorter exposures to higher concentrations. Health effects, diseases which may result from exposure to different types of dust, include pneumoconioses, cancer, systemic poisoning, hard metal disease, irritation and inflammatory lung injuries, allergic responses (including asthma and extrinsic allergic alveolitis), infection, and effects on the skin. The same agent can cause a variety of adverse health effects, for example, certain wood dusts cause such impairment as eye and skin irritation, allergy, reduced lung function, asthma, and nasal cancer.

Pneumoconioses

"pneumoconiosis is the accumulation of dust in the lungs and the tissue reaction to its presence". The lung changes in pneumoconiosis range from simple deposition of dust, as in the case of siderosis (deposition of iron dust in lungs, clearly observed by X-ray examination but with no clinical manifestations), to conditions with impairment of lung function, such as byssinosis (caused by cotton and flax dust) and to the more serious fibrotic lung diseases such as silicosis (caused by free crystalline silica dust).

Coal-miners' pneumoconiosis may be a serious problem in countries where coal mining is appreciable.

Asbestosis may be a very serious problem wherever asbestos is mined and/or processed, but the cancers it causes are a problem at low exposures also.

Other pneumoconioses may be produced by inhalation of excessive amounts of the following dusts: beryllium (berylliosis); kaolin (kaolinosis); barium (baritosis); tin (stannosis); iron oxide (siderosis); talc; graphite; and mica. With the exception of berylliosis, these other pneumoconioses are relatively benign.

Silicosis

Silicosis is a fibrotic lung disease that is caused by overexposure to dusts composed of or containing free crystalline silica. It is irreversible, progressive, incurable, at later stages disabling and eventually fatal. The silicosis risk depends on the amount of free crystalline silica inhaled and actually deposited in the alveolar region

The first symptom of silicosis is dyspnoea (breathing difficulty), which may become increasingly serious. In view of the restrictive nature of this lung disease, compensatory emphysema (destruction of the alveolar walls) may occur. The most usual complication of silicosis, and a frequent cause of death in silicotic persons, is tuberculosis (silico-tuberculosis). Respiratory insufficiency due to the massive fibrosis and emphysema, sometimes accompanied by enlargement of the heart due to the continued effort to breathe with a restrictive lung disease is another cause of death.

Silicosis, like most pneumoconioses, is a chronic disease, taking many years to appear. However, if exposure is massive enough, it may occur in the accelerated (acute) form.

Byssinosis

Byssinosis is an obstructive lung disease, usually characterized in the initial stages by shortness of breath, chest tightness and wheezing on the first day after returning to work, but with symptoms increasing and becoming more permanent as the disease progresses. Byssinosis (also called "brown lung") is caused by overexposure to dusts from cotton (mainly in operations such as ginning, carding and spinning), flax, sisal and soft hemp.

Cancer

Many dusts are confirmed carcinogens (Cancerous) for example: asbestos .which may cause lung cancer and mesothelioma, free crystalline silica hexavalent chromium and certain chromates, arsenic (elemental and inorganic compounds), particles containing polycyclic aromatic hydrocarbon (PAH) and certain nickel-bearing dusts. Certain wood dusts have been recognized as causing nasal cancer. Deposited radioactive particles expose the lungs to significant doses of ionizing radiation, which may cause carcinoma of the lung tissue, or they may be transported from the lungs and damage other parts of the body. Soluble carcinogens may pose a risk to both lungs and other organs. It should be mentioned that, in the case of lung cancer, cigarette smoke constitutes a confirmed non-occupational causal agent. Moreover, there is a strong synergistic effect between cigarette smoke and certain airborne dusts, for example asbestos, by which the potential risk is enormously increased.

Cancers due to asbestos, particularly mesothelioma, have been clearly linked to occupations such as building maintenance, where exposure is incidental, and would be expected to be low. This has clear implications for 'recognition': there may be an asbestos-cancer risk where people are working with asbestos-containing materials in building maintenance.

The establishment of cause-effect between chemicals in the work environment and cancer is complicated by factors which include: the lapse of time between exposure and disease (latency period); exposure to multiple agents; and the fact that cancers from occupational and non-occupational causes are often pathologically identical.

Ischemic Heart Disease

Dusts may have health effects on organs other than lungs. Recently studies have found effects on the cardiovascular diseases related to dust exposure. There is a possible association between occupational exposure to dust and ischaemic heart disease.

Systemic Poisoning

Some chemical dusts can enter the organism and pass to the bloodstream, thus being carried through the organism and exerting toxic action on one or more organs or systems, e.g., kidneys, liver, blood. Systemic intoxication can be acute (i.e., of rapid onset and short duration), or chronic (of long duration and usually slow onset), depending on the type of chemical and degree of exposure. Toxic metal dusts - such as lead, cadmium, beryllium and manganese - may cause systemic intoxications, affecting blood, kidneys or the central nervous system (CNS) "Although less usual, certain toxic dusts may also enter the organism by absorption through the skin, e.g. pentachlorophenol crystals may dissolve in sweat and easily penetrate through intact skin.

There are some wood dusts which can also be toxic if inhaled or ingested . Wood toxins are usually alkaloids.

Hard Metal Disease

Overexposure to certain hard metal dusts (e.g., cobalt and tungsten carbide) or hard metal-containing dusts may lead to a diffuse pulmonary fibrosis, with increasing dyspnoea (shortness of breath) Severe cases may progress even after cessation of exposure. This disease is often complicated with occupational asthma.

Irritation and Inflammatory Lung Injuries

Although most widely associated with gases and vapours, irritation to the respiratory system may be caused by airborne particles. Certain dusts have irritant effects upon the upper respiratory tract and can produce chronic bronchitis from continuous irritation, which can lead to chronic emphysema. Exposure to irritants may also lead to tracheitis and bronchitis, pneumonitis, and pulmonary oedema. Airborne irritant particles include: beryllium (acute chemical pneumonitis), vanadium pentoxide, zinc chloride, boron hydrides, chromium compounds, manganese, cyanamide, phthalic anhydride, dusts of some pesticides, and some vegetable dusts.

Vegetable dusts such as tea, rice and other grain dusts may cause lung disorders, such as chronic airways obstruction and bronchitis. Some of these conditions are called 'mill fever'.

Allergic Responses

Some dusts may cause allergic reactions, either in the respiratory system (asthma-like), or skin (rashes and eruptions). Most sensitizers have a gradual effect, which appears only weeks or even years after exposure started. The sensitizer induces certain specific cellular changes so that, after a period of latency, further contact results in an acute allergic reaction. Cobalt, for example, can cause asthmatic effects, which may be crippling.

The two main respiratory diseases of allergic type caused by occupational exposure to particles are occupational asthma and extrinsic allergic alveolitis. **Occupational asthma** may be caused by certain grain dusts, flour and wood dusts, and metals (e.g., cobalt, platinum, chromium, vanadium, nickel). **Extrinsic allergic alveolitis** is caused by moulds (and their spores) that grow on other materials, particularly under damp conditions. This is the case of farmer's lung, bagassosis, suberosis and other types, as given below in the table

Examples of extrinsic allergic alveolitis

S.No.	Disease caused	Disease causing Agents/dusts
1.	Farmers' Lung	Mouldy grains, straw, hay (*Micropolyspora faeni, Thermoactinomyces vulgaris)*
2.	Suberosis	Cork dust
3.	Bagassosis	Mouldy sugar cane (*Thermoactinomyces vulgaris)*
4.	Malt Workers' Lung	Mouldy barley (Aspergillus)
5.	Wheat disease	Wheat flour (*Sitophilus granarius*)

Infection (biological hazards)

Inhalation of particles containing fungi, viral or bacterial pathogens may cause transmission of infectious diseases. For example, pulmonary anthrax - a serious and often fatal disease - results from the inhalation of dusts from animal products (e.g., bones, wool or hides) contaminated with the anthrax bacillus. The highly dangerous pulmonary form is rather rare, the most usual form of anthrax being through skin contact.

Exposures to heavy concentrations of organic dusts (contaminated with microorganisms) may lead to serious respiratory and systemic illness, such as *organic dust toxic syndrome.* Examples of health effects resulting from exposure to a number of airborne dusts are presented in the table below:

Human health effects and organs affected by exposure to various types of dust

S.No.	Types of dust	Main health effect	Target organ affected	dust Fraction of interest
1.	Free crystalline silica	Silicosis (lung fibrosis); progressive and irreversible restrictive lung disease; also carcinogenic	Lungs, gas-exchange region, alveoli	Respirable fraction
2.	Coal dust	Coal workers' pneumoconiosis; restrictive lung disease	Lungs; gas-exchange region; alveoli	Respirable fraction
3.	Asbestos	Asbestosis; lung cancer; mesothelioma	Lungs; bronchial and gas-exchange region;	Thoracic and respirable fraction
4.	Lead dust	Systemic intoxication (blood and central nervous system)	Through respiratory system into the bloodstream	Inhalable fraction
5.	Manganese	Systemic intoxication (blood and central nervous system)	Through respiratory system into the bloodstream	Inhalable fraction
6.	Wood dusts	Certain hard woods cause nasal cancer	Nasal airways	Inhalable fraction

S.No.	Types of dust	Main health effect	Target organ affected	dust Fraction of interest
7.	Cotton dust	Byssinosis; obstructive lung disease	Lungs	Thoracic fraction
8.	Dried sugar cane (Baggase) dust	Bagassosis (extrinsic allergic alveolitis)	Lungs	Respirable fraction
9.	Cement dust	Dermatoses	Skin	Any particle size
10.	Pentachlorophenol	Systemic poisoning	Through skin into blood stream	Any particle size

Appendix 1.51: Mechanism of Formation of Acid Rain components

S.No.	Air pollutants	Combinations/reactions	Products formed
1.	Sulfur dioxide	i. SO_2 with water	i. Sulfurous acid
		ii. Oxidation of sulphurous acid	ii. Sulfuric acid
		iii. Oxidation of SO_2	iii. Sulfurous trioxide
		iv. SO_2 with water	iv. Sulfuric acid
2.	Nitrogen dioxide	i. NO_2 with water	i. Nitrous and nitric acids
		ii. HNO_2 with water	ii. Nitric acid
3.	Carbon dioxide	CO_2 with water	Carbonic acid
4.	Hydrogen fluoride	HF with water	Hydrofluoric acid

Note:

The Effect of Acid Deposition on Materials

Iron, **steel and steel alloys.** The rate of corrosion is related to the amount of SO2 in the atmosphere. The rate of corrosion was also found to depend on the deposition rate of $S0_2$. A study showed that corrosion of steel samples under natural meteorological conditions was highly correlated with long-term SO_2 concentrations and only minimally related to nitrogen oxides. Stainless steels, incorporating chromium, molybdenum and nickel, are highly corrosion resistant because of the protective properties of the oxide corrosive film.

Aluminum: Aluminum is generally considered corrosion-resistant.

Copper: long-term corrosion of copper not controlled by deposition of pollutants, but rather, it was likely controlled by availability of copper to react with deposited pollutants. The patina, that is mostly basic sulfate, is not readily dissolved by acids and thus provides significant protection for the substrate metal. However, this patina can take as long as 5 years to form and varies with meteorological conditions. In bronze, as with many metals, dry deposition between rain events dominate soluble corrosion. For copper and copper alloys, though, if the patina color has aesthetic value, and $S0_2$ accelerates the formation, then the presence of $S0_2$ may be beneficial.

The effects of dust alone: The effect of dust lessened as the rust layer was-is formed. particles, by forming nuclei for the concentration of active ionic species, increase the corrosion rate of $S0_2$. A laboratory study of the synergistic effects of different types of particles and SO_x on the corrosion of aluminum, iron and zinc showed that four most aggressive species of particles were salt and salt/sand from marine and de-iced locations, ash from iron smelters, ash from municipal incinerators and coal mine dusts.

Paints: Paints, opaque film coatings, are the dominant class of manmade materials exposed to air pollutants in both indoor and outdoor environments. Paints are used as decorative coverings and protective coatings against environmental elements on a variety of finished including woods, metals, cement and asphalt. Paints primarily consist of two components: the film forming component and the pigments. Paints undergo natural weathering processes from exposure to environmental factors such as sunlight, moisture, fungi and varying temperatures. pollutants affect the durability of paint. Paint films permeable to water are also susceptible to penetration by $S0_2$ and $S0_4^{2\prime}$ aerosols (PM). Unpigmented polymer films have a large range of permeability's but polymers used in paint formulations do not generally form barriers to $S0_2$ either in the gaseous state or in solution as sulfurous acid. There is little degradation to the polymer itself front $S0_2$ at low concentrations. $S0_2$ and relative humidity markedly affect the erosion of oil-based house paint. The presence of NO_2 increase the weight of the paint film. Blisters are formed on acrylic latex house paint at high $S0_2$ levels. The vinyl and acrylic coating are resistant to SO_2. In addition, the weathering of wood prior to painting decreases paint adhesion.

Automobile finishes: Particles can cause damage to automobile finishes. The formulation of the paint will affect the paint's durability under exposure to varying environmental factors and pollution. However, failure of the paint system results in the need for more frequent repainting and additional cost.

Stone and cement: In general, stone containing lime is especially susceptible to the effects of acid deposition. There can be facilitating effects between different constituents of particulate matter. For instance, marble in cemeteries showed that SO_2 is more reactive with the calcium in marble under high $N0_2$ conditions.

Soiling: An additional, significant and detrimental effect of particle pollution is the soiling of man-made surfaces. The black crust found in the protected areas of buildings is formed from a hard crust of gypsum mixed with dust, aerosols and carbonaceous particles. Increased frequency of cleaning, washing or repainting over soiled surfaces becomes an economic burden and can reduce the life usefulness of the material soiled.

Appendix 1.52: Forms of Damage to the Leaves of the Plants/Tree due to Exposure of Air Pollutants

1	Necrosis	Necrosis is the killing or collapse of tissue
2	Chlorosis	Chlorosis is the loss or reduction of the green plant pigment, chlorophyll. It is similar to anaemia in animals
3	Abscission	Leaf abscission is dropping of leaves
4	Epinasty	Leaf epinasty is a downward curvature of the leaf due to higher rate of growth on the upper surface.

Note:

Air pollutions Effects on Plants

Plants are often the first to show damage associated with increased ambient levels of air pollution. The most severe damage seems to have been associated with high levels of SO+ and heavy metal particulates associated with mining and smelting, but many other pollutants such as NO, Cl_2 and NH_3 are now known cause direct toxic effects on plant tissue.

The problems of acid rain and photochemical smog were first suggested as major environmental problems by their actions on plants.

Air pollutants may injure plants in several ways– with some effects being subtle, but others clearly visible. Visible effects normally involve identifiable changes in the leaf structure such as chlorophyll destruction (chlorosis), tissue death (necrosis) and pigment formation. Subtle effects include inhibition of growth and lowered rates of photosynthesis.

Plant tissue damage may be either acute or chronic. Acute exposures to high levels of pollutants generally lead to **necrosis**. Chronic injuries result from intermittent or long term exposures to lower levels of pollutants, with chlorophyll destruction or chlorosis being the most common symptoms of injury. Different pollutions produce differing symptomatic effect on plants.

Effects of Sulfur Dioxide (SO_2) on plants

Atmospheric sulfur dioxide is harmful to certain plants at levels below 1 ppm, causing tissue damage and destruction of chlorophyll. However, it is through chemical reactions in the atmosphere which produce sulfuric acid, that sulfur dioxide emissions have caused their most significant damage. In conjunction with the nitric acid produced from nitrogen oxide emissions, the phenomenon known as acid rain has transformed ecosystems destroying forests and increasing acidity levels in lakes.

Sulfur dioxide causes damage to plant tissues by entering through stomatose^where it immediately comes into contact with the spongy mesophyll cells of the leaf. Initially the damage is restricted to the near vicinity of the pore, but as exposure continues, there is a progressive expansion of injury and tissue collapse (see Figure) As the injury develops, affected tissue may have a greyish green water-soaked appearance, which on drying becomes ivory-to-white, red, brown or even black, depending on the species. The injury extends from the bottom to the top of the leaf and is visible on both surfaces.

The severity of the injury, that is, the amount of leaf tissue affected, is dependent on the dose to which the plant has been exposed. The greater the dose or exposure, the more severely injured are individual leaves as well as the whole plant. The entire leaf, or in some extreme cases the entire plant, may be killed.

The severity of observed injury on individual leaves is also dependent on leaf maturity. Young, fully expanded leaves are most sensitive to SO_2 Older leaves and those not fully expanded are much less sensitive.

In dicots, acute SO_2 injury is usually manifested as an interveinal necrosis, although necrosis in some species may appear on the margins. In some instances, chlorosis may be associated with the necrotic areas. In monocots, such as grains, lilies, and gladioli, injury may appear as irregular necrotic streaks between the larger veins on both surfaces of the leaf. At low to moderate concentrations, SO_2 may cause leaf tip injury on monocots. Sulfur dioxide injury on conifers is observed as reddish brown to brown tip necrosis that may have a banded appearance from repeated exposures.

Different species and varieties vary in their sensitivity to SO_2 exposure. Alfalfa is the most sensitive species, with acute injury observed at a dose of 1 ppm for 1 hr. Because of its hypersensitivity, alfalfa has been used as a bio indicator of SO_2levels in the ambient atmosphere.

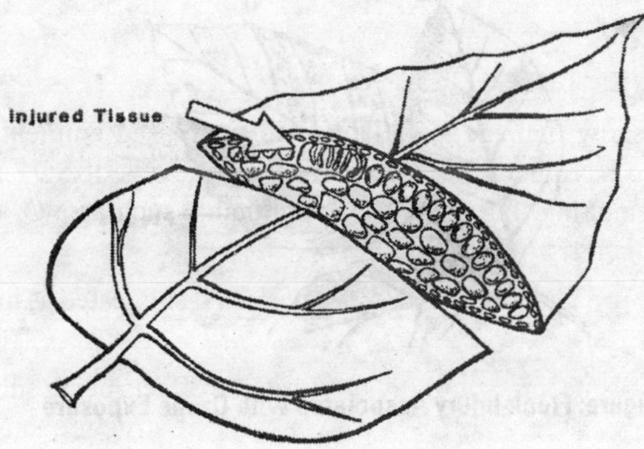

Figure: Tissue Injury Symptoms Associated with Exposure to High Levels of SO_2

Effects of Ozone (O_3) on plants

Ozone also enters the leaf through the stomates on the lower leaf surface. In dicots it bypasses the spongy cells and injures the palisade mesophyll cells preferentially (see Figure). Due to this the symptoms of acute injury are visible on the upper leaf surface. In monocots, conifers, and dicots without a palisade mesophyll, cells near the stomates are injured, with symptoms appearing on either or both surfaces. Developing leaves ranging from 65-90% of their full size are most sensitive to O_3 injury. As with SO2 the maturity of the plant determines sensitivity, with younger plants being more sensitive and older plants more resistant.

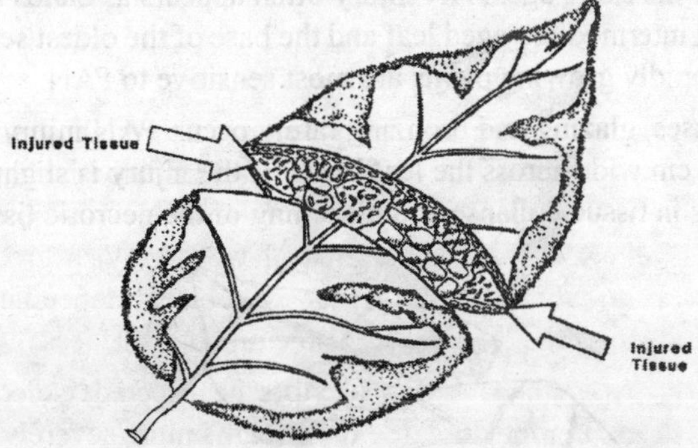

Figure : Leaf Tissue Injury Associated with Ozone Exposure

The most common 03-induced symptom patterns observed on dicots are upper surface flecks (see Figure) These flecks are produced when groups of palisade cells are killed or injured, resulting in chlorotic or necrotic lesions. The adjacent epidermal cells may in some cases also collapse. The flecks may be white, tan or yellow in color. If flecking is extensive the upper leaf surface may appear bronzed or chlorotic. A variation of this symptom is called stipple. In stipple, the injured palisade cells may become intensely pigmented, giving the leaf surface a red to purple to black to brown appearance. The collapse of the overlying epidermal cells is generally not observed with the stipple symptom. Flecking is generally a more severe reaction than stipple, but it is apparent that these symptoms are more related to plant species than to ozone dose. In both the fleck and stipple symptom patterns, injury is confined to the interveinal tissues, the veins being uninjured.

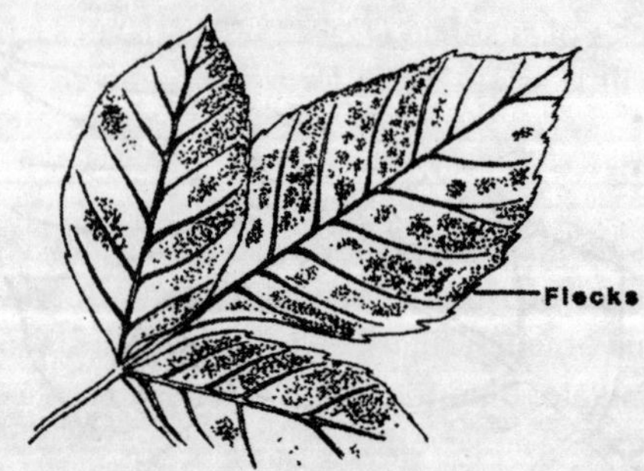

Figure: Fleck Injury Associated with Ozone Exposure

Ozone sensitivity varies from species to species and from variety to variety within species. Ozone is more toxic to plants than SO2. Symptoms may be observed on sensitive plants from exposures of as little as 0.10 - 0.30 ppm for a few hours.

Ozone is considered to be the most important of all of the phytotoxic air pollutants

Effects of Peroxyacyl Nitrate (PAN) on Plants

Peroxyacyl nitrate is the causal agent of "smog injury" observed on vegetable crops. PAN enters the leaf through the stomate, then reacts immediately with the spongy mesophyll around the stomate. This causes the stomate to collapse, forming large air pockets that lead to a glazing/browning appearance on the lower surface of the leaf. Patterns of banding may also be observed with PAN exposures. These bands occur in the same location on leaves of the same age. PAN injury often appears as bands at the apex of the youngest sensitive leaf, the middle of an intermediate-aged leaf and the base of the oldest sensitive leaf. Young, rapidly developing leaves on young rapidly growing plants are most sensitive to PAN.

In monocots, such as grasses, glazing and bronzing rarely occur. PAN injury usually appears as distinct bands several millimeters to 2 cm wide across the leaf blade. If the injury is slight, the injured tissue may be chlorotic. Severe injury results in tissue collapse and bleaching of the necrotic tissue.

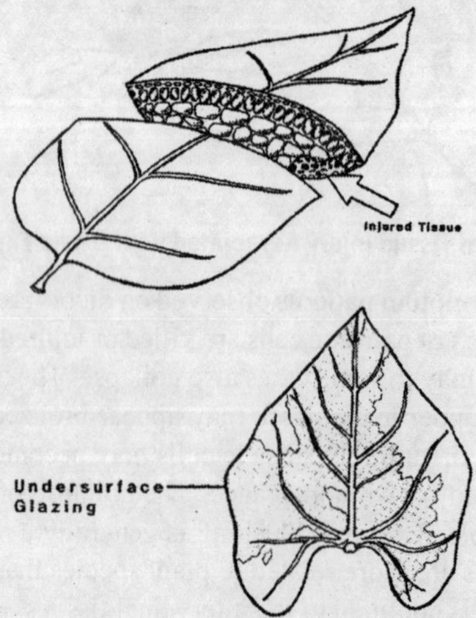

Figure: Tissue Damage and Symptoms Associated with Leaf Exposure To PAN's

Peroxyacyl nitrate is more toxic to plants than O_3. Sensitive plants such as petunia, tomato, and lettuce may be injured by an exposure to as little as 15-20ppb for a few hours

Effects of Fluorides on Plants

Fluoride can have substantial effects on growing plant tissue. Injury may result from the uptake of gaseous hydrogen fluoride through the stomates or from soluble particulate fluorides absorbed through the leaves and/or roots. Regardless of the means of entry, fluorides enter the veins and are transported to leaf margins and/or the leaf tip, where they accumulate. Fluoride injury generally results from the accumulation of small quantities of fluoride over long periods. The severity of the injury is directly related to the dose absorbed level of exposure.

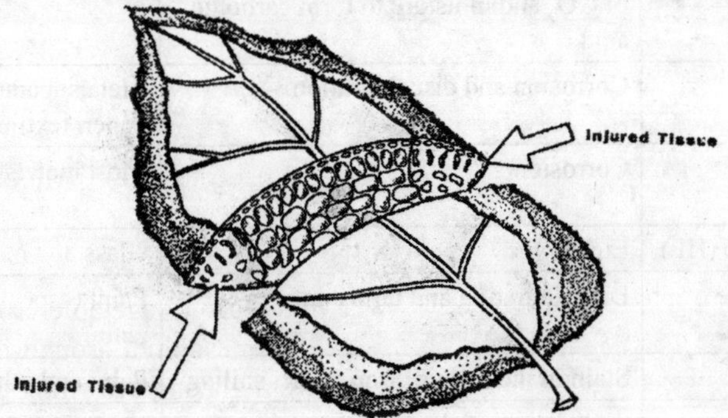

Figure: Tissues of a Leaf Affected by Fluoride Exposure

Fluoride injury normally appears as a marginal or tip necrosis (commonly called "tip bum"). The injury occurs in both the mesophyll and the epidermis. There is often a narrow reddish band between the living and dead tissue.

Typical fluoride-induced injury on selected species is illustrated in Figures. Some sensitive plant species (such as iris, wheat, tulip and gladiolus) can be injured by fluoride exposures of only a few ppb for a 1 week period.

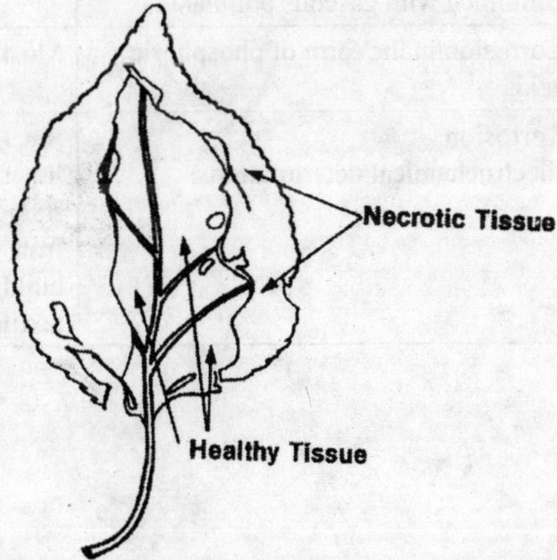

Figure: Symptoms of a Typical Leaf Exposed to Fluoride for an Extended Period

Appendix 1.53: Air Pollutants Effects on Various Materials

S.No.	Air Pollutants	Type of effect on materials	Materials affected by the pollutants
1.	Ammonia (NH_3)	Deterioration associated with SO_2 and moisture	Damage to varnish and paint surfaces, discolours fabrics
2.	Biological aerosols/ Saprophytic bacteria	Deterioration of surfaces and food spoilage	All material surfaces in contact with air, most unprotected food
3.	Fungi	Deterioration of surfaces	Electronic circuits, leather
4.	Carbon dioxide (CO_2)	Deterioration caused by combination of CO_2 and moisture to form carbonic acid	Building stone
5.	Chlorine (Cl_2)	Corrosion and discolouration	Metals, paints, building materials, paper, textiles
6.	Hydrochloric acid (HCl)	Corrosion	Most metals and alloys
7.	Hydrogen fluoride (HF)	Etching	Glass and metals
8.	Hydrogen sulphide (H_2S)	Discolouration and tarnishing	Paint (especially lead based), copper, zinc
9.	Iron (Fe)	Stain in the form of iron oxide, soiling	Paint and other materials, textiles
10.	Manganese (Mn)	Soilant, especially near ferro-manganese plant	Most materials, textiles
11.	Nitrogen oxides (NO_x)	Cause dyes to fade and whites to turn yellow	Textile
12.	Odour-producing pollutants	Odours that cling to skin, hair, clothing (especially bad near soap plant)	Requires more frequent laundering and drycleaning of clothes
13.	Ozone-organics oxidants (O_3)	Deterioration, fading of dyes, cracking	Textiles, rubber
14.	Particulates	Abrasion and corrosion when combined with gaseous pollutants	Most metals, paint, textiles
15.	Phosphorus (P)	Corrosion in the form of phosphoric acid	Most materials
16.	Sulphur oxides (SO_x)	Corrosion Electrochemical deterioration	Steel, zinc, electrical equipment, limestone, roofing state, mortar, statues, tiles, leather, book bindings Iron, aluminium, copper, silver, building materials, leather, paper, textiles.

Appendix 1.54: Heavy Metals Found in the Effluents from some Major Industries

S.No.	Industry generating effluents	Heavy metals present in the effluents generated
1.	Chlor-alkali plants	Cd, Cr, Cu, Pb, Zn, Hg, Se
2.	Electroplating units	Cu, Cr, Zn, Ni
3.	Leather industries	Cr
4.	Batteries manufacture	Pb, Cd, Hg
5.	Paints and Dyes industry	Cd, Cr, Cu, Pb, Hg, Se
6.	Textiles industries	Cr
7.	Paper and pulp plant	Hg
8.	Petroleum refining plant	Zn, Cd, Cu, Cr, Pb
9.	Fertilizers plant	As, Cd, Cr, Cu, Pb, Mn, Hg, Zn
10.	Motor Vehicles	Cd, Cr, Pb, Hg, Se, Zn
11.	Pharmaceuticals industry	As, Hg
12.	Mining and Metallurgy industries	As, Cd, Cr, Cu; Hg, Se, Zn
13.	Explosives industries	As, Hg, Pb, Cu
14.	Pesticides manufacture	As, Hg, Se, Pb

Appendix 1.55: Target Organs Affected by Heavy Metal Pollutants

S.No.	Heavy Metal Pollutants	Target Organs affected
1.	(As), Hg, Mo, Se	Liver
2.	(As), Cd, Hg, Pb	Blood
3.	(As), Pb, Hg	Brain
4.	(As), Cd, Hg	Lungs
5.	(As), Pb, Hg, Cd	Kidney
6.	Cd, Se	Bones and teeth

Note:

Heavy Metals: Factsheet

Human health and the aquatic environment are at significant risk from excessive heavy metal exposure.

Mercury, Cadmium, Copper, Iron, Lead and Zinc are the environment important heavy metals

- Toxic metals, including "heavy metals," are individual metals and metal compounds that negatively affect people's health.
- **In very small amounts, many of these metals are necessary to support life. However, in larger amounts, they become toxic.**

- **They may build up in biological systems and become a significant health hazard.**
- **Many heavy metals tend to be sequestered at the bottom of water bodies.**
- **The aquatic environment is more susceptible to the harmful effects of heavy metal pollution because aquatic organisms are in close and prolonged contact with the soluble metals**
- **The potential effects of exposure to excessive heavy metal contamination on human health and marine life is given below**

Mercury: Readily absorbed via respiratory tract. causes Inflammation of mouth and gums, swelling of salivary glands, excessive flow of saliva, loosening of teeth; kidney damage; muscle tremors, jerky gait, spasms of extremities; personality changes, discouragement, depression, irritability, nervousness, dementia, loss of coordination. Mercury accumulation in fish may harm the fish and other animals that consume them. Birds and mammals that eat fish are then exposed to. Effects on wildlife can include reduced fertility, damaged kidneys, slower growth and development, abnormal behaviour and even death. Whales and dolphins may also be at high risk from mercury exposure.

Cadmium: Target organs are the liver, placenta, kidneys, lungs, brain, and bones. Renal dysfunction. Biopersistent and once absorbed by an organism, remains resident for many years. Cadmium is also toxic to plants, animals and micro-organisms. Cadmium accumulates mainly in the kidney and liver of vertebrates and in aquatic invertebrates and algae. Acute toxic effects on fish, birds and other animals may include death or fetal malformations. Cadmium can affect plants resulting in decreased growth rate and even death.

Copper: Copper toxicity is a contributor to many health problems; including anorexia, fatigue, premenstrual syndrome, depression, anxiety, liver and kidney damage, migraine headaches, allergies, childhood hyperactivity and learning disorders.

Iron: Ingestion accounts for most of the toxic effects of iron because iron is absorbed rapidly in the gastrointestinal tract. The corrosive nature of iron seems to further increase the absorption. Target organs are the liver, cardiovascular system, and kidneys. Elevated blood pressure; cognitive and neurobehavioral effects in children and adults. Lead exposure *in utero,* in infancy and childhood may result in low birth rate, anaemia, neurological impairment, IQ deficits, renal alterations, colic, growth retardation or impaired metabolism of vitamin D.

Lead: Target organs are the bones, brain, blood, kidneys, and thyroid gland. Accounts for most of the cases of paediatric heavy metal poisoning. Lead poisoning is the leading environmentally induced illness in children. At greatest risk are children under the age of six because they are undergoing rapid neurological and physical development. In the environment, lead bioaccumulates in most organisms and is toxic to plants, animals and micro-organisms. Young fish are more susceptible to lead poisoning than mature fish or eggs. Symptoms of lead toxicity in fish include spinal deformity and blackening of the caudal region (rear part of the fish).

Appendix 1.56: Effects of Some Non-Biological Air Pollutants on Human Beings

S.No.	Pollutants	Pollutants Sources	Effects of pollutants
1.	Aldehyde (formaldehyde and acrolein)	Autoexhaust, waste incineration, fuel combustion, photo-chemical reactions	Odours irritant to eye, skin and respiratory system.
2.	Ammonia (NH_3)	Chemical industries, coke ovens, refineries, stocks yards and fuel incineration.	Corrosive to mucous membranes, damaging to eyes and respiratory tract.
3.	Arsenic (As)	Metal smelters, arsenical pesticides and herbicides	If inhaled, ingested or absorbed through skin, it causes dermatibis, mild bronchitis and nasal irritation, suspected carcinogen.
4.	Asbestos dust	Asbestos factory or mine construction sites	Pulmonary fibrosis, pleural calcification, lung cancer.
5.	Barium (Ba)	Industries, mining, refining, barium based chemicals, smoke suppressant, additives in diesel fuels	Affects heart muscles, gastrointestinal tract and central nervous system and respiratory tract.
6.	Beryllium (Be)	Industrial usage, production of fluorescent lamps, motor fuels	Pulmonary damage and damage to skin membranes from handling and mucous soluble salts of beryllium.
7.	Boron (B)	Industry producing boron, petroleum fuel, or additives, present in coal	Toxic through ingestion of inhalation, as dust causes irritation and inflammation; boron hydrides can cause damage to central nervous system which may result in death.
8.	Cadmium (Cd)	Metal industries engaged in extraction, refining, machining, el-electroplating and welding materials. Byproduct of rcfining lead, zinc and copper. Pesticides, fertilizer, cadmium, nickel batteries, reactor poison in nuclear fission plants, used in production of tetraethyl lead gasoline.	Chronic and acute poisoning. Inhalation of fumes and vapours, causes damage to kidney, emphysema, bronchitis, cancer, gastric and intestinal disorders, diseases of heart, liver and brain.
9.	Chlorine (Cl_2)	Process industries using chlorine, accidental breakage during storage and transportation.	Irritates eye, nose and throat. Large doses damage lungs, producing edema, pneumonitis, emphysema and bronchitis.

S.No.	Pollutants	Pollutants Sources	Effects of pollutants
10.	Chromium (Cr)	Metallurgical and chemical industries, products employing chromate compounds, cement, asbestos.	Irritative, corrosive and toxic to body tissues. Believed to cause carcinogenic action. Develops dermatitis and ulcers of the skin and perforation of nasal septum.
11.	Ethylene (C_2H_4)	Motor vehicle emissions, chemical industries, incineration of agricultural wastes, emitted by growing plants.	Eye irritation as a result of photochemical reaction with nitrogen oxide and ozone.
12.	Hydrochloric acid (HCl)	Byproduct from chlorination of organic compounds, burning of chlorinated plastics and paper, combustion of gasoline containing ethylene chloride.	Coughing and choking from inhalation, inflammation and ulceration of upper respiratory tract, clouding of the cornea.
13.	Hydrogen sulphide (H_2S)	Biological decay of protein material in stagnant water, kraft paper mills, industrial waste disposal ponds, sewage treatment plant, refineries, coke ovens.	Headaches, conjunctivitis, sleeplessness, and pain in eyes. In high concentrations, can lead to blockage of O_2 transfer, it acts as a cell poison and damages nerve tissues.
14.	Iron (Fe)	Iron and steel plants, fly ash from combustion of coal and fuel oil, municipal waste incineration, use of welding rods.	Benign pneumoconiosis and siderosis (iron pigmentation of the lungs). Iron oxides act as a vehicle for transporting carcinogens and sulphur dioxide, deep into the lungs.
15.	Lead (Pb)	Automobile emissions, lead smelters, combustion of coal and fuel oil, lead arsenate pesticides.	Absorbed through gastro-intestinal and respiratory tract and deposited in mucous membrane of nose, throat and in the lungs.
16.	Manganese (Mn)	Blast furnaces producing ferromanganese compounds, organic manganese fuel additives, use of welding rods, incineration of manganese containing products.	Manganese poisoning of central nervous system, manganic pneumonia caused by absorption, inhalation, ingestion or through skin.
17.	Mercury (Hg)	Mining and refining of mercury, use of mercury in laboratories. Pesticides containing mercury.	Vapours inhaled may cause intoxication or protoplasmic poisoning.

S.No.	Pollutants	Pollutants Sources	Effects of pollutants
18.	Nickel (Ni)	Metallurgical plants using nickel, engines burning fuels containing nickel additives, burning coal and oil, nickel plating facilities, incineration of nickel products.	May cause cancer of lungs and sinus, other respiratory disorders, dermatitis.
19.	Phosphorous (P)	Plants producing phosphate fertilizers, phosphoric acid, phosphorous pentoxide. Emission from vehicles and aeroplanes using phosphorous as corrosion inhibitors in fuel.	Skin irritation to systematic poisoning. High concentrations affect nervous system.
20.	Radioactive substances	Direct contamination by radioactive gases or suspended dust from natural or artificial sources. Indirect contamination when radio-nuclides are ingested from food chain as a result of contact with contaminated ground, water, plants or animals.	Somatic effects such as leukemia and other types of cancer, cataracts and reduction in life expectancy. Genetic effects include mutation in human gametes that show up only in future generations.
21.	Selenium (Se)	Combustion of industrial and residual fuels, refinery waste gases and fumes, incineration of waste including paper products.	Irritation to eyes, nose, throat, respiratory tract, and gastrointestinal tract, chronic intoxication from industrial toxicants. Effects on kidney, liver and lungs.
22.	Vanadium (V)	Vanadium refining industries, alloy industries, power plants and utilities using vanadium-rich residual oil.	Physiological effect of varying severity on the gastrointestinal and respiratory tracts. Inhibition of chloesterol synthesis. Chronic exposure related to heart disease and cancer.
23.	Zinc (Zn)	Zinc refineries, manufacture of brass, zinc-galvanizing process.	Fumes have a corrosive effect on the skin which irritate and damage the mucous membrane.

Appendix 1.57: Major Anatomical Features of Human Beings Affected by Air Pollutants

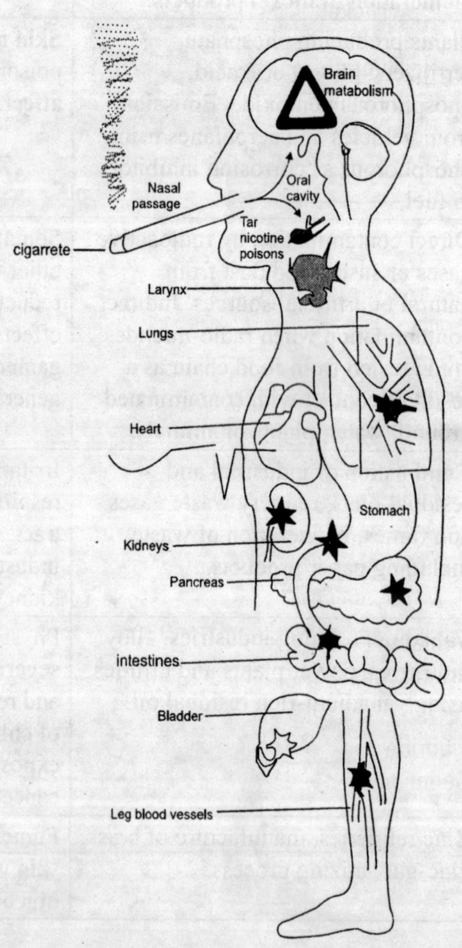

Appendix 1.58: The Site of Effects of Some Major Air Pollutants in Human Beings

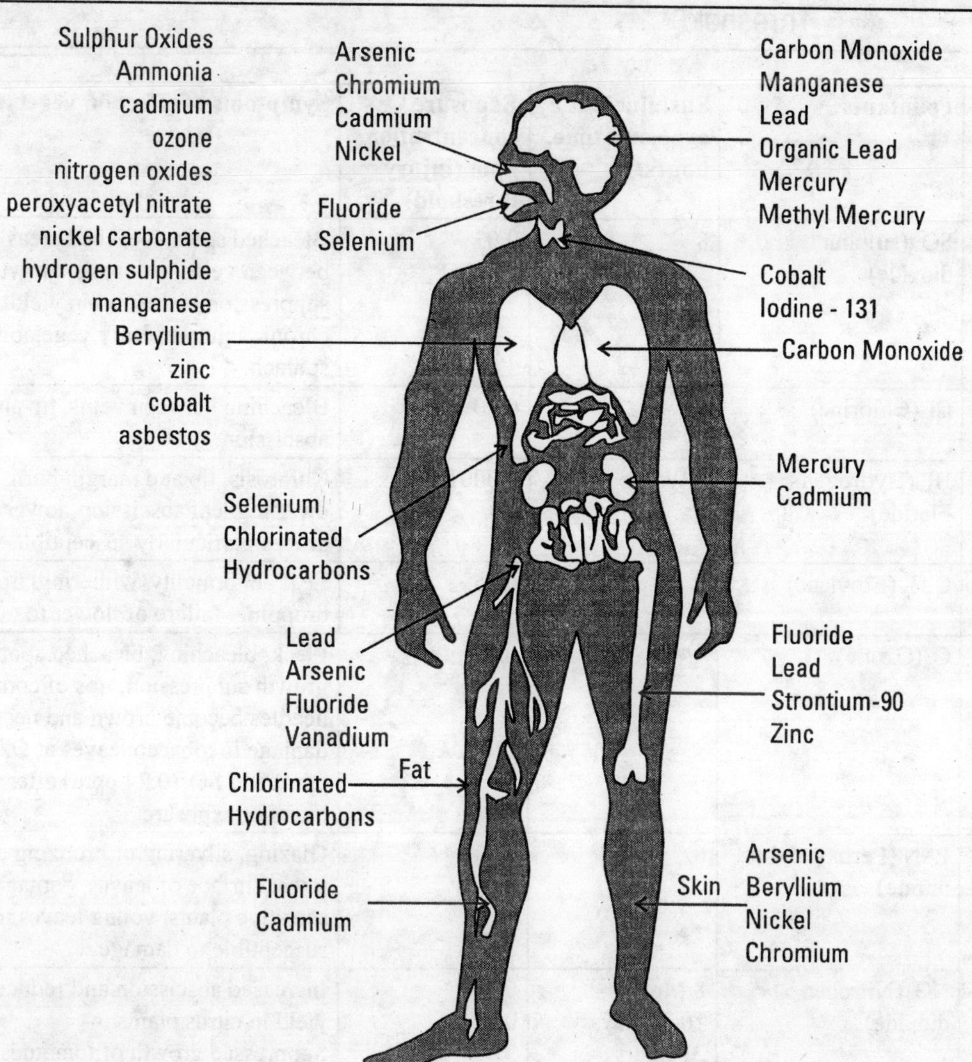

Appendix 1.59: Air Pollution Damage to Vegetation – Symptoms and Injury Threshold

S.No.	Pollutants	Sustained exposure time, hours	Exposure concentration, ppm (injury threshold)	Symptoms of effect on vegetation.
1.	SO_2 (sulphur dioxide)	8	0.03	Bleached spots, bleached areas between veins, chlorosis, growth suppression, reduction in yield, chronic injury to leafy vegetables/ spinach.
2.	Cl_2 (Chlorine)	2	0.10	Bleaching between veins, tip and leaf abscission.
3.	HF (Hydrogen Floride)	5 Weeks	0.0001	Chlorosis, tip and margin burn, dwarfing leaf abscission, lower yield, grapes particularly susceptible.
4.	C_2H_4 (Ethylene)	6	0.05	Leaf abnormality, withering, flower dropping, failure of flower to open.
5.	O_3 (Ozone)	4	0.03	Fleck, bleaching, bleached spotting, growth suppression, tips of conifer needles become brown and necrotic, damage to tobacco leaves at O_3 (0.027 ppm) and SO_3 (0.24 ppm) after two hours of exposure.
6.	PAN (Peroxyacyl nitrate)	6	0.01	Glazing, silvering or bronzing on lower surface of leaves, damage to sensitive plants, young leaves more susceptible to damage.
7.	NO_2 (Nitrogen dioxide)	8 Months 10-12 days 21 hours 1 hour	0.25 0.50 3.5 2.5	Increased abscission and reduced yield in citrus plants. Suppressed growth of tomatoes Spots of mild necrosis on cotton and bean plants. Acute leaf injury.

Appendix 1.60: Carbon dioxide (Green House Gas) Emission from Various Power Generating Sources

S.No.	Power generating sources	CO_2 emission rate, kg CO_2/Kwh
1	Coal	0.9 – 1.2
2	Oil	0.75 – 0.80
3	Gas (Open cycle)	0.58 – 0.60
4	Combined cycle	0.43
5	Hydropower, wind power, solar power, nuclear power	Negligible

NOTES

a. Total global emission = 23 billion tons of CO_2 = 6.5 billion tons of carbon.
b. 1 kg of coal (60% fixed carbon) on combustion produces 2 kg of CO_2.
c. 100 m^3 of Blast furnace gas/coke oven gas/mixed gas on combustion produces 40 -42 m^3 of CO_2.

Appendix 1.61: Various Approaches to Control The Health Effects of Air Pollutants

A. Political Approach

1. "Command and Control" method that legally limits or forbids certain emissions, a common approach of state or federal implementation plan under the clean air act.
2. "Emission trading"methed that limts the quantity of emission permits distributed among sources through market mechanism, favoured by economists to maximise efficienay of pollution controls.
3. Direct taxes on measurable emissions.
4. Taxes on goods and services associated with pollution.

B . Technological Approch

1. Source modification
   a. Discontinue or solate processes that emit pollutants.
   b. Substitution of less polluting for more polluting processes.
   c. Emission control on sources.
2. Ventilation / dispersion
   a. Dilution eg, by tall stock (trade off of high regional versus low loval pollution).
   b. Exhaust from work areas with unternal soutrces.

3. Removal from human-occupied space
   a. Physical filbation
   b. Adsorption
   c. Catalytic destruction
   d. Electrostatic preciptation

C. Personal Approach

1. Avoidance by changing activity patterns, job location.
2. Medical management eg, asthmatics inhale bronchodilator prior to exercising in SO_2 rich atmosphere.
3. Education of others

Appendix 1.62: Toxic Effects of Various Gases/Vapour on Man

S.No.	Toxic effects	Pollutant gases/vapor	Nature of pollutant	Action
1.	Irritants	H_2SO_4, HCl, NH_3 SO_2, Br_2, Cl_2 NO- NO_X	Highly soluble Reasonably soluble Poorly soluble	Nasal passage irritation Lower lung irritation Bronchitis
2.	Asphyxiants	N_2, CH_4, CO_2 CO	Simple asphyxiant Chemical asphyxiant	Causes oxygen deficiency Causes oxygen deficiency in blood
3.	Anaesthetics	Petrol vapour, C_2H_5OH, $C_2H_5OC_2H_5$, C_6H_6	Simple anaesthetic avtion Short exposure	Depressant action Action on blood

Appendix 1.63: Pathological Effects of Common Air Pollutants on Human Beings.

S.No.	Air pollutants	Sources	Pathological effect on man
1.	Aldehydes	Thermal decomposition of fats, oils or glycerols	Nasal & respiratory tract irritation
2.	Ammonia	Chemical processes, dye making, explosives, fertitiser	Inflammation of upper respiratory passages.
3.	Arsenic	Processes involving metals or acids containing arsenic soldering	Breakdoan of red cells in blood, kidney damage, jaundice.
4.	Carbon monoxide	Burning of fossil fuels, automobile exhaust	Reduces oxygen carraying capacity of blood leading to asphyxiation.

5.	Chlorine	Chlor-alkali cell, many other chemical processes	Attacks entire respiratory tract s mucous membranes of eyes, causes pulmonary edema.
6.	Hydrogen cynide	Fumigation, iron blast furnaces, chemical manufacturing, metal plating .	Interferes with nerve cells, . produces dry throat, indistinct vision, headache .
7.	Hydrogen fluoride	Petroleum refining, glass etching, aluminium s fertitiser production	Irritates s corrodes all body passages .
8.	Hydrogen sulphide	Refineries s chemical, bituminous fuels	Smells like rotten eggs, causes nausea, irritates eyes s throat .
9.	Nitrogen oxides	Coal combustion, motor vehicle exhaust	Inhabit cilia action so that soot s dust penetrate for into lungs.
10.	Phosgene (carbonyl chloride)	Chemical s dye manufacturing	Induces coughing, irritation and sometimes fatal pulmonary edema.
11.	Sulphur dioxides	Coal s oil combustion	Causes chest constriction, headache, vomiting and death from respiratory ailments .
12.	Suspended particles (ash, soot, smoke)	Incinerators, almost any manufacturing plants .	Causes emphysema, eye irritation and possibly cancer

Appendix 1.64: Health Hazards Present in Various Chemical Industries/Processes

S.No.	Chemical industries/processes.	Health hazards encountered.
1.	Coal and zinc mining (cadmium), refining of metal, tobacco smoke, incineration of plastic containers.	Pneumoconiosis, Cardiovascular disease shypertension, kidney damage.
2.	DDT & Parathion Manufacturing.	Insecticide poisoning.
3.	Printing press workers, accumulator.	Lead toxicity.
4.	Rayon industry.	Toxicity due to CS_2 and Zn, Narcotic poison.
5.	Dichromate exposure.	Ulceration of wounds
6.	Ferromanganese, Steel processing & refining.	Toxicity of Mn & V.
7.	Cotton fibre dust.	Byssionsis.
8.	Textile & jute mill workers.	Chronic bronchitis.
9.	Dyestuff industry/ intermediates.	Carcinogenic, bladdercancer (Compounds like β-naphthylamine, benzidine etc.)
10.	Carbon black & graphite plants.	Blackening of lung tissues.

S.No.	Chemical industries/processes.	Health hazards encountered.
11.	Rubber, refractory s abrasive manufacturing Limestone manual mining, cement plants (Silica dust inhalation).	Interfering with passage of air to lungs due to production of fibrous tissues in it, shortness of breath, tuberculosis.
12.	Manufacture of (chromate dust).	Lung cancer
13.	Benzol distillation plant (benzene).	Narcotic effect, damage to bone marrow, fatigue, Insomania, weakness, haemorrhage from gums s under the skin.
14.	Phenol recovery plant (phenol).	Headache, dizziness, respiratory paralysis, diarrhea. Loss of appetite, kidney s liver damage.
15.	Lead storage battery manufacturing plant, pipes, auto- exhaust (petrol engine)	Lead affects the blood, bowel s nervous system. It causes anemia, colic s constipation .By entering the nervous system, it paralyses muscles of hand s foot .It also causes depression s mania. It is nevrotoxin.
16.	Caustic-chlorine plant (Mercury poisoning) evaporation from Hg ore mining	Hg vapour inhalation or its absorption through skin causes skin Irritation. It also affects nervous system leading to irritability, depression s loss of confidence. Its effect is noted by excessive salivation, gum inflammation and loss of teeth.
17.	Manganese ore mining & handling, Fe-Mn alloy making plant	Mn dust acts as a poison by entering through respiratory system. It is absorbed and conveyed by blood stream through the lungs to various organs. Manganese poisoning develops slowly and affects the brain causing tremor of hands, salivation etc.

Appendix 1.65: General Toxicological Effects of Metal Pollutants

S.No.	Name of metal/ metallic compound pollutants	Toxicological effects.
1	C_dO, B_eO, NO_2	Irritation effects.
2	HCN, CO .	Asphyxiation.
3	$CHCl_3$, CCl_4	Narcosis.
4	As, AsH_3	Hepatic & renal disorder.
5	Mn, Hg.	Neurtoxic effects.
6	Cr(VI), $Ni(Co)_4$, Pt, As.	Carcinogenic effects.
7	Pb, organolead compounds.	Hemotoxic effects.

8	Be compounds.	Pneumoconiosis.
9	Alkalis, Phenol.	Dermatitis.
10	Organic compounds.	Systemic intoxication.

Common Toxic manifestation: (Ingestion- - swallowing through the mouth, inhalation – Breathing in through the nose)

Pulmonary: Disorder of lung is termed as pulmonary.

Cardiac disfunction: Any problem with heart is called cardiac disfunction.

Renal disorder: Any problem with kidney is termed as renal disorder.

Hepatic disorder: Any problem with lever is called hepatic disorder.

Central nervous system: It is mainly operated by brain.

Problems Associated with Blood Vascular System, Digestive Systems, Skin Ailments

General Causes of Origin of Various Toxicological Effects

a. **Asphyxiation:** It means suffocation. Absence of oxygen causes cessation of breathing and may result in death. Gases like Co s HCN react chemically and produce asphyxiation.

   Simple asphyxants (N_2, CO_2 etc) act by displacing oxygen of the air and reducing it to a level where it does not support life and causes suffocation chemical asphyxants (co, hydrocynic acid etc) have an asphyxiating effect by toxic chemical action they damage the oxygen carraging capaoily of the blood

b. **Irritation effects:** Irritation of eyes, skin s respiratory passages are caused by exposure to corrosive acids s caustic alkalis .Harm caused to lower respiratory passage (by No_2, PH_3, Cdo, Beo etc.) are more dangerous pulmonary irritants which affect the lungs are d_2 $cocl_2$, N_2o, SO_3, O_3, Br_2, F etc. Upper respiratory passage s eye irritants are NH_3, SO_2, HCHO, aldehydes, styrene etc

c. **Pneumoconiosis:** this is a problem related to lung caused by inhalation of dust, quqrtz, asbestos, beryllium compounds etc.

d. **Narcosis:** Most of the organic solvents are narcotic in nature eg, halogenated hydrocarbons, esters, Ketones, ethers, benzene etc. Intoxication effects are not of permones nature but at the same time many effects are irreversible. Halogenated hydrocarbons includes chlorofrom, carbon tetrachloroethane, trishloroethylene, chlorinated aromatics etc

E. **Hepatic And Renal Disorder:** All problems of liver (which forms blood) s kidney (which filters s removers toxins from the body) fall under this category and is caused by compounds of Hg, Se, U, As, TNT, halogenated hydrocarbons etc.

f. **Neurotoxic Effects:** It is an indirect fall out of renal or hepatic disorder causing tremor, paralysis, mental disturboance due to CH_3Br, CS_2, Mn or Hg compounds. Bile fluid getting mixed with blood causes jaundice, overdrinking of wine causes of cirrhosis of liver, problem with urinary infection cause cystitis s nephritis.

g. **Carcinogenic (Cancerous) Effects:** Cancer causing compounds leads to uncontrolled multiplication of calls – β – naphthylamine, benzidine, benzopyrene, nickel carbonyl, asbestos, Cr(VI)compounds s As compounds cause cancer.

h. **Systemic Intoxication/Hemotoxic Effects:** It causes irreversible injury with slow recovery of vital organs, nervous system or of hemoropolectic system hemotoxic effects are caused by benzene and can lead to leekemia. Anemia is commonly caused by lead poisoning carbon monoxide converts haemoglobin in blood to carboxyhaemoglobin leading to death due to asphyxiation.

i. **Dermatitis:** It is an occupational disease caused by contact with corrosive chemicals .this means skin inflammation caused by chemical burns, irritant action of chemical or allergic reaction. Strong acid s alkalis have a corrosive effect on the skin causing deep burn and sever tissue damage chemical, nickel salt, lime creates skin. Tar, pitch and dinitrobenzene cause coloration of skin. Use of chemical solvent for cleaning hands/skins should be avoided as they cause defatting action on skin causing dermatitis.

   Amines used for hardening epoxy resins, formaldehyde, rubber accelarators s antioxidants and dinitrochlorobenzene affects the skin

j. **Central Nervous System Poisons:** They have narcotic effect on the body eg, alcohols, toluene, xylene CS_2 etc.

Some more terminologies related with occupational disease caused by air pollutants/chemicals:-

1. **Rales:** It means rattling sound heard in an unhealthy lung i.e., bubbling of chest.
2. **Hepatitis:** It is irritation and inflammation of liver caused by certain chemical substances.
3. **Nausea:** Feeling of sickness/loathing.
4. **Cirrtiosis**: A condition in which frequent or sustained damage or irritation to the liver promote scarring.
5. **Dyspnoe:** Difficulty of breathing.
6. **Hepatomegaly:** Swelling of lever caused when it can not act as an efficient filter resulting in building up of fluid inside it.
7. **Mucous:** Slimy substance secreted by an organ of the body (cough).
8. **Methemoglobinemia:** It is the stoppage of hemoglobin (blood iron) from binding with oxygen provoking a form of internal suffocation; oxidized hemoglobin forming methaemoglobin is not able to carry oxygen to different body parts. Methaemoglobin in blood (concentration 71.5 gm/cc) produces bluish grey discoloration or cyanosis of skin and mucous membrane of body. It its concentration is 2.5-4.5 gm/cc, then exertion leads to severe headache, dizziness, breathlessness, tiredness not impairment of mental ability and a rapid heart rate. Methemoglobinemia is caused by many chemicals and drugs notable among them are nitrites, nitrates, anailine, dyes, analgesic drugs etc.
9. **Anorexia:** Shortage/hunger/dearth.
10. **Thrombocytopenia:** It is the depression of bone marrow function which causes stoppage bone marrow from producing new blood cells to replace those circulating in the blood.
11. **Necrosis**: Mortification of bones (decay of bones).
12. **Plumonary edema:** Filling up of the lung with a fluid.
13. **Photophobia:** Intolerancy of light.
14. **Pulmonary fibrosis:** It is the lung loosing its breathing area to a great deal and becoming in elastic.
15. **Pharyngitis, Anthropogenic & natural sources, background concentration, estimated life time and removal mechanisms of trace gases present in atmosphere** Disease of pharynx (food pipe disease).

16. **Cancer:** It is a malignant tumor of a body organ or tissue; a sort of runaway growth.
17. **Cutaneous:** Pertaining to skin.
18. **Embryotoxic or embryolethal effects:** It is the poisoning or killing of the embryo carried by a pregnant woman.
19. **Plumbism:** Lead poisoning.
20. **Shavers disease:** It is a ling disorder found in aluminum abrasive industry.
21. **Colic**: Griping pain in belly.
22. **Glomerulitis:** Loss of filtration capacity of Kidney leading to inability to retain protein in the system and promotion of high-blood pressure and edema.
23. **Meningitis**: Inflamation of membrane of brain.
24. **Hemangiothelioma:** It is a type of liver cancer prevalent among workers working in smelters or mines with arsenical ores.
25. **Encephlitis**: Brain fever.
26. **Berylliosis:** Caused by exposure to beryllium dust, it is a pneumonia like disease characterized by coughing, shortness of breath, bloody phelgum, loss of light, inflammation of lings and frequently death.
27. **Gingivitis:** Gum disease.
28. **Magnesium tetany:** Caused by exposure to magnesium results in digestive upsets and muscular paralysis.
29. **Pneumonia:** Inflammation of lungs.
30. **Erethism:** Caused by mercury vapor poisoning, it results is severe behavioral disorder/psychic disturbance.
31. **Cancer:** It is a malignant tumor.
32. **Leukamia:** A progressive, malignant disease of the blood forming organs, marked by distorted proliferation and development of leucocytes and their precursors in the blood and bone marrow.
33. **Emphysema:** Severe form of asthma, enlargement of air vesicles of lungs.
34. **Angina Pectoris:** Pressure pain in chest.
35. **Edema:** Lung water.
36. **Keratitis of eyes:** It is the inflammation of the cornea of the eyes, eye water and is very painful; vision is blurred; it is caused by abrasion, dust, vapour, ultra-violet light etc.
37. **Lymphocytosis:** It means increase in the number of lymphocytes (a variety of white blood cell) in the blood. It occurs im many diseases such as leukemia and infection due to virus.
38. **Blood dyscrosia:** Dyscrosia means abnormal state of the body or part of the body, especially one due to the abnormal development or metabolism.
39. **Platinosis:** Caused by platinum dust, it is characterized by sneezing, running nose, shortness of breath, wheezing, itching of skins inflammation of eyes.
40. **Hyperplasia of bone marrow:** Hyperplasia means increased production and growth of normal cells in a tissue or organs. Affected part becomes larger but retains its normal form e.g., breath growth during pregnancy of women.

41. **Aplastic anemia** - Aplasia means total or partial failure of development of an organ or tissue.
42. **Wheezing:** Low pitch breathing sound associated with bronchospasm suchas occurs in asthma s byssinosis.
43 **Argyria:** caused by silver dust, it results in the skin and mucous lining beginning to turn grey. Also caver lung irritation and bronchitis.
44. **Myocardial infraction:** Death of a segment of heart muscle followed by interruption of is blood supply. It results in heart attack and severe chest pain spreading to arms s throat.
45. Anuria: It is failure of kidney to produce urine from an obstruction to the flow of urine from the kidney.
46. **alterations Mutagenic:** Mutagene is an external agent that when applied to calls or organism can increase the rate of mutation.
47. **Byssinosis:** It is a lung disease caused by inhalation of dust of cotton, flex or hemp, It results in chest tightness s wheeze after the weekend break which wears off during the working week.
48. **Protoplasmic poison:** Protoplasm means the material of which the living cells are made which includes the cytoplasm s nucleus.
49. **Erythema attropy:** Erythema means abnormal flushing of skin caused by dilation of blood capillaries. Erythema nodosum is a disease of sudden onset characterised by fever, joint pain and an eruption of painful swelling on the legs. Erythema multiforme is a disease caused by toxic in blood, eruptions consisting of circular or irregular red patches occurring on the back of arms s hands.
50. **Ataxia:** Ataxia means shaky movements and unsteady gait resulting from brains failure to regulate body postures and the strength s direction of limb movements. It is caused by the disease of sensory nerves or the cerebellum.
51. **Stupor:** Stupor is a condition of near unconsciousness with apparent mental inactivity and reduced ability to respond to stimulation.
52. **Caisson disease:** Also called compressed air illness. It is a syndrome occurring in people working under high pressure in diving bells or at great depths or attitudes with breathing apparatus. chronic compressed air illness may cause damage to bones, heart s lungs.
53. **Cyanosis of blood:** It means turning skin blue because the blood is not adequately transporting oxygen to the cells It is caused by cadmium bearing welding rods.
54. **Scelerosis of bone:** It means hardening s breathless of bones caused by calcium fluoride (Caf_2) used in welding rods and steelmaking additive fluorspar.
55. **Insomania:** Inability to sleep.
56. **Nausea:** Anunpleasant sensation with a tendency to vomit.
57. **Leptospirosis:** It is an infections disease caused by pathogenic (disease causing) bacteria called leptospires. Man acquires this disease through contact with the contaminated water symptoms include fever, anorexia, headacle, nausea, vomitting, profuse sweating, diarrhoea, constipation .
58. **Metal fume fever:** It is an industrial disease caused by inhalation of high concentration of metal oxide fumes eg, due to exposure of copper fumes in rolling mills, iron fumes from welding operation, magnesium oxide s cadmium fumes, molten zinc handling in galvanising industry, prass foundaries etc

Symptoms of metal fume fever include foul taste in the mouth, upper respiratory tract irrition, coughing, dryness of mucous membranes, muscular pain, weakness, fatigue general malaise, nausea, vomittings, headache, exaggerated mental activity.

59. **Mycoses:** It is caused by variety of fungi and include dermotophytosis, aspergillosis, blastocomysis, histoplasmsis, actinomycosis, and nocardiosis.

    Symptoms include skin infection.

60. **Neurosis:** It is pyschoneurotic changes resulting from an occupational injury. Symptoms include changes in central nervous system, headache, nightmare, dizziness, shortness of breath, weakness, instability of mood, slight irritability, low working capacity.

61. **ornithosis and psittacosis:** It is an infection caused by birds s animals to man.

    Symptoms include pain in the head, Limbs s small of the back, coughing, diarrhoea, stupor or confusion .

62. **Peptic ulcer:** caused by physival or emotional stress prevalent among drivers moter mechanic, railway employees, tram car operators etc. Symptoms include the relief of distress on ingestion of food or alkalis or gastrointestinal bleeding dietry management help in preventing it foods which stimulate gastric secretion eg, meat extracts, beverages containing caffiens s alcohol should not be taken by ulcer patient.

63. Pneumoconiosis: It is a lung disease caused by inhalation of dust symptoms include breathlessness from chronic bronchitis or emphysema, inflammation of airways. It includes silicosis, asbestosis etc.

64. **Q fever:** Q (Query) fever is an infections disease caused by a rickettsia (coxiella burnetti) which is a tiny organism. Human infection arises from direct /indirect association with cattle, sheep or goat. Symptoms include fever (19 to 40°c)severe headache, palpable spleen, liver damage S jaundice. It is diagonised by fever with severe headache, slow pulse rate s cough.

65. **Rheumatic diseases:** It is disorders of musculoskeleted systems and includes rheumatic fever, rheumatoid arthritis, spondylitis, joint diseases, gout, fibrositis, muscular rheumatism.

66. **Schistosomiasis:** It is an infection disease caused by bacteria to human beings May be picked up from the animals contect.

67. **Siderosis:** It is a special type of pneumoconiosis caused by inhalation of iron dust this disease is observed in ironore miners, metal polishers, are welders and ochre workers.

    Symptoms include chronic Bronchitis, dyspnoea, tubercular complications. Bronchial cancer is frequently found in iron miners.

68. **Silicosis/Silicotubercolosis:** It is caused by free silica containing dust inhalation.It is a nodular fibrosis of the lung. Symptoms include dyspnea on exertion, emphysema. It is diagonised radiologically.

Appendix 1.66: Characteristics of Two Major Functions of Leaf of a Plant i.e., Photosynthesis and Respiration

$$Co_2 + H_2o \underset{\text{Respirotion}}{\overset{\text{Photosynthesis}}{\rightleftharpoons}} (CH_2o)_n + o_2$$

S.No.	Photosynthesis	Respiration
1.	Oxygen is given off.	Carbon dioxide is given off.
2.	Raw materials are H_2O and CO_2	Raw materials are sugars O_2
3.	Carbohydrates are synthesized.	Carbohydrates are consumed
4.	Energy is stored.	Energy is transformed.
5.	Weight is increased.	Weight is decreased.
6.	Goes on only in cells containing chlorophyll in presence of light.	Goes on at all time in all living cells.

Appendix 1.67: Physiological Role of some Essential Trace Metals present in Plants and Animals

S.No.	Element	Physiological function
1.	Zinc (Zn)	Activator For several mammalian enzymes
2.	Copper(Cu)	Activator for several enzymes
3.	Chromium (Cr)	Phosphoglytamase
4.	Cobalt (Co)	Methymalonyl G A mutase in livers, constituent of Vitamin B12
5.	Manganese(Mn)	Activator for pyruvate corboxylase and arginase in liver
6.	Molybdenum (Mo)	Found in metalloflavin enzymes and in xanthine oxidase in liver.

Appendix 1.68: Sources and Health Effects of Some prominent Heavy Metals

S.No.	Elements	Sources of elements	Health Effects of elements
1.	Mercury (Hg)	Coal, electrical batteries, other industrial processes	Kidney damage, nerve damage and death
2.	Lead (Pb)	Auto exhaust, paints	Brain, liver, and kidney damage; convulsions, behavioral disorders, death

3.	Cadmium (Cd)	Coal, zinc mining, water mains and pipes, tobacco smoke, burning plastics	High blood pressure and cardiovascular disease, interferes with zinc and cooper metabolism
4.	Nickel (Ni)	Diesel oil, residual oil, coal, tobacco smoke, chemicals and catalysts, steel and nonferrous alloys	Lung cancer
5.	Arsenic (As)	Coal, petroleum, detergents, pesticides, mine tailings	Hazard disputed, may cause cancer
6.	Germanium (Ge)	U Coal	Little innate toxicity
7.	Vanadium (V)	Petroleum (Venezuela, Iran), chemicals and catalysts, steel and nonferrous alloys	Probably no hazard at current levels.
8.	Antimony (Sb)	Industry	Shortened life span in rats

Appendix 1.69: Increasing Order of Pollutants Emission from Various Plants/Units

Increasing orders of pollutants emission	Hydro Electric power plants, wind, solar, renewable-**Minimum Pollution**
	Nuclear power plants
	Gas turbine power plants
	Integrated coal gasification combined standard power plant.
	Fluidised bed combustion power plant.
	Coal fired power plant with emission controls
	Conventional coal fired power plants *without* emission controls. **-Maximum pollution.**

Appendix 1.70: Carboxyhemoglobin (COHb) levels in Blood and Related Health Effects on Human Beings

% COHb in blood	**Effects Associated with this COHb Level**
80	Death[a]
60	Loss of consciousness; death if exposure continues[a]
40	Confusion; collapse on exercise[a]
30	Headache; fatigue; impairedjudgment[a]

% COHb in blood	Effects Associated with this COHb Level
7-20	Statistically significant decreased maximal oxygen consumption during strenuous exercise in healthy young men[b]
5-17	Statistically significant diminution of visual perception, manual dexterity, ability to learn, or performance in complex sensorimotor tasks (such as driving)[b]
5-5.5	Statistically significant decreased maximal oxygen consumption and exercise time during strenuous exercise in young healthy men[b]
Below 5	No statistically significant vigilance decrements after exposure to CO[b]
2.9-4.5	Statistically significant decreased exercise capacity (i.e., shortened duration of exercise before onset of pain) in patients with angina pectoris and increased duration of angina attacks[b]
2.3-4.3	Statistically significant decreased (about 3-7%) work time to exhaustion in exercising healthy men[b]

Note:

Carbon Monoxide (CO) poisoning is the most common of all poisonings in the industry

Carbon Monoxide is colour less and odourless and, therefore, gives no warning of its presence under any circumstances. Nearly all other industrial toxins have a characteristic odour, taste, or appearance, or produce some sort of irritation that warns the potential victim, a factor not present in carbon monoxide poisoning. Another reason that makes this form of poisoning particularly insidious is that the early symptoms can resemble other disorders. Nausea, headache and dizziness are easily confused with the onset of a cold, flu or some other common disorders.

It is to be noted that the amount of CO produced usually do not create problems, CO poisoning may strike without warning if either the source of CO increases for some reason or ventilation is decreased.

Fact sheet about carbon monoxide (CO) poisoning

- Carbon Monoxide is absorbed through Lungs and combines reversibly with Hemoglobin and to a lesser extent with myoglobin, cytochrome oxidase, and other compounds with Iron containing porphyrin.
- The relation between equilibrium concentrations of Oxygen and Carbon Monoxide in inspired air and Oxyhaemoglobin (HBO_2) and Carboxyhaemoglobin (COHb) is expressed by

  $COHb/HbO_2 = M \times P_{CO} / P_{O2}$

  Where M is an affinity constant (208 to 235 at 37°C) and P_{CO} and PO_2 are the partial pressures of CO and O_2 respectively.
- This relationship is affected by the concentration of COHb before exposure, normally about 0.5% to 0.8% from endogenous breakdown of haemoglobin plus that produced by the tobacco smoke.
- COHb also disassociates in accordance with this equation, a process that is accelerated by increased concentration of oxygen in inspired air and increased pulmonary ventilation by exercise.
- Under conditions of cessation of exposure, excretion of CO begins rapidly but becomes slower with time and with lowered levels of Carboxyhaemoglobin. Hyperbaric oxygen will accelerate the lowering of high levels of Carboxyhaemoglobin.

- The action of carbon monoxide is favoured by heat, humidity and muscular exertion.
- Young men are more susceptible than older men in industrial experience, probably
- because they breathe more deeply.
- Men with chronic bronchitis or asthma handle CO poorly and the course of CO poisoning
- is unfavorably influenced by alcoholism, obesity and chronic heart disease / chronic *> vascular disease.
- The asphyxia brought about by CO is caused by the affinity of haemoglobin for this gas, which is 240 times greater than its affinity for oxygen, and the resulting . U carboxyhaemoglobin is much more stable than oxyhaemoglobin.
- It is to be noted that the CO is physiologically inert except for one property – it i drives OXYGEN out of red cells of the blood and thus DEPRIVES the body of its normal supply of oxygen.
- In simple asphyxia, there are pleural haemorrhages and respiratory symptoms appear first followed by nervous symptoms; while in CO poisoning this order is reversed.
- Clinically, there is a wide difference between a sudden, rapid and excessive exposure to CO and one that is less great but more prolonged.
- This difference is shown by the complete recovery of the victims in an accident of severe sudden exposure to CO. If, however, the inhalation of CO proceeds slowly and if the victim stays for sometime in an atmosphere of CO, there may be cell destruction from the prolonged anoxaemia.
- Fire Fighters may have unusual intermittent exposures to CO from smoke inhalation. Apart from this fire-smoke, there is a significant increase in mean COFb levels after building fires and Fighters should avoid smoking after attending a fire in order to prevent further increases in COFIb levels.
- Intermittent use of self-contained compressed air breathing apparatus offers no advantage over non-use and continuous use of this equipment is required for adequate protection.
- In fighting fires, the exertion level. is very high and therefore the ventilation rates should be so high that even in low or moderate concentrations of CO, the COHb does not can rise to dangerous levels within minutes.

Concentration of CO necessary to produce symptoms is tabulated below:

Concentration of CO in Air (%)	**CO ppm in air**	**Effects on human health on exposure to CO of various-concentration**
0.02	200	Possibly headache, mild frontal in 2 to 3 hrs.
0.04	400	Headache frontal and Nausea after one or 2 hrs followed by Occipital Headache.
0.08	800	Headache, Dizziness and Nausea in hour: Collapse and possibly unconsciousness in 2 hours.
0.16	1600	Headache, Dizziness and Nausea in 20 minutes; collapse, unconsciousness, possibly death in 2 hrs.
0.32	3200	Headache, Dizziness in 5 to 10 minutes, unconsciousness and danger of death in 30 minutes.
0.64	6400	Headache, Dizziness in 1 to 2 minutes, unconsciousness and danger of death in 10 to 15 minutes.

1.28	12, 800	Immediate effect; Unconsciousness and danger of death in 1 to 3 minutes.

1% = 10, 000ppm

Permissible Limits of Exposure of CO (Carbon Monoxide)				
	TWA (8 hours)		STEL (15 minutes)	
	50 ppm	55 mg/m^3	400 ppm	440 mg/m^3

TWA-Time weighted average, STEL- short exposure limit

Toxic Effects of CO on human system

Blood

Carbon monoxide, like oxygen, is reversibly bound by haemoglobin, it is consequently in competition with oxygen for binding sites on the haemoglobin molecule. CO not only diminishes the total amount of available oxygen by direct replacement in the alveolar air but it also alters the binding of oxygen to haemoglobin so that it is held more tenaciously.

Since Myoglobin may act as a reservoir for oxygen in the muscle fibers, it too can combine with CO to form carboxyhaemoglobin, although the affinity constant, M, is much lower in this case - approximately 40. But this value may be higher for CARDIAC muscle so that when blood COHb is 10%, 30% of the cardiac myoglobin may be saturated with CO.

Cardiovascular System

a. Cardiac symptoms are not at all unusual immediately after severe exposure, and they may be persistent or even permanent.

b. Palpitation, breathlessness on exertion, and pre-cordial pain are fairly common complaints in CO poisoning.

c. Carbon monoxide asphyxia may exacerbate pre-existing disease of the Coronary Arteries.

d. Evidences were have found for the effect of CO on aggravation of Angina Pectoris . (after exposure of 50 ppm for 90 minutes), aggravation of intermittent claudication (after exposure of 50 ppm for 120 minutes), and alteration of ECG in normal subjects(after exposure of 100 ppm for 4 hours).

e. Smokers and Alcoholics are more prone to all such effects. (Blood analysis show carboxyhaemoglobin higher in smokers, averaging 4.1% in light smokers and 5.4% in those smoking at least 20 cigarettes a day. Non-smokers averaged 1.7% COHb.)

Central Nervous System (CNS)

a. Marked hyperaemia of the cerebral vessels is the first effect of CO induced anoxiemia. Ophthalmoscopy revealed bilateral venous engorgement but normal optic discs. Blindness - complete or incomplete, transient or permanent is extremely rare.

b. The first stage of symptoms: a sense of pressure in the head or a band like constriction or throbbing, a feeling of weakness in the knees, mental confusion, headache, nausea, roaring in the ears, and perhaps vomiting.

c. The second stage: characterizes by increasing weakness and confusion - the headache, dizziness, inability to think clearly, to decide and act with energy - increase, and all of these render the victim peculiarly helpless.

d. Symptoms of brain involvement may come on some time after the accident, in form of either of an apoplectic seizure or of an increasing psychosis. Epileptoid convulsions may be a symptom, not of an early stage, but of recovery.

e. Attacks of headache and dizziness may persist for many weeks after recovery from severe symptoms.

Pre-employment medical examinations should be performed with special attention to cardiovascular diseases, anaemia, respiratory insufficiency or any other disease or medical condition which could be exacerbated by the hypoxic effect of carbon monoxide. Pregnant women should not be employed in tasks where there is a risk of dangerous concentrations of CO in the air.

Treatment of CO poisoning

- "First Aid", in case of any gas poisoning, plays an extremely important role in determination of prognosis and immediate treatment. Therefore, any treatment should be preceded by the following First Aid management:
- Removal on to fresh air, loosening of garments.
- Keep the victim lying on hard surface. Total rest.
- Never splash cold water on his face.
- Cover with dry blanket. Keep him warm.
- Do not give him alcoholic stimulant.
- Assess the condition of the victim.
- If required - start immediate artificial breathing continue it until normal breathing starts.
- If available, may start O_2 inhalation
- Arrange to transfer the victim to hospital / clinic.
- If condition improves by First Aid only, do not let him go home on his own efforts.

A. Biochemical repair starts as soon as the CO poisoned worker is exposed to FRESH AIR, if the victim is still breathing. Therefore the victim should immediately be moved from the exposure area to uncontaminated air. If breathing has stopped, resuscitation should be started without delay. Pure oxygen should be administered to hasten the carbon monoxide excretion. Hyperbaric oxygen treatment at three atmospheres of pressure reduces the biological half-life of carboxyhaemoglobin to 24 minutes. The toxicity of oxygen itself should not be forgotten. The addition of 5% carbon dioxide to oxygen to serve as a respiratory stimulant entails some risk of compounding the mild metabolic acidosis that arises from tissue hypoxia. Stimulant drugs should be avoided. Absolute bed rest for at least 48 hours should be ensured. After recovery the patient may be watched for late neurological and cardiac complications.

If a hyperbaric chamber is available, this is by the best method of treatment in severe CO poisoning. Hyperbaric chambers may be either of the large walk-in type or small, mono- place chambers that are pressurized with pure oxygen. Under hyperbaric conditions, the CO is eliminated much more rapidly. The rationale of such hyperbaric treatment is as follows:

A. CO is apparently displaced from cytochrome A_3 oxidase by a surfeit of oxygen, and electron transport is restored. Lipid peroxidation in the brain is halted.

B. If the patient is given oxygen to breathe at 3 ATA, the half-time for CO elimination from the blood is reduced to 24 minutes.

C. As soon as the patient breathes oxygen at 3 ATA, 6.4 vol % of completely available oxygen are physically dissolved in plasma - enough to support metabolism even in complete absence of functioning haemo globin. Thus patient's problem with oxygenating the brain and other tissues is immediately over on reaching 3ATA.

D. Intracranial pressure secondary to cerebral oedema is reduced of the order of 50% within one minute of the commencement of oxygen breathing.

Recovery from severe exposure of CO depends on several factors. If the rescue from the ? poisoned atmosphere is delayed and if there is pre-existing disease - especially disease of the blood vessels or the respiratory tract - recovery is doubtful unlikely, or may be abnormally delayed.

Appendix 1.71: Global Sources of Emission of Hydrocarbon Pollutants

S.No.	Sources of hydrocarbon emissions	% emission
1.	Petrol engines	38.5
2.	Incinerators and refuse burning	28.5
3.	Solvent evaporation	11.5
4.	Storage and transfer evaporation	8.5
5.	Petroleum refining	7.0
6.	Wood burning (fuel and forest fires)	2.0
7.	Industrial uses of coal	1.0
8.	Coal carbonization and heating	2.0
9.	Coal based power generation plant	0.5
10	Fuel oils and petroleum distillates	0.5
	Total	**100%**

Appendix 1.72: Harmful Effects of Some Air Pollutants on Human Health

S.No.	Air Pollutants	Harmful effects on human health
1.	Asbestos	Disease of lungs (Fibriosis, asbestosis) tumors, lung cancer
2.	Arsenic (As)	Bronchitis, dermatitis, skin cancer
3.	Aldehydes (RCHO)	Irritation of eyes and respiratory track
4.	Beryllium (Be)	Systemic poison causes berilliosis, skin damage
5.	Chlorine (Cl_2)	Irritation to eye, nose, throat. Bronchitis, pneumonitis
6.	Cadmium (Cd)	Highly toxic, causes bronchitis, fibriosis of lungs hypertension, carcinogenic

7.	Lead (Pb)	Systemic poison, causes anemia (reduction of haemoglobin) affects kidney function, damages brain
8.	Mercury (Hg)	Highly toxic, enzyme poison, nervous system disorder
9.	Nickel (Ni)	Respiratory disorder
10.	Pesticides	Various insecticides, herbicides, fungicides, nematicides and repellants are toxic and produce headache, stomach disturbances and irritations.
11.	Radioactive substances	Somatic (on body cells) effects including lukemia (blood cancer), and other cancers. Genetic effects are mainly mutations in gem cells
12.	Selenium (Se)	Irritation to eyes, nose, respiratory tract and gastrointestinal tract, badly affects kidney, lungs and liver
13.	Vanadium (V)	Affects gastrointestinal tract and respiratory tract. Cholesterol synthesis inhibition, carcinogenic

Appendix 1.73: Characteristics of Chronic Obstructive Lung Diseases Caused by Air Pollutants

S.No.	Names of diseases:	Characteristics
1.	Chronic bronchitis	The clinical image of this disease is that excessive mucus is secreted in the bronchus and that phlegm is cough out chronically or repeatedly. In most cases, these symptoms continue almost every day for at least three months each year for at least two consecutive years.
2.	Asthma	The main characteristic of this disease is exacerbation of reaction to various stimuli of the trachea and the bronchus and its most important symptom is stenosis of extensive parts of the airway, whose intensity change spontaneously or can by changed through treatment. The most recent definition of the disease, which recognizes that "the existing definition of asthma is unsatisfactory because it is impossible to fully understand the nature of the disease from its viewpoint of obstructed airway and airway hyperreactivity." is: "chronic inflammatory injury to the airway in which many cells such as mast cells and eosinophils are implicated."
3.	Pulmonary emphysema	In this disease an anatomical change to a morbid state is presented: this disease is mainly characterized by enlargement of air chambers from the ends of the tracheae to the farthest ends of the whole respiratory organ, which is accompanies by destruction of some cases do not reveal the disease. In many cases, the disease is accompanies by usually coughing, which is turn is accompanied by wheeze. Its patients find it difficult to breathe and feel very much fatigued.

Appendix 1.74: Sources and Health Effects of indoor Air Pollutants

S.No.	Pollutant	Major Indoor Sources	Potential Health Effects*
1.	Asbestos	Damaged or deteriorating insulation, fireproofing, and acoustical materials	Asbestosis, lung cancer, mesothelioma, and other cancers
2.	Biological Agents (House Dust Mites, Amimal Dander, Mold, bacteria, Viruses)	House dust; pets; bedding; poorly maintained air conditioners, humidifiers and dehumidifiers; wet or moist structures (e.g., due to plumbing leaks)	Allergic reactions; asthma symptoms; eye, nose, and throat irritation; humidifier fever, influenza, and other infectious diseases
3.	Carbon Monoxide (CO)	Gas stoves, malfunctioning gas appliances, wood stoves, tobacco smoke, car or truck exhaust from attached garages	Headache; nausea; angina; difficulty concentrating; death at high concentrations
4.	Environmental Tobacco Smoke (ETS)	Cigarettes, cigars, and pipes	Respiratory irritation, bronchitis and pneumonia in children, emphysema, lung cancer, and heart disease
5.	Formaldehyde	Pressed wood products such as plywood and particleboard, furnishings; wallpaper; durable press fabrics; personal care products	Eye, nose, and throat irritation; headache; allergic reactions; cancer
6.	Lead	Sanding or open-flame burning of lead paint; house dust	Nerve and brain damage, particularly in children; anemia; kidney damage; cardiovascular effects; growth retardation
7.	Nitrogen Oxides (NO_x)	Gas stoves, malfunctioning gas appliances	Eye, nose, and throat irritation; lung irritation and damage; increases respiratory infections in children
8.	Organic Chemicals	Solvents, glues, cleaning agents, pesticides, paints, moth repellents, air fresheners, dry-cleaned clothing, and treated water	Eye, nose, and throat irritation; headaches; loss of coordination; damage to liver, kidney and brain; various types of cancer
9.	Ozone (O_3)	Ozone-generating indoor air cleaners, hobbies involving soldering or welding	Respiratory tract (lung) irritation and inflammation, serious breathing difficulty including asthma, permanent lung damage

S.No.	Pollutant	Major Indoor Sources	Potential Health Effects*
10.	Particulate Matter (PM)	Cigarettes, wood stoves, fireplaces, cooking, vacuuming, burning candles and incense, products of reactions of ozone with fragrances	Eye, nose and throat irritation; worsening of asthma; increases respiratory disease; lung cancer; cardiovascular disease; premature death
11.	Phthalates	Used to soften plastics and add flexibility [i.e., polyvinyl chloride (PVC)]; plastic children's toys; cosmetics and personal care products (i.e., to make fragrances last longer); automotive (e.g., air filters, battery covers) and building/construction products; and food and beverage packaging	Associated with asthma and respiratory symptoms, allergies, and rhinitis in some studies; reproductive and developmental problems, especially in children under 3 years of age
12.	Polybrominated Diphenyl Ethers (PBDE)	Flame retardants in foams in furniture and automobiles, electronic printed circuit boards, electronics casings, carpet backing, upholstery	Disrupt thyroid hormones, may cause developmental deficits, may act as a reproductive toxin, and may cause cancer
13.	Radon	Soil under buildings, some earth-derived construction materials, and groundwater	Lung cancer

* Depends on factors such as the amount of pollutant inhaled, the duration of exposure and susceptibility of the individual exposed.

Appendix 1.75: Sources and Impacts of Selected Pollutants on Aquatic Organisms and Human Health & Welfare

S.No.	Pollutants	Sources	Impact on Aquatic Organisms	Impact on human Health and Welfare
1.	Sediment	Agricultural fields, pastures, and livestock feed lots; logg hill sides; degraded streambanks; road con-struction.	Reduced plant growth and diversity, reduced prey for predators; clogging of gills and filters; reduced survival of eggs and young, smothering of habitats.	Increased water treatment costs; transport of toxics and nutrients; reduced availability of fish, shellfish, and associated species; shortened lifespan of lakes streams and artificial reservoirs and harbours

S.No.	Pollutants	Sources	Impact on Aquatic Organisms	Impact on human Health and Welfare
2.	Nutrients	Agricultural fields pastures and livestock feed lots; landscaped urban areas; raw and treated sewage discharges; industrial discharges.	Algal blooms resulting in depressed oxygen levels and reduced diversity and growth of large plants, release of toxins from sediments; reduced diversity in vertebrate and invertebrate communities fish kills.	Increased costs of water treatment; reduced availability of fish, shellfish, and associated species.
3.	Organic materials	Agricultural fields and pasture landscaped urban areas; combined sewers; logged areas; chemical manufacturing and other industrial processes.	Reduced dissolved oxygen in affected waters; fish kills; reduced abundance and diversity of aquatic life	Increased costs of water treatment; reduced availability of fish, shellfish, and associated species.
4.	Disease causing agents	Raw and partially treatment sewage; animal wastes; dams that reduce water flows.	Reduced survival and reproduction in fish, shellfish, and associated species.	Increased costs of water treatment; river blindness, elephantiasis, schistosomiasis, cholera, typhoid, dysentery reduced availability and contamination of fish, shellfish, and associated
5.	Heavy metals	Atmospheric deposition; road runoff; industrial discharges; sludge and discharges from sewage treatment plants; creation of reservoirs; acidic mine effluents.	Declines in fish populations due to failed reproduction lethal effects on invertebrates leading to reduced prey for fish.	Increased costs of water treatment; lead poisoning, and kidney dysfunction, reduced availability and healthfulness of fish, shellfish and associated species.
6.	Toxic chemicals	Urban and agricultural runoff municipal and industrial discharges; leachate from landfill.	Reduced growth and survivability of fish eggs and young; fish diseases.	Increased costs of water treatment; Increased, risk of rectal, bladder, and colon cancer; reduced availability and healthfulness of fish, shellfish, and associated species.

S.No.	Pollutants	Sources	Impact on Aquatic Organisms	Impact on human Health and Welfare
7.	Acids	Atmospheric deposition mine effluents; degrading plant materials	Elimination of sensitive aquatic organisms; release of trace metals from soils, rocks, and metal surfaces such as water pipes.	Reduced availability of fish, shellfish and associated species
8.	Chlorides	Roads treated for removal of ice or snow; irrigation runoff brine produced in oil extraction; mining.	At high levels; toxic to fresh-water life.	Reduced availability of drinking water supplies; reduced availability of fish, shellfish, and associated species.
9.	Elevated temperatures	Urban landscapes; unshaded stream; impounded waters; reduced discharges from dam; discharges from power plants and industrial facilities.	Elimination of cold-water species of fish and shellfish reduced dissolved oxygen due to increased plant growth; increased vulnerability of some fishes to toxic wastes, parasities, and diseases.	Reduced availability of fish, shellfish and associated species.

Appendix 1.76: Possible Respiratory Hazards Present in some Typical Industrial/ Manufacturing Processes

S.No.	Industrial/manufacturing processes	Possible respiratory hazards present
1.	Brake shoe lining	Asbestos dusts
2.	Coke ovens	Coal tar, pitch dust, smoke
3.	Core blowing	Phenol and/or formal dehye vapor, triethylamine vapor
4.	Degreasing of oxygen carrying pipelines	Methyl chloroform vapors or, other organic vapor
5.	Foundry core knock out	Silica containing dust
6.	Grey iron mill room –grinding	Silica containing dust
7.	Machining	Oil mist
8.	Mulling, core knock out, shake out	Silica containing dust
9.	Oil quenching	Smoke
10.	Pickling of steel sheet by HCl /H_2So_4	Acid mists

11.	Plating	Vapors/mists containing acids and/or chromates
12.	Sanding and gridding on surfaces painted red, orange or yellow	Dusts containing lead or chromium
13.	Solvent handling	Organic vapor
14.	Spray painting	Airborne mist containing lead and chromium
15.	Welding through surfaces painted yellow, orange or red	Fume containing lead and chromium

NOTES

Oxygen deficiency in air is also a respiratory hazard. Atmospheric air contains 21% oxygen (by volume in environment) in which oxygen level is < 16% immediately hazardous to life.

Appendix 1.77: Physico- Chemical Properties of Benzene, Toluene and Xylene (BTX)

Benzene is a naturally occurring colourless liquid at room temperature (25 °C) and pressure (760 mm Hg). It has a characteristic aromatic odour, a relatively low boiling point (80.1 °C) and high vapour pressure, which causes it to evaporate rapidly at room temperature. Benzene is released into the environment from both natural and man-made sources, although the later is the most significant source. Benzene in air exists predominantly in the vapour phase, with residence time varying between few hours to few days depending on the environment and climate. Degradation of benzene in air occurs mainly by reaction with hydroxyl, alkoxy and peroxy radicals. Toluene and Xylene are also present in most of the sources along with Benzene.

	Benzene	**Toluene**	**o-Xylene**	**m-Xylene**	**p-Xylene**
Physical form (20 °C)	Clear colourless liquid	Clear colourless liquid	Colourless liquid	Colourless liquid	Colourless liquid
Flash Point ,°C	-11.1	4.4	30	25	25
Flammable limits, %	1.3-7.1	1.17-7.1	-	-	-
Melting/Freezing point, °C	5.5	-95	-25.2	-47.9	13.3
Boiling point (760 mmHg), °C	80.1	110.6	144.4	139.1	138.3
Density (20 °C), g/ml	0.878	0.8669	0.876	0.860	0.857
Vapour pressure, mmwg (26 °C)	13.3 kPa	28.7	0.66	0.79	0.86

Solubilities	Water. 1800mg/lt at 25 °C Non aq. Solvents miscible with most	Fresh water: 535mg/lt at 25 °C Sea water 380mg/lt at 25 °C	142mg/lt in water	146mg/lt in water	185mg/lt in water

Note:

Benzene is present in both exhaust and evaporative emissions. Motor vehicles account for approximately 85% of the total benzene emissions. The remaining is attributed to stationary sources (15%). Benzene is a component of gasoline.

Another important source of benzene in air is active smoking of tobacco which accounts for roughly half of the total population exposure to benzene, which is over and above that from motor vehicles.

Benzene is quite stable in atmosphere and the only important reaction in the lower atmosphere in the reaction with OH radical. This reaction too is very slow. Benzene is one of the ingredients of petroleum product.

Benzene hydrocarbons can be condensed and washed out to the coke-oven gas. Crude oil contains benzene and its homologues account for 3.9%-4.8%. Benzene is also produced from olefins. Benzene in the motor fuel is infact burnt in the combustion process, but at the same time there is an additional amount of benzene produced through dealkylation which is emitted. Benzene is mainly used as raw material for the production of substituted aromatic hydrocarbons. The major source is emission from motor vehicles and evaporation losses during the handling, distribution and storage of petrol.

Sources of BTX Emission in the Environment

Transport	I. **Vehicular Pollution:** a. Vehicular exhaust b. Fuel filling station c. Fuel adulteration II. **Railways** III. **Airways**
Industries:	I. Major industrial units (Refinery, Petrochemical etc): II. Industrial estates: III. Medium scale chemical industries
Domestic and commercial Emissions:	I. Domestic combustion units II. Commercial combustion units

While people working with petrol, in its manufacture and distribution, derive their main exposure from this source, a major source of exposure for the general population is from vehicle exhausts, and here benzene is mostly produced by chemical reactions occurring during combustion of petrol in the engine. While petrol exhaust is the major source of benzene in the air we breathe outdoors especially in urban areas, there are other sources of human exposure, some of which contribute a significant proportion of an individual's total intake. Cigarette smoke contains benzene, and may be the main source of exposure for a heavy smoker; passive smoking may make a small contribution to benzene intake. Benzene may also be present in low concentrations in foods and in drinking water, and diet may be the main source of benzene for non-smokers living in unpolluted, rural areas.

Benzene Emission from Passenger Car

Benzene is a major constituent of evaporative emission due to its high volatility. In the gasoline engine exhaust against benzene is a significant component of hydrocarbon fraction due to its refractory nature in the combustion process.

Benzene emission in cars are correlated by the following equations.

For noncatalyst cars:

BEE = 1.515 + 0.765.BF + 0.0414. NBAF

For catalyst cars:

BEE = 1.237 + 0.599.BF + 0.0602.NBAF

Where, BEE – wt% benzene in total hydrocarbons in engine exhaust,

BF – wt% benzene in feed and

NBAF – wt% nonbenzene aromatics in feed.

BEE is very sensitive to benzene content in gasoline and also depends on aromatic content in the fuel.

Sources of Benzene Emission

A. **Emissions from Benzene Production:**
- Catalytic reforming/separation process
  - Toluene dealkylation and toluene disproportionation process
  - Ethylene production
  - Coke oven and coke by product recovery plants

B. **Emissions from major Uses of Benzene:-**
- Ethylbenzene and Styrene production
- Cyclohexane production
- Cumene production
- Phenol production
- Nitrobenzene production
- Aniline production
- Chlorobenzene production
- Linear alkylbenzene production
- Other organic chemical production
- Benzene use as a solvent

C. **Emission from other Sources**
- Oil and gas wellheads
- Glycol dehydration units
- Petroleum refinery processes
- Gasoline marketing
- Publicly owned treatment works
- Municipal solid waste landfills
- Pulp, paper and paper board industry
- Synthetic graphite manufacture

- Rayon – based carbon fiber manufacture
- Aluminium casting
- Asphalt roofing manufacturing
- Consumer products/building supplies.

D. **Emissions from combustion sources**

- Medical waste incinerators
- Sewage sludge incinerators
- Hazardous waste incinerators
- External combustion solid, liquid and gaseous fuels in stationary sources heat and power generation.
- Stationery internal combustion
- Secondary lead smelting
- Iron and Steel Foundries
- Portland cement production
- Hot – mix ashalt production
- Open burning of biomass, scrap tires

Benzene Emission Factor From Different Sources

Emissions from Refuelling losses

Refelling losses: These depend on the ambient temperature at the time of refueling, and are given by the following equation

$$Bz_e = 0.23\ BZ_d + BZ_f$$

where,

Bz_e Is the mass fraction of benzene in total hydrocarbons emitted
BZ_d Is the fraction of benzene in the petrol distributed
BZ_f Is the fraction of benzene in the petrol in the tank

Emissions from Evaporative losses from the Carburetor and the Fuel Tank

These results from the dissipation of heat when the engine is turned off and from variations in temperature between day and night. The following equations are used:

Engine stopped: % weight of benzene in vapour = 0.45 of the weight of benzene in petrol

Respiration: % weight of benzene in vapour = 0.89 of the weight of benzene in petrol

Emissions from Exhaust

It is given by the following equation:

$$Bz_{exh} = 0.50 + 0.44\ BZ_{gas} + 0.04\ Ar$$

where,

Bz_{exh} = percent weight of benzene in the exhaust
BZ_{gas} = percent weight of benzene in petrol (gasoline)
Ar = percent weight of other aromatics in petrol (gasoline)

For Benzene exhaust from petrol vehicles, separate equations are used for three way plus oxidation catalysts, and other catalyst types. For vehicles **with a three- way catalyst**, running on baseline gasoline, the following equation is used:

3-way + Bz%THC = 1.077+0.7732* (Volume % benzene) + 0.0987* (Volume % aromatics – volume % benzene)

For vehicles with a **three-way plus oxidation catalyst**, running on baseline petrol, the equation used is:

3-way + oxBz%THC = 0.6796* (Volume % benzene)

+ 0.0681* (Volume % aromatics – volume % benzene)

For vehicle **with no catalyst or an oxidation catalyst**, the equation used is:

Other Bz%THC = 0.8551* (Volume % benzene)

+0.12198 * (Volume % aromatics) – 1.1626

Benzene Emission from gasoline Loading Racks at Bulk Terminals and Bulk Plants

Table 5: Benzene Emission Factors for Gasoline Loading Racks at Bulk Terminals and Bulk Plants

Loading Method	Gasoline Vapour Emission Factor[a] (mg/litre)	Benzene Emission Factor[b] (mg/litre)
Splash loading – normal service	1430	12.9
Submerged loading[c] normal service	590	5.3
Balance service[d]	40	0.36

a. Gasoline factors represent emissions of non-methane VOC. Factors are expressed as mg gasoline per liter transferred.

b. Based on an average benzene/VOC ratio of 0.009.

c. Submerged loading is either top or bottom submerged.

d. Splash and submerged loading. Calculated using a Stage 1 control efficiency of 95 percent

Gasoline Vapour and Benzene Emission from a Typical Service Station

Emission Source	Gasoline Vapour Emission Factor[a] (mg/litre)	Benzene Emission Factor[b] (mg/liter)
Underground storage Tanks – Tank Filling Losses – splash Fill	1, 380	12.4
Underground storage Tanks – Tank Filling Losses – Submerged Fill	880	7.9
Underground storage Tanks – Tank Filling Losses – Submerged Filling[c]	40	0.4
Underground storage Tanks – Breathing Losses	120	1.1
Vehicle Refueling[d] – Displacement Losses	1.320	11.9
• Uncontrolled	132	1.2
• Controlled		
Vehicle Refueling[d] – Spillage	84	0.76

a. Typical service station considering a gasoline throughout of 19, 000 liters/day

b. Based on gasoline emission factor and an average benzene/VOC ratio of 0.009

c. Calculated using a Stage 1 control efficiency of 95 percent.

d. Vehicle refueling emission factors can also be derived for specific geographic locations and for different seasons of the year using EPA's mobile source emission factor computer model.

Gasoline Vapour and Benzene Emission from a Typical Bulk Plant

Emission Source	Gasoline Vapour Emission Factor[a] (mg/litre)	Benzene Emission Factor[b] (mg/litre)
Storage Tanks – Fixed Roof – Breathing Loss	600	5.4
Storage Tanks – Fixed Roof – Working Loss Filling	1150	10.3
Emptying	460	4.1
Gasoline Loading Racks	1430	12.9
Splash Loading (normal service)	590	5.3
Submerged Loading (normal service)	40	0.4
Splash and Submerged Loading (normal service)[c]		

a. Typical bulk plant with gasoline throughout of 19, 000 liters/day

b. Based on gasoline emission factors and an average benzene /VOC ratio of 0.009

c. Calculated using a Stage 1 control efficiency of 95 percent.

Benzene Emissions from Locomotives and Aircraft Loading and Take Off

Benzene Emissions Factors for Locomotives and Benzene Content In Aircraft Loading and Take Off Emission's

Source	Toxic Emission Fraction	Emission Factor (lb/gal)	Weight Percent Benzene
LineHaul Locomotive	0.0106[a]	0.00022	–
Yard Locomotive	0.0106[a]	0.00054	–
Military Aircraft	–	–	2.02
Commercial Aircraft	–	–	1.94
Air Taxi Aircraft	–	–	3.44
General Aviation	–	–	3.91

a. These fractions represent toxic emission fractions for heavy-duty diesel vehicles. Toxic fractions for locomotive are assumed to be the same, since no fractions specific for locomotive are available. The toxic emission fractions are multiplied by the HC emission factors to obtain the toxic emission factors.

Toluene and Xylene are used as solvent in rubber and plastic cement and in manufacturing of some compounds. During the production process toluene and Xylene get vaporized and find their way into the air.

Health Effects of Benzene Exposure

About 50% of benzene is absorbed by inhalation and absorption of it via skin is limited. The high lipophilicity and low water solubility of benzene favour its distribution to fat-rich tissues. Benzene distributed by blood accumulates in fat-rich tissues like adipose tissue bone marrow and liver.

Benzene is readily absorbed by the body during inhalation or ingestion and is less extent in the bone marrow, producing intermediates, which account for the toxicity of benzene. In human, half-life of benzene is 1–2 days. Accumulation is not expected for benzene or its metabolites. Benzene is primarily exhaled through the lungs unchanged or excreted as metabolites in the urine.

Benzene is oxidized by cytochrome P–450- dependent mixed-function oxidase system. In humans voluntarily exposed to 100ppn (320 mg/m^3) benzene for 5 hours, 61% absorbed benzene was metabolized to phenol, 6.4% to catechol and 2% to hydroquinoue while 26% was exhaled unmetabolized. The major part of the metabolites was excreted as sulfate or glucoronic acid conjugates.

Metabolites of benzene are responsible for haematotoxicity. The bio-transformation path ways are the formation of phenol via epoxide and catcehol via benzene dihydrodiol.

Benzene is ubiquitous in the environment, resulting in the exposure of most humans to trace levels (or more) of this chemical. Exposure in the general population is primarily to air borne benzene and derives from active and passive tobacco smoke, industrial activities, and use of the automobile (gasoline fumes from refilling, etc. and exhaust emissions). Estimates of the daily amounts of benzene consumed in drinking – water and food-stuffs vary considerably and are of the order of µg/day. Depending upon the assumptions made with respect to levels of benzene from tobacco products and foodstuffs, estimates for the exposure of the general smoking population in industrialized countries range from 2000 to 3500 µg benzene/day. Adult (70kg) non-smokers are considered to be exposed to about 200 to 1700 µg benzene/day (about 3 to 25 µg/kg body weight per day).

About 50% of inhaled Benzene in air is absorbed. Benzene intake based on 24 hour exposure volume of 20 m^3 at rest will be 10mg/day for each 1 mg/m^3 benzene in air. The daily adult intake at a typical benzene level of 16 µg/m^3 will, therefore, be about 160 µg. Together with other pollutants, benzene also participate in photochemical process which result in formation of oxidants and smog. Exposure to high level of benzene causes neurotoxic symptoms. Persistent exposure to high level of toxic level of benzene may cause injury to human bone-marrow. Early manifestation of toxicity are anaemia, leucocytopania or thromo-cytopania. Benzene is a known human carcinogen.

Occupational Exposure

The major factors controlling industrial exposure to benzene are process technology, worker practices and the efficiency and sophistication of engineering controls. When appropriate engineering controls are in place, available monitoring data indicate that exposures of workers involved in the production, handling and use of benzene and benzene containing materials which vary from non-detectable levels to approximately 15 mg/m^3 (8-h TWA), in addition to the amounts estimated for the general population. In developing countries the exposure can be several times higher. Due to the nature of the processes involved, a small percentage of workers may be exposed to more than 320 mg benzene/shift. In some developing countries, benzene exposure may be sufficiently high to cause acute toxicity. Dermal exposure to benzene has generally not been included in these estimates.

As regards to permissible level of benzene concentration for occupational exposure, American Conference of Government Industrial Hygienists prescribes a threshold limit value (TLV) of 0.5 ppm (1622 µg/m^3) (TWA). OSHA regulations also call for human exposure limit of less than 0.5 ppm. National Institute of Occupational Safety and Health (NIOSH) suggests Recommended Exposure Limit (REL) of 320 µg/m^3 (0.1 ppm)

Thus, a non- smoker living in an unpolluted rural area may be exposed to as little as 120 µg benzene daily, while a 20 cigarettes per day smoker living in a city may be exposed to as much as 1, 250 µg daily.

Toxicological and Carcinogenic Effects

Long term exposure to benzene in air causes leukaemia in human beings. In animal studies, leukaemias, lymphomas and other types of tumours are observed. Exposure to benzene is linked to genetic changes, increased proliferation of bone marrow cells and occurrence of certain chromosomal aberrations in humans and animals. US EPA has classified benzene as Group A human carcinogen. The International Agency for Research on Cancer (IARC), Lyon (France) has listed benzene as carcinogenic to humans. In addition a number of non cancer health effects are associated with benzene exposure such as disorders of blood, harmful effects on bone marrow, anaemia and reduced abilities of blood to clot, damage to immune system and a reproductive and developmental toxicant.

High levels of benzene exposure produce haematotoxic effects like leucopenia, lymphopenia and anemia in laboratory animals. Exposure to high levels of benzene causes neurotoxic symptoms.

Substances that can induce benzene –metabolizing enzymes are likely to modify the haematotoxicity of benzene. It has been found that benzene itself, Phenobarbital, Toluene ethanol can modify the metabolism and haematotoxicity of benzene, if animals an pretreated with these substance. Toluene has been found to inhibit the metabolism of benzene and decrease its haemotototoxicity. Ethanol inhances the haemotototoxicity of benzene in mice.

Acute lethal doses of benzene in experimental animals cause narcosis, ventricular tachycardia and respiratory failure.

European Commission Limit Value for Benzene

The limit value must be expressed in µg/m³, standardized at a temperature of 293 K (20 °C) and a pressure of 101.3 kPa.

	Averaging period	Limit Value	Margin of tolerance	Date by which limit value is to be met
Limit value for the protection of human health	Calendar year	5 µg/m³	5 µg/m³ (100 %) on 13 December 2000, reducing on 11 January 2006 and every 12 months thereafter by 1 µg/m³ to reach 0% by 1 January 2010	1 January 2010

Air Quality Standards for Benzene in Different Countries

Country	Standard Concentration measured as	To be achieved by
UK	16 µg/m³ – running annual mean	31st Dec., 2003
European Commission	5µg/m³ – annual average	31st Dec., 2010
Japan	3 µg/m³ annual average*	Existing Standard
WHO Guidelines	5-20 µg/m³	–

* Prefer GC-MS (sample to be collected with canister or tube or its equivalent method).

Appendix 1.78: All about Persistent Organic Pollutants (POPs)

Of all the pollutants released into the environment by human activity. Persistent Organic Pollutants (POPs) are among the most dangerous. They are ubiquitous: found in the most remote areas, far from any industrial activities. Even though they are present in the environment at extremely low levels, they have been linked to many health and environmental effects. With the evidence of their long- range transport to regions where they have never been used or produced, and the consequent threats they pose to the environment of the entire planet, the international community has called for urgent global action to reduce and eliminate releases of these POPs.

POPs: The 'dirty dozen'

Stockholm Convention are DDT, aldrin, dieldrin, endrin, chlordane, heptachlor, mirex. toxaphene, hexachlorobenzene (HCB), polvchlorinated biphenyls (PCBs), dioxins and furans. Of these, nine are organochlorine pesticides. PCBs are compounds with varied industrial applications, while HCB, PCBs, dioxins and furans are produced unintentionally, industrially as well as otherwise.

Global action against POPs

In 1997, efforts were made on a global level to deal with the problem of POPS under the aegis of United Nations Environmental Programme (UNEP)

Long-Range Transport and Deposition of POPs

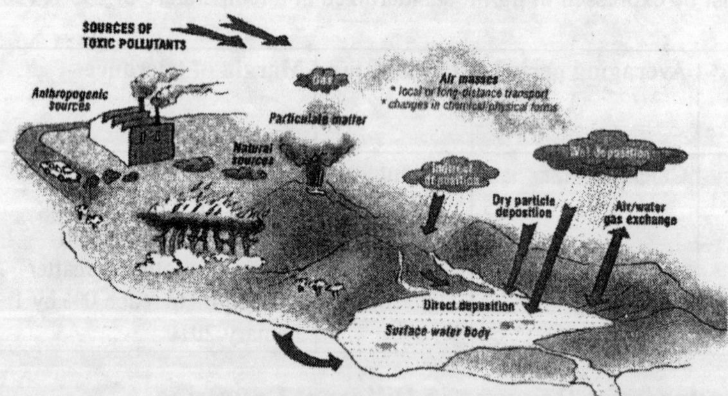

- POPs are most hazardous of all pollutants released into the environment every year by humans.
- Among these, dioxins and furans are unintentionally produced by processes such as incineration and are of most concern because of their significant toxicity.
- POPs travel globally on wind and marine currents to regions far from where they are produced.
- Collective global action is the only way to stop the spread of POPs. The Stockholm Treaty lays guidelines for their phase-out.
- POPs are transported via convection currents to colder regions, where they persist for longer periods. The Himalayas could possibly be a major sink for POPs to reenter the ecosystem through rivers.

A multilateral, internationally legally binding treaty to ban POPs was signed on May 23, 2001, in

Stockholm by 87 countries. It has come to be known as the Stockholm Convention.

The Stockholm Convention sets out control measures covering the production, import, export, disposal and use of POPs. Govern ments are to promote the best available technologies and practices for replacing existing POPs while preventing the development of new POPs.

While the control measures will apply to an initial list of 12 chemicals, a POPs Review Committee will consider additional candidates for the POPs list on a regular basis.

Pathways of POPs is shown below in figure:

(POPs always find a way to reach us. They travel on a variety of pathways to eventually land up in our bodies.)

Sources of POPs:

Sources of POPs can be classified according to their type. Pesticide POPs have similar source characteristics. PCBs have specific uses in different industrial sectors and have source characteristics accordingly Similarly, unintentionally produced POPs have commonalities in terms of the kinds of sources and release mechanisms associated with them.

Pesticide POPs

Sources of pesticides can be typically characterised as point and non-point.

Point sources of POPs pesticides are in the form of pesticide manufacturing facilities (both technical grade manufacturers as well as formulators) and stockpiles of obsolete, unwanted or date-expired pesticides.

Non-point sources arise due to the general application of pesticides (such as in agriculture) resulting in crop run-offs or leaching into ground water reserves. POPs pesticides have been banned for use in agriculture, though they are present in food, water, etc. The use of DDT as part of the malaria programme also constitutes a non- point source.

Production of POPs

Potential hotspots exists in closed factory premises which still house stocks of manufactured POPs It is expected that the environment around such sites might also be contaminated due to exposure to toxics during the period of manufacture.

Unintentional POPs

A variety of industrial and non-industrial processes (typically involving high temperatures') result in the release of POP chemicals dioxins, furans, PCBs and HCB) into the environment. Among these POPs, dioxins and furans are of most concern because of their significant toxicity. In most cases, dioxins

Sources identified by the Stockholm Convention for unintentionally produced POPs

a. **Industrial sources with potential tor comparatively high formation and release of POPs :**
   - Waste incinerators, including co-incinerators of municipal, hazardous or medical waste or sewage sludge
   - Cement kilns firing hazardous waste
   - Production of pulp using elemental chlorine or chemicals generating elemental chlorine for bleaching
   - Thermal processes in the metallurgical industry, including secondary copper production, sinter plants in the iron and steel industry, seçondary aluminum production, secondary zinc production, etc.

b. Other sources listed
   - Open burning of waste including burning of landfill sites
   - Thermal processes in the metallurgical industry not mentioned in Part II
   - Residential combustion sources
   - Firing installations for wood and other biomass fuels
   - Specific chemical production processes releasing unintentionally formed POPs, especially production of chlorophenols and chloranil
   - Crematoria
   - Motor vehicles, particularly those burning leaded gasoline
   - Destruction of animal carcasses
   - Textile and leather dyeing (with chloranil) and finishing (with alkaline extraction)
   - Shredder plants for the treatment of end-of- life vehicles
   - Smoldering of copper cables
   - Waste oil refineries and furans released are accompanied with PCBs and HCB as well.

POPs pervade the environment through a variety of media. Pesticides, though intended for the target pest species, end up in the food chain, water and air and into non-target species and ecological systems. Unintentional and industrial POPs are released in an unregulated manner and are assimilated into environmental systems owing to their properties of persistence, lipophilicity and volatility.

Exposure of humans to pesticide POPs can be intentional (suicides and homicides) and unintentional (accidental, occupational and non-occupational exposure from water, air and

Personal action on POPs to reduce risk

POPs that are already in the environment will be around for decades. Still, we can reduce our exposure to POPs and help stop more POPs from getting into circulation. To reduce the risks of POPs, the following steps should be considered:

- **Try to eat lower on the food chain or avoid fats**

  This will reduce lifetime accumulation of POPs and is especially relevant for children.

- **Choose unbleached paper products.**

  Chlorine bleaching processes unintentionally generate POPs by-products such as dioxins. furans and others.

- **Avoid polyvinyl chloride (PVC or vinyl) plastics.**

  This might be an impossible task given that there is an endless list of common vinyl items that include packaging material, utility items, wall coverings, blood and infusion bags, medical equipment, credit cards, office supplies, etc. Don't burn these items.

- **Avoid using weed killers containing POPs chemicals.**

  A useful hint for the customers is to check the label for the active ingredient 2, 4-D; they may contain dioxins and other POPs byproducts food). Amongst unintentional exposures in the environment are the direct toxic effects during application to non-target groups such as pollinators, predators, wildlife, etc, as well as post application hazards due to pesticide residues in food, air and water.

Pesticide POPs can get into the ecosystem at various junctures such as during production, transport, storage and application.

Once in the environment, POPs are transported within the region mainly through natural media like fresh water systems (rivers and streams), atmospheric currents (usually adsorbed to suspended sediment) and marine currents.

Air transport can also occur through precipitation and the movement of dust particles with adsorbed POPs. The persistence of POPs in soil/sediments may result in their re-suspension and re-mobilisation back to the atmosphere under favourable high temperature conditions through volatilisation and to the surface waters through run-offs during monsoon/flood periods.

Most inflows into the marine environment occur due to the outflows of rivers into the oceans and seas.

Factsheet of POPs

- POPs are highly stable compounds and can circulate globally through a process known as the "grasshopper effect" According to UN, the breast milk of the average Inuit mother has five times as much dioxins as that of her counterpart in the industrialised world. However, breast milk is still recommended for infants.

- On entering the environment, POPs can alter the normal bio-chemical and physiological functions of plants and animals, including humans. Effects may include molecular changes (enzyme activity, DNA damage), cellular changes (tumour formations), tissue changes (organ functioning), individual changes (behaviour, deformities), population changes (mortality, abundance, distribution), and community changes (numbers of species and their interactions).

- POPs concentrate in living organisms through another process called bioaccumulatlon. POPs are readily absorbed in fatty tissue, where concentrations can become magnified by up to 70, 000 times the background levels.

- Fish, predatory birds, mammals and humans are high up the food chain and so absorb the greatest concentrations.
- Human health impacts may be felt most acutely in populations that consume large amounts of fish (for example, subsistence fishermen), since fish have a high fat content and thus can contain high concentrations of POPs.
- Though most of the 12 POPs have been banned or subjected to severe use restrictions in many countries for more than 20 years, many are still in use and stockpiles of obsolete POPs exist in many parts of the world.
- Humans can be exposed to POPs through diet, occupational accidents and the environment,
- Exposure to very low doses of certain POPs can lead to cancer, damage to the central and peripheral nervous systems, diseases of the immune system, reproductive disorders, and interference with infant and child development.
- Shifting from POPs to chemical and non- chemical alternatives is the key to reducing the impact of these hazardous substances. A high priority is finding alternatives to hazardous chemicals for insect control.
- There are many safer chemical and non- chemical alternatives, but their development and dissemination will require time, money and training.
- Many countries face barriers to identifying and controlling releases of POPs. These include high prices of some alternatives, the need for training on the hazardous nature of POPs, a lack of information on alternatives, a lack of reliable data about the current uses of POPs and the need for regulations/infrastructure to manage their use.

Dietarv exposure of POPs take precedence over other pathways like air, water and dermal exposure. Primary exposure by dietary pathways is, however, not applicable occupational and accidental exposure where dermál, atmospheric and water pathways can assume significance.

The presence of POPs residues in human and animal samples is the most effective way to confirm the exposure and magnitude of POPs in the population. POPs exposures even at low doses can cause undesirable health effects, especially in the more vulnerable sections of the population, such as pregnant mothers and young children.

POPs problem in India

Despite the fact that most short-listed POPs are banned from use, production, import or export in India, a combination of factors has led to their continued production, trade, use and release into the environment.

Chlordane has been imported in India subsequent to its ban, while aldrin, chlordane and heptachlor have been exported subsequent to their ban. It is suspected that DDT which is allowed for vector control but banned for agricultural purposes is pilfered and used as an insecticide.

Though stockpiles of obsolete POPs exist, there is no known government programme to monitor them.

Electrical equipment, such as transformers containing PCBs, are indiscriminately disposed into the waste stream.

Some developed nations, are pushing for the expansion of POPs producing technologies such as waste incinerators, waste-to-energy and PVC manufacturing plants

Extensive use of DDT for malaria control has led to a high level of contamination of all quarters of the environment and food commodities.

Collective action to he taken on POPs

The problems associated with POPs are too large, complicated and expensive for any individual nation to tackle. Suggested ways of doing so are as follows:

- **Elimination of all POPs sources**

  The sources of POPs must be properly identified, and then phased out, replaced or removed. This is true for POPs that are produced intentionally as well as unintentionally, such as dioxins.

- **Technological and other resource assistance**

  The global POPs treaty includes measures to discourage wealthy countries from promoting and exporting POPs-generating technologies, processes and materials to newly industrialised nations. At the same time, measures must be included to encourage wealthy countries and aid programmes to assist newly industrialised countries in developing and implementing non- POP technologies and materials.

- **No incineration**

  Since incineration itself has been identified as a maior source of dioxins and furans, it cannot be used as a disposal route for existing POPs stockpiles. While so-called state-of-the- art incinerators can gready reduce stack emissions, they cannot stop the formation of dioxins and other POPs which may be present in fly ash and bottom ash. Those would still require further treatment and disposal.

  Alternative destruction and de-toxification methods and technologies which do not generate POPs should be used instead.

"The Dirty Dozen" POPs

These 12 POPs are often referred to as the "dirty dozen" information about them is given below:

1. **Aidrin** - an organochlorine insecticide;Acts as stomach and contact poison
   - A pesticide used to control soil insects termites and ants 'It has been widely used to protect crops such as corn and potatoes, to protect wooden structures from termites.
   - It is oxidized in the insect to form dieldrin, a neurotoxin.
   - It is a carcinogen as well as a mutagen.it is most toxic to aquatic invertebrates
   - It binds strongly to soil particles, is resistant to leaching into ground water. It is released from soil by volatilization.
   - Due to its persistent nature, aidrin is known to bio-concentrate.
   - Aidrin is toxic to humans causing headache, dizziness, nausea, general malaise, and vomiting, followed by muscle twitchings,
   - myoclonic jerks, and convulsions. Lethal dose is 80 mg/kg if adult body weight MRLs range from 0.006 mg/kg meat fat and 0.1 to 180 mg/kg for water quality
   - Occupational exposure to aidrin, in conjunction with dieldrin and endrin, has been associated with a significant increase in liver and biliary cancer. -Aldrin may affect immune responses.
   - Many countries, including India have ended its usage manufacture, import and export
   - It is a carcinogen as well as a mutagen.
   - **Half life in soil ranges from 20 days to 1.6 years**
2. **Chlordane**
   - Chlordane is a contact insecticide that has been used on agricultural crops as well as being used extensively in the control of termites, household pests. Uses as fumigant, acaricide, most effective on termites

- Banned in US in 1988 and India alos used manufacture, import and export
- Chlordane sticks strongly to soil particles at the surface and is not ' likely to enter groundwater. As a result it can stay in the soil for over 20 years and breaks down very slowly. Its half life in soil is 4 years.
- Bio-concentrates in fish, mammals, and birds;
- In both humans and animals, it can damage nervous and digestive systems, and liver - Has caused convolutions and death;
- Recent human studies have linked chlordane exposure with prostate and breast cancers.
- MRLs are 0.002 mg/kg milk fat and 0.5 mg/kg poultryfat for water - 1.5 to 6mg/litre

3. DDT- a synthetic pesticide, was earlier used on agriculture crops.
   - DDT is toxicant; used for vector control of malaria, dengue and kala azar.
   - It has a half life of 2-15 years^and is immobile in most soils. Half life in air is 7 days
   - Breakdown products in the soil environment are DDE and DDD, which are also highly persistent and have similar chemical and physical properties.
   - Banned in the US for most uses in 1972. In India, It is banned for agriculture uses but allowed for restrictive use in the pblic health sector
   - Subsequently banned for agricultural use worldwide, but is stiil used to a limited extent in mosquito control in certain parts of the world
   - -concentrates significantly in fish and other aquatic species, leading to long-term DDT and its metabolic products DDE and DDD magnify through the food chain -
   - DDT bio exposure to high concentrations.
   - At the chronic level, individuals who consumed contaminated fish increased diabetes occurrences;
   - DDT is a probable human carcinogen. It exhibits estrogen like activity
   - The Cows milk samples found to contain detectable levels of DDE.
   - MRLs are 0.02 mg/kg if milk fat, 5 mg/kg of meat fat and 1 mg/litre of drinking water
4. Dieldrin- an insecticide for soil pests, used in wool processing industries and for termite control
   - Closely related to aidrin which itself breaks down to form dieldrin
   - it accumulates as it is passed along the food chain.
   - MRLs are 0.006 mg/kg milk fat, 0.2 mg/kg meat fat and 0.1 to 1.8 mg/litre of drinking water.
   - It acts as non-systematic stomach and contact poison.
   - Long-term exposure is toxic to many animals (humans included) far greater than to the original insect targets;
   - Currently banned in most of the world including India (except use for locust control in the desert areas).
   - Linked to Parkinson's disease, breast cancer and immune, reproductive, and nervous system damage.
   - It has a high half life of 3-4 years in soil in temperate climates. Persistence in air for 4 to 40 hours.
5. Endrin- an insecticide/rodenticidein cotton, rice, sugarcane and other crops.
   - Banned in many countries including India.
   - It is likely to adsorb onto the sediments in surface water.
   - An insecticide used on cotton, maize, and rice; a rodenticide used to control mice. It is a contact and stomach poison, has a delayed neurotoxic action.
   - It can bio-concentrate in the fatty tissues, of organisms living in water. It is very toxic to aquatic

organisms, namely fish, aquatic invertebrates, and phytoplankton. Bioconcentration up to 1800 has been reported in fish.

- Half-life in soil estimated 12 years.
- endrin poisoning in humans primarily affects the nervous system. Food contaminated with endrin has caused several clusters of. poisonings worldwide, especially affecting children. It suppresses the immune system in humans.

6. **Heptachlor** - an insecticide for soil pests, termites, household pests; also used for seed treatment.
   - Similar to the insecticide chlordane; toxic to wildlife ; in birds induces behavioural changes red uced reproductiva success and mortality.
   - The U.S. has banned the sale of heptachlor products and virtually eliminated its use for any purpose. In India also, it is completely banned.
   - Has a very stable structure, thus it can remain in environment for decades; Its soil half life is 0.75 to2 years in temperate climate
   - Possible human carcinogen (Group 2B)
   - MRLs are 0.006 mg/kg of milk fat, 0.2 mg/kg meat or poultry fat.

7. **Hexachlorobenzene (HCB)** - a fungicide;
   - A fungicide formerly used as a seed treatment, especially on wheat.
   - Known animal carcinogen (liver, kidney, thyroid);
   - Probable human carcinogen; also causes urinary, arthritic, neurological and metabolic disorders in human.
   - After its introduction as a fungicide in 1945, for crop seeds, this toxic chemical was found in all food types.
   - Banned in many countries. It is not registered in India.
   - In humans it can cause liver disease, skin lesions, ulceration, hair loss, thyroid damage colic, severe weakness, kidneydamage, CNS effects and respiratory depression.
   - Human and animal studies have demonstrated that HCB accumulate in fetal tissues and is transferred in breast milk.
   - Extremely toxic to aquatic creatures. Risk of bioaccumulation in an aquatic species is high and has a long half-life in biota.
   - *Estimated half life in soil is 2.7 to 5.7 years in soil and 0.5 to 4.5 years in air.*

8. **Mirex**
   - An insecticide; flame retardent
   - Listed as a persistent, accumulative, and toxic pollutant by EPA in the U.S. Capable of under going long range transport due to volatility.
   - It was used to control fire ants and as a flame retardant in plastic, rubber, paint, paper and electronics.
   - Mirex is transported across the placenta and can be passed from mother to child through breast milk;
   - Most affected in animals in the. It has moderate acute toxicity for mammals.
   - Proven to cause cancer in mice and rats and is a carcinogenic risk to humans It has potential of endocrine disruption in humans.
   - toxic for a range of aquatic organisms, with crustacea being particularly sensitive.
   - Mirex induces pervasive long-term physiological and biological disorders in vertebrates.
   - There is evidence of accumulation of mirex in aquatic and terrestrial food chains to harmful levels

- Mirex is one of the most stable of the organochlorine insecticides and is it is widespread in the environment.
- All uses of mirex as a pesticide were banned in the U.S. in 1978.It is never registered in India.
- mirex is still used in the USA mainly as a flame-retardant in plastics, rubber, paint, paper and electronics.
- Effects on organisms combined with its persistence suggest that mirex presents a long-term hazard for the environment. Its half life is 10 years.
- *It has a bioconcentration factor of 2600 for pink shrimp and 51400 for fathead minnows.*

9. **PCBs** (polychlorinated biphenyls)

- Used as coolants/insulating fluids, also used in flexible PVC coatings of electrical wiring and electronic components, pesticide extenders, cutting oils, flame retardants, hydraulic fluids, sealants (used in caulking, etc), adhesives, wood floor finishes, paints, and in carbonless copy paper.
- PCBs are very stable compounds and do not degrade readily.
- Despite being banned in the 1970s due to their high toxicity, PCBs still persist in the environment
- PCBs have been detected globally in the atmosphere, from the most urbanized areas that are the centers for PCB pollution, to regions north of the Arctic Circle, carried by wind currents.
- Bio-concentrates in animals;
- PCBs can be transmitted to children via breast milk as well as through the placenta.
- In humans, PCBs can cause liver disease, ocular lesions and lessened immune response.
- Babies born to women exposed to PCBs have been shown to weigh less, have problems with motor skills, a decrease in short-term memory, and compromised immune systems.
- PCBs alter estrogen levels in the body and contribute to reproduction problems.
- Endocrine Disrupting Chemicals (EDC's) such as PCBs pose a serious threat to reproduction in top-level predators.
- Biological magnification of PCBs has led to polar bears and whales that have both male and female sex organs and males that cannot reproduce.
- Effects on animals are liver, stomach, thyroid damage, plus immune system changes, changes in behavior and impaired reproduction
- Some studies indicate that PCBs are associated with cancer in humans, such as cancer of the liver and biliary tract.
- PCBs are known to induce Vitamin A deficiency in mammals, an affect that may be associated with impairment of the immune system, reproduction and growth.

10. **PCDDs (Polychlorinated dibenzodioxins)^- commonly referred to as DIOXINS;**

- Dioxins are teratogens (cause birth defects), mutagens, potential human carcinogens;
- Known to accumulate in humans and wildlife due to dioxins' ability to dissolve in fats and oils and their tendency to be water insoluble. This means that even small amounts in contaminated water can bio-concentrate up the food chain to dangerous levels.
- In humans, they can cause effects in reproductive/sexual development, plus immune system damage, thyroid disorders, nervous system disorders, endometriosis and diabetes.
- In animals and fish, studies have shown dioxin exposure to cause cancer birth defects, liver damage, endocrine damage, and immune system suppression.
- Studies have shown that exposure to dioxin increases the ratio of female births to male births among a population.
- Concentrations of dioxins are found in all humans today, with higher levels found in persons living in more industrialized countries.

- The estimated elimination half-life for dioxins in humans ranges from 8 to 132 years.
- Dioxin enters the general population almost exclusively from ingestion of food, specifically through the consumption of fish, meat, and dairy products since dioxins are fat-soluble and readily climb the food chain
- Because Dioxins are lipophilic, breast fed children usually have substantially higher dioxin body burdens than non-breast fed children until they are about 8 to 10 years old.
- According to the EPA, 80% of Dioxin emissions are caused by coal burning plants, municipal waste incinerators, metal smelting, diesel trucks, land application of sewage sludge, burning treated wood and trash burn barrels. Dioxins are also generated in bleaching fibers for paper and textiles.

11. **Polychlorinated dibenzofurans (PCDFs)**
    - highly toxic;
    - Properties and chemical structure similar to dioxins;
12. **Toxaphene - a non-systemic contactand stomach insecticide**: Used extensively for cotton pests, vegetables, fruits and cereal grains.
    - highly toxicto fish also used to combat livestck parasites.
    - Banned in US in 1990, outlawed in 1991 by the Stockholm Convention. It has half life of upto 12 years; bioconcentrates in aquatic orgarisms.
    - Exposure can cause damage to lungs, nervous system, kidneys and can be fatal. It has potential of endocrine disruption and is a possible human carcinogen.

Additional POPs

a. **PBDEs - (Polybrominated** diphenyl ethers)-flame **retardants**
   - "PBDEs are used in plastic, foam and textiles in clothing, computers, televisions, furniture and cars.
   - PBDEs have been found at high levels in indoor dust
   - Research has linked some chemicals in the flame retardants to effects on thyroid function on brain function, reduced male fertility and damaged ovarian development and the development of the embryonic nervous system, impacting motor skills and behavior.
   - These chemicals were banned in Europe in 2004/2005.
   - levels of concentration in humans and marine mammals continue to increase.
   - Studies have found that PBDEs accumulate in human blood, fat tissue and breast milk. It has been found that a woman can pass these chemicals to her unborn child through the placenta.
   - In the United States levels in human breast milk are 40 times higher than in Europe and are steadily rising.
   - Marine mammals like bottlenose dolphins, harbor porpoises, pilot and beluga whales have been found to be contaminated with PBDEs. Birds including cormorants and glaucous gulls, and popular food fish including salmon and tuna have also been found to carry high concentrations of PBDEs

b. **PFCs** - perflorinated compounds;
   - PFCs are used as industrial and commercial surfactants - wetting agents that lower the surface tension of liquid.
   - In contrast to the majority of organic pollutants that are deposited in fatty tissue, perflorinated contaminants circulate in the blood and accumulate primarily in the liver.

- PFCs are synthetic molecules that are thought to bioaccumulate and are believed to be extremely resistant to physical degradation, biodegradation, and biotransformation.
- Clear links exist between chemicals such as phthalates, bisphenol A, and perflorinated compounds found in consumer products including baby bottles, toys, and cosmetics to reproductive disorders.

Appendix 1.79: Typical Diameters of Aerosol Particles

Aerosols	Diameter, μ (micron)
Tobacco smoke	0.25
NH_4Cl smoke	0.1
H_2SO_4 mist	0.8-5.5
ZnO smoke	0.05
Coal mine air	10
Flour-mill air	15-20
Cement mill (kiln exhaust)	10
Grain elevator air	15
Fog	50
Talc dust	10
Pigments	1-4

Appendix 1.80: Particle Sizes of Airborne Contaminants (Note: The Logarithmic Scale)

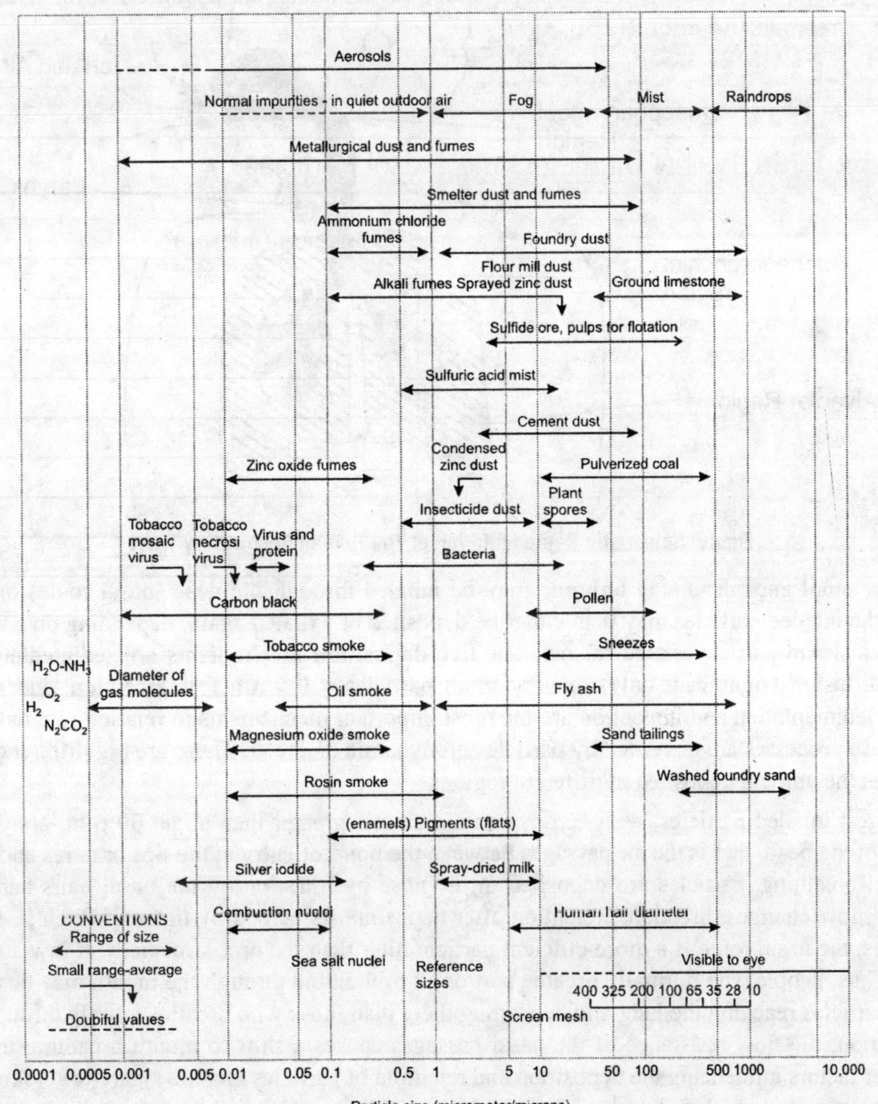

NOTES

Penetration and Deposition of Particles in the Human Respiratory Tract

A schematic representation of the respiratory system is presented in Figure indicating the different regions, namely, nasopharyngeal, tracheobronchial region and alveolar region.

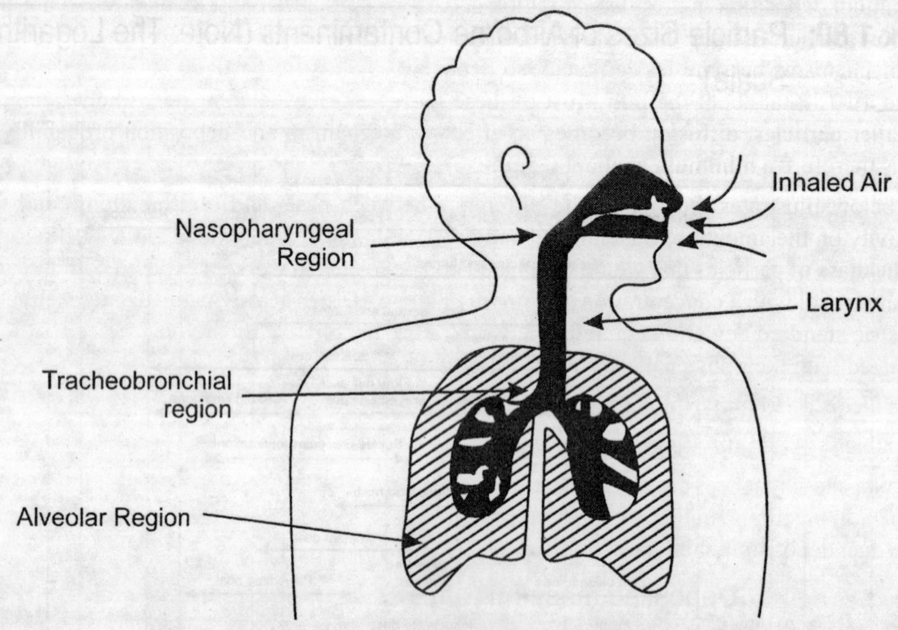

Figure: Schematic Representation of The Human Respiratory Tract

Particles small enough to stay airborne may be inhaled through the nose (nasal route) or the mouth (oral route. the inhaled particles may than either be deposited or exhaled again, depending on a whole range of physiological and particle-related factors. The five **deposition mechanisms** are sedimentation, inertial impaction, diffusion (significant only for very small particles < 0.5 μ m), interception, and electrostatic deposition. Sedimentation and impaction are the most important mechanisms in relation to inhaled airborne dust, and these processes are governed by particle aerodynamic diameter. There are big differences between individuals in the amount deposited in different regions

The largest inhaled particles, with aerodynamic diameter greater than about 30 p m, are deposited in the airways of the head, that is the air passages between the point of entry at the lips or nares and the larynx. During nasal breathing, particles are deposited in the nose by filtration by the nasal hairs and impaction where the airflow changes direction. Retention after deposition is helped by mucus, which lines the nose. In most cases, the nasal route is a more efficient particle filter than the oral, especially at low and moderate flow rates. Thus, people who normally breathe part or all of the time through the mouth may be expected to have more particles reaching the lung and depositing there than those who breathe entirely through the nose. During exertion, the flow resistance of the nasal passages causes a shift to mouth breathing in almost all people. Other factors influencing the deposition and retention of particles include cigarette smoking and lung disease of the particles which fail to deposit in the head, the larger ones will deposit in the tracheobronchial airway region and may later be eliminated by mucociliary clearance or - if soluble - may enter the body by dissolution. The smaller particles may penetrate to the alveolar region the region where inhaled gases can be absorbed by the blood. In aerodynamic diameter terms, only about 1% of 10-μm particles gets as far as the alveolar region, so 10 μ m is usually considered the practical upper size limit for penetration to this region. Maximum deposition in the alveolar region occurs for particles of approximately 2-μ m aerodynamic diameter. Most particles larger than this have deposited further up the lung. For smaller particles, most deposition mechanisms become less efficient, so deposition is less for particles smaller than 2 μm until it is only about 10-15% at about 0.5 μm. Most of these particles are exhaled again without being deposited. For still smaller particles, diffusion becomes an effective mechanism and deposition probability is higher. Deposition is therefore a minimum at about 0.5 μm.

Figure below illustrates the size of the difference between nasal and oral breathing, and the role of physical activity on the amount of dust inhaled and deposited in different regions of the respiratory airways. It presents the mass of particles that would be inhaled and deposited in workers exposed continuously, during 8 hours, to an aerosol with a concentration of 1 mg/m³, a mass median aerodynamic diameter equal to 5.5 μm and a geometric standard deviation equal to 2.3. Respiratory parameters Itidal volume, V_t, and frequency, f) -were associated with their physical activity as follows:

V_t = 1450 cm³ f= 15 min⁻¹ (moderate physical activity)

V_t = 2150 cm³ f= 20 min⁻¹ (high physical activity)

The results show that oral breathing increases dust deposit in the alveolar (gas-exchange) region when compared to nasal breathing, indicating the protective function of the nasal airways. A higher physical activity can increase dust deposition in all parts of the respiratory airways.

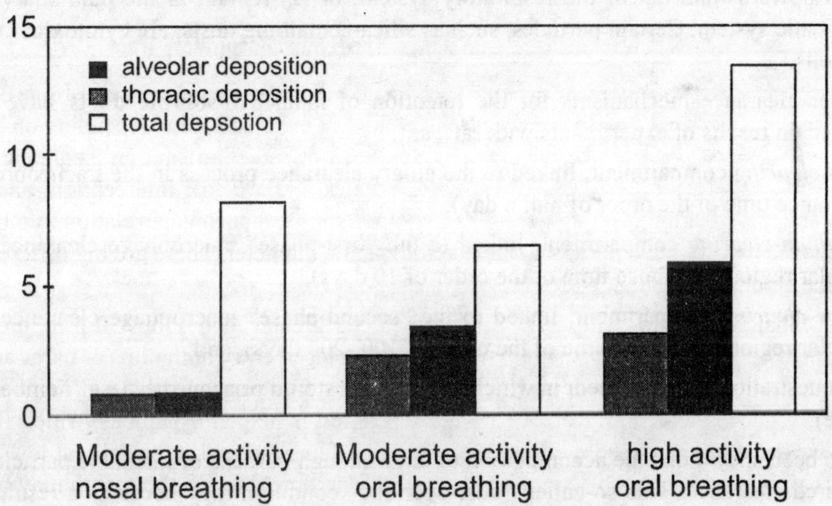

Figure: Difference between nasal and oral breathing and the role of physical activity on the amount of dust inhaled and deposited in different regions of the respiratory airways

Fibres behave differently from other particles in their penetration into the lungs, fine fibres even as long as 100 μ m have been found in the pulmonary spaces of the respiratory system. This is explained by the fact that the aerodynamic diameter of a fibre, which governs its ability to penetrate into the lung, is primarily a function of its diameter and not its length. However, for longer fibres, deposition by interception becomes increasingly important.

Clearance of particles from the respiratory tract

After deposition, the subsequent fate of insoluble particles depends on a number of factors. *(Soluble* particles depositing anywhere may dissolve, releasing potentially harmful material to the body.)

Mucociliary clearance

The trachea and bronchi, down to the terminal bronchioles, are lined with cells with hairlike cilia (the ciliated epithelium) covered by a mucous layer. The cilia are in continuous and synchronized motion, which causes the mucous layer to have a continuous upward movement, reaching a speed in the trachea of 5-10 mm per minute. Insoluble particles deposited on the ciliated epithelium are moved towards the epiglottis, and then swallowed or spat out within a relatively short time. The rate of clearance by the mucociliary mechanism may be significantly impaired by exposure to cigarette smoke.

Bronchiole movement

Intermittent peristaltic movements of the bronchioles, and coughing and sneezing, can propel particles in the mucous lining towards the larynx and beyond.

Phagocytosis

The epithelium of the alveolar region is not ciliated; however, insoluble particles deposited in this area are engulfed by macrophage cells (phagocytes), which can then (1) travel to the ciliated epithelium and then be transported upwards and out of the respiratory system; or (2) remain in the pulmonary space; or (3) enter the lymphatic system. Certain particles, such as silica-containing dusts, are cytotoxic; i.e. they kill the macrophage cells.

Defence or clearance mechanisms for the retention of inhaled **insoluble dusts h**ave been broadly classified, based on results of experiments with rats, as

- a *fast-clearing* compartment, linked to the ciliary clearance process in the tracheobronchial region (clearance time of the order of half a day);
- a *medium-clearing* compartment, linked to the "first-phase" macrophage clearance action in the alveolar region (clearance time of the order of 10 days),
- a *slow-clearing* compartment, linked to the "second-phase" macrophage clearance action in the alveolar region (clearance time of the order of 100 -200 days), and,
- a "sequestration" compartment in which particles are stored permanently (e.g., "embedded" in fixed tissue).

It has also been shown that the accumulation of large enough burdens of insoluble particles in the lungs leads to impaired clearance. This so-called "dust overload" condition may occur as a result of prolonged occupational exposures, even at relatively low levels.

Wherever the particles are deposited, either in the head or in the lung, they have the potential to cause harm either locally or subsequently elsewhere in the body. Particles that remain for a long time have increased potential to cause disease. This is why inhaled particles are important in relation to environmental evaluation and control.

Appendix 1.81: Particle Size Ranges for Aerosols, Dusts and Fumes

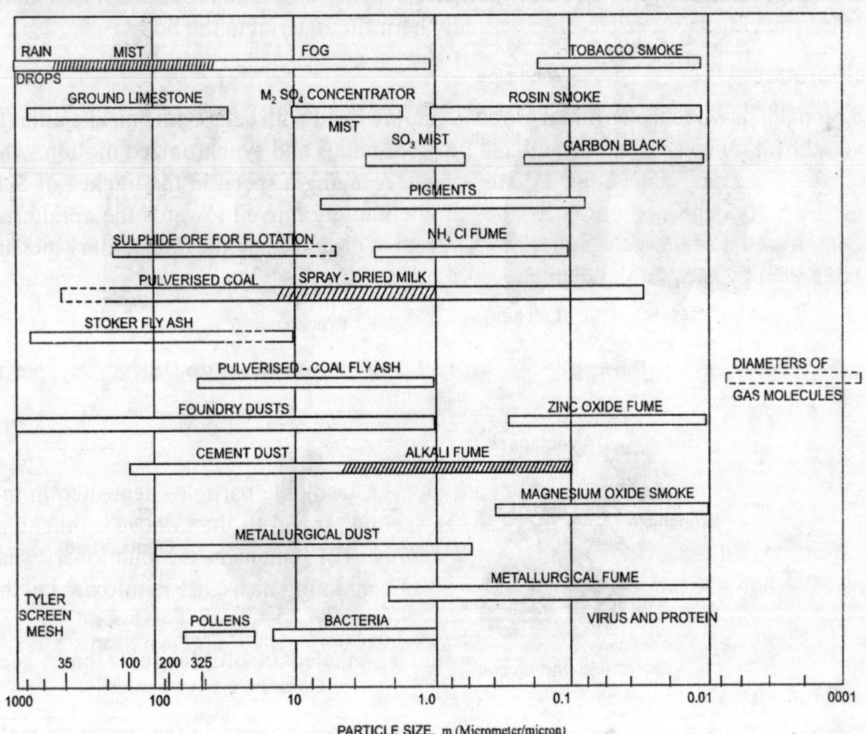

Appendix 1.82: Features of Human Respiratory System Exposed to Air Pollutants

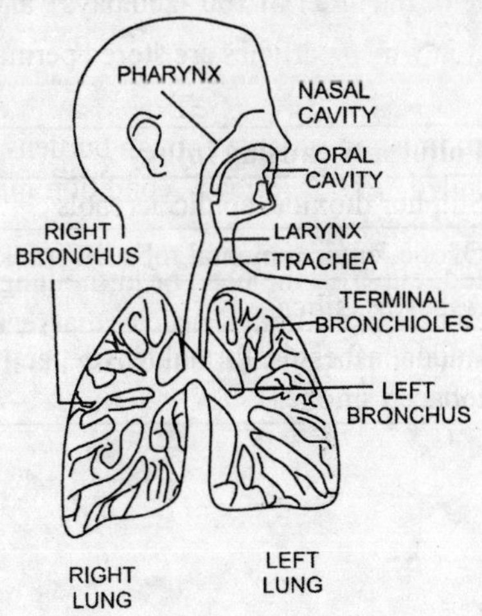

Note:

Anatomy of the lungs have been shown below in figure:-

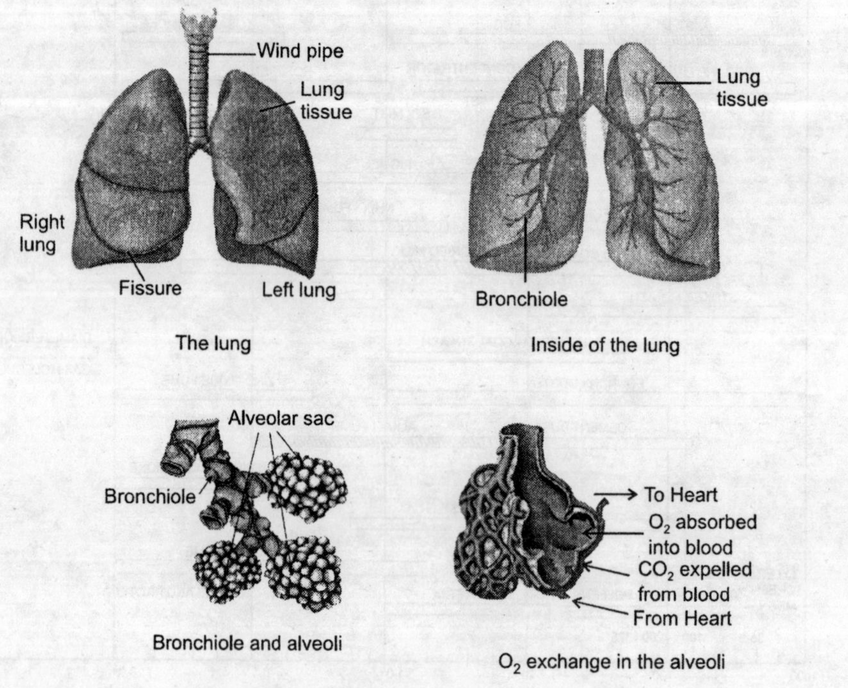

Anatomy of the lungs

Appendix 1.83: Injury Mechanism by which Deposited Inhaled Pollutants Injure The Airways

S.No.	Injury mechanisms	Pollutants causing injury
1.	Bronchoconstriction	Sulphur dioxide, acidic aerosols
2.	Inflammation	Ozone, environmental tobacco smoke
3.	Fibrosis	Asbestos, silica
4.	Carcinogenesis	Radon, asbestos, formaldehyde, active smoking, environmental tobacco smoke.

Appendix 1.84: Sizes of Typical Stationary Source Particulate Emissions

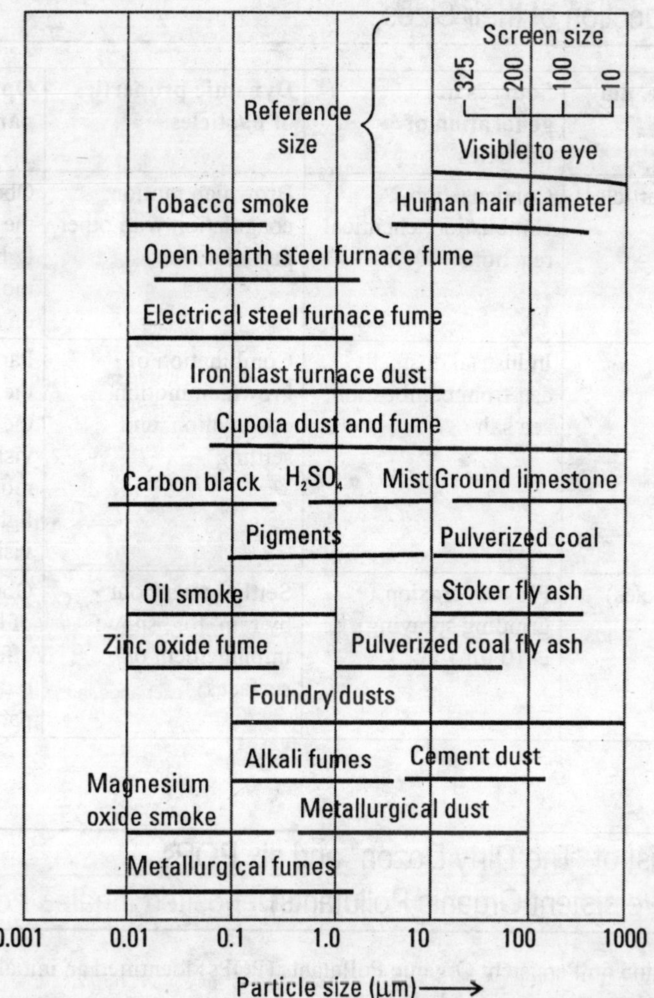

Appendix 1.85: Dynamic & Optical Properties and Sources of Particles as a Function of their Sizes

S.No.	Size of particle, μm	Sources of generation of particles	Dynamic properties of particles	Optical properties of particles
1.	<0.1 (Aitken particles)	Condensation, combustion, chemical reactions	Brownian motion, coagulation with other particles	Obey roughly the same laws of light scattering as molecules, little effect on visibility.
2.	0.1 – 1 (Large particles)	Industrial dusts, fly ash from combustion, sea salt	Combination of brownian motion, coagulation, and settling	Particle sizes of the same order as the wavelength of visible radiation; most effective in light scattering and visibility reduction.
3.	>1 (Giant particles)	Fly ash, erosion, grinding spraying (>10 μm)	Settling; washout by rain, fog, snow; impingement on surfaces	Obey scattering laws of large bodies; not as effective in visibility reduction as smaller particles.

Appendix 1.86: List of "The Dirty Dozen" and six POPs (Persistent Organic Pollutants)

The Stockholm Convention on Persistent Organic Pollutants (POPs) identified an initial twelve chemicals or chemical groups for priority action as given below.

"The Dirty Dozen POPs" Designated for International Action

S.No.	Dirty Dozen POPs identifies for priority action
1.	**Aldrin:** (organochlorine pesticide) A pesticide applied to soils to kill termites, grasshoppers, corn rootworm, and other insect pests. chemical name: 1, 2, 3, 4, 10, 10 Hexachloro 1, 4, 4a, 5, 8, 8a hexahydro 1, 4:5, 8 dimethanonaphthalene. molecular formula: $C12H8Cl6$; formula weight: 364.92 Appearance: White, odourless crystals when pure; technical grades are tan to dark brown with a mild chemical odour.

S.No.	Dirty Dozen POPs identifies for priority action
2.	**Chlordane:** (organochlorine pesticide) Used extensively to control termites and as a broad-spectrum insecticide on a range of agricultural crops. Chemical Name: 1, 2, 4, 5, 6, 7, 8, 8 octachloro 2, 3, 3a, 4, 7, 7a hexahydro 4, 7 methano 1H indene Appearance: colourless to yellowish brown viscous liquid with an aromatic, pungent odour similar to chlorine;
3.	**DDT:** (organochlorine pesticide) DDT was widely used during World War II to protect soldiers and civilians from malaria, typhus, and other diseases spread by insects. Also used as insecticide on agricultural crops, especially cotton. It continues to be applied against mosquitoes in several countries to control malaria. Chemical Name: 1, 1 (2, 2, 2Trichloroethylidene)bis(4 chlorobenzene) molecular formula: C14H9C15: formula weight: 354.49 Appearance: Odourless to slightly fragrant colourless crystals or white powder.
4.	**Dieldrin:** (organochlorine pesticide) Used principally to control termites and textile pests, dieldrin has also been used to control insect-borne diseases and insects living in agricultural soils. Chemical Name: 3, 4, 5, 6, 9, 9 Hexachloro 1a, 2, 2a, 3, 6, 6a, 7, 7a octahydro 2, 7:3, 6 dimetanonapth[2, 3 b]oxirene. 1; molecular formula: C12H8CI60; formula weight: 380.91. Appearance: A stereo isomer of endrin, dieldrin may be present as white crystals or pale tan flakes, odourless to mild chemical odour.
5.	**Endrin:** (organochlorine pesticide) This insecticide is sprayed on the leaves of crops such as cotton and grains. It is also used to control mice, voles and other rodents. Chemical Name: 3, 4, 5, 6, 9, 9, Hexachloro 1a, 2, 2a, 3, 6, 6a, 7, 7a octahydro 2, 7:3, 6 dimethanonaphth[2, 3 b]oxirene. molecular formula: C12H8CI60; formula weight: 380.92. Appearance: White, odourless, crystalline solid when pure, light tan colour with faint chemical odour for technical grade.
6.	**Heptachlor:** (organochlorine pesticide) Primarily employed to kill soil insects and termites, heptachlor has also been used more widely to kill cotton insects, grasshoppers, other crop pests, and malaria-carrying mosquitoes. Chemical Name: 1, 4, 5, 6, 7, 8, 8-Heptachloro-3a, 4, 7, 7a-tetrahydro-4, 7-methanoindene (C10H5CI7). Appearance: It is white crystalline solid and soluble in most organic solvents. The substance has a Camphor-like odor.

S.No.	Dirty Dozen POPs identifies for priority action
7.	**Hexachloroberizene (HCB):** (organochlorine pesticide and by-product chemical) Fungicide used for seed treatment of wheat, onions, sorghum. Also found as impurity in several pesticides formulations. An industrial chemical used to make fireworks, ammunition, synthetic rubber. It is also released as a by-product of the manufacture of industrial chemicals including carbon tetra-chloride, perchlorethylene, trichloroethylene and pentachlorobenzene and as a result of the processes that give rise to dioxins and furans. Chemical Name: hexachlorobenzene Appearance: White monoclinic crystals or crystalline solid
8.	**Mirex:** (organochlorine pesticide) This insecticide is applied mainly to combat fire ants and other types of ants, leaf cutters, harvester termites, mealbug and yellow jacket wasps. It has also been used as a fire retardant in plastics, rubber, and electrical goods. chemical name: 1, 1 a, 2, 2, 3, 3a, 4, 5, 5a, 5b, 6 dodecachloroacta hydro 1, 3, 4 metheno 1H cyclobuta[cd]pentalene Appearance: White crystalline, odourless solid;
9.	**Texaphene:** (organochlorine pesticide) This insecticide, a mixture of more than 670 chemicals, is also called camphechlor and is applied to cotton, cereal grains, fruits, nuts, and vegetables. It has also been used to control ticks and mites in livestock. Chemical Name: Toxaphene molecular formula: $C10H10CI8$; formula weight: 413.82 Appearance: Yellow, waxy solid with a chlorine/terpene like odour.
10.	**Polychlorinated Biphenyls (PCBs):** (compounds with varied industrial applications) These compounds are employed in industry for a variety of industrial applications, including as heat exchange fluids, in electric transformers and capacitors, and as additives in paint, carbonless copy paper, sealants and in plastics as a flame retardant. In PCBs the presence of Chlorine atoms which attached to biphenyl ring can vary from 2 to 10. PCBs show isomerism. There are a lot of homolog PCB molecules that have different isomeric structures. PCBs are generally inert, they are resistant to alkalies, acids and they have high thermal stability. They are industrially valuable compound used generally in dielectric fluids, transformers, capacitor, lubricants etc.
11.	**Dioxins:** (unintentionally formed compounds) These chemicals are produced unintentionally due to incomplete combustion, as well as during the manufacture of certain pesticides and other chemicals like polyvinyl chloride, chlorinated solvents and other chlorinated solvents. In addition, certain kinds of metal recycling and pulp and paper bleaching can release dioxins. Dioxins have also been found in automobile exhaust, tobacco smoke and wood and coal smoke. Dioxins are a class of hydrocarbons that contain two oxygen atoms generally attached to the para position of two benzene rings. Chlorinated dioxins are very stable which are less soluble in water and are less volatile. They particulate easily with air, water and soil. They are very toxic. eg:-Pentachlorodibenzo-p-dioxins ($C12H3CI502$).

S.No.	Dirty Dozen POPs identifies for priority action
12.	**Furans:** (unintentionally formed compounds) These compounds are produced unintentionally from the same processes that release dioxins, and they are also found in commercial mixtures of PCBs. Dibenzofuran contains two benzene rings fused to a central furan ring. Choloro-Dibenzofurans(CDFs) are a class of compunds contain 1 to 8 chlorine atoms attached to the benzene rings of Dibenzofurans. It is nearly insoluble in water, relatively stable towards acids and alkalies and it withstands high temperatures.

The UNECE Convention negotiations for a protocol on POPs named following six additional compounds:

1.	**Hexabromobiphenyl:** Hexabromobiphenyl is a flame retardant suitable for use in thermosetting resins and thermoplastics.
2.	**Pentachlorophenol:** Pentachlorophenol has been used as herbicide, insecticide, fungicide, algaecide, disinfectant and as an ingredient in antifouling paint. Some applications were in agricultural seeds (for nonfood uses), leather, masonry, wood, cooling tower water, rope and paper mill system.
3.	**Chlordecone:** Kepone (also known as Chlordecone) is an insecticide related to mirex used for ant and roach baits.
4.	**Lindane:** Lindane is an insecticide, also known as gamma-hexachlorocyclohexane (HCH) and benzene hexachloride (BHC). It has been used in agriculture and in pharmaceutical products for the treatment of headlice and scabies.
5.	**Short chain chlorinated parffins**
6.	**Polyaromatic hydrocarbons**

Status of Dirty Dozen POPs' in India

Aldrin : Used on crops like corn and to keep away birds. It has since been banned in India.

Chlordane : A carcinogenic pesticide widely used to control termites and since has been banned in India.

Dichlorodiphenyl trichloroethane (DDT): Small amounts of DDT are today used on crops; however, DDT is still widely employed as the cheapest means of controlling mosquitoes that carry malaria and other diseases. It is banned in 34 countries including India because of toxicity to mammals, including humans.

Dieldrin : It is an insecticide with probably the strongest carcinogenic effect amongst all organochlorine pesticides and since has been banned in India.

Endrin : Used mainly on field crops such as cotton and grains. Also used as a rodenticide and to keep away birds. It has since been banned in India.

Heptachlor : Used in Mexico, Bulgaria and the US to kill insects in seed grains and on crops, and to fight termites. It is now banned in India.

Hexachlorobenzene (HCB): Once widely used as a fungicide to protect seeds, HCB is also released into the environment as a by-product during the manufacture of chlorine gas and some chlorinated pesticides, and during waste incineration. It causes porphyria turcica, a metabolic disorder, and damages reproductive and immune systems.

Mirex : Used as an insecticide and fire retardant in the US till 1987 and also known to cause cancer in laboratory animals. It has not been produced and not used in India.

Toxaphene : It took over from DDT in 1970s as the world's most popular insecticides, particularly used in cotton-growing. It is now banned in 37 countries including India. Toxaphene is still manufactured in China, Nicaragua and Pakistan.

Polychlorinated biphenyls (PCBs) : PCBs are the leading industrial POPs. Being chemically stable and heat-resistant, they are widely used in electrical equipment oils such as transformers and capacitors as a heat-exchange fluid, and as a lubricant and plasticiser. Acutely toxic in high concentrations, they damage immune and reproductive systems and cause birth defects. Though only small quantities are now produced, they are still being released into the environment due to the break-up of old electrical equipment.

Dioxins and furans . These chlorine based compounds are produced in high-temperature incinerators used for burning organic materials including plastics, as an unwanted by-product in industrial processes ranging from metallurgy to the bleaching of paper, and as a trace contaminant in chlorophenol compounds (such as wood preservatives) and herbicides (such as Agent Orange). They are more potent than PCBs, but the effects are similar.

Status of Some Other POPs in India

Hexachlorocyclohexane (HCH) : This insecticide is widely used on cotton and on locust control. There are three chemical forms of HCH - alpha, beta and gamma. The crude manufactured from, known as technical HCH contains all the three forms. Only gamma-HCH is a useful pesticide, modern manufacturers make a pesticide formulation called lindane, which is exclusively gamma-HCH.

Endosulfan : Widely used as an insecticide on many crops, including rice, and to control these flies. Endosulfan is being manufactured in Germany, Brazil, India, Israel, Indonesia and South Korea. Change of CCl_4 by toluene in endosulfan production could be environmental relevance of the considered synthesis.

Pentachlorophenol : An insecticide mainly used as a timber preservative, especially for power line poles, railway sleepers and fence posts.

Note:

Factsheet about POPs

- Persistent organic pollutants (POPs) are organic compounds that resist environmental breakdown via biological, chemical, and photolythic processes, some taking as long as a century to degrade.
- POPs exposed to the environment are proven to travel long distances from their origin via wind and ocean currents. Precipitation carry PCBs (polychlorinated biphenyls).
- POPs therefore can be found globally, even in areas such as the Arctic and Antarctica, far from their source. Human exposures in certain Arctic areas are among the highest worldwide.
- POPs bio-concentrate as they move up through the marine food chain and accumulate in the fatty tissues of living organisms at higher trophic levels.
- Marine mammals around the world carry high burdens of POPs.
- Although many countries have banned these chemicals, they remain stockpiled, are produced or used illegally, or, because of lengthy half-lives, they continue to exist in soil, or other environmental media.

- The United Nations is currently considering the elimination or reduction of twelve of some of the most damaging POPs through the formulation of an international treaty. Nine of the POPs chemicals under consideration are pesticides that have been extensively used in both developed and developing countries.

General Health Effects

- POPs can disrupt the endocrine, reproductive, and immune systems The developing brain and nervous system may be most vulnerable.
- POPs are capable of causing behavioral problems, cancer, diabetes and thyroid problems.

Babies whose mother ate large amounts of highly contaminated fish from Lake had lower birth weights, poor memory smaller head circumferences and shorter attention spans than babies whose mothers did not eat fish.

Marine mammals are affected with diseases associated with immune system dysfunction.

Marine mammals present a metabolic imbalance, so they are considered one of the most vulnerable organisms with respect to long-term toxicity of man-made chemicals such as organochlorines. (Whales and dolphins) have a poor ability to detoxify organic pollutants because they lack isozymes that are required to detoxify DDT and PCBs.

Appendix 1.87: List of Human Carcinogens (cancer causing agents)

S.No.	Carcinogens	Characteristics	Uses
1.	Aldrin	Brown to white crystalline solid, mild chemical odour	Pesticides
2.	Asbestos	Fibres	Fireproof fabrics, brake linings, gaskets, roofing, composition, electrical and thermal insulation, paint filler, chemical filters, reinforcing agent in rubber and plastics, component of paper dryers felts.
3.	Arsenic	Silver grey brittle crystalline solid which darkness in moist air Also may be a black amorphous or yellow crystalline solid	Non-ferrous metal industries (except aluminum), paint pigments, pesticides, medicine.
4.	Anthracene oil	Oil	Component of coal tars used in chemical industry to make antracene and other polycyclic aromatic hydrocarbons, also used as wood preservatives and pesticides.

S.No.	Carcinogens	Characteristics	Uses
5.	Alpha- Naphthylamine	White crystals that become redon exposure to air, soluble in alcohol slightly soluble in water	Dye dye intermediates.
6.	2- acetylamino fluorene	Cystal; soluble into glycols, alcohols, and fats solvents: insoluble in water	
7.	Bis-chloromethyl ether	Colorless liquid or vapour	Formed when formaldehyde and hydrochloric acid combine.
8.	Benezene	Colorless liquid or vapour	Formed when formal dehyde and hydrochloric acid combine.
9.	Beta-naphthylamine	While to reddish lustrious leaflets, faint odour	Dyes
10.	Benzidine's and its salts	Greyish yellow, white or reddish grey crystals or crystalline powder	Organic synthesis, manufacture of dyes especially congo red, Detection of blood stains agent in rubber compounding.
11.	Colorless liquid, pungent odour	Colorless liquid, pungent odour	Organic synthesis, vapor sterilant, disinfectant.
12.	Chromates	Various colors	Chromate or processing, chromium plating industry, chemical industry, textile industry dyes
13.	Creosote oil	Oily liquid	Wood preservative, disinfectants
14.	Coal soot	-	Chimney cleaners
15.	Coal tar-lignite, bitumen, anthractic	Fume	Chemical industry, carbon electorate industry, coke ovens
16.	Coal tar-lignite, bitumen, anthracite	Black, viscous liquid(or semi-solid), Naphthalene like odour	Roofer, pavers, carbon electrode industry, coke ovens and by-product plant.
17.	3-3 dichlorobenzidine	Grey purple crystals; insoluble in water, soluble in alcohol, benzene and glacial acetic acid.	Intermediate for drugs and pigments, curing agent for some risins
18.	Dieldrin	Light flaked solids; soluble in acetone and benzene.	Pesticides
19.	4-dimethylaminoazobenzene	Yellow crystalline leaflets; soluble in alcohol, ether; strong mineral acids and oils, insoluble in water	Indicators in volumetric analysis; also in test for peroxidised fats; coloring agents

S.No.	Carcinogens	Characteristics	Uses
20.	Ethyleneimine	Clear colorless liquid with an ammonia smell (however, smell is not strong enough to act as a warning)	Intermidiate and monomer for fuel oil and lubricant refining, ion exchange, protective coatings, including paper and textile, pharmaceuticals, adhesive, polymers stabilizers, surfactants.
21.	Hydrogenated coal oils(Bergius process)	-	Hydrogenation is a possible means of converting coal oil into gasoline.
22.	Isoprophyl oil	-	Chemical industry, isopropyl alcohol markers.
23.	Lubricating oil	–	Machine shop, metal fabricating industry, garages, maintenance workers, petroleum industry.
24.	Methyl chloromethyl ether	Colorless liquid or vapor	Formed when formal dehyde and hydrochloric acid combine
25.	4-4 methylene bis (2 choloraniline) (MOCA) (cyanaset)	Tan-colored pallets; soluble in hot methyl ethyl ketone, acetone, esters and aromatic hydrocarbons	Curing agent for materials which become ingradients in the protective coatings on tanks, chemical reactors, acid handling equipments and solid synthetic rubber products.
26.	Mustard gas	-	Chemical industry war+gas, medicine
27.	4-nitrodiphenyl	Crystalline	-
28.	N-nitrosodimethylamine	Yellow to brown or orange powder or flakes; soluble in alcohol and acetone	Retarder of vulcanization of rubber; pesticides
29.	Nickel	-	Nickel smelting operartions-Mond processs stainless steel welding, petroleum industry
30.	Nickel carbonyl	Colorless liquid, soluble in alcohol, many organic solvents and concentrated nitric acid; not soluble in water	-do-
31.	p-xenylamine(4-aminodiphenyl)	Colorless crystals	-

S.No.	Carcinogens	Characteristics	Uses
32.	Polycylic aromatic hydrocarbons (PAH)	-	Component of tobacco smoke, coal tar and pitch fumes, anthracene oil, coal soot, creosote oil, lubricating oil, paraffin oil, petroleum coke etc.
33.	Petroleum coke	-	Petroleum industry, carbon electrode industry electro thermal chemical industry
34.	Paraffin oil	Oily	Floor treatment, lubricant, medicine-pure.
35.	Radioactive materials	-	Medical profession, air craft gages, non-destructive testing, radioactive material minning.
36.	Tobacco smoke	-	Everywhere found
37.	Vinyl chloride(chlorothene)	Compressed gas, easily liquefied, smells faintly sweet like ether; usually handled as liquid; slightly soluble in water; soluble in alcohol and ether .	Manufacture of chemicals and PVC (polyvinyl chloride), significant amount may be found in plastic resin all during processing to form finished plastic products and adhesive for plastics
38.	Vinylidene chloride	Colorless liquid, readily polymerises	Plastic industry, chemical industry, synthetic fibres
39.	Wood-dust-oak Beech	-	Oak wood products industry, carpenters, lumber industry.
40.	X-rays	-	Medical profession, non-destructive testing.

Appendix 1.88: Occurrence and Site of Attack of Substances/Chemicals Causing Occupational Cancer

S.No.	Substances/Chemical causing occupational cancer	Occurrence/Industries	Site of attack
1.	Mineral oil, paraffin pitch, tar	Petroleum and petrochemical industry	Skin
2.	Chromium/Chromates Nickel	Alloysteel foundries electroplating, chemical plants	Lungs, bladder, nasal passage

S.No.	Substances/Chemical causing occupational cancer	Occurrence/Industries	Site of attack
3.	Asbestos	Asbestos units	Lungs, larynx, (Gastro-- Intestinal)
4.	Organic chemicals, Chlorinated hydrocarbons	Chemical and petrochemical industry	Liver
5.	U.V. Rays, Ionising radiation	All industry (Inspection, NDT) (NDT-Non destructive Testing)	Eyes, parts of body exposed
6.	Benzole		Blood forming cells
7.	Benzene	Organic chemical industry	Blood forming cells, Leukemia (acute nonlymphocytic)
8.	Arsenie and trioxide compounds	Chemical plants	Skin, lung, angiosarcoma, liver
9.	Benzidine	Chemical plants	Bladder
10.	Bis (chloromethyl) ether (BCME) and chloromethyl ether	-	Lung
11.	Coke oven emissions	Integrated steel plant's coke overs	Skin, lung, urinary tract
12.	Hematite underground mining	Iron ore mines	Lung
13.	Isopropyl alcohol	Chemical plants	Lung, larynx, paranasal sinuses
14.	Mustard gas	-	Lung
15.	Chemical plants	Chemical plants	Bladder
16.	Tar distillation and recovery units	Tar distillation and recovery units	Skin, lung, bladder, G.I.
17.	Chemical plants	Chemical Plants	Liver (angiosarcoma), lung, brain
18.	Multiple sites	Multiple sites	-
19.	Shoe manufacturing units	Shoe manufacturing units	Nasal sinuses, lung, bladder, leukemia
20.	Furniture making units	Furniture making units	Nasal sinuses

NOTES

Information on occupational cancer is received from following three sources.

a. Environmental monitoring and co-relating data with epidemiological findings (i.e. how many people in the exposed category and unex-posed category have symptoms of disease).
b. Animal experiments – i.e. studying effect on animals (eg: rats, rabbits.) exposed to controlled levels and deducing the effects in human beings with extrapolation.
c. Laboratrry experiments based on mutation in bacteria.

Summary of Environmental and Occupational Links with Cancer is tabulated below:

S.No.	Category	Carcinogenic Agent	Sources/Uses	Strong evidence of causal link (IARC Group 1)	Suspected evidence of causal link
1.	Aromatic Amines	Benzidine, 2-naphylamine, 4, 4'-methylenebis 2-chloraniline chlornaphazine, Heterocyclic Aromatic Amines	Rubber and cutting oil production, azo dye manufacturing, and used as pesticides. Common in chemical and mechanical industries and aluminum transformation. Found in tobacco smoke and widely used in textile industry and as hair dyes.	Bladder (Benzidine, 2-naphylamine, 4, 4'-methylenebis 2-chloraniline chlornaphazine)	Prostate (Heterocyclic Aromatic Amines)
2.	Chlorination By-Products	Trihalomethanes	Chloroform, bromodichloromethane and bromoform. (formed by interaction of chlorine and bromine with organic chemicals)	Bladder	Colorectum, Esophagus
3.	Environmental Tobacco Smoke	Contains more than 50 known carcinogens	Combined release from burning end of tobacco products and exhaled smoke.	Lung, Breast	–
4.	Metals	Arsenic	By-product of non-ferrous metal production (mainly copper). Greater than 10% of dust in some smelter operations.	Bladder, Kidney, Lung, Skin, Soft Tissue Sarcoma (angiosarcoma of the liver)	Brain/Central Nervous System, Liver/Biliary, Prostate, Soft Tissue Sarcoma
		Beryllium	Nuclear, aircraft and medical devices industries. Alloy or in specialized ceramics for electrical and electronics applications. Contaminant of coal and fuel oil combustion.	Lung	–

S.No.	Category	Carcinogenic Agent	Sources/Uses	Strong evidence of causal link (IARC Group 1)	Suspected evidence of causal link
		Cadmium	Naturally occurring in zinc-, lead-, and copper-bearing ores. Stabilizer in polyvinyl chloride products, colour pigment, common in rechargeable nickel-cadmium batteries, pollutant in phosphate fertilizers.	Lung	Pancreas, Kidney, Prostate
		Chromium	Steel and other alloy production. Chromium III and Chromium IV used in chrome plating, manufacture of dyes and pigments, leather tanning, wood preserving.	Lung, Nasal/ Nasopharynx	–
		Lead	Batteries, ammunition, solder, pipes and X-ray shielding devices. Still in some gasoline, paints, ceramic products, caulking, and pipe solder. Common in costume jewellery.	–	Brain/Central Nervous System, Lung, Kidney, Stomach
		Mercury	Production of chlorine gas and caustic soda. Thermometers, dental fillings, batteries. In some skin lightening creams and antiseptic creams and ointments. Transformed by microorganisms to methylmercury in water and soil.	–	Brain/Central Nervous System (CNS)
		Nickel	Primarily as alloy in stainless steel. Nickel platting and batteries.	Lung, Nasal/ Nasopharynx	Larynx, Pancreas, Stomach
5.	Metalworking Fluids and/or Mineral Oils	Straight Oils, Soluble Oils, Synthetic and Semisynthetic Fluids	Various industries including metal machining, print press operating, cotton and jute spinning	Bladder, Larynx, Lung, Nasal/ Nasopharynx (mineral oils), Rectum, Skin, Stomach	Esophagus, Pancreas, Prostate
6.	Natural Fibres/ Dust	Asbestos	Acoustical and thermal insulation.	Larynx, Lung, Mesothelioma, Stomach	-
		Silica	Foundries, brick-making and sandblasting	Lung	-
		Talc containing asbestiform fibres	Manufacturing of pottery, paper, paint and cosmetics	Lung	Ovary
		Wood dust	Carpentry, joinery, furniture and cabinetry	Lung, Nasal/ Nasopharynx	Larynx
7.	Pesticides	Herbicides, Fungicides, Insecticides	Preventing, destroying, repelling, mitigating pests. Plant regulators, defoliants, dessicants. Majority used in agriculture. Also residential applications.	-	Brain/Central Nervous System, Breast, Colon, Lymphoma, Leukemia, Lung, Multiple Myeloma, Ovary, Pancreas, Kidney, Soft Tissue Sarcoma, Stomach, Testicle

S.No.	Category	Carcinogenic Agent.	Sources/Uses	Strong evidence of causal link (IARC Group 1)	Suspected evidence of causal link
8.	Petrochemicals and Combustion By-Products	Petroleum Products, Motor Vehicle Exhausts (including diesel) Polycyclic Aromatic Hydrocarbons (PAHs), Soot, Dioxins	Petrochemicals derived from natural gas or petroleum and used to produce pesticides, plastics, medicines and dyes. Substances produced as building blocks for other products but mainly from incomplete combustion of coal, oil, gas (including diesel), household waste, tobacco and other organic substances. Dioxins produced as by-products of combustion of chlorine and carbon-based chemicals such as PVC plastics, and via chlorine-bleaching process for whitening paper and wood pulp.	Lung (PAHs, air pollution, including diesel exhaust, soot, dioxin), Non-Hodgkin Lymphoma (dioxin), Soft Tissue Sarcoma (dioxin), Skin (PAHs)	Bladder (PAHs, diesel exhaust), Breast (dioxin, Larynx (PAHs), Multiple Myeloma (dioxin), Prostate (dioxin, PAHs)
9.	Radiation	Ionizing Radiation	Several particle types released by radioactive material, high-voltage equipment, nuclear reactions, and stars. Particles of concern include alpha and beta particles, X-rays and gamma rays.	Bladder, Bone, Brain/ Central Nervous System, Breast, Colon, Leukemia, Liver/Biliary, Lung, Multiple Myeloma, Nasal and Nasopharynx, Ovary, Soft Tissue Sarcoma, Skin, Stomach, Thyroid.	-
	Non-ionizing radiation	Non-ionizing	Microwaves and electromagnetic frequencies, including radiowaves and extremely low-frequency electromagnetic fields	-	Brain, Breast, Leukemia, Salivary Gland
		Ultraviolet Radiation	Solar radiation from the sun	Skin	-
10	Reactive Chemicals	Butadiene	Production of polymers for manufacture of styrene- butadiene rubber for tires; nitrile rubber for hoses, gaskets, adhesives, footwear; styreme-butadiene latexes for paints and carpet backing; acrylonitrile-butadiene-styrene polymers for parts, pipes and various appliances.		Leukemia
		Ethylene Oxide	Sterilant, disinfectant and pesticide. Ingredient in making resins, films and antifreeze.	Leukemia	Breast
		Formaldehyde	Production of urea, phenol or melamine resins for molded products such as appliances, electric controls and telephones. Also used in particle-board and in surface coatings.	Nasal/Nasopharynx	Leukemia
		Sulphuric Acid	Production of isopropanol, ethanol, treatment of metals, manufacture of soaps, detergents and batteries.	Larynx	Lung
		Vinyl Chloride	In polyvinyl resins for production of plastic pipes, floor coverings, and in electrical and transportation applications.	Liver Soft Tissue Sarcoma (angio-sarcoma of the liver)	-

S.No.	Category	Carcinogenic Agent	Sources/Uses	Strong evidence of causal link (IARC Group 1)	Suspected evidence of causal link
11.	Solvents	Benzene	Intermediate in production of plastics, resins and some synthetic and nylon fibres. Used to make some types of rubbers, lubricants, dyes, detergents, drugs and pesticides. Also in crude oil, gasoline and cigarette smoke.	Leukemia, Multiple Myeloma, Non-Hodgkin Lymphoma	Brain/Central Nervous System, Lung, Nasal/ Nasopharynx
		Carbon Tetrachloride	Industrial applications. Before being banned, used in the production of refrigeration fluid and propellants for aerosol cans, as a pesticide, as a cleaning fluid and degreasing agent, in fire extinguishers, and in spot removers.	-	Leukemia
		Methylene Chloride	Solvent in industrial applications and as a paint stripper. In some aerosol and pesticide products and in production of photographic film.	-	Brain/Central Nervous System, Liver
		Styrene	Production of rubber, plastic insulation, fiberglass, pipes, automobile parts, food containers and carpet backing.		Lymphoma
	Solvents (cont'd)	Toluene	Production of paints, paint thinners, fingernail polish, lacquers, adhesives and rubber. Some printing and leather tanning processes.		Brain/Central Nervous System, Lung, Colorectum
		Trichloro-ethylene	Degreasing metal parts. Previously used in dry cleaning. May be found in printing inks, varnishes, adhesives, paints, and lacquers.	Liver	Cervix Lymphoma, Kidney, Leukemia, Lymphona
		Tetrachloro-ethylene	Degreasing metal parts and industrial solvent. Dry cleaning.	-	Bladder, Cervix, Kidney, Lymphona
		Xylene(s)	Cleaning agent, thinner for paint and in paints and varnishes. Used in printing, rubber and leather industries. Low levels in gasoline and airplane fuel.	-	Brain/Central Nervous System, Colorectum
12	Other	Creosotes	Includes coal tar and coal tar pitch formed via high temperature treatment of wood, coal, or from the resin of creosote bush. Coal tar products used in medicine, animal and bird repellants and as pesticides. Coal tar creosote widely used as wood preservative. Coal tar, coal tar pitch, and coal tar pitch volatiles using in roofing, road paving, aluminum smelting and coking.	Bladder (coal tars), Lung, Skin	-

S.No.	Category	Carcinogenic Agent	Sources/Uses	Strong evidence of causal link (IARC Group 1)	Suspected evidence of causal link
		Endocrine Disruptors	Several natural and synthetic chemicals capable of mimicking the body's natural hormones	Breast Cervic	Breast (bisphenol A), Prostate (bisphenol A), Testicle (chlorinated insecticides)
		Hair Dyes	Hair colouring products - chemicals agents vary according to the colour and degree of dye permanency (temporary, semi-permanent, demi, and permanent).	-	Bladder, Brain/ Central Nervous System, Leukemia, Multiple Myeloma, Lymphoma
		Nitrosamines and N-nitroso Compounds	Formed when amines and nitrosating agents chemically react. Found in the rubber, metal and agricultural industries, and in cosmetics and foods such as fried bacon and cured meats.	-	Brain/Central Nervous System, Kidney
		Poly-chlorinated Biphenyls (PCBs)	Banned but still circulating in the environment - used as coolants and lubricants in transformers, capacitors, and other electrical equipment.	Liver	Lymphoma

Appendix 1.89: Excess Cancer Incidence Sites in Persons Engaged in Various Occupations

S.No.	Occupational groups/industrial processes	Main target cancer sites
1.	Boot and shoe manufacturers/repairers	Nose
2.	Coal mines	Stomach
3.	Chemists	Pancreas, lymphatic system
4.	Coke ovens and by-product workers (in steel plants)	Large intestine, pancreas
5.	Cadmium production workers	Lung, prostrate
6.	Foundary workers	Lung
7.	Furniture making workers	Nasal cavities and sinuses
8.	Leather workers	Bladder
9.	Metal miners	Lung
10.	Manufacturer of auramine	Bladder
11.	Manufacture of isopropyl alcohol by strong acid process	Nose and nasal sinuses
12.	Nickel refining	Lung, nose and nasal sinus

S.No.	Occupational groups/industrial processes	Main target cancer sites
13.	Printing pressmen (newspapers)	Mouth and pharynx
14.	Rubber industry: a) Processing of rubber b) Tyre building c) Tyre curing	Stomach leukemia Bladder, brain Lung
15.	Shoe workers	Nasal cavity and sinuses, blood marrow (Leukemia)
16.	Underground mining of hematite	Lung
	Cancerous target sites of some chemicals:	
17.	4-Aminodiphenyl	Bladder
18.	Asbestos	Lung, pleura, peritoneum, larynx, gastrointestinal tract
19.	Diethylstilboestrol	Female genital tract, breast
20.	Melphalan	Haematolymphopoietic system
21.	Conjugated oestrogens	Uterus
22.	Cyclophosphamide	Bladder

Appendix 1.90: Classification Groups for Carcinogenic Substances

Group 1: carcinogenic to human (common called "know") carcinogen (examples : abestos, arsenic) (evidence supports the chemical or substance as human carcinogen)

Group 2A: probably carcinogenic to human (examples : diethyl sulfate vinl bromide) (limited evidence in humans and sufficient evidence in experimental animals)

Group 2B: **possibly** carcinogenic to humans (examples: bracken term chlordane) (limited evidence in humans and less than sufficient evidence in experimental animals)

Group 3: unclassified or not classified as carcinogenic to humans (examples : Adrian, aniline) (inadequate evidence in humans and inadequate or limited evidence in animals)

Group 4: **probably not** carcinogenic to humans (example: capolactain) (evidence suggesting lack of carcinogenicity in human and animals)

Appendix 1.91: Various Entry Routes of Environment Chemicals in Our Body and their Health Effects

In order for a chemical to become hazardous to a person's health, it must first contact or enter the body and the chemical must have some biological effect on the body. There are five major routes of entry of environment chemicals in our body:

- Inhalation (breathing)
- Skin contact (dermal route)
- Digestive system (ingestion or eating)
- Injection
- Eye contact

Breathing of contaminated air is the most common way that workplace chemicals enter the body. Some chemicals, when contacted, can pass through the skin into the blood stream. Less commonly, workplace chemicals may be swallowed/ingested accidentally if food or cigarettes (or hands) are contaminated. For this reason we should not drink, eat, or smoke in areas where they may be presence of toxic chemicals.

Injection is the another way chemicals may enter the body. While uncommon in most workplaces, it can occur when a sharp object (e.g., needle) punctures the skin and injects a chemical (or virus) directly into the bloodstream.

The eyes may also be a route of entry. Usually, however, only very small quantities of chemicals in the workplace enter through the mouth or the eyes.

Regardless of the way the chemical gets into the body, once it is in the body it is distributed to anywhere in the body by the blood stream. In this way, the Chemicals can attack and harm organs which are far away from the original point of entry as well as where they entered the body.

Fate of contaminated air breathe in

Contaminated air in the workplace can be inhaled. Air is drawn through the mouth and nose and then into the lungs. An average person will breath in and out about 12 times a minute. Each of the 12 inhalations brings in about 500 mL of air, corresponding to 6 litres of air per minute, together with any contaminants that the air contains.

Involved in hard physical work will breathe harder and take in more than 6 litres a minute. Over an 8-hour working day, more than 2, 800 litres of air will be breathed in and out of the lungs. In conditions of hard physical work, up to 10, 000 litres may be exchanged. Air breathed in through the nose is filtered by the nasal hairs so that large, solid particles (>10 microns) in the atmosphere are prevented from going any further. Inside the nose there are small bones and cartilages that cause the inhaled air to swirl around. This swirling air can cause some large contaminating articles to be deposited in the nose and trapped by the moisture of the mucus lining.

Air coming in from the nose and the mouth reaches the back of the throat and enters an area known as the pharynx. The pharynx, which is the entrance to the airways, divides into two tubes, one called the esophagus, which carries food to the stomach, and one called the trachea, which leads down towards the lungs. Contaminated air passes into the trachea which itself divides into two large tubes, each called a bronchus. Each bronchus enters a lung. Once inside its lung, each bronchus starts to branch. The tubes of the bronchus get thinner and thinner as they spread, rather like the branches of a tree. Eventually, the tiniest tubes, which are called bronchioles, end in thin-walled air sacs. Each of these sacs is called an alveolus. Collectively, they are called alveoli and there are many thousands of these in each lung. The walls of the alveoli are very thin and are richly supplied with tiny blood vessels (capillaries). Waste carbon dioxide from the body, carried in the blood inside the veins, can pass out of the veins through the walls of the alveoli to become a part of the air which is exhaled.

Oxygen in the inhaled breath crosses the alveolar walls to enter the blood within the capillaries. Once oxygen has become attached to the blood inside the veins, it is then distributed throughout the body. Chemical vapours, gases and mists which reach the alveoli in the lungs can also pass into the blood and be distributed around the body.

Sometimes, the concentration of chemicals reaching the alveolar air sacs is lower than in the workplace air. This is because the airways contain a lining of sticky, thick fluid called mucus. Tiny hairs, known as cilia, on the inside of the tubes constantly carry this mucus upwards towards the back of the throat. In some instances, a portion of the gases, vapours and mists may be dissolved in this mucus before they reach the alveolar sacs.

Solid, visible particles found in dusts, fumes and smoke that have escaped the filtering mechanisms of the nose may also be trapped by the mucus. The mucus is wafted by the tiny cilia hairs until it reaches the back of the throat where it is either expelled through the mouth or swallowed and passed to the stomach. In this latter case, the contaminating chemicals will enter the body in the same way as contaminated food or drink. This is dealt with in more detail in the section on the digestive system.

Much smaller particles (so small that they cannot be seen by the eye) may not be stopped by the mucus in the trachea and bronchiole tubes. They travel through the various branches of the airway and eventually reach the alveoli. Solid particles which cannot pass through the thin wall of the air sacs may lodge and stay where they are. Some may dissolve, others may be attacked and destroyed by the scavenger cells of the body's defence system. Others may prove too big or too insoluble to be disposed of in this way and simply stay in the air sacs. Some of these particles, if they are present only in small quantities, do no apparent harm. Other types of dusts may damage the surrounding alveolar walls. The damage may permanent and may cause scars to form, which eventually interfere with the lung's ability to pass oxygen into the blood stream.

Some acids, caustics or organic chemicals, when inhaled in sizable amounts, can cause serious and irreparable, "burn" damage to the mouth, nose, trachea, bronchi and lungs.

Examples of workplace chemicals that can be inhaled

Gases and Vapours

Workplace chemicals can enter the air in a number of different ways. Simple evaporation is probably the most common way. Organic solvents, such as toluene, methyl ethyl ketone (MEK) or alcohols, generally evaporate more rapidly than water, acids or caustics, though this is not always the case. Evaporation produces vapours. Vapours are formed from substances that exist as solids or liquids under normal temperature and pressure conditions. Substances that do not exist as solids or liquids at normal temperatures and pressures are called gases. Gases as well as vapours can contaminate the workplace air.

Mists

In some instances, an industrial process might produce tiny liquid droplets that are able to float in the air. These droplets are called mists. Mists are formed by gases that condense into small liquid droplets in the air. Alternatively, mists may form by breaking up, splashing or atomizing a liquid. Examples include acid mists from electroplating, oil mists from cutting and grinding, or paint spray mists from painting operations.

Dusts, fumes and smoke

Some workplace processes can generate tiny solid particles which are light enough to float in the air, and these are referred to as dusts, fumes and smoke. Dusts are solid particles often generated by some mechanical or abrasive activity. They are usually heavy enough to settle slowly to the ground. Fumes are very tiny solid particles which can remain airborne that are formed when a heated metal has evaporated in the air and then condensed back to a solid form. This occurs in welding operations. Smoke is carbon or soot from burning. Smoke particles can settle or remain airborne depending on their size.

Entry of chemicals in the body through skin

Chemicals which pass through the skin are nearly always in liquid form. Solid chemicals and gases or vapours do not generally pass through the skin unless they are first dissolved in moisture on the skin's surface.

The skin is the second most common route by which occupational chemicals enter the body which consists essentially of two layers, a thin, outermost layer called the epidermis and a much thicker underlayer called the dermis. The epidermis consists of several layers of flat, tightly-packed cells which form a barrier against infections, water, and some chemicals. This barrier is the external part of the epidermis. It is called the keratin layer, and is largely responsible for resisting water entry into the body. It can also resist weak acids but is much less effective against organic and some inorganic chemicals. The keratin layer contains fat and fat- like substances which readily absorb chemicals which are solvents for fat, oil, and grease.

Examples of chemicals that can enter through skin

Organic and caustic (alkaline) chemicals can soften the keratin cells in the skin and pass through this layer to the dermis, where they are able to enter the veins and hence the blood stream. Areas of the body such as the forearms, which may be particularly hairy, are most easily penetrated by chemicals since they can enter down the small duct containing the hair shaft. Chemicals can also enter through cuts, punctures or scrapes of the skin since these are breaks in the protective layer. Contact with some chemicals such as detergents or organic solvents can cause skin dryness and cracking. There can also be hives, ulcerations or skin flaking. All these conditions weaken the protective layer of the skin and may allow chemicals to enter the body.

Chemicals can vary enormously in the degree to which they penetrate the skin. Some solvents such as trichloroethylene, naphtha and toluene may soften the keratin layer but are not believed to penetrate much further unless there is prolonged skin contact. On the other hand, chemicals such as benzene, carbon tetrachloride, carbon disulfide and methyl alcohol can readily pass through the epidermis and subsequently enter the blood stream. Some chemicals are so corrosive they burn holes in the skin, allowing entry for infection or other chemicals.

In some instances, chemicals may enter by accidental injection through the skin. This may occur in hospital settings or in industrial hole-punching or injection processes. Once in the blood stream, the chemicals can be transported to any site or organ of the body where they may exert their effects.

Entry of chemicals to digestive system by ingestion/swallowing

Chemicals can enter the stomach either by swallowing contaminated mucus which has been expelled from the lungs, or by eating and drinking contaminated food. Food and drink are most frequently contaminated by contact with unwashed hands, gloves or clothing, or by being left exposed in the workplace. Nail-biting and smoking also contribute.

Once inside the mouth, workplace chemicals pass down the esophagus and then into the stomach. Food in the stomach is digested with a strong acid produced by the stomach. A few chemicals, such as alcohol, may pass across the stomach wall and enter the veins and the blood stream here, but most chemicals move from the stomach into a long, twisting tube known as the small intestine. The inside of the small intestine has many hundreds of tiny finger-like projections called villi. The villi have very thin walls and are filled with tiny blood vessels. This allows the digested food to pass from the small intestine across the walls of the villi and enter the veins. The food is then carried around in the blood stream to the parts of the body that need it.

Some workplace chemicals which contaminate food or drink can also pass across the thin walls of the villi and into the blood stream in this manner. Other workplace chemicals, which are not soluble or whose basic units (molecules) are too big to pass across the villi walls, will stay in the gut and pass out through the feces without being absorbed into the blood stream to any extent.

Some acids, caustics and organics may cause severe "burn" damage to the digestive system if ingested in high concentrations.

Examples of chemicals that can enter our body by ingestion (swallowing)

In workplaces, dusts, smoke or fumes can enter the digestive system by accidental ingestion (e.g., swallowing contaminated mucus which has been expelled from the lungs) or by eating something with contaminated hands (e.g., lead paint on unwashed hands).

Chemicals entry to body through eyes

Although eye splashes or eye contamination by workplace chemicals is fairly common, large quantities of chemicals probably do not enter the body this way. Small amounts of chemicals may enter by dissolving in the liquid surrounding the eyes and larger, but probably not significant amounts, may enter the eyes if they are splashed with chemicals.

The eyes are richly supplied with blood vessels and many chemicals can penetrate the outer tissues and pass into the veins. The eye may or may not be damaged during this process, depending on the corrosive nature of the chemical and its ability to penetrate the outer tissues.

Examples of chemicals that can enter the body through the eye

Any chemical, whether in the form of a liquid, dust, vapour, gas, aerosol or mist, can contact and enter the eye. For example, the organic solvent toluene can pass through the outer layers of the eye and probably enter the blood stream. During this process it can cause keratitis, an inflammation of the outer layer of the eye.

Appendix 1.92: Available Recovery and Abatement Techniques for Treatment of Waste Gases from Chemical Industries in Relations to Type of Contaminants/Pollutants

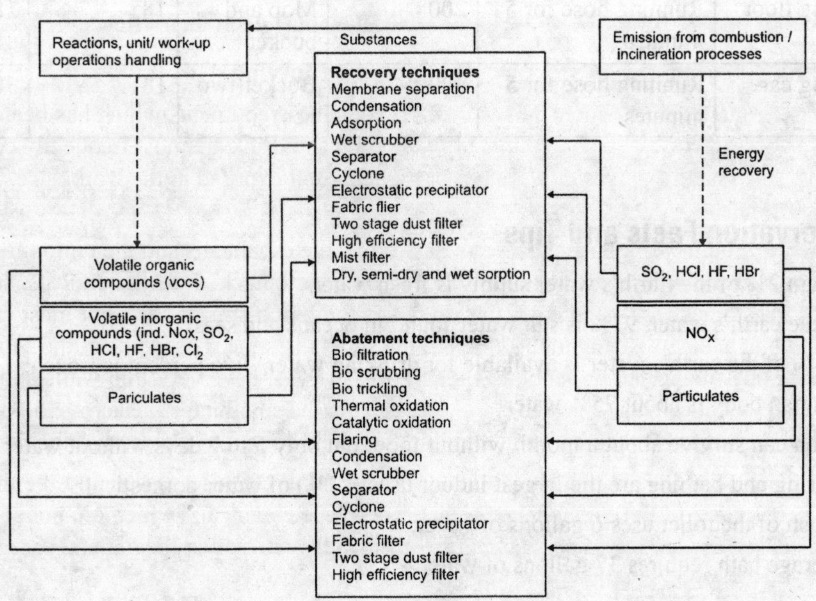

Appendix 1.93: Unintentional Wastage of Water

S.No	Activity	Method adopted	Quantity used in litres	Method to be adopted	Quantity used, litre	Quantity saved litre
1.	Brushing teeth	Running water tap for 5 minutes	45	Tumbler or glass	0.5	44.5
2	Washing hands	Running water tap for 2 minutes.	18	Half filled wash basin	2	16
3.	Shaving	Letting shower run while soaping, staying under shower too long.	18	Shaving mug	0.25	17.75
4.	Shower	Using old Fashioned large capacity cisterns.	90	Wet down-tap off, soap up rise off.	20	70
5.	Flushing toilet	Running hose for 5 minutes.	13.5 or more	Dual system short flush for liquid waste, full flush for solid waste.	4.5	4.5
6.	Watering plants	Running hose for 5 minutes.	120	Water can	5	115
7.	Washing floor	Running hose for 5 minutes.	200	Mop and bucket	18	182
8.	Washing car	Running hose for 5 minutes.	400	Bucket(two nos)	18	382

Note:

Water Conservation Facts and Tips

- Less than 2% of the Earth's water supply is fresh water.
- Of all the earth's water, 97% is salt water found in oceans and seas.
- Only 1% of the earth's water is available for drinking water. Two percent is frozen.
- The human body is about 75% water.
- A person can survive about a month without food, but only 5 to 7 days without water.
- Showering and bathing are the largest indoor uses (27%) of water domestically.
- One flush of the toilet uses 6 gallons of water.
- An average bath requires 37 gallons of water.

- The average 5-minute shower takes 20 gallons of water – around 40 gallons are used in 10 minutes.
- If you water your grass and trees more heavily, but less often, this saves water and builds stronger roots.
- Water your lawn only when it needs it. If you step on the grass and it springs back up when you move, it doesn't need water. If it stays flat, it does need water.
- Run your dishwasher and washing machine only when they are full.
- When washing a car, use soap and water from a bucket. Use a hose with a shut-off nozzle for rinsing.
- Never put water down the drain when there may be another use for it such as watering a plant or garden, or cleaning.
- When washing dishes by hand, fill one sink or basin with soapy water. Quickly rinse under a slow-moving stream from the faucet.
- An automatic dishwasher uses 10 gallons of water while hand washing dishes can use up to 20 gallons.
- Store drinking water in the refrigerator rather than letting the tap run every time you want a cool glass of water.
- Water lawns during the early morning hours, or evening when temperatures and wind speed are the lowest. This reduces losses from evaporation.
- Do not hose down your driveway or sidewalk. Use a broom to clean leaves and other debris from these areas. Using a hose to clean a driveway wastes hundreds of gallons of water.
- Don't leave the water running when brushing your teeth or shaving. Get in the habit of turning off the water when it's not being used.
- Use of bowl of water to clean fruits and vegetables rather than running water over them. You can reuse this for your house plants.
- A dairy cow must drink four gallons of water to produce one gallon of milk.
- One inch of rainfall drops 7, 000 gallons or nearly 30 tons of water on a 60' by 180' piece of land.
- There is the same amount of water on Earth as there was when the Earth was formed.
- Water is composed of two elements, Hydrogen and Oxygen. 2 Hydrogen + 1 Oxygen = H_20.
- Nearly 97% of the world's water is salty or otherwise undrinkable. Another 2% is locked in ice caps and glaciers. That leaves just 1% for all of humanity's needs – all its agricultural, residential, manufacturing, community, and personal needs.
- Water regulates the Earth's temperature. It also regulates the temperature of the human body, carries nutrients and oxygen to cells, cushions joints, protects organs and tissues, and removes wastes.
- 75% of the human brain is water and 75% of a living tree is water.
- Water is part of a deeply interconnected system. What we pour on the ground ends up in our water, and what we spew into the sky ends up in our water.
- Water expands by 9% when it freezes. Frozen water (ice) is lighter than water, which is why is floats in water.

Appendix 1.94: Key Facts about Seas and Oceans

- Oceans cover 70% of the earth surface.
- 40% of the earth's population lives within 60kms of a sea coast.
- Three quarters of the world's mega cities are by the sea.
- 80% of the people will live within 100Kms of a sea coast by the year 2020.
- Average sea level has risen between 10 and 25 cms in the past 100 years.
- If all the world's ice melts, then the ocean would rise by 66 metres.
- 60% of the Pacific shoreline and 35% of the Atlantic shoreline are receding at the rate of one meter a year.

NOTES

Ocean Facts Sheet

Area: about 362 million sq km or nearly 71% of the Earth's surface.
Average Depth:: 3, 720 m
Deepest point: 10033 min the Mariana Trench in the western Pacific.
Mountains: The ocean ridges form a great mountain range, almost 64, 000 km long, that weaves its way through all the major oceans. It is the largest single feature on Earth.
Highest Mountain: Mauna Kea, Hawaii, rises 10, 203 m from its base on the ocean floor; only 4, 170 m/are above sea level.

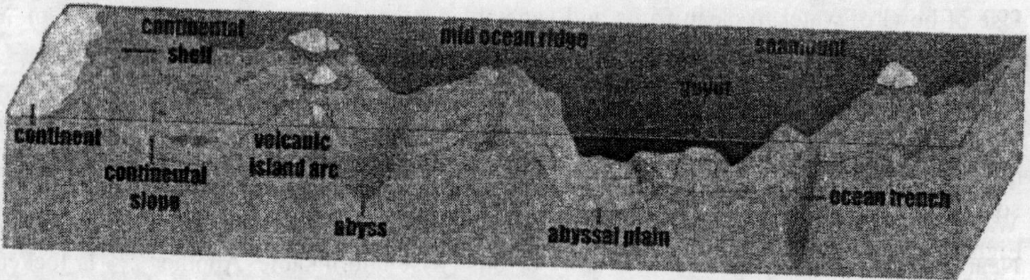

Some more facts about our oceans are listed below:

- The oceans occupy nearly 70% of our planet's surface
- More than 97% of all our planet's water is contained in the ocean
- The top 3 metres of the ocean hold as much heat as our entire atmosphere
- The average depth of the ocean is about 4 kms
- The oceans provide 99 percent of the Earth's living space- the largest space in our universe known to be inhabited by living organisms
- More than 90% of this habitat exists in the deep sea known as the abyss
- Mount Everest (the highest point on the Earth's surface 8.8 kms is more than 1.6 kms shorter than the Challenger Deep (the deepest point in the ocean at 11 kms)

- The longest continuous mountain chain known to exist in the Universe resides in the ocean at more than 64000 kms long
- The Antarctic ice sheet that forms and melts over the ocean each year is nearly twice the size of the United States
- Water pressure at the deepest point in the ocean is more than 8 tons per square inch, the equivalent of one person trying to hold 50 jumbo jets.
- The Gulf Stream off the Atlantic seaboard of the United States flows at a rate nearly 300 times faster than the typical flow of the Amazon river, the world's largest river
- The worlds oceans contain nearly 20 million tons of gold
- The color blue is least absorbed by seawater; the same shade of blue is most absorbed by microscopic plants, called phytoplankton, drifting in seawater
- A new form of life, based on chemical energy rather than light energy, resides in deep-sea hydrothermal vents along mid-ocean ridges
- A swallow of seawater may contain millions of bacterial cells, hundreds of thousands of phytoplankton and tens of thousands of zooplankton
- The blue whale, the largest animal on our planet ever (exceeding the size of the greatest dinosaurs) still lives in the ocean
- The gray whale migrates more than 16000 kms each year, the longest migration of any mammal
- The Great Barrier Reef, measuring 12000 kms, is the largest living structure on Earth. It can be seen from the Moon.
- More than 90 percent of the trade between countries is carried by ships and about half the communications between nations use underwater cables
- Lot of oil reaches the oceans each year as a result of leaking automobiles and other non-point sources
- Fish supply the greatest percentage of the world's protein consumed by humans
- Most of the world's major fisheries are being fished at levels above their maximum sustainable yield; some regions are severely overfished
- Eighty per cent of all pollution in seas and oceans comes from land-based activities.
- Three-quarters of the world's mega-cities are by the sea.
- By 2020, 80 per cent of people will live within 100 kms of the coast.
- Death and disease caused by polluted coastal waters costs the global economy heavily and. the annual economic impact of hepatitis from tainted seafood alone is
- Plastic waste kills up to 1 million sea birds, 100. 000 sea mammals and countless fish each year. Plastic remains in our ecosystem for years harming thousands of sea creatures everyday.
- Over the past decade, an average of 600, 000 barrels of oil a year has been accidentally spilled from ships.
- Tropical coral reefs border the shores of 109 countries, the majority of which are among the world's least developed. Significant reef degradation has occurred in 93 countries.
- Although coral reefs comprise less than 0.5 per cent of the ocean floor, it is estimated that more than 90 per cent of marine species are directly or indirectly dependent on them.

- There are about 4, 000 coral reef fish species worldwide, accounting for approximately a quarter of all marine fish species.
- Nearly 60 per cent of the world's remaining reefs are at significant risk of being lost in the next three decades.
- The major causes of coral reef decline are coastal development, sedimentation, destructive fishing practices, pollution, tourism and global warming.
- Less than one half a per cent of marine habitats are protected – compared with 11.5 per cent of global land area.
- The High Seas – areas of the ocean beyond national jurisdiction – cover almost 50 per cent of the Earth's surface. They are the least protected part of the world.
- Although there are some treaties that protect ocean-going species such as whales, as well as some fisheries agreements, there are no protected areas in the High Seas.
- Protecting critical marine habitats – such as warm-and cold-water coral reefs, seagrass beds and mangroves – can dramatically increase fish size and quantity.
- More than 3.5 billion people depend on the ocean for their primary source of food. In 20 years, this number could double to 7 billion.
- Populations of commercially attractive large fish, such as tuna, cod, swordfish and marlin have declined by as much as 90 per cent in the past century.
- As many as 100 million sharks are killed each year for their meat and fins, which are used for shark fin soup. Hunters typically catch the sharks, de-fin them while alive and throw them back into the ocean where they either drown or bleed to death.
- The annual global by-catch mortality of small whales, dolphins and porpoises alone is estimated to be more than 300, 000 individuals.

Coastlines

The total length of the world's coastlines is about 5, 00, 000 kms, enough to circle the Equator 12 times.

As coastal zones become more and more crowded, the quality of coastal water will suffer, the wildlife will be displaced, and the shorelines will erode. 60% of the Pacific and 35% of the Atlantic Coast shoreline are eroding at a rate of a meter every year.

More than half the world's population live within a 100 km distance from the coast. This is more than 2.7 billion people. Rapid urbanization will lead to more coastal mega-cities containing 10 million or more people. By the end of the millennium 13 out of 15 of the world's largest cities will be located on or near the coast. Growing population in coastal areas leads to more marine pollution and distribution of coastal habitats. Some 6, 5 million tons of litter finds its way into the sea each year.

Fisheries

The sea provides the biggest source of wild or domestic protein in the world. Each year 75 million tons of fish are caught in the ocean. Of this amount around 29 million tons is for human consumption. The global fish production exceeds that of cattle, sheep, poultry or eggs. Fish can be produced in two ways: by capture and by aqua culture.

The largest numbers of fish are located in the Southern Hemisphere due to the fact that these waters are not largely exploited by man.

Fifteen out of seventeen of the world's largest fisheries are so heavily exploited that the reproduction can't keep up. With the result that many fish populations are decreasing rapidly.

Species of fish endangered by overfishing are: tuna, salmon, haddock, halibut, and cod.

Rising Sea Level

The sea level has risen with an average of 10 to 25 cm over the past 100 years and scientists expect this rate to increase. Sea levels will continue rising even if the climate has stabilized, because the ocean reacts slowly to changes.

10, 000 years ago the ocean level was about 110 mtr lower than it is now.

If all the world's ice melted, the oceans would rise mtr

Volcanic Activity

90% of all volcanic activity on Earth occurs in the ocean. The largest known concentration of active volcanoes (approximately 1, 133) on the sea floor is located in the South Pacific

Density

The density of ocean water varies. It becomes more dense as it becomes colder, right down to $_{o}$its freezing point of -1.9 °C. *(This is unlike fresh water, which is most dense at 4 °C, well above its freezing point.)*

Water Temperature

Under the enormous pressures of the deep ocean, sea water can reach very high temperatures without boiling. A water temperature of 400 °C has been measured at one hydrothermal vent.

The average temperature of all ocean water is about 3.5 °C.

Almost all of the deep ocean temperatures are only a little warmer than freezing

Ice

Antarctica has as much ice as the Atlantic Ocean has water.

10% of the earth's surface is covered with ice.

The Arctic Ocean is the smallest ocean, holding only one percent of the Earth's seawater. This is still more than 25 times as much water as all rivers and fresh water lakes.

The average thickness of the Arctic ice sheet is about 3 metres, although there are some areas as thick as 20 metres.

In the unlikely event that all the polar ice were to melt, the sea level all over the world would rise 180 meters. As a result, 85 to 90% of the Earth's surface would be covered with water as compared to the current 71%.

The Arctic produces 10, 000 to 50, 000 icebergs annually. The amount produced in the Antarctic regions is inestimable. Icebergs normally have a four-year life-span; they begin entering shipping lanes after about three years.

Carbon Dioxide Absorption

Oceans absorb between 30% and 50% of the carbon dioxide produced by burning fossil fuel. Carbon dioxide is transported downwards by plankton. Any change in the temperature of the ocean water, influences the ability of plankton to take up carbon dioxide. This has consequences for the ecosystem, because plankton form the base of the food web.

Reefs

Over 60% of the world's coral reefs are threatened as a result of pollution, sedimentation and bleaching due to rising water temperatures caused by global warming. Currently 27% of all coral reef worldwide has disappeared and around 2050 only 30% will be left.

Rubbish/Contamination

In one year, three times as much rubbish is dumped into the world's oceans as the weight of fish caught.

Gold

If all the gold suspended in the world's seawater were mined, each person on Earth could have about 4 kgs of gold.

Oil

Oil is one of the ocean's greatest resources. It gives us heat for our homes, endless consumer products, and the ability to run the engines of cars, planes, and boats for auto transport all over the world. Nearly one-third of the world's oil comes from offshore fields in our oceans which, as can have devastating effects on our ocean's ecosystems. The transport of ocean oil from the Arabian Gulf, The North Sea, and the Gulf of Mexico reaches all corners of the globe on a daily basis.

Oil was also borne from the sea. Millions of years ago, countless marine microscopic plants (phytoplankton) and animals (zooplankton) lived in the ancient seas as they do today. As they died, the skeletal remains of these tiny organisms settled to the sea floor, mixed with mud and silt, and over millions of years, formed organic-rich sedimentary layers. Other sediments continued to be deposited and further buried the organic-rich sediment layer to depths of thousands of feet, compressing the layers into a rock that would become the source for oil.

Over the years, as the depth of the burial increased, pressure increased, along with the temperature. Under such conditions, and over long periods of time, the original skeletal remains of phytoplankton and zooplankton changed, breaking down into simpler substances called hydrocarbons - compounds of hydrogen and carbon. This process still continues, although it will be millions of years before the next batch of oil is done cooking.

Salinity

The oceans contain as much as 50 quadrillion tons (50 million billion tons=50, 000, 000, 000, 000, 000) of dissolved solids. If the salt in the ocean could be removed and spread evenly over the Earth's land surface it would form a layer more than

The ocean's principal dissolved solids are sodium salts (sodium chloride or common salt), calcium salts (calcium carbonate or lime, and calcium sulfate), potassium salts (potassium sulfate), and magnesium salts (magnesium chloride, magnesium sulfate, and magnesium bromide).

calcium salts (calcium carbonate or lime, and calcium sulfate), potassium salts (potassium sulfate), and magnesium salts (magnesium chloride, magnesium sulfate, and magnesium bromide).

Atlantic sea water is heavier than Pacific sea water due to its higher salt content.

The freezing point of sea water depends on its salt content. Typical ocean water has about 35 grams of salt per liter and freezes at -19 °C.

Appendix 1.95: Elemental Analysis of Sea Water

S.No.	Elemental constituents of sea water	Concentration in sea water, mg/kg sea water
1.	Chlorine	18980
2.	Sodium	10561
3.	Magnesium	1277
4.	Sulphur	884
5.	Calcium	400
6.	Potassium	380
7.	Bromine	65
8.	Caron	28
9.	Strontium	13
10.	Boron	4.6
11.	Fluorine	4.6
12.	Silicum	4.0
13.	Other 28 elements	<1

Note:

The ocean remains one of the most expansive, mysterious and diverse places on Earth. Unfortunately, it is being threatened by pollution from people on land and from natural causes. Marine life is dying, and as a result the whole oceanic ecosystem is threatened simply by various sources of pollution.

Before, it was assumed that because the ocean was so big, vast and deep, that the effects of dumping trash and litter into the sea would only have minimal consequences, this has proven to not be the case. While all four oceans have suffered as a result of human consequence, it has accelerated in the past few decades. Oil spills, toxic wastes, floating plastic and various other factors have all contributed to the pollution of the ocean.

Fact Sheet of Ocean Pollution

Plastic is the most common element that is found in the ocean. It is harmful for the environment as it does not get break down easily and is often considered as food by marine animals.

The biggest source of pollution in the ocean is directly from land based sources, such as oil, dirt, septic tanks, farms, ranches, motor vehicles, among larger sources. Thousands of tons of waste and trash are dumped into the ocean on a daily basis.

Over one million seabirds are killed by ocean pollution each year. Three hundred thousand dolphins and porpoises die each year as a result of becoming entangled in discarded fishing nets, among other items. One hundred thousand sea mammals are killed in the ocean by pollution each year.

Even though much the trash and waste dumped into the ocean is released hundreds of miles away from land, it still washes up on beaches and coastal areas, and affects everything in between. Every marine animal is affected by man-made chemicals released in the water.

There is an island of garbage twice the size of Texas inside the Pacific Ocean: the North Pacific Gyre off the coast of California is the largest oceanic garbage site in the entire world. There, the number of floating plastic pieces outnumbers total marine life six to one in the immediate vicinity.

Oil is the fastest source of deterioration to the ocean, being far more harmful than trash and waste. However, only a small percentage of oil (around 12%) dumped in the ocean comes as a result of actual oil spills. Most oil causing harm in the ocean is a result of drainage from land. Oil spills suffocate marine life to death, and leads to behavioral changes and a breakdown in thermal insulation to those that do survive.

It essentially changes the entire ecosystem of an affected area, such as a long coastline or deep ocean.

Toxic metals can destroy the biochemistry, behavior, reproduction, and growth in marine life.

Plastic debris can absorb toxic chemicals from ocean pollution, therefore poisoning whatever eats it. In facf, plastic pollution is one of the most serious threats to the ocean. Plastic does not degrade; instead, it breaks down into progressively smaller pieces, but never disappears. They then attract more debris. It poses a significant health threat to the various sea creatures, and to the. entire marine ecosystem. Overall, plastic is the pumber one source of pollution in the ocean.

Not all sources of contamination in the ocean come from just oil, trash and solid wastes. The dumping of radioactive waste from nuclear reactors, industrial race (such as heavy metals and acids), and drained sewage are also heavy contributors to pollution. Billions of tons of litter end up in the ocean each year, and it is substantially more than the 250 million tons of trash generated. This has led to a gradual loss in marine life and an increase in the number of endangered species. Littering causes pollution in the ocean, which also causes a substantial loss of life beneath the seas.

Sewage leads to the decomposition of organic matter that in turn leads to a change in biodiversity. Even if the ocean's ecosystem isn't destroyed entirely, it is still changed drastically, and usually not for the better.

Fertilizer runoff creates eutrophication that flourishes algal bloom (rapid increase or accumulation in the population of algae in aquatic systems) which depletes the oxygen content in the water that affects marine life.

Small animals at the bottom of food chain absorb the chemicals as part of their food. These small animals are then eaten by larger animals that again increases the concentration of chemicals. Animals at the top of hierarchy of food chain have contamination levels millions times higher than the water in which they live.

People get contaminated easily by eating contaminated seafood that can cause serious health problems, from cancer to damage to immune system.

The garbage like plastic bottles, aluminium cans, shoes, packaging material - if not disposed correctly, can reach the sea and the same garbage can again reach the sea shore where it pollutes beaches and affects local tourism industry.

Salty water of ocean has the capability to move pollutants from the ocean into coastal freshwater making wells and groundwater contaminated.

Chemicals from industries and mines can also enter ocean through land based activities. They can seep through soil, water or land either during their manufacture, use or accidental leaks. From soil, water or land, they can enter into ocean currents and can travel longer distances.

As 70% of the earth is covered with water, people actually assumed that all pollutants would be diluted and get disappeared. But in reality, they have not disappeared and their effects can be easily seen as they have entered the food chain.

Until 1970's, the chemicals and garbage was deliberately dumped into the oceans and became as common practice for disposing everything including pesticides and radioactive waste, assuming that it would get dissolved to safe levels.

In several parts of the world including Gulf of Mexico and the Baltic Sea, Eutrophication has created enormous dead zones.

Till today, in many parts of the world, sewage water is discharged in the ocean - untreated or under-treated. This can cause serious effect on marine and human life and can also lead to eutrophication.

Appendix 1.96: Breakdown of Global Water Reserves and Share of Fresh Water

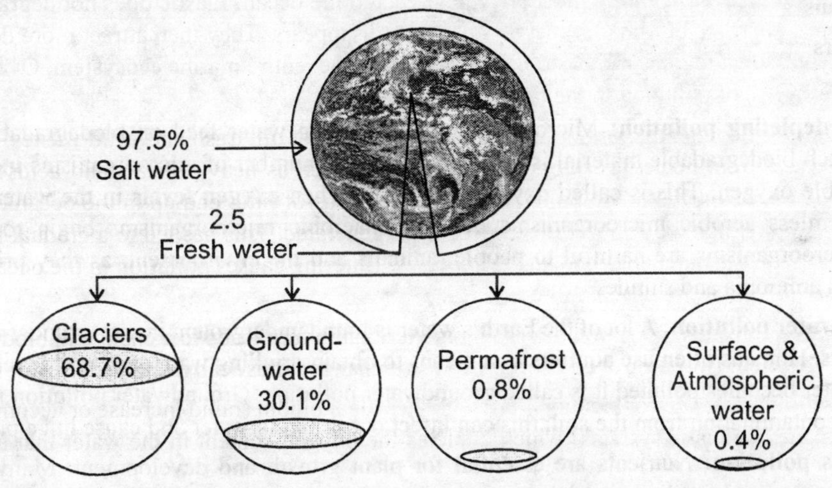

Appendix 1.97: Examples and Sources of Major Categories of Water Pollutants

Category	Examples of water pollutants	Sources of water pollutants
	A. Causes health problems	
1. Infectious agents	Bacteria, viruses, parasites	Human and animal excreta
2. Organic chemicals	Pesticides, plastics, detergents, oil, and gasoline	Industrial, household, and farm use
3.Inorganic chemicals	Acids, caustics, salts, metals	Industrial effluents, household cleansers, surface runoff.
4. Radioactive materials	Uranium, thorium, cesium, iodine, radon	Mining and processing of ores, power plants, weapons production, natural sources

B. Causes ecosystem disruption		
1. Sediment	Soil, silt	Land erosion
2. Plant nutrients	Nitrates, phosphates, ammonium	Agricultural and urban fertilizers, sewage, manure
3. Oxygen-demanding wastes	Animal manure and plant residues	Sewage, agricultural runoff, paper mills, food processing
4. Thermal	Heat	Power plants, industrial cooling

Note:

Types of Water Pollution

Surface water pollution: Surface waters are the **natural water** resources of the Earth. They are found on the exterior of the Earth's crust and include:

- Oceans
- Rivers
- Lakes

Oxygen depleting pollution: Microorganisms that live in water feed on biodegradable substances. When too much biodegradable material is added to water, the number of microorganisms increase and use up the available oxygen. This is called oxygen depletion. When **oxygen levels** in the water are depleted, relatively harmless aerobic microorganisms die and anaerobic microorganisms begin to thrive. Some anaerobic microorganisms are harmful to people, animals and the environment, as they produce harmful toxins such as ammonia and sulfides

Groundwater pollution: A lot of the Earth's water is found underground in soil or under rock structures called aquifers. Humans often use aquifers as a means to obtain drinking water, and build wells to access it. When this water becomes polluted it is called groundwater pollution. Groundwater pollution is often caused by **pesticide** contamination from the soil; this can infect our drinking water and cause huge problems.

Nutrients pollution: Nutrients are essential for plant growth and development. Many nutrients are found in wastewater and fertilizers, and these can cause excess weed and algae growth if large concentrations end up in water.

- This can contaminate drinking water and clog filters.
- This can be damaging to other aquatic organisms as the algae use up the oxygen in the water, leaving none for the surrounding marine life.

Microbiological pollution: Microbiological water pollution is usually a natural form of water pollution caused by microorganisms.

Many types of microorganisms live in water and cause fish, land animals and humans to become ill.

Microorganisms

- Bacteria Viruses
- Protozoa

Serious diseases such as cholera come from microorganisms that live in water. These diseases usually affect the **health of** people in poorer countries, as they do not have the facilities to treat polluted water.

Suspended matter pollution: Some pollutants do not dissolve in water as their molecules are too big to mix between the water molecules. This material is called particulate matter and can often be a cause of water pollution.

- The suspended particles eventually settle and cause thick silt at the bottom. This is harmful to marine life that lives on the floor of rivers or lakes.
- Biodegradable substances are often suspended in water and can cause problems by increasing the amount of anaerobic microorganisms present.
- Toxic chemicals suspended in water can be harmful to the development and survival of aquatic life.

Chemical pollution: Industrial and agricultural work involves the use of many different chemicals that can run-off into water and pollute it.

- Metals and solvents from industrial work can pollute rivers and lakes. These are poisonous to many forms of aquatic life and may slow their development, make them infertile or even result in death
- **Pesticides** are used in farming to control weeds, insects and fungi. Run-offs of these pesticides can cause water pollution and poison aquatic life. Subsequently, birds, humans and other animals may be poisoned if they eat infected fish.
- Petroleum is another form of chemical pollutant that usually contaminates water through **oil spills** when a ship ruptures. Oil spills usually have only a localized effect on wildlife but can spread for miles. The oil can cause the death of many fish and stick to the feathers of seabirds causing them to lose the ability to fly.

Causes of Water Pollution

Sewage and wastewater: Domestic households, industrial and agricultural practices produce wastewater that can cause pollution of many lakes and rivers.

- **Sewage** is the term used for wastewater that often contains feces, urine and laundry waste.
- There are billions of people on Earth, so treating sewage is a big priority.
- Sewage disposal is a major problem in developing countries as many people in these areas don't have access to sanitary conditions and clean water.
- Untreated sewage water in such areas can contaminate the environment and cause diseases such as diarrhea.
- Sewage in developed countries is carried away from the home quickly and hygienically through sewage pipes.
- Sewage is treated in water **treatment plants** and the waste is often disposed into the sea.
- Sewage is mainly biodegradable and most of it is broken down in the environment.
- In developed countries, sewage often causes problems when people flush chemical and pharmaceutical substances down the toilet. When people are ill, sewage often carries harmful **viruses and bacteria** into the environment causing health problems.

Marine dumping of industrial waste: Dumping of litter in the sea can cause huge problems. Litter items can get caught in marine animals and may result in death. Different items take different lengths of time to degrade in water:

- Cardboard: Takes 2 weeks to degrade.
- Newspaper: Takes 6 weeks to degrade.

- Photodegradable packaging: Takes 6 weeks to degrade.
- Foam: Takes 50 years to degrade.
- Styrofoam: Takes 80 years to degrade.
- Aluminum: Takes 200 years to degrade.
- Plastic packaging: Takes 400 years to degrade.
- Glass: It takes so long to degrade that we don't know the exact time.

Radioactive waste dumping: Nuclear waste is produced from industrial, medical and scientific processes that use radioactive material. Nuclear waste can have detrimental effects on marine habitats. Nuclear waste comes from a number of sources:

- Operations conducted by nuclear power stations produce radioactive waste. Nuclear-fuel reprocessing plants in northern Europe are the biggest sources of man-made nuclear waste in the surrounding ocean. Radioactive traces from these plants have been found as far away as Greenland.
- Mining and refining of uranium and thorium are also causes of marine nuclear waste.
- Waste is also produced in the nuclear fuel cycle which is used in many industrial, medical and scientific processes.

Oil spills: Oceans are polluted by oil on a daily basis from oil spills, routine shipping, run-offs and dumping. Oil spills make up about 12% of the oil that enters the ocean. The rest come from shipping travel, drains and dumping. An oil spill from a tanker is a severe problem because there is such a huge quantity of oil being spilt into one place. Oil spills cause a much localized problem but can be catastrophic to local marine wildlife such as fish, birds and sea otters. Oil cannot dissolve in water and forms a thick sludge in the water. This suffocates fish, gets caught in the feathers of marine birds stopping them from flying and blocks light from photosynthetic aquatic plants.

Underground storage tanks(UST's) leaks: A tank or piping network that has at least 10 percent of its volume underground is known as an **underground storage tank** They often store substances such as

petroleum, that are harmful to the surrounding environment should it become contaminated. Many UST's constructed before 1980 are made from steel pipes that are directly exposed to the environment. Over time the steel corrodes and causes leakages, affecting surrounding soil and groundwater.

Atmospheric deposition: Atmospheric deposition is the pollution of water caused by air pollution.

- In the atmosphere, water particles mix with **carbon** dioxide sulfur dioxide and nitrogen oxides, this forms a weak acid.
- Air pollution **means** that water vapor absorbs more of these gases and becomes even more acidic.
- When it rains the water is polluted with these gases, this is called acid rain.
- When acid rain pollutes marine habitats such as rivers and lakes, aquatic life is harmed *

Global warming: An increase in **water** temperature can result in the death of many aquatic organisms and disrupt many marine habitats. For example, a rise in water temperatures causes coral bleaching of reefs around the world. This is when the coral expels the microorganisms of which it is dependent on. This can result in great damage to coral reefs and subsequently, all the marine life that depends on it. The **rise** in the Earth's water temperature is caused by global warming.

- Global warming is a process where the average global temperature increases due to the greenhouse effect.
- The burning of fossil **fuel** releases greenhouse gasses, such as carbon dioxide, into the atmosphere.

- This causes heat from the sun to get 'trapped' in the Earth's atmosphere and consequently the global temperature rises.

Eutrophication: Eutrophication occurss when the environment becomes enriched with nutrient (eg. N, P, K, etc.) This can be a problem in marine habitats such as lakes as it can cause algal blooms.

- Fertilizers are often used in farming, sometimes these fertilizers run-off into nearby water causing an increase in nutrient levels.
- This causes phytoplankton to grow and reproduce more rapidly, resulting in algal blooms.
- This bloom of algae disrupts normal ecosystem functioning and causes many problems.
- The algae may use up all the oxygen in the water, leaving none for other marine life. This results in the death of many aquatic organisms such as fish, which need the oxygen in the water to live.
- The bloom of algae may also block sunlight from photosynthetic marine plants under the water surface.
- Some algae even produce toxins that are harmful to higher forms of life. This can cause problems along the food chain and affect any animal that feeds on them.

Dangerous Effects of Water Pollution

On Health: Virtually all types of water pollution are harmful to the health of humans and animals. Water pollution may not damage our health immediately but can be harmful after long term exposure. Different forms of pollutants affect the health of animals in different ways:

- Heavy metals from industrial processes can accumulate in nearby lakes and rivers. These are toxic to marine life such as fish and shellfish, and subsequently to the humans who eat them. Heavy metals can slow development; result in **birth defects** and some are carcinogenic.
- Industrial waste often contains many toxic compounds that damage the health of aquatic animals and those who eat them. Some of the toxins in industrial waste may only have a mild effect whereas other can be fatal. They can cause immune suppression, reproductive failure or acute poisoning.
- Microbial pollutants from sewage often result in infectious diseases that infect aquatic life and terrestrial life through drinking water. Microbial water pollution is a major problem in the developing world, with diseases such as cholera and typhoid fever being the primary cause of infant mortality.
- **Organic matter** and nutrients causes an increase in aerobic algae and depletes oxygen from the water column. This causes the suffocation of fish and other aquatic organisms.
- Sulfate particles from acid rain can cause harm the health of marine life in the rivers and lakes it contaminates, and can result in mortality.
- Suspended particles in freshwater reduces the quality of drinking water for humans and the aquatic environment for marine life. Suspended particles can often reduce the amount of sunlight penetrating the water, disrupting the growth of photosynthetic plants and microorganisms.

On Economy: Water pollution can be damaging to the economy as it can be expensive to treat and prevent contamination. Waste that does not break down quickly accumulates in the Earth's waters and eventually makes its way to the oceans.

Water pollution can be prevented by stopping pollutants from contaminating nearby waters. There are a **number** of water treatments to prevent pollution-

These simple techniques cost money to maintain, but **prevention** is much cheaper than cleaning up water pollution that has already occurred. The cost of a pollution clean-up depends on many factors:

- The location of the pollution is important in determining how much the clean-up will cost. If the contamination is in an area that is easy to get to, then the clean-up cost will be cheaper.
- The contamination size also needs to be considered, the larger the area of contamination, the more expensive the cost of the clean-up.
- The type of pollutant may also have an effect on the clean-up cost, some pollutants are more difficult to clean up than others, and therefore more expensive.

On Environment: There are many different types of water pollution and all have a different adverse effect on the environment.

- Heavy metals from industrial processes can accumulate in nearby lakes and rivers. These are toxic to marine life such as fish and shellfish, and can affect the rest of the food chain. This means that entire animal communities can be badly affected by this type of pollutant.
- Industrial waste often contains many toxic compounds that **damage** the health of aquatic animals and those who eat them. Some toxins affect the reproductive success of marine life and can therefore disrupt the community structure of an aquatic environment.
- Microbial pollutants from sewage often result in infectious diseases that infect aquatic life and terrestrial life through drinking water. This often increases the **number** of mortalities seen within an environment.
- Organic matter and nutrients causes an increase in aerobic algae and depletes oxygen from the water column. This is called eutrophication and causes the suffocation of fish and other aquatic organisms.
- Sulfate particles from acid rain change the pH of water making it more acidic, this damages the health of marine life in the rivers and lakes it contaminates, and often increases the numberof mortalities within an environment.
- Suspended particles can often reduce the amount of sunlight penetrating the water, disrupting the growth of photosynthetic plants and micro-organisms. This has subsequent effects on the rest of the aquatic community that depend on these organisms to survive.

Pervention of Water Pollution

We can prevent water pollution of *nearby* rivers and lakes as well as groundwater and drinking water by following some simple guidelines in our everyday life.

- Conserve water by turning off the tap when running water is not necessary. This helps prevent water shortages and reduces the **amount** of contaminated water that needs treatment.
- Don't throw paints, oils or other forms of litter down the drain.
- Use environmentally household products, such as **washing powder**, household cleaning agents and toiletries.

Do not overuse pesticides and fertilizers. This will prevent runoffs of the material into nearby water sources.

- By having more plants in our garden we are preventing fertilizer, pesticides and contaminated water from running off into nearby water sources.
- Don t throw litter into rivers, lakes or oceans. Help clean up any litter you see on beaches or in rivers and lakes, make sure it is safe to collect the litter and put it in a nearby dustbin.

Appendix 1.98: Utilization of Solid Waste Generated by Industries

S.No.	Waste generated by industries	Areas of application of solid waste
1.	Flyash (fine coal ash generated by combustion of coal in power plants etc. One portion is the bottom ash another is the one collected in the separators from the flue gases.)	i. As raw material in manufacturing of cement ii. As binding material with cement iii. As filler in mines iv. As plasticizer v. As an aggregate in cellular concrete bricks and blocks vi. For stabilization of soil
2.	Blast furnace slag	i. Manufacturing slag cement ii. Making expansive cement, coloured cement and high early-strength cement iii. In refractory and ceramic industry iv. As a structural fill v. As aggregate in concrete.
3.	Ferro-alloy and other metallurgical slag	i. As structural fill ii. In making pozzolana metallurgical cement
4.	Byeproduct gypsum	i. Gypsum plaster ii. As retarder in cement setting iii. As mineraliser
5.	Lime sludge (phos-phochalk, paper and sugar sludges)	i. As a sweetener for lime in cement manufacture ii. Manufacturing of lime pozzolana brick/ binders iii. For recycling in parent industry iv. Manufacturing of building lime
6.	Chromium sludge	i. As a raw material component in cement sludge

Appendix 1.99: Key Routes of Chemical Absorption, Distribution and Excretion in Human Body

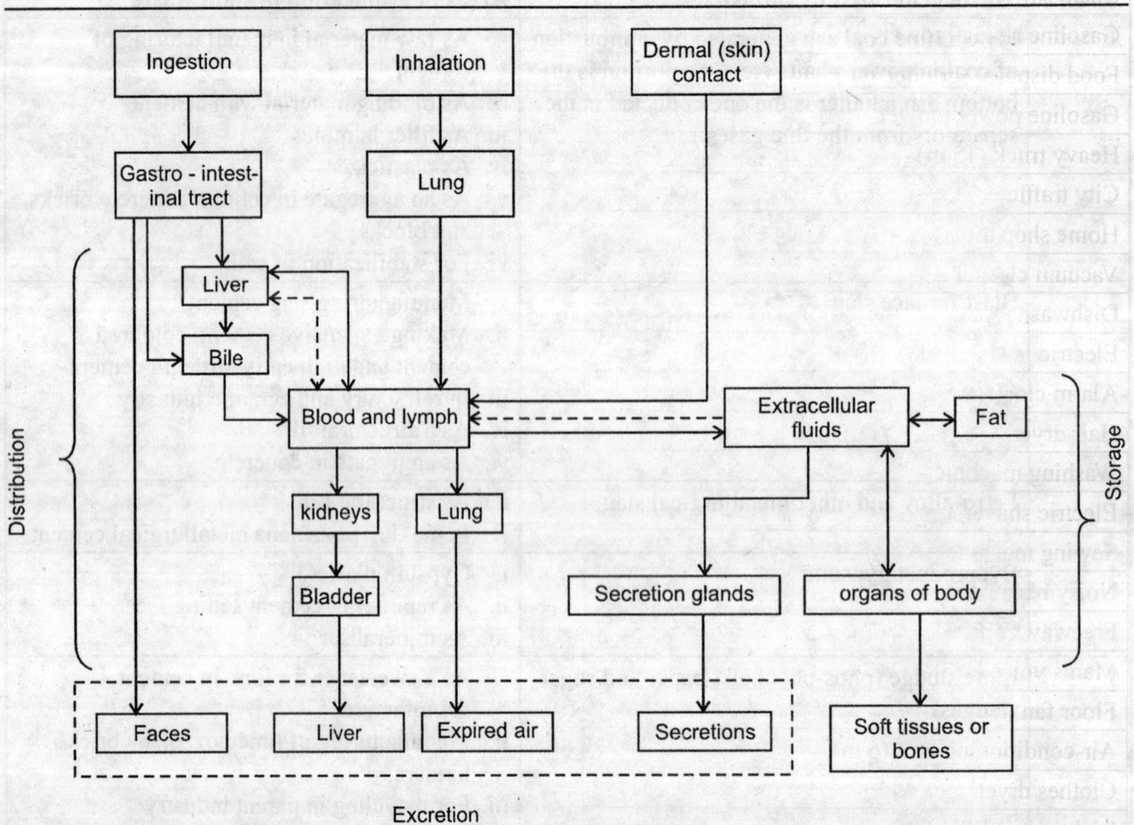

Appendix 1.100: Sound Levels and Human Response to Various Sound Levels

Common Sounds encountered	Noise Level (dB)	Effect on human being
Carrier deck jet operations	140	Painfully loud
Air raid siren	130	
Jet takeoff (65m)	120	Maximum vocal effort
Thunderclap		
Discoteques		
Auto horn (1m)		
Stereo	Up to 120	
Pile drivers	110	

Common Sounds encountered	Noise Level (dB)	Effect on human being
Garbage truck	100	
Chain saw		
Gasoline riding mower	90-95	
Food disposal (grinder)	67-93	
Gasoline power mower	87-92	
Heavy truck (15 m)	90	Very annoying
City traffic		Hearing damage (8 h)
Home shop tools	85	
Vacuum cleaner	62-85	
Dishwasher	54-85	
Electric lawn edger	81	
Alarm clock (0.5 m)	80	Annoying
Hair dryer		
Washing machine	47-78	
Electric shaver	75	
Sewing machine	64-74	
Noisy restaurant	70	Telephone use difficult
Freeway traffic		
Man's voice (1m)		
Floor fan	38-70	Intrusive
Air-conditioning unit (6 m)	60	
Clothes dryer	55	
Light auto traffic (30 m)	50	Quiet
Living room	40	
Bedroom		
Quiet office		
Refrigerator		
Library	30	Very quiet
Soft whisper (15 ft)		
Broadcasting studio	20	
	10	Just audible
	0	Hearing begins

Appendix 1.101: Typical Sound Pressure Level (SPL) in Decibels (DB) in various Environments

Pressure sound level in dB (decibel)	Environment
140	Threshold of pain
130	Pneumatic drill
120	Loud motor car horn (distance 1 m)
110	Rush hour traffic freely flowing
100	Inside tube train
90	Inside motor bus
80	Average traffic on street corner
70	Conversational speech
60	Typical business office
50	Living room, suburban area
40	Library
30	Bedroom at night
20	Broadcasting studio
10	Threshold of hearing
0	

Appendix 1.102: Sound Pressure Levels (Decibels) of Various Industrial Noise Sources

Industrial Noise Sources	Sound level (db)	Speaking Effort Required
Pneumatic chipper (at 5 ft)	115	Nearly impossible to communicate by voice
Chain saw (at ear) Teenage rock n roll band Riveting machine Nail machine Casting shakeout area Punch press Forging hammer	110	Very difficult to communicate by voice
	105	Shout with hands cupped between mouth and other person's ear
Pueumatic air hoist: 4000 lb		
	100	Shout at 1/2 foot
Tumbler 6in × 3 in, small castings		
	95	Shout at 1 foot
Automatic screw machine Nut blanking	90	normal voice at 1/2 foot, shout at 2 feet
Boiler room Arc welder	85	Normal voice at 1 foot, shout at 4 feet
Milling machine (at 4 ft)	80	Normal voice at 1-1/2 feet, shout at 6 feet
Inside a car (50 mph)	75	Normal voice at 2 feet, shout at 8 feet

Telephone use impossible (80 dB and above)

Appendix 1.103: OSHA's Table of Permissible Exposure Level (PEL) For Continuous Noise (in decibels)

A-Weighted Sound Level, dB (A)	Maximum permissible duration of exposure Time (hr)
80	32
81	27.9
82	24.3
83	21.1
84	18.4
85	16
86	13.9
87	12.1
88	10.6
89	9.2
90	8
91	7.0
92	6.2
93	5.3
94	4.6
95	4
96	3.5
97	3.0
98	2.6
99	2.3
100	2
101	1.7
102	1.5
103	1.4
104	1.3
105	1
106	0.87
107	0.76
108	0.66
109	0.57
110	0.50
111	0.44

A-Weighted Sound Level, dB (A)	Maximum permissible duration of exposure Time (hr)
112	0.38
113	0.33
114	0.29
115	0.25
116	0.22
117	0.19
118	0.16
119	0.14
120	0.125
121	0.110
122	0.095
123	0.082
124	0.072
125	0.063
126	0.054
127	0.047
128	0.041
129	0.036
130	0.031

OSHA: Occupational safety and Health Administration, U.S.A.

Appendix 1.104: Classification of Welding Fumes and Gases by Hazards (OSHA – U.S.A)

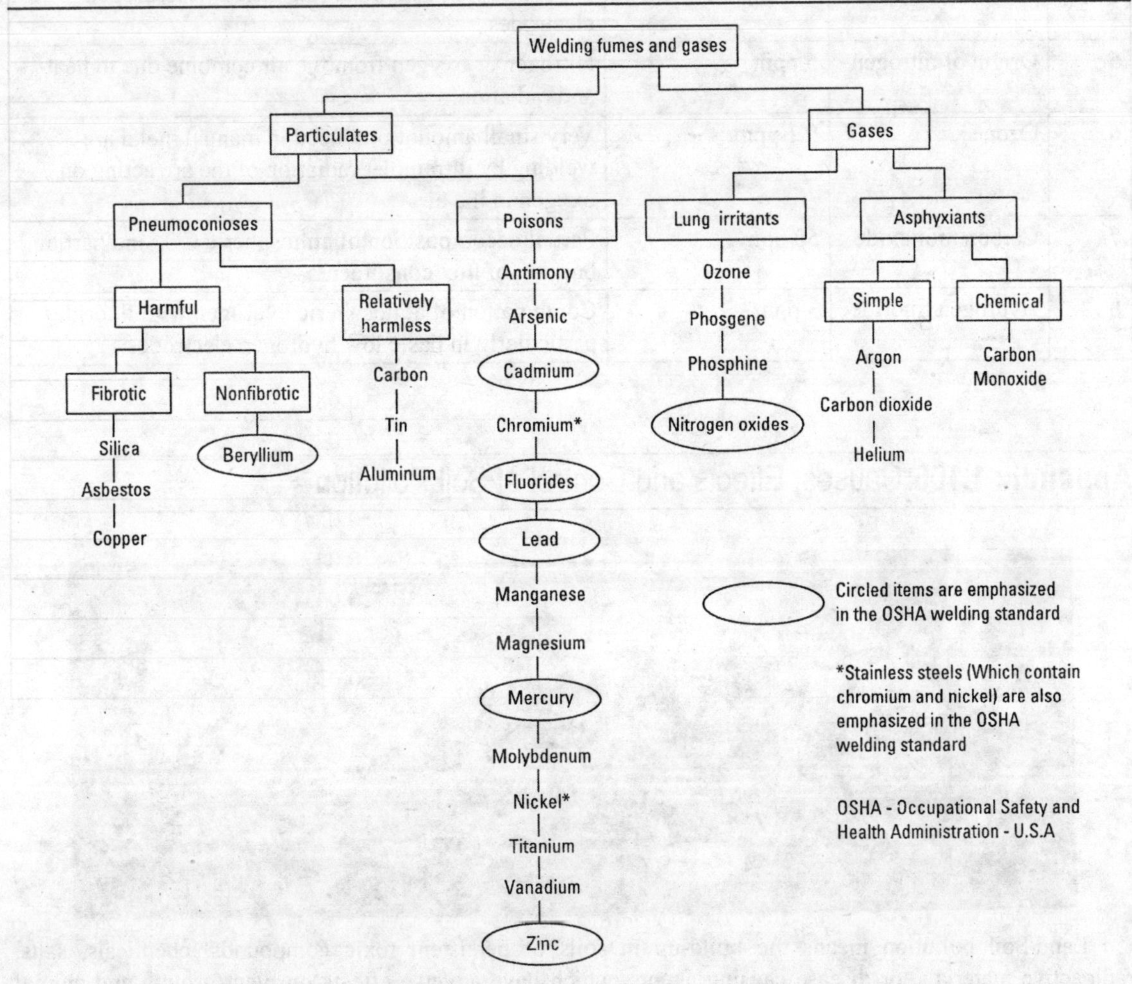

Appendix 1.105: Toxic Components of Welding Fumes – their TLV & origin.

S.No.	Toxic components of welding fumes	Threshold limit values (TLV)	Origin of the toxic constituents
1.	Origin of iron	10mg/m^3	Molten iron from electrode care wire, iron particbs in the flux s from the workplace.
2.	Origin of zinc	5 mg/m^3	Molten zinc from surface coatings s from electrode flux.

3.	Origin of manganese	5 mg/m³	Present in small quantities in mild steel and in largor quantity in higher tensile s alloy steels.
4.	Inorganic fluorides	2.5 mg/m³	Present as flux constituent in basic low hydrogen electrodes .
5.	Origin of nitrogen	5 ppm	Nitrogen & oxygen from the air combine due to heat and radiation.
6.	Ozone	0.1 ppm	Very small amount produced in manual metal are welding by ultrainolet radiation of the are acting on oxygen in the air.
7.	Carbon monoxide	50 ppm	Partial decomposition of atmospheric CO_2 and partial burning of flux constituents .
8.	Hydrogen fluoride	3 ppm	Combination of atmospheric hydrogen with fluorides particularly in basic low hydrogen electrodes.

Appendix 1.106: Causes, Effects and Control of Soil Pollution

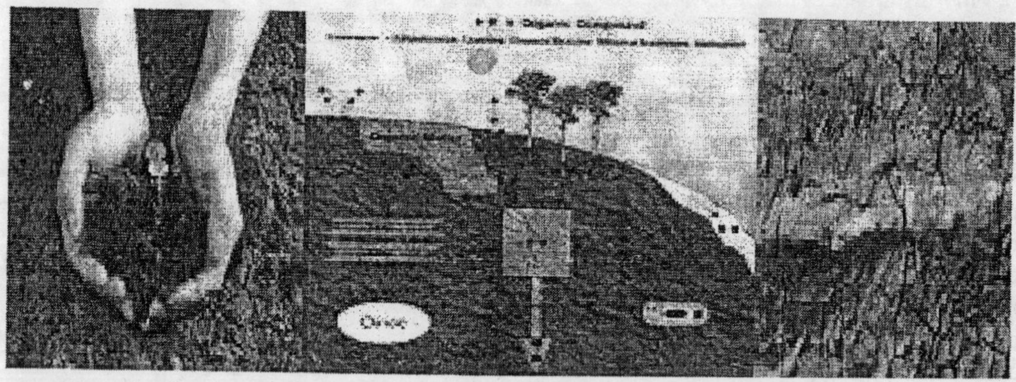

Land/Soil pollution means the build-up in soils of persistent toxic compounds, chemicals, salts, radioactive materials, or disease causing agents, which have adverse effects on plant growth and animal health. Soil pollution is the degradation of earth's land surfaces caused by human activities and their misuse of land resources. Urbanisation and industrialisation are major causes of soil pollution.

Soil is the thin layer of organic and inorganic materials that covers the Earth's rocky surface. The organic portion, which is derived from the decayed remains of plants and animals, is concentrated in the dark uppermost topsoil. The inorganic portion made up of rock fragments, was formed over thousands of years by physical and chemical weathering of bedrock. Productive soils are necessary for agriculture to supply the world with sufficient food.

Different ways by which soil can become polluted are:

- Improper disposal of waste and exploitation of minerals
- Seepage from a landfill
- Discharge of industrial waste into the soil

- Percolation of contaminated water into the soil
- Rupture of underground storage tanks
- Excess application of pesticides, herbicides and fertilizers
- Solid waste seepage
- Improper use of soil by inadequate agricultural practices

The most common chemicals involved in causing soil pollution are:

- Petroleum hydrocarbons
- Heavy metals
- Pesticides
- Solvents

Types of Soil Pollution:

a. Agricultural Soil Pollution
   i. pollution of surface soil
   ii. pollution of underground soil
b. Soil pollution by industrial effluents and solid wastes:
   i. pollution of surface soil
   ii. disturbances in soil profile
c. Pollution due to urban activities:
   i. pollution of surface soil
   ii. pollution of underground soil

Causes of Soil Pollution

Soil pollution is caused by the presence of man-made chemicals or other alteration in the natural soil environment. This type of contamination typically arises from the rupture of underground storage links, application of pesticides, percolation of contaminated surface water to subsurface strata, oil tanks and fuel dumping, leaching of wastes from landfills or direct discharge of industrial wastes to the soil. The most common chemicals involved are petroleum hydrocarbons, solvents, pesticides, lead and other heavy metals. The occurrence of this phenomenon is correlated with the degree of industrialization and intensities of chemical usage.

A soil pollutant is any factor which deteriorates the quality, texture and mineral content of the soil or which disturbs the biological balance of the organisms in the soil. Pollution in soil has adverse effect on plant growth.

Pollution in soil is associated with:

A. Indiscriminate use of fertilizers.
B. Indiscriminate use of pesticides, insecticides and herbicides.
C. Dumping of large quantities of solid waste.
D. Deforestation and soil erosion.
E. Pollution due to industrial activities.
F. Pollution due to Urbanisation.

A. Indiscriminate use of fertilizers:

Soil nutrients are important for plant growth and development. Plants obtain carbon, hydrogen and oxygen from air and water. But other necessary nutrients like nitrogen, phosphorus, potassium, calcium, magnesium, sulphur and more must be obtained from the soil. Farmers generally use fertilizers to correct soil deficiencies. Fertilizers contaminate the soil with impurities, which come from the raw materials used for their manufacture. Mixed fertilizers often contain ammonium nitrate (NH_4NO_3), phosphorus as P_2O_5, and potassium as K_2O. For instance, As, PB and Cd present in traces in rock phosphate mineral get transferred to super phosphate fertilizer. Since the metals are not degradable, their accumulation in the soil above their toxic levels due to excessive use of phosphate fertilizers, becomes an indestructible poison for crops.

The over use of fertilizers reduce quantity of vegetables and crops grown on soil over the years. It also reduces the protein content of wheat, maize, grams, etc., grown on that soil. The carbohydrate quality of such crops also gets degraded. Excess potassium content in soil decreases Vitamin C and carotene content in vegetables and fruits. The vegetables and fruits grown on over-fertilized soil are more prone to attacks by insects and disease.

B. Indiscriminate use of pesticides, insecticides and herbicides:

Plants on which we depend for food are under attack from insects, fungi, bacteria, viruses, rodents and other animals, and must compete with weeds for nutrients. To kill unwanted populations living in or on their crops, farmers use pesticides. The first widespread insecticide use began at the end of World War II and included DDT (dichlorodiphenyltrichloroethane) and gammaxene. Insects soon became resistant to DDT and as the chemical did not decompose readily, it persisted in the environment. Since it was soluble in fat rather than water, it biomagnified up the food chain and disrupted calcium metabolism in birds, causing eggshell to be thin and fragile. As a result, large birds of prey such as the brown pelican, ospreys, falcons and eagles became endangered. DDT has been now been banned in most countries. Ironically many of them including USA, still produce DDT for export to other developing nations whose needs outweigh the problems caused by it.

The most important pesticides are DDT, BHC, chlorinated hydrocarbons, organophosphates, aldrin, malathion, dieldrin, etc. The remnants of such pesticides used on pests may get adsorbed by the soil particles, which then contaminate root crops grown in that soil. The consumption of such crops causes the pesticide remnants to enter human biological systems, affecting them adversely.

An infamous herbicide used as defoliant in the Vietnam War called Agent Orange (dioxin), was eventually banned. Soldiers' cancer cases, skin conditions and infertility have been linked to exposure to Agent Orange.

Pesticides not only bring toxic effect on human and animals but also decrease the fertility of the soil. Some of the pesticides are quite stable and their bio-degradation may take weeks and even months.

Pesticides problems such as resistance, resurgence, and health effects have caused scientists to seek alternatives. Pheromones and hormones to attract or repel insects and using natural enemies or sterilization by radiation have been suggested.

C. Dumping of Solid Wastes:

Solid waste includes garbage, domestic refuse and discarded solid materials such as those from commercial, industrial and agricultural operations. They contain increasing amounts of paper, cardboards, plastics, glass, old construction materials, packaging materials and toxic or otherwise hazardous substances. Since a significant amount of urban solid waste tends to be paper and food waste, the majority is recyclable or biodegradable in landfills. Similarly, most agricultural waste is recycled and mining waste is left on site.

The portion of solid waste that is hazardous such as oils, battery metals, heavy metals from smelting industries and organic solvents are the ones we have to pay particular attention to. These can in the long run, get deposited to the soils of the surrounding area and pollute them by altering their chemical and biological properties. They also contaminate drinking water aquifer sources. More than 90% of hazardous waste is produced by chemical, petroleum and metal-related industries and small businesses such as dry cleaners and gas stations contribute as well.

Solid Waste disposal was brought to the forefront of public attention by the notorious Love Canal Case in USA in 1978. Toxic chemicals leached from oozing storage drums into the soil underneath homes, causing an unusually large number of birth defects, cancers and respiratory, nervous and kidney diseases.

D. Deforestation and Soil Erosion:

Soil Erosion occurs when the weathered soil particles are dislodged and carried away by wind or water. Deforestation, agricultural development, temperature extremes, precipitation including acid rain, and human activities contribute to this erosion. Humans speed up this process by construction, mining, cutting of timber, over cropping and overgrazing. It results in floods and cause soil erosion.

Forests and grasslands are an excellent binding materials that keeps the soil intact and healthy. They support many habitats and ecosystems, which provide innumerable feeding pathways or food chains to all species. Their loss would threaten food chains and the survival of many species. During the past few years, lot of vast green land has been converted into deserts. The precious rain forest habitats of South American, tropical Asia and Africa are coming under pressure of population growth and development (especially timber, construction and agriculture) as many believe that a wealth of medicinal substances including a cure for cancer and aids, lie in these forest. Deforestation is slowly destroying the most productive flora and fauna areas in the world, which also form vast tracts of a very valuable sink for CO_2.

E. Pollution due to Industrial Activities:

Industrial activities also are a contributing factor to land pollution. Industrial wastes are the effluents discharged from chemical industries, paper and pulp mills, tanneries, textile mills, steel industries, distilleries, refineries, pesticides and fertilizer industries, pharmaceutical industries, food processing industries, cement industries, thermal and nuclear power plants, mining industries etc.

Thermal power plants generate a large quantity of 'Fly ash'. Huge quantities of these wastes are dumped on land which cause land pollution. Mining also affects ground and surface waters, the aquatic life, vegetation, soils, animals, and the human health. Acid mine drainage can cause damage to streams which in return can kill aquatic life. Mining gas and petroleum also pollutes the land. Petroleum extraction and manufacturing contaminates the soil with bitumen, gasoline kerosene and mining brine solutions. Open cast mining, which is a process where the surface of the earth is dug open to bring out the underground mineral deposits, destroys the topsoil and contaminates the area with toxic metals and chemicals.

F. Pollution due to Urbanisation

Urban activities generate large quantities of city wastes including several Biodegradable materials (like vegetables, animals wastes, papers, wooden pieces, carcasses, plant twigs, leaves, cloth wastes as well as sweepings) and many non-biodegradable materials (such as plastic bags, plastic bottles, plastic wastes, glass bottles, glass pieces, stone/cement pieces). On a rough estimate Indian cities are producing solid city wastes to the tune of 80, 000 metric tons every day. If left uncollected and decomposed, they are a cause of several problems such as:

- Clogging of drains: Causing serious drainage problems including the burst/leakage of drainage lines leading to health problems.
- Barrier to movement of water: Solid wastes have seriously damaged the normal movement of water thus creating problem of inundation, damage to foundation of buildings as well as public health hazards.
- Foul smell: Generated by dumping the wastes at a place.
- Increased microbial activities: Microbial decomposition of organic wastes generate large quantities of methane besides many chemicals to pollute the soil and water flowing on its surface
- When such solid wastes are hospital wastes they create many health problems: As they may have dangerous pathogen within them besides dangerous medicines, injections etc.

Pollution of Underground Soil

Underground soil in cities is polluted by:

- Chemicals released by industrial wastes and industrials wastes.
- Decomposed and partially decomposed materials of sanitary wastes.

Many dangerous chemicals like cadmium, chromium, lead, arsenic, selenium products are likely to be deposited in underground soil. Similarly underground soil polluted by sanitary wastes generate many harmful chemicals.. These can damage the normal activities and ecological balance in the underground soil.

Causes of soil pollution can be summarised as follows

- Polluted water discharged from factories.
- Runoff from pollutants (paints, chemicals, rotting organic material) leaching out of landfill.
- Oil and petroleum leaks from vehicles washed off the road by the rain into the surrounding habitat.
- Chemical fertilizer runoff from farms and crops.
- Acid rain (fumes from factories mixing with rain).
- Sewage discharged into rivers instead of being treated properly.
- Over application of pesticides and fertilizers.
- Purposeful injection into groundwater as a disposal method.
- Interconnections between aquifers during drilling (poor technique).
- Septic tank seepage.
- Sanitary/hazardous landfill seepage.
- Lagoon seepage.
- Cemeteries.
- Scrap yards (waste oil and chemical drainage).
- Leaks from sanitary sewers.

Various sources of land/soil pollution are tabulated below:

S.No.	Sources	Methods of land pollution
1.	Agricultural sources	accumulation of animal manures. excessive input of chemical fertilizers. illicit dumping of tainted crops on land
2.	Mining and Quarrying	using of explosives to blow up mines. use of machineries which emits toxic by products and leaks to the ground.
3.	Sewage sludge	improper sanitation system causes sludge to leak at surrounding soil.
4.	Dredged spoils	improper method of dredging at fertile land causes soil infertility, leaving the soil more prone to external pollution.
5.	Household	improper waste disposal system causes waste accumulation. improper sanitation system.
6.	Demolition and construction	non-biodegradable rubbles/debris which are not cleared settles in the soil.
7.	Industrial sources	Poisonous/toxic gaseous emissions undergo chemical reactions and increase soil toxicity.

Effects of Soil Pollution

a. **Agriculture:**
- Reduced soil fertility.
- Reduced nitrogen fixation.
- Increased erodibility.
- Larger loss of soil and nutrients.
- Deposition of silt in tanks and reservoirs.
- Reduced crop yield.
- Imbalance in soil fauna and flora.
- Destruction of vegetation that provides food and shelter

b. **Industrial:**
- Dangerous chemicals entering underground water.
- Ecological imbalance.
- Release of pollutant gases.
- Release of radioactive rays causing health problems.
- Increased salinity.
- Reduced vegetation.

c. **Urban:**
- Clogging of drains.
- Public health problems.
- Inundation of areas.
- Public health problems.

- Pollution of drinking water sources.
- Foul smell and release of gases.
- Waste management problems.

Long Term Effects of Soil Pollution on Environment

Soil that has been contaminated should not be used to grow food, because the chemicals can leech into the food and harm people who eat it.

If contaminated soil is used to grow food, the land will usually produce lower yields than it would if it were not contaminated. This, in turn, can cause even more harm because a lack of plants on the soil will cause more erosion, spreading the contaminants onto land that might not have been polluted before.

In addition, the pollutants will change the makeup of the soil and the types of microorganisms that will live in it. If certain organisms die off in the area, the larger predator animals will also have to move away or die because they've lost their food supply. Thus it's possible for soil pollution to change whole ecosystems.

Effects of soil pollution are summarised below:

- pollution runs off into rivers and kills the fish, plants and other aquatic life.
- crops and fodder grown on polluted soil may pass the pollutants on to the consumers.
- polluted soil may no longer grow crops and fodder.
- soil structure is damaged (clay ionic structure impaired).
- corrosion of foundations and pipelines.
- impairs soil stability.
- may release vapours and hydrocarbon into buildings and cellars.
- may create toxic dusts.
- may poison children playing in the area.

Pesticides can damage crops; kill vegetation; and poison birds, animals, and fish. Most pesticides kill or damage life forms other than those intended. For example, pesticides used in an effort to control or destroy undesirable vegetation and insects often destroy birds and small animals. Some life forms develop immunity to pesticides used to destroy them.

- Acid rain kills trees and other plants
- Exterminates wild life.

Control of Soil Pollution

The following steps have been suggested to control soil pollution. To help prevent soil erosion, we can limit construction in sensitive area. We will need less fertilizer and pesticides, if we could adopt the three R's: Reuse, and Recycle. This would give us less solid waste.

a. **Reducing chemical fertilizer and pesticide use:**

   Applying bio-fertilizers and manures can reduce chemical fertilizer and pesticide use. Biological methods of pest control can also reduce the use of pesticides and thereby minimize soil pollution.

b. **Reusing of materials:**

   Materials such as glass containers, plastic bags, paper, cloth etc. Can be reused at domestic levels rather than being disposed, reducing solid waste pollution.

c. **Recycling and recovery of materials:**
This is a reasonable solution for reducing soil pollution. Materials such as paper, some kinds of plastics and glass can and are being recycled. This decreases the volume of refuse and helps in the conservation of natural resources. For example, recovery of one tonne of paper can save 17 trees.

d. **Reforesting**
Control of land loss and soil erosion can be attempted through restoring forest and grass cover to check wastelands, soil erosion and floods. Crop rotation or mixed cropping can improve the fertility of the land.

b. **Solid waste treatment**
Proper methods should be adopted for management of solid waste disposal. Industrial wastes can be treated physically, chemically and biologically until they are less hazardous. Acidic and alkaline wastes should be first neutralized; the insoluble material if biodegradable should be allowed to degrade under controlled conditions before being disposed.

As a last resort, new areas for storage of hazardous waste should be investigated such as deep well injection and more secure landfills. Burying the waste in locations situated away from residential areas is the simplest and most widely used technique of solid waste management. Environmental ad aesthetic considerations must be taken into consideration before selecting the dumping sites.

Incineration of other wastes is expensive and leaves a huge residue and adds to air pollution. Pyrolysis is a process of combustion in absence of oxygen or the material burnt under controlled atmosphere of oxygen. It is an alternative to incineration. The gas and liquid thus obtained can be used as fuels. Pyrolysis of carbonaceous wastes like firewood, coconut, palm waste, corn combs, cashew shell, rice husk paddy straw and saw dust, yields charcoal along with products life tar, methyl alcohol, acetic acid, acetone and a fuel gas.

Anaerobic'aerobic decomposition of biodegradable municipal and domestic waste is also being done and gives organic manure. Cow dung which releases methane into the atmosphere, should be processed further in 'gobar gas plants' to produce gobar gas' and good manure.

Natural Land Pollution

Land pollution occurs massively during earth quakes, land slides, hurricanes and floods. All cause hard to clean mess, which is expensive to clean, and may sometimes take years to restore the affected area. These kinds of natural disasters are not only a problem in that they cause pollution but also because they leave many victims homeless.

Appendix 1.107: Sources, Effects and Control of Marine Pollution

The Ocean covers approximately 70% of the earth's total surface area. In the total water content of earth, 97% is present in the Oceans.

Oceans are the main regulatory agent of earth's climate. About 60% of the world's population live within 60KM if coastline and use the coastine for their livelihood. Environmental pollution of the coast, inshore water and deep ocean is one of the important topical issue in the context of human health and global warming.

The major marine pollutants are oil, sewage, garbage, toxic chemicals, pesticides, heavy metal, radioactive waste, thermal pollution, and eutrophication.

Marine pollution means introduction by man, directly/ indirectly of substances / energy into the marine environment resulting in deleterious effects such as harm to living resources, hazards to quality for use of seawater and reduction of amenities.

The continuing environmental degradation of our oceans and costal areas is a matter of concern because it is detrimental to human health, economic development, climate and our planet's store of biodiversity. It is interfering with the sustainability of environment and its resources.

"Biodiversity" means the variability among living organisms from all sources including terrestrial, marine and other aquatic ecosystems and the ecological complexes of which they are part; this includes diversity within species, between species and of ecosystems".

Sustainability is defined as the development that meet the need of the present, without compromising the ability of future generation to meet their own needs.

Sources of Marine Pollution

The main sources are atmosphere, river runoff, agriculture, livestocks, urban runoff, automobiles, land clearing, sewage outfall, industrial waste etc. In general the coastal zone receives pollutants from major ways viz., atmosphere, reverine and glacier. Also anthropogenic activities like discharging the effluents through piped outfalls, direct dumping, operation of ships etc also cause marine pollution.

In the earth's surface, 18% is represented by coastal zone, which provide space for 60% of the world's human population, because, 70% of the world cities with population exceeding 1.6 million are located in this zone. 90% of the world fish catch is obtained from coastal zone. The coastal zone is only 8% in the hydrosphere, but responsible for 18-33% of the total primary production. The biological wealth of this zone is very high as it served as feeding, nursery and spawning grounds with rich biodiversity.

Pollution may arise from point source like a single sewer pipe or factory wastewater outfall, or it can arise from a variety of geographic points called as non point source.

In the point source pollution the concentration of the substance or the intensity of the effect (e.g. temperature near cooling water outfalls of power plants) declines with increasing distance from the point source.

The non-point source effects cannot be attributed to any single. Runoff arises due to rain is an example of a non point source. The toxic substance and other fertilizer based nutrients may spread over the coast.

Indicators of Marine Pollution

Many organisms serve as indicators of marine pollution. For example, the ecological shift in the succession of phytoplankton communities from diatoms to dinoflagellates due to metal pollution and thermal pollution. The poor hatching of eggs and soft calcium shell of birds and mass mortalities of birds due to failure in the earlier metabolites by the interference of pesticides,

Large scale mass killing of birds due to casting, asphyxiation and smothering by oil pollution, reproductive failure in mammal are also some of the indicators of severe pollution.

Usually the pollutants exceeding threshold limits combined with environmental variables such as temperature, salinity, PH, dissolved oxygen, hydrogen sulphide etc., give stress to the organisms.

In most cases the migratory organisms such as fishes could avoid the polluted area.

Whereas the residential or sedentary organisms are exposed to the pollutants

Experience the stress conditions.

Effect of Pollutants on Marine Organisms

Pollutants enters into to the oceans are mostly diluted, however the organisms living in the oceans tend to concentrate the pollutants into their body by various mechanisms, like adsorption, absorption, ingestion etc. The concentration of pollutant Increases with higher trophic levels, i.e. from primary producers to the tertiary consumer. The threshold limit of organisms to the pollutant concentration increase with increasing trophic levels. Various mode of pollution transport in organisms includes bioconcentration, bioaccumulation and biomagnifications.

Bioconcentration

The bioconcentration id the process by which a contaminant such as oil is directly taken up (i.e., by absorption only) from water and is accumulated to levels greater than those found in the surrounding water.

Bioaccumulation

Marine organisms can accumulate chemical species in amounts far exceeding their surrounding (sea water or sediment) concentrations. The bioaccumulation is the process by which a contaminant is taken up by organisms directly through the physical exposure pathway or through consumption fo food containing the contaminated substance,

Bioaccumulation incorporates the concepts of bioconcentration and biomagnifications. The term 'concentrator factor' has been used to quantify the bioaccumulation and is defined as the concentration of the chemical species in the Organisms or one of its components divided by the seawater or surrounding concentration the environment from which the organism was living. Organisms may take up the food, concentration factors can vary from one environment to another for a specific organisms with a given chemical.

Biomagnification

Biomagnification is the increase in tissue concentrations of a bioaccumulated chemical substance such as oil as it passes up though the trophic levels.

Biodegradation

Biodegradation is the process of degradation of a contaminant such as oil because of its use as a food a source for certain microorganisms. This process is limited to a great extent by temperature, nutrients and oxygen availability.

Types and Effects of Various Marine Pollutants

There are two types of marine pollutants–biodegradable and non-biodegradable pollutants. Most of the materials, which reach the sea, disintegrate either through simple chemical reactions or because of the activities of bacteria and some larger organisms. There are some substances which are either extremely stable or else they have a very slow rate of degradation. The biodegradable pollutants are not persistent in the environment (e.g. oil, organic compound in sewage). But the non-biodegradable pollutants are persisting in environment for longer time (e.g. Heavy metals, plastics, nuclear wastes etc.).

Sewage

Sewage is discharged into the oceans all over the world mostly from urban settlement. Sewage adds to the amount of small particles suspended in the water and contributes large amounts of nutrients.

The raw sewage contains higher concentrations of organic particulate matter, home washings, detergents small amount of oil, higher concentration of bacteria and viruses, which includes harmful pathogenic forms. Sewage has bad odour, irritant and harmful to human and other organisms.

The main sources of sewage are the coastal outfall located near the cities. Also the increasing shipping activities also add higher concentration of sewage to the harbor and shipping routes. Moreover many rivers transport sewage from the inner regions of the land.

In open coast, the effect of sewage is difficult to detect due to higher mixing up of tidal, wave and current actions. However in semi-enclosed areas like small bays, harbours etc., the effects are devastating such as higher Biochemical Oxygen Demand(BOD) which expresses the organic load concentration. BOD is the amount of oxygen required by the microorganisms for the decomposition of the organic compounds dissolved in a given volume of sewage.

Consequently high BOD values indicate high concentrations of organic material and therefore low quantites of dissolvedoxygen (DO). Near sewage outflow areas in temperate waters, the benthic invertebrate communities have degraded, kelp debs have disappeared and diseased fish have become more prevalent. Sewage input in the coastal environment normally increases the nutrient concentration of the outfall area. In tropical waters, outflows near coral reefs have caused a bloom of algal species that grows over the coral, and eventually it leads to the death of coral and its associated forms. This some times leads to the change in the ecological succession, that is the coral reef may be replaced by an algal reef.

Oil

Oil is discharged in to the sea in various forms as crude oil and as separate fractions. Most of the oil fractions are biodegradable. Oil and its fractions are used in various ways from household needs to automobiles and industries. The spilled oil cause more devastation in the marine environment. Petroleum oils are very complex mixtures of large numbers of hydrocarbons, containing also small amounts of sulphur, nitrogen, oxygen and different metals combined with the hydrocarbons.

The main ways in which oil enters in to the sea are,

- Natural release
- Tanker operations
- Oil tanker and other ship accidents
- Operation of ships other then tankers
- Offshore oil drilling and production platforms
- Ship-shore oil terminal operation
- Refinery operations
- Discharges of oil products on land and subsequent discharge by runoff

As the oil enters into sea it is exposed to processes, which modify it both physically and chemically. This process is called weathering. As the oil tends to float on the surface of sea, the prevailing wind, wave, tide and current conditions make the oil to spread to a wider area from the point of release. The majority of crude oil forms sticky layers on the surface, which prevents free diffusion of gasses, clogs adult organisms feeding structures and decreases the sunlight available for photosynthesis. Volatile components of an oil spill eventually evaporate into the air, leaving heavier tars behind which forms into tar balls, which fall to the bottom and may be assimilated by bottom organisms or incorporated into sediments.

The alteration of physical and chemical properties of discharged oil through a series of natural processes, which begin when the discharge occurs and continue until the oil is removed. Major processes, which contribute to weathering, include evaporation, dissolution, photooxodation and emulsification.

Very little fraction of miscible fraction of oil is dissolved in the seawater and other fractions are immiscible. Evaporation is the single most important weathering process for the first several days of an oils discharge. Evaporation results in the loss of lighter fractions through volatization from petroleum products (e.g. benzene, naphthalene).

The process where by oil is incorporated into water in the form of small droplets is called emulsification. Emulsions in water can either be oil-in-water or water-in-oil, formed as a result of wave actions. There is some solution of hydrocarbons in the water and this also is more marked in the more volatile constituents. The oil-in-water leading to the wide dispersion of fine oil droplets in the sea. The water-in-oil composition leading to the formation of semisolid masses known as 'chocolate mousse'.

If silt or sediment is present, oil may be absorbed on to it and may fall with it to the seabed. This effect can be seen as tar ball formation, which is mostly observed over the seabed of oil spoiled area.

Some of the oil constituents are naturally attached by bacteria present in the sea. The simpler compounds, the normal paraffin, are most easily broken down and consumed. The other higher and complex molecular weight fractions are broken down very slowly.

When there is a major oil spill the degree of effect depends on many factors such as

- Type of oil
- Quantity of oil
- Phsiography of the area
- Weather conditions at the time of spill
- Biodiversity of the area
- Season of the spill
- Previous exposure of the area to oil
- Exposure to other pollutants and
- Treatment of oil spill.

As the oil is of floating type it tend it spread and eventually reach the coastline. In many coastal areas, animals and plants situated between high and low water marks are heavily coated with oil and smothered. The exposure of organism to oil adds foreign smell in its body and tissues and the process is called tainting. The shellfish can be tainted either by deposition of oil on their shell or by ingestion of naturally emulsified oil during normal feeding. The local fish population in the immediate vicinity of oil discharge can acquire a marked taint from oil components dissolved in the water.

Almost all groups of organisms like invertebrates, vertebrates like fishes, reptiles, mammals etc., are affected by oil. When the plumage of a diving birds get oiled, the water repellent properties of capacity. This leads to a condition called hypothermia, subsequently the bird may die.

Birds may also ingest the oil. All these cause large scale death of marine birds.

The tainting can be cleared by rearing the affected organisms in clean water for a considerable period. In shellfishes, during the process of moulting, the tainted shells are removed. Also, the birds's feathers and some animal's body can be cleaned with detergent. In this type of clean up, the within about five years, as crude oil is not as highly toxic as it is biodegradable.

The polynuclear aromatic hydrocarbons (PAH) in particular 3, 4 benzopyrene are well known carcinogens (cancer producing). These compounds are present in crude oil in small amounts. The PAH assimilated by marine organisms can be passed upwards through food chain towards man. This will lead to serious concern of human health.

The environmental degradation of oil includes the destruction of algal beds, spreading of oil over rocky shores, incorporation in beach sediment and subsequent loss on scenic beauty.

Heavy Metals

Normally, the metals having the atomic weight more than twenty is considered as heavy metals. The term 'heavymetals' is used to denote elements with specific gravity higher than those of Iron (Fe) and mainly include Lead (Pb), Mercury(Hg), Copper(Cu), Cadmium(Cd) and Chromium (Cr).

Coal combustion, electric utilities, steel and iron manufacturing, fuel oils, fuel additives, and incineration of urban refuse are the more sources of oceanic and atmospheric contamination by heavy metals. Heavy metal contaminated runoff from the land, rain of pollutants from the air, and fallout from ship wrecks pollute the ocean with dangerous metals.

In general, the effects of heavy metals toxicity can be classified into three main categories:

- The blocking of essential functional groups of the biomolecule (e.g. proteins, enzymes).
- The displacement of a metal ion from a biomolecule.
- The modification of the structure of biomolecules in space which is vey important for their function.

Some Metal Pollutants and their Effect in Human Beings and Marine Organisms

Lead

Affects human central nervous system. A poison by ingestion and moderately irritating. A common air contaminant due to use of bad fuel in the auto industry, now being phased out through the introduction of lead-free petrol; also in air in the vicinity of industrial plants using lead. It is a carcinogen of the lungs and kidneys. Lead is flammable in the form of dust when exposed to heat or flame. From the point of view of industrial poisoning, inhalation of lead is much more important than is ingestion. Lead is a cumulative poison. Increasing amounts build up in the body and eventually a point is reached where symptoms and disability occur. Lead can cause irreversible behavioral disturbances, neurological damages and other developmental problems in young children and babies.

Cadmium

This metal is toxic to humans by inhalation and other routes. It can enter through ingestion, intraperitoneal, subcutaneous, intramuscular and intravenous routes. Has no known biochemical or nutritional function. Also cadmium is highly toxic to freshwater and marine organism. It is bioaccumulative through the food chain. Increased exposure can increase risk of lung cancer, cadmium was responsible for the 'Itai-itai' disease in japan.

Chromium

The metal exists in two forms, i.e. trivalent and hexavalent. Hexavalent chromium in high dosages has been implicated as the cause of digestive tract cancers, cutaneous and nasal mucous membrane ulcers and dermatitis. Certain chromate salts, i.e. calcium chromate are carcinogenic, at least when inhaled.

Zinc

Zinc is human skin irritant and produces pulmonary system effects. The difficulty arises from oxidation of zinc fumes prior to inhalation or presence of impurites such as cadmium, antimony, arsenic and lead. Zinc however is a needed micronutrient for humans. It is needed in agriculture. It is used in galvanizing industry. The real problem is of its association with heavy metals like lead.

Arsenic

Toxic by subcutaneous, intramuscular and intraperitoneal routes and reported to produce systemic, skin and gastrointestinal effects. It is also a human carcinogen and experimental teratogen.

Mercury

The silvery-white liquid metal is a potent neurotoxin, capable of causing severe brain damage in developing fetuses and mild tremors and emotional disturbances in exposed asults.

Implicated in the occurrence of Minamata disease in Japan, largest source of mercury contamination is through aquatic animals in whom it accumulates and gets bio-transferred to organocompounds, e.g. methy! mercury. Mercury and lead poisoning cause brain damage and behavioral disturbances in children.

Copper

It is also dangerous to marine organisms and it has bee used in marine anti-corrosive paints.

Pesticides

Pesticides are toxins (chemical compounds) used in agricultural farm to kill the unwanted pest organisms attaching the cultivates plants. A pesticide is any substance or mixture of substances intended for preventing, destroying, repelling, or mitigating any pest. Pests can be insects, mice and other animals, unwanted plants (weeds), fungi, or microorganisms like bacteria and viruses.

There are natural and artificial pesticides. The natural pesticides are biodegradable and are mostly derived from the extracts of plants and animals. However, artificial pesticides are man made and produced extensively only after the industrial revolution. These pesticides belong to four main categories.

1. Organochlorine compounds (e g, DDD- Dichlorodiphenyldichloroeathane)
   - Persistent in the environment or decompose very slowly.
   - Half life is 9-11 years.
   - Accumulate many organisms
   - Higher concentrations on higher organisms-mammals like dolphins, whales, porpoise etc.
   - Detected in antartic penguins.
2. Organophosphate esters (e.g., parathion)
   - Very toxic
   - Much toxic than previous one.
3. Carbamates (Baygon)
   - Decompose rapidly through hydrolysis.
   - Half-life period approximately one week.
4. Chlorophenyl-acils 2, 4, 5-T
   - Hazardous compounds.
   - Used in Vietnam War for defoliation.

Pesticides are the examples of non-point source pollutants. These toxic materials used in the highland for agricultural and other purposes end up into the sea either through runoff into streams and rivers or having been transported by the wind(dust and aerial spray).

Depending on the type of the chemical, its partial and vapour pressure, evaporation, condensation and deposition occur at various places of the globe leading to their global partitioning. Because of such conditions these chemicals, which originate from lower latitudes, move to higher latitudes.

Depending on various conditions, these chemicals migrate in a series of relatively short jumps termed as "grasshoffer effect". Because of this grasshopper effect, Concentration of such chemicals

Even in the remote parts of the world (Arctic and Antarctic) may rise even if the use of these chemicals are curtailed in source countries.

Existence of contamination by these chemicals in the environmental media leads to potential impact on almost all species including human. All the organisms, obtain their first dosage of these chemical from their food and water, wile the mammalian species derive these toxins from theirs mothers through milk. While the plants and lower trophic level organisms in not openly seen. Organochloines have been found to cause microsomal enzyme induction, accelerating body steroids such as estrogens and also to have an effect on the normal reproductive activities of many experimental animals. It is generally known that 5% eggshell thinning occurs at a concentration 5 mg/g of DDE in the bird egg. It is also found that concentrations of 20-1000 mg/g of DDE in the liver of birds are considered to pose a threat to individual bird reproduction and therefore on the population as a whole.

Polychlorinated Biphenyls (PCBS)

PCBs are a series of technical mixtures consisting of many isomers and compounds that vary from mobile oily liquid to white crystalline solids and hard noncrystalline resins.

PCB's usually come from older electrical equipment like transformers, insulations etc.

Toxic by ingestion, inhalation and skin contact. A suspected human carcinogen, affects skin and liver. Usual signs of systemic intoxication include nausea, vomiting, weight loss, jaundice, edema and abdominal pain. Severe liver damage may cause coma and death. PCB's typically cause reproduction problems in most marine organisms. The reduction in sperm motility in human with increasing PCB concentration was observed indication the possible effect of these chemicals on the reproductive capability of mammalian species.

Radioactive Materials

The world's oceans have been a sink for radioactive waste from the production waste enters the ocean from nuclear weapon testing (this is banned by international conventions), the releasing or dumping of wastes from nuclear fuel cycle systems, nuclear accidents (chernoby! Accident in USSR in may 1985), ongoing pollution from smaller but more or less continuous leaks from nuclear power stations (sellafield) or nuclear submarines and carries etc.

Dumping of high-level radioactive waste is no longer permitted in the ocean, but dumping of low-level wastes is still permitted. Low-level waste contains less radioactivity per gram than high-level waste. High-level wastes usually have longer half-lives. For example, one common high-level waste that is produced by spent nuclear fuel has a half-life of 24, 100 years.

It has been suggested that contained nuclear waste should be disposed in the deep sea.

The time required for half the atoms in a given quantity of a radioactive element to disintegrate is called its half life.

So little is known about the deep sea environment or the consequences of containment leakage and failure, that the effects could be devastating. The bioaccumulation process is not associated with the radioactivity properties of a molecule but with its chemical properties. If, however, a bioacumulated molecule happens to be radioactive, its effects on the organism are much more severe.

Radioactive materials bring about two principal sorts of consequences at the organism level:

- Direct toxicity caused by the ionization atoms and molecules of living materials (and especially water) resulting in the production of strong oxidizing agents, which are harmful to living tissues.
- The mutagen activity: here radiation is a factor, which causes errors during the DNA transcription leading to genome alteration. These errors, called mutations, are in most cases lethal or else lead to teratogenesis when they occur during reproduction. They can also result in cancer when they take place during the division of somatic cells.

Nutrients

The nutrients are mainly nitrogen, phosphorus and silicates. Nutrients are used by plants and animals in a complicated cycle, returned to the environment and then re-used. They are essential for the growth of marine primary producers like phytoplankton, seaweed and seagrass. Studies have shown that too much nutrient in the sea and abnormal algal blooms can upset the balance of plants and animals living near the seabed.

Agriculture in the land, aquaculture practices in the coastal areas and release of sewage as outfall are some the major sources. Fertilizers used on land are washed into the ocean via rivers and streams. Also the aquaculture farms locates near the coastal zone also release the wastewater in to the streams and estuaries or directly to the marine environment. Apart from this the sewage released into the coastal areas also contains higher concentration of nutrients.

The higher nutrient concentrations are always utilized by phytoplankton. The release of excess nutrients into coastal waters leads to the condition called Eutrophication. Eutrophication leads to higher production of phytoplankton called bloom. The death and decay of phytoplankton lead to higher oxygen consumption, which is detrimental to the organisms.

High nutrient concentration cause phytoplankton blooms such as, red tides, various yellow and green foams, slimes, and slicks. Although algal blooms are natural, a higher frequency of their occurrence in the past twenty years indicates an unhealthy ecosystem. The toxicity of recent blooms is increasing, which can have a direct effect on the organisms that fees on them. Also, phytoplankton naturally contains DMS (dimethy! Sulfoxide), which is released from dead phytoplankton into the atmosphere and can be changed to sulfuric acid, which eventually contribute to acid rain.

Thermal Pollution of Sea

In this type, the normal temperature of marine source water is increased to 7-8^{0}C from the ambient temperature.

Mostly the power generating plants along the ocean coastlines use the marine waters for cooling purpose, which leads to heated water expelled into the marine environment.

The excess temperature usually gets reduced by various ways. As the water has higher specific heat, it takes prolonged time for cooling unless there is an external mechanism to accelerate this, The dilution by surrounding water, evaporation and precipitation may accelerate cooling within a reasonable time.

Thermal pollution seems to only effects the communities immediately adjacent to the discharge. In continuous discharge the exposure time is also continuous, which generally result in community shift that is more heat tolerant species may replace the normal biota. The longtime exposure to the heat conditions may be detrimental to the normal biota. Many studies have been done on the effects of thermal pollution on the marine environment. Thermal discharge is most noted in the tropical areas, where organisms are near their thermal maximum. For example, mangrove trees in a thermal heated bay no longer reproduce and no new seedlings can be found in the lagoon.

Solid Wastes

Plastic is produced in large amount and has danger of solid waste into the marine environment.

Plastic is not biodegradable and therefore effecting the oceans for long periods of time. When compared to other types of pollutants the biological consequences are low.

Solid wastes are dumped into the sea mainly by the anthropogenic activities.

Mostly of no biodegradable plastics made materials such as used bottles, cups, plastic bags, plastic fishing nets, metal pieces, abandoned wreckage ships, etc. are dumped into the sea. This is also called Ocean dumping.

- Six-pack holders will not decompose for 400 years
- 100, 000 marine mammals and 2 million sea birds die every year after ingesting or being caught in plastic debris.
- Adding ingredients to plastics that would hasten decomposition would add 5-7% to costs, but this price increase in unacceptable to industries.

The solid waste on the other hand acts as a substratum for many fouling marine organisms. The initial biofilm formation on a new substratum and subsequent macro fouling is useful in some ways.

If the solid waste is not releasing any toxic substances then it can be used to build artificial reefs and has been considered as a very good way to increase the environmental heterogeneity (biodiversity) in areas which function as nursery or feeding grounds for important fishing stocks.

There have been cases that fish having been trapped inside plastic bags and other plastic objects are dying of suffocation. Also some cases have been reported of fish getting entangled in plastic fishing nets which have been torn off from fishing boats and drift in the water column, causing a phenomenon known as "ghost fishing". Sea turtles mistake plastic bags for jellyfish, eat and die from internal blockages. Seals and sea lions starve after being muzzled by six-pack rings or entangled by nets.

Birds get entangled in discarded fishing line and drowned, while lost drift nets are particularly dangerous to marine forms.

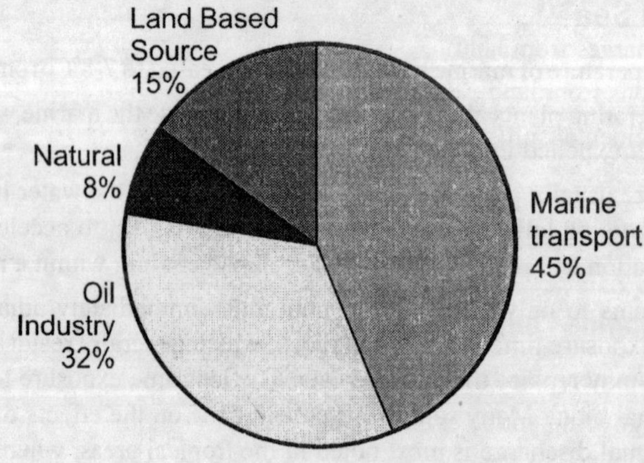

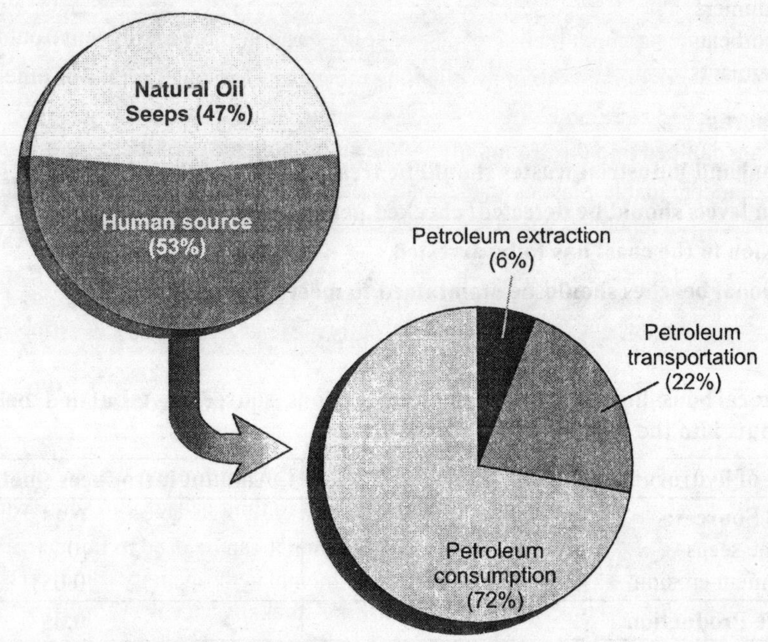

Control Measures for Oil Pollution

- Natural process of emulsification of oil by use of chemical dispersants: can be sprayed on the oil.
- Slick-lickers: continuous belt of absorbent material dips through the oil slick and is passed through rollers to extract oil.
- Rocks can be cleaned with high pressure steam.

Share of Various Sources of Marine Pollution

S.No.	Sources	Share of total(percent)
1.	Runoff and discharges from land	44
2.	Airborne emissions from land	33
3.	Shipping and accidental spills	12
4.	Ocean dumping	10
5.	Offshore mining, oil and gas drilling	1
	All sources	100

Control Measures for Marine Pollution

- Local communities near sea must protect sea
- Management by government be enforcing the laws.
- Incentives must be offered for conservation.
- Industrial units should be equipped with pollution control equipments.
- Oil floating on the sea can be removed by using:-

a. Skimmers
b. Absorbents
c. Coagulants

Protective Measures

- **Municipal and industrial wastes should be treated well and then disposed to sea.**
- **Pollution levels should be detected/ checked periodically.**
- **Soil erosion in the coast has to be arrested.**
- **Recreational beaches should be maintained to meet hygienic stanard.**

Note:

Sources of hydrocarbons inputs into the seas by various sources is tabulated below., Sources of hydrocarbon inputs into the sea

S.No.	Sources of hydrocarbon inputs	Amount introduces (metric tons/year)
1.	**Natural Sources:-** • Maine seeps • Sediment erosion	**0.25** (0.2) (0.05)
2.	**Offshore Production**	**0.05**
3.	**Transportation:-** • Tanker operations • Dry- docking • Marine terminals • Bilge and fuel oils • Tanker accidents • Non tanker accidents	**1.47** (0.7) (0.03) (0.02) (0.3) (0.4) (0.02)
4.	**Atmospheric Transport**	**0.3**
5.	**Municipal and Industrial:-** • Municipal wastes • Refineries • Non refining industrial wastes • Urban runoff • River runoff • Ocean dumping	**1.18** (0.7) (0.1) (0.2) (0.12) (0.04) (0.02)
	Total	**3.2**

Pollutants and Their Effect on the Marine Environment/is Tabulated below

S.No.	Pollutants Sources	Impact on marine environment
1.	Municipal and domestic waste	Reduced dissolved Oxygen(DO); increased hydrogen sulphide levels; incidence of faecal coli form and faecal streptococci; high biological oxygen demand (BOD)
2.	Industrial waste	Affect DO, temperature, turbidity, PH, ammonia values; increases BOD, suspended solids

S.No.	Pollutants Sources	Impact on marine environment
3.	Toxic metals	Causes change in chemical and biochemical processes, increase in turbidity, lethal and sublethal effects on marine life
4.	Oil pollution	Cause smothering, clogging and toxicity
5.	Fertilisers	Affect nutrient levels and may cause eutrophication
6.	Dredging and reclamation siltation	Affects habitats of marine organisms; lethal and sublethal effects; affects flushing capacity of the water body increases in nutrient levels and can cause excessive algal bloom; may also cause damage to coral reefs and coastal nurseries
7.	Discharge of coolant waters	Raises the temperature of the water; can cause the growth of the blue-green algae
8.	Toxic chemicals	Causes lethal and sublethal effects on marine organisms
9.	Offshore mining	Increases particulate loading which can lead to loss of light and reduced primary productivity; smothering and clogging of benthic communities
10.	Radionuclides	Bioaccumulation in fish and other benthick community

Appendix 1.108: Classification of Disorder caused by Exposure to High Level of Environmental Heat

S.No.	Classification of disorder caused by exposure to heat	Effects of heat disorders
1.	Systemic disorders	Heat stroke (hyperphyrexia), heat exhaustion (from circulating deficiency: heat syncope), water, and salt deficiency, heat cramps or sweating deficiency
2.	Skin disorders	Prickly heat (miliaria, rubra), skin cancer (rodent alcer)
3.	Psychoneurotic disorders	Mild chronic (tropical) heat fatigue, acute loss of emotional control

Note:

Heat - Related Health and Safety

Warm temperatures combined with high humidity levels can pose a risk of heat-related illness and death. Extreme heat is described as temperatures that hover 10 degrees or more above the average high temperature for the region and last for several weeks. Humid or muggy conditions, which add to the discomfort of high temperatures, occur when a "dome" of high atmospheric pressure traps hazy, damp air near the ground.

Although anyone at any time can suffer from heat-related illness, some people are at greater risk than others.

- Infants and children up to four years of age are sensitive to the effects of high temperatures and rely on others to regulate their environments and provide adequate liquids.

- People 65 years of age or older may not compensate for heat stress efficiently and are less likely to sense and respond to change in temperature.
- People who are overweight may be prone to heat sickness because of their tendency to retain more body heat.
- People who overexert during work or exercise may become dehydrated and susceptible to heat sickness.
- People, who are physically ill, especially with heart disease or high blood pressure, or who take certain medications, such as for depression, insomnia, or poor circulation, may be affected by extreme heat.

when temperatures approach or go above 90° F, the following actions are recommended;

- Avoid strenuous activity during the hottest part of the day. If such activity is unavoidable, drink plenty of fluids and take, frequent breaks in air-conditioned, or shaded, areas. Consider monitoring body weight and oral temperature. A weight loss of more than 1 Kg. or an oral temperature above 99° F is cause for concern.
- Do not leave anyone - children, disabled individuals, pets - in cars for even brief periods. Temperatures can rise to life-threatening levels in a matter of minutes.
- Make frequent checks on the status of elderly or ill relatives or neighbors. If necessary, move them to an air-conditioned environment during the hottest part of the day.
- To avoid dehydration, a conscious effort should be made to drink more fluids during hot weather. Rapid weight loss may be a sign of dehydration! Don't drink liquids, that contain caffeine, alcohol, or large amounts of sugar–these actually cause you to lose more body fluid.
- Use fans to increase ventilation. If the temperatures exceed 90° F, instead of having a fan blow hot air in from a window, have the fan blow the hot air to the outside. At extreme high temperatures, a fan loses its ability to effectively reduce heat-related illness.
- Cool showers, baths, and sponge baths can be used to reduce body temperatures; In addition, wet clothing has a cooling effect. Preventing Heat-Related Illness

Individuals who are medically compromised, elderly, very frail or very young are at high risk for heat-related illness.

Risk Factors for Heat-Related Illness

Conditions that can increase the risk for heat-related illness include:

- **Age:** the very young (under 4) and the elderly (oyer 65) are more vulnerable.
- **Obesity:** people who are overweight have greater difficulty regulating body temperature.
- **Medical conditions:** conditions such as cardiovascular disease, respiratory disease, or renal diseases may increase a person's susceptibility to heat-related illness. Medications for these conditions may cause dehydration.
- **Medications:** use of psychotropic medications and other medications that affect the body's ability to perspire and/or regulate the heat response can greatly decrease the person's ability to cope in hot weather.
- Medications of particular concern include the following:
  - Antipsychotic agents ;
  - Tranquilizers;

– Anticholinergic such as Benadryl; and
– Lithium, especially in combination with excessive alcohol consumption, can . cause dehydration and may produce serum drug levels that are toxic.

- **Alcohol abuse:** people who are prone to acute alcohol intoxication are at greater risk for sudden death when intoxication is combined with extreme heat.
- **Excessive exposure;** people who are outdoors for long periods can be overcome easily with heat and humidity.
- **Dehydration:** people prone to body fluid loss due to extreme exertion, lack of sufficient fluid intake, medications, or a medical condition.

Heat Exhaustion and Heat Stroke Comparision:

Heat Exhaustion	Heat Stroke
I. Cause. Depletion of body fluids and electrolytes due to exposure to intense ; heat or the inability to acclimatize to : heat, resulting in prolonged or severe diaphoresis. May progress to heat stroke	**I. Cause.** Failure of temperature- regulating mechanism of the body due to ; prolonged exposure to high temperature.
II. Onset. May develop slowly after exposure to heat for several days and inadequate or unbalanced replacement of fluids and electrolytes.	**II. Onset.** May develop quickly (within minutes).
III. Similarities • Headache • Vomiting • Dizziness • Muscle cramps (arms, legs, abdomen) • Fatigue • Rapid pulse (tachycardia) • Nausea	**III. Similarities** • Headache j • Vomiting • Dizziness • Muscle cramps (arms, legs, abdomen) • Fatigue • Rapid pulse (tachycardia) • Nausea
IV. Differences • Profuse perspiration • Cool, moist skin • Rapid respiration • Body temperature may be normal, or slightly below normal, or as high as 102.2 degrees F. • Possible giddiness	**IV. Differences** • Absence of perspiration • Hot, dry, red or mottled skin • Slow deep respiration • Extremely high temperatures • (104 degrees F/40 degrees C or above, rectally) • Mental confusion, disorientation, delirium, irrational behavior, feeling of euphoria or impending doom, diminished level of • consciousness, loss of consciousness may be abrupt • Confusion may occur early or late • Signs of shock • Coma

Heat Exhaustion	Heat Stroke
V. How to Intervene. Cool fluids, loosen tight clothing, recumbent position, cool shaded environment, elevate legs, transport to medical facility for medical follow up.	**V. How to Intervene.** Call for transport to medical facility immediately; cool; water, sponging, and fluids by mouth if still alert; reduce body temperature to 102 degrees F as rapidly as possible. **Transport to hospital for temperature stabilization and prevention/ treatment of complications**

Appendix 1.109: Typical Working Environment/Situations involving Heat Stress

S.No.	Working Situations involving heat stress
1.	Working in very hot weather.
2.	Working in direct sun-light in summer, as in the case of construction jobs and open stock yards.
3.	Working in hot confined spaces as in cellar of coke even batteries
4.	Working near furnaces, heat treatment tanks and sources of heat.
5.	Foundries, forge shops and rolling mills.
6.	Working near or around boilers, steam pipes, and hot processes.
7.	Operations like welding, gas cutting.
8.	Working in hot processes like glass manufacture and hot metal
9.	Working in sheds/ workshops which are not due to radiant heat.
10.	Working in poorly ventilated rooms.
11.	Working in east houses of irons blast furnaces, soking pits. Reheating furnaces of steel plant
12.	Working on top of coke oven batteries.

Appendix 1.110: Effects of over Exposure to Heat Stress on Human Beings

S.No.	Effect of over exposure to heat stress
1.	When the body had to cope up with high temperature, blood circulation has to increase which necessitates the heart to work harder. Consequently, the pulse rate and blood pressure increases. Pulse rate is one of the indications of heat stress.
2.	At higher temperatures, there is increased secretion of sweat, which results in loss water and salt form body. The human body needs a minimum quantity of water and salt. Depriving the body of salt and water leads to exhaustion and impairment of body functions.

3.	Blood plays and important part in the dissipation of heat and the heat regulating process. Since blood and the cardiovascular system are linked with the most of the body functions, over exposure to heat will result in many physiological disorders, (for example, it affects digestion resulting in vomiting and diarrhea.)
4.	In process of thermo-regulation, physiological changes take place to prevent muscles from producing heat. These changes are associated with tiredness, tendency to sleep and fatigue. The person also causes to be alert and his capacity for work gets affected. Continuing to work at this stage will lead to and accidents.

Appendix 1.111: The Electromagnetic Spectrum, Encompassing The Ionizing & Nonionizing Radiation (Expanded Portion and Right). Top Portion Expands Spectrum Between 10^{-7} and 10^{-6} m.

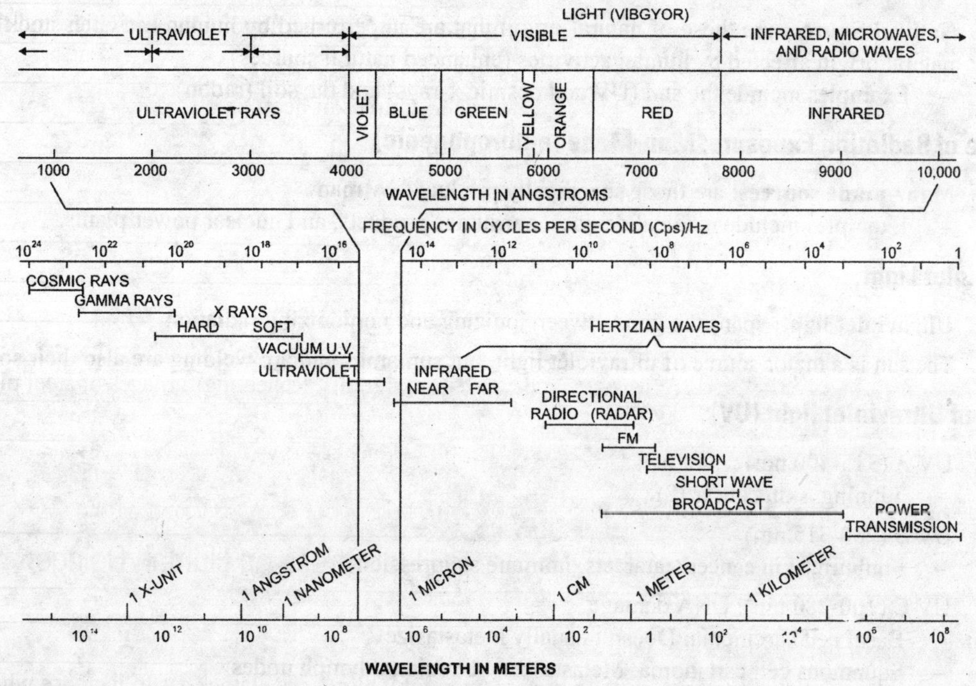

NOTES

[1°A = 1°Angstrom = 10^8 cm = 10^{10} metre, 1Nanometer = 1nm = 10^9 metre, 1hertz = 1 cycle/ second]

Note:

- **Ionizing radiation:** is radiation with sufficient energy to eject electrons from atoms; this process is called
- **Non – ionising radiation:** is radiation without sufficient energy to produce ionization

Types of Radiation

- **Electromagnetic radiation:** UV, visible light, x-rays, EMFs: Electric field and magnetic field traveling at right angles; no mass, no charge
- **Particulate radiation:** Alpha particle, beta particle, neutrons: Have mass and (alphas and betas) charge
- **Ionizing radiation(~>14 eV):** Particulate: Alpha (2p + 2n), beta (e-), positron (e+) Electromagnetic: X-rays, gamma rays
- **Non-ionizing radiation(~<14 eV):** Electromagnetic: UV, microwaves, EMFs

Source of Radiation Exposure: Natural

- **Natural Source:** are those of natural origin that are unperturbed by human activities and those of natural origin affected by human activities (enhanced natural sources)
  - Examples include the sun (UV and cosmic x-rays) and the soil (radon)

Source of Radiation Exposure: Man-Made (anthropogenic)

- **Man –made sources:** are those specifically produced by man
  - Examples include medical devices, consumer products, and nuclear power plants.

Ultraviolet Light

- **Ultraviolet light:** spans the gap between ionizing and nonionizing radiation
- The sun is a major source of ultraviolet light, but sunlamps and arc welding are also their sources.

Types of Ultraviolet light (UV):

- UV-A (315-400 nm)
  - Tanning, skin cancer (?)
- UV-B (280-315 nm)
  - Sunburn, skin cancer, cataracts, immune suppression
- UV-C (100-280 nm): DNA damage
  - Basal cell carcinoma: Doesn't usually metastasize
  - Squamous cell carcinoma: Metastasizes to regional lymph nodes
  - Malignant melanomas: Rapid overall metastasis

Extremely Low-Frequency EMFs

- 1 -3, 000 Hz (cycles per second)
  - High-voltage transmission lines, lower voltage distributor lines, radar and communication equipment, electric blankets
- EMFs probably impact biological systems by interfering with bio-electric processes at the molecular level–not by direct heat, chemical bond breakage, or ion formation

- No evidence of direct DNA damage, but EMFs produce changes in DNA synthesis and RNA transcription
- ELF EMFs are possibly carcinogenic to humans
- Public health approach
  - "Prudent avoidance"
  - ALARA–as low as reasonably achievable

Ionizing Radiation Exposure

Average annual effective dose equivalent (U.S.) from natural sources in thousandths of a rem (mrem)	Annual dose from all sources (percentage from each)
Radon–200 mrem In the body–39 mrem Terrestrial–28 mrem Cosmic–27 mrem	Radon–55% In the body–11% Terrestrial–8% Cosmic–8% Medical–15% Products–2% Other–1%
	Total – 100%

Ionizing Radiation and Cancer Causation

- Leukemia
  - Cancers in A-bomb survivors
  - Patients irradiated for ankylosing spondylitis
- Thyroid cancer
  - Children irradiated for enlarged thymus.
  - Children irradiated for tinea capitis
- Lung cancer
  - Pitchblende (Uranium ore)miners are affected with it.
- Bone tumors
  - Radium dial painters
- Liver tumors
  - Patients who received thorotrast
  - Children irradiated for tinea capitis
- Skin cancer
  - X-ray workers

Three Types of Ionizing Radiation

Three types of ionizing radiation and their penetrating power is shown below:

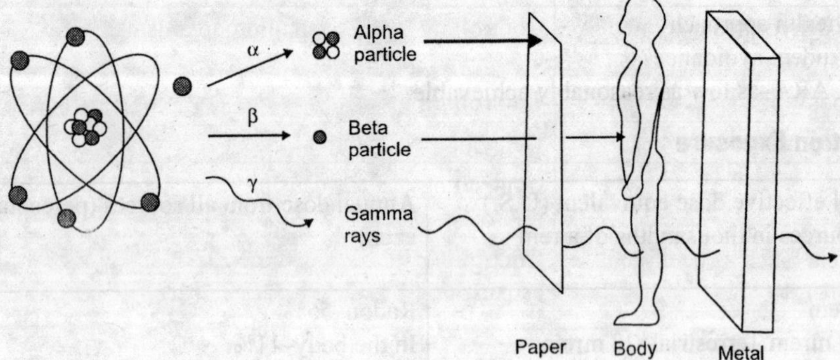

Radon Gas

- Alphas lose all their energy over a short distance; they thus have a short "range" in tissue (~ 2-4 cell diameters)
- If a radioactive atom decays in the airways by alpha decay, the alpha particle will give a very high dose to radiosensitive epithelial stem cells
- The alphas emitted by radon progeny are thus particularly "dangerous"

Radioactive Sources

- Radioisotopes are unstable and emit radiation (in the form of gamma rays) as they disintegrate or decay to a stable state
  - Cobalt-60 or cesium-137 radionuclides
  - The market is dominated by Co-60 sources (emit 1 MeV photons)

Radiation Machines

- Electron accelerators are a source of high-energy radiation that use no radioactive materials
- Accelerators produce 10 MeV electrons
  - Penetration is limited compared to gamma rays from Co-60 or Cs-137
- Electrons can be converted to more-penetrating x-rays
  - Conversion efficiency is low

Pros and Cons of Sources and Radiations

- Radioactive source vs. machine
  - Source is always "active"; machine can be turned on and off-
  - Source produces gamma rays (like x-rays); machine can produce either electrons or x-rays
- Electrons vs. x-rays or gamma rays
  - Electrons deliver a large dose near the surface; x-rays are more penetrating (for thicker packages)
  - Gamma rays from radioactive sources are widely available; x-rays from machines are relatively inefficient to produce

S.No.	Food	Approved use	Dose (kGy)
1.	Spices and dry vegetable seasoning	Decontaminating and controlling insects and microorganisms	30
2.	Dry or dehydrated enzyme preparations	Controlling insects and microorganisms	10
3.	All foods	Controlling insects	1
4.	Fresh foods	Delaying maturation	1
5.	Poultry	Controlling disease-causing microorganisms	3
6.	Red meat (such as beef, lamb, and pork)	Controlling spoilage and disease-causing microorganisms	4.5 (fresh) 7 (frozen)

K – Kilo (10^3), Gy – Gray

Nuclear Device Detonation Physical Effects

- Air blast
- Thermal radiation
- Initial nuclear radiation
- Residual nuclear radiation
- Crater formation
- Ground shock

Radiation Poisoning

- Small area contamination (localized release of radioactivity)
- Larger release in air or water
- (Must consider single event vs. ongoing poisoning)

Medical Issues

- Acute
  - "Usual" medical problems for bomb-type attack (injuries, burns)
  - Acute radiation syndromes (unlikely for radiation poisoning)
  - Patient internal contamination
- Delayed
  - Radiation carcinogenesis
  - (Must consider both acute and delayed effects of in utero irradiation)

Psychosocial Issues

- Radiation as an "invisible toxin"–some attacks may be explicitly designed more for psychosocial than physical impact.
- Acute and chronic psychosocial reaction is a central concern.
- Risk perception is high-risk–toxic hazard per se plus terrorist (i.e., non-accidental) event.
- High-risk groups for psychosocial harm–children, mothers with young children, emergency workers, clean-up workers etc.

Appendix 1.112: Environment and Occupational Hazards of Electromagnetic Radiation

We are under continual attack from electromagnetic fields (EMFs) radiating from power lines, household wiring, microwave ovens, computers, televisions, clock radios, cellular phones, electric blankets, and other appliances. Researchers have correlated electro-pollution with increase in cancer, birth defects, depression, learning disabilities, chronic fatigue syndrome, Alzheimer's disease, and sudden infant death syndrome.

Interaction between electricity and magnetic fields produces electromagnetic radiation. Electromagnetic radiation (EMR) r electromagnetic fields (EMFs) are the terms that broadly describe exposures created by the vast array of wired and wireless technologies that have altered our lives in beneficial ways. Though these technologies were designed to maximize energy efficiency and convenience; there is growing evidence about possible health risks associated with these technologies based on new studies. This is also called EMR hazards or RADHAZ.

EMF radiation passes through space at the speed of light at about 300×10^6 meters per second. We cannot see it, taste it or smell it, but it is one of the most pervasive environmental exposures in industrialized countries today. Human beings are bioelectrical systems. Our hearts and brains are regulated by internal bioelectrical signals. Environmental exposures to artificial EMFs can interact with fundamental biological process in the human body. In some cases, this can cause discomfort and disease.

EMFs are not blocked or weakened by the trees, walls, buildings, or other structures. Like X-ray, they pass through these objects and into our bodies- disrupting normal cellular function and biological processes.

Sources of Exposure

Environmental and Natural

Our greatest exposure to EMFs comes from within our homes and offices. All household and office appliances emit EMFs. But we cannot tell which are most dangerous by their size or function. Often the small devices like electric can openers and hair dryers are much more dangerous than the larger ones such as electric ovens or refrigerators.

The existence of non-thermal effects of EMFs on living matter, occur at every level from molecular to epidemiological.

Environmental exposure to man-made electromagnetic fields has been steadily increasing as growing electricity demand and ever-advancing technologies. Everyone is exposed to a complex mix of weak electric and magnetic fields, both at home and at work, from the generation and transmission of electricity, domestic appliances and industrial and industrial equipment, to telecommunications and broadcasting.

We are fundamentally electro-magnetic beings with micro electrical currents being generated to control our bodily functions such as growth, metabolism, thought, movements, etc. Disturbances to his electrical network in our body can cause malfunctioning of our internal organs, especially to the brain. Exposure to consistent external frequency for more than a few minutes can cause our body's electrical functionality to be disrupted. This is true even with exposures to very weak EMFs.

Occupational Exposure

At work, there are similar dangers: radiation sources and radiation enhancers. Certain occupations present more of a risk than others. The workers listed below have the greatest radiation exposure daily:

Occupations	Typical Exposure (in milligauss)
Cable splicers	upto 15 mG
Power Distribution Substation Operators	upto 34 mG
Electronics	upto 34 mG
Line workers	upto 35 mG
Machinists	upto 28 mG
TV Repair Workers	upto 08 mG
Welders	upto 96 mG

Office workers are not safe either. They have to contend with fluorescent lighting, computers and other electronic devices that people rarely, hesitate to use.

Classification Of EMFs

According to their frequency and energy, electromagnetic waves can be classified as either ionizing radiations or non-ionizing radiations (NIR).

Ionizing radiations are extremely high frequency electromagnetic waves (X-ray and gamma rays), which have enough photon energy to produce ionization by breaking the atomic bonds that hold molecules in cells together.

Non-ionizing radiations (NIR) is a term for that part of the electromagnetic spectrum which has photon energies too weak to break atomic bonds. They include ultraviolet radiation, infrared radiation, infrared radiation, radiofrequency and microwave fields. NIR cannot cause ionization however they have been shown to produce other biological effects, for instance by heating, altering chemical reactions or inducing electrical currents in tissues and cells.

There are four subgroups of electromagnetic radiation fields with frequency and intensity. This electromagnetic spectrum begins at a frequency of 1 Hertz (Hz), which is 1 wave/sec.

Subgroups of Electromagnetic Radiation

Type	Frequency range	Source
Static	0Hz	• Natural • Video
Extremely low frequency (ELF)	$(0<f\leq 300$ Hz)	• Power-lines • Domestic distributions • Electric engines in cars, train, tramway
Intermediate frequency (IF)	(300Hz$<f\leq$100kHz)	• Monitors • Anti-theft devices • Hands free access control systems • Card readers • Metal detectors

Type	Frequency range	Source
Radio frequency (RF)	(100kHz<f<300GHz)	• Broadcasting and TV • Mobile telephony • Microwave oven • Radar • Portable and stationary radio transceivers • Personal mobile/radio

Hazards of EMFs

Radiation Damage

An electromagnetic radiation hazard exists when electronic equipment generates a strong enough electromagnetic field to fall in a category listed below:

- Causes harmful or injurious effects to humans and wildlife.
- Induces or otherwise couples currents and/or voltages of magnitudes large enough to initiate electro-explosive devices or other sensitive explosive components of weapons systems, ordnance, or other explosive devices.
- Creates sparks large enough to ignite flammable mixtures or materials that must be handled in the affected areas. Electromagnetic fields above certain levels can trigger biological effects.
- Short-term exposure at the levels present in the environment or in the home do not cause any apparent detrimental effects.

The EMF are not in themselves dangerous, they are not devastating our body cells like X-Rays, but our immune system is recognizing them as an enemy like a virus or bacteria. Fighting with them surrounding us everywhere makes our immune system overtired which are not able to handle all tasks and are unable to completely defend our bodies e.g. catching cancer cells.

Cellular Effects

Even though RF/microwaves don't have the energy to directly break chemical bonds, unlike ionizing radiation such as X-ray, there is scientific evidence that this energy can cause DNA damage indirectly leading to cancer by a combination of biological effects.

Radiation can damage the DNA in the cells of living things. Damaged DNA can make the cell stop working or unable to reproduce. It can also cause the cell to grow out of control, causing cancer.

Health Effects

Human bodies are very sensitive to EMFs. When we interact with natural energies, we reinforce the natural balance within our energy system. But when we are exposed to man-made EMFs that are unnatural for our bodies, they create a chaotic, situation that is harmful. Our bodies absorb and store the energy fields that weaken our immune system, resulting to various illnesses and diseases. Some diseases linked to constant exposure to EMFs are: Headaches, disturbance in sleep patterns, shortened attention span, increased blood pressure, damage to eyes especially when glaucoma medication are also taken chronic fatigue syndrome, memory loss, miscarriages, birth defects, leukemia, lymphoma, brain tumor and even cancer.

- Studies have shown EMF exposure above 2 mG will start to develop biological stress.
- Evidence links prolonged exposure to levels between 2 and 12 mG+ with cancer and possible immune system effects.

- Exposure to 12 mG suppresses the human hormone melatonin (critical for sleep, mood regulation, and overall health), powerful artificial EMFs overwhelm body's electrical fields, changing their frequency and distorting the balance of the body's electromagnetic field and its communication systems. This causes physical, mental and emotional chaos.

The danger doesn't always pass once you get away from the strong electromagnetic field. That's because biological systems have been proven to store electromagnetic radiation within the cells in the form of electromagnetic oscillations. These oscillations can stubbornly remain inside you, wreaking havoc with your body's most important processes.

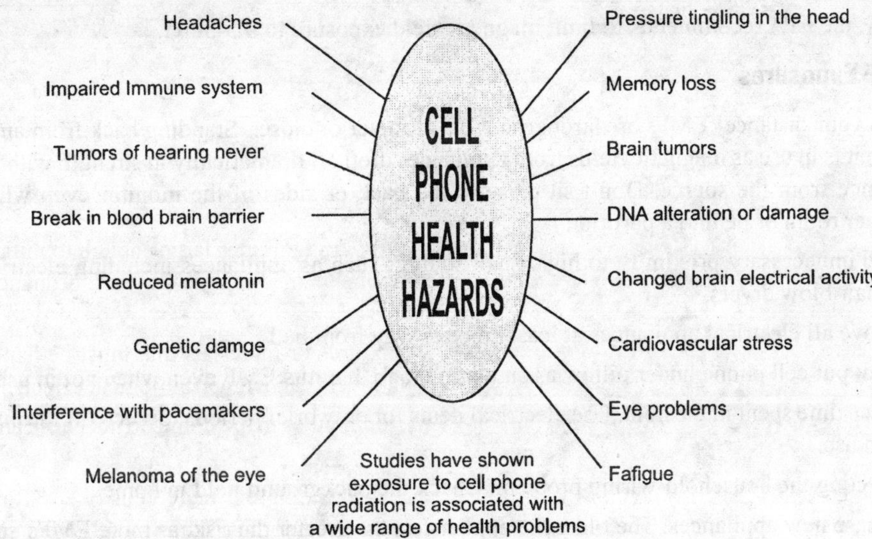

Effects on Blood Brain Barrier (BBB)

Environmental heat in excess of the mammalian thermoregulatory capacity can increase the permeability of the BBB to macromolecules. Neuronal albumin uptake in various brain regions was shown to be dose dependently related to brain temperature, with effects becoming apparent with temperature increases of 1°C or more. Thus, albumin bounded drugs uptake increases.

Electromagnetic Hypersensitivity

Despite regulatory protection against thermal effects, a small percentage of people have reported health problems that are believed to be caused by exposure to electromagnetic fields (EMF); for example from cell phone towers, WiFi and cordless phones Authorities have labeled this group of people as having "electromagnetic hypersensitivity" or EHS. Common symptoms reported by people with EHS include heart palpitations, dizziness, nausea, digestive disturbances, fatigue, tiredness, difficulty concentrating, and skin irritation.

Prevention of EMFs Hazards

Measurement

1 mG (milligauss) is the common unit of measurement for magnetic fields. One milliGauss is 1/1000 of a Gauss. These fields are measured using an instrument called a milligauss meter or Electrosmog Detector.

Safety Limits

Today's public exposure limits for telecommunications are based on the presumption that heating of tissue (for RF) or induced electric currents in the body (for ELF) are the only concerns when living organisms are exposed to RF. These exposures can create tissue heating that is well known to be harmful in even very short-term doses.

Bio-effects and some adverse health effects occur at far lower levels or RF and ELF exposure where no heating (or induced currents) occurs at all; some effects are shown to occur at several hundred thousand times below the existing public safety limits where heating is an impossibility.

For safety, the EPA recommends to limit magnetic field exposure to 0.5-5mG.

Limiting EMF Exposures

- Keep your distance! EMFs are strongest at 0.5–1 meter or closer. Standing back from an appliance when it is in use as magnetic fields from appliances drop off dramatically in strength with increasing distance from the source. Don't sit close to the back or sides of the monitor even when it is in another room or behind a partition.
- Avoid unnecessary proximity to high EMF sources such as appliances including electric blankets and hair blow dryers.
- Remove all electrical appliances at least 2metres way from bed
- Do not put cell phone under pillow as an alarm clock. It emits EMF even when not in use.
- Reduce time spent in the field. Use electrical items for only brief periods. Switch off the power when not in use.
- Correcting the household-wiring problem. Check the background field in home
- Purchase new appliances. The older the appliances, the greater the risk, as more EMFs are emitted.
- Limit the time spent around home's electrical appliances.
- Place all clocks, cell phones, cordless phones and other electrical devices at least 2 meters from the place of sleep.
- If one wears glasses, go for plastic frames or non-metal frames. Good conductor material can serve as an antenna to channel radio waves directly into brain.
- Grow some plants indoors. Plants are natural eco-friendly air purifiers and their leaves can help absorb some infrared radiation and produce negative ions that are necessary for bodies.
- Avoid prolonged exposure of children to television and computers.

Radiation Barriers

The effects of EMFs on the health of individuals and environments is of concern to people. Those who want to do something may consider the following suggestions:

- Use for all members of family the Peace Ball, a personal device. Be sure, to wear it 24 hours every day.
- Install a Neutralizer in house, car and in workplace

Existing Legislations

WHO's International EMF Project was launched to provide scientifically sound and objective answer to public concerns about possible hazards of low level electromagnetic fields.

Knowing that there is danger inherent in owning and operating cellular devices has not curbed the manufacture or selling strategies of cellular providers. This knowledge has not encouraged the government to stop cellular providers from raising the acceptable emissions allowances. This information hasn't even inspired a cellular provider to voluntarily include a frequency-absorbing chip to protect their customers.

Cellular towers continue to be erected on office buildings sides and tops. Only monetary consideration between the building's owners and the cellular providers determines the feasibility. The towers harm people by emitting very high levels of the RF radiation sometimes exceeding the guidelines. The harm is done to anyone in close proximity to the tower and certainly to the workers in the facility.

Radio-Frequency Radiation Hazard Warning Symbol

The warning symbol for radio frequency radiation hazards shall consist of a red isosceles triangle above an inverted black isosceles triangle, separated and outlined by an aluminum color border. The words "Warning-Radio–Frequency Radiation Hazard" shall appear in the upper triangle.

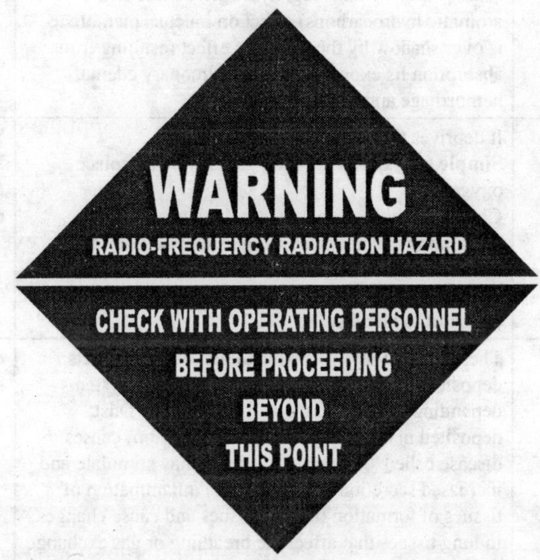

Some Interesting Facts about EMF

- Low dose exposure doesn't mean low risk. In fact, low-level EMF exposures increases health risks!!
- Everything generates an EMF. The natural earth creates one, and so does the human body. Human body creates an EMF of about 10 Hz.
- There is no evidence to conclude that exposure to low level EMFs is harmful to human health.
- Scientists say that chances of developing Alzheimer's, Multiple sclerosis and Parkinson's is enhanced by two minutes of exposure to emissions from mobile phones which can disable a safety barrier in blood causing proteins and toxins to leak into the brain.
- The eyes and testes are the most vulnerable body organs to EMR.

Appendix 1.113: Classification and Health Hazards of Toxic Chemicals

S.No.	Classification of toxic/ hazardous chemicals	Health hazards involved	Typical examples
1.	Irritants	Its contact inflaming of mucous membrane of upper respiratory tract. It can also cause changes in the mechanics of respiration and lung functions. Long terms of exposure to irritants results in increased mucous secretions and chronic bronchitis. A primary irritants (eg, HCL) exerts no systemic toxic action either because the products formed on the tissue of the respiratory tract are nontoxic or because the irritant action is far in excess of any systemic toxic action. A secondary irritants (eg, hydrogen sulfide and aromatic hydrocarbons) effect on mucous membrane is over shadow by the systemic effect resulting from absorption Its exposure causes pulmonary edema, hemorrhage and tissue necrosis.	NH_3, HCl, HF, NO_2, O_3, SO_2, H_2SO_4, halogens, alkaline dust and mist, diethyl/dimenthy sulphate, phosphorus chloride phosgene, acetic acid, acrolein, formaldehyde, formic acid.
2.	Asphyxiants	It deprives the tissues of oxygen supply. **Simple ashyxiants** are inert gases which displace oxygen. **Chemical asphyxiants** render the body incapable of utilizing and adequate oxygen supply. They can cause damage to very low concentration eg co, forms carboxyhaemoglobin leading to lack of oxygen transport to blood cells.	**Simple asphyxiants** –CO_2, H_2, He, N_2, N_2O etc. **Chemical asphyxiants**-CO, HCN etc.
3.	Pulmonary agents	These are mainly dusts which upon inhalation gets deposited in various sites in the respiratory systems depending on their aerodynamic diameter. Dust deposited in deep lungs (ie. Alvedar region) causes disease called "Pneumoconisis and may stimulate and increased secretion of the mucous, inflammation of tissues of formation of fibre tissues and cause changes in lung tissues that affect the breathing or gas exchange functions leading to lung damage or even lung cancer. Silica dust inhalation causes a lung disease called silicosis	dusts of asbestos, coal, cotton, silica, and wood.
4.	Anaesthetics	These chemical act on the central nervous system in the brain causing stupor and unconsciousness by reaching the brain and producing effects depending on their practical pressure in the blood.	Chloroform, diethyl, ether, hexane, acetylene, olefin hydrocarbons, ether, ketones, alcohols etc.
5.	Skin and ingestion (corrosive)toxins	These toxins cause contact allergies/acute irritation corrosion. A sensitisor causes a substained proportion of exposed people to develop an allergic reaction in normal tissues after repeated exposure to the chemical.	**Contact allergies**-Formaldehye Benzocaine, neomycin **Sensitizer** – chlorinated hydrocarbons, expoxies chromium compounds nickel compounds.

S.No.	Classification of toxic/ hazardous chemicals	Health hazards involved	Typical examples
6.	Systemic poisons	These chemicals causes damage to one or more system of the body either by direct contact of the skin or mucous membrane or by systemic chemicals. Antimalarial drug (quinacrine and chloroquine) affects eye cornea after oral administration. It also causes conjunctivitis. Inhalation/ingestion of phenols, naphthalene damages blood system and CS_2 damages nerve system	Quinacrine, chloroquine, phenol, naphthalene, CS_2
7.	Carcinogens	These chemicals cause cancer which is characterized by unrestricted growth of tissue cells leading to eventual death. It is a multistage process with distinct biological features namely initiation, promotion and progression. These are five types of chemical carcinogens: i) confirmed human carcinogens. ii) suspected human carcinogens iii) Animal carcinogens iv) Chemicals not classified as human carcinogens v) Chemicals not suspected as human carcinogens	Ex:- arsenic, asbestos, benzene, beryllium, chromium (vi) compounds, **Nickle compounds (insoluble)** Ex:- ehthylene oxides Ex:- hydrazine Ex:-Nitrogen dioxide Ex:- phosphorus trichloride

NOTES

Some chemicals belong to more than one category. As for example, benzene is carcinogenic as well as irritant. Hydrogen peroxide is an explosive, heat sensitive and an oxidizing agent. Phosphorus trichloride is corrosive, toxic and water sensitive.

Health Effects of Some Chemicals on Body Systems

The Respiratory System's function is to supply oxygen to the body and remove carbon dioxide. It includes the nasal passages, pharynx, trachea, bronchi, and lungs. Possible health effects of the respiratory system include asbestosis, lung cancer, chronic bronchitis, fibrosis, emphysema, and decreased oxygen supply in blood.

Possible Contaminants	Source
Asbestos	Old insulation
Radon	The ground
Cadmium	Old batteries
Benzene	Degreasers
Carbon monoxide	Car exhaust, unvented or faulty furnaces
Soot	Furnaces, wood burning stoves

The Renal System's function is to rid the body of waste, to regulate the amount of body fluids, and to regulate the amount of salts in the body. It includes the kidneys, the urethra, the bladder, and the ureter. Possible health effects of the renal system include decreased formation of urine, decreased blood flow to kidney, decreased ability to filter the blood, prevented urine flow, kidney tissue damage, and kidney cancer.

Possible Contaminants	Source
Cadmium Lead Mercury Uranium Chlorinated hydrocarbon solvents (TCE, PCE, PCT)	Old batteries, cigarette smoke Old paint, outdated plumbing Thermostats, thermometers, some fish Food and water, proximity to nuclear testing sites Degreasers, paint removers, dry cleaning solutions

The CARDIOVASCULAR SYSTEM'S function is to move nutrients, gases, and wastes to and from the body, to help stabilize body temperature, and to fight diseases and infections by transporting white blood cells to important areas. It includes the heart, blood, arteries, veins, and capillaries. Possible health effects include heart failure and the inability of blood to carry the necessary oxygen to the body.

Possible Contaminants	Source
Carbon monoxide Carbon disulfide Nitrates Methylene chloride	Car exhaust, unvented or faulty furnaces Industrial production Fertilizers, Auto part cleaners, paint removers

The Reproductive System's function is to produce hormones. For males it includes females it includes the uterus, bladder, vagi of the reproductive system include decreas defects, and infertility (the inability to have to produce egg and sperm cells, to nurture a developing fetus, and the testicles, seminal vesicles, prostate gland, and the penis. For na, Fallopian tubes, ovaries, and the cervix. Possible health effects ad ability to have a baby, increased baby deaths, increased birth

Possible Contaminants	Source
Methyl mercury Carbon monoxide Lead	Some fish, coal-burning power Car exhaust, unvented or faulty furnaces /Old paint, outdated plumbing

The Nervous System's function is to tranmit messages from one part of the body to another . It includes the central nervous system (the brain and spinal cord) and the peripheral nervous system. Possible health effects of the nervous system include inability to move, loss of feeling, confusion, and decreased speech, sight, memory, muscle strength, or coordination.

Possible Contaminants	Source
Arsenic	Pressure treated wood
Cadmium	Discarded batteries
Carbon monoxide	Car exhaust, unvented or faulty furnaces
Cyanide	Rat poison

The Immune System's function is to protect the body from tumor cells, environmental substances, and invading viruses or bacteria. It includes the lymph system, bone marrow, white blood cells, and the spleen. Possible health effects of the immune system include overreaction to environmental substances (allergy), immune system slow down or failure, and auto immunity (autc immunity causes the body to attack itself - which fl .makes it more likely to have an over-reaction or infection).

Possible Contaminants	Source
Mercury Lead | Pesticides Polychlorinated biphenyls (PCBs) Polycyclic aromatic hydrocarbons (PAHs)	Thermostats, thermometers, some fish Old paint, outdated plumbing Unwashed fruits and vegetables Industrial waste, fish from contaminated water Cigarette smoke, vehicle exhaust, asphalt roads

The Skin serves as a barrier to germs and other substances, prevents dehydration, and regulates body temperature. Possible health effects of the skin include irritation, rash, redness or discoloration, dermatitis, and health effect related to other systems and organs due to contamination through the skin.

Possible Contaminants	Source
Nickel Mercury Arsenic Chromium Polychlorinated biphenyls (PCBs) VOC (volatile organic compounds)	Cement Thermostats, thermometers, some fish Pressure treated wood Paints, industrial production Industrial waste, fish from contaminated water Fumes from gasoline, paint, adhesives, building supplies

The Hepatic System's function is to break down food and store nutrients, to make proteins which are essential for blood to clot, and to purify the body of drugs, contaminants, or chemicals. It includes the liver and its veins. Possible health effects of the hepatic system include liver damage, tumors, accumulation of fat, (steatosis), and death of liver cells.

Possible Contaminants	Source
Carbon tetrachloride Methylene chloride Vinyl chloride	Adhesives Auto part cleaners, paint removers Pipe sealer ‘

Appendix 1.114: Permissible Exposure Limit of Some Chemicals in the Work Environment

(Second schedule under Indian Factories Act, 1948)

S.No.	Chemical substances	Permissible limit of exposure			
		Time weighted average (TWA) concentration (8 hours)		Short term exposure limit (STEL) (15 minutes)	
		ppm	mg/m³	ppm	mg/m³
1.	Acetaldehyde	100	180	150	270
2.	Acetic acid	10	25	15	37

S.No.	Chemical substances	Permissible limit of exposure			
		Time weighted average (TWA) concentration (8 hours)		Short term exposure limit (STEL) (15 minutes)	
		ppm	mg/m³	ppm	mg/m³
3.	Acetone	750	1780	1780	2375
4.	Acrolein	0.1	0.25	0.3	0.8
5.	Acrylonitrile-skin (S.C)	2	4.5	-	-
6.	Aldrin-skin	-	0.25	-	-
7.	Allyl chloride	1	3	2	6
8.	Ammonia	25	18	35	27
9.	Aniline-skin	2	10	-	-
10.	Anisidine (0, p -isomers)-skin	0.1	0.5	-	-
11.	Arsenic and soluble compounds (As)	-	0.2	-	-
12.	Benzene(S.C)	10	30	-	-
13.	Beryllium and compounds (as Be) (S.C)	10	30	-	-
14 .	Boron trifluoride-c	1	3	-	-
15.	Bromine	0.1	0.7	0.3	2
16.	Butane	800	1900	-	-
17.	2-butanone(methyl ethyl ketone-MEK)	200	590	300	885
18.	n-Butyl acetate	150	710	200	950
19.	n-butyl alcohol-skin-c	50	150	-	-
20.	Secondary/tertiary butyl acetate	200	950	-	-
21.	Butyl mercapton	0.5	1.5	-	-
22.	Cadmium dust and salts (as Cd)	-	0.05	-	-
23.	Calcium oxide	-	2	-	-
24.	Carbonyl (sevin)	-	5	-	-
25.	Carbofuran (Furadan)	-	0.1	-	-
26.	Carbon- disulfide-skin	10	30	-	-
27.	Carbon monoxide	50	55	400	440
28.	Carbon tetrachloride	5	30	-	-
29.	Chlordane-skin	-	0.5	-	2
30.	Chlorine	1	3	3	9

S.No.	Chemical substances	Permissible limit of exposure			
		Time weighted average (TWA) concentration (8 hours)		Short term exposure limit (STEL) (15 minutes)	
		ppm	mg/m^3	ppm	mg/m^3
31.	Chlorobenzene (Monochlorobenzene)	75	350	-	-
32.	Chloroform (S.C)	10	-	-	-
33.	Bis(cholormethyl)ether (H.C)	0.001	0.005	-	-
34.	Chromic acids and chromates (as Cr)	-	0.05	-	-
35.	Chromus salt (as Cr)	-	0.5	-	-
36.	Copper fume	-	0.2	-	-
37.	Cotton dust, raw	-	0.2	-	-
38.	Arosol, all isomers, skin	5	22	-	-
39.	Cyanide (as CN)- skin	-	5	-	-
40.	Cyanogens	10	10	-	-
41.	DDT (dichlorodi phenyl tirchlorothane)	-	1	-	-
42.	Demeton-skin	0.01	0.1	-	-
43.	Diazonon-skin	-	0.1	-	-
44.	Dibutylphthalane	-	5	-	-
45.	Dischlorus (DDVP)-skin	0.1	1	-	-
46.	Dieldrin –skin	-	0.25	-	-
47.	Dinitrobenzene(all isomers)-skin	0.15	1	-	-
48.	Dinitrobenzene- skin	-	1.5	-	-
49.	Diphenyl (Biphenyl)	0.2	1.5	-	-
50.	Endosulofan(thidan-skin)	-	0.1	-	-
51.	Endrin-skin	-	0.1	-	-
52.	Ethyl acetate	400	1400	-	-
53.	Ethyl amine	10	18	-	-
54.	Ethyl amine	10	18	-	-
55.	Fluorides (as F)	-	2.5	-	-
56.	Fluorine	1	2	2	4
57.	Formaldehyde (S.C)	1	1.5	3	3
58.	Formic acid	5	9	-	-
59.	Gasoline	300	900	500	1500

S.No.	Chemical substances	Permissible limit of exposure			
		Time weighted average (TWA) concentration (8 hours)		Short term exposure limit (STEL) (15 minutes)	
		ppm	mg/m^3	ppm	mg/m^3
60.	Hydrazine-skin(S.C)	0.1	0.1	-	-
61.	Hydrogen chloride-c	5	7	-	-
62.	Hydrogen cyanide-skin-c	10	10	-	-
63.	Hydrogen fluoride(as F)-c	10	10	-	-
64.	Hydrogen peroxide	1	1.5	-	-
65.	Hydrogen sulfide	10	14	15	21
66.	Iodine-c	0.1	1	-	-
67.	Iron oxide fume (Feo) (as Fe)	-	5	-	-
68.	Isoamyl acetate	100	525	-	-
69.	Isoamyl acetate	100	360	125	450
70.	Isobutyl alcohol	50	150	-	-
71.	Lead, inorganic dust and fumes (Pb)	-	0.15	-	-
72.	Lindane-skin	-	0.15	-	-
73.	Malathion-skin	-	10	-	-
74.	Manganese (as Mn) dust and compounds –c	-	5	-	-
75.	Manganese (as Mn) dust and compounds –c	-	5	-	-
76.	Mercury (as Hg)-skin- (L) i. alkyl compounds (ii) all forms except alkyl vapor	- -	0.01 0.05	- -	- -
77.	Amyl and inorganic compounds	-	0.1	-	-
78.	Methyl alcohol (methanol)-skin	200	260	250	310
79.	Methyl cellosolve(2-methoxyethanol)-skin	5	16	-	-
80.	Methyl isocyanate –skin	0.02	0.35	-	-
81.	Methyl isobutyl ketone	50	205	75	300
82.	Naphthalene	10	50	15	75
83.	Nickel carbonyl (as Ni)	0.05	0.35	-	-
84.	Nitric acid	2	5	5	10
85.	Nitric acid	25	30	-	-
86.	Nitrobenzene –skin	1	5	-	-

S.No.	Chemical substances	Permissible limit of exposure			
		Time weighted average (TWA) concentration (8 hours)		Short term exposure limit (STEL) (15 minutes)	
		ppm	mg/m^3	ppm	mg/m^3
87.	Nitrogen dioxide	3	6	5	10
88.	Oil mist mineral	-	5	-	10
89.	Ozone	0.1	0.2	0.3	0.6
90.	Parathion –skin	-	0.1	-	-
91.	Phenol –skin	5	19	-	-
92.	Phorate (Thimate)- skin	-	0.05	-	0.2
93.	Phosgene(carbonyl chloride)	0.1	0.4	-	-
94.	Phosphine	0.3	0.4	1	1
95.	Phosphoric acid	-	1	-	3
96.	Phosphorus (yellow)	-	0.1	-	-
97.	Phosphorus pentachloride	0.1	1	-	-
98.	Phosphorous trichloride	0.2	1.5	0.5	3
99.	Picric acid-skin	-	0.1	-	0.3
100.	Pyridine	5	15	-	-
101.	Silane (silicone tetrahydride)	5	7	-	-
102.	Sodium hydroxide-c	-	2	-	-
103.	Styrene monomer (phenlthylene)	50	215	100	425
104.	Sulfur dioxide	2	5	5	10
105.	Sulphur hexafluoride	1000	6000	-	-
106.	Sulphuric acid	-	1	-	-
107.	Tetraethyl lead (as Pb)-skin	-	0.1	-	-
108.	Toluene (toluol)	100	375	150	560
109.	o- Toluidine- skin (s.c)	2	9	-	-
110.	Tributyl pyhosphate	0.2	2.5	-	-
111.	Trichloroethylene	50	270	200	1080
112.	Natural uranium	-	0.2	-	-
113.	Vinyl chloride (H.C)	5	10	-	-
114.	Welding fumes	-	5	-	-
115.	Xylene (o, m, p-isomers)	100	435	150	655
116.	Zinc oxide- i) Fume ii) dust (total dust)	 - -	 5 10	 - -	 10 -
117.	Zirconium compounds(as Zr)	-	5	-	10

Note:

i. C: denotes ceiling limit
ii. S.C: denotes suspected human carcinogens
iii. H.C: denotes confirmed human carcinogens
iv. Skin: denotes potential contribution to the overall exposure by the cutaneous route including mucous membrane and the eye.
   a. **TLV-TWA**- is the value for 8 hours work-day and 40 hours work-week
   b. **TLV-STEL** indicates the concentration to which the workers can be continuously for not more than 15 minutes without suffering form (i) irritation (ii)chronic or irreversible to tissue damage or, (iii) narcosis that can cause accident or reduce work efficiency. The duration of exposure at STEL level should not exceed 15 minutes and not be repeated more than four times per day with a gap of at least 60 minutes in between successive exposure at the STEL STEL is recommended for chemicals that have toxic effects on exposure only to high concentration.
   c. **TLV-C:** indicates the concentration of chemicals that should not exceed even once in a shift "c" value (ie, ceiling limit) are given to the chemicals which are fast acting. Where, there is a hazard of absorption of chemical through the skin, the fact is indicated in the TLV document.

Appendix 1.115: Systemic Toxicological Effects of Various Chemical Substances

S.No.	Nomenclature of toxic chemical substances	Examples	Site of action	Signs and symptoms	Remark
1.	Hepatotoxic	Carbon tetrachloride, nitrosomines, tetrachlorothene.	Cause damage to liver	Jaundice and liver enlargement	Tetrachloro ethane is the most toxic of chlorinated hydrocarbons cccl4 produces severe diffuse central necrosis of liver.
2.	Nephrotoxic	Halogenated hydrocarbons, uranium compounds.	They affect kidneys	Edema and proteinurea.	Some halogenated hydrocarbons produce damage to the both kidneys and liver.
3.	Neurotoxins	Mercury, acrylamide, carbon disulfide, organomettalic compounds, certain sulfide compounds, manganese, methyl mercury, organo phosphate insecticides tetra ethyl lead, thallium trialkyl tin compounds.	They produce primary toxic effects on central nervous systems	Narcosis, behavioural changes, decrease in motor function	Its toxic effects produces nervous system damage. Central nervous system is sensitive to organometallic compound Neurological damage is caused by methyl mercury and tetraethyl lead. Trialkyl tin compounds causes edema of central nervous system. Carbon disulfide acts on the nervous system mainly

S.No.	Nomenclature of toxic chemical substances	Examples	Site of action	Signs and symptoms	Remark
4.	Hematotoxic	Aniline, benzene, nitirites, nitrobenzene, toluidine	They act on blood, bone narrow or hematopoietic systems.	Causes cyanosis (a bluish coloration to the skin) and loss of consciousness	Benzene damages bone marrow which can lead to leukemia nitrobenzene forms methemoglobin and lower blood pressure arsine produces hemolysis.
5.	Carcinogens	Alkylating agents (alphahalo-ether, sulfonates epoxides, electrophilic alkenes and alkynes), Acylating agents, organohalogen compounds, hydrazines, N-nitro compounds, aromatic amines, aromatic hydrocarbons(benzene), cadmium compounds, asbestos	Lung and other vital organs	They initiate or speed up the development of malignant or potentiall malignant tumours of other malignant neoplastic proliferation of cells It induces or enhances neoplasia.	The terms tumorigen, oncogen and blastomogen are synonymous with carcinogen. Naturally occurring carcinogens include aflatoxins, some alkaloids, parascarbon acid, safrol and isosafro
6.	Reproductive toxins	Lead, ethylene oxide, 1-2 dibroma-3-chloroporpane (DBCP)	It damages reproductive system directly or to the fetel tissue	Causes infertility in (azoospermia) males and females	-
7.	Teratogens	Acrylic acid, ailine cadmium, carbon disulfide hexachlorobenzene, lead compounds, nitrobenzene, DBCP etc. Vitamin A and iodine deficiency, ionizing nuclear radiation, rubella and herpes virus, methyl mercury, aspirion, thalidomide.	They cause growth abnormalities in embryos and genetic modification in cells they interface with normal embryonic development without damage to the mother or lethal effects of the fetus	Sterility and birth defects, deviation during prenatal development is, congenital malformations, development of conception and may result even in death.	Effects are not heraditory. It inhibits normal level of growth and development of feuts, during some gestational period. It is harmful to the embryo and not to the mother.
8.	Immune system effects causing chemicals	Arsenic, benzene, cadmium, lead, methyl, mercury, nitrous oxide, polycylic aromatic hydrocarbons (PAH)	-	Causes in immunousuppression	-

S.No.	Nomenclature of toxic chemical substances	Examples	Site of action	Signs and symptoms	Remark
9.	Cardiovascular system damaging chemicals	a. Aliphatic alcohols, aldehydes, glycols. b. Lead and cadimium (heavy metals)	a. Causes direct damage to the heart (cardiotoxic b. Causes damage to the vascular system	-	-
10.	Mutagens	Diethylstibestro(DES) diethyl sulfate, nuclear radiation, benzene (a)pyrene, aflatoxin, aeorlein, DDT	It causes changes in chromosomes of living cells resulting in inherited birth defects.	They are able to induce mutations in DNA and in living cells.	It causes birth defects ie, abnormality in the development of and embryo.
11.	Cummulative	Heavy metals	These materials tend to build up in the body as a result of chronic exposure	The effects are not seen until a critical body burden is reached	-
12.	Synergistic or potentially effect causing chemicals	Exposure to alcohol and chlorinated solvents	-	The resulting effect is greater than the effects predicted based on the individual chemical substances.	This occurs when two or more hazardous chemical are present in combination.

Appendix 1.116: Fate of PAH'S in the Environment

- PAH's enter the air mostly as releases from burning coal, automobiles exhaust wood fire, and forest fire.
- PAH's remains in air attached to dust particles.
- Some PAH's particles can readily evaporate into air from soil or surface waters.
- Most PAH's do not dissolve easily in water, they stick to solid particles and settle to the bottom of lakes or rivers.
- Microorganisms break down PAH's in soil or water after a period of weeks to months.
- In soil, PAH's are most likely to stick tightly to particles, certain PAH's move through soil to contaminate ground water.

Note:

Factsheet about Polycyclic Aromatic Hydrocarbons (PAHs)

Health Effects

Acute: EPA has found polycyclic aromatic hydrocarbons (PAHs) similar to benzo(a)pyrene to potentially cause the following health effects from acute exposures at levels above the MCL: red blood cell damage, leading to anemia; suppressed immune system. MCL - 0.0002 mg/litre

Drinking water levels which are considered "safe" for short-term exposures have not been established at this time.

Chronic: Benzo(a)pyrene has the potential to cause the following health effects from long-term exposures at levels above the MCL: developmental and reproductive effects.

Cancer: There is some evidence that benzo(a)pyrene has the potential to cause cancer from a lifetime exposure at levels above the MCL.

Benzo(a)pyrene is one of a group of compounds called polycyclic aromatic hydrocarbons (PAHs), or polynuclear aromatic hydrocarbons (PNAs). They are not produced or used commercially but are ubiquitous in that they are formed as a result of incomplete combustion of organic materials.

PAHs are found in exhaust from motor vehicles and other gasoline and diesel engines, emission from coal-, oil-, and wood-burning stoves and furnaces, cigarette smoke; general soot and smoke of industrial, municipal, and domestic origin, and cooked foods, especially charcoal-broiied; in incinerators, coke ovens, and asphalt processing and use.

There are two major sources of PAHs in drinking water: 1) contamination of raw water supplies from natural and man-made sources, and 2) leachate from coal tar and asphalt linings in water storage tanks and distribution lines. PAHs in raw water will tend to adsorb to any particulate matter and be removed by filtration before reaching the tap.

PAHs in tap water will mainly be due to the presence of PAH-containing materials in water storage and distribution systems. There are reports that levels can reach 0.01 mg/L with optimum leaching conditions.

Released benzo (a) pyrene is largely associated with particulate matter, soils, and sediments. Although environmental concentrations are highest near sources, its presence in places distant from primary sources indicates that it is reasonably stable in the atmosphere and capable of long distance transport.

When released to air it may be subject to direct photolysis, although adsorption to particulates apparently can retard this process. It may also be removed by reaction with ozone (half-life 37 min) and N02 (half- life 7 days), and an estimated half-life for reaction with photochemically produced hydroxyl radicals is 21.49 hr.

If released to water, it will adsorb very strongly to sediments and particulate matter. It will not hydrolyze. It has been shown to be susceptible to significant metabolism by microorganisms in some natural waters without use as carbon or energy source, but in most waters and in sediments it is stable towards biodegradation. BaP will be expected to undergo significant photodegradation near the surface of waters. Evaporation may be significant with a predicted half-life of 43 days. However, adsorption to sediments and particulates may significantly retard biodegradation, photodegradation, and evaporation.

If released to soil it will be expected to adsorb very strongly and will not be expected to leach to the groundwater. However, its presence in some groundwater samples indicates that it can be transported there by some mechanism. It will not hydrolyze, and evaporation from soils and surfaces is not expected to be significant. Biodegradation tests in soils have resulted in a wide range of reported half-lives: 2 days to 1.9 yr. Based on these values and the apparent lack of a significant competing fate process, biodegradation may be an important process in soils.

Benzo(a)pyrene is expected to bioconcentrate in aquatic organisms that can not metabolize it. The presence of humic acid in solution has been shown to decrease bioconcentration. Those organisms which lack a metabolic detoxification enzyme system, tend to accumulate polycyclic aromatic hydrocarbons.

Human exposure will be from inhalation of contaminated air and consumption of contaminated food and water. Especially high exposure will occur through the smoking of cigarettes and the ingestion of certain foods (eg smoked and charcoal broiled meats and fish).

Chemical/ Physical Properties

Color/ Form/Odor: Pale yellow needlelike crystals, faintly aromatic

M.P: 179-179.3°C B.P.: >360°C

Vapor Pressure: >1 mm Hg at 20°C

Density/Specific Gravity: 1.35 at 15°C

Solubility: 0.0038 mg/L of water at 25°C; very low solubility in water Soil adsorption: very low mobility in soil

Bioconcentration Factor: BCFs range from <1 to 2675 in fish; expected to bioconcentrate in aquatic organisms which are unable to metabolize it.

Trade Names/Synonyms: 3, 4-Benz(a)pyrene; BaP; BP

Appendix 1.117: Modes of Exposure from PAH'S

- Breathing air containing PAH's in the workplace of coke ovens, tar and asphalt production plants, smoke houses and municipal incinerators.
- Breathing air containing PAH's from cigarette smoke, wood smoke, vehicle exhaust and asphalt.
- Coming in contact with air water or soil near hazardous waste site.
- Eating contaminated cereals flour bread vegetables, fruits, meats.
- Drinking contaminated water or milk.
- Nursing infants of mothers living near hazardous waste sites may be exposed to PAH;s through their mother's milk.

PAH- Polycyclic Aromatic Hydrocarbons.

Note:

The term Polycyclic Aromatic Hydrocarbons (PAH's) refer to a large group of organic compounds widely distributed in the atmosphere. Whose molecular structure contains two or more aromatic rings fused together. They are mainly formed during the incomplete burning of coal. Oil and gas, garbage and tobacco. Because of their low vapour pressure, some PAH's are presents both in the form of gases as well as associated with particles. The lighter PAH's such as phenanthrene, are found exclusively in the gas phase, the heavier PAH's such as Benzo (α) pyrene are almost totally absorbed into particles. PAH's are relatively insoluble in water but dissolve in fats and oils.

Humans are always exposed to environmental mixture of PAH's. Population exposed to high concentrations of PAH mixture depict an increased incidence of tumours. Lung cancer is obviously linked to PAH's through inhaled air irrespective of their physical properties and initial routes of exposure, the tumor inducing effects of carcinogenic PAH's depend upon their conversion in the body into compounds which, unlike parent substances are capable of reacting with and damaging the genetic material (DNA) in the nuclei of cells, thus defined as genotoxic carcinogens.

In 1987, the international Agency for Research and Cancer (IARC) classified 48 PAH's according to their likely human carcinogencity.

Classification of Selected PAH's According to Carcinogencity (IARC, 1987)	
Classification	**PAH Compounds**
IARC Group 2A ***'Probably Carcinogenic to humans'***	Benzo (α) pyerene Benzo (α) anthracene Dibenz (ah) anthracene
IARC Group 2B ***'Possibly carcinogenic to humans'***	Benzo (k) fluoranthene Benzo (b) fluoranthene Indeno (123 cd) pyrene
IARC Group 3 ***'not classifiable'***	Benzo (glin) perylene Chrysene, Coronene, Fluorene, Anthrace, Dibenzanthraceone.

Polycyclic Aromatic Hydrocarbons (PAHs) Fact Sheet

PAHs, (polycyclic aromatic hydrocarbons) are a group of chemicals consisting of numerous carbon atoms joined together to form multiple rings. There are at least 10, 000 different PAH compounds. Most are formed from the incomplete combustion of plant or animal matter, or carbon fuels, such as coal or petroleum. PAHs in the environment and familiar as the sooty part of smoke or ash. PAHs can be the result of natural (forest fires) or man-made practices or industrial processes. Many useful products contain PAHs, such as mothballs, blacktop, and creosote wood preservatives. They are also found at low concentrations in some special-purpose skin creams and anti-dandruff shampoos that contain coal tars.

Automobile exhaust, industrial emissions and smoke from burning wood, charcoal and tobacco contain high levels of PAHs. In general, more PAHs form when organic materials burn at low temperatures, such as in wood fires or cigarettes. High-temperature furnaces produce fewer PAHs. Smoke from fires contains tiny particles of PAHs and other chemicals. These are known as particulate matter (PM) which is microscopic in size, It remains suspended in air and can move long distances.

Most PAHs do not dissolve in water but, instead, bind to sediments. When sediments become suspended in water, PAHs can be transported with the sediment. PAHs can enter groundwater from ash, tar, or creosote that is improperly disposed in landfills.

PAHs are everywhere. Urban environments, in particular, have many sources of PAHs.

Low-level PAH exposures: PAHs are extremely common in the environment. Ail of us are exposed to PAHs every day in the air we breathe and the food we eat. Our metabolism is adapted to handle these frequent low-level exposures.

Most people are exposed to PAHs when they breathe smoke, auto emissions or industrial exhausts. Most exhausts contain many different PAH compounds. Frequent exposure over many years may lead to health problems, particularly to the lungs and heart. People with the highest exposures are smokers and, people who live or work with smokers, as well as, roofers, road builders and people who live near major highways or industrial sources.

Grilled, smoked and charbroiled foods, especially meats, are a source of some PAH exposure. PAHs may be in soil near disposal sites where construction wastes or ash is buried. Due to metabolism of PAHs by vertebrates, there is no significant accumulation of PAHs in fish meat. However, eating shellfish living in contaminated water may be another major source of exposure. Vegetables do not take up significant amounts of PAHs that are in soil. PAHs may be in groundwater near disposal sites where construction wastes or ash is buried; people may be exposed by drinking this water.

Concentrated PAHs can affect the skin, particularly when dissolved in oily solvents to form tar.

Contact with oily or tarry materials, combined with sunlight, can cause redness or irritation to the skin. In contaminated waterways, exposures can come from contact with oily sheens in the water or contact with oil on fish or driftwood. Most of the exposure is to the hands, with a lesser probability from splashes to limbs or face. In addition to exposure related to fishing, there is a potential for skin contact via wading. Skin that comes in contact with PAHs should be washed immediately with soap.

Drinking water standards for five PAHs: Anthracene - 3, 000 parts per billion (ppb), Benzo(a)pyrene - 0.2 ppb, Benzo(b)fluoranthene - 0.2 ppb, Fluoranthene - 400 ppb, and Fluorene - 400 ppb. Stop drinking water containing more than these amounts.

PAHs in air are indirectly regulated by the National Ambient Air Quality Standard for PM2.5. PAHs are often an important component of PM.

In general, chemicals affect the same organ systems in all people who are exposed. However, the seriousness of the effects may vary from person to person. A person's reaction depends on several things, including individual health, heredity, previous exposure to chemicals and medicines, and personal habits, such as smoking or drinking.

It is also important to consider the length of exposure to the chemical, the amount of chemical exposure, and whether the chemical was inhaled, touched, or eaten.

Short-Term Exposures

When we breathe high concentrations of PAHs, it is usually in the form of smoke. The harmfulness of breathing smoke comes from the combined effect of many reactive chemicals in smoke. We can also have skin contact with high concentrations of PAHs in the form of tar, such as blacktop or roofing material. When these products get on the skin and are exposed to sunlight, they may cause redness, blistering, and peeling.

Long-term exposures: The following health effects can occur after several years of exposure to PAHs:

Cancer: Occupational and chronic exposure to PAHs may cause cancer. Several PAHs have been shown to cause lung and skin cancer in laboratory animals. People who have worked in industries where they had regular exposure to very high levels developed tar warts and skin cancer. Extracts of various types of smoke containing PAHs caused lung tumors in laboratory animals.

Reproductive Effects: Reproductive problems and problems in unborn babies' development have occurred in laboratory animals that were exposed to benzo(a)pyrene. Other PAHs have not been studied enough to determine whether they cause reproductive problems.

Organ Systems: A person's lungs, liver, skin, and kidneys can be damaged by exposure.

Awareness of potential PAH exposure through hobbies, recreational, and home/outdoor scenarios, and taking action to minimize or avoid exposure may decrease the risk of PAH overexposure. Cigarette smoke contains PAHs and other carcinogenic substances. Exposure to PAHs by smoking or second hand smoke may increase the risk of overexposure to PAHs and PAH-related disease.

Additionally, there are some foods that contain PAHs. Minimizing consumption of charbroiled, chargrilled and smoked meats and fish can all reduce exposure to PAHs.

You can decrease your exposure to PAHs by wearing gloves when working with cutting oils, washing your skin immediately after coming in contact with products or contaminated soils containing PAHs, and avoiding smoke from campfires. Produce grown in contaminated soil should be washed before consumption and root vegetables should be washed and peeled.

Many PAHs can be detected in blood or urine soon after exposure. Tests for these compounds are not routine and can only be performed using special equipment not usually found in doctor's offices. People who think they may have been exposed to PAHs for a long time should contact their physician. Blood tests of liver and kidney function are available. People exposed to PAHs in air should get their lung function tests done.

Appendix 1.118: Environmental and Occupational Hazards of Chemicals

The physical form of a chemical can affect how it enters your body and to some extent, the damage it causes. The main physical forms of chemicals are solids, dusts, liquids, vapours and gases.

Solids

- Solids are the least likely of the chemical forms to cause chemical poisoning. However, certain chemical solids can cause poisoning if they get onto your skin or food and you then ingest them.
- The greatest danger with solids is that some work processes can change them into a more dangerous form. For example, wood that is being cut can turn into wood dust which can then be inhaled. Welding rods can decompose into fumes and gases. Polyurethane foam is safe in its normal solid form but gives off deadly gases, if it bums.
- Chemicals in solid form can give off toxic vapours which can be inhaled, and solids can be flammable and explosive, and corrosive to the skin.
- Effective control measures should be used with chemical solids, particularly during work processes which may change them into more hazardous materials.

Dusts

- You can be exposed to dust in the workplace from materials that normally exist in dust form (for example, bags of cement), or from work processes that create dust (for example, handling glass fibre can produce toxic dust).
- The main danger from harmful dusts is that you can breathe (inhale) them into your lungs. When breathed in, the larger dust particles are usually trapped by nasal hairs and mucus and then removed by the body. Smaller particles, however, are more dangerous because they can get deep inside the lungs where they can have damaging effects, or they can be absorbed into the bloodstream and travel to other parts of the body where they can cause damage. They can also cause eye damage.
- Dusts can be hard to see – you often cannot even see a cloud of tiny dust particles except with special lighting. Particles less than 40 microns are not visible to naked eye.
- Under certain conditions dusts can explode. An example of this is an explosion in a grain silo, flour mill or coal grinding ball mills.

- Effective control measures should be used to keep dust in the workplace at "safe" levels.

Liquids

- Many hazardous substances, such as acids and solvents, are liquids when they are at normal temperature.
- Many liquid chemicals give off vapours which you can inhale and which may be highly toxic, depending on the chemical.
- Liquid chemicals can be absorbed by skin. Some liquid chemicals may cause immediate skin damage (they may or may not be absorbed into the bloodstream as well). Other liquids pass directly through the skin into the bloodstream, where they can travel to different parts of the body and cause damaging effects.
- Effective control measures should be used with liquid chemicals to eliminate or reduce the possibility of inhalation, skin exposure and eye damage.

Vapours

- A vapour is the gas phase of a material which is normally liquid under standard conditions.
- Tiny droplets of liquid which are suspended in the air are called mists.
- Many liquid chemicals evaporate at room temperature, which actually means that they form a vapour and stay in the air.
- The vapours from some chemicals can irritate your eyes and skin.
- There can be a variety of serious health effects from inhaling certain toxic chemical vapours.
- Vapours can be flammable or explosive. To avoid fire or explosion, it is important to keep chemicals that vaporize away from any sparks, sources of ignition or incompatible chemicals.
- Controls should be used to prevent worker exposure to vapours from liquids, solids or other chemical forms.
- Gasoline and water are two examples of liquids which generate vapour under standard conditions.

Gases

- Some chemical substances are in the form of a gas when they are at a normal temperature. However, some chemicals in liquid or solid form become gases when they are heated.
- You can detect some gases easily by their colour or smell, but there are other gases that you cannot see or smell at all - you can only detect them with special equipment.
- Gases can be inhaled.
- Some gases produce irritant effects immediately. The health effects of other gases may be noticeable only after your health has already been seriously damaged. Gases may be flammable or explosive. Extreme caution should be used when working around flammable or explosive gases.
- Workers should be protected from the potential harmful effects of chemical gases with effective control measures in the workplace.
- Some examples of gases are: nitrogen, nitrogen dioxide, carbon monoxide, carbon dioxide, and oxygen.

The physical form, route of entry, affected organ and type of toxicity of some common industrial chemicals.

S.No.	Chemicals	Physical form	Route of entry to the body	Organ(s) that can be affected	Class of toxicity	Symptoms	Examples of industries
1.	Cadmium metal and some of its compounds	Dusts, vapours.	Inhalation	Lungs, throat, kidneys	Poisonous, causing damage to lungs, kidneys on chronic exposure.	Dry burning throat, chest pain, vomiting, headaches	Metal industries, welding processes, heavy chemicals
		Dusts	Ingestion				
2.	Toluene diisocyanate	Vapour	Inhalation	Lungs	Allergenic	Industrial asthma due to lung effects	Industrial processes involving polyureth ane manufact ure, nts and inks
		Solid	Spillage on skin	Skin	Allergenic	Dermatitis	
3.	Mercury and many of its compounds	Vapour (mercury itself) dust	Inhalation	Brain and nervous system, kidneys	Poisonous. Often irreversible damage to nervous system	Loss of muscular coordination, loss of mental ability.	Heavy chemicals, laboratory workers, engineering
		Liquids dusts	Spillage on skin, ingestion				
4.	Cloroform, Carbon tetrachloride Trichloro-ethylene	Vapour	Inhalation	Brain, liver, kidneys skin	Poisonous. Carcinogenic? Chronic exposure may lead to liver and kidney failure	Drowsiness	Light engineeri ng, 1 heavy chemical s, cleaning, ice workers
		Liquid	Spillage on skin			Dermatitis	
5.	Auramine	Dusts, vapour	Inhalation	Bladder	Carcinogenic to bladder.	Blood in urine	Dyes industry, pottery and glazing industries
		Dusts	Spillage on skin	Skin	Irritant	Inflammation, bums	

Concepts of Toxicology Everything is Toxic; It all Depends on the Dose

How Well the Body Accepts a Substance Depends on:

- The Type of Substance,
- The Amount (Dose) Absorbed.
- The Period of lime Over Which It Is Absorbed.
- The Susceptibility/Sensitivity of the Person Exposed.
  - Acute Effects
  - Chronic Effects
- **Target Organs**
  - Etiologic agents
  - Hepatotoxic
  - Nephrotoxins
  - Neurotoxins
  - Pulmonary Toxir
  - Mutagens
  - Teratogens
  - Carcinogens
- **Hazardous Atmospheres**
  - Flammable
  - O_2 deficient
  - Toxic
- **Hazardous Concentrations**
  - IDLH
  - LD50
  - LC50
  - TLV
  - TWA
  - STEL
  - PEL

Routes of Entry of Chemicals into the Body of Human Beings

- Inhalation by breathing
- Absorption by skin
- Ingestion/swallowing
- Injection

Chemical Hazards

- Routes of entry-Inhalation, Ingestion, skin absorption. (inhalation is the main route of entr)
- Chemical agents can be classified into-

1. Metals – Lead, TEL, As, Hg, Cd, Ni, Co etc.
2. Aromatic Hydrocarbons- Benzene.
   Toluene, phenol etc.
3. Aliphatic Hydrocarbons- Methyl alcohol

4. Gases *Simple asphyxiants: N_2, CH_4, CO_2

*Chemical asphyxiants : CO, H_2S, HCN
*Irritant gases: Ammonia, SO_2, Cl_2,
*Systemic poison: CS_2

Entry Routes of Chemicals in the Body

Inhalation (Breathing)

Chemicals in the air are breathed in through the mouth or nose.
Gases and vapors are absorbed through the lungs directly into the bloodstream.
The size of dust particles or mist droplets can affect where the chemical settles in the respiratory tract.

Ingestion (Swallowing)

Chemicals that are swallowed are absorbed in the digestive tract.

Chemicals can rub off dirty hands and contaminate food, drinks or tobacco products.

Chemicals in the air can settle on food or drink and be swallowed.

Skin Absorption

Some chemicals can pass through the skin into the body.
These chemicals can then cause various health effects.

Biochemical Actions of Chemicals

There are a number of factors that determine the type of toxic effect a chemical can have on you. These factors include:

- the chemical composition of the hazardous substance (certain substances are more harmful than others because of their chemical structure);
- the physical form of the chemical (dust, vapour, liquid, etc.);
- the route of entry by which the chemical gets into the body (chemicals have different routes of entry. Some chemicals can enter the body in more than one way. Different health effects can occur depending on the route of entry);
- the particular tissues and organs in which the chemical collects or localizes;
- the frequency, concentration, and length of exposure; and
- the worker's individual response to the chemical, which can vary a great deal from person to person.

Table below shows some of the types of toxic effects that certain industrial chemicals can cause. The table includes the toxic property of the chemical (for example, whether it is carcinogenic, poisonous, causes an allergic reaction, etc.), the part of the body affected by the chemical, how long it takes until disease develops, the type of effect the chemical causes, and some examples of chemicals that cause those effects.

Dangerous Toxic Properties of Chemicals Found in the Work Environment

S.No.	Toxic Property of Chemical	Part of body affected	Time scale of appearence	Effects	Examples of chemicals
1.	Irritant or corrosive	Any, but usually the eyes, lungs and skin	A few minutes to several days	inflammation, bums and blisters of exposed area. Frequently healed after acute exposure. Chronic exposure may lead to permanent damage.	Ammonia, sulphuric acid, nitrogen oxides, caustic soda
2.	Fibrogenic	Generally lungs	Years	Gradual cumulative loss of lung function leading to disability and death if there is chronic exposure.	Bauxite dust, asbestos, bagasse
3.	Allergic	Any, but frequently lungs and skin	Days to years	In lungs may lead to chronic asthma-like disease and permanent disability. In skin may produce industrial dermatitis.	Toluene, di-isocyanate (TDI), amine hardeners for epoxy resins.
4.	Dermatitic	Skin	Days to years	Inflamed, peeling skin rashes. May result from chronic exposure to irritants, allergenic agents, solvents or detergents.	Strong acids, alkalis, detergents, carbon tetrachloride, trichloroethylene.
5.	Carcinogenic (Cancerous)	Any organs, bu frequently skin, lungs, bladder	10 to 40 years	Cancer in affected organ or tissue. Ultimately this may cause premature death.	2-Naphthylamine, certain tars and oils, benzidine, asbestos
6.	Poisonous	Any organs but frequently liver, brain, kidney	A few minutes to many years	Death of cells in vital organis with eventula failure of organ to carry out important biological functions. Ultimately can cause death.	Carbon tetrachloride, mercury, cadmium, carbon monoxide, hydrogen cyanide.
7.	Asphyxiants (suffocants)	Lungs	Minutes	Gases replace normal oxygen content of air	Acetylene, carbon dioxide

The Chemicals Produce a Variety of Health Effects Including

- **irritant or corrosive**: As the name implies, these chemicals can cause inflammation and/or blistering of the skin. Short-term exposure frequently heals whereas long-term exposure can lead to permanent damage (chemicals examples- ammonia, sulphuric acid, and caustic soda);
- **allergenic:** for susceptible workers, these chemicals produce asthma-like symptoms (lungs) or industrial dermatitis (skin) [examples include toluene diisocyanate- TDI, epoxy resins, formaldehyde];
- **fibrogenic**: these chemicals result in the gradual cumulative loss of lung function as "elastic" alveolar tissues are damaged by dust. Virtually any dust can have this effect
- **asphyxiant:** these are chemicals which stop the normal absorption of oxygen by the body either by replacing the oxygen or inhibiting its uptake; The gases, carbon dioxide and carbon monoxide, are examples;
- **narcosis:** chemicals such as a number of solvents which depress the central nervous system and produce "light headedness" which can be a factor in accidents;
- **poisonous**: these chemicals cause damage/death of cells in vital organs leading to the failure of that organ to function effectively which can ultimately lead to death;
- **carcinogenic**: these chemicals cause cancer in exposed workers usually over a long period of time. Examples include - arsenic, asbestos, chromium, nickel, 2- napthylamine, etc.
- **teratogenic and mutagenic**: these chemicals affect the foetus in the womb or eggs/sperm for future generations.

Carcinogens

Carcinogens are cancer-causing compounds.

Some chemicals are known human carcinogens, others are only suspected as carcinogens.

DOSH has regulations covering the general use of carcinogens, and has specific regulations for several known human carcinogens.

DOSH has specific regulations on the following carcinogens

- Vinyl Chloride
- Acrylonitrile
- l, 2, -Dibromo-3-chloropropane (DBCP)
- Arsenic
- Ethylene Oxide
- Cadmium
- Butadiene
- Methylene Chloride
- Benzene
- Hexavalent Chromium

Other Groups of Toxic Chemicals

Teratogens are compounds that can harm the developing fetus, causing birth defects or death.

Mutagens cause genetic mutations or changes. These mutations can cause' birth defects or other problems in following generations or may lead to cancer in the exposed person.

Sensitizers can "switch on" a reaction in an individual worker.

The reaction to a sensitizer depends upon the individual worker.

Once a worker becomes sensitized to a compound, smaller and smaller exposures can cause a reaction, and the reactions can become more severe.

Toxicity:
how poisonous are chemicals?

Dose: The effects of any toxic chemical depends on the amount of a chemical that actually enters the body.

Acute Toxicity: the measure of how toxic a chemical is in a single dose over a short period of time.

Chronic Toxicity: the measure of the toxicity of exposure to a chemical over a long period of time.

Some chemicals will only make you sick if you get an `acute" or high dose all at once.

Example: ammonia

Some chemicals are mainly known for their chronic or long-term effects. Example - asbestos

Most chemicals have both acute and chronic effects. Example carbon monoxide

The health risk is dependent upon the toxicity of the chemical, the types of effects and the various routes of entry.

Toxicity, vs. Hazard of Chemicals

a. Toxicity is ability of a chemical to act as a poison or cause injury to tissues.

b. Hazard is likelihood that a chemical will cause injury in a given environment or situation; degree of hazard depends on how toxic the substance is, how it is absorbed, etc.

Acute vs. Chronic Exposures

a. Acute Exposures means exposure of short duration, usually to relatively high concentrations or amounts of material.

b. Chronic Exposures means continuous or intermittent exposure extending over a long period, usually to relatively low material amounts or concentrations.

Local vs. Systemic Effects

a. Local effects of the chemical may be localized on a specific area of the body such as nose or throat.

b. Systemic Effects means entire body system and organs are affected.

Exposure to toxic chemicals can also lead to higher rates of accidents at work. For example, chemicals such as solvents and asphyxiants may slow your reaction time by affecting nervous system or limiting the amount of oxygen that gets to lungs. A slow reaction can be very serious (or even fatal) if you are in a dangerous situation that requires an immediate response. Unfortunately, when accidents occur in the workplace, management often blames the worker, claiming he or she was careless. This tendency to "blame the victim" is yet another reason to learn about the substances you work with, to make sure the proper control measures are in place, and to know your rights

Chemical Exposure Limits

Many chemicals have exposure limits, or allowable amounts of a chemical in the air.

These limits are often called "permissible Exposure Limits (PEL)" or Threshold limit Values (TLV)" or "Recommended exposure Limits(RELs)", or, "Time weighted Averages (TWA)" based on 8-hour average exposure – Short Term Exposure Limits(STEL)" or ceiling or peak levels.

Chemicals exposure Levels must be kept below these limits for safety.

The NIOSH recommended exposure limits (RELs). For NIOSH RELs, "**TWA**" indicates a time-weighted average concentration for up to a 10-hour workday during a 40-hour workweek. A short-term exposure limit (STEL) is designated by "**ST**" preceding the value; unless noted otherwise the STEL is a 15-minute TWA exposure that should not be exceeded at any time during a workday. A ceiling REL is designated by "**C**" preceding the value; unless noted otherwise, the ceiling value should not be exceeded at any time. Any substance that NIOSH considers to be a potential occupational carcinogen is designated by the notation "**Ca**".

Keep them away from heat, sparks, or flames. Friction from rubbing objects together may produce enough heat or sparks to ignite these materials. If you need to heat flammable solvents, use Standard Methods of Prevention:

1. **Incompatible materials.** Keep incompatible chemicals away from each other. For example, don't place strong acid solutions near strong bases, peroxides near organic solvents, bleach near chlorine, etc. When mining materials, such as oxidizers and reducers, or acids and bases, dilute them first.
2. **Avoid contact**: Protect your five routes of entry. Wear suitable PPE. Eye protection must be worn when you handle chemicals. Avoid skin contact. Wear compatible gloves if you are going to use organic solvents or flammable or combustible liquids. Wear an apron when handling corrosive materials. Use appropriate engineering controls, such as a hood, if there is a risk of inhaling fumes or dust. Keep containers tightly sealed.
3. **Hygiene**: Keep yourself clean of chemicals:

   Keep everything away from your face to prevent contact with your mouth and eyes. No smoking, eating, drinking, or applying makeup or lip balm. No mouth pipeting: attach approved bulbs or other devices to the pipet. Keep pens, pencils, and your fingers away from your face.

   Assume your hands are contaminated, even if you wear gloves. Wash your hands frequently and upon leaving the laboratory.

   - Clean up chemical spills promptly. They tend to migrate if left alone, and they can contaminate other areas.
   - Remove soiled gloves, lab coats, aprons, etc., before leaving the laboratory.
   - Tie back long hair, clip or remove neckties, arid avoid wearing dangling jewelry. You don't want these items to fall into chemicals or touch an open flame. They may also get tangled in vacuum pump motors and other moving parts.
4. **Transporting chemicals**: When transporting chemicals between the storage area and your workplace, put containers in protective carrying devices. Glass bottles can break if they bump against a wall, door, or other obstruction. Carry only what you can hold in your hands. Don't hug

several jars or bottles near your body; you may drop something, or someone may walk into you and break a container. If you have a lot to carry, use a cart with a rim on each shelf. Keep containers in protective devices so they won't rattle or tip. Don't transport incompatible chemicals together.

Chemical Handling

Review the following list from time to time to check your own knowledge of the safe handling of chemicals:

Be careful: Watch where you place containers, and close them when you have finished. Read the label to be sure that you're working with the right materials. For example, there's a big difference between sodium thiosulfate and sodium thiosulfite. Don't rush – that's when you're more likely to spill or make a mistake.

Stay alert: Fatigue affects judgment. You need to focus your attention on a number of different procedures and assignments. Often you're rushing to meet deadlines. If you need to work extra hours, take periodic breaks. Be careful not to leave anything unattended if it creates a hazardous situation. If you take prescription medication that causes fatigue, avoid handling hazardous chemicals or equipment. If you are unable to concentrate, or if you feel fatigued or confused, speak to your supervisor immediately-«You may be affected by an overexposure and not be aware of it!

Follow directions: Make sure you know the correct procedures, and any associated hazards, for any analysis you perform or technique you use. Sometimes larger quantities of materials may require additional precautions or different procedures. Some exothermic reactions may generate too much heat or may explode if scaled-up. You may need different equipment or more engineering controls. When working on a new technique or synthesizing a new compound, discuss any new activities with your supervisor.

Monitor Heating: Monitor any reactions you are heating. They may boil over, or the solvent may evaporate. Check water or oil levels in temperature baths. Many liquids, solids and gases are flammable.

Know what you're working with: Read material safety data sheets (MSDSs) and container labels for all chemicals. Know how the chemicals can hurt you, how to avoid harmful conditions in your work area, and how to respond to an accident.

Think about what you're doing: It's easy to develop a routine and habits in your work. Even though you've used a particular procedure for years, you should still pay close attention to what you're doing. Accidents are more likely to occur when your level of concentration is low especially early in your career, they stay with you, and you're more likely to stay healthy.

Follow all safety procedures: Obey the warning signs posted in your lab and follow the directions and precautions on chemical labels. Use engineering controls as they are intended. Use a hood if fumes or dust will be released. Use a shield for protection against splashes or explosion. Always wear approved eye protection and the PPE required for the chemicals you are handling.

Practice good housekeeping and personal hygiene: Keep your work area clean and free of chemical residues. Wash your hands frequently, even if you wear gloves while handling chemicals. Remove protective clothing when you leave the lab (including lab coats and aprons). Make sure all contaminated clothing is properly cleaned.

Report dangerous activities or situations: Pay attention to the way other workers handle hazardous materials. Someone else's mistake can hurt you. Take care of any frayed wires, spills, non-labeled containers, or malfunctioning equipment. If you can't handle the situation alone, talk to your supervisor. These hazards won't take care of themselves and could lead to injury.

Review the following list from time to time to check your own knowledge of the safe handling of chemicals:

Be careful: Watch where you place containers, and close them when you are finished. Read the label to be sure you're working with the right materials. For example, there's a big difference between sodium thiosulfate and sodium thiosulfite. Don't rush - that's when you're more likely to spill or make a mistake.

Stay alert: Fatigue affects judgment. You need to focus your attention on a number of different procedures and assignments. Often you're rushing to meet deadlines. If you need to work extra hours, take periodic breaks. Be careful not to leave anything unattended if it creates a hazardous situation. If you take prescription medication that causes fatigue, avoid handling hazardous chemicals or equipment. If you are unable to concentrate, or if you feel fatigued or confused, speak to your supervisor immediately You may be affected by an overexposure and not be aware of it!

Follow directions: Make sure you know the correct procedures, and any associated hazards, for any analysis you perform or technique you use. Sometimes larger quantities of materials may require additional precautions or different procedures. Some exothermic reactions may generate too much heat or may explode if scaled-up. You may need different equipment or more engineering controls. When working On a new technique or synthesizing a new compound, discuss any new activities with your supervisor.

TWA concentrations for OSHA PELs must not be exceeded during any 8-hour work shift of a 40-hour workweek. A STEL is designated by **"ST"** preceding the value and is measured over a 15-minute period unless noted otherwise. OSHA ceiling concentrations (designated by **"C"** preceding the value) must not be exceeded during any part of the workday; if instantaneous monitoring is not feasible, the ceiling must be assessed as a 15-minute TWA exposure. In addition, there are a number of substances from e.g., beryllium, ethylene dibromide that have PEL ceiling values that must not be exceeded except for specified excursions. For example, a "5-minute maximum peak in any 2 hours" means that a 5-minute exposure above the ceiling value, but never above the maximum peak, is allowed in any 2 hours during an 8-hour workday.

Concentrations are given in ppm, mg/m^3, mppcf (millions of particles per cubic foot of air as determined from counting an impinger sample), or $fibers/cm^3$ (fibers per cubic centimeter). The **"[skin]"** designation indicates the potential for dermal absorption; skin exposure should be prevented as necessary through the use of good work practices, gloves, coveralls, goggles, and other appropriate equipment. The **"(total)"** designation indicates that the REL or PEL listed is for "total particulate" versus the **"(resp)"** designation which refers to the "respirable fraction" of the airborne particulate.

Chemical Health Risk Assessment (CHRA)

Chemical Health Risk Assessment (CHRA) is corner stone on which compliance with the Use and Standard of Exposure of Chemicals Hazardous to Health (USECHH) Regulations 2000 is achieved. USECHH Regulations require employer to make a comprehensive assessment of the risk of employees exposure to chemical hazardous to health in the workplace for the purpose of enabling decision to be made on appropriate control, measure, further training of employee, and monitoring health surveillance activities as may be required to protect the health of employees who may be exposed to chemical hazardous to health at work.

NIOSH offer comprehensive CHRA services in providing guidance on improving current control measure Complete with report and the recommendation in accordance to USECHH 2000.

Common Laboratory Safety Practices

Your laboratory will have policies for your specific activities but all laboratories follow these common safety practices:

Think safety first: Review the procedures you will use, and look for potential hazards from chemicals and equipment Plan your work. Determine what protection you need, such as engineering controls and

equipment, personal protective equipment, and safe procedures and gases are flammable. Keep them away from heat, sparks, or flames. Friction from rubbing objects together may produce enough heat or sparks to ignite these materials. If you need to heat flammable solvents, use Standard Methods of Prevention.

Incompatible materials: Keep incompatible chemicals away from each other. For example, don't place strong acid solutions near strong bases, peroxides near organic solvents, bleach near chlorine, etc. When mixing materials, such as oxidizers and reducers, or acids and bases, dilute them first.

Avoid contact: Protect your five routes of entry. Wear suitable PPE (personal protective equipments) Eye protection must be worn when you handle chemicals. Avoid skin contact. Wear compatible gloves if you are going to use organic solvents or flammable or combustible liquids. Wear an apron when handling corrosive materials. Use appropriate engineering controls, such as a hood, if there is a risk of inhaling fumes or dust. Keep containers tightly sealed.

Hygiene: Keep yourself clean of chemicals:

- Keep everything away from your face to prevent contact with your mouth and eyes. No smoking, eating, drinking or applying makeup or lip balm. No mouth pipeting; attach approved bulbs or other devices to the pipet. Keep pens, pencils, and your fingers away from your face.
- Assume your hands are contaminated, even if you wear gloves. Wash your hands frequently and upon leaving the laboratory.
- Clean up chemical spills promptly. They tend to migrate if left alone, and they can contaminate other areas.
- Remove soiled gloves, lab coats, aprons, etc., before leaving the laboratory.
- Tie back long hair, clip or remove neckties, and avoid wearing dangling jewelry. You don't want these items to fall into chemicals or touch an open flame. They may also get tangled in vacuum pump motors and other moving parts.
- **Transporting chemicals:** When transporting chemicals between the storage area and your workplace, put containers in protective carrying devices. Glass bottles can break if they bump against a wall, door, or other obstruction. Carry only what you can hold in your hands. Don't hug several jars or bottles near your body; you may drop something, or someone may walk into you and break a container. If you have a lot to carry, use a cart with a rim on each shelf. Keep containers in protective devices so they won't rattle or tip. Don't transport incompatible chemicals together.

Hazardous Concentrations of chemicals are represented by following parameters

- IDLH (Immediate danger to life and health)
- LD50 (Lethal dose 50)
- LC50 (Lethal concentration 50)
- TLV (Threshold Limit Value)
- TWA (Time weighted Average)
- STEL (Short term Exposure Limit)
- PEL (Permissible Exposure Limit)
- REL (Recommended Exposure Limit)

Control Methods for Chemical Hazards

Control of chemicals in the workplace should be according to the following hierarchy of priorities:

a. Elimination of the most toxic materials,

b. Substitution of hazardous materials by less hazardous ones.

c. Containment, or total enclosure of toxic processes or substances.

d. Adoption of safe handling procedures.

e. Exhaust ventilation, using efficient hoods, ducting, filters, and fume cupboards, and noise-controlled fans or suction devices.

f. Environmental monitoring – fixed point, grab sampling and/or personal monitoring - should always accompany containment and ventilation control measures, to monitor their effectiveness.

A. Designated Area

This is an area assigned for the usage of either a particularly hazardous substance or purpose. For example, if carcinogens are being used in the lab, a "designated area" should be assigned, and warning label should be posted.

There are no longer clean areas allowed in an active research laboratory.

B. Engineering Controls

This is the most effective and desirable method for minimizing risk of exposure either to toxic chemicals or to mechanical equipment. Examples of engineering controls: guards, remote controls, or interlock systems. However, for toxic fumes, mists, and vapors, ventilation systems are the best approaches to help reduce personal exposure to acceptable levels. Generally, there are two types of ventilation systems:

a. Dilution Ventilation

In most buildings a certain percentage of the building air is recirculated periodically through the building ventilation systems but in laboratories all air is exhausted directly to the outside. This "single pass" system is a type of "dilution ventilation" system for controlling low risk airborne contaminants. They are simply exhausted to the outside before they can build to hazardous levels.

b. Local Exhaust Ventilation

Used for moderate to high-risk contaminants. Local exhaust systems capture the airborne contaminants much more effectively than dilution systems such as chemical fume hoods.

- Local exhaust ventilation (e.g. an exhaust hood) should be used to reduce solvent exposures during the spotting process. An exhaust hood should be located near the source of the spotting chemicals to prevent the vapors from entering the workers' breathing zone. A slot hood at table level or a downdraft table exhausting to the outside of the building provides good protection for workers. Spotting tables equipped with an exhaust hood are now commercially available
- General ventilation should include both supply and exhaust air. Make-up air supply should be designed and used in conjunction with the exhaust system. An inadequate volume of make-up air may result in negative pressure areas, perhaps resulting in drafts that interfere with the exhaust hood. When make-up air is introduced, it should enter the opposite side of the room from the exhaust to prevent short circuiting. Ventilation systems in dry cleaning shops should provide a minimum of 30 cubic feet per minute (cfm) of outside air per person.

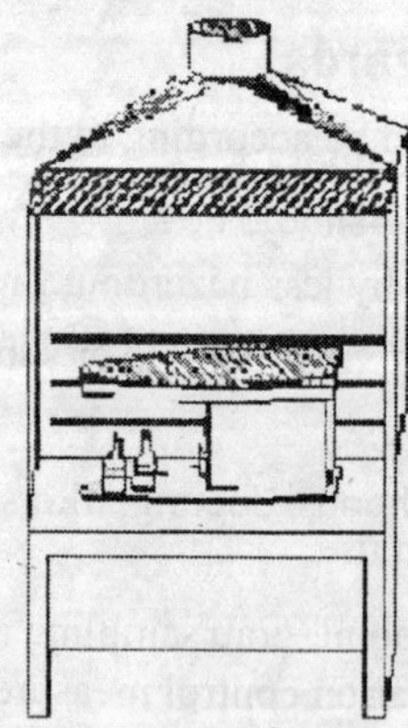

Diagram showing spotting table with local exhaust ventilation hood.

c. **Fume Hood**

The fume hood is designed to contain and disperse gases, vapors, and aerosols to the external environment. It does not provide absolute containment or protection from the materials in the hoods, however, a properly designed hood in a properly designed room can provide adequate protection if the following practices are observed:

- Inspect and ensure that the hood is working.
- Do not store chemicals and equipment in the hood
- Remove unnecessary chemicals and equipment. 15
- All equipments should be at least 1.5cms back from the front sash
- Position the sash no higher than the approved working height that is designated by a fluorescent yellow sticker.
- When evaporating or distilling perchloric acid, special perchloric acid fume hoods MUST be used.

C. Work Practice Controls

In most cases, a well-designed set of work practices is the best risk management tool.

a. **Chemical Transportation:** Assure that an unbreakable secondary container is being used, and that transport carts are designed for this purpose.

b. **Eating. Drinking and Smoking:** There should be no eating, drinking, smoking, chewing of gum or tobacco, application of cosmetics, storage of utensils, food, or food containers in the laboratories.

c. **Pipetting:** Mechanical pipetting aids should always be used for all pipetting procedures. Oral pipetting is prohibited.

d. **Personal Hygiene:** All personnel should wash their hands immediately after the completion of any procedure in which chemicals have been used and when they leave the laboratory. If hazardous chemical exposures occur to skin, immediately shower or wash affected areas.

e. **Housekeeping:** Keeping the working area clean and orderly reduces the frequency and severity of accidents. Here is some common sense tips:

- Keep aisles, exits, stairs and hallways free of obstructions.
- Avoid slip hazards by keeping the floor clean of ice, stoppers, glass beads or rods, other small

items and spilled liquids.

- Keep drawers and cabinet doors closed.
- Never store chemicals on the floor.
- Workspaces and storage areas should be kept clear of broken glassware, leftover chemicals and scraps of paper.
- Place all non-contaminated broken glass in rigid containers clearly marked "Broken Glass".

Chemical Safety: Ten Basic Rules

1. Know tie hazards of chemicals in use.
2. Label all chemicals and their waste properly.
3. Use PPE while handling hazardous chemicals.
4. Work with volatile and hazardous chemicals in a fume hood.
5. Store flammables properly.
6. Do not work alone with hazardous chemicals.
7. Maintain clear access to exits, showers and eyewashes
8. Keep work areas free to clutter.
9. Wash promptly when chemical contacts skin.
10. Do not eat, drink, and apply cosmetics in laboratory.

1. Recognition of Chemical Hazards

It may sound obvious but how do you know if you have a chemical problem in your workplace? Some chemical hazards are obvious as in the case of an acid bum, accordingly you can come up with various control strategies for handling the acid. However, many chemical hazards are more insidious in their effect.

They do not produce any immediate, obvious effects on the workers and you may not be aware of the slow onset of any occupational disease. You may have a suspicion that something is wrong but be unsure about what to do. In such cases, there a number of ways of ways to look into the problem:

- ***Direct health indicators*** - Are specific workers off sick more than others?

  Has the company nurse or doctor seen an increase in ill health in certain groups of workers? Do any workers have obvious health effects from working with a specific chemical? Have the workers complained when using a specific chemical?.
- ***Experience/knowledge of a specific chemical***- Managers should have approved Material Safety Data Sheets (MSDSs) for all chemicals used in their workplaces. These MSDSs are produced by the chemical manufacturers and, in many countries of the **world, are required by law**

Warning Properties

Warning properties, such as those shown below, are examples of characteristics that could alert you to the presence of a chemical in the air.

- Odor threshold
- Color of product
- Other senses

Odor Threshold

The odor threshold is the airborne concentration at which a hazardous chemical can be detected by smell. This concentration of gas or vapor is expressed as parts per million parts (ppm) of air.

This odor threshold is considered to be one of the warning properties of gases and vapors but must always be used with caution because of variations between individuals and their senses of smell. Allergies, head colds, and olfactory fatigue also can reduce one's ability to smell.

Color of product

Color can present a visual indication of the presence of a potentially hazardous chemical. Two examples are fuming nitric acid, which creates a red cloud, and chlorine gas, which has a greenish color. Both of these materials are highly toxic.

Other Senses

Mild irritation of the eyes, nose, or throat can be a useful warning property if it occurs at a concentration that does not produce other harmful or toxic effects. Some chemicals will produce a taste before, or instead of, odor or irritation.

- Ammonia is a good example of a chemical with a useful odor threshold. Some individuals can detect ammonia at 5 ppm which is below the average of 17ppm.
- It is also 10 times lower than its OSHA Permissible Exposure Limit. (PPL) Since the odor threshold can warm you well before the Threshold Limit Value (TLV) and PEL are reached, ammonia is said to have adequate warning properties.
- Another reference indicates that Methyl Formate has an odor threshold of 2, 000 ppm, Which is 20 times higher than its PEL of 100 ppm. Thus, it is dangerous to rely on methyl formates odor as a warning property.

Odour threshold and Permissible exposure Limits (PEL) of some chemicals are given below:

S.No.	Chemical	Odour Threshold (ppm)	PEL (ppm)
1.	Acetone	62	1000
2.	Ammonia	5	50
3.	Hydrazine	3.7	1
4.	1, 1, 2-Tricholoroethelyene	82	100
5.	Methyl Formate	2000	100
6.	Hydrogen sulfide(H_2S)	0.3	10(ceiling Limit)

2. Evaluation of Chemical Hazards

Evaluating or measuring the risk of chemicals is difficult. This normally involves taking air samples and comparing the level with recognized exposure limit standards. This is often time consuming, expensive and open to different interpretations. Often the equipment gives only a "qualitative" snap-shot value. Evaluating the risk means ranking them as low, medium or high depending upon the toxicity of the chemical, duration of exposure, type of exposure, use of PPE etc. Even though there are legal requirements in Cambodia relating to Air Pollution Control which specify the maximum allowable concentration of hazardous substances in ambient air, the enforcement agencies do not have the requisite equipment to measure chemicals in the workplace.

3. Control of Hazards

Once you have recognised that you have a problem with a chemical hazard in the workplace, what can you do about it? For all health and safety problems, including chemical hazards, there is a logical, systematic strategy or sequence for dealing with them ranging from *elimination* to the use of *personal protective equipment* (PPE). As the name implies, the most effective way of controlling chemical hazards, is by eliminating them from the process or by finding a safe substitute. Obviously if you can remove the chemical hazard from the workplace, you have solved the problem. This however rarely occurs as the chemical in question is usually a vital part of the process or a natural by-product. The next best method is to use a safer substitute for the original chemical (this is best done by seeking advice from the competent authorities and by comparing the information for the different chemicals on their specific MSDSs Some examples of safe substitution could include:

- Use less hazardous solvents instead of toxic ones;
- Use water and detergents instead of solvents for cleaning;
- Use leadless glazes and paints.

Using safer substitutes does not mean the new chemical is safe - all it means is that it is safer than the original hazard.

There are a number of engineering measures (including ***enclosure, isolation*** and ***ventilation***) that can be used to control chemical hazards either by partially or totally enclosing the process. Highly toxic materials that may be released into the workplace atmosphere should be totally enclosed, usually by using a mechanical handling device or a closed glove system that can be operated from the outside. If the process is particularly hazardous, it can also be isolated to other parts of the factory or separate rooms where there are fewer workers. When using enclosure and isolation it is important to consider all aspects of ventilation. As we saw in the section on ventilation, general ventilation can be used to "dilute" chemical hazards in the workplace and local exhaust ventilation/partial enclosures can be used to remove more toxic chemical hazards to the outside of a factory.

Only as a last resort should workers be required to wear personal ***protective equipment*** – after all, by the use of PPE, we have admitted that we cannot control the hazard at source or along the pathway between the source and the worker. The use of PPE is the least efficient, but often the cheapest, method of control. PPE is often uncomfortable to wear, especially in the hot, humid conditions found in many workplaces and, it is not uncommon to find workers only wearing the PPE at the times of a visit by a Labour Inspector Officer. Workers sometimes complain that wearing PPE inhibits their work performance and, in the case of hearing protection, can limit communication and prevent workers from hearing any warning signals. Examples of PPE include - safety glasses; ear muffs and plugs; respirators (different types for dusts, fumes, vapours etc.); gloves; safety shoes; helmets and hard hats; aprons and overalls.It is important that workers are provided with the correct type of PPE for the specific hazard and given training in its correct use, namely, how to ensure the best fit; how to avoid leaks; how to tell when it needs replacing; how to keep it clean and maintain the PPE etc. All too often however, you can see the workers wearing totally the wrong type of PPE or wearing it incorrectly e.g. dust masks only being worn over the mouth and not the nose as well. Another of the most common mistakes is to find workers wearing simple dust masks when dealing with a chemical vapour. This can actually make them situation worse as the vapour is actually absorbed into the fabric of the dust mask (rather, like tissue absorbs a liquid or blotting paper absorbs ink) and then it remains in worker's breathing zone for as long as he/she is wearing the mask.It must be emphasised again that the choice of PPE depends upon the nature of the hazard as well as the route and duration of exposure. The PPE must be the best available and not necessarily the cheapest– and it should not be improvised.

Other control methods can include ***administrative controls*** which means limiting the amount of time workers spend at a hazardous job thereby reducing exposure. Job rotation, changing work schedules are typical examples. It must be remembered that the hazard still exists so that any administrative measures must be used with other forms of control. Linked to administrative controls are the ***general cleanliness*** of the workplace and ***personal hygiene,*** both of which can reduce the exposure to chemical hazards.

Chemicals Hazards Communication should be done by informing the workers about

- What are hazardous chemicals,
- how hazardous chemicals affect the body,
- what are the different types of hazardous chemicals,
- what is on product labels,
- what are material safety data sheets,
- how to protect yourself from hazardous chemicals.

Monitoring of Chemical Hazardous to Health

Monitoring is to be conducted on employees' exposure to chemical hazardous to health. Workplace exposure monitoring also may be required for other reason such as:

- To ensure the employees' exposure level are maintained below the Permissible Exposure Limits.
- To ensure the maintenance of adequate control measure.
- To quantify exposure during new process set-up.
- To assess the effects of a change in process specification.
- To ensure that the risk assessment is still valid.

Control of basic Chemicals Hazards

Substitution of Chemicals

One means of preventing exposure is ***substitution with a safer chemical.***

1, 1, 1-trichloroethane, which was widely used by small engineering companies in metal degreasing has been replaced with trichloroethylene(TCF). However TCE is more acutely toxic and carcinogenic, and has a significantly lower occupational exposure limit than 1, 1, 1-trichloroethane.Substitution is not about "shifting risks between workers, consumers or parts of the environment."

The Montreal Protocol also banned CFC-114, which was used as a refrigerant. In one Belgian smelting plant, CFC-114 was replaced with an HCFC mixture in the air-conditioning unit of an overhead gantry cabin. After nine gantry drivers developed hepatitis, an occupational health doctor traced the cause to the new, more environmentally friendly coolant. The new mixture contained HCFC-123, which had already been found to cause liver damage in rats. The drivers had been exposed to HCFC-123 at levels much higher than the occupational exposure limit because of a leaking hose in the air-conditioning system. The problem might have been noticed sooner if someone had realised that the unit must have been leaking because it needed to be refilled so often.

Personal Exposure Monitoring

Biological monitoring is the measurement of the amount of a substance or its metabolite, or a biochemical effect, from which exposure can be assessed, Biological monitoring measures the amount of a substance that has been absorbed into the body, rather than measuring the amount that is present in the workplace air. A variety of specimens can be used, such as urine, blood or exhaled air. For example, exposure to an organophosphate pesticide can be monitored either by measuring the amount of a metabolite in urine, or by measuring the activity of an enzyme (acetylcholinesterase) in the blood. Workplace exposure to over 100 different chemicals can be estimated by biological monitoring.

Health surveillance, which includes biological monitoring, is a collective term for a variety of procedures designed to protect workers health by early identification of exposure or disease. As well as biological monitoring, health surveillance includes biological effect monitoring, medical surveillance, examinations, and inspections and review of records.

Health surveillance should be triggered by the risk assessment process and cover workers exposure to industrial chemicals by all routes including inhalation, ingestion and skin absorption.

Biological Monitoring

Biological monitoring is the assessment of worker exposure to a hazardous agent through the measurement of a biomarker which results from contact with the agent. The biomarker typically is the agent or its metabolite in a biological specimen derived from the worker; examples are styrene in expired air, styrene in blood, and mandelic and phenylglyoxylic acids (metabolites of styrene) in urine. The biomarker also can be an effect of the agent, such as elevated levels of zinc protoporphyrin in blood, caused by exposure to lead. Industrial hygiene professionals use biological monitoring to assess the risk to workers from exposure hazards and to demonstrate the adequacy of control technologies and intervention strategies. Biological monitoring has the potential to assess worker exposure to industrial chemicals by allroutes, including inhalation, skin absorption, and ingestion.

Monitoring Goals. Air monitoring (or workplace environmental monitoring) and biological monitoring have complementary goals and frequently are applied simultaneously in industrial hygiene investigations.

1. ***Air monitoring:*** Air monitoring provides an estimate of the potential for exposure to an agent. The presence of a health hazard is estimated, by reference to environmental exposure limits such as the NIOSH recommended exposure levels, the Occupational Safety and Health Administration (OSHA) permissible exposure levels, or the threshold limit values ($TLVs_{TM}$) of the American Conference of Governmental Industrial Hygienists (ACGIH). Compared with biological monitoring, air monitoring offers advantages in certain situations. If the agent has acute toxic effects on the respiratory tract or the eyes, air monitoring is the logical tool for controlling the exposure. Air monitoring can be conducted continuously and, thus, can detect peak exposures to dangerous chemicals.
2. ***Biological monitoring of exposure***: A biomarker of exposure represents uptake of the agent through all routes of exposure. Thus, compared to air monitoring, biological monitoring offers a better estimate of the health risk in situations where routes of exposure other than inhalation are significant.
   a. **The rate of disappearance of a biomarker** determines the period of time after exposure during which the level of the biomarker is still affected by the exposure. The levels of rapidly disappearing biomarkers primarily reflect exposures during the previous several hours. On the other hand, biomarkers which disappear over the course of several weeks reflect one, several, or numerous exposure incidents occurring anytime during a period of several weeks previous to the measurement.
   b. **Some toxicants accumulate** in one or several parts of the body and are in dynamic equilibrium with the sites of toxicity. In the case of polychlorinated biphenyl (PCB), which accumulates in fatty tissue, the blood level of PCB reflects the amount stored in the body.

c. **When the site of critical action** for a toxicant is known, the concentration of the biomarker at that site can be used as a measure of the biologically effective dose. Carboxyhemoglobin is such a biomarker for carbon monoxide poisoning. In this case, the biomarker level is correlated with the health effect.

Biological monitoring of effect: This term is defined as monitoring reversible biochemical changes resulting from exposures. The degree of change is less than that which leads to injury and is not associated with a known irreversible pathological effect Biological monitoring of effect is not health surveillance through which individuals with early signs of adverse health effects are identified. Some examples of biomarkers of effect are:

a. Zinc protoporphyrin in blood, levels of which increase with lead exposure, because lead inhibits the biosynthesis of heme
b. Protein and DNA adducts of aromatic amines in blood. These adducts can both reflect the intensity of exposure and be correlated with the biologically effective dose.
c. Antibodies produced against low-molecular-weight molecules. Some chemicals, while not immunogenic in their own right because of small size and other limitations, may bind to constitutive polymers (such as host proteins) and become immunogenic, causing the production of specific antibodies. Alternatively, such exposures may lead to production of new antigenic determinants, through nonadduct-forming reactions of the agent with selected protein-carrier molecules. Antibodies can be made to these modified proteins or to the parent hapten-conjugate In both cases, the antibodies may remain in the human system much longer than the toxicant which initiated their development.

Biological Matrices. The most common matrices used for biological monitoring are exhaled air, blood, and urine.

- Monitoring exhaled air is limited to volatile chemicals. Exhaled air monitoring is not suitable for chemicals inhaled as aerosols or for gases and vapors, which decompose upon contact with body fluids or tissues, or which are highly soluble in water, such as ketones and alcohols.
- Blood is the medium which transports chemicals metabolites in the body.

  Therefore, most biomarkers present in the body can be found in the blood during some period of time after exposure.

  a. A chemical in the blood is in dynamic equilibrium with various parts of the body: the site of entry, tissues in which the chemical is stored, and organs in which it is metabolized or from which it is excreted. Thus, the concentration of a biomarker in the blood may differ between regions of the circulatory system. This would be the case during pulmonary uptake or elimination of a solvent, which would cause differences in concentration between capillary blood (mainly arterial blood) and venous blood.
  b. Two advantages of blood monitoring are: (1) The gross composition of blood is relatively constant between individuals. This eliminates the need to correct measured biomarker levels for individual differences. (2) Obtaining specimens is straightforward and with proper care can be accomplished with relatively little risk of contamination.
  c. An important consideration in blood monitoring is that obtaining blood specimens requires an invasive procedure and should be performed only by trained persons.
- Urine is more suitable for monitoring hydrophilic chemicals, metals, and metabolites than for monitoring chemicals poorly soluble in water. The concentration of the biomarker in urine usually is correlated to its mean plasma level during the period the urine dwells in the bladder.

a. In some instances the urine concentration is affected by the amount of the biomarker stored in the kidneys. Examples are cadmium and chromium.

b. The accuracy of the exposure estimate, using urine monitoring, depends upon the sampling strategy. The most influential factors are time of collection and urine output.

c. Measurements from 24–hour specimens are more representative than from spot samples and usually correlate better with intensity of exposure. However, collection, stabilization, and transportation of 24-hour specimens in the field are difficult and often not feasible.

Sources of Biomonitoring Action Levels

a) ***Biological Exposure Indices (BEI)*** adopted by the American Conference of Governmental Industrial Hygienists (ACGIH)

b) ***Biological Tolerance Values for Working Materials*** (BAT) published by the Deutsche Forschungsgemeinschaft's (DFG) Commission for the Investigation of Health Hazards of Chemical Compounds in the Work Area

c) ***Lauwerys' and Hoet's "Summary of Recommendations"*** in Industrial Chemical Exposure. Guidelines for Biological Monitoring

d) Occupational Safety and Health Administration (OSHA) standards

Appendix 1.119: Environmental Effects of Some Nuclear Fission Reactor Emission

S.No.	**Radioisotopes**	**Half Life**	**Environmental Effects**
1.	Strontium-90	25 years	May cause leukaemia, collects in bone
2.	Iridium-131	8 days	Concentrated in thyroid
3.	Krypton-85	10.4 years	Not concentrated because it is chemically unreactive
4.	Cessium-137	30 years	Irradiates the whole body, deposited in the soft tissue

Appendix 1.120: The Pesticide Cycle in The Environment

The pesticide cycle

Sun

Atmosphere

Degraded by ultraviolet light

Deposited by fainfall

Pesticides

Vaporized to atmosphere

Absorbed by crop

Surface run off to lakes and river's

Adheres to soil particles

Leached below root zone by rain or irrigation

Degraded by bacterial oxidation or chemical hydrolysis

Leached to water courses

NOTES

The term "pesticide" is a composite term that includes all chemicals that are used to kill or control pests.

Classification of pesticides

- herbicides (weeds, plants),
- insecticides (insects),
- fungicides (fungi),
- nematocides (nematodes),
- rodenticides (rodents),
- molluscicides (snails and slugs),
- avicides (repel birds),
- piscicides (fish control),
- bactericides (bacteria),
- slimicides (slime-causing organisms in water),
- algicides (algae).

A general diagrammatic classification of pesticides based on pests they kill and based on chemical groups is given below:

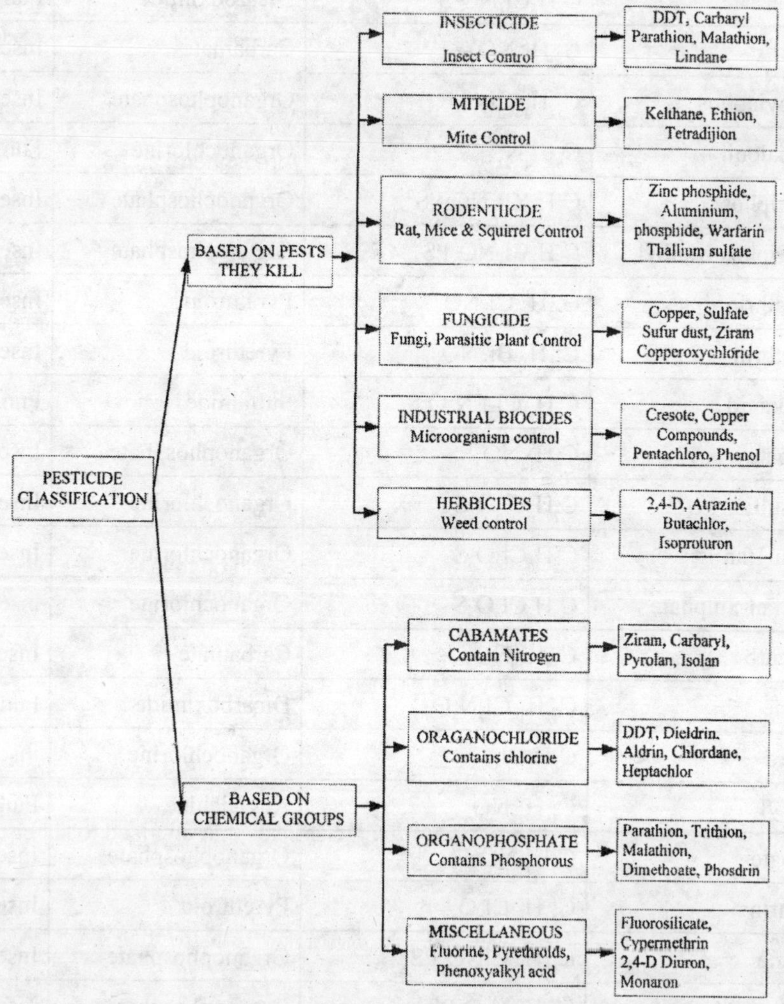

Diagram of Classification of Pesticides

Features of major pesticides are tabulated below:

Major Pesticides

S.No.	Pesticide name	Chemical Formulae	Chemical group	Use
1.	Bifenthrin	$C_{23}H_{22}ClF_3O_2$	Pyrethroid	insecticide
2.	Bromoprapylate	$C_{17}H_{18}Br_2O_3$	Benzilate	Acaricide
3.	Bupirimate	$C_{13}H_{24}N_4O_3S$	Pyrimidine	Fungicide

S.No.	Pesticide name	Chemical Formulae	Chemical group	Use
4.	Captan	$C_6H_8Cl_3NO_2S$	Dicarboximide	Fungicide
5.	Carbaryl	$C_{12}H_{11}NO_2$	Carbamate	Insecticide
6.	Chlorfenvinphos	$C_{12}H_{14}Cl_3O_4P$	Organophosphate	Insecticide
7.	Chlorothalonii	$C_6Cl_4N_2$	Organochlorine	Fungicide
8.	Chlorpyriphos	$C_6H_{11}Cl_3NO_3PS$	Organophosphate	Insecticide
9.	Chtorpyriphos-Methyl	$C_7H_7Cl_3NO_3PS$	Organophosphate	Insecticide
10.	Cypermethrin	$C_{22}H_{19}Cl_2NO_3$	Pyrethroid	Insecticide
11.	Deltamethrin	$C_{22}H_{19}Br_2NO_3$	Pyrethroid	Insecticide
12.	Dichlofiuanid	$C_9H_{11}Cl_2FN_2O_2S_2$	Sulfamide	Fungicide
13.	Dimethoate	$C_5H_{12}NO_3PS_2$	Organophosphate	Insecticide
14.	α-Endosulfan	$C_2H_6Cl_6O_2S$	Organochlorine	Insecticide
15.	β-Endosu1fan	$C_2H_6Cl_6O_2S$	Organochlorine	Insecticide
16.	Endosulfan-Sulphate	$C_2H_6Cl_6O_4S$	Organochlorine	insecticide
17.	Fenoxycarb	$C_{17}H_{19}NO_4$	Carbamate	Insecticide
18.	Iprodsone	$C_{13}H_{13}Cl_2N_3O_3$	Dicarboximide	Fungicide
19.	Lindane	$C_6H_6Cl_5$	Organochlorine	insecticide
20.	Melalaxyl	$C_{15}H_{21}NO_4$	Acylalanine	Fungicide
21.	Omethoate	$C_5H_{12}NO_4PS$	Organophosphate	Insecticide
22.	Permethrin	$C_{21}H_{20}Cl_2O_3$	Pyrethroid	Insecticide
23.	Phosalone	$C_{12}H_{15}ClNO_4PS_2$	Organophosphate	Insecticide
24.	Pirimiphos- Methyl	$C_{11}H_{20}N_3O_3PS$	Organophosphate	insecticide
25.	Tetradiphon	$C_{12}H_5Cl_4O_2S$	Organochlorine	Acaricide
26.	Tolylfluanid	$C_{10}H_{13}Cl_2FN_2O_2S_2$	Sulfamide	Fungicide
27.	Triazophos	$C_{12}H_{16}N_3O_3PS$	Organophosphate	Insecticide

List of various technical grades pesticides manufactured in India is tabulated below:

S.No.	Pesticides	Name of Pesticides manufactured in India
1.	Insecticides	Acephate, BHC, chlorpyriphos, Cypermethrins, DDT, Dichlorvos, Dimethoate, Deltamethrin, Dichloropropane and dichloropropanes, mixture (D.D. mixture), Dicofol, Ethion, Endosulfan, Ethylene dibromide and carbon tetra chloride mixture, Fenthion, Fenitrothion, Fenvalerate, Lime-sulfur, Lindane, Malathion, Methyl parathion, Monocrotophos, Nicotine sulphate, Oxydemetonmethol, Phosphamidon, Pyrethrum extract, Phorate, Phosalone, Quinalphos, Temephos, Triazaphos.
2.	Fungicides	Aureofungin, Copper oxychloride, Copper sulphate, Carbendazim, Captafol, Ferbam, Mancozeb, Nickel Chloride, Organomercurials (MEMC and PMA), Sulphur (Colloida, wettable and dust), Streptocycline, Thiram, Tridemorph, Ziram, Zineb.
3.	Rodenticides	Barium carbonate, Comafuryl, Sodium cyanide, Warfarin, Zinc phosphide.
4.	Weedicides	Anilophos, Butachlor, Diuron, 2, 4-D, Fluchlorallin, Glyphosate, Isoproturon, Paraquat dichloride.
5.	Plant Growth Regulants	Alpha naphthalene acetic acid, Chloromequat chloride.
6.	Fumigants	Aluminiumphosphide, Ethylene bromide, Methyl bromide

Properties and health effects of some major pesticides is given below.

Properties and health effects of some major pesticides

S.No.	Pesticides	Properties	Characteristics/health effects
1.	D.D.T (di-chloro, difluro, trichloro ethane)	Insecticide, colorless white crystal, odorless or slight aromatic odor BP=109°C, TLV=1mg/m^3 of air	Vomiting, numbness, partial paralysis, mild convulsions
2..	Parathion	Insecticide, yellowish liquid BP=375°C	Headache, blurred vision, weakness, nausea, cramps, sweating, loss of reflexes, coma
3.	2, 4-D	Herbicide, white powder, mol wt.-221	Nausea, vomiting, kidney and liver failure
4.	Carbaryl	Insecticide, solid, melting point=142 C, TLV=5mg/m^3 of air,	Symptoms same as parathion but to a lesser extent
5.	Dieldrin	Crystalline, melting point=150°C	Headache, nausea, vomiting, general malaise and drowsiness
6.	B.H.C (Benzene Hexachloride)	White crystalline powder, melting point=157°C	Irritation, excitation, hyperirritability, loss of equilibrium, depression

S.No.	Pesticides	Properties	Characteristics/health effects
7.	Lindane	White crystalline powder, melting point=157°C, Mol wt. =290.84 TLV=0.5 mg/m^{3}of air	Acute toxicity may be excitation, hyperirritability, and loss of equilibrium, depression, headache, nausea, irritation of eyes, nose and throat no chronic toxicity as such. It emits highly toxic fumes of phosgene when heated.
8.	Atrazine	Crystalline, soluble in water, alcohol and chloroform	Relatively low toxicity but a possible carcinogen. It emits dangerous fumes when strongly heated.
9.	Guthion	Crystals, slightly water soluble but soluble in organic solvents, mol wt. =317.3, MP =74°C	Highly toxic: toxicity is almost same as that of parathion.
10.	Malathion	Yellow liquid, BP=156-157°C, slightly soluble in water, miscible with most polar organic solvents, combustible, TLV=10 mg/m^3 of air	Toxic by ingestion and inhalation, absorbed by skin
11.	Copper oxychloride	Fungicde, bluish green powder, soluble in acids, ammonia. Insoluble in water	Toxic by ingestion and inhalation
12.	Ziram	Fungicide, white and odorous when pure, sp gr.– 1.71, MP=246°C, insoluble in water, soluble in some solvents like acetone and carbon disulfide.	Moderately toxic, strong irritant to eyes and mucous membrane
13.	Zinc phosphide	Rodenticide, Dark grey gritty powder, sp gr.– 4.55, MP>246°C, insoluble in alcohol, soluble in acids, decomposes in water.	Highly toxic by ingestion, reacts violently with oxidizing agents, Phosphine is produced when reacted with the acids.
14.	Aldrin	Insecticide, brown to white crystalline solid, insoluble in water, soluble in most organic solvents, MP=104 - 105 °C, TLV=0.25mg/ m3of air	Highly toxic by ingestion and inhalation, carcinogen, use restricted to non-agricultural application.
15.	Dimethoate	Insecticide, white solid, MP=51 -52 °C, moderately soluble in water, soluble in most organic solvents except hydrocarbons.	Highly toxic, a cholinesterase inhibitor, restricted use.

Pesticide manufacturing consists of the chemical synthesis of active ingredients for crop protection, which is very often the synthesis of complex organic chemical compounds, and subsequent formulation of these active ingredients (usually mixing and grinding processes).

Major chemical reactions involved in production of technical grade pesticides are: alkylation, carboxylation, acetylation, condensation, cyclization, dehydration, halogenations, oxidation, sulfonation, nitration and amination. Main physical "unit operations" include Liquid/liquid extraction, liquid/liquid separation, liquid/solid separation, gas/solid separation, distillation, crystallization, gas absorption, drying, grinding and mixing.

Banned or Severely Restricted Chemicals /Pesticides

Central Insecticides Board and Registration Committee (CIBRC) have been constituted under the insecticides Act. CIBRC have banned certain pesticide formulations, refused registration of some products and also prepared a list of pesticides for restricted use in country, which are tabulated below:

List of Pesticides / Pesticides Formulations in India

A.	**Pesticides Banned for manufacture, import and use (25 Nos.)**			
	1.	Aldrin	14	Pentachloro Nitrobenzene
	2.	Benzene Hexachloride	15	Pentachlorophenol
	3.	Calcium Cyanide	16	Phenyl Mercury Acetate
	4.	Chlordane	17	Sodium Methane Arsonate
	5.	Copper Acetoarsenite	18	Tetradifon
	6.	Dibromochloropropane	19	oxafen
	7.	Endrin	20	Aidicarb
	8.	Ethyl Mercury Chloride	21	Chlorobenzilate
	9.	Ethyl Parathion	22	Dieldrin
	10.	Heptachlor	23	Maleic Hydrazide
	11.	Menazone	24	Ethylene Dibromide
	12.	Nitrofen	25	TCA (Trichloro acetic acid)
	13.	Paraquat Dimcthyl Sulphate		
B.	**Pesticides / Pesticide formulations banned for use but their manufacture is allowed for export (2 Nos.)**			
	1.	Nicotin Sulphate		
	2.	Captafol 80% Powder		
C.	**Pesticide formulations banned for import, manufacture and use (4 Nos.)**			
	1.	Methomyl 24% L	3	Phosphamidon 85% SL
	2.	Methomyl 12.5% L	4	Carbofuran 50% SP
D	**List of Pesticides Refused Registration**			
	1.	Calcium Arsonate	10	Azinphos Ethyl

	2.	EPM	11	Binapacryl
	3.	Azinphos Methyl	12	Dicrotophos
	4.	Lead Arsonate	13	Thiodemeton / Disulfoton
	5.	Mevinphos (Phosdrin)	14	Fentin Acetate
	6.	2, 4, 5-T	15	Fentin Hydroxide
	7.	Carbophenothion	16	Chinomethionate (Morestan)
	8.	Vamidothion	17	Ammonium Sulphamate
	9.	Mephosfolan	18	Leptophos (Phosvel)
E	**Pesticides Restricted for Use in India**			
	1.	Aluminium Phosphide	5	Methyl Parathion
	2.	DDT	6	Sodium Cyanide
	3.	Lindane	7	Methoxy Ethyl Mercury Chloride (MEMC)
	4.	Methyl Bromide		

Air Pollutants from Pesticide Industry

In general, process emissions can be classified into channelised and fugitive emissions. The channelised emission is a point source emission from process operations and the fugitive emission is an uncontrolled emission from storage tanks/drums, spills, leaks, overflows etc.

The identified pollutants associated with pesticides manufacture are tabulated below:–

Pesticides and associated pollutants emitted during their manufacture

S.No.	Pesticides	Pollutants emitted
1.	Acephate	HCl
2.	Aluminium phosphide	P_2O_5 fumes (as H_3PO_4 mist)
3.	Captafol	Cl_2 and HCl
4.	Captan	Cl_2 and HCl
5.	Cypermethrin	Cl_2, HCl and SO_2
6.	Dimethoate	H_2S
7.	2, 4 – D Acid	HCl and Cl_2
8.	Dichlorvos (D.D.V.P)	CH_3Cl
9.	Ethion	H_2S and C_2H_5SH
10.	Endosulphan	HCl
11.	Fenvalerate	HCl, Cl_2 and SO_2
12.	Isoproturon	NH_3
13.	Malathion	H_2s
14.	Monocrotophos	HCl and CH_3Cl
15.	Phosalone	NH_3, HCl and H_2S

S.No.	Pesticides	Pollutants emitted
16.	Phorate	H_2S and C_2H_5SH (ethyl mercarptan)
17.	Phosphamidon	HCl and CH_3Cl
18.	Zinc Phosphide	P_2O_5 as H_3PO_4 mist

Besides the air pollutant listed above in Table the pesticide manufacturing processes use various types of solvents for separation of desired product from other chemicals. The list of solvents for various pesticides manufacturing process is given below in Table. Spent solvent are recovered by either recycle in the same system or reuse for other purposes. The unrecovered solvents are generally incinerated in the pesticide industries. The loss of solvent is not so significant. A well designed system can prevent loss due to fugitive emissions.

List of Solvents used in manufacturing process of pesticides

S.No.	Name of the Solvents	Pesticides/Product list in which the solvent is used
1.	Acetone	Quinalphos, Acetamiprid, Triazophos, Pendimethalin
2.	Acetonitrile	Divenyl acid chloride, Acetamiprid, Imidachlorprid, Tetrachloro butyronitrile, Cypermethrin, Alphamethrin, Permethrin
3.	Acetic acid	Acephate, 2, 4-D acid, Mancozeb
4.	Acetic anhydride	Acepate, 2, 4-D acid
5.	Benzene	Malathion, Diethyl Aniline, Dimdethoate, Lindane
6.	Butanol	Aureofungin
7.	Carbon tetrachloride	Endosulfan, 2, 6, Dichloro pyridine, Divenyl acid chloride, isoproturan, Tetrachloro buty ronitrile, Cypermethrin, Alphamethrin, Permethrin
8.	Dioxane	Metconazole
9.	Diethyl Amine	Cypermethrin, Alphamethrin, Permethrin, Devrinol
10.	Di Methyl Formamide (DMF)	Fenvalerate, Hexaconazole, Propiconazole, Hexaconazole
11.	Ethanol	Malathion, Phorate, Ethion, Terbufos, Diethyl Aniline, Ethyl chloride, Phenthoate, Lindane
12.	Ethyl acetate	Acepate
13.	Ethylene Di Chloride	Captan, Dimethote, Monocrotophos, Meta Bromo Benzaldehyde, Fenvelerate, Cypermethrin, Phosphamidon
14.	Formaldehyde	Phorate, Tebufos
15.	Hexane	Cypermethrin, Alphamethrin, Permethrin, Deltamethrin, Fenvalerate, Cypermethrin, Divenyl acid chloride, Iso proturon
16.	Isopropyl alcohol	Delatamethrin, Phosphamidon, Imidachlorprid
17.	Methanol	Dimethote, Malathion, Methyl Bromide, hexaconazole, Quinalphos, Metamitron, Phenthoate, Methyl Parathion, Temephos
18.	Methylene Chloride	Acephate, Dichiorvos, Mepiquate chloride, Hexaconazole

S.No.	Name of the Solvents	Pesticides/Product list in which the solvent is used
19.	Methylene dichloride	Acephate, Deltamethrin, Propiconazole
20.	Piperidine	Mepiquate chloride
21.	Phenol	2, 4-D acid, Meta Phenoxy Denzaldehyde
22.	Pyridine	D-Allethrin, Carbendazin
23.	Renine	For formulations,
24.	Toluene	Captafol, Captan, Cypermethrin, Ethion, Malathion, Dimethote, Fenvalerate, 5-Amino Salicylic acid, Andpa, Meta Phenoxy Denzaldehyde, Permethrin, Benfuresate, d-Allethrin, Penpropathrin, Tenephos, Metconazole
25.	Triethyl amine	Cypermethrin, Divenyl acid chloride, Benfuresate, Permethrin, Deltamethrin
26.	Xylene	Quinalphos, Triazophos, Cymoxanil, Melalxyl, Marcozeb, Anilofos, DMTA, Oxadiargyl, Benfuresate, Phosphamidon, Pendimethalin,

Physical and chemical properties and health effects of some priority pollutants generated during pesticides manufacture is tabulated below:

Properties and Health effects of some pollutants produced during pesticides manufacture

S.No.	Pollutants	Physical and Chemical Property	Exposure, odour and Corrosiveness	Health effects
1.	Ammonia (NH_3)	Colourless, stable at room temp, Anhydrous ammonia will react exothermically with acids and water. Anhydrous ammonia decompose to hydrogen and nitrogen gases above 450° C. Decomposition temperature may be lowered by contact with certain metals.	TWA 50ppm, Ammoniacal strong, high corrosive in presence of Cu and its alloys. Slightly corrosive in presence of Al and Zn. Very slightly corrosive in presence of Mild Steel. Non Corrosive in presence of glass or Stainless steel (304 or 316)	Exposure can cause coughing, chest pains difficulty in breathing. Repeated significant overexposure can cause permanent lung function damage, edema and chemical pneumonitis. May cause serious damage to eyes
2.	Chlorine (Cl_2)	Greenish yellow gas, extremely reactive can react violent with many combustible materials and other chemicals including water	TLV 0.5 ppm, pungent suffocating bleach like odour, strongly corrosive to most metals in the presence of moisture	Can cause itching and burning of the eyes, nose, throat. At high concentrations chlorine is respiratory poison.

S.No.	Pollutants	Physical and Chemical Property	Exposure, odour and Corrosiveness	Health effects
3.	Hydrogen Bromide (HBr)	Colourless gas, stable incompatible with strong oxidising agents, strong bases. Hazardous decomposition products: Bromine and Hydrogen	TWA 3 ppm, Pungent suffocating odour, corrosive to most common metals and alloys	Toxic- causes serious burns
4.	Hydrochloric Chloric Acid (HCl)	Colourless to slightly yellow gas, stable incompatible with alkalis	TWA 5 ppm, irritating pungent odour, odour threshold 0.77 ppm, and dry gaseous hydrogen chloride can be handled in many metals and alloys, although at elevated temperature the corrosion rates increase. Commonly used metals are Carbon steel alloy 400, 300, series stainless steels, alloy 600 and Nickel 200	Toxic may be fatal if inhaled, severe irritant, very harmful by inhalation, ingestion or through skin contact.
5.	Hydrogne Sulphide (H_2S)	Colourless gas, stable, highly inflammable, may form explosive mixture with air incompatible with oxidising agents, oxides and metals	TWA 10 ppm, smell of rotten eggs, corrosive to most metals	High toxic may be fatal if inhaled. Skin contact may cause burns. Asphyxiant.
6.	Sulphur Dioxide (SO_2)	Colourless gas, stable, incompatible with strong reducing or oxidising agents, moisture Zinc and its alloys	TWA 2ppm, irritating pungent odour, non corrosive to common materials except Zn when dry; corrosive when wet	Toxic-high concentrations are fatal
7.	Ethyl Mercaptan (C_2H_5SH)	Colourless gas, stable under normal storage condition, decomposition of chemical can emit carbon monoxide, hydrogen sulphide and sulphur dioxide	TWA 0.5ppm, extremely strong and repulsive smell. Odour threshold is 0.001 ppm	Highly toxic, affects the central nervous system

TLV– Threshold limit value

TWA– Time weighted average

Pollution control technologies adopted in pesticides industries in India

Arising waste gases are treated by techniques where:

- The waste gas content is recovered and either recycled to the original process or used in another process as raw material or energy carrier.
- The contaminants are abated.

Compounds normally worth to recover comprise

- VOCs recovered from, solvent vapours or vapours of low-boiling products.
- VOC used as energy carrier in incinerators or boilers.
- Hydrogen chloride, transferred into hydrochloric acid.
- Ammonia to recycle into the production process.
- Sulphur dioxide, transferred into sulphuric acid, sulphur or gypsum.
- Dust containing higher amounts of solid raw products or end products.

The control technologies adopted to control the identified gaseous pollutants are given in Table below:

Gaseous Pollution Control technologies adopted in Indian pesticides industries

S.No	Gaseous Pollutants	Pollution Control Systems used
1.	HCl	• Water Scrubber, • Caustic Scrubber, • Water / Caustic Scrubber
2.	Cl_2	• Water Scrubber, • Caustic Scrubber, • Water / Caustic Scrubber
3.	CH_3Cl	• Charcoal Bed Scrubber, • Liquification, filling and /or Combustion
4.	H_2S	• Scrubber with NaOCl media • Scrubber with NaOH media • Charcoal Bed Scrubber
5.	SO_2	• Caustic Scrubber • Water/Caustic Scrubber
6.	P_2O_5 (as H_3PO_4)	• Water Scrubber • Water Scrubber (Ring jet scrubber) + Mist Eliminator + Demister • Mist Eliminator
7.	NH_3	• Two Stage Water Scrubber • Recovery System
8.	CH_3OH	• Channelised Emission was not observed, because this chemical (solvent) is used in less quantity in the manufacturing
9.	HBr	• Caustic Scrubber
10.	Odours compounds, non recovered solvents	• Incinerator / chemical reaction

The efficiency of gaseous pollutants cleaning devices is expressed in a variety of ways, including control efficiency, penetration, and decontamination factor. The most common means for expressing the efficiency of performance is in terms of the control efficiency, which is defined as the ratio of the quantity of pollutant prevented from entering the atmosphere by the control device to the quantity that would have been emitted (inlet quantity to the gas cleaning device) to the atmosphere had there been no control device.

Efficiency (n) - Collected / Inlet = C/I = (Inlet – Outlet)/Inlet = (I-O)/I

Penetration is defined as the ratio of the amount of pollutant escaping (penetrating without control) the gas cleaning device to the amount entering.

Penetration (P) = Outlet/Inlet

Hence penetration focuses attention upon the quality of the emission stream, but in reality it is actually another way of looking at control efficiency. Since

Efficiency (n) = 1 – [outlet / inlet], then = *1–P*

For a control efficiency of 99.999% (i.e., $E = 0.99999$), the penetration is 0.00001, or 10–5. Alternatively, efficiency can be expressed as the decontamination factor (*DF*), which is defined as the ratio of the inlet amount to the outlet amount.

Decontamination Factor (DF) = Inlet / Outlet = 1 / (1 – E)

For a percentage control efficiency of 99.999%, DF is 10^5. The logarithm to the base 10 of the decontamination factor is the decontamination index. In the numerical example above, this index is 5.0. These efficiency terms are used to represent the overall efficiency of control of a single device or any combination of control devices.

Based on the data obtained from the industry and also monitored / measured it is found that the control system as presented in table below is most suitable for identified priority pollutants (from efficiency point of view) control in Pesticide industries.

Most Suitable gaseous pollutants Control Systems for Priority Pollutants of Indian Pesticide industries.

S.No.	Gaseous Pollutants	Most suitable Pollutants Control System
1.	HCl	Water/Caustic Scrubber
2.	Cl_2	Water/Caustic Scrubber
3.	CH_3Cl	Liquification system and/or Incinerator
4.	H_2S	Scrubber with NaOH media
5.	SO_2	Water/Caustic Scrubber
6.	P_2O_5 (as H_3PO_4)	Water Scrubber (Ring jet scrubber) + Mist Eliminator + Demister
7.	NH_3	Two Stage Water Scrubber
8.	CH_3OH	Adsorption Bed (Charcoal or Molecular Sieve)
9.	HBr	Caustic Scrubber
10.	Mercaptan	Incinerator

Technologies based on absorption and chemical reaction are being commonly used to control the gaseous pollutants e.g. HCl, H_2S, SO_2, P_2O_5 (as H_3PO_4), Cl_2, NH_3 and HBr. In case of H_2S, adsorption and incineration technology is also adopted by some pesticide industry. To control the emission of CH_3Cl, condensation (liquification) and incineration system is used in large scale pesticide industry.

Indian National Emission Standards of pollutants for Pesticide Manufacturing and Formulation Industry

S.No.	Pollutant	Not to exceed, mg/Nm³
1.	HCl	20
2.	Cl_2	05
3.	H_2S	05
4.	P_2O_5 (as H_3PO_4)	10
5.	NH_3	30
6.	Particulate matter with pesticide compounds	20
7.	CH_3Cl	20
8.	HBr	05

Guidelines for Fugitive Emission Control in Indian Pesticide Industries

i. Fugitive emissions over reactors, formulation areas, centrifuges, chemical loading, transfer areas etc., are yet to be collected through hoods and ducts by induced draft and controlled by scrubber / dust collector.

ii. Usually scrubbers installed for channelised emissions are used for fugitive emissions to control also and sometimes dedicated scrubbers are provided. This practice may be permitted as long as tail gas concentrations are within the prescribed limit.

iii. In addition, organic gaseous emissions (odorous and toxic) be routed to activated carbon beds (adsorption) or to thermal oxidiser, and for dust emissions cyclones / bag filters are to be provided.

iv. Emphasis to be given to solvent management / solvent loss prevention.

v. Enclosures to be provided for chemical storage area and collection of emissions from loading of raw materials, in particular, solvents through hoods and ducts by induced draft, and control by scrubber / dust collector to be ensured.

vi. Vapour balancing, nitrogen blanketing, ISO tanks etc. to be provided; special care needs to be taken for control in respect of odorous chemicals.

Appendix 1.121: The Mercury Cycle in the Environment

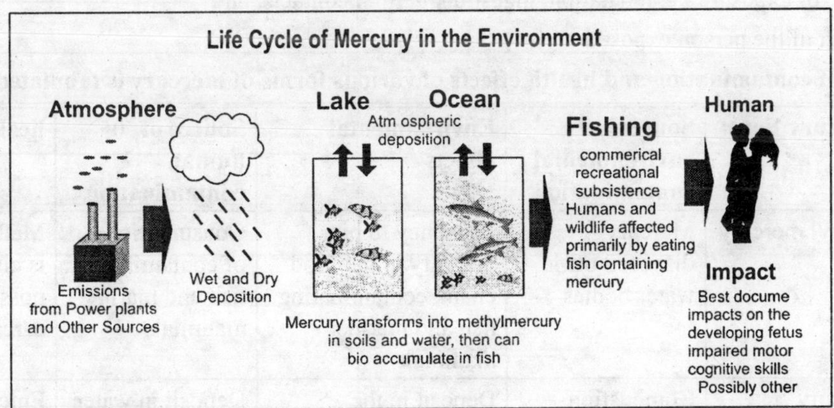

Note:

Mercury is a natural trace element which exists as a liquid at room temperature in its metallic form. Its vapour pressure is sufficiently high to yield hazardous concentrations of vapour at temperatures normally encountered both indoors and outdoors under most climatic conditions.

Typically, mercury is released into the atmosphere in one of three forms:

- elemental mercury: can travel a range of distances, may remain in the atmosphere up to one year and may travel globally before undergoing transformation
- particle-bound mercury: can fall out of the air over a range of distances
- oxidized mercury (sometimes called ionic or reactive gaseous mercury (RGM)): found predominantly in water-soluble forms, which may be deposited at a range of distances from sources depending upon topographic and meterological conditions downwind of a source.

Exposure to Mercury occurs when elemental mercury is spilled or products that contain elemental mercury break and expose mercury to the air, particularly in warm or poorly ventilated indoor spaces.

Hg is released into the environment by nearly all coal power plants, municipal and medical incinerators. Mercury is also emitted to the air from natural sources, such as volcanoes.

It is also released from chloralkali plants, electrical equipment, paints, measurements and control systems, such as thermometers and blood pressure meters, agriculture, dental, laboratory and other uses including military uses as detonators.

Effects of Mercury on Human Health

Mercury is very toxic and linked with a wide range of health effects including irreversible damage to the human nervous system. The severity of health effects from mercury exposures is influenced by the following factors:

i. Chemical form of mercury i.e. inorganic or organic mercury,

ii. Dose,

iii. Age of the person exposed (the foetus is the most susceptible),
iv. Duration of exposure,
v. Route of exposure i.e. inhalation, ingestion, dermal contacts, and
vi. Health of the person exposed

Sources of contamination and health effects of various forms of mercury is tabulated below

S.No.	Mercury Form	Sources of environmental contamination	Environmental effects	Source of human contamination	health effect
1.	Methyl mercury	Mercury disposed off in water bodies	Tendency to bio-magnify in the food chain, contaminating fish and marine mammals	Consumption of contaminated fish and marine mammals	Methyl mercury is classified as a possible human carcinogens
2.	Mercury	Combustion of coal, incineration of mercury bearing waste	Deposit in the environment and sometimes travel long distances	Deposit in water bodies and also falls with rain water	Enters body via food
3.	Mercury vapours	–		Ambient air, dental amalgam	Dangerous if inhaled or ingested

Routes of Mercury Exposure

Followings are the routes of mercury exposure:

Inhalation

Mercury vapor is highly toxic via inhalation. It can cause severe respiratory tract damage. Symptoms of mercury toxicity include sore throat, coughing, pain, tightness in chest, breathing difficulties, shortness of breath, headache, muscle weakness, anorexia, and gastrointestinal disturbance, ringing in the ear, liver changes, fever, bronchitis and pneumonitis. Mercury can be absorbed through inhalation with symptoms similar to those appearing after ingestion.

Ingestion

Ingestion (through mouth) of mercury may cause burning of the mouth and pharynx, abdominal pain, vomiting, corrosive ulceration, bloody diarrhea. Ingestion may be followed by a rapid and weak pulse, shallow breathing, paleness, exhaustion, tremors and collapse.

Skin Contact

Contact of mercury with skin causes irritation and burns to skin. Symptoms include redness and pain. It May cause skin allergy and sensitization. Mercury can be absorbed through the skin with symptoms parallel to ingestion.

Eye Contact

Contact of mercury with eye causes irritation and burns to eyes. Symptoms include redness, pain, blurred vision; may cause serious and permanent eye damage, also depending upon exposure.

Chronic Exposure

Chronic exposure of mercury through any route can produce central nervous system disorders. It may cause muscle tremors, personality and behavior changes, memory loss, metallic taste, loosening of the teeth, digestive disorders, skin rashes, brain damage and kidney damage. Mercury can cause skin allergies and accumulate in the body. Repeated skin contact with mercury can cause the skin to turn gray in color. Mercury is a suspected reproductive hazard; may damage the developing foetus and decrease fertility in males and females.

Levels of mercury in air are found to be in the range of 20 – 60 ng/ m^3.

Measurement method: The following flowchart depicts the procedure used to analyse mercury.

Flowchart of Measurement Method of Mercury (EPA method)

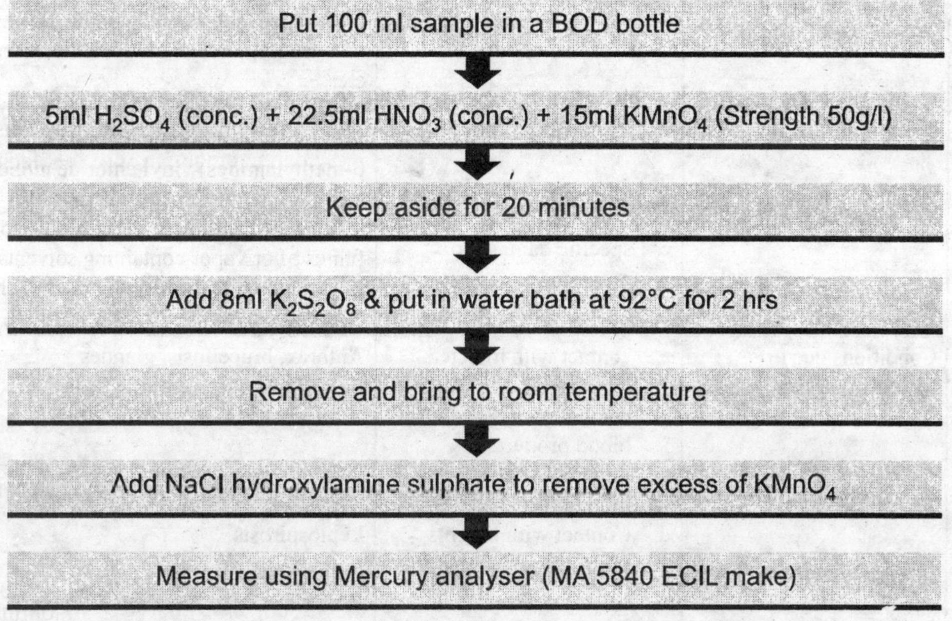

The maximum permissible value of Hg as given by the WHO is 0.05 mg/ m^3.

Appendix 1.122: Principal Conditions and Agents of Occupational Diseases

S.No.	Conditions causing Occupational diseases	Agents causing occupational diseases	Examples of typical occupational diseases caused
1.	Conditions due to physical agents	a. Dust b. Lighting c. Manual work d. Noise e. Pressure f. Radiation g. Repetitive movement h. Temperature i. vibration	Pneumoconiosis, silicosis, asbestosis, occupational asthma cancer. Miner's nystagmus Beat elbow, beat knee, beat hand Noise induced hearing loss Decompression sickness Radiation sickness, arc eyes, burns Writer's cramp Heat strokes, heat cataracts Vibration induced white finger
2.	Conditions due to chemical agents	a. poisoning	Caused by the use or handling of or exposure to the fumes dust or vapour of a wide range of chemical substance including lead, manganese, phosphorus, arsenic, mercury etc.
		b. occupational cancers	Associated with exposure tar, pitch, bitumen, minerals oils, aromatic amines (e.g. I and β-methylamines) vinyl chloride monomer etc.
		c. specific damage to organs and systems	The use or handling of or exposure to the fumes of or vapor containing solvents such as carbon tetrachloride, and methyl chloride (chloromethane)
3.	Conditions due to biological agents	a. contact with infected animals b. contact with blood/ blood products. c. Vegetable borne infections d. Contact with rodents	Anthrax, brucellosis, glandes Viral hepatitis Farmer's lung (asperillosis) Leptospirosis
4.	Conditions due to related agents	a. Friction and pressure b. Job movements	Cellulitis, bursitis's Writer's cramp, 'beat' conditions e.g. beat hand, beat elbow, beat knee, traumatic inflammation of the tenders sheaths of the hand or forearmed tenosynovitis, work related upper limb disorders (repetitive strain injury)

Appendix 1.123: Occupational Disease Prevalent in Various Industries /Processes

S.No.	Occupational diseases	Prevalent in industries/process
1.	Byssinosis	Textile industry
2.	Dermatitis	Mineral oil handling
3.	ECG abnormalities	Miner oil handling
4.	Lead poisoning	Lead handlers, printing, battery manufacturing
5.	Mercury poisoning	Chloro-alkali plants
6.	Noise induced hearing loss	Industries/process where noise levels are high
7.	Pneumoconiosis	Coal mining industry
8.	Silicosis	Coal mines workers, stone quarrying and slate pencil manufacturing units.

Appendix 1.124: Occupational Disease Caused by Inhalation of Various Dust Particles

S.No.	Dust particles inhaled	Name of disease caused
1.	Asbestos dust	Asbestosis
2.	Coal dust	Anthracosis/black lung
3.	Cotton dust	Byssinosis
4.	Hay or grain dust	Farmer's lung
5.	Iron dust	Siderosis
6.	Silica dust	Silicosis
7.	Sugar cane dust	Bagassosis
8.	Tobacco	Tobaccosis

Note:

Particulate matter (PM) is the term used for a mixture of solid particles and liquid droplets suspended in the air. These particles originate from a variety of sources, such as power plants, industrial processes, and diesel trucks, ana they are formed in the atmosphere by transformation of gaseous emissions. Particulate matter is composed of both coarse and fine particles.

Coarse particles (PM_{10}) have an aerodynamic diameter between 2.5μ m and 10μ m. They are formed by mechanical disruption (e.g. crushing, grinding, abrasion of surfaces); evaporation of sprays, and suspension of dust. PM_{10} is composed of aluminosilicate and other oxides of crustal elements, and major sources including fugitive dust from mads, industry, agriculture, construction and demolitions activities and fly ash from fossil fuel combustion. The lifetime of PM_{10} is from minutes to hours, and its travel distance varies from <1km to 10 km.

Fine particles have an aerodynamic diameter less than 2.5μ m ($PM_{2.5}$). They differ from PM_{10} in origin and chemistry These particles are formed from gas and condensation of high-temperature vapors during combustion, and they are composed of various combinations of sulfate compounds, nitrate compounds, carbon compounds, ammonium, hydrogen ion, organic compounds, metals (Pb, Cd, V, Ni, Cu, Zn, Mn, and Fe), and particie bound water. The major sources of $PM_{2.5}$ are fossil fuel combustion, vegetation burning, and the smelting and processing of metals. Their lifetime is from days to weeks and travel distance ranges from 100 to >1000 km.. In addition, fine particles are associated with decreased visibility (haze) impairment in many cities.

Comparisons of fine and coarse particle matter is tabulated below

S.No.	Parameters	Fine Particle	Coarse Particle
1.	Formed from	Gases	Large solids/droplets
2.	Formed by	Chemical reaction; nucleation; condensation, coagulation; evaporation of fog and cloud droplets in which gases have dissolved and reacted	Mechanical disruption (e.g. crushing, grinding, abrasion of surfaces); evaporation of sprays; suspension of dusts
3.	Composed of	Sulfate (SO_4^{-2}); Nitrate (NO 3); ammonium (NH4+); hydrogen ion (H+); elemental carbon; organic compounds (e.g. PAHs, PNAs); metals (Pb, Cd, V, Ni, Cu, Zn, Mn, Fe); particle bound water	Resuspended dust (e.g. soil dust, street dust); coal and oil fly ash; metal oxides of crustal elements (Si, Al, Ti, Fe); $CaCO_3$, NaCl, sea salt; pollen, mould spores; plant, animal fragments; tyrewear debris
4.	Solubility	Largely soluble, Hygroscopic and deliquescent	Largely insoluble and non-hygroscopic
5.	Sources	Combustion of coal, oil, gasoline, diesel, wood; atmospheric transformation products of NOx, SO_2 and organic compounds including biogenic species (e.g. terpenes); high temperature processes, smelters, steel mills, etc.	Resuspension of industrial dust and soil tracked onto roads; suspension from disturbed soil (e.g. farming, mining, unpaved roads); biological sources; construction and demolition; coal and oil combustion; ocean spray
	Lifetimes	Days to weeks	Minutes to Hours
	Travel distance	100s to 1000s of kilometers	<1 to 10s of kilometers

Physical Properties and Physiological Effects

The capacity of particulate matter to produce adverse health effects in humans depends on its deposition in the respiratory tract. Particle size, shape, and density affect deposition rates. The most important characteristics influencing the deposition of particles in the respiratory system are size and aerodynamic properties. The aerodynamic diameter of a particle is the diameter of a unit density sphere having the same settling velocity as the particle in question, whatever its size, shape or density. Particles between 2.5 and 10 micrometer in aerodynamic diameter correspond to the inhalable particles capable to be deposited, in the upper respiratory tract. Particles with aerodynamic diameter smaller than 2.5 micrometer ($PM_{2.5}$), called fine particles, correspond to the respirable particle fraction capable of penetrating the alveolar region of the lung. Inhaled particles come in contact with surface of the respiratory system. These particles pass the proximal airway (throat and larynx) of the respiratory tract, and deposit in the tracheobronchial conductive airway of the lungs (bronchial and bronchiolar airway) or in the gas exchange region (respiratory bronchioles, alveolar ducts, and alveoli of the lung parenchyma).

There are five mechanisms that influence particle deposition within the respiratory tract. The primary mechanisms are gravitational settling, impaction, and Brownian diffusion. Secondary mechanisms are electrostatic attraction and interception. These last two processes have minimal importance for inhalation and deposition of particulate matter. Deposition by gravitational settling occurs as a result of the influence of gravity on particles suspended in the air. The settling rate of particles is directly proportional to particle size. This process is most important in the distal region of the bronchial airway and in proximal portions of the gas exchange region. Another mechanism of particle deposition is impaction. Due to inertia airborne particles do not follow changes in direction or speed of airflow and they may impact on the wall of the airway. This mechanism occurs primarily in the throat and larynx with particles larger than 3µ m and increases with increasing particle size. Brownian diffusion involves collision between gas molecules and micrometer-sized particles, which push the particle in an irregular manner. Brownian diffusion increases with decreasing particle size. This mechanism is predominant in the gas exchange alveolar region of the lung for particles smaller than 0.5µ m.

There are other factors that also influence particle deposition, including mode of breathing (oral breathing permit the passage of particles greater than 10µ m to the lung), physical activity (exercise), age, lung diseases (chronic obstructive lung disease), and ambient conditions (increase in temperature or the presence of other pollutants).

The ability of the lung trying to protect itself against inhaled particles, clearance, will determine the adverse health effects of particulate matter. There are two clearance mechanisms: the mucociliary system and the alveolar macrophages. Particles deposited in the ciliated region of the tracheobronchial airway, rise on the mucociliary ladder to be expelled by coughing or swallowing. Particles deposited on the terminal bronchioles are cleared by lung macrophages. An early cellular response to an acute particulate exposure is damage to epithelial cells of respiratory tract, which also produce many different types of inflammatory mediators. The local pulmonary inflammation induced by PM_{10} could impact on the cardiovascular system via the local production of procoagulant factors in the lung or as a result of the effects of mediators released from the lungs which act on the liver, to increase the levels of procoagulant factors which could promote myocardial infarction.

Health Effects of Particle Pollution

The size of particles is directly linked to their potential for causing health problems. Small particles less than 10 micrometers in diameter pose the greatest problems, because they can get deep into lungs, and some may even get into blood stream.

Exposure to such particles can affect both lungs and heart. Small particles of concern include "inhalable coarse particles" (such as those found near roadways and dusty industries), which are larger than 2.5 micrometers and smaller than 10 micrometers in diameter; and "fine particles" (such as those found in smoke and haze), which are 2.5 micrometers in diameter and smaller.

Health Effects

Particle pollution – especially fine particles – contains microscopic solids or liquid droplets that

Are so small that they can get deep into the lungs and cause serious heath problems particle pollution exposure causes a varity health problems, including:

- increased respiratory symptoms, such as irritation of the airways, coughing, or difficulty breathing, for example;
- decreased lung function;
- aggravated asthma;
- development of chronic bronchitis;
- irregular heartbeat;
- nonfatal heart attacks; and
- premature death in people with heart or lung disease.

People with heart or lung diseases, children and older adults are the most likely to be affected by particle pollution exposure. However, even if you are healthy, you may experience temporary symptoms from exposure to elevated levels of particle pollution.

Acute Particle Pollution Exposure Causes

- Irritation to the eyes, nose, throat and lungs.
- Neurological effects such as lightheadedness
- Cough
- Exacerbate asthma and difficulty in breathing

Chronic Particle Pollution Exposure Causes

- Decreased lung function
- Development of chronic bronchitis
- Irregular heartbeat
- Nonfatal heart attacks
- Premature death in people with heart or lung disease
- Bladder Cancer
- Heart Disease
- Nervous System impairment
- Stroke
- DNA Damage

Particle size	Respiratory rate	Deposition process	Deposition site
Coarse (2.5- 10μm)	High	Impaction	Large airways
Coarse (2.5- 10μm)	Low	Sedimentation	Large and smaller airways
Fine (2.5μm)	High	Diffusion	Large and smaller airways
Fine (2.5μm)	Low	Diffusion	Alveoli

Appendix 1.125: Maximum Permissible Process Emission Standards for Various Chemicals (as per CPCB)-India

(CPCB-Central Pollution Control Board of India)

S.No.	Chemicals	Maximum Permissible concentration, ppm
1.	Sulphur penta fluoride, Dimethyl sulfide	0.1
2.	Fluorine, Bromine, mercaptans, phosgene, Trimethyl amine,	0.5
3.	Total Amines	1
4.	Phosphorous trichloride	2.5
5.	Hydrazine	3
6.	Chlorine, benzyl chloride, sulphur mono chloride	5
7.	Ethanolamine, formal dehyde, hydrogen bromide, hydrogen fluoride, hydrogen cynide, nitric acid, pyridine,	10
8.	Creosol, formic acid, nitrogen oxides, sulphur dioxide	15
9.	Acetic anhydride, aniline, hydrogen chloride, hydrogen sulfide, hydrocarbons,	25
10.	Dimethyl amine, benzene, carbon tetrachloride	30
11.	Acetic acid, carbon disulphide	50
12.	Ethylene dichloride	150
13.	Ammonia	250
14.	Vinyl chloride	500

Appendix 1.126: Summary of Indian Legislations related to Environmental Health

Various environmental health protection legislations of India have been described below.

A. The Environment (Protection) Act, 1986

"An Act to provide for the protection and improvement of environment and for matters connected therewith.

Subject to the provisions of this Act. the Central Government, shall have the power to take all such measures as it deems necessary or expedient for the purpose of protecting and improving the quality of the environment and preventing controlling and abating environmental pollution. in particular, and without prejudice to the generality of the provisions of subsection:

1., such measures may include measures with respect to all or any of the following matters, namely:-

i. coordination of actions by the State Governments, officers and other authorities-

   a. under this Act, or the rules made

   thereunder, or

   b. under any other law for the time being in force which is relatable to the objects of this Act,

ii. planning and execution of a nation-wide programme for the prevention, control and abatement of environmental pollution;

ii. laying down standards for the quality of environment in its various aspects;

iv. laying down standards for emission or discharge of environmental pollutants from various sources whatsoever:

   Provided that different standards for emission or discharge may be laid down under this clause from different sources having regard to the quality or composition of the emission or discharge of environmental pollutants from such sources;

v. restriction of areas in which any industries, operations or processes or class of industries, operations or processes shall not be carried out or shall be carried out subject to certain safeguards;

vi. laying down procedures and safeguards for the prevention of accidents which may cause environmental pollution and remedial measures for such accidents,

vii. laying down procedures and safeguards for the handling of hazardous substances;

viii. examination of such manufacturing processes, materials and substances as are likely to cause environmental pollution:

ix. carrying out and sponsoring investigations and research relating to problems of environmental pollution;

x. inspection of any premises, plant, equipment, machinery, manufacturing or other processes, materials or substances and giving, by order, of such directions to such authorities, officers or persons as it may consider necessary to take steps for the prevention, control and abatement of environmental pollution;

xi. establishment or recognition of environmental laboratories and institutes to carry out the functions entrusted to such environmental laboratories and institutes under this Act;

xii. collection and dissemination of information in respect of matters relating to environmental pollution;

xiii. preparation of manuals, codes or guides relating to the prevention, control and abatement of environmental pollution;

xiv. such other matters as the Central Government deems necessary or expedient for the purpose of securing the effective implementation of the previsions of this Act.

Persons carrying on industry operation, etc., not to allow emission or discharge of environmental pollutants **excess of the standards** - No person carrying on any industry, operation or process shall discharge or emit or permit to be discharged or emitted any environmental pollutants in excess of such standards as may be prescribed.

Persons handling hazardous substances to comply with procedural safeguards - No person shall handle or cause to be handled any hazardous substance except in accordance with such procedure and after complying with such safeguards as may be prescribed.

Environmental **laboratories:**

i. The Central Government may, by notification in the Official Gazette, -

   a. establish one or more environmental laboratories;

   b. recognize one or more laboratories or institutes as environmental laboratories to carry out the functions entrusted to an environmental laboratory under this Act.

ii. The Central Government may, by notification in the Official Gazette, make rules specifying-

   a. the functions of the environmental laboratory;

   b. the procedure for the submission to the said laboratory of samples of air, water, soil or other substance for analysis or tests, the form of the laboratory report thereon and the fees payable for such report;

   such other matters as may be necessary or expedient to enable that laboratory to carry out its functions.

B. Water (Prevention and Control of Pollution) Act, 1974

An Act to provide for the prevention and control of water pollution and the maintaining or restoring of wholesomeness of water for the establishment, with a view to carrying out the purposes aforesaid, of Boards f or the prevention and control of water pollution, for conferring on and assigning to such Boards powers and functions relating thereto and for matters connected therewith :

i. **Power to take samples of effluents and procedure to be followed in connection there with**

   A State Board or any officer empowered by it in this behalf shall have power to take for the purpose of analysis samples of water from any stream or well or samples of any sewage or trade effluent which is passing from any plant or vessel or from or over any place into any such stream or well.

ii. **Prohibition on use of stream or well for disposal of polluting matter, etc.**

   Subject to the provisions of this section-

   a. no person shall knowingly cause or permit any poisonous, noxious or polluting matter determined in accordance with such standards as may be laid down by the State Board to enter (whether directly or indirectly) into any [Stream or well or sewer or on land]; or

   b. no person shall knowingly cause or permit to enter into any stream any other matter Which may tend, either directly cr in combination with similar matters, to impede the proper flow of the water of the stream in a manner leading or likely to lead to a substantial aggravation of pollution due to other causes or of its consequences.

iii. Restrictions or- new outlets and new **discharges**

   Subject to the provisions of this section, no person shall, without the previous consent of the State Board

   a. establish or take any steps to establish any industry, operation or process, or any treatment and disposal system cr any extension or addition thereto, which is likely to discharge sewage cr trade effluent into a stream or well or sewer or on land (such discharge being hereafter in this section referred to as discharge of sewage); or

   b. bring into use any new or altered outlet for the discharge of sewage; or

   c. begin to make any new discharge of sewage: Provided that a person in the process of taking any steps to establish any industry, operation or process immediately before the commencement of the Water (Prevention and Control of Pollution) Amendment Act, 1988, for which no consent was necessary prior to such commencement, may continue to do so for a period of three months from such commencement or, if he has made an application for such consent, within the said period of three months, till the disposal of such application.

C. The Air (Prevention and Control of Pollution) Act, 1981

An Act to provide for the prevention, control and abatement of air pollution, for the establishment, with a view to carrying out the aforesaid purposes, of Boards, for conferring on and assigning to such Boards powers and functions relating thereto and for matters connected therewith.

a. **Power to declare air pollution control areas:**

The State Government may, after consultation with the State Board, by notification in the Official Gazette declare in such manner as may be prescribed, any area or areas within the State as air pollution control area or areas for the purposes of this Act.

If the State Government, after consultation with the State Board, is of opinion that the use of any fuel, other than an approved fuel, in any air pollution control area or part thereof, may cause or is likely to cause air pollution, it may, by notification in the Official Gazette, prohibit the use of such fuel in such area or part thereof with effect from such date (being not less than three months from the date of publication of the notification) as may be specified in the notification.

b. **Powers to give instructions for ensuring standards** for emission **from automobiles**

With a view to ensuring that the standards for emission of air pollutants from automobiles laid down by the State Board tinder clause (g) of subsection (1) of section 17 are complied with, the State Government shall, in consultation with the State Board, give such instructions as may be deemed necessary to the concerned authority in charge of registration of motor vehicles under the Motor Vehicles Act, 1939 (Act 4 of 1939), and such authority shall, notwithstanding anything contained in that Act or the rules made thereunder be bound to comply with such instructions.

c. **Restrictions on use of certain industrial plants**

Subject to the provisions of this section, no person shall, without the previous consent of the State Board, establish or operate any industrial plant in an air pollution control area:

Provided that a person operating any industrial plant in any air pollution control area, immediately before the commencement of section 9 of the Air (Prevention and Control of Pollution) Amendment Act, 1987, (47 of 1987) for which no consent was necessary prior to such commencement, may continue to do so for a period of three months from such commencement or, if he has made an application for such consent within the said period of three months, till the disposal of such application.]

Every person to whom consent has been granted by the State Bool'd under sub-section (4), shall comply with the following conditions, namely:

- the control equipment of such specifications as the State Board may approve in this behalf shall be installed and operated in the premises where the industry is carried on or proposed to be carried on;
- the existing control equipment, if any, shall be altered or replaced in accordance with the directions of the State Board;
- the control equipment referred to in clause (i) or clause (ii) shall be kept at all times in good running condition;

- chimney, wherever necessary, of such specifications as the State Board may approve in this behalf shall be erected or re-erected in such premises;
- such other conditions as the State Board, may specify in this behalf; and
- the conditions referred to in clauses (i), (ii) and

shall be complied with within such period as the State Board may specify in this behalf

d. **Persons carrying on industry, etc., to allow emission** of air pollutants in **excess of the standard laid down** by State Board

No person operating any industrial plant, in any air pollution control area shall discharge or cause or permit to be discharged the emission of any air pollutant in excess of the standards laid down by the State Board.

e. **Power of entry and inspection**

Subject to the provisions of this section, any person empowered by a State Board in this behalf shall have a right to enter, at all reasonable times with such assistance as he considers necessary, any place for the purpose of examining and testing any control equipment, industrial plant, record, register, document or any other material object or for conducting a search of any place in which he has reason to believe that an offence under this Act or the rules made has been or is being or is about to be committed and for seizing any such control equipment, industrial plant, record, register, document or other material object if he has reasons to believe that it may furnish evidence of the commission of an offence punishable under this Act or the rules made thereunder.

f. **Power to take samples of air or emission and procedure to be followed in connection therewith**

A State Board or any officer empowered by it in this behalf shall have power to take, for the purpose of analysis, samples of air or emission from any chimney, flue or duct or any other outlet in such manner as may be prescribed.

g. **State air laboratory**

The State Government may, by notification in the Official Gazette:

1. establish one or more State Air Laboratories; or
2. specify one or more laboratories or institutes as State Air Laboratories to carry out the functions entrusted to the State Air Laboratory under this Act.

The State Government may, after consultation with the Board, make rules prescribing:

h. the functions of the State Air Laboratory;

i. the procedure for the submission to the said Laboratory of samples of air or emission for analysis or tests, the form of the Laboratory's report thereon and the fees payable in respect of such report;

D. The Noise Pollution (Regulation and Control) Rules, 2000:

a. **Ambient air quality standards in respect of noise for different areas/zones**

The State Government may categorize the areas into industrial, commercial, residential or silence areas/zones for the purpose of implementation of noise standards for different areas.

The State Government shall take measures for abatement of noise including noise emanating from vehicular movements and ensure that the existing noise levels do not exceed the ambient air quality standards specified under these rules.

All development authorities, local bodies and other concerned authorities while planning developmental activity or carrying out functions relating to town and country planning shall take into consideration all aspects of noise pollution as a parameter of quality of life to avoid noise menace and to achieve the objective of maintaining the ambient air quality standards in respect of noise.

An area comprising not less than 100 metres around hospitals, educational institutions and courts may be declared as silence area/zone for the purpose of these rules.

b. Restrictions on the use of loud speakers/public address system

1. A loud speaker or a public address system shall not be used except after obtaining written permission from the authority.
2. A loud speaker or a public address system shall not be used at night (between 10.00 p.m. to 6.00 a.m.) except in closed premises for communication within, e.g. auditoria, conference rooms, community halls and banquet halls.

c. Consequences of any violation in silence zone/area

Whoever, in any place covered under the silence zone/area commits any of the following offence, he shall be liable for penalty under the provisions of the Act:

i. whoever, plays any music or uses any sound amplifiers,
ii. whoever, beats a drum or tom-tom or blows a horn either musical or pressure, or trumpet or beats or sounds any instrument, or
iii. whoever, exhibits any mimetic, musical or other performances of a nature to 44raq crowds.

Ambient Air Quality Standards in respect of Noise

Area Code	Category of Area/Zone	Limits in dB(A) Leq *	
		Day (6am-10pm)	Night (10pm -6am)
A.	Industrial area	75	70
B.	Commercial area	65	55
C.	Residential area	55	45
D.	Silence Zone	50	40

Silence zone is defined as an area comprising not less than 100 metres around hospitals, educational institutions and courts. The silence zones are zones which are declared as such by the competent authority.

E. Ozone Depleting Substances (Regulation and Control) Rules, 2000

"Ozone depleting substance" means the ozone depleting substances specified in column (2) of Schedule I, whether existing by itself or in a mixture, excluding any such substance or mixture (blend) which is in a manufactured product other than a container used for the transportation or storage of such substance.

- Regulation of production and consumption of ozone depleting substances
- Prohibition on export to or import from countries not specified in Schedule VI
- Ozone depleting substances are to be exported to or imported from countries specified in Schedule VI under a licence
- Regulation of the sale of ozone depleting substances
- Regulation on the purchase of ozone depleting substances

- Regulation on the use of ozone depleting substance
- Prohibition on new investments with ozone depleting substances
- Regulation of import, export and sale of products made with or containing ozone depleting substances

F. Hazardous Wastes (Management and Handling) Rules, 1989

Cegories of Hazardous Wastes

Waste Category	Type of wastes	Regulatory Quantities
1.	Cyanide Wastes	1 kgs/year calculated as cyanide
2.	Metal Wastes Finishing	10 kgs/year the sum of the specified substance calculated as pure metal
3.	Waste containing water solublechemical compounds of Pb, Cu, Zn, Cr and Sb.	10 kgs/year the sum of the specified substance calculated as pure metal
4.	Hg, As, Ti and Cd bearing wastes.	5 kgs/year the sum of the specified substance calculated as pure metal
5.	Non-halogenated hydrocarbons including non solvents	200 kgs/year calculated as halogenated hydrocarbons
6.	Halogenated hydrocarbons including solvents	50kgs/ year calculated as halogenated hydrocarbons
7.	Wastes from paints, pigments, glue, varnish and printing ink.	250 kgs/year calculated as oil or oil emulsions.
8.	Wastes from Dyes and Dye intermediate containing inorganic chemical compounds	200 kgs/year calculated as inorganic chemicals.
9.	Wastes from Dyes and Dye intermediate containing organic chemical compounds	50 kgs/year calculated as organic chemicals.
10.	Waste oil and oil emulsions.	1000 kgs/year calculated as oil and oil emulsions.
11.	Tarry wastes from refining and tar residues from distillation or Prolytic treatment.	200 kgs per year calculatcd as tar.
12.	Sludgesarising from treatment of waste containing heavy metals, toxic organics, oils emulsions and spend chemical and incineration ash.	Irrespective of any quantity.
13.	Phenols	5 kgs/year calculated as phenols.
14.	Asbestos	200 kgs/year calculated asbestos.
15.	Wastes from manufacturing of pesticides and herbicides and . residues from ; pesticides and herbicides formulation units.	5 kgs per year calculated as pesticides and their intermediate products.
16.	Acid/Alkaline/ Slurry	200 kgs/year calculated as Acids/Alkalies.

Waste Category	Type of wastes	Regulatory Quantities
17.	Off-specification and discarded products	Irrespective of any quantity.
18.	Discarded containers ; **and** Containers liners i of **hazardous** and toxic **i wastes.**	Irrespective of any quantity.

Responsibility of The Occupier for Handling of Wastes

The occupier generating hazardous wastes listed in quantities equal to or exceeding the limits given, shall take ail practical steps to ensure that such wastes are properly handled and disposed of without any adverse effects which may result from such wastes and the occupier shall also be responsible for proper collection, reception, treatment, storage and disposal of these wastes either himself or through the operator of a facility.

G. Radiation Protection Rules, 1971

In exercise of the powers conferred by section 30 of the Atomic Energy Act, 1962 (33 of 1962) and all other powers enabling it in this behalf, the Central v Government hereby makes the following rules :

i. **Prohibition of employment of persons below the age of 18 years:** No person under the age of eighteen years shall be employed as a radiation worker.

ii. **Radiological Safety Officer**

Every employer shall designate, with the approval of the competent authority, either himself or a person under his employ as Radiological Safety Officer.

iii. **Duties and functions of the Radiological Safety Officer-** The duties and functions of the Radiological Safety Officer in any radiation installation shall be as follows; namely-

a. to take all necessary steps aimed at ensuring that the operational limits are not normally exceeded;
b. to instruct the radiation workers under his charge on the hazards of radiation and on suitable safety measures and work practices aimed at minimizing exposure to radiation and contamination;
c. to carry out leakage tests on sealed sources as specified in rule;
d. to regulate the safe movement of all radioactive materials (including wastes containing radioactive materials) within the area under his charge;
e. to investigate and initiate prompt and suitable remedial measures in respect of any situation that could lead to radiation hazards;
f. to ensure that reports on all hazardous situations (including situations of the type referred to in rule 47 or as laid down in the notifications issued by the competent authority regarding operational limits) along with details of any immediate remedial measures that may have been initiated, are made available immediately to his employer;
g. to ensure that the ultimate disposal of wastes containing radioactive materials is done in a manner approved by the competent authority.

iv. **Radiation surveillance:**

The competent authority shall, by notification specify appropriate radiation surveillance procedures and the employer shall comply with the same.

Without prejudice to the generality of the above power such procedures may provide that:

a. in any radiation installation, the planning, design, construction and operation of facilities involving the use of radiation shall be done with the prior approval of and in accordance with the specifications laid down by the competent authority;
b. the working conditions adopted and the monitoring and personnel protective equipment provided, shall be in accordance with the specifications laid down by the competent authority;
c. the radiation workers shall be subjected to personnel monitoring procedures specified by the competent authority;
d. the radiation workers shall be subjected to medical examinations as provided in rule 19 or rule 20;
e. the appropriate records shall be maintained on the results of any routine measurements and investigations such as radiation and radioactivity level measurements, personnel monitoring measurements and medical examinations that may be stipulated by the competent authority.
f. an employer shall comply with any other special or general radiation surveillance procedures that may be specified by the competent authority.

v. **Radiation symbol**

a. The radiation symbol shall be prominently displayed in areas where exposure to radiation or contamination is likely and on all containers (including those used for transport) which contain
b. The radiation symbol shall not be used for any purpose other than those mentioned in these rules.
c. The colours for the radiation symbol shall be such as may be specified, by notification, by the competent authority.

Medical examination of radiation workers - Every radiation worker, prior to commencing radiation work, and subsequently at intervals not exceeding 12 months, shall be subjected to the following medical examinations, and any other examinations which may be specified by the competent authority, namely

d. X-ray examination of the chest;
e. all general laboratory investigations such as the examination of blood and excreta;
f. special investigations such as examinations of the skin, hands, fingers, finger nails and eyes.

Appendix 1.127: Environmental and Occupational Hazards of 'MERCURY'

Mercury is a toxic metal and is the only metal which is liquid at room temperatures; it is liquid at 298 Kelvin. Mercury is some-times called quicksilver because of its silvery white appearance it rarely occurs free in nature and is found mainly in cinnabar ore (HgS) in Spain and Italy. It is a heavy, odourless, lustrous liquid metal that sinks in water. Mercury can volatilize and enter into the air, and through air, it can enter water and the soil system. If a liquid effluent contains mercury it can directly deposit it in water, and soil. Some bacteria fund in soil and water can change mercury to its organic form, which is called methyl mercury, which is then taken in by live form.

Human are exposed to mercury from various man made and natural sources. A substantial number of human are exposed to low doses of mercury through dental amalgam and professionals of density are also exposed while preparing and filling up amalgam containing mercury. There are three form of mercury: 1) organic 2) Inorganic and 3) elemental mercury. Elemental mercury is absorbed mainly through inhalation. The inorganic salts are absorbed in our body through injection and they are highly irritating causing mucosal damage. Organic mercury compounds are readily absorbed in the body through all the available routes. Mercury is encountered in the workplace in two major forms inorganic mercury, which includes both elemental and inorganic salts and as organic mercury compounds, also known as alkyl mercury compounds. Occupational exposure to inorganic mercury compounds is much more common than the organic mercury compounds.

Elemental mercury gas is nearly insoluble in water and rather inert chemically, so it can be transported up to thousands of miles while gradually being converted to other forms and deposited. The ionic form, Hg(II) or Hg^{2+}, is soluble in water and thus incorporated into rain, for, and snow. Also, Hg(II) is both physically and chemically active and is known as "reactive gaseous mercury" or RMG. Most of the Hg(II) emitted is deposited via both precipitation and dry gases within about 50 to 100 kms from the stack/source. In many cases, this "local" deposition can be the most important impact of mercury from combustion sources. The fate of particle-bound mercury depends on the size of the particles, though generally they deposit on earth within a few hundred kms of the emitting stack.

India is one of largest user of mercury in the world. Occupational uses of mercury include manufacture of amalgams, lab equipments, valves, switches and rectifiers; as catalyst and pigments; and as medicines, biocides and antifouling agents.

Metallic mercury is used to produce chlorine gas and caustic soda, and is also used in thermometers, dental fillings, and batteries. Mercury salts are sometimes used in skin lightening creams and as antiseptic creams and ointments.

Major consumers of mercury in India

Form of mercury used	Consumers/Ingredients
Elemental mercury	Chlor-alkali industry (Mercury cell process) **Instruments manufacturing unit** 1. Clinical thermometers 2. Laboratory thermometers 3. Sphygmomanometer 4. Barometers 5. Other Instruments
Mercury oxide	**Health set-ups** 1. Drugs, pharmaceuticals 2. Dental amalgams 3. Others **Electrical apparatus,** Electric switches **Fungicides** 1. Phenyl mercury acetate 2. Methoxy ethyl mercury chloride 3. Others

Mercury Compounds	**Electrical apparatus** 1. Electric lamps a. Fluorescent lamps b. Mercury vapour lamps 2. Batteries **Paints** **Cosmetics**

The consumption of fish is a common route by which mercury enters the human body; freshwater fish tends to have slightly higher mercury levels than manna species.

Health Effects: Elemental (metallic) mercury primarily causes health effects when it is breathed as a vapor where it can be absorbed through the lungs. Symptoms include tremors; emotional changes (e.g., mood swings, irritability, nervousness, excessive shyness); insomnia; neuromuscular changes (such as weakness, muscle atrophy, twitching); headaches; disturbances in sensations; changes in nerve responses; performance deficits on tests of cognitive function. At higher exposures there may be kidney effects, respiratory failure and death. High exposures to inorganic mercury may result in damage to the gastrointestinal tract, the nervous system, and the kidneys. Both inorganic and organic mercury compounds are absorbed through the gastrointestinal tract and affect other systems via this route. However, organic mercury compounds are more readily absorbed via ingestion than inorganic mercury compounds. Symptoms of high exposures to inorganic mercury include: skin rashes and dermatitis; mood swings; memory loss; mental disturbances; and muscle weakness. USEPA has concluded that mercury chloride and methylmercury are possible human carcinogens.

Sources of Exposure

The exposure to mercury can results either from naturally occurring mercury or through atmospheric condition. Natural mercury arises from degassing of the earth's crust through volcanic gases and, probably, by evaporation from the oceans. Local levels in water derived from mercury ores may also be high (up to 80 µg/litre). Atmospheric pollution from industrial production is probably low, but pollution of water by mine tailings is significant.

The major sources of mercury pollution in India are chlorakali industries, industrial processes in thermal power plants, steel and cement industries, coal fired power and heat production mercury-containing products such as Sphygmomanometer, blood pressure equipment, pesticides, dental amalgam, waste incineration processes etc.

Bioaccumulation and Bio-Magnification

Two properties that make mercury extremely unmanageable are bioaccumulation and bio-magnification. Bioaccumulation is the retention of the toxic substance in the tissues, especially muscles. Bio-magnification is the process by which the toxic metal increases in concentration as it moves up the food chain (up to 100, 000 times the original levels, in some cases). Further organic mercury, mostly methyl mercury (MeHg) the most toxic one is bio-accumulating in the biota and subsequently biomagnified in the aquatic food chain, especially in fish.

Major Emitters of Mercury in India

Mercury	**Consumers**
Mercury in effluent (mercury cell process)	Chloralkali industry

Mercury in effluents and soil	Instrument manufacturing industries
Mercury in air and fly ash	Coal-based thermal power plants
Mercury in air and dust	Cement manufacturing plants
Mercury in air	Burning of mineral oil
Mercury in air and effluents	Disposal of municipal solid waste
Mercury in air	Disposal of medical wastes

Human exposure to mercury can happen through:

- Eating fish contaminated with organic mercury.
- Release of mercury from dental and medical treatments. Amount of mercury released from dental amalgams range from 5 to 20 g/day.
- Breathing contaminated workplace air or skin contact during use in the workplace (dental health services, chemical, and other industries that use mercury)
- Breathing vapours in air from spills, incinerators, and industries that burn mercury-containing fuels.
- From a pregnant woman/nursing mother exposed to mercury to their child.

The major target organs for the mercury toxicity are kidney and CNS. All forms of mercury are able to cause kidney damage if large amount enter the body.

Pregnant women, breast-feeding women, the fetus new born and young children residing in the hot spot area of the country are the potential vulnerable groups. The health effect of the mercury depends upon the compound to which a person is exposed.

Mercury (Hg) and its Health Effects

Chemical form of Hg	Health Effects
Methyl mercury	For fetuses, infants and children the primary health effect of methylmercury is impaired neurological development Methylmercury exposure in the womb, resulting from a mother's consumption of fish and shellfish containing methylmercury, adversely effects a baby's growing brain and nervous system. Impacts on cognitive function, memory attention language and fine motor and visual spatial skills have been seen in children expose to methy mercury in the womb
Elemental Mercury	Elemental (metallic) mercury primarily causes health effects when it is breathed as a vapor. Inhaled mercury vapor may cause headaches, cough, chest pains, chest rightness and difficulty in breathing. It may also cause chemical pneumonits. Other symptoms include tremors; emotional changes (e.g. mood swings, irritability, nervousness, excessive shyness); insomnia; neuromuscular changes (such as weakness, muscle atrophy, twitching); headaches, disturbances in sensations; changes in nerve responses; performance deficits on tests of cognitive function.
Inorganic Mercury	High exposures to inorganic mercury may result in damage to the gastrointestinal tract, the nervous system and the kidneys symptoms of high exposure to inorganic mercury include; skin rashes and dermatitis; mood swings memory loss; mental disturbance; and muscle weakness.

Minamata Disease

Minamata disease (Minamata-byo), sometimes referred to as Chisso-Minamata disease (Chisso-Minamata-byo), is a neurological syndrome caused by severe mercury poisoning Symptoms include ataxia, numbness in the hands and feet, general muscle weakness, narrowing of the field of vision and damage to hearing and speech In extreme cases, insanity, paralysis coma and death follow within a few weeks after the onset of symptoms. A congenital from of the disease can also affect fetuses in the womb Minamata disease was first discovered in Minamata city in Kumamota prefecture, Japan in 1956. It was caused by the release of methyl mercury in the industrial wastewater from the Chisso Corporation's chemical factory, which continued from 1932 to 1968. This highly toxic chemical bioaccumulated in shellfish and fish iin Minamata Bay and the Shiranul Sea, which when eaten by the local population resulted in mercury poisoning. while cat, dog and human deaths continued over more than 30 years, the government and company did little to prevent the pollution.

Other Occurrences of mercury poisoning.

- The phrase "Mad as a hatter" is likely a reference to mercury poisoning, as mercury based compounds were once used in the manufacture of felt hats in the 18th and 19th century.
- An early scientific study on mercury poisoning was in 1923 by the German. Alfred Stock who himself became poisoned, together with his colleagues, by breathing mercury vapour that was being was being released in his laboratory from the equipments diffusion pumps, float valves, and manometers all these contained mercury, and also due to accidentally mercury spilt which remained in the cracks of the linoleum floor covering.
- The term Hunter-Russell syndrome derives from a study of mercury poisoning among workers in a seed packing factory in England in the late 1930s because they breathed vapour of meth mercury that was used a seed disinfectant and preservative.
- Widespread mercury poisoning occurred in rural Iraq in 1971-1972 when grain treated with a methyl mercury based fungicide that was intended for planting and subsequently used by the rural population for the preparation of bread. The incident caused at least 6530 cases of mercury poisoning including 450 deaths.
- On August 14, 1996 professor working at Dartmouth college spilled as small amount of dimethyimercury on her later glove. She began experiencing the symptoms of mercury posining five months later from brain malfunction due to mercury intoxication.
- In April of 2000 Alan Chmurny attempted to kill a former employee Marta Bradley by pouring mercury into the ventilation system of her car.
- The first emperor of unified China, Qin Shi Huang Di was driven insane and killed by mercury pills intended to give him eternal life.

Prevention of Mercury Poisioning

1. Environmental monitoring

Regularly monitor mercury levels in air, water soil and food materials.

Tolerance Guidelines

Title	Details
Max Permissible limit in water	1 g/1

Max concentration in compost	0.15 mg/kg dry basis
Max concentration in treated Leacheate	10 g/1 in land surface water and public
Max allowed/recommended levels of mercury in fish	0.5 ppm total mercury

Permissible Limits

Permissible concentration of mercury as recommended by the National institutes of safety and Health, USA for occuptation exposure in 0.05 mg/m^3. The current OSHA standard for mercury is a ceiling level of 0.1milligram of recommended that the mean permissible exposure limit may be changed to 0.05 mg/m^3 over an eight-hour work shift.

2. Medical surveillance

The following medical procedures should be made available to each employee which is exposed to mercury at potentially hazardous levels.

Initial Medical Examination

A complete history and physical examination: The purpose is to detect pre existing conditions that might place the exposed employee at increased risk, and to establish a baseline date for further health monitoring. Persons with a history of allergies or known sensitization to mercury, chronic respiratory disease, nervous system disorders or kidney disease would be expected to be at increased risk from exposure.

Urine Analysis

Since kidney damage has been observed in humans exposed to mercury, a urine analysis should be carried out at least for specific gravity, albumin, glucose and a microscopic examination of centrifuged sediment. Determination of mercury level in urine may be helpful in assessing the extent of absorption.

3. Minimisation of Exposure Risk

- Advise children not to play with shiny, siiver liquids.
- Medications containing mercurial state should be stored safely and properly disposed.
- Pregnant women and women of child bearing age should limit their consumption of fish with high levels of MeHg e.g. shark and sword fish.

4. Safer alternatives

There is an urgent need to bring in some policy for gradual shift from mercury containing equipments to safer alternatives.

5. Proper handling and disposal.

Healthcare staff needs to be trained to handle this toxic mental safely and the disposal and emission issues need to be addressed.

6. Personal protective equipment

Employees should be provided with and required to use impervious clothing, gloves, face shields (eight-inch minimum) and other appropriate protective devices/clothing necessary to prevent repeated or prolonged skin contact with liquid mercury. Non-impervious clothing which becomes contaminated with mercury should be removed promptly and not re-use until the mercuric removed from the clothing.

7. First Aid

If liquid mercury gets into the eyes, wash eyes immediately with large quantity of water, lifting the lower and upper lids occasionally. Contact lenses should not to worn when working with this chemical.

Skin Exposure

If liquid mercury gets on the skin, promptly wash the contaminated skin with soap or mild detergent and water. If liquid mercury penetrates through the clothing, remove the clothing promptly and wash the skin using soap or mild detergent and water.

Breathing

If a person breathes in large volume of mercury, move the exposed person to fresh air at once. If breathing has stopped, perform artificial respiration. Keep the affected person warm and the keep him at rest.

Swallowing

When large quantities of mercury have been swallowed or mercury has been swallowed repeatedly and the person is conscious given the person large quantities of water immediately. After the water has been swallowed, try to get the person to vomit by having him touch the back of his throat with his finger. Do not apply the same practice with an unconscious person.

8. Generate data on the environmental release of mercury and its origin and speciation

Legislations

Two rules that deal with hazardous substances are: The Hazardous Waste Management and Handling Rules (1989), which list mercury and mercury containing waste as hazardous waste. Another rule is the Manufacture, Storage and Import of Hazardous Chemical Rules, 1989, which includes a few mercury compounds.

Ingested elemental mercury is only 0.01 per cent absorbed, but methyl mercury is nearly 100 percent absorbed from the gastrointestinal tract.

There is consumer legislation in place the Prevention of Food Adulteration Act, 1955, which aims to protect consumers from adulterated food and ensure food safety. The Rules declare mercury as a poisonous metal and limit its concentration in fish to 0.5 ppm and in other food items upto 1.0 ppm, Methyl mercury concentration in fish is limited to 0.25 ppm.

The WHO have also limited the concentration of methyl mercury to 0.5 mg/kg for all fishes except predatory fishes and mg/kg for predatory fishes such as shark, swordfish, tuna, pike etc.

A draft notification was issued by the Ministry of Environment and Forest in 2000 for a phased elimination of mercury from consumer products including thermometer, fluorescent tube, batteries, electrical thermostat and switches, medical instruments, certain pharmaceutical and agricultural products with an exemption for essential products. Government of India is trying to ensure that potentially harmful mercury waste are recycled or disposed off under strict standards to prevent the emission into the environment.

Although individual thermometers only contain a small amount of mercury, hospitals, break hundreds every year. This has contributed to healthcare becoming the third largest source of mercury contamination in the environment.

Appendix 1.128: Environmental and Occupational Hazards of 'ARSENIC'

Arsenic a metalloid element is a natural part of the earth's crust in some part of the world and is found in water that has flowed through arsenic rich rocks. Arsenic is also emitted into air by burning of coal & vegetation and volcanic eruptions. High concentration of arsenic is found in drinking water in various parts of the world including India (in West Bengal)

Souces of Arsenic

Natural Sources

Concentrations of arsenic in open ocean water are typically 12 μg/litre. The concentrations of arsenic in unpolluted surface water and groundwater are typically in the range of 110 μg/litre. Elevated concentrations in surface water and groundwater up to 100-5000 μg/litre can be found in areas of sulfide, mineralization.

Elevated concentrations (> 1 mg As/litre) in groundwater of geochemical origins have also been found in Taiwan, India (West Bengal) and more recently in most districts of Bangladesh.

Elevated arsenic concentrations were also found in the drinking-water in Chile, North Mexico and several areas of Argentina. Arsenic-contaminated groundwater was found in parts of PR China and the USA (California, Utah. Nevada, Washington and Alaska). More recently, arsenic concentrations of < 0.98 mg/litre have been found in wells in south-western Finland. Levels as high 35 mg As/litre and 25.7 mg As/litre have been reported in areas associated with hydrothermal activity.

Arsenic is the main constituent of more than 200 mineral species of which about 60% are arsenate, 20% sulfide and sulfosalts and the remaining 20% include, arsenides, arsenites, oxides & elemental arsenic.

Man-made(anthropogenic) sources

Industry

Smelting of non-ferrous metals and the production of energy from fossil fuels are the two major industrial processes for anthropogenic arsenic contamination of air water and soil other sources of contamination are the manufacture and use of arsenical pesticides and wood preservatives.

Tailings from metal-mining operations are a significant source of contamination, and can lead to contamination of the surrounding topsails, and. because of leaching, sometimes the groundwater too could be contaminated As sulfur is often present in these tailings, exposure to the atmosphere in the presence of water leads to the production of an acid solution that cart teach many elements including arsenic

Arsenic is also present in the rock phosphate used to manufacture fertilizers and detergents

Past Agricultural Use

In 1983, arsenical pesticides were one of the largest classes of biocontrol agent. From the 1960s there was a shift in herbicide use, from inorganic compounds (including lead calcium arsenate and copper acetoarsenite) to inorganic and organic compounds (arsenic acid and sodium arsenate)

Sewage Sludge

The levels of arsenic sewage sulfate reflect the extent of industrialization of the area served by the local sewage system. Significant quantities may be a mod by arsenic contaminated wastewater runoff derived from sources including atmospherically deposited arsenic residues from pesticide usage, phosphate detergents and. industrial effluent, particularly from the metal-processing industry

Exposure to Arsenic

Occupational Exposure

There is the potential for significant occupational exposure to arsenic in several industries, eg. non-ferrous smelting, electronics, wood preservation, wood joinery shops, arsenic production, glass manufacturing, and the production and application of arsenical pesticides .Exposure primarily through inhalation of arsenic-containing particulate, but ingestion and dermal exposure may be significant in particular situations eg preparation of CCA-treated timber.

It is extremely rare for workers to be exposed to arsenic alone the exposure is usually to arsenic in combination with other element.

Arsenic (a naturally occurring element) is a notoriously poisonous metalloid that has many allotropic forms; yellow, black and grey are a few that are regularly seen. Pure arsenic is a gray metal-like material usually found combined with other elements such as oxygen, chlorine, and sulphur. Arsenic and its compounds are used as pesticides, herbicides, insecticides and various alloys. Arsenic in animals and plants combines with carbon and hydrogen to form organic arsenic compounds. Most inorganic and organic arsenic compounds are white or colorless powders that do not evaporate. They have no smell, and most have no special taste. Most arsenic compounds can also dissolve in water. Arsenic cannot be destroyed in the environment. It can only change its form. Arsenic in air will settle to the ground or is washed out of the air by rain. Many arsenic compounds can dissolve in water.

Lead hydrogen arsenate has been used, well into the 20th century, as an insecticide on fruit trees (resulting in neurological damage to those working the sprayers), and in the 19th century as a coloring agent in sweets. Inorganic arsenic occurs naturally in many kinds of rock, especially in ores that contain copper or lead. When these ores are heated at smelters to get the copper or lead, most of the arsenic enters the air as a fine dust. Smelters collect this dust and purify the arsenic for several uses. But once in the air, the arsenic particles travel with the wind for a while and then settle back to the ground. Thus, arsenic can get into lakes, rivers, or underground water by dissolving in rain or snow, or through the discharge of industrial wastes. Some of the arsenic will stick to the sediment on the bottom of the lake or river, and some will be carried along by the water. The main use is as a preservative for wood to make it resistant to rotting and decay. Inorganic arsenic compounds are also used to make several types of insect killers and weed killers are mainly used to preserve wood. Organic arsenic compounds are used as pesticides, primarily on cotton plants. Fish and shellfish can accumulate arsenic, but the arsenic in fish is mostly in a form that is not harmful. Eating food, drinking water, or breathing air containing arsenic, working in Arsenic using industries, smoke from wood treated with arsenic, living near uncontrolled hazardous waste sites containing arsenic, living in areas with unusually high natural levels of arsenic in rock are a few important sources of the metal. In the past, several kinds of products used in the home (rat poison, ant poison, weed killer, some types of medicines) had arsenic in them.

Health effects of inorganic arsenic are sore throat or irritated lungs, nausea, vomiting, decreased production of red and - white blood cells, abnormal heart rhythm, damage to blood vessels, a sensation of "pins and needles" in hands and feet, redness and swelling of the skin and even death, central and peripheral nervous system disorders are a common occurring amongst workers in Arsenic rich environment. Long term exposure to inorganic arsenic are peripheral neuropathy, darkening of the skin and the appearance of small "corns" or "warts" on the palms, soles, and torso and liver or kidney damage in humans. Inorganic arsenic can increase the risk of lung cancer, skin cancer, bladder cancer, liver cancer, kidney cancer, and prostate cancer. Organic arsenic compounds are less toxic than inorganic arsenic compounds. Arsenic and many of its compounds are especially potent poisons.

Arsenic kills by inhibition of the metabolic enzyme lipothiamide pyrophosphatase, leading to death from multi-system organ failure. The IARC recognizes arsenic and arsenic compounds as group 1 charcinogens, and the EU lists arsenic trioxide, *arsenic pentoxide and arsenate salts as category 1 carcinogens.* Chronic arsenic poisoning from drinking water causes a disease known as *Arsenicosis.*

Measured concentrations of Arsenic in Ambient Air at Various places

S.No.	Location	Sampling period	Particle size and/ or species	Concentration (ng/m^3)
1.	Antarctica	NS		0.019
2.	Antarctic Ocean	1988–1989	AS_i	0.05 (0.01–0.2)
		1988–1989	As_o	0.002 (single sample)
3.	North Pacific Ocean	1981–1987	AS_i	0.1 (0.01–0.95)
		1981–1987	As_o	0.008 (0.001–0.03)
4.	North Atlantic Ocean	1989		0.1 (0.01–0.45)
		1989	As_o	0.007 (0.001–0.3)
5.	Baltic Sea	1985		1.1 (0.3–3.7)
6.	Mid-Atlantic coast, USA	1985–1986		1.05
7.	Continental shelf waters, south-eastern USA	1975–1976	particulate	1.7 (0.2–4.3)
8.	Northern Chesapeake Bay, USA	1990–1991	<10 μm	0.66 (0.11–1.96)
9.	Rural US sites (National Parks)	1979–1981	0.45 μm	< 1.6–2.3 (range of means)
10.	Midwestern USA	1990		1.6 (0.7–2.5)
11.	Natural geysers, northern California, USA	1989	As (III)	0.22 and 0.54 (0.06–3.08)
			As (V)	0.46 (0.08–1.3) and 2.29 (0.7–6.54)
12.	Bagauda, Nigeria	1976		0.6
13.	Pelindaba, South Africa	1976		1.7
14.	Chilton, United Kingdom	1976		4.2
I5	Rural sites, United Kingdom	1972–1973		1.5–2.5 (range of means)
16.	Rural area near Thessaloniki, Greece	1989–1990		2.7
17.	Birkenes, Norway	1978–1979	particulate	1.2 (0.02–12)
		1985–1986	particulate	0.63 (< 0.04-4.6)

As_i=inorganic As; As_o = organic As; NS = not stated

Non-Occupational Exposure

Non-occupational human exposure to arsenic in the environment is primarily through the food and water. Of these, food is generally the principal contributor to the daily intake of total arsenic. In some areas, arsenic in drinking-water often constitutes the principal contributor to the daily arsenic intake. Contaminated soils such as mine tailings are also a potential source of arsenic exposure. The daily intake of total arsenic from food and beverages is generally between 20 and 300 µg/day. Approximately 25% of the arsenic present in food is inorganic, but this depends highly on the type of food ingested inorganic arsenic levels in fish and shellfish are low (< 1%). Foodstuffs such as meat, poultry, dairy products and cereals have higher levels if inorganic arsenic. Pulmonary exposure may contribute up to approximately 10 µg day in a smoker and about 1 µg/day in a non-smoker and more in polluted areas. The concentration of metabolites of inorganic arsenic in urine (inorganic arsenic) reflects the absorbed dose of inorganic arsenic by an individual. Generally, it ranges from 5 to 20 µg As/litre, but may even exceed 1000 µg litre.

Health Effects of Arsenic

Because arsenic is a recognized carcinogen, focus has been on cancer risks. Studies have also examined effects of arsenic on atherosclerotic diseases diabetes mell.tus and peripheral vascular diseases in different geographical areas and under varying exposure conditions following are the principal health effects of arsenic exposure:-

Acute Effects

Inorganic arsenic is acutely toxic and ingestion of large doses leads to gastrointestinal symptoms, disturbances of cardiovascular and eventually nervous system functions, multiorgan failure and eventually death. In survivors, tone marrow depression, haemolysis, hepatomegaly, melanosis, polyneuropathy and encephalopathy may be observed.

Vascular Effects

Vascular effects include heart disease and high blood pressure, which are caused by a progressive narrowing of blood vessels that diminishes me normal flow of blood. The flow through small vessels in the extremities ran be reduced so severely that tissue starves and dies. Untreated can lead to gangrene of the feet or blackfoot disease, a condition seen in people from Bangladesh and other regions of the world where arsenic concentration in drinking water die above 500parts per billion (500 micrograms per liter). In contrast, people drinking water with less than 500 parts per billion may be at a greater risk for hypertension and heart attacks

Diabetes Mellitus

Non-occupational exposure occurs mainly through drinking water. Exposure to levels of arsenic in drinking water well above 100 ppb has been associated with an increased risk of type 2 diabetes in the high- arsenic areas Taiwan and Bangladesh. The biological mechanisms for an association between chronic arsenic exposure and increased diabetes risk are not known.

Neurological Effects

Human exposure to arsenic causes nervous system disturbances such as polyneuropathy, EEG abnormalities and in extreme cases, hallucination, disorientation and agitation.

Skin Changes

Skin changes, including hyperpigmentation and generalized hyperkeratosis, are the most common signs of chronic at ingestion from drinking water.

An elevated risk of arsenic-associated skin lesions hyperpigmentation and/or keratosis) has been found in west Bengal where the lowest concentration of arsenic in drinking (<50µg/litre) is reported

Arsenic Rule at a Glance

Maximum Contaminant Level (MCL)

MCL-0.010 ppm

Maximum Contaminant Level ground (MCLG)

MCLG-0.005 ppm

Health-Effects

Skin damage or problems with circulatory systems. and may haw increased risk of getting cancer

Sources of Contamination

Erosion of natural deposits run off from orchards. run off from glass and electronic production wastes.

Arsenic Cancer

Ingestion of arsenic is known to cause skin cancer. The evidence indicates that arsenic can also cause liver, lung, kidney and skin cancer.

Occupation exposure to arsenic, primary by inhalation, is associated with lung cancer. Increased risks have been observed at cumulative exposure levels >0.75(mg/m^3)/year (e.g.15 year exposure to a work room air concentration of 50 µg/m$^{3)}$

Smokers are exposed to arsenic by the inhalation of mainstream cigarette smoke. It has been estimated that someone smoking 40 cigarette per day would inhale about 10 µg of arsenic.

Herbal asthma medicines contain up to 107mg/g of inorganic arsenic.

Biomarkers of Arsenic exposure

Blood

Inorganic arsenic is rapidly cleared from blood, hence blood arsenic is typically used only as an indicator of very recent or relatively high-level exposure.(e.g. in cases of poisoning), or chronic stable exposure (e.g. to drinking water). The limitation of blood arsenic levels as indicators of low-level exposure of drinking water is that it is difficult to distinguish the contributions of inorganic arsenic from water and organic arsenic from food.

Hair and Nail

As arsenic accumulates in keratin-rich tissues such as skin, hair and nail as a consequence of its affinity for sulfhydryl groups, arsenic levels in hair and nails may be used as an indicator of past arsenic exposure. When exposed to water containing high arsenic levels, hair can bind arsenic extremely and may not be removed readily by washing.

The arsenic content of fingernails and toenails has also been used as a bioindicators of past arsenic exposure and fingernail arsenic has been reported to be significantly correlated with hair arsenic content.

Urine

Since arsenic is rapidly metabolized and excreted through the urine, total arsenic, inorganic arsenic and the sum of arsenic metabolites in urine have all been used as biomarkers of recent arsenic exposure. Levels of arsenicals in urine may be a consequence of inhalation exposure or ingestion of arsenic from drinking water, beverages, soil or foodstuffs.

Purification of Water

Aresnic Removal Units (ARUs)

The arsenic removal units are designed to both remove arsenic from the ground-water making it safe for human consumption and provide for safe containment of the removed arsenic without long-term adverse ecological impact. The ARU lowers the arsenic concentration from as high as 500 parts per billion to less than 50 parts per billion More than 150, 000 villagers in Bangladesh and West Bengal are drinking and using arsenic safe potable water from ARUs attached to ground well pumps which do not require chemical additions, pH adjustments or electricity, Water is pumped from the well prior to installation of the well-head unit where it passes through beads of activated alumina that remove the arsenic.

Only indigenous durable materials are used in constructing the ARUs Many such units have been providing safe water for well over five years.

To combat the arsenic crisis .following steps should be taken:

- In most of the villages surveyed in West Bengal and Bangladesh, an average of 35% of the tube wells contain water that is safe to drink. The mouths of the safe tube wells must be colored green and the unsafe ones red so that the villagers can use water from green tube wells for drinking and cooking, and red tube wells water for bathing washing toiled, etc .The safe tube wells should be tested for arsenic every 3-6 months.
- The effective preventive measures for preventing chronicarsenic toxicity include safe water and nutritious food.

Note:

Arsenic Fact Sheet

- Arsenic is naturally present at high levels in the groundwater of a number of countries.
- Arsenic is highly toxic in its inorganic form.
- Contaminated water used for drinking, food preparation and irrigation of food crops poses the greatest threat to public health from arsenic.
- Long-term exposure to arsenic from drinking-water and food can cause cancer, skin lesions developmental effects, cardiovascular disease, neurotoxicity and diabetes.
- The most important action in affected communities is the prevention of further exposure to arsenic by provision of a safe water supply.

Sources of Exposure

Arsenic is a natural component of the earth's crust and is widely distributed throughout the environment in the air, water and land. It is highly toxic in its inorganic form.

People are exposed to elevated levels of inorganic arsenic through drinking contaminated water, using contaminated water in food preparation and irrigation of food crops, industrial processes, eating contaminated food and smoking tobacco.

Long-term exposure to inorganic arsenic, mainly through drinking of contaminated water, eating of food prepared with this water and eating food irrigated with arsenic- rich water, can lead to chronic arsenic poisoning. Skin lesions and skin cancer are the most characteristic effects.

The greatest threat to public health from arsenic originates from contaminated groundwater. Inorganic arsenic is naturally present at high levels in the groundwater of a number of countries, including Argentina, Bangladesh, Chile, China, India, Mexico, and the United States of America. Drinking-water, crops irrigated with contaminated water and food prepared with contaminated water are the sources of exposure.

Fish, shellfish, meat, poultry, dairy products and cereals can also be dietary sources of arsenic, although exposure from these foods is generally much lower compared to exposure through contaminated groundwater. In seafood, arsenic is mainly found in its less toxic organic form.

Arsenic is used industrially as an alloying agent, as well as in the processing of glass, pigments, textiles, paper, metal adhesives, wood preservatives and ammunition. Arsenic is also used in the hide tanning process and, to a limited extent, in pesticides, food additives and pharmaceuticals.

People who smoke tobacco can also be exposed to the natural inorganic arsenic content of tobacco because tobacco plants essentially take up arsenic naturally present in the soil. Also, in the past, the potential for elevated arsenic exposure was much greater when tobacco plants used to be treated with lead arsenate insecticide.

Health Effects

Arsenic occurs in inorganic and organic forms. Inorganic arsenic compounds (such as those found in water) are highly toxic while organic arsenic compounds (such as those found in seafood) are less harmful to health.

Acute Effects

The immediate symptoms of acute arsenic poisoning include vomiting, abdominal pain and diarrhoea. These are followed by numbness and tingling of the extremities, muscle cramping and death, in extreme cases.

Chronic Effects

The first symptoms of long-term exposure to high levels of inorganic arsenic (e.g. through drinking-water and food) are usually observed in the skin, and include pigmentation changes, skin lesions and hard patches on the palms and soles of the feet (hyperkeratosis). These occur after a minimum exposure of approximately five years and may be a precursor to skin cancer.

In addition to skin cancer, long-term exposure to arsenic may also cause cancers of the bladder and lungs. The International Agency for Research on Cancer (IARC) has classified arsenic and arsenic compounds as carcinogenic to humans.

Other adverse health effects that may be associated with long-term ingestion of inorganic arsenic include developmental effects, neurotoxicity, diabetes and cardiovascular disease. In China (Province of Taiwan), arsenic exposure has been linked to "blackfoot disease", which is a severe disease of blood vessels leading to gangrene. However, this disease has not been observed in other parts of the world, and it is possible that malnutrition contributes to its development.

Arsenic contamination of groundwater is widespread and there are a number of regions where arsenic contamination of drinking-water is significant.

Arsenic in Bangladesh has attracted much attention since recognition in the 1990s of its wide occurrence in well-water in that country. People in Bangladesh are at risk of being exposed to arsenic concentrations that are greater than the national standard of 50 microgram/litre and the WHO guideline value of 10 μg/1itre respectively

The symptoms and signs caused by long-term elevated exposure to inorganic arsenic. differ between individuals, population groups and geographical areas.

There is no method to distinguish cases of cancer caused by arsenic from cancers induced by other factors.

Prevention and Control of Arsenic Poisoning

There are a number of options as discussed below to reduce levels of arsenic in drinking-water, food preparation water and water used for irrigation of food crops.

- Substitute high-arsenic sources such as groundwater, with low-arsenic, microbiologically safe sources such as rain water and treated surface water. Low- arsenic water can be used for drinking, cooking and irrigation purposes, whereas arsenic-rich water can be used for other purposes such as bathing and washing clothes.
- Discriminate between high-arsenic and low-arsenic sources. For example, test water for arsenic levels and paint tube wells or hand pumps in different colours.This can be an effective and low-cost means to rapidly reduce exposure to arsenic.
- Blend low-arsenic water with higher-arsenic water to achieve an acceptable arsenic concentration level.
- Install arsenic removal systems - either centralized or domestic - and ensure the appropriate disposal of the removed arsenic. Technologies for arsenic removal include oxidation, coagulation-precipitation, absorption, ion exchange and membrane techniques. There is an increasing number of effective and low-cost options for removing arsenic from small or household supplies.

Long-term actions are also required to reduce occupational exposure from industrial processes.

High-risk populations should also be monitored for early signs of arsenic poisoning usually skin problems

Arsenic is one of WHO's 10 chemicals of major public health concern. The current recommended limit of arsenic in drinking-water is 10μg/liter, although this guideline value is designated as provisional because of measurement difficulties and the practical difficulties in removing arsenic from drinking-water.

Appendix 1.129: Environmental and Occupational hazards of 'CHROMIUM'

Chromium (Cr) is a lustrous, brittle, hard metal having silver-gray colour and can be highly polished. When heated it burns and forms the green chromic oxide. Chromium is unstable in oxygen, it immediately produces a thin oxide layer that is impermeable to oxygen and protects the metal.

Chromium is naturally found in rocks, animals, plants, and soil. It can exist in several different forms. Depending on the form it takes, it can be a liquid, solid, or gas. The most common forms are chromium(O), chromium(III), and chromium(VI). No taste or odor is associated with chromium compounds. Chromium (III) is an essential nutrient that can be toxic in large doses. The National Academy of Sciences (USA) has established a safe and adequate daily intake for Cr(III) in adults of 50 -200 micrograms per day. Major factors governing the toxicity of chromium compounds are oxidation state and solubility. Since Cr(III) is poorly absorbed by any route, the toxicity of chromium is mainly attributable to the Cr(VI) form. It can be absorbed by the lung and gastrointestinal tract, and even to a certain extent by intact skin. Any of these species could attack DNA, proteins, and membrane lipids, thereby disrupting cellular integrity and functions.

Chromium is known to enhance the action of insulin, a hormone critical to the metabolism and storage of carbohydrate, fat, and protein in the body. Chromium also appears to be directly involved in carbohydrate, fat, and protein metabolism.

Occurrence of Chromium in Nature

World Wide

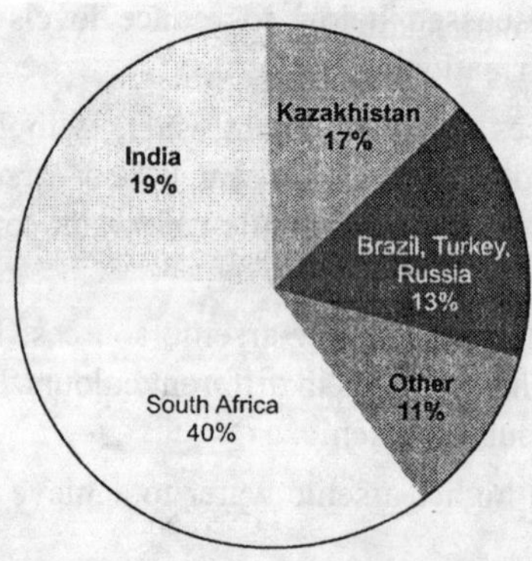

Chromium in Food

Food items	Cr (mcg)
Broccoli, 1/2 cup	11
Grape juice, 1 cup	8
Potatoes, mashed, 1 cup	3
Garlic, dried, 1 teaspoon	3
Beef cubes, 3 ounces	2
Orange juice, 1 cup	2
Whole wheat bread, 2 slices	2
Red wine, 5 ounces	1-13
Apple, unpeeled, 1 medium	1
Banana, 1 medium	1
Green beans, ½ cup	1

mcg--microgram

Chromium

Sources of Chromium Exposure

Occupational: Occupational exposures to chromium occurs to workers through

- Chromate added in anti-algae
- Antifreeze agents
- Cement, Chrome alloy production

- Chrome electroplating (Cr[VI])
- Copier servicing, Glassmaking
- Leather tanning (soluble Cr[lll])
- Paints/pigments (insoluble Cr[VI])
- Photoengraving, Porcelain and ceramics manufacturing
- Production of high-fidelity magnetic audiotapes
- Tattooing, Textile manufacturing
- Welding of alloys or steel
- Wood preservatives, i.e. acid copper chromate

Environmental: Environmental exposures to chromium occurs to workers through

- Air borne emissions from chemical plants and incineration facilities
- Cement dust, contaminated landfill
- Effluents from chemical plants
- Asbestos lining erosion
- Road dust from catalytic converter erosion and asbestos brakes
- Tobacco smoke
- Top soil and rocks

Top Six Toxic Threats

Toxics	Estimated population at risk at identified sites (million people)	Estimated global impact** (million people)
1. Lead	10	18.22
2. Mercury	8.6	15.19
3. Chromium	7.3	13.17
4. Arsenic	3.7	5.9
5. Pesticides	3.4	5.8
6. Radionuclides	3.3	5.8

* Population estimates are preliminary and based on an ongoing assessment of polluted sites
** Estimated global impact is extrapolated from current site research and assessment surveys

Routes of Exposure

Chromium exposure can occur by breathing in breathing in dust or by skin contact with solutions or solids or by eating, drinking or smoking in areas where chromium is used. Members of the public could be exposed to it due to a spillage or inadequate safety measures during disposal/incineration, most often by ingestion through chromium content in soil, food, and water.

Enviromental Effects

Chromium enters the air, water and soil in the chromium (III) and chromium (VI) form through natural processes and human activities.

Chromium Concentrations in the Environment

Sample	Concentration
Fresh Water	0-117 μgL^{-1}
Sea Water	0-0.5 μgL^{-1}
Air	1-545, 000 ng m^{-3}
Plants	0.006-18mg kg^{-1}
Animals	0.03-1.6mg kg^{-1}
Sediments	0-31000 mg kg^{-1}

In India, about 20000-32000 tons of elemental Cr annually escape into the environment from tanning industries. Even if the recommended limit for Cr concentration in water are set differently for Cr(III) (8 μg/L) and Cr(VI) (1 μg/L), it ranges from 2 to 5g/L in the effluents of these industries.

Chromium Deficiency

Only 10% of Americans have an adequate amount of chromium in his/her diet.

There is no specific recommended daily allowance for chromium. Average daily intake may be about 80-100 mcg. Even mild deficiencies of chromium can produce problems in blood sugar metabolism, such as anxiety or fatigue. Abnormal cholesterol metabolism and increased progress of atherosclerosis are associated with chromium deficiency, and deficiency may also cause decreased growth in young people and slower healing time after injuries or surgery.

Chromium Toxicity

S.No.	Biological effects	Symptoms
1.	Respiratory	Asthma, chronic bronchitis, chronic irritation, chronic pharyngitis, chronic rhinitis, congestion and hyperemia, polyps of the upper respiratory tract, tracheobronchitis, and ulceration of the nasal mucosa/ septal perforation
2.	Skin	Dryness, erythema, fissuring, papules, scaling, small vesicles, and swelling
3.	Carcinogenic	Occupational exposure to Cr (VI) compounds has been associated with increased risk of respiratory system cancers.
4.	Renal	Elevated urinary β2-microblobulin levels (an indicator of renal tubular damage) have been found in chrome platers, and higher levels have generally been observed in younger persons exposed to higher Cr(VI) concentrations
5.	Hepatic	Liver effects include derangement of the liver cells, necrosis, lymphocytic and histocytic infiltration, and increases in Kupffer cells, liver failure in acute cases
6.	Gastro-intestinal	Ulcer formation, hypertrophic gastritis
7.	Cardio-vascular	Early hypoxic changes in the myocardium, cardiogenic shock that may complicate by pancreatitis and gut mucosal necrosis and hemorrhage

S.No.	Biological effects	Symptoms
8.	Hematological	Decreased Hb content and increased total WBC and reticulocyte counts, and plasma Hb
9.	Genotoxic and mutagenic studies	DNA oxidative damage from the urinary samples, chromosome aberrations in mammalian cells and has been associated with increased frequencies of chromosome aberrations in lymphocytes
10.	Other effects	Dizziness, headache, weakness

Management of Chromium Toxicity

Irrespective of the route of exposure, the initial approach to an affected individual includes a brief assessment of clinical status followed by support of basic cardiopulmonary functions. Once the airway has been stabilized and cardiopulmonary support has been instituted as indicated, further measures can be considered.

Management of Acute Chromium Toxicity

- No proven antidote is available for chromium poisoning. Acute poisoning is often fatal regardless of therapy. Treatment in cases of acute high-level chromium exposure is usually supportive and symptomatic.
- Fluid and electrolyte balance is critical.
- Affected patients should be monitored carefully for evidence of gastrointestinal bleeding, hemolysis, coagulopathy, seizures and pulmonary dysfunction. Appropriate supportive measures may include ventilatory support, cardiovascular support, and renal and hepatic function monitoring.
- When renal function is compromised, maintenance of adequate urine flow is important. Progression to anuria is associated with poor prognosis.
- Induction of vomiting is contraindicated, owing to the potential corrosive effects of the chromium compounds and the potential for rapid deterioration of the patient.
- Gastric lavage with magnesium hydroxide or another antacid might be useful in cases of chromium ingestion.
- If renal failure ensues, hemodialysis may be necessary for management of the renal failure itself.
- If the eyes and skin are directly exposed, flush with copious amounts of water.
- The ulcers heal in several weeks without specific treatment.
- EDTA ointment 10% might facilitate removal of chromate scabs.
- Weeping dermatitis can be treated with 1% aluminum acetate wet dressings, and chrome ulcers can be treated with topical ascorbic acid.

Management of Chronic Chromium Toxicity

- In most patients with chronic low-dose exposure, no specific treatment is needed.
- The mainstay of management is removing the patient from further exposure and relying on the urinary and fecal clearance of the body burden.
- Dermatitis, liver and renal injury will not progress after removal from exposure, and, in most cases, the patient will recover.
- If the exposure has been to high levels or lengthy, the increased risk of lung cancer should be examined

- Although no reliable tests are currently available to screen patients for lung cancer, the physician can provide advice and patient education regarding smoking cessation, avoiding or minimizing exposure to other known pulmonary carcinogens, and general preventive health measures.
- Annual chest radiographs advisable in selected cases.

Prevention

Self-Care

- Wear proper personal protective equipment such as respiratory protection, protective clothing, eye protection, and gloves.
- Maintain work area free of dust.
- Shower and change clothes immediately on completion of work.
- Leave or dispose of contaminated clothing at the worksite.
- Do not track dust from the work area to the rest of the home.
- Do not smoke, eat, or drink in the work area.
- Wash hands well before eating, drinking, or smoking.
- Head employer should provide patient and worker education.
- If a smoker, stop smoking.

Avoid Tobacco Smoke

Chromium is a component of tobacco smoke. Avoid smoking in enclosed spaces like inside the home or car in order to limit exposure to children and other family members.

Launder Clothing from Worksites

Clothing or items removed from the workplace may contain chromium if employed in a setting where occupational exposure is significant. Therefore, common sense hygiene and laundry practices may help avoid unnecessary exposures.

Clinical Follow-Up

Patients should be advised to call if they develop any of the following symptoms:

- Symptoms of kidney dysfunction such as lower extremity swelling, reduced urine volume, changes in color of urine, etc.
- Dyspnea, cough, wheezing, nausea, vomiting, diarrhea
- Yellowing of the teeth
- Altered sense of smell

Remedial Method for Environmental Contamination

For *in situ* remediation of soil and groundwater, a variety of geochemical fixation or anaerobic biological treatment methods have been designed which make use of Cr's ability to change the valence state of the oxidized, toxic, and highly mobile nature of Cr (VI) and convert it into the more stable, nontoxic, and immobile Cr(III). Cr (III) ultimately precipitates out as Cr(III) hydroxide [$Cr(OH)_3$]. In these technologies, Cr is generally not removed from the environment, but becomes less toxic and immobile.

Many treatment methods use sulfur-based reductants; anaerobic biodegradation enhancements such as lactic acids, molasses, or other chemical reductants; or elemental iron [Fe(O)] technologies to create a reducing environment.

A variety of technologies are used in the extraction and treatment of surface water or groundwater (known as pump and treat).

Once on the surface, water containing Cr(VI) can be reduced by Fe(II) compounds followed by several procedures including alkaline precipitation, ion exchange with regenerant treatment, or disposal.

Electrochemical reduction is another method used where Cr (VI) reduction is followed by alkaline precipitation in which Fe(II) forms electrochemically, instead of being added as a purchased chemical, and acidic reduction of the Cr at $pH < 3.0$ with sulfur dioxide, sodium sulfite, sodium bisulfite, or sodium meta bisulfite completes the conversion to Cr(III).

If reduction with a sulfite compound is used, there is a greater potential for incomplete conversion of Cr(VI) to Cr(III). Consequently, these reactions must be monitored carefully to ensure complete conversion to Cr(III), which ultimately precipitates out of solution as $Cr(OH)_3$

Regulations and Guidelines

Workplace Guidelines

Occupational Safety and Health Administration (OSHA): OSHA has established an 8-hour time- weighted average (TWA) exposure limit of 5 μ/m^3 of air for Cr (VI) Workers exposed to Cr (VI) are at an increased risk of developing lung cancer occupational exposure to Cr(VI) may result in asthma and damage to the nasal epithelia and skin.

The final rule contains ancillary provisions for worker protection such as requirements for exposure determination, preferred exposure control methods, including a compliance alternative for a small sector for which the new PEL is infeasible, respiratory protection, protective clothing and equipment, hygiene areas and practices, medical surveillance, record keeping, and start-up dates that include four years for the implementation of engineering controls to meet the PEL

The National Institute for Occupational Safety and Health (NIOSH): NIOSH has recommended a 10-hour TWA exposure limit for all Cr (VI) compounds of 1 µg Cr (VI)/m^3 For chromium metal and Cr (II) and Cr (III) compounds, there commended exposure limit is 500 µg/m^3 as an10-hour TWA. On the basis of current evidence, NIOSH considers all Cr (VI) compounds potential occupational carcinogens.

Environmental Guidelines

Air: The **U.S. Environmental Protection Agency (EPA)** regulates chromium emissions uses technology-based standard for categories of industries, rather than numerical emissions standards, to reduce chromium levels in ambient air. These maximum achievable control technology standards are based on emissions levels already achieved by the best-performing similar facilities.

Drinking Water: EPA has an enforceable maximum contaminant level of total chromium in drinking water of 100µg/L (100 ppb) for public water systems.

Regulations and Guidelines for Chromium

Agency	Focus Area	Level	Comments Remarks
ACGIH	Air, workplace	10 μg/m³ as Cr	TWA to avoid carcinogenic risk from insoluble Cr(VI) compounds
		50μ g/m³ as Cr	TWA for water- soluble Cr(VI) compounds
		500 (1μg/m³ as Cr	TWA for Cr metal and Cr(III) compounds
NIOSH	Airworkplace	1 μg/m³as Cr	TWA (10-hour) for chromic acid and all Cr(VI) compounds
		500 μg/m³ as Cr	TWA (10-hour) for Cr metal and Cr(II) and Cr(III) compounds
OSHA	Air, workplace	5 μg/m³ as Cr_2O_3/m³	Regulation; PEL for chromic acid and chromates, (8-hour TWA)
		500μg/m³ as Cr	PEL for Cr(II) and Cr(III) compounds (8 hour TWA)
		1, 000 μg/m³ as Cr	PEL for Cr metal and insoluble compounds (8-hour TWA)
EPA	Air, environment	Not Available	Cr is listed as a hazardous pollutant
	Drinking water	100 μg/L	Regulation; current MCL for total Cr

TWA: Time weighted average
PEL: Permissible Exposure Level
MCL: Maximum Concentration Level

Appendix 1.130: Environmental and Occupational Hazards of 'SILICA'

The chemical compound silicon dioxide, also known as silica or silox is the oxide of silicon (chemical formula SiO_2) and has been known for its hardness. Silica is commonly found in nature as sand or quartz. It is a principal component of most types of glasses and some other substances such as concrete.

Free silica exists in two main forms: amorphous silica and crystalline silica. Crystalline silica is one of the most common minerals found on the surface of the earth. It may be found in a variety of forms the most common of which is quartz. Other types of crystalline silica include tridymite and cristobalite.

Amorphous silica is the term used to describe forms, which have no regular crystal structure. It includes a wide range of fossilized skeletal remains of marine organisms. While there is some scientific evidence that amorphous silica may occasionally cause adverse health effects, most silica-related disease has been a result of breathing of crystalline silica.

Chronic exposure to crystalline silica is associated with increased incidences of tuberculosis, bronchitis, emphysema, chronic obstructive pulmonary disease, renal diseases, silicosis and lung cancer. Of these potential health effects silicosis and lung cancer are the effects of most concern associated with occupational exposure to reparable quartz dust.

Crystalline silica in the form of quartz or cristobalite from occupational sources is (carcinogenic to humans. The severity and incidence of silicosis and silica induced cancer increases with intensity of dust exposure, with increased cumulative duration of exposure, with increasing peak concentrations of silica and with increased percentage of silica within the respirable dust.

Main industries/activities causing silica exposure to man

S.No.	Industry or activity	Operations and tasks	Source materials
1.	Agriculture	Ploughing, harvesting, using machinery, burning agricultural waste, processing agricultural products	Soil
2.	Mining and related milling operations	Most occupations (underground, surface, mill) and mines (metal and nonmetal, coal), rock drilling, dredging	Ores, associated rock
3.	Quarrying and related milling operations	Crushing stone, sand and gravel processing, stone monument cutting and abrasive blasting, slate work (e.g., pencil manufacturing), diatomite calcination	Sandstone, granite, flint, sand, gravel, slate, diatomaceous earth
4.	Construction activities	Abrasive blasting of structures and buildings, highway and tunnel construction, excavation and earth moving and digging, masonry, concrete work, demolition, dry sweeping and brushing, pressurized air blowing, jack hammering, laying railroad track, removing rust or paint, sanding and scaling, replacement of asphalt roofing, and hauling, pouring, mixing, or dumping silica containing materials	Sand, concrete, rock, soil, mortar, plaster, shingles
5.	Glass, including fiberglass	Raw material processing, refractory installation and repair	Sand, crushed quartz, refractory materials
6.	Cement	Raw material processing	Clay, sand, limestone, diatomaceous earth
7.	Abrasives	Silicon carbide production, abrasive products fabrication	Sand, tripoli, sandstone
8.	Ceramics, including bricks, tiles, sanitary ware, porcelain, pottery, refractories, vitreous enamels	Mixing, molding, glaze or enamel spraying, finishing, sculpting, firing	Clay, shale, flint, sand, quartzite, diatomaceous earth
9.	Iron and steel plants	Refractory preparation and furnace repair	Refractory material
10.	Silicon and ferro-silicon foundries (ferrous and nonferrous)	Raw materials handling, casting, molding and shaking out, abrasive blasting, fettling, furnace installation and repair	Sand, refractory material

S.No.	Industry or activity	Operations and tasks	Source materials
11.	Metal products, including structural metal, machinery, transportation equipment	Abrasive blasting	Sand
12.	Shipbuilding and repair	Abrasive blasting	Sand
13.	Rubber and plastics	Raw materials handling	Fillers (tripoli, diatomaceous earth)
14.	Paint	Raw materials handling, site preparation	Fillers (tripoli, diatomaceous earth, silica flour)
15.	Soaps and cosmetics	Manufacturing or occupational use of abrasive soaps and scouring powders	Silica flour
16.	Roofing asphalt felt	Riling and granule application	Sand and aggregate, diatomaceous earth
17.	Agricultural chemicals	Raw material crushing, handling, bagging; or dumping products or raw materials	Phosphate ores and rock
18.	Jewellery	Cutting, grinding, polishing, buffing, etching, engraving, casting, chipping, sharpening, sculpting	Semiprecious gems or stones, abrasives, glass
19.	Arts, crafts, sculpture	Pottery firing, ceramics, clay mixing, kiln repairs, abrasive blasting, sand blasting, engraving, cutting, grinding, polishing, buffing, etching, casting, chipping, sharpening, sculpting	Clays, glazes, bricks, stones, rocks, minerals, sand, silica flour
20.	Dental material	Sana blasting, polishing	Sand, abrasives
21.	Boiler scaling	Coal-fired boilers	Ash and concretions
22.	Automobile repair	Abrasive blasting, sanding, removing paint and rust	Sand, metals, priming putty

Silicosis

Silicosis is the chronic fibrosing disease of the lungs produced by the prolonged and extensive exposure to free crystalline silica dust. When workers inhale crystalline silica (dust), the lung tissue reacts by developing fibrotic nodules and scarring around the trapped silica particles. This fibrotic condition of the lung is called silicosis. If the nodules grow too large, breathing becomes difficult and death may result. Silicosis victims are also at high risk of developing active tuberculosis. Silicosis treatment is very limited as there is no cure for the disease, but it is 100 percent preventable if employers, workers, and health professionals work together to reduce exposure

Types of Silicosis

Chronic Silicosis: Results from long-term exposure (more than 20 years) to low amounts of silica dust. Nodules of chronic inflammation and scarring provoked by the silica dust form in the lungs and chest lymph nodes. This disease may feature breathlessness and may resemble chronic obstructive pulmonary disease (COPD) Chronic silicosis can be either simple silicosis or complicated silicosis (also called progressive massive fibrosis), a distinction based on the chest radiographic appearance.

Simple Silicosis may be asymptomatic or may present with exertional dyspnea and cough with sputum production.

Complicated Silicosis, dyspnea and productive cough often are accompanied by constitutional symptoms of malaise and weight loss.

Accelerated Silicosis: Occurs after exposure to larger amounts of silica over a shorter period of time (4-8 years). Inflammation, scarring, and symptoms progress faster in accelerated silicosis than in simple chronic silicosis.

Acute Silicosis: Results from short-term exposure to very large amounts of silica mainly in unregulated environments. The lungs become very inflamed and may fill with fluid, causing severe shortness of breath and low blood oxygen levels. Acute silicosis causes symptoms of severe dyspnea, cough, fever, and weight loss. On chest radiograph, ground-glass appearances, as well as coarse linear or rounded opacities are seen. A miliary picture with very small round opacities may also occur in the lower lung fields. The diagnosis of alveolar proteinosis may be readily made by the appearance of the bronchoalveolar fluid, which has a milky appearance.

Particles of silica dust get trapped in the tiny sacs (alveoli) in the lungs where air exchange takes place. White blood cells called macrophages in the alveoli ingest the silica and die. The resulting inflammation attracts other macrophages to the region. The nodule forms when the immune system forms fibrous tissue to seal off the reactive area. The disease process may stop at this point, or speed up and destroy large areas of the lung. The fibrosis may continue even after the worker is no longer exposed to silica.

Early symptoms of silicosis include shortness of breath after exercising and a harsh, dry cough. Patients may have more trouble in breathing and cough up blood as the disease progresses.

Congestive heart failure can give their nails a bluish tint. Patients with advanced silicosis may have trouble in sleeping and experience chest pain, hoarseness, and loss of appetite

Other Silica Related Diseases

Chronic Obstructive Pulmonary Disease (COPD)

Chronic obstructive Pulmonary Disease is a term used for a number of conditions, including chronic bronchitis and emphysema COPD. leads to damaged airways in the lungs, causing them to become narrower and making it harder for air to get in and out of the lungs.

Exposure to silica and coal mine dusts may result in chronic obstructive pulmonary disease, including emphysema and chronic bronchitis that appears indistinguishable from obstructive lung disease caused by exposure to tobacco smoke. This may cause significant respiratory dysfunction in sensitive individuals.

Pulmonary Tuberculosis

Exposure to crystalline silica dust causes multiple diseases, but silicosis and silica dust-associated tuberculosis (TB), in particular, are the two diseases that remain high on the list of occupational health priorities. The prevalence of silica-related TB is exacerbated by the human immunodeficiency virus (HIV) epidemic in low-income countries.

The relative risk for tuberculosis is more for men with silicosis compared with that in the men without silicosis.

Autoimmune Disorders

There have been reports linking silica exposure to a variety of autoimmune diseases (systemic sclerosis, rheumatoid arthritis, lupus, chronic renal disease). Evidence of this association in larger epidemiologic studies has been increasing in the last decade.

Renal Diseases

Silica exposure has been associated with kidney disease (chronic renal disease and glomerulonephritis). Excess risk of end-stage renal disease due to a lifetime of occupational exposure at currently recommended limits is above a background end-stage renal disease risk.

Caplan's Syndrome

Caplan's syndrome is characterized by the presence of seropositive rheumatoid arthritis associated with a specific form of pneumoconiosis, consisting of multiple, well-defined homogeneous rounded opacities on chest X-ray. It develops especially in miners working in anthracite coal-mines and in persons exposed to silica and asbestos. In the development of the disease, genetic factors are considered to play an important role by influencing immunological reactivity of the organism exposed to various heteroantigens.

Silica exposed men had increased risk of rheumatoid arthritis, with an odds ratio (OR), adjusted for age, residential area, and smoking.

Prevention and Control

Employers and workers should take appropriate actions to reduce silica flour exposure to below permissible limits. This can be accomplished by following measures:

1. Exposure Monitoring

Workers exposure surveys should be made by competent industrial hygienist and engineer. Surveys are necessary to determine the extent of worker exposure and the effectiveness of engineering controls. In the collected air samples, crystalline silica is measured by:

XRD (filter redeposition) (X Ray Diffraction) FTIR (KBr pellet) (Fourier Transform infrared spectro photometry

2. Engineering Controls

The most effective control of airborne concentrations of silica flour is at the source of contamination by enclosure of the operation and/or use of local exhaust ventilation.

3. Medical Surveillance

Pre-placement and annual medical examinations should be made available to all workers who manufacture, use or handle silica flour or materials containing silica flour.

Workers with radiographic evidence of silicosis should be given the opportunity to get transferred to other jobs/department having no silica exposure (defined as exposure at concentrations less than half of the recommended standard).

4. Work Practices

The following work practices are recommended:

a. Work procedures should be developed so they do not produce dust.

b. Work clothes should be vacuumed before removal.

c. General housekeeping duties should be intensified so that there is no dust accumulation on machinery, beams, corners, and other surfaces. Dustless methods of cleaning such as vacuuming or wetting down should be used. Dry sweeping or blowing with com pressed air should be avoided

d. Emphasis should also be given to cleanup of spills, preventive maintenance, and timely repair of equipment.

5. Personal Protective Equipment

Personal protective equipment is not recommended as a primary means of control. Exposure of workers to airborne silica flour should not be controlled with the use of respirators except:

a. During installation and implementation of engineering or work practice controls;

b. In work situations in which engineering and work practice controls are technically not feasible;

c. During major overhaul and repair of equipment, if exposure to silica flour is possible;

d. In operations that require entry into tanks or closed vessels; or

e. In emergencies.

6. Worker Education

Worker education is a vital aspect of a good control program. Workers should be informed of the hazardous nature of silica flour, the results of workplace monitoring and medical tests, and the correct usage and maintenance of respirators.

7. Labelling of Silica Flour Containers

Packaged silica flour should be labelled correctly, and health warnings should be placed on each container to alert users, handlers and producers to the hazards of silica flour.

Limit values					
Agency/ Country	**Exposure limit and interpretation**	**Crystalline silica form**	**Dust fraction**	**Limit value [mg/m³]**	**Value range in reference to 100% SiO_2**
ACGIH, USA (2006)	TLV - Threshold Limit Value (TWA-8h, 40-h week)	Crystalline silica: a - quartz cristobalite	Respirable	0.025	-
NIOSH, USA (1974)	REL - Recommended Exposure Limit (TWA-10h, 40-h week)	Quartz cristobalite tridymite	Respirable	0.05	-

Limit values					
Agency/ Country	**Exposure limit and interpretation**	**Crystalline silica form**	**Dust fraction**	**Limit value [mg/m³]**	**Value range in reference to 100% SiO_2**
OSHA, USA	PEL - Permissible Exposure Limit (TWA-8h, 40-h week)	Dust containing quartz (1971)			0.03-0.1
			Total dust		0.09-0.3
		Dust containing cristobalite, tridymite (1978)	Respirable	of the value for quartz	0.015-0.05
			Total dust	of the value for quartz	0.045-0.15
India (1987)	PLE -Permissible Limit of Exposure	Dust containing free silica	Respirable		0.03-0.1
			Total dust		0.075-0.3

ACGIH has mentioned that free silica is a suspected human carcinogen while NIOSH has identified it as a potential occupational carcinogen.

Indegemously Developed Control Devices

Agate Grinding Units

Local exhaust system containing air-cleaning device was developed and fitted over grinding machine along with bag fitter especially designed for agate grinding. The exhaust system did not interfere with the work and the workers found it very useful as it facilitated the work by removing the dust, which hampered their sight. To reduce electricity consumption, the emery wheel and blower were rotated with the same motor by attaching it to a pulley. Dust reduced by about 93% in the work environment.

Silica Flourmills (Hammermill)

The crushing process involves jaw crusher, disintegration, screening and bagging. Dust control device was developed with bag filter and installed. Since jaw crusher is outside the shed, separate dust control device was developed and installed with motors. Dust control measures like, enclosure of dust source, use of powerful exhaust and humidification of work environment, reduced dust by about 75-85%.

Stone Quarry

Developed a protective shield of dust hood made of an iron frame covered by an acrylic shield at its food and cloth on its sidewalls. The dust hood could be fitted to a chisel. The height of the hood could be adjusted. Workers satisfied with the device as it reduced the dust exposure.

Silica Flourmill (Ballmill)

NIOH designed Dust control system using ACGIH ventilation guidelines which involves hood, ducting, centrifugal fan and bag filters with reverse pulse jet system. Owners were encouraged to install dust control system for feeding and bagging operations. The work is in progress for testing the efficacy of dust control system.

Appendix 1.131: Environmental and Occupational Hazards of 'ABSESTOS'

The name asbestos comes from the word asbeston, which in Greek meant unquenchable, inextinguishable, or indestructible. The Romans, called asbestos by the name of amiantus, used it in their building materials, cremation cloths, lamp wicks, and even made napkins.

Asbestos clothing was also used by the Egyptians to dress their pharaohs at the time of burial, by the ancient Finnish to strengthen pottery, and in the middle ages as insulation for armor.

New applications began to expand the usage of asbestos. It was used in steam locomotives to insulate their boilers, pipes, and fireboxes, as well as line their brakes and clutches, it was made into packings, gaskets and pipe lagging materials and was even used for protective suits by firemen.

In the 1900s the use of asbestos began to increase dramatically. Military began to make use of it in ships, flares, and fireproof suits among other things. Since reaching its peak in 1973 asbestos production and consumption has been on a constant decrease and many countries including Germany, Italy, Netherlands, Sweden, Norway, Denmark, and Switzerland have passed comprehensive bans on asbestos mining, production, sale, use, import and export.

Asbestos is not a single substance, but is the generic name for a family of six related polysilicate fibrous minerals of which one (chrysotile) belongs to the serpentine family and five (actinolite, amosite, anthophyllite, crocidolite, and tremolite) belong to the amphibole family. These minerals differ from each other in physical and chemical properties, and each mineral can exist in a wide range of fiber sizes. These differences between fiber type and, more importantly, fiber size (length and diameter) are believed to be important determinants of the health risks posed by asbestos.

Sources of Exposure

Mining Activities

Exposure to asbestos can occur in workers involved in mining, milling and handing of other ores and rocks containing asbestos. Residents who live near mining, milling, or manufacturing sites that involve asbestos-containing material may be exposed to higher levels of airborne asbestos.

Demolition and Maintenance Work

Amphibole asbestos can be found in a variety of building materials, such as insulation, ceiling or floor tiles, and cement pipes. Workers or homeowners involved in demolition work, maintenance, repair, or remodeling of buildings containing these products can be exposed to higher airborne levels than in ambient air.

Some of the work environments or occupations in which workers are exposed include:

Work Environments

Asbestos product manufacturing (insulation, roofing, building, materials)	**Automotive repair (brakes and clutches)**
Construction Sites	Maritime operations
Mining Operations	Offshore rust removals
Oil refineries	Power plants
Shipyards/ships/ shipbuilders	Sand or abrasive manufacturers
Railroads	Steel mills

Occupations

Plumbers	Plasterers
Workers at exfoliation facilities where vermiculite ore are processed	Asbestos removal workers
Demolition workers	Roofers
Workers at asbestos products manufacturing plants	Floor covering manufacturers or installers
Auto mechanics	Grindrs
Boilermakers	Hod carriers
Bricklayers	Insulators
Building inspectors	iron workers
Carpenters	Laborers
Drywallers	Maintenance workers
Electricians	Merchant marines
Furnace workers	
Glazers	Operating engineers
	Painters

Uses of Asbestos

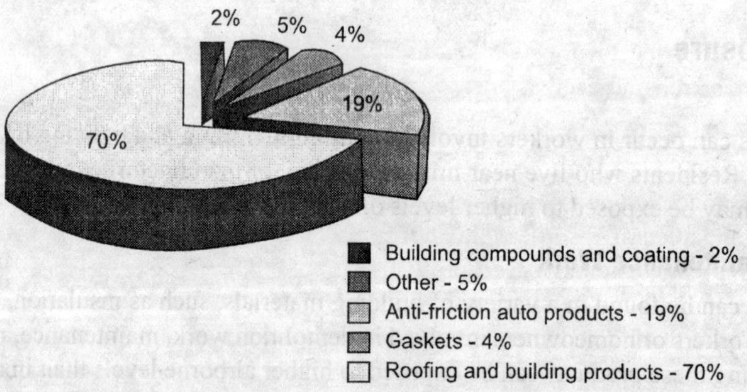

Asbestos Facts

- When asbestos fibers are inhaled, most fibers are expelled, but some can become lodged in the lungs and remain there throughout life. Fibers can accumulate and cause scarring and inflammation. Enough scarring and inflammation can affect breathing, leading to disease.
- The term "naturally occurring asbestos" refers to the mineral as a natural component of soils or rocks as opposed to asbestos in commercial products, mining or processing operations. Naturally occurring asbestos can be released from rocks or soils by routine human activities, such as construction, or natural weathering processes.
- People are more likely to experience asbestos-related disorders when they are exposed to high concentrations of asbestos, are exposed for longer periods of time, and/or are exposed more often.

- Exposure to asbestos can increase the likelihood of lung cancer, mesothelioma, and non-malignant lung conditions such as asbestosis (restricted use of the lungs due to retained asbestos fibers) and changes in the pleura (lining of the chest cavity, outside the lung).
- Changes in pleura such as thickening, plaques, calcification, and fluid around the lungs (pleural effusion) may be early signs of asbestos exposure. These changes can affect breathing more than previously thought. Pleural effusion can be an early warning sign for mesothelioma (cancer of the lining of the lungs).
- Most cases of asbestosis or lung cancer in workers occurred 15 years or more after the person was first exposed to asbestos.
- Most cases of mesothelioma are diagnosed 30 years or more after the first exposure to asbestos.
- Asbestos-related disease has been diagnosed in asbestos workers, family members, and residents who live close to asbestos mines or processing plants.
- Health effects from asbestos exposure may continue to progress even after exposure is stopped.

Risk Factors

- Exposure concentration - what was the concentration of asbestos fibers?
- Exposure duration - how long did the exposure time period last?
- Exposure frequency - how often during that time period was the person exposed?
- Size, shape and chemical makeup of asbestos fibers:
- Individual risk factors, such as a person's history of tobacco use (smoking) and other pre-existing lung disease, etc.

Cigarette smoke and asbestos together significantly increase chances of getting lung cancer. Stopping smoking can improve health and decrease risk of cancer.

Health Effects

Significant exposure to any type of asbestos will increase the risk of lung cancer, mesothelioma and nonmalignant lung and pleural disorders, including asbestosis, pleural plaques, pleural thickening, and pleural effusions. This conclusion is based on observations of these diseases in groups of workers with cumulative exposures ranging from about 5-1, 200 fiber-year/mL. Such exposures would result from 40 years of occupational exposure to air concentrations of 0.125-30 fiber/mL.

Diseases from asbestos exposure take a long time to develop. Most cases of lung cancer or asbestosis in asbestos workers occur 15 or more years after initial exposure to asbestos. Tobacco smokers who have been exposes to aspects have a tar greater than additive risk for lung cancer than do nonsmokers who have been exposed meaning the risk is greater than the individual risks from asbestos and smoking added together. The time between diagnosis of mesothelioma and the time of initial occupational exposure to asbestos commonly has been 30 years or more. Cases of mesotheliomas have been reported after household exposure of family members of asbestos workers and in individuals without occupational exposure who live close to asbestos mines.

Chronic exposure to asbestos may increase the risk of lung cancer, mesothelioma, and nonmalignant lung and pleural disorders.

Asbestosis is a serious progressive, long-term disease of the lungs. Asbestosis is not a cancer. Inhaling asbestos fibers that irritate and inflame lung tissues, causing the lung tissues to scar, causes asbestosis.

The scarring makes it hard to breathe and difficult for oxygen and carbon dioxide pass through the lungs. Asbestosis generally progresses slowly. The latency period for the onset of asbestosis is typically 10-20 years after the initial exposure. The disease can vary from asymptomatic (no symptoms) to disabling and potentially fatal.

Signs and Symptoms of asbestosis include:

- Shortness of breath (primary symptom)
- A persistent and productive cough (A cough that experts mucous)
- Chest tightness
- Chest pain
- Loss of appetite
- A dry, crackling sound in the lungs while inhaling

Pleural Abnormalities: Persons with significant exposure to asbestos are at risk for developing various types of pleural (lining of the chest cavity, outside the lungs) abnormalities. These abnormalities include pleural plaques, pleural thickening, pleural calcification, and pleural mesothelioma.

Mesothelioma is a rare cancer, which may affect the lining of the chest cavity, outside the lung (pleura) or the abdominal contents (peritoneum). Most mesotheliomas are caused by exposure to asbestos.

Lung cancer is a malignant tumor that invades and obstructs the lung's air passages. Cigarette smoking greatly increases the likelihood of a person developing lung cancer as the result of asbestos exposure. The most common symptoms of lung cancer are cough, wheezing, unexplained weight loss, coughing up blood, and labored breathing. Other symptoms of lung cancer include shortness of breath, persistent chest pain, hoarseness, and anemia. People who develop these symptoms do not necessarily have lung cancer, but they should consult a physician for advice.

Gastrointestinal and Colorectal Cancers

Studies of asbestos workers suggest that asbestos exposure might be associated with gastrointestinal (esophagus and stomach) and colorectal (colon and rectum) cancers. However, the evidence is unclear.

Some evidence shows that short-term (acute) oral exposure to asbestos might bring on precursor lesions of colon cancer, and that long-term (chronic) oral exposure might increase the incidence of gastrointestinal tumors.

Kidney, Brain, Bladder, Laryngeal, and Other Cancers

Results of studies of cancers at other sites are also inconclusive. One reason is that relatively few studies have tried to evaluate the relation between asbestos exposure and non-respiratory cancers.

Some studies have noted excess deaths from, or reported cases, of certain cancers such as the kidneys brain, and bladder. Several epidemiologic studies have reported an increased risk of laryngeal cancer in workers exposed to asbestos.

Mechanisms of Various Health Effects of Asbestos

When asbestos mineral fibres are inhaled, many of them are deposited on the epithelial surface of the respiratory tree. Entry and deposition of the fibres depend upon the type of fibres and more importantly fibre size (length and diameter) which are important determinants of the health risk posed by asbestos. The number of fibres that are deposited and the location within the airway where deposition occurs is a function of the aerodynamic properties of the fibres. For typical fibres of chrysotile, amosite and crocidolite, about 30-40% of all the fibres in inhaled air are retained with most of these (about 60%) being deposited in the upper air ways (nose, throat, trachea). The fibres in the upper airway consist mainly of relatively thick fibres (greater than 3μm) with thinner fibres being carried deeper into the airways. Most fibres deposited into the airways are removed from the lung by mucociliary transport or by alveolar macrophages (AM), but a small fraction remains in the lung for long periods. In addition, some fibres pass from the lung to the pleura, although the precise mechanism of transport is not known. Those fibres that enter the lymph are presumably able to reach other organs of the body.

Epidemiological surveys and experimental studies established that asbestos is a carcinogen as well as co-carcinogen. Chronic inhalation of airborne asbestos results in the fibrotic lung disease. Deposition of these fibres/particles in the lung is followed by a sequence of events, which starts with change in the free cell population, which includes AM and polynuclear inflammatory cells via their influx.

There have always been debates about the nature of interaction between multiple environmental pollutants in causing diseases to human. Predisposing factors like food preservatives, exposure to cigarette smoke, kerosene soot and bio-fuels at indoor levels would accelerate the disease processes. Cigarette smoke alone has been shown to cause lung cancer but the risk of lung cancer increases substantially due to cigarette smoke in conjugation with exposure to asbestos.

The development of the asbestos industries has always been linked with the recognition of health risk involved. Owing to the growing activities in mining, grinding and manufacturing of asbestos-products, the risk of health hazards has also received wide attention world over.

Asbestos - Mediated Toxicity and Diseases

Inflammation

The sequence of events in the lung, following deposition of fibres includes modulation in the free cell population, primarily characterized by an increase in AM and polymorpho nuclear inflammatory cells. Further, a change in the composition of the lung lining fluids has also been reported. The inflammatory response to these fibres stimulates the release of a variety of inflammatory cell mediators and growth factors which play an important role in the fibrogenesis of the lung.

Mesothelioma

Approximately 80% of mesotheliomas occur in men exposed to mineral fibres at workplaces and sometimes in their family members or in persons who lives near asbestos sources. Mesothelioma may develop in pleural and peritoneal cavity of the lung.

Peritoneal Mesothelioma

Peritoneal mesothelioma involves the abdominal cavity, infiltrating the liver and spleen and the bowels pain is the most common presenting complaints. In addition, fluid accumulation in the abdominal cavities, the abdomen appears enlarged, the patient experience nausea, vomiting, swelling of their feet, fever and difficulty in moving their bowels.

Pleural Plaques

The pleuron is a set of thin membrane that lines the chest cavity. Pleura are co-important as they provide lubrication, friction free surface to lung for easily expand and contract against. After an exposure of asbestos for minimum 10 years, pleural changes may begin to appear and these changes may be some times called as pleural thickening, pleural calcification and more commonly pleural plaques. Hyaline plaques of the parietal pleura occur in association with exposure to all types of asbestos. The majority occurs after 20 years or more of the exposure.

Asbestosis

This is a typical asbestos-related disease. Asbestos fibres when inhaled and reach in the lung start to damage the lung cells and result is asbestosis (formation of scar tissue in the lung), and /or lung cancer. The risk of lung cancer among people exposed to asbestos is increased by 7 times compared with the general population. Asbestosis is an interstitial pulmonary fibrosis, which reduces the lung capacity to deliver the oxygen in proper way to the whole body because the lung tissue loses its ability to function. It is characterized by the airway obstruction and air trapping, reducing vital capacity. This disease has relatively long latency period of about 40 years.

Clinically, asbestosis is very similar to interstitial pulmonary fibrosis (IPF). Most patients with well-established asbestosis characterized with shortness of breath, dry cough, and physical examination typically reveals dry rales at the base on inspiration. The usual function changes in the fully developed case are a restrictive defect and decreased diffusing capacity.

Bronchogenic Carcinoma

Bronchogenic carcinoma is a tumor, arising in tracheobronchial epithelial or alveolar epithelial cells. The average latency period of the disease i.e. the diagnosis of the disease from the time of first exposure to asbestos ranges from 20 to 30 years. The degree of association varies with the type of mineral fibre, morphology, concentration, exposure regimen, and other predisposing factors like smoking habits or the presence of certain other chemicals, but there is usually a dose- response relation (fiber per cubic centimeter of air times the number of years of exposure). Lung tumor is rare among the mineral fibre workers who do not smoke; although effect of mineral fibres and smoking combines in a multiple fashion to produce lung cancers.

Prevention and Control

Employers and workers should take appropriate actions to reduce asbestos exposure to below permissible limits. This can be accomplished by following measures:

1. Exposure Monitoring

Worker exposure surveys should be made to determine the extent of worker exposure and the effectiveness of engineering controls.

Detecting Asbestos

The combined use of detection methods called light microscopy, electron microscopy, and energy dispersive X-ray analysis offer the most accurate approach to identify asbestos and to estimate concentrations that may become airborne upon disturbance. For the purposes of counting asbestos fibers in these samples, regulatory agencies commonly count as "fibers" those particles of asbestos minerals at least 5 micrometers in length and/with length width ratios of 3:1. For other purposes, such as detecting fibers in bulk building materials, asbestos particles with length: width ratios of 5:1 are counted.

2. Engineering Controls

The most effective control of airborne concentrations of asbestos fibres is at the source of contamination by enclosure of the operation and/or use of local exhaust ventilation.

Ventilation equipment should be checked to ensure adequate performance. System effectiveness should be checked soon after any change in production, process, or control, which might result in significant, increases in airborne exposure to asbestos fibres.

3. Medical Surveillance

Pre-placement and annual medical examinations should be made available to all workers who manufacture, use, or handle asbestos. These examinations should include at least:

a. Comprehensive work and medical histories to evaluate exposure and signs and symptoms of respiratory disease:

b. Pulmonary function tests (PFT) including forced vital capacity (FVC) and forced expiratory volume in 1 second (FEV_1), with calculation of the FEV_1/FVC ratio.

Workers with radiographic evidence of asbestosis should be given the opportinity to transfer to jobs without exposure.

4. Work Practices

Work practices involve both the design of work procedures and the actions of workers. The following work practices are recommended:

a. Work procedures should be developed so they do not produce dust.

b. Work clothes should be vacuumed before removal.

c. General housekeeping duties should be intensified so that there is no dust accumulation on machinery, beams, comers, and other surfaces. Dustless methods of cleaning such as vacuuming or wetting down should be used. Dry sweeping or blowing with compressed air should be avoided.

d. Emphasis should also be given to cleanup of spills preventive maintenance, and timely repair of equipment.

5. Personal Protective Equipment (PPE)

Personal protective equipment is not recommended as a primary means of control. Exposure of workers to airborne asbestos should not be controlled with the use of respirators except:

a. During installation and implementation of engineering or work practice controls.

b. In work situations in which engineering and work practice controls are technically not feasible:

c. During major overhaul and repair of equipment, if exposure to asbestos is possible;

d. In emergencies.

6. Record keeping

The employer must keep an accurate record of all measurements taken to monitor employee exposure to asbestos. This record is to include: the date of measurement, operation involving exposure, sampling and analytical methods used, and evidence of their accuracy; number, duration, and results of samples taken; type of respiratory protective devices worn; name, and the results of all employee exposure measurements. This record must be kept for 30 years.

7. Worker Education

Workers should be informed of the hazardous nature of asbestos, the results of workplace monitoring and medical tests, and the correct usage and maintenance of respirators.

8. Labelling of Asbestos Containers

Packaged asbestos should be labelled correctly, and health warnings should be placed on each container to alert users and handlers as well as producers to the hazards of asbestos.

Regulations and Guidelines Applicable to Asbestos

Agency	Description	Information
Guidelines:		
IARC	Carcinogenic classification	Group 1[b]
Regulations and guldeines:		
a. Air:		
ACGIH	TLV-TWA	0.1 f/mL
EPA	Carcinogenic inhalation unit risk Carcinogenic classification	23×10-1 (f/mL)-1 Group A[c]
NIOSH	REL(100-minuite TWA in a 400 L	0.1 f/mL
OSHA	PEL (8-hour TWA)	0.1 f/ml
	PEL (8-hour TWA) for construction	0.1 f/mL
	PEL (excursion limit) averaged over a 30-minute sampling period for construction	1.0 f/mL
	PEL (8-hour TWA) for shipyard	1.0 f/mL
	PEL (excursion limit) averaged over a 30-minute sampling period for shipyard	1.0 f/mL
IFA	PEL (and -hour TWA)	1.0 f/ml
b. Water:		
EPA	Concentration at cancer risk of 10^{-4} for drinking water	700 MFL
	Human heath for consumption of: Waters organism	7 MFL
	MCL (f>10 μm In length)	7 MFL
	MCLG (10 μm in length)	7 MFL

[*] Includes: Actinolite, amosite, anthophyllite, chrysotile, crocidolite, and tremolite

[b] Group 1: Carcinogenic to humans

[c] Group A: Human carcinogen

[d] A 1: Confirmed human carcinogen ACGIH=American Conference of Governmental Industrial Hygienists;

EPA=Environmental Protection Agency; IARC = International Agency for Research on Cancer; NIOSH = National Institute of Occupational Safety and Health; OSHA= Occupational Safety and Health Administration; PEL= permissible exposure limit; REL=recommended exposure limit; TLV = threshold limit value; TWA = time-weighted average; IFA=Indian Factories Act, 1948
f = fibres

9. Hygiene Facilities and Practices

Clean change rooms must be furnished by employers for employees who work in areas where exposure is above the TWA and/or excursion limit. Showers must be furnished so that employees may shower at the end of the work shift. Lunchroom facilities for those employees must have a positive pressure, filtered air supply and should be readily accessible to employees. Employees must wash their hands and face prior to eating, drinking or smoking.

Asbestos Vaccine

Researchers from Netherlands have demonstrated the safety of a potential vaccine against mesothelioma, a rare cancer associated primarily with asbestos exposure. The vaccine, which infuses uses a patient's own dendritic cells (DC) with antigen from the patient's tumor, was able to induce a T-cell response against mesothelioma tumors.

Appendix 1.132: Environmental and Occupational Hazards of 'PESTICIDES'

Besides being beneficial for increased crop yield as well as in vector control programme, a pesticide has resulted in several health-related problems. Unintended exposure to pesticides can occur during their manufacturing, formulation and application or from environmental residues after application. Pesticides have been an integral part of modern agriculture and until very recently, not much attention was given to the potential problem of environmental and groundwater contamination. Indiscriminate and injudicious use of chemical pesticides, in agriculture has resulted in several associated adverse effects such as environmental pollution, ecological imbalances, pesticides residues in food, fruits and vegetables, fodder, soil and water, pest resurgence, human and animal health hazards, destruction of biocontrol agents, development of resistance in pests etc. Pesticides, which are toxic to pests, might produce some adverse health effect. Nine of the twelve most unwanted persistent organic pollutants (POP's) are pesticides used in agriculture crops and for public health vector control programme. These twelve POP's have been identified as a powerful threat to the human and wildlife health on a global basis.

Definition and Classifiction

A pest has characteristics that are regarded by human as injurious or unwanted. The term pesticide covers a wide range of compounds including insecticides, fungicides, herbicides, rodenticides, molluscicides, nematocides, plant, growth regulators and others, Ideally a pesticide must be lethal to the targeted pests, but not to non target species, including man.

Pesticides can be classified on the basis of type of pest control toxic property and chemical composition.

On the basis of pest control:

- Insecticides (against insect pest)

- herbicides (for killing and controlling weeds)
- Fungicides (against fungal diseases)
- Others (e.g. molluscicides, nematocides, etc.)

On the basis of their acute toxicity

- Class la: Extremely hazardous, demarcated in red.
- Class lb: Highly hazardous, symbolized by an yellow triangle.
- Class II: Moderately hazardous, marked by a blue triangle
- Class III: Slightly hazardous

On the basis of chemical composition:

- Organophosphate compounds
- Organochlorines
- Synthetic pyrethroids
- Carbamates
- Biopesticides

Production and Consumption

Use of pesticides in India began in 1948 when DDT was imported for malaria control and BHC for locust control. India started pesticide production with manufacturing plant for DDT and benzene hexachloride (BHC) in the year 1952.

The introduction of other synthetic insecticides organophosphate (OP) insecticides in the 1960s, carbonates in 1970 and pyrethroids in 1980s and the introduction of herbicides and fungicides in 1970s-1980s contributed greatly in pest control and agricultural output.

The following working condition of farmers makes them susceptible to high exposure to pesticides.

- Safety gears costly and also not suitable for tropical climate.
- Practice Poor to dangerous.
- Barefoot and barehanded, wearing minimum cloth, leaking spray tank etc.
- Mixing of concentrated chemicals and refilling spraying tanks and spraying even tasting the mixture.
- Multiple pesticides used mix and match subsequent medical management difficult.
- Unintended exposure of women and children to pesticides while mixing and spraying.

List of Pesticides / Pesticides formulation banned in India

I. Pesticides banned for manufacture, import and use:	
1. Aldrin	2. Pentachloro Nitrobenzene
3. Benzene Hexachloride	4. Pentachlorophenol
5. Calclum Cyanide	6. Phenyl Mercury Acetate
7. Chlordane	8. Sodium Methane Arsonate
9. Copper Acetoarsenite	10. Tetradifon
11. Cibromochloropropane	12. Toxafen
13. Endrin	14. Aldicarb

15. 15. Ethyl Mecury Chloide	16. Chlorobenzilate
17. Ethyl Parathion	18. Dieldrin
19. Heptachlor	20. Maleic Hydrazide
21. Menazone	22. Ethylene Dibromide
23. Nitrofen	24. TCA (Trichloro acetic acid)
25. Paraquat Dimethyl Suiphate	
II. Pesticide / Pesticide formulation banned for use but their manufacture is allowed for export:	
1. Nicotin Sulfate	2. Captafol 80% Powder
III. Pesticide formulations banned for import, manufacture and use:	
1. Methomyl 24% L	2. Phosphamidon 85% SL
3. Methomyl 12.5% L	4. Carbofuron 50% SP
IV. Pesticide Withdrawn	
1. Dalapon	2. Paradichlorobenzene (PDCB)
3. Ferbam	4. Simazine
5. Formotin	6. Warfarm
7. Nickel Chloride	

Pesticides restricted for use in India

S.No.	Name of Pesticides
1.	Aluminium Phosphide
2.	DDT
3.	Lindane
4.	Methlyl Bromide
5.	Methyl Parathion
6.	Sodium Cyanide
7.	Methoxy Ethyl Mercuric Chloride (MEMC)
8.	Monocrotophos (ban for use on vegetables)
9.	Endosulfan (Banned in the state of Kerala)
10.	Fenitrothion

Health Effects

Organophosphorus Insecticides

Orgnophosphorus insecticides are normally esters, amides or thiol derivatives of phosphoric, phosphonic, phospnorothioic, or phosphonothioic acids. Most organophosphorus insecticides are more stable in the pH range that may be encountered in the environment (pH 3.6) than at neutral pH. Different climatic conditions, especially temperature and humidity, before, during, and after spraying may influence the survival time of insects.

The metabolic fate of organophosphorus insecticides is basically the same in insects, animals, and plants. Uptake in animals and insects may occur through the skin, respiratory system, or gastrointestinal tract. Metabolism occurs principally by oxidation, hydrolysis or by conformational changes by transfer of a group of the molecule.

The signs and symptoms of acute organophosphate poisoning are an expression of the effects caused by excess acetylcholine (cholinergic syndrome). According to the degree of the severity of poisoning, the following signs and symptoms can occur:-

Mild anorexia, headache, dizziness, weakness, anxiety, substernal discomfort, fasciculation of the tongue and eyelids. miosis and impairment of visual acuity.

Moderate nausea, salivation, bronchorrhoea, lacrimation, abdominal cramps, diorrhoea, vomiting.

Sweating, hypertension or hypotension, and muscular fasciculation.

Severe miosis or mydriasis, non - reactive pupils, dyspnoea, respiratory depression, pulmonary oedema, cyanosis, loss of sphincter control, convulsions coma, bradycardia or tachycardia, cardiac ischaemia, cardiac dysrhythmias, hypokalaemia, and hyperglycaemia. Acute pancreatitis has also occurred. Muscular paralysis may involve the respiratory muscles. Some organophosphorus pesticides have causes delayed peripheral neuropathy.

Organochlorines (OC)

Organochlorines are compounds that contain carbon, chlorine and hydrogen. Their chlorine - carbon bonds are very strong which means that they do not break down easily. They are highly insoluble in water, but are lipophlic.

The organochlorines compound exhibit an ability to resist degradation, associate with sediments and to accumulate in the tissue of invertebrates, fish and mammals. Most of the organochorine compounds have properties that contribute to their ability to concentrate in biota and magnify in the foodchain.

Organochlorines contribute to many acute and chronic illnesses. Symptoms of acute poisoning can include tremors, headache, dermal irritation, respiratory problems, dizziness, nausea, and seizures. Organochlorines are also associated with many chronic diseases. Studies have found a correlation between organochlorine exposure and various types of cancer, neurological damage (several organoclorines are known neurotoxins). Parkinson's disease, birth defects, respiratory illness, and abnormal immune system function.

Many organochlorines are known or suspected hormone disruptors, and recent studies shows that extremely low levels of exposure in the womb can cause irreversible damage to the reproductive and immune systems of the developing fetus.

Carbamate

Carbamates, or urethanes, are a group of organic compounds sharing a common functional group with the general structure - NH(CO)O-. Carbamates are esters of carbamic acid, NH_2COOH, an unstable compound. These insecticides can cause cholinesterase inhibition poisoning by reversibly inactivating the enzyme acetylcholinesterase.

Health hazards for man occur mainly from occupational over-exposure to carbamate insecticides resulting in poisoning characterized by cholinergic symptoms caused by inhibition of the enzyme ACHe. The main routes of exposure are inhalation and skin. Apart from the symptoms indicative of ChE poisoning, other signs and symptoms induced by certain carbamates are skin and eye irritation, hyperpigmentation, and influence on the function of testes (sight increase of sperm abnormalities).

Pesticides as Endocrine disruptors

An endocrine disruptor is an exogenous substance or mixture that after functions(s) of the endocrine system and consequently causes adverse health effects in an intact organism or its progeny or sub populations. Exposure to low doses of endocrine-disrupting chemicals may have little effects on the exposed adult organism, but the offspring of that organism may suffer drastic repercussions. Another group of pesticide i.e. pyrethroid compounds, which are considered to be safe till recently as compared to organochlorine and organochlorine compounds have also been reported to be able to disrupt estrogen function in vitro.

* Exposure to some of the pesticides during infancy, even at very low levels, can lead to serious life-long consequences if the pesticides disrupt hormone - driven development processes.

Preventive Measures

Integrated Pest Management (IPM)

IPM is an eco-friendly approach, which encompasses cultural, mechanical, biological and need based chemical control measures. The IPM approach is being disseminated through various schemes/projects at national and state level with the following objectives:

- To maximize crop production with minimum input costs.
- To minimize environmental pollution soil, water and air due to pesticides.
- To minimize occupational health hazards due to chemical pesticides.
- To preserve ecosystem and maintain ecological equilibrium.
- To no or less use of chemical pesticides for minimum pesticide residues.

Curtailing Pesticide Misuse

The pesticide misuse should be minimized to reduce the exposure. The pesticide misuse includes:

- Failing to follow label or permit instructions
- Using pesticides in a way that is likely to injure people or damage property
- Harming a non-target plant or animal.
- Storing pesticides in container that do not have that approved label attached
- Disposing of a pesticide or its container illegally e.g. pouring pesticide waste down a drain.
- Spraying pesticides from an aircraft without a relevant license.
- Spraying pesticide against the wind direction.

Safety Precautions

- When using pesticides always read the pesticide container label before use and strictly follow the directions for use provided by the manufacturer or supplier.
- Prepare pesticides in a well - ventilated area, wear appropriate protective clothing (such as gloves, facemask and goggles) and keep children and animals away from preparation and application area.
- Prepare only enough pesticide for the particular application, use all the prepared pesticide on the application site (but avoid excess application), and avoid using pesticides on windy days.
- Persistent organochlorine pesticides should be not be as these chemicals are not readily broken down in the environment and accumulate in humans and animals and therefore have the potential to harm human health and the environment many years after application.

Follow Pesticide 'NEVERS'

- NEVER store pesticides in non-pesticide containers, e.g. containers used or labeled for food and drink.

- NEVER re-use pesticide containers for any purpose other than to hold the original pesticide.
- NEVER dispose of pesticide wastes down drains, toilets, sinks, gully traps or into bodies of water (such as rivers, lakes, streams or dams).
- NEVER dispose of pesticide wastes or contaminated or uncleaned pesticide containers in public jitter bins or private garbage bins, or leave them out for curbside collection.
- NEVER dispose of surplus pesticides in the garden.
- NEVER deliver used glass or plastic pesticide containers to bottle collection depots.
- NEVER burn pesticide wastes, including pesticide containers.

First Aid Measures

If pesticides are spilled on the skin:

- Follow the first-aid directions on the label - or immediately wash with soap and water.
- Remove pesticide contaminated clothes immediately wash them separately from other clothes.
- Monitor for symptoms of poisoning if in any doubt, seek immediate medical attention.
- If pesticides come in contact with the eyes, wash with copious amounts of water.

Regulation of Pesticide Use

Central Insecticides Act, 1968

The Central insecticides Act 1968 [Act 46 of 1968 As amended by Acts 46 of 1972, 24 of 1977 and 23 of 2000]. This Act is made to regulate the import, manufacture, storage, transport, distribution and use of pesticides with a view to prevent risk to human beings, animals and the environment. It came into force with effect: from 1st August, 1971. Through this Act, a Central Insecticides Board has been set up to advise the state and central governments on technical matters and for inducing insecticides into the Schedule of the Act.

Prevention of Food Adulteration [PFA] Act, 1954

This act governs the production and use of insecticides in India This Act and its Rules lay down standards for different food articles as well as provisions for their storage, distribution and sale. The Maximum Residue Limits [MRLs] for different pesticides are regulated through this PFA Act.

Use of biopesticides

Biopesticides are a group of pesticides derived from natural materials like animals, plants, bacteria, and certain minerals. For example, canola oil and baking soda have pesticidal applications and are considered biopesticides. Biopesticides fall into three major classes

1. **Microbial pesticides** consist of a microorganism (e.g., a bacterium, fungus, virus or protozoon) as the active ingredient.
2. **Plant-incorporated-Protectants (PIPs)** are pesticidal substances that plants produce from genetic material that has been added to the plant.
3. **Biochemical pesticides** are naturally occurring substances that control pests by non-toxic mechanisms.

Advantages Of Biopesticides

Biopesticides are usually less toxic than conventional pesticides.

- Biopesticides generally affect only the target pest and closely related organisms, in contrast to broad spectrum, conventional pesticides that may affect a group of organisms such as birds, insects, and mammals.
- Biopesticides often are effective in very small quantities and often decompose quickly, thereby resulting in lower exposures and largely avoiding the pollution problems caused by conventional pesticides

Appendix 1.133: Sources of Environmental Lead Exposure

Home

Cosmetics
Burning of contaminated firewood or kerosene
Indigenous remedies
Tobacco and Tobacco products
Drinking water (contaminated)
Toys

Environment

Automobile exhaust
Industrial effluents
Lead acid batteries
Ammunition (acid rain leachates)
Paints and varnishes
Water pipes

Appendix 1.134: Biological Effects of Hydrogen Sulfide

S.No.	H_2S concentration, ppm	Biological effects on human beings
1.	1.3×10^{-3}	Olfactory threshold
2.	10	Maximum acceptable concentration; no danger after prolonged inhalation
3.	50-150	Slight damage after a few hours of inhalation
4.	170-250	Maximum concentration which causes no damage after 60 minutes of inhalation

5.	300-600	Dangerous affects after 30 minutes
6.	600-1000	Acute poisoning
7.	2000	Death within a few minutes

Note:

Hydrogen sulfide is colorless, flammable extremely hazardous gas with a "rotten egg" smell. Some common names for the gas include sewer gas, stink damp, swamp gas and manure gas. It occurs naturally in crude petroleum, natural gas, and hot springs. In addition, hydrogen sulfide is produced by bacterial break-down of organic materials and human and animal wastes (e.g., sewage). Industrial activities that can produce the gas include petroleum/natural gas drilling and refining, wastewater treatment, coke ovens, tanneries, and kraft paper mills. Hydrogen sulfide can also exist as a liquid compressed gas.

Hazardous properties of H_2S gas

Hydrogen sulfide is heavier than air and may travel along the ground. It collects in low-lying and enclosed, poorly-ventilated areas such as basements, manholes, sewer lines, underground telephone vaults and manure pits.

For work within confined spaces, use appropriate procedures for identifying hazards, monitoring and entering confined spaces.

The primary route of exposure is inhalation and the gas is rapidly absorbed by the lungs. Absorption through the skin is minimal. People can smell the "rotten egg" odor of hydrogen sulfide at low concentrations in air. However, with continuous low-level exposure, or at high concentrations, a person loses his/her ability to smell the gas even though it is still present (olfactory fatigue). This can happen very rapidly and at high concentrations, the ability to smell the gas can be lost instantaneously. Therefore, **DO NOT** rely on your sense of smell to indicate the continuing presence of hydrogen sulfide or to warn of hazardous concentrations.

In addition, hydrogen sulfide is a highly flammable gas and gas/air mixtures can be explosive. it may travel to sources of ignition and flash back. If ignited, the gas burns to produce toxic vapors and gases, such as sulfur dioxide.

Contact with liquid hydrogen sulfide causes frostbite. If clothing becomes wet with the liquid, avoid ignition sources, remove the clothing and isolate it in a safe area to allow the liquid to evaporate.

Hydrogen sulfide is both an irritant and a chemical asphyxiant with effects on both oxygen utilization and the central nervous system. Its health effects can vary depending on the level and duration of exposure. Repeated exposure can result in health effects occurring at levels that were previously tolerated without any effect.

Low concentrations irritate the eyes, nose, throat and respiratory system (e.g., burning/ tearing of eyes, cough, shortness of breath). Asthmatics may experience breathing difficulties. The effects can be delayed for several hours, or sometimes several days, when working in low-level concentrations. Repeated or prolonged exposures may cause eye inflammation, headache, fatigue, irritability, insomnia, digestive disturbances and weight loss.

Moderate concentrations can cause more severe eye and respiratory irritation (including coughing, difficulty breathing, accumulation of fluid in the lungs), headache, dizziness, nausea, vomiting, staggering and excitability.

High concentrations can cause shock, convulsions, inability to breathe, extremely rapid unconsciousness, coma and death. Effects can occur within a few breaths, and possibly a single breath.

Protection against H_2S exposure

Before entering areas where hydrogen sulfide may be present:

Air must be tested for the presence and concentration of hydrogen sulfide using air monitoring equipment, such as hydrogen sulfide detector tubes or a multi-gas meter that detects the gas.

Testing should also determine if fire/ explosion precautions are necessary.

If the gas is present, the space/area must be ventilated continually to remove the gas.

If the gas cannot be removed, the person entering the space/area must use appropriate respiratory protection and any other necessary personal protective equipment, rescue and communication equipment.

Entering dangerous H_2S atmosphere

A level of H_2S gas at or above 100 ppm is Immediately Dangerous to Life and Health (IDLH). Entry into IDLH atmospheres can only be made using: 1) a full facepiece pressure demand self-contained breathing apparatus (SCBA) with a minimum service life of thirty minutes, or 2) a combination full facepiece pressure demand supplied-air respirator with an auxiliary self-contained air supply.

If H_2S levels are below 100 ppm, an air-purifying respirator may be used, assuming the filter cartridge/ canister is appropriate for hydrogen sulfide. A full facepiece respirator will prevent eye irritation.

If air concentrations are elevated, eye irritation may become a serious issue. If a halfmask respirator is used, tight fitting goggles must also be used.

Workers in areas containing hydrogen sulfide must be monitored for signs of overexposure.

A rescue in an area that may contain hydrogen sulfide using appropriate respiratory protection should be done.

Appendix 1.135: Properties, Toxicity and Health Effects of Chlorine (Cl_2) Gas

Properties of Chlorine

Physical Properties

Molecular Formula = Cl_2

Molecular Weight = 70.914

Atomic Weight = 35.457

Boiling Point : /-34.05°C/

It is the temperature at which liquid Chlorine vaporizes at one atmosphere, pressure

Melting Point: /-101.0°C/

It is the temperature at which liquid chlorine solidifies at one atmosphere, pressure

Density :

Dry Gas - 3.209 g/L at 0°C and 1 Atmosphere

Liquid - 1.468 kg/L (Pressure of liquid at 0°C is 3.617 atm.)

Density represents the mass of a unit volume at specified temperature and pressure conditions. Density decreases rapidly with increasing temperature. Chlorine gas is about 2.5 times heavier than air at the same condition.

Heats of Fusion and Vaporization :

Heat of Fusion (at melting point) = 22.9 g cal/g

Heat of Vaporization (at boiling point) = 68.7 g cal/g

It is the heat required to change one mass of solid Chlorine to liquid (heat of fusion) or to change one mass of liquid to vapour (heat of vaporization) without change of temperature.

Viscosity : The viscosity of Cl_2 gas is nearly the same as that of saturated steam between 1 and 10 atmosphere : that of liquid Cl_2 is about one third that of water at the same temperature. Hence, its friction loss in pipelines is definitely lower than that of water.

Specific Volume :

Dry gas : at 0°C and 1 atm. pressure (0.31098 m^3/kg)

Liquid : at 0°C and 3.617 atmospheric pressure

It is the volume of unit mass of Cl2 at specified condition of temperature and pressure.

Liquid-Gas Volume Relationship : Liquid Cl_2 vaporizes rapidly at Standard atmospheric, conditions. One volume of liquid vaporises into about 457.6 Volume of gas, or 0.453 kg of liquid forms about 0.1415 m^3 of gas.

Volume and Temperature Relationship : The volume of liquid Cl_2 increases rapidly as its temperature increases. The volume-temperature relationship of liquid Cl_2 in a container loaded to its authorized limit is as follows:

Temperature								
°F	10.00	0.00	20.0	60.0	100.0	120.0	160.0	200.0
°C	23.00	-32.0	-6.6	15.55	37.7	48.88	70.11	93.33
% Volume (as liquid)	81.4	82.1	84.4	87.7	2.3	94.8	97.7	100

Vapour Pressure:

It is the pressure of Cl_2 gas above liquid Cl2 when they are in equilibrium. The relation between vapour pressure and temperature is as follows:

Temperature								
°F	29.29	10.00	0.00	20.0	60.0	100.0	160.0	200.0
°C	34.05	23.00	-32.0	-6.6	15.55	37.7	70.11	93.33
Vapour Pressure								
Kg/cm^2	0	0.56	0.93	1.89	4.82	9.54	21.12	32.7

Critical Properties

- The critical density (573 g/L) is the mass of a unit volume of Chlorine at the critical pressure and temperature.
- The critical pressure (76.1 atmosphere) is the vapour pressure of liquid Chlorine at the critical temperature.

- The critical temperature (144°C) is the temperature above which chlorine exists only a gas despite the pressure.
- The critical volume (1745 ml/g) is the volume of a Unit mass of Chlorine at the critical pressure and temperature. It is the reciprocal of critical density.

Chemical properties of Chlorine

Chlorine is neither flammable nor combustible but it is highly reactive material. It reacts rapidly with most elements, many inorganic and organic compounds, in some cases with explosive violence and frequently with evolution of heat.

Reactions with water : When it is dissolved in water between 9.6°C and 100°C, it reacts to form hypochlorous acid (HOCl) and hydrochloric acid (HCl).

$$Cl_2 + H_2O \rightarrow HOCL + HCL$$

$$HOCL \rightarrow H^+ + OCL$$

The formation of HOCl and HCl acids explain why moist chlorine has a corrosive effect on metals which ordinarily are unaffected by gas or liquid Chlorine.

Reaction with gases :

Ammonia : It reacts with NH3 as follows :

$$12\ NH_3 + 6\ Cl_2 \rightarrow 9\ NH_4Cl + NCl_3 + N_2$$

It may react in aqueous ammonia solution to form various Chloramines or mixtures of Chloramines.

$$NH_3 + Cl_2 \rightarrow NH_2Cl + HCl$$

$$NH_3 + 2Cl_2 \rightarrow NHCl_2 + 2HCl$$

$$NH_3 + 3Cl_2 \rightarrow NCl_3 + 3HCl$$

$$NH_3 + XCl_2 \rightarrow N_2 + H_2O + HCl$$

Hydrogen: It reacts with H2 slowly in dark but explosively in sunlight or at high temperature to form Hydrogen Chloride ($H_2 + Cl_2 \rightarrow 2\ HCl$).

The lower explosive limit of Hydrogen - Chlorine mixture varies from 3.1% to 81% depending upon pressure and other variables.

Oxygen : Chlorine does not react directly with O2 but several Oxides are formed by using indirect means.

Hydrogen sulfide: Cl_2 reacts with H_2S, and Chloroxidation processes for odour control are frequently employed.

$$H_2S + 4Cl_2 + 4H_2O \rightarrow H_2SO_4 + 8HCl$$

$$H_2S + Cl_2 \rightarrow 2HCl + S$$

Sulfur Dioxide : Dechlorination with SO_2 is based on this reaction :

$$SO_2 + Cl_2 + 2H_2O \rightarrow H_2SO_4 + 2HCl$$

Reactions with Metals and other Elements : Under certain conditions, Cl_2 reacts with nearly all metals to form metal Chlorides which are usually quite soluble in water (Cuprous, Silver, Plumbous, Mercurous, platinous, Thallous are only slight soluble). Below 110°C, Copper, iron, lead, nickel, platinum, silver, Steel are chemically resistant to dry Chlorine, gas or liquid. The reaction rate of Cl_2 with most metals increases with extreme rapidity above certain temperatures. Dry Cl_2 reacts with Aluminium, Arsenic, Gold, Mercury, Selenium, Tellunium, Tin and Titanium.

$$\text{e.g. } Sn + 2\,Cl_2 \rightarrow SnCl_4$$

At certain temperature potassium and sodium burn in Cl_2 gas. Carbon steel ignites at 250°C. Antimony, Arsenic, Bismuth, Boron, Copper, Iron, Phosphorus, and certain of their alloys, in powdered, sponge or wire in finely divided form, ignite spontaneously in Cl_2.

Moist chlorine, primarily because of HCI and HOCI formed through Hydrolysis is very corrosive to all common metals. Gold, platinum, Silver and Titanium are resistant. At temperature below 140°C, Tantalum is totally inert to wet and dry Chlorine. Moist and dry chlorine reacts readily with Mercury.

Reactions with Inorganics : Chlorine reacts with many inorganic compounds like Calcium Carbonate, Calcium Borate, Ammonium Chloride, Carbon Mono Oxide, H_2O_2, $FeSO_4$, $MgSO_4$, HgO, NO, KClO, KCN, KOH, NaOH, AgF, $AgNO_3$, $NaHSO_3$, $NaClO_2$ etc.

Reactions with Organics : In the organic and Petrochemicals field Cl_2 is most widely used as a chemical intermediate. There are thousands of Chlorine containing organic compounds of commercial importance are made by chlorinating hydrocarbons. In most instances, the reaction is very rapid and gives off heat.

Solubility in Water : Chlorine is only slightly soluble in H_2O, reaching its maximum solubility of approximately one per cent at 9.6°C. Below (9.6°C, it forms crystalline. Hydrate with H_2O ($Cl_2.8H_2O$), referred to as "Chlorine ice". Above 9.6°C its solubility decreases when temperature increases upto boiling point of water at which it is completely insoluble.

Toxicity of Cl_2 : Chlorine is corrosive. It reacts with body moisture to form corrosive acids. Cl_2 gas is primarily a respiratory irritant. It is irritating at concentration above 3 to 5 ppm by volume in air or readily detectable by a normal person. Inhalation of Cl_2 can cause respiratory injury ranging from irritation to death depending upon its concentration and duration of inhalation.

Acute effect of Cl_2 : The first symptom of exposure to Cl_2 is irritation to the mucous membrane of eyes, nose and throat with increase in burning pain. If the duration of exposure of the concentration of Cl_2 is excessive, it will cause restlessness, throat irritation and sneezing. In extreme cases lung tissues may be attacked resulting in pulmonary adema, fall of blood pressure.

Health Effect of Cl_2 at Various Concentrations

Cl_2 Concentration in air (ppm)	Health Effects
0.2–0.5	No noxious long term effect
0.5	Slight odour
1 to 3	Definite odour, irritation to eyes
6	Irritation of throat
30	Intense cough fits
40–60	Exposure for 30 to 60 minutes may cause serious damage
100	May cause lethal damage
1000	Danger to life even after few deep inhalations

Chronic effect: In small concentration men rapidly lose their ability to detect the smell of Cl_2 gas therefore concentration higher than TLV (Threshold Limit Value) of 1 ppm may occur without immediate notice. Due to prolonged exposure to concentration more than 5 ppm bronchial problem and T.B may result and enamel of teeth may get damaged.

Appendix 1.136: Maximum Permitted Dose of Radiation (rem) for Human Beings

S.No.	Type of radiation	Body part exposed to radiation	Three months exposure limit	One year exposure limit	Coordinative dose
1.	Any ionizing radiation	a. Extremities eg, feet, hands b. Eyes c. Other body parts	40 8 15	75 15 30	– – –
2.	X-rays, gamma rays, neutron	Whole body	3 (1.3 for women)	5	5x(Age–18)

Appendix 1.137: General Standards *for Discharge of Environmental Pollutants (CPCB- India)

Part A: Effluents

S.No.	Parameter	Inland surface water	Public sewers	Land for irrigation	Marine/castal areas
		a.	b.	c.	d.
1.	Colour and odour	See 6 of annexure-11		See 6 of annexure-11	See 6of annexure -11
2.	Suspended solids mg/1 max	100	600	200	a. For process waste water b. For cooling water effluent 10 percent above total suspended matter of influent.

S.No.	Parameter	Inland surface water	Public sewers	Land for irrigation	Marine/castal areas
		a.	b.	c.	d.
3.	Particle size of suspended solids	Shall pass 850 micron IS sieve	-	-	a. Floatable solids, solids max. 3mm b. settleable solids max 856 microns
4.	pH value	5.5 to 9.0	5.5 to 9.0	5.5 to 9.0	5.5 to 9.0
5.	Temperature	Shall not exceed 5°c above the receiving water temperature	-	-	Shall not exceed 5°c above the receiving water temperature
6.	Oil and grease, mg/l max,	10	20	10	20
7.	Total residual chlorine, mg/l, max	1.0	-	-	1.0
8.	Ammonical nitrogen (as N), mg/l, max	50	50	-	50
9.	Total kjeldahl nitrogen (as N); mg/l, max mg/l max	100	-	-	100
10.	Free ammonia (as NH_3), mg/l, max	5.0	-	-	5.0
11.	Biochemical oxygen demand(3 days at 27°c), mg/l, max	30	350	100	100
12.	Chemical oxygen demand mg/l, max	250	-	-	250
13.	Arsecic (as As)	0.2	0.2	0.2	0.2
14.	Mercury (as Hg), mg/l max	0.01	0.01	-	0.01
15.	Lead (as Cd) mg/l, max	0.1	0.1	-	2.0

S.No.	Parameter	Inland surface water	Public sewers	Land for irrigation	Marine/castal areas
		a.	b.	c.	d.
16.	Cadmium (as Cd) mg/l, max	2.0	1.0	-	2.0
17.	Hexavalent chromium(as Cr + 6), mg/l max	0.1	2.0	-	1.0
18.	Total chromium (as Cr) mg/l, max	2.0	2.0	-	2.0
19.	Copper (as Cu) mg/l max	3.0	3.0	-	3.0
20.	Zinc (as Zn) mg/l max	5.0	15	-	15
21.	Selenium (as Se)	0.05	0.05	-	0.05
22.	Nickel (as Ni mg/l)	3.0	3.0	-	5.0
23.	Cyanide (as CN) mg/l, max	0.2	2.0	0.2	0.2
24.	Fluoride (as F mg/l), max	2.0	15	-	15
25.	Dissolved phosphates (as P, mg/l, max)	5.0	-	-	-
26.	Sulphide (as S) mg/l, max	2.0	-	-	5.0
27.	Phenolic compounds (as S) mg/l, max	1.0	5.0	-	5.0
28.	Radioactive materials: a. Alpha emitters micro curie mg/l, max. b. Beta emitters micro curie mg/l	10-7 10-6	10-7 10-6	10-8 10-7	10-7 10-6

S.No.	Parameter	Inland surface water	Public sewers	Land for irrigation	Marine/castal areas
		a.	b.	c.	d.
29.	Bio-assay test	90%h survival of fish after 96 hours in 100% effluent	90% suival of fish after 96 hours in 100% effluent	90% survival of fish after 96 hours in 100% effluent	90% survival of fish after 96 hours in 100% effluent
30.	Manganese	2 mg/l	2 mg/l	-	2 mg/l
31.	Iron (as Fe)	3 mg/l	3 mg/l	-	3mg/l
32.	Vanadium (as V)	0.2 mg/l	0.2 mg/l	-	0.2 mg/l
33.	Nitrate nitrogen	10 mg/l	-	-	20 mg/l

* These standards shall be applicable for industries, operations or processes other than those industries, operations or process for which standards have been specified in Schedule of the Environment Protection Rules, 1989.

Part B: Wastewater Generation Standards

S.No.	Industry	Quantum
1.	Integrated Iron and steel	16 m^3/ tonne of finished steel
2.	Sugar	0.4 m3/tone of cane crushed
3.	Pulp and paper industries a. Large Pulp and paper (i) Pulp and paper (ii) Rayon grade pulp b. Small pulp and paper (i) Agro- residue based (ii) Waste paper based	175m^3/tone of paper produced 50m^3/tone of paper 150m^3/tone of paper produced 50m^3/tonne of paper produced
4.	Fermentation industries a. Maltry b. Brewer c. Distillery	3.5m^3 /tonne of grain processed 0.25 m^3/KL of beer produced 12 M^3/KL of alcohol produced
5.	Caustic soda a. Membrane cell process b. Mercury cell process	1m^3/tonne of caustic soda produced excluding cooling tower blowdown 4 m^3/tonne of caustic soad produced (mercury bearing) 10% below down permitted for cooling tower .

6.	Textile industries : Man-made fibre i. Nylon and polyster ii. viscose staple Fibre iii. viscose filament yarn	 $120m^3$/tonne of fibre produced $150m^3$/tonne of product $500m^{3/}$ tonne of product
7.	Tanneries	$28m^3$/tonne of raw hide
8.	Starch Glucose and related products	$8m^3$/tonne of maize crushed
9.	Dairy	3 m^3/kl of Milk
10.	Natural rubber processing industry fertilizer a. Straight nitrogenous fertilizer b. Straight phosphatic fertilizers (SSp and TSP) Excluding manufacture of any acid c. Complex fertilizers	$4m^3$/tonne of rubber
11.	Biochemical oxygen demand (3 days at 27°c) mg/l, max a. Straight nitrogenous fertilizer b. Straight phosphatic fertilizer (SSP and TSP) excluding manufacture of any acid c. Complex fertilizer	 $5m^3$/tonne of urea orequivalent produced 0.5 m^3 tonne of SSP/TSP Standards of nitrogenous and phospathic fertilizers are applicable depending on the primary product

Part C: Load Based Standards

1. Oil Refinery Industry

Parameter	Quantum in Kg/1000 tonnes of crude Processed
Oil and grease	10.00
Phenol	0.70
BOD	10.50
Suspended solids	14.00
Sulphide	0.35

2. Large Pulp and Paper, News Print/Rayon grade plants of capacity above 24, 000 Tonne/annum

Parameter	quantum
Total organic chloride (TOCI)	2 Kg/tonne of product

Part D: Concentration Based Standards

1. General Emission Standards

S.No.	Parameter	concentration not to exceed (in mg/Nm^3)
1.	Particulate matter (PM)	150

2.	Total fluoride	25
3.	Asbestos	4 Fibres/cc and dust should not be more than mg/Nm^3
4.	Mercury	0.2
5.	chlorine	15
6.	hydrochloric acid vapour And mist	35
7.	sulphuric acid mist	50
8 .	carbon monoxide	1%
9 .	Lead	10

2. Equipment based standards

For dispersion of sulphur dioxide; a minimum stack height limit is accordingly prescribed as below:

S.No.	Power generation capacity/steam generation capacity	stack height (metre)
1.	power generation capacity:	275
	• 500 MW and more	220
	• 200/210 MW and Above to less than 500 MW	H=(Q)0.3
	• less than 200/210 MW	
2.	steam generation capacity	
	• less than 2 tonne/hr	09
	• 2 to 5 tonne/hr	12
	• 5 to 10 tonne/hr	15
	• 10 to 15 tonne/hr	18
	• 15 to 20 tonne/hr	21
	• 20 to 25 tonne/hr	24
	• 25 to 30 tonne/hr	27
	• more than 30 tonne/hr	30 or as per formula $H=14\ (q)^{0.3}$which ever is more

NOTES

H = Physical height of the stack in metre
Q = Emission rate of SO_2 in Kg/hr.

3. Load/ Mass Based standards

S.No.	Industry	Parameter	Standard
1.	Fertilizer (urea) • commissioned prior to 1.1.82 • commissioned after 1.1.82	Particulate Matter Particulate Matter	2Kg/tonne of product 0.5 Kg/tonne of product
2.	Copper, lead & zinc smelter	Sulphur dioxide	4Kg/ tonne of concentrated (100%) acid produced

S.No.	Industry	Parameter	Standard
3.	Nitric acid	Oxides of nitrogen	3kg/tonne of weak acid (before concentration produced)
4.	Sulphuric acid	Sulphur dioxide	4Kg/ tonne of concentrated (100%) acid produced
5.	Coke oven	Carbon monoxide	3kg/tonne of coke produced
6.	Oil refineries –distillation (atmosphere + vacuum)	Sulphur dioxide	0.25kg/tonne of feed in this process
	• catalytic cracker	-do	0.25kg/tonne of feed in this process
	• sulphur recovery unit	-do	120kg/tonne of sulphur in the feed
7.	Aluminium plants: i. anode bake oven	Total fluoride	0.3 Kg/tonne of aluminium
	(ii) pot room		
	(a) VSS	-do-	4.7 kg/tonne of aluminium
	(b) HSS	-do-	6kg/tonne of aluminium
	(c) PBSW	-do-	2.5kg/tonne of aluminium
	(d) PBCW	-do-	1.0kg/tonne of aluminium
8.	Glass industry	-do-	2kg/hr
	a. Furnace capacity (i) up to the product draw capacity Particulate Matter of 60 tonne/day (ii) product draw capacity more than 60 tonne/day		0.8kg/tonne of product drawn

NOTES

VSS = vertical stud soderberg; HSS=horizontal stud soderberg;
PBSW= pre backed side work and PBCW= pre backed centre work

Part E: Noise Standards

A. Noise limits for automobiles (from at 7.5 metre in dB(A) at the manufacturing stage)

1. Motorcycle, scooters and three wheelers 80
2. Passengers car 82
3. Passenger of commercial vehicles upto 4 tonne 85
4. Passenger of commercial vehicles above 4 tonne and Upto 12 tonne 89
5. Passenger or commercial vehicles exceeding 12 tonne 91

B. Domestic appliances and construction equipments at the manufacturing stage to be achieved by 31st December, 1993.

1. Window air conditioners of 1-1.5 tonne 68
2. Air coolers 60
3. Refrigerators 46

4. Diesel generator for domestic purpose 85-90
5. Compactors (rollers), front loaders, Concrete mixers, cranes (movable), virators and saws 75

Annexure I

(For the purpose of parts – A, B and C)

The state boards shall fallow the following guidelines in enforcing the standards specified under schedule IV

1. The waste water and gases are to be treated with the best available technology (BAT) in order to achieve the prescribed standards.
2. The industries need to be encouraged for recycling and reuse of waste materials as far as practicable in order to minimize the discharge of wastes into the environment.
3. The industries are to be encouraged for recovery of biogas, energy and reusable materials.
4. While permitting the discharge of effluents and emissions into the environment, State Boards have to taken into account the assimilative capacities of the receiving bodies, especially water bodies so that quality of the intended use of the receiving waters is not affected. Where such quality is likely to be affected, discharge should not be allowed into water bodies.
5. The central and state boards shall put emphasis on the implementation of clean technologies by the industries in order to increase fuel efficiency and reduce the generation of environmental pollutants.
6. All efforts should be made to remove colour and unpleasant odour as far sa practicable.
7. All efforts should be mentioned in this Schedule shall also apply to all other effluents discharged such as mining, and mineral processing activities and sewage.
8. The limit given for the total concentration of mercury in the final effluent of caustic soda industry, is for the total concentration of mercury in the final effluent of caustic soda handling; (d) hydrogen handling and (e) hydrochloric acid plant.
9. All effluents discharged including from the industries such as cotton textile, composite woolen mills, synthetic rubber, small pulp and paper, natural rubber, petro? Chemicals, into surface waters shall conform to the BOD limit specified above, namely, 30 mg/l. For discharge of an effluent having a BOD more than 30mg/l, the standards shall conform to those given above for other receiving bodies, namely, sewers, coastal waters and land irrigation.
10. Bio-assay shall be made compulsory for all industries, where toxic and non biodegerable chemicals are involved.
11. In case of fertilizer industry, the limits in respect of chromium and fluoride shall be complied with at the outlet of chromium and fluoride removal units respectively.
12. In case of pesticides.
    a. The limits should be complied with at end of the treatment plant before dilusion.
    b. Bio-assay test should be carried out with the available species of fish in the receiving water, the COD limits to be specified in the content conditions should be correlated with the BOD limits.
    c. In case metabolites and isomers of the pesticides in the given list are found in significant concentrations, standards should be prescribed for these also in the same concentration as the individual pesticides.
    d. Industries are required to be to analyse pesticides in waste water by advanced analytical methods such as GLC/HPLC.

13. The chemical oxygen demand (COD) concentration in a treated effluent, if observed to be persistently greater than 250 mg/l before disposal to any receiving body (public sewer, land for irrigation, inland surface water and marine coastal areas,) such industrial units are required to indentify chemicals causing the same. In case these are found to be toxic as defined in the Schedule- I of the Hazardous Rules, 1989, the state boards in such cases shall direct the industries to install tertiary treatment stipulating time limit.
14. Standards specified in Part A of Schedule – VI for discharge of effluents into the public sewer shall be applicable only if such sewer leads to secondary treatment including biological, treatment system otherwise the discharge into sewers shall be treated as discharge into inland surface waters.

Annexure II

(For the purpose of Parts D)

1. The states boards shall fallow the following guidelines enforcing the standards specified under Schedule –VI
   a. In case of cement plants, the total dust (from all sections) shall be within 400 mg/Nm^3 and 250mgNm^3 for the plants upto 200 tpd and more than 200 tpd capacities respectively.
   b. In respect of calcinations process (e.g. aluminium plants), kilns and step grate begasse fired boilers, particulate matter emissions shall be within 250mg/Nm^3.
   c. In case of thermal power plants commissioned prior to 1/1/1982 and having generation capacity less than 62.5 MW, the particulate matter emission shall be within 350 mg/Nm^3.
   d. In case of the Lime Kilns of capacity more than 5 tpd and 40 tpd, the particulate matter emission shall be within 500mg/Nm^3.
   e. In case of horse shoe/pulsating grate and spreader stroker bagasse fired boilers, the particulate matter emission shall be within 500 (12% CO_2) and 800 (12% CO_2) mg/Nm^3 respectively. In respect of these boilers, if more than one is attached to single stack, the emission standard shall be fixed, based on added capacity of all the boilers connected with the stack.
   f. In case of asbestos dust, the same shall not exceed 2 mg/Nm^3.
   g. In case of the urea plants commissioned after 1.1.1982 coke ovens and lead glass units, the particulate matter emission shall be within 1600 and 1200 mg/Nm^3.
   h. In case of all small boilers of capacity less than 2 tonne/hr and between 2 to 5 tonne/hr, the particulate matter emissions shall be within 1600 and 1200 mg/Nm^3.
   i. In case of intergrated iron and steel plants, particulate matter emission upto 400 mg/Nm^3 shall be allowed during oxygen lancing.
   j. In case of stone crushing units, the suspended particulate matter contribution value at a distance of 40 meter from a controlled, isolated as well as from a unit located in cluster should be less than 600 micrograms/Nm^3.

   These units must be adopt the following pollution control measures:

   a. Dust containment cum suppression system for the equipment;
   b. Construction of wind breaking walls;
   c. Construction of the metalized roads within the premises;
   d. Regular cleaning and wetting of ground within the premises;
   e. Growing of a green belt along the periphery.
   f. In case of ceramic industry, from the other sources of pollution, such as basic raw materials and

processing operations, heat recovery dryers, mechanical finishing operation all possible prevention measures should be taken to control particulate matter emissions as far as practicable.

1. The total fluoride emissions in respect of glass and phosphatic fertilizers shall not exceed 5mg/Nm3 and 25mg/Nm^3 respectively.
2. In case of copper, lead and zinc smelting, the off? Gases may as far as possible be utilized for manufacturing sulphric acid.
3. In case of cupolas (foundries) having capacity (melting rate) less than 3 tonne hour, the particulate matter emissions shall be within 450 mg/Nm^3. In these cases it is essential that stack is constructed over the cupolas beyond the charging door and the emissions are directed through the stack, which should be at least six times the diameter of cupola. In respect of arc furnaces and induction furnaces. Provisions has to be made collecting the fumes before discharging the emissions through the stack.

Appendix 1.138: Maximum Permissible Exposure Limits of Fibrogenic Dusts

S.No.	Group	Fibrogenic dusty substances	Maximum permissible concentration, mg/m^3
1.	I	a. Silica which has been heated above 870°C. b. Cryodolite (or blue asbestos).	50
2.	II	a. Silica in the form of quartz, ganister, gritstone or any rock containing not less than 80% total silica. b. Fire clay dust with total silica in excess of 60%.	100
3.	III	a. Non crystalline silica (including unheated Keisselguhr). b. Aluminous fire clays (with total silicate as "silica" not more than 40% c. Carbides d. Graphite (Natural graphite may contain more than 20% silica)	1000

NOTES

Fibrogenic dust comprise those respirable dust which contain a proportion of quartz (crystalline SiO_2) or asbestos (magnesium silicate) fibres, which can cause diffuse fibrosis in the lung tissue and lymph-nodes. Fibrogenic dust cause scar tissue to be formed in lungs when inhaled in excess quantity.

PBSW= pre backed side work and PBCW= pre backed centre work

Appendix 1.139: Permissible Exposure Limits (PEL) for various Air Pollutants as per OSHA

Air Pollutants	Permissible Exposure Limit (PEL)[1)]
Acetaldehyde	200 ppm
Acetic Acid	10 ppm
Acetone	1000 ppm
Asbestos[2]	0.1 fibre/cubic centimetre
Benzene[2]	1 ppm
Bromine	0.1 ppm
Cadmium	5 $\mu g/m^3$
Carbon dioxide	5000 ppm
Carbon disulphide	200 ppm
Carbon monoxide	50 ppm
Carbon tetrachloride	10 ppm
Chlorine	1 ppm
Chloroform	50 ppm
Cresol	5 ppm
Ethanol (Ethyl alcohol)	1000 ppm
Ethyl chloride	1000 ppm
Fluorine	0.1 ppm
Formaldehyde	0.75 ppm
Gasoline	300 ppm
Hydrogen cyanide	10 ppm
Iodine	0.1 ppm
Iron oxide (fume)	10 mg/m^3
Isopropyl alcohol	400 ppm
Lead	50 $\mu g/m^3$
Manganese compounds	5 mg/m^3
Mercury	$mg/10\ m^3$
Methanol (Methyl alcohol)	200 ppm
Nitric Oxide	25 ppm
Nitrogen dioxide	5 ppm
Propane	1000 ppm
Selenium compounds	0.2 mg/m^3
Sulphur dioxide	5 ppm
Sulphur acid	1 mg/m^3
Tellurium compounds	0.1 mg/m^3
Tetraethyl lead	0.075 mg/m^3
Toluene	200 ppm
Turpentine	100 ppm

Air Pollutants	Permissible Exposure Limit (PEL)[1)]
Vinyl chloride[2]	1 ppm
Zinc oxide, fume Zinc oxide, dust	5 mg/m^3 15 mg/m^3

NOTES

1) PEL-Permissible Exposure Limit – 8-hour time weighted averages – An employee's exposure to any substance listed, in any 8-hour work shift of a 40-hour work week, shall not exceed the 8-hour time weighted average limit given for that substance – OSHA, Occupational Safety and Health Administration of U.S.A.

2) Human carcinogen - confirmed

Appendix 1.140: Source, Effect and Control of Thermal Pollution of Water

Thermal pollution of water means the addition of excess of undesirable heat to water thereby making it harmful to aquatic life in river bodies, lakes, ponds, oceans etc.

Waste water/ cooling water generated by the thermal power plants, nuclear plants etc. has higher temperature than the normal temperature of the surface water (More than 8-10°C). This causes thermal pollution of water.

Thermal pollution is defined as sudden increase or decrease in temperature of natural body of water (e.g., ocean, lake, river, pond) by human influence. This normally occurs when a plant or facility takes in water from a natural resource and puts back with an altered (high or low) temperature.

Thermal Pollution occurs when an industry draws in water from a natural source and either cools it down or heat up. Then it ejects this hot/cold water back into the natural source (river/ocean/lake/pond etc.) which changes the dissolved oxygen levels and have disastrous effects on local ecosystem and community.

- The basic definition of thermal water pollution is the deterioration of water quality as a result of any process that changes ambient water temperature
- It is usually associated with the increases of water temperatures in a stream, lake, or ocean due to the discharge of heated water from industrial processes, such as the generation of electricity from the thermal power plants
- However this rise in the ambient water temperature can also occur in streams where shading vegetation along the banks is removed or where sediments have made the water more turbid. Both of these effects allow more energy from the sun to be absorbed by the water and thereby increase its temperature
- instances in which the effects of colder-than-normal water temperatures can be also observed. For example, the discharge of cold bottom water from deep-water reservoirs behind large dams changed the downstream biological communities in systems such as the River.

Causes of Thermal Pollution

- **Water as Cooling Agent in Power, Manufacturing and Industrial plants:** Production and Manufacturing plants are biggest source of thermal pollution. These plants draw water from nearby source to keep machines cool and then release back to the source with higher temperature. When heated water returns to the river or ocean, the water temperature rises sharply. When oxygen levels are altered in the water, this can also degrade the quality and longevity of life in wildlife that lives underwater. This process can also wipe away streamside vegetation, which constantly depends on constant levels of oxygen and temperature. By altering these natural environments, industries are essentially helping decrease the quality of life for these marines based life forms and can ultimately destroy habitats if they are not controlled and careful about their practices.

- **Soil Erosion:** Soil erosion is another major factor that causes thermal pollution. Consistent soil erosion causes water bodies to rise, making them more exposed to sunlight. The high temperature could prove fatal for aquatic biomes as it may give rise to anaerobic conditions.
- **Deforestation:** Trees and plants prevent sunlight from falling directly on lakes, ponds or rivers. When **deforestation** takes place, these water bodies are directly exposed to sunlight, thus absorbing more heat and raising its temperature. Deforestation is also a main cause of the higher concentrations of **greenhouse gases** i.e. global warming in the atmosphere.
- **Runoff from Paved Surfaces:** Urban runoff discharged to surface waters from paved surfaces like roads and parking lots can make water warmer. During summer seasons, the pavement gets quite hot, which creates warm runoff that gets into the sewer systems and water bodies.
- **Natural Causes:** Natural causes like volcanoes and geothermal activity under the oceans and seas can trigger warm lava to raise the temperature of water bodies. Lightening can also introduce massive amount of heat into the oceans. This means that the overall temperature of the water source will rise, having significant impacts on the **environment**.

Source of Thermal Pollution

Heat producing industries i.e.,

- thermal power plants, nuclear power plants, refineries, steel mills etc. are the major sources of thermal pollution. Power plants utilize only 1/3 of the energy provided by fossil fuels for their operations. Remaining 2/3 is lost in the form of heat generally to cooling water. Cold water, generally, is drawn from some nearby water-body. passed through the plant and returned to the same water body with temperature about 10°C higher than the initial temperature. Excess of heat reaching such water bodies causes thermal pollution of water.

The following source contribute to thermal pollution:

- Nuclear power plants
- Coal fired plants
- Industrial effluents
- Domestic sewage
- Hydro-electric power

Nuclear power plants: Nuclear power plants including drainage from hospitals, research institutions, nuclear experiments and explosions, discharge a lot of heat that is not utilized along with traces of toxic radio nuclides into nearby water streams. Emissions from nuclear reactors and processing installations are also responsible for increasing the temperatures of water bodies. The operations of power reactors ant) nuclear fuel processing units constitutes the major contributor of heat in the aquatic environment. Heated effluents from power plants are discharged at 10°C higher than the receiving waters that, affects the aquatic flora and fauna.

Coal-fired power plants: Coal fired power plants constitute a major source of thermal pollution. The condenser coils in such plants are cooled with water from nearby lakes or rivers. The resulting heated water is discharged into streams thereby raising the water temperature by 15C. Heated effluent decreases the dissolved content of water resulting in death of fish and other aquatic organisms. The sudden fluctuation of temperature also leads to *'thermal shock'* killing aquatic life that have become acclimatized to living in a steady temperature.

Industrial effluents: Industries like textile, paper, pulp and sugar manufacturing release huge amounts of cooling water along with effluents Into nearby natural water bodies. The waters polluted by sudden and heavy organic loads result in severe drop in levels of dissolved oxygen leading to death of several aquatic organisms.

Domestic Sewage: Domestic sewage is discharged into rivers, lakes, canals or streams with minimal treatment or without any treatment. These wastes have a higher organic temperature and organic load. This leads to decrease in dissolved oxygen content in the receiving waters resulting in the set-up of anaerobic conditions causing release of foul and offensive gases in water. Eventually, this leads to development of anoxic conditions resulting in rapid death of aquatic organisms.

Hydro-electric power: Generation of hydroelectric power sometimes leads to negative thermal loading in water systems. Apart from electric power industries, various factories with coding requirement contribute to thermal loading.

Thermal Pollution in Streams by Human Activities

- Industries arid power plants use water to cool machinery and discharge the warm water into a stream
- Stream temperature rises when trees and tall vegetation providing shade are cut.
- Soil erosion caused due to construction also leads to thermal pollution
- Removal of stream side vegetation
- Poor farming Practices also lead to thermal pollution

Effects of Thermal Pollution

- With rise in water temperature in river bodies, density and viscosity of water reduces and its vapor pressure in creases increasing the evaporation loss.
- Warmer temperatures of the discharged effluents from the thermal power plants lower the dissolved oxygen content of the lake or stream by decreasing the solubility of oxygen in water. High temperature becomes barrier for oxygen penetration in deep cold water. The toxicity of pesticides, detergents and chemical in the effluent increases with increases in temperature. Warning of air reduces its air saturation value and its capacity to dissolve oxygen.
- These warmer water also enhance the respiration rates to the aquatic organisms thus forcing them to consume oxygen faster. Warning of water accelerates the decomposition process taking palace in water. If water is chemically polluted, then its warning increases the toxicity to fish warmer water allows algae to flourish on surface of water and over the long term, growing algae can decrease oxygen level in the water.
- With the decrease in the dissolved oxygen level and increase in the respiration rate of the aquatic species, berating breathing problems arise in the aquatic organisms which may prove to be final. Temperature rise of water affects physiology controlling reproductive cycle, digestion rates,

respiration rate and many chemical activities. Dissolved oxygen in water for fish to survive is 5-7 mg/litre (ppm). In fishes, several activities like nest building, spawning, hatching and reproduction depend on optimum temperature

- It also increase their susceptibility to disease, parasites, and toxic chemicals
- Temperature rise of water causes rise in metabolic rates which implies that an aquatic animal will require more food to maintain body weight. With reduction in temperature to 5°C fishes require less food compared to 15°C (natural food). Higher temperature implies faster growth rate and shorter life span temperature is of vital signification to physical metabolic and biochemical processes that control respiratory rates, digestion, excretion and overall development of aquatic organisms. Temperature changes cause total disruption to ecosystem.
- Water temperature of 35°C is dangerous for fishes survival at higher temperature (a) bacterial growth rate and rate of physiological process increases, (b) decomposition of organic and other oxygen demanding waste is speeded up and (c) rate of oxygen depletion increases and dissolved oxygen content decrease
- Thermal discharge is favorable to bacteria and pathogens.
- Discharge of heated water into shallow water near the shore of a lake may also damage spawning and kill young fish. Fishes do not survive above 30°C. Salmon fry sockeye fish die at 22.5°C while large mouth bass fish die at 26°C., Thermal death of fish is due to thermal shock of water
- The adaptability of the fishes and other aquatic organisms do not safeguard them form sudden and sharp changes in temperature of the effluent water from the thermal power plants.
- These effect of sharp changes in water temperature when new power plants open up or when plants shut down for repair is known as thermal shock. Abrupt changes in temperature alters the seasonal variation in the type and abundance of lower organisms leading to shortage of right food for fish at the right time.
- Fish and other organisms adapted to a particular temperature range can also be killed from thermal shock. Thermal pollution is directly responsible for mortality of aquatic organisms. Increase in water temperature leads to exhaustion of micro-organisms thereby shortening the life span of fishes. Above a certain temperature, fishes die due to failure of respiratory system and nervous system failure.
- Many fish do die on the intake screens which are employed to prevent the fishes from clogging the heat exchanger pipes.
- Growth of green algae is good at 30 – 35°C and that of blue green algae at 35 – 40°C. Thermal discharges in rivers/ lakes/ ocean favors their resulting in damage to eco-system as blue green algae is poorer food source and toxic to fish.
- The composition of flora and fauna changes because the species sensitive to increased temperature due to thermal shock will be replaced by temperature tolerant species
- Metabolic activities of aquatic organisms increase requiting more oxygen and food
- Fish migration is affected due to formation of various thermal zones
- Warning of earth due to nuclear plants/fossil fuel fired power plants and geothermal energy increases heat load on the earth added by heat radiation from earth due to green house effect, the average earth temperature increase causing rise in water body temperature. Increase in per capita energy consumption increases the thermal load on earth.

- **Increase in Toxins:** With the constant flow of high temperature discharge from industries, there is a huge **increase in toxins** that are being regurgitated into the natural body of water. These toxins may
- contain chemicals or radiation that may have harsh impact on the local ecology and make them susceptible to various diseases.
- **Loss of Biodiversity:** A dent in the biological activity in the water may cause significant **loss of biodiversity**. Changes in the environment may cause certain species of organisms to shift their base to some other place while their could be significant number of species that may shift in because of warmer waters. Organisms that can adapt easily may have an advantage over organisms that are not used to the warmer temperatures.
- **Ecological Impact:** A sudden thermal shock can result in mass killings of fish, insects, plants or amphibians. Hotter water may prove favorable for some species while it could be lethal for other **species**. Small water temperature increases the level of activity while higher temperature decreases the level of activity. Many aquatic species are sensitive to small temperature changes such as one degree Celsius that can cause significant changes in organism metabolism and other adverse cellular biology effects.
- **Affects Reproductive Systems:** A significant halt in the reproduction of marine wildlife (although this may be true, reproduction can still occur between fish - but the likelihood of defects in newborns is significantly higher) can happen due to increasing temperatures as reproduction can happen with in certain range of temperature. Excessive temperature can cause the release of immature eggs or can prevent normal development of certain eggs.
- **Increases Metabolic Rate:** Thermal pollution increases the metabolic rate of organisms as increasing enzyme activity occurs that causes organisms to consume more food than what is normally required, if their environment were not changed. It disrupts the stability of food chain and alter the balance of species composition.
- **Migration of organisms:** Warm water may also cause particular species of organisms to migrate to suitable environment that would cater to its requirements for survival resulting is loss for those species that depend on them for their daily food as the food chain is interrupted.
- **A one megawatt nuclear power plant**, may require five million m^3 cooling water each day. The water used in such a plant has its temperature increased by about 17°C during the cooling process. For this reason, such plants are usually built very close to an abundant water supply such as a lake, a large river, or the ocean.

When thermal pollution drives water temperatures up, most aquatic and marine wildlife cannot survive. Immobile organisms, such as plants and shellfish, simply die. The amount of any gas that can be dissolved in water varies inversely with the temperature. As water is warmed, therefore, it is capable of dissolving less and less oxygen. Organisms that need oxygen to survive will, in such cases, cant be able to survive.

When heated water is released from a plant or factory, it does not readily mix with the cooler water around it. Instead, it forms a stream like mass known as a **thermal plume** that spreads out from the outflow pipes. It is in this thermal plume that the most severe effects of thermal pollution are likely to occur. Only over an extended period of time does the plume gradually mix with surrounding water, producing a mass of homogenous temperature.

Invasion of Destructive Organism

Water temperatures can have other, less expected effects also. As an example, trout can swim less rapidly in water above 19°C making them less efficient predators. Organisms may become more subject to disease in warmer water too. The bacterium *Chondrococcus columnaris* is harmless to fish at temperatures of less than 10°C. Between temperatures of 10 – 21°C, however, it is able to invade through wounds in a fish's body and at temperatures above 21°C it can even attack healthy tissue.

Urban Heat dome

Another example of thermal pollution is the development of urban heat islands. An urban heat island consists of a dome of warm air over an urban area caused by the release of heat in the region. Since more human activity occurs in an urban area than in the surrounding rural areas, the atmosphere over the urban area becomes warmer than it is over the rural areas.

It is not uncommon for urban heat islands to produce measurable climate changes. For example, the levels of pollutants trapped in an urban heat island can reach 5 to 25 percent greater than the levels over rural areas. Fog and clouds may reach twice the level of comparable rural areas, wind speeds may be produced by up to 30 per cent, and temperatures may be 0.5 - 2°C higher than in surrounding rural areas. Such differences may cause both personal discomfort and, in some cases, actual health problems for those living within an urban heat island.

Undesirable changes in Algal population: Excess Nutrients from the washout water from farm lands combined with thermal pollution cause an excessive algal growth with consequent changes. High Temperature promotes blue green algal blooms which disrupts the aquatic food chain

Thermal Enrichment of Water

- It is the concept of using heated water released from the thermal power plants for beneficial purposes
- Warm water released from the power plants can be used for irrigation purpose to prolong the growing season of some species in frost-prone areas.
- Cycled through aquaculture pens it can be also used to enhance the growth of commercially valuable fish and shellfish. For example, waste hot water is used to cultivate oysters in aquaculture lagoons in Japan and in New York's Long Island Sound and to cultivate catfish and redfish in Texas.
- Heated water can be also used to heat buildings and greenhouses.
- Can be run under sidewalks to melt snow in the cold countries.
- To desalinate ocean water.

Contra! measures for thermal pollution

The following methods can be adapted to control high temperature of water caused by thermal discharges:

- **Cooling towers**: Use of water from water systems for cooling systems for cooling purposes, with subsequent return to the water way after passage through a condenser, Is called cooling process. Cooling towers transfer heat from hot water to the atmosphere by evaporation. Cooling towers are of two types:

  i. **Wet cooling tower**: Hot water coming out from the condenser (reactor) is allowed to spray over baffles. Cool air. with high velocity, is passed from sides, which takes away the heat and cools the water.

  ii. Dry cooling **tower**: Here, hot water is allowed to flow In long spiral pipes. Cool air with the help of a fan is passed over' these hot pipes, which cools down hot water. This cool water can be recycled.

- **Cooling ponds:** Cooling ponds are the best way to coot thermal discharges. Heated effluents on the surface of the water in cooling panels maximise dissipation of heat to the atmosphere and minimise the water area and volume. The warm water wedge acts like a cooling pond.
- **Spray ponds:** The water coming out from condensers is allowed to pass into the ponds through sprayers. Here water is sprayed through nozzles as fine droplets. Heat from the fine droplets gets dissipated to the atmosphere.
- **Artificial lakes:** Artificial takes are man made water bodies that offer once-through cooling. The heated effluents can be discharged into the lake at one end and water for cooling purposes may be withdrawn from the other end. The heat is eventually dissipated through evaporation. Working principles and the diagram of the above methods of thermal pollution control is given below

Cooling Tower

a. **Wet cooling tower:** Hot water is sprayed over baffles. Cool air entering from sides takes away the heat and cools water.

   This cool water can be recycled or discharged. Large amount (2%)of water is lost through evaporation and in the vicinity of wet cooling tower extensive fog is formed which is not good for environment and causes damage to vegetation.

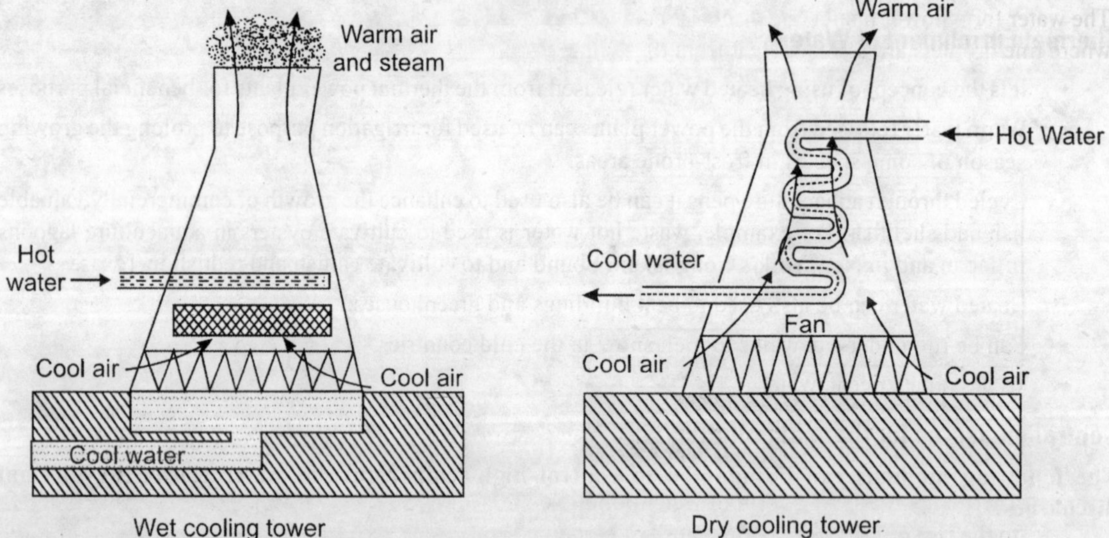

b. **Dry cooling tower:** The heated water flows in a system of spiral pipes. Air is passed over these hot pipes with fans. There is no water loss in this method and installation and operation of dry cooling tower is many times higher than wet cooling tower.

Cooling Ponds/ Artificial lakes

Outgoing hot water from power plant condensers is stored in cooling ponds or artificial lakes natural evaporation cools water which can be recirculated or discharged in nearby water body.

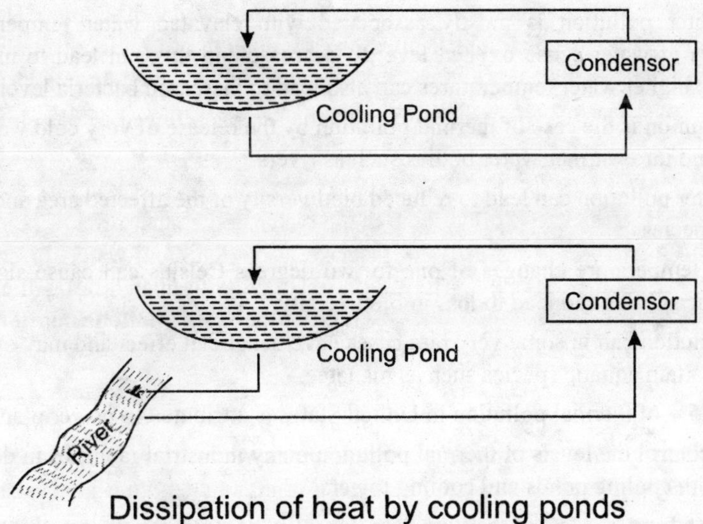

Dissipation of heat by cooling ponds

Spray Towers

The water form power plant condensers is received in spray ponds. Here the water is sprayed through nozzles where fine droplets are formed. Heat from these fine droplets is dissipated to the atmosphere.

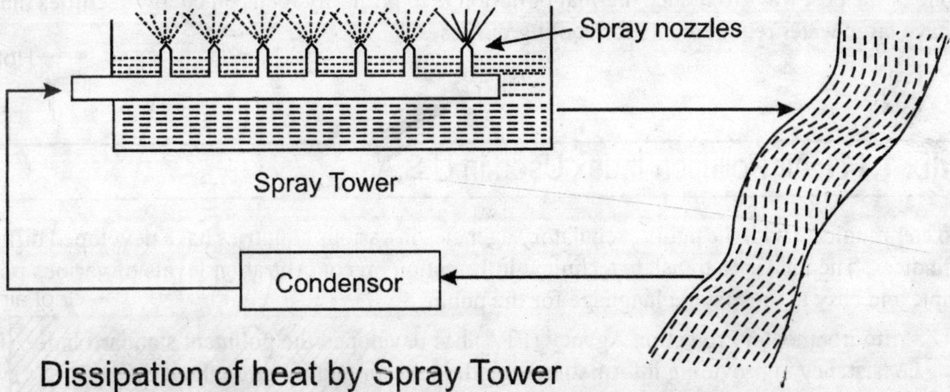

Dissipation of heat by Spray Tower

- Thermal pollution is a form of water pollution characterized by the degradation of water quality due to the rise or fall in the temperature of certain water body.
- Thermal pollution of water bodies is almost always caused by humans, and is mostly the result of use of water as a coolant in power plants and different industrial facilities,
- Unlike chemical pollution of water, thermal pollution primarily results in a change in the physical properties of water,
- Thermal water pollution can cause huge damage to many aquatic ecosystems because there are many fish, and other aquatic organisms that have adapted to particular temperature of the water and once this temperature changes they often fail to adapt and die in the process as a result of thermal shock.

- Thermal water pollution is mostly associated with elevated water temperatures. The higher temperatures of water cause oxygen level to decrease and this can lead to major decline of fish species. The higher water temperatures can also lead to increased bacteria levels.
- The less common is the case of thermal pollution by the release of very cold water from the base of reservoirs and into warmer water bodies such as rivers.
- Thermal water pollution can lead to reduced biodiversity of the affected area and cause migration of many fish species.
- Even small temperature changes of one to two degrees Celsius can cause significant changes to many aquatic species and lead to loss in biodiversity.
- Thermal pollution can in some very rare cases have beneficial effect and may even, lead to positive impact on certain aquatic species such as manatee.
- More than 75% of thermal pollution in United States is attributed to power plants.
- In order to control the levels of thermal pollution many industrial facilities in developed world' are equipped with cooling ponds and cooling towers'.
- Urban runoff from roads and parking lots can also become major source of thermal pollution, particularly during warm weather.
- Thermal pollution of water can also lead to increased levels of chemical pollution because toxicity of many harmful chemicals often increases with the increase in temperature.
- one of the best ways to reduce thermal pollution Is to primarily focus on energy facilities that do hot have large water requirements for cooling purposes.

Appendix 1.141: Air Pollution Index Used in U.S.A

In order to communicate with the public, regulatory agencies in various countries have developed different air pollution indices. The idea is to translate technical information on concentration levels of various pollutants into a simple and easy to understand language for the public.

The U.S. Environmental Protection Agency (EPA) has developed the pollutant standard index (PSI) for introducing consistency in providing information regarding the air quality throughout the U.S. The system is based on a scale of 0-500. The computed index below 100 indicates that the air quality is within acceptable range. A value over 100 implies potential health problems. The alerts are issued at 200, 300 and 400 levels. Five pollutants (carbon monoxide, sulfur dioxide, total suspended particulate, ozone and nitrogen dioxide) are included in the index. The reported index is based on the highest index value of any of the five pollutants.

The concentration associated with the index value is given in Table below and the health effects are discussed in the table following next.

U.S. Pollutant Standard Index

Pollutants Index Value	Air Quality Level	TSP (24 hr) 10^{-6} gm/m³	SO_2 (24 hr) 10^{-6} gm/m³	CO (8 hr) 10^{-3} gm/m³	O_3 (24 hr) 10^{-6} gm/m³	NO_2 (24 hr) 10^{-6} gm/m³
500	Significantly Harmful level	1000	2620	57.5	1200	3750
400	Emergency Level	875	2100	46	1000	3000
300	Warning Level	625	1600	34	800	2260
200	Alert Level	375	800	17	400c	1130
200	NAAQS	260	365	10	235	a
50	50 % of NAAQS	75[b]	80[c]	5	80	a
0	–	0	0	0	0	a

NAAQS – National Ambient Air Quality Standards, TSP–Total Suspended Particles

Associated Health Effects With U.S. Pollutant Standard Index

Index Value	Health Effect Descriptor	General Health Effects	Cautionary measures
400 - 500	Hazardous	Premature death of ill and elderly. Healthy people will experience adverse symptoms that affect their normal activity.	All persons should remain indoors, keeping windows and doors closed. All persons should minimize physical exertion and avoid automobile traffic
300 - 400	Hazardous	Premature onset of certain diseases in addition to significant aggravation of symptoms and decreased exercise tolerance in healthy persons	Elderly and persons with existing diseases should stay indoors and avoid physical exertion. General population should avoid outdoor activity.
200 - 300	Very Unhealthy	Significant aggravation of symptoms and decreased exercise tolerance in persons with heart or lung disease with widespread symptoms in the healthy population	Elderly and persons with existing heart or lung disease should stay indoors and reduce physical activity
100-200	Unhealthy	Mild aggravation of symptoms with susceptible persons, with irritation symptoms in the healthy population	Persons with existing heart or respiratory ailments should reduce physical exertion and outdoor activity
50-100	Moderate	–	–
0-50	Good	–	–

NOTES

a : No index values reported at concentration levels below those specified by " Alert Level" criteria.
b : Annual primary NAAQS
c : $400*10^{-6}$ gm/m³ was used instead of the O_3 Alert Level of $200*10^{-6}$ gm/m³

Appendix 1.142: Environmental Legislations in India

S.No.	Environmental Legislations
1.	The water (Prevention and Control of Pollution) Act, 1974 as amended upto 1988
2.	The water (Prevention and Control of Pollution) rules, 1975
3.	The Water (Prevention and Control of Pollution) cess act 1977 as amended upto 1991
4.	The Water (Prevention and Control of Pollution) cess rules 1978 as amended upto 1992
5.	The Air (Prevention and Control of Pollution) rules 1982 and 1983
6.	The Environment (Protection) rules, 1986
7.	The Environment (Protection) rules, 1986
8.	The Hazardous wastes (Management and handling) rules, 1989
9.	Manufacture, storage and import of hazardous chemicals rules, 1989
10.	Manufacture, use, import, export and storage of hazardous micro-organisms, genetically engineered micro-organiser or cells rules, 1989
11.	The Public liability insurance act, 1991
12.	The Public liability insurance rules, 1991
13.	Environmental (Protection) rules, 1992 and 1993 – "Environmental Statement"
14.	Environmental (Protection) rules 1994 – "Environmental Clearance"

Note:

In the Constitution of India it is stated that it is the duty of the state to 'protect Improve the environment and to safeguard the forests wildlife of the country'. It imposes a duty on every citizen 'to"protect and improve the natural environment including forests, lakes, rivers, and wildlife. The Department of Environment was established in India in 1980 to ensure a healthy environment for the country. This later became the Ministry of Environment and Forests in 1985.

The constitutional provisions are backed by a number of laws - acts, rules, and notifications. The EPA (Environment Protection Act), 1986 came into force soon after the Bhopal Gas Tragedy and is considered an umbrella legislation as it fills many gaps in the existing laws. Thereafter a large number of laws came into existence as the problems began arising, for example, Handling and Management of Hazardous Waste Rules in 1989.

Following is a list of the environmental legislations that have come into effect:

- **General**
- **Forest and wildlife**

- **Water**
- **Air**

General

1986: The Environment (Protection) Act authorizes the central government to protect and improve environmental quality, control and reduce pollution from all sources, and prohibit or restrict the setting and /or operation of any industrial facility on environmental grounds.

1986: The Environment (Protection) Rules lay down procedures for setting standards of emission or discharge of environmental pollutants.

1989: The objective of Hazardous Waste (Management and Handling) Rules is to control the generation, collection, treatment, import. storage. and handling of hazardous waste.

1989: The Manufacture, Storage, and Import of Hazardous Rules define the terms used in this context, and sets up an authority to inspect, once a year, the industrial activity connected with hazardous chemicals and isolated storage facilities,

1989: The Manufacture, Use, Import, Export, and Storage of hazardous Micro-organisms/ Genetically Engineered Organisms or Cells Rules were introduced with a view to protect the environment, nature, and health, in connection with the application of gene technology and microorganisms.

1991: The Public Liability Insurance Act and Rules and Amendment, 1992 was drawn up to provide for public liability insurance for the purpose of providing immediate relief to the persons affected by accident while handling any hazardous substance.

1995: The National Environmental Tribunal Act has been created to award compensation for damages to persons, property, and the environment arising from any activity involving hazardous substances.

1997: The National Environment Appellate Authority Act has been created to hear appeals with respect to restrictions of areas in which classes of industries etc. are carried out or prescribed subject to certain safeguards under the EPA.

1998: The Biomedical waste (Management and Handling) Rules is a legal binding on the health care institutions/hospitals/nursing home to streamline the process of proper handling of hospital waste such as segregation, disposal, collection, and treatment.

1999: The Environment (Siting for Industrial Projects) Rules, 1999 lay down detailed provisions relating to areas to be avoided for siting of industries, precautionary measures to be taken for site selecting as also the aspects of environmental protection which should have been incorporated during the implementation of the industrial development projects..

2000: The Municipal Solid Wastes (Management and Handling) Rules, 2000 apply to every municipal authority responsible for the collection, segregation, storage, transportation, processing, and disposal of municipal solid wastes.

2000: The Ozone Depleting Substances (Regulation and Control) Rules have been laid down for the regulation of production and consumption of ozone depleting substances.

2001: The Batteries (Management and Handling) Rules, 2001 rules shall apply to every manufacturer, importer, re-conditioner, assembler, dealer, auctioneer, consumer, and bulk consurger involved in the manufacture, processing, sale, purchase, and use of batteries or components so as to regulate and ensure the" environmentally safe disposal of used batteries,

2002: The Noise Pollution (Regulation and Control) **(Amendment)** Rules lay down such terms and conditions as are necessary to reduce noise pollution, permit use of loud speakers or public address systems during night hours (between 10:00 p.m. to 12:00 midnight) on or during any cultural or religious festive occasion •

2002: The Biological Diversity Act is an act to provide for the conservation of biological diversity, sustainable use of its components, and fair and equitable sharing of the benefits arising out of the use of biological resources and knowledge associated with it.

Forest and Wildlife Related Legislations

1927: The Indian Forest Act and Amendment, 1984, is one of the many surviving colonial statutes. It was enacted to 'consolidate the law related to forest, the transit of forest produce, and the duty leviable on timber and other forest produce'.

1972: The Wildlife Protection Act, Rules 1973 **and Amendment** 1991 provides for the protection of birds and animals and for all matters that are connected to it whether it **be** their habitat or the waterhole or the forests that sustain them.

1980: The Forest (Conservation) Act and Rules, 1981, provides for the protection of and the conservation of the forests.

Water Pollution Related Legislations

1882: The Easement Act allows private rights to use a resource that is, groundwater, by viewing it as an attachment to the land. It also states that all surface water belongs to the state and is a state property.

1897: The Indian Fisheries Act establishes two sets of penal offences whereby the government can sue any person who uses dynamite or other explosive substance in any way (whether coastal or inland) with intent to catch or destroy any fish or poisonous fish in order to kill.

1956: The River Boards Act enables the, states to enroll the central government in setting up an Advisory River Board to resolve issues in inter-state cooperation.

1970: The Merchant Shipping Act aims to deal with waste arising from ships along the coastal areas within a specified radius.

1974: The Water (Prevention and Control of Pollution) Act establishes an institutional structure for preventing and abating water pollution. It establishes standards for water quality and effluent. Polluting industries must seek permission to discharge waste into effluent bodies.
The CPCB (Central Pollution Control Board) was constituted under this act.

1977: The Water (Prevention and Control of **Pollution)** Cess **Act** provides for the levy and collection of cess or fees on water consuming industries and local authorities.

1978: The Water (Prevention and Control of Pollution) Cess Rules contains the standard definitions and. indicate the kind of and location of meters that every consumer of water is required to affix.

1991: The Coastal Regulation Zone Notification puts regulations on various activities(including construction) are regulated. It gives some protection to the backwaters and estuaries.

Air Pollution Related Legislations

1948: The Factories Act **and Amendment in** 1987 was the first to express concern for the working environment of the workers. The amendment of 1987 has sharpened its environmental focus and expanded its application to hazardous processes.

1981: The Air (Prevention and Control of Pollution) Act provides for the control and abatement of air pollution. It entrusts the power of enforcing this act to the CPCB .

1982: The Air (Prevention and Control of Pollution) Rules defines the procedures of the meetings of the Boards and the powers entrusted to them.

1982: The Atomic Energy Act deals with the radioactive waste.

1987: The Air (Prevention ana control or pollution) Amendment Act empowers the central and state pollution control boards to meet with grave emergencies of air pollution.

1988: The Motor Vehicles Act states, that all. hazardous waste is to be properly packaged, labelled, and transported.

The above laws have been sourced from;

- Environmental policy-making in India - The process and its pressure, TER! report.
- Indian Environmental Legislations, list from the MOEF web site.
- Strengthening Environmental Legislations in India, document by Centre for Environmental Law, WWF

Appendix 1.143: Key Functions of Various Environmental Pollution Control Agencies in India

S.No.	Name of Agency	Key Functions
1.	Ministry of environment and forest	• Environmental policy planning • Ensure effective implementation of legislation • Monitoring and control of pollution • Eco-development • Environmental clearances for industrial and development projects • Environmental research • Promotion of environmental education, training and awareness • Co-ordination with concerned agencies at the national and international level • Forest conservation, development and wildlife protection • Biosphere reserve programme

2.	Central Pollution Control Board (CPCB)	• Promotes cleanliness of streams and wells • Advise the central government on matters concerning prevention, control and abatement of water and air pollution • Co-ordinate and provide technical and research assistance to state boards Information dissemination, training and awareness • Lay down, modify or annual the standards for a stream or well and for air quality • Planning and execution of nationwide programme for the prevention, control or abatement of water and air pollution • Ensure compliance with the provisions of the Environment (Protection) act 1986
3.	State Pollution Control Board	• Planning and execution of state wide programmes for the prevention, control or abatement of water and air pollution • Advise the state government on prevention, control and abatement of water and air pollution and sitting of industries • Information dissemination, training and awareness • Ensure compliance with the provisions of the relevant acts • Lay down, modify or annul the effluent and emission standards • Ensure legal action against defaulters • Evolve techno-economic methods for treatment, disposal and utilisation of the effluent

NOTES

Govt. of India has addressed the concerns of environment protection by enacting enacted. "The Water (Prevention and contral of Pollution) Act, 1974; Air pollution (Prevention and Control of Pollution) Act, 1981 and the Environment (Protection) Act of 1986, with corresponding Rules and Regulations. Summary of provisions and the salient features of these there acts have been tabulated in the table

Table: Summary of Provisions in the Water (Prevention and Control (Act), Air (Prevention and Control Act) and Environment (Protection) Act

S.No.	Title Chapter/ section	The Water (Prevention and Control) Act, 1974	The Air (Prevention and Control) Act, 1981	The Environment (Protection) Act, 1986
1.	Enactment date	March 23, 1974 Has Eight Chapters and 64 Sections	March 29, 1981 Has Seven Chapter and 54 Sections	May 23, 1986 Has Four Chapters and 26 Sections

S.No.	Title Chapter/ section	The Water (Prevention and Control) Act, 1974	The Air (Prevention and Control) Act, 1981	The Environment (Protection) Act, 1986
2.	Chapter 1 Preliminaries - Title and Definitions	Section (1) Water (Prevention and Control) Act, 1974 It applies to the States of Assam, Bihar, Gujarat, Haryana, Himachal Pradesh, Jammu and Kashmir, Karnataka, Kerala, M.P., Rajasthan, Tripura, W. Bengal and in Union Territories. It shall apply to such other States which adopts this Act by resolution passed in this behalf under clause (1) of article 252 of constitution. The date on which the Act comes into force.	Section (1) Air (Prevention and Control) Act, 1981 It extends to the whole of India. Has come into force w.e.f. 16.05.1981 vide notification GSR 351 (E) Dt. 15.05.1981	Section (1) Environment (Protection) Act, 1986 It extends to the whole of India. Has come into force w.e.f. 19.11.1986 vide notification no. GSR 1198 (E) Dt. 12.11.1986
		Section (2) Definitions of Control Board, member (Chairman and others), occupier, pollution, prescribed, sewage effluent, State Boards, stream, inland water, subterranean water, sea, tidal waters, trade effluent, have been made.	Section (2) Definitions of Act (air pollution) approved appliance, approved fuel, automobile, Control Board Chimney, control equipment, emission, industrial plant, member, occupier, State Board,	Section (2) Definitions: environment, environment pollutant, environmental pollution, hazardous substance, occupier, prescribed are presented.

S.No.	Title Chapter/ section	The Water (Prevention and Control) Act, 1974	The Air (Prevention and Control) Act, 1981	The Environment (Protection) Act, 1986
3.	Chapter II The Central and State Boards for Prevention and Control of Water Pollution.	Section (3) The Constitution of Central and State Boards: Central Board a. full time Chairman b five officials nominated by Central Govt, to represent Govt. c. not exceeding five to be nominated by Central Govt, from among members of State Boards d. three non officials to represent the interests of industry, fishery and agriculture e. two persons to represent the companies or corporations, owned controlled or managed by Central Govt. f. full time member secretary (1+5 + 5 + 3+2+1 = 17)	Section (3) The Central Board Constituted under Water (Prevention and Control) Act. will perform functions of the Central Board for the Prevention and Control of Air Pollution under this Act. Same as for Water Act	section (31 The Central Govt. shall have the power to take all such measures for protecting the environment including measure to co-ordinate with State Govt., laying down standard, declaration of sensitive areas, laying down procedures and safeguards of accidents causing environmental pollution and hazardous substances and such related measures of inspection of premises, establishment of laboratories. May constitute an authority to exercise powers and functions of the EPA.
		Section (4) State Board a..full time Chairman b..five officials of State Govt, nominated by State Govt. c..five persons nominated by State Govt, from local bodies	Section (4 and 5) State Board State Board constituted under water (prevention and control) Ate, shall be deemed to be State Board for the prevention and control of air pollution under this functioning within the state.	Section (4) Central Govt, may constitute authority to exercise powers and functions Section (5) Central Govt, may issue directions to any person/officer/ authority in exercise of its power and performance of its functions.

S.No.	Title Chapter/ section	The Water (Prevention and Control) Act, 1974	The Air (Prevention and Control) Act, 1981	The Environment (Protection) Act, 1986
		a. three non officials to be nominated by State Govt, to represent the interests of agriculture, industry and fishery. b. two persons to represent company or corporation controlled, owned, managed by State Govt. c. a full time member secretarv 11 + 5 + 5 + 3 + 2 + 1 = 17) Section (5) Terms and conditions of service of members Section (6) Provisions of disqualifications Section (7) Vacation of seats by members Section (8) Meeting of Boards Section (9) Constitution of Committees Section (10) Temporary association of persons with boards for particular purpose	Act. Section (6) Central Board to exercise the powers and functions of State Boards in the union territories Section (7) Terms and conditions of service of members. Section (8) Vacation of seat by members Section (9) Meeting of Boards Section (10) Constitution of Committees. Section (11) Temporary association of persons with boards Section (12) Same as in Water Act. Section (13)	Section (6) Central Govt, may make rules in respect of all matters referred in Section (3) viz. standards of air, water or soil for various areas, maximum allowable limits of various pollutants (including noise) for different areas, procedures and safeguards of hazardous, substances, location of industries, accidents and remedial measures.
		Section (11) Vacancy in Board Section (12) Member Secretary and officers and other employees of Boards	Similar to Water Act.	
4.	Chapter III Joint Boards	Powers and Functions of Joint Boards Under Section (13, 14, 15), there are provisions of Joint Boards, with full time Chairman, members from participating States and a full time Member Secretary	Same as Water Act	Prevention, control and Abatement of Environmental Pollution. Section (7) No person carrying on any industry, operation, process shall discharge/emit any pollutant in excess of such standards.

S.No.	Title Chapter/ section	The Water (Prevention and Control) Act, 1974	The Air (Prevention and Control) Act, 1981	The Environment (Protection) Act, 1986
5.	Chapter IV	Powers and Functions of Boards Section (16) Central Board may • advise Central Govt, on any matter concerning prevention and control of water pollution • co-ordinate activities of State Boards • provide technical assistance to State Boards, Carryout/Sponsor research Section (17) State Board may • plan a comprehensive programme for the prevention of water pollution • advise State Govt, on any matter concerning prevention and control of water pollution • collect, disseminate information • encourage, conduct and participate in research Section (18) Alter any water pollution, prevention and control area and define a new water pollution area	Chapter III Powers and Functions of Boards Section (16) As in Water Act Section (16) for air pollution prevention and control. Section (17) Same as in Water Act but for air pollution prevention and control Section (18) Power to give directions	Section (8) No person shall handle any hazardous substance except in accordance with procedures and safeguards Section (9) In case of accidents or apprehension of accident, the person shall be bound to prevent and mitigate environmental pollution and intimate to authorities of such occurrence. Section (10) Empowered persons by Central Govt, will have the right of entry for examining and listing any equipments plant, record, document at all places and all times. Section (11) Central Govt, or an officer empowered shall have the power to take samples for analysis of air, water-soil or other substance from a factory promise.

S.No.	Title Chapter/ section	The Water (Prevention and Control) Act, 1974	The Air (Prevention and Control) Act, 1981	The Environment (Protection) Act, 1986
6.	Chapter V	Prevention and Control of Water Pollution (Sectons 19-33) Section (19) • Power of the State Govt, to restrict the application of Act to certain areas • Such areas may be defined by referring to a map or geographical boundaries • Alter any pollution area by extension or reduction or by defining a new area in merging with another area Section (20) State Board or an officer designated may make survey of any area and keep record of flow and characteristics Give direction on abstraction/ addition of waste from a stream/well Give direction to industry to furnish details about construction/operation or disposal system Section (21) Board or an officer designated to collect and analyze samples of sewage and trade effluent as per procedures prescribed. Section (22) Report of results of an analysis on samples as per details stipulated	Chapter IV Prevention and Control of Air Pollution (Sections 19-31) Section (19) Power State Govt, to declare air pollution areas Section (20) Power to give instructions to concerned authorities (under Motor Vehicles Act, 1939) for ensuring emissions from automobiles. Section (21) Restricting on the use of industrial units in an air pollution control area. Industries already operating before the commencement of section (9) of Air Act (may continue for 3 months and apply for consent, subsequently as per norms specified Section (22) No person operating an industrial plant in any pollution control area shall discharge emission in excess of standards laid down by State Boards Board may make application to court in restraining persons from causing air pollution.	Section (12) Central Govt, may establish laboratories or recognize laboratories/institutes for the purpose of environmental pollution analysis. Under Sections 13, 14, 15, 16, 17 Central Govt, may recognize persons as Govt, analysts and any document signed by them may be used as evidence in court. Whoever fails with or contravenes the provisions of this act shall be punishable with imprisonment for five years with fine which may extend to Rs. 1.0 lac or both. If the contravention continuous beyond a period of one year the offender, shall be punishable with imprisonment upto 7 years. If the offence is proved to be committed by a Govt. Deptt. Head of the Deptt. shall be deemed to be guilty and liable to be punished accordingly.

S.No.	Title Chapter/ section	The Water (Prevention and Control) Act, 1974	The Air (Prevention and Control) Act, 1981	The Environment (Protection) Act, 1986
		Section (23) An empowered person by State Board to enter for inspection any premise Section (24) Prohibition on use of stream or well for disposal of polluting matter (poisonous noxious or polluting matter) Restrict persons in constructing a water regulating structure, adding/ depositing material on bank Section (25) Restraining persons on constructing new outlets and new discharges Section (26) Restraining a person from altering old drains as per clause (25) Sections 27, 28, 29, 30 and 31 Define restraints on sampling, appeals and revisions. Section (32) Defines measures for discharge of polluting matter by accident. Section (33) Power of the Board to make application to court for restraining pollution in streams and wells.	Section (23) Furnishing information to State Board of apprehension due to accident in exceeding limits of air pollution standards. Section *(24)* Provision of entry and inspection concerning functions, testing and control equipment/ documents and any other offence and any other information (Section 25) Section *(26)* State Board or an officer designated can take samples of air, emission (Chimney, flue or duct or any other outlet) as per procedures outlined. Section 27, 28, 29, 30, 31 Samples so collected will be analyzed in State Laboratory or a laboratory recognized by Analysts of State Board or recognized by State Govt, submit report and accept appeals concerning sampling and analysis	

S.No.	Title Chapter/ section	The Water (Prevention and Control) Act, 1974	The Air (Prevention and Control) Act, 1981	The Environment (Protection) Act, 1986
7.	Chapter VI	Funds Accounts and Audit Sections (34-401 Funds, Accounts and Audit Contributions to Central Board, State Boards by Central, State Govts, will be made. Accounts will be maintained as per budget, and annual accounts will be audited.	Will be maintained as under Water Act (Sections 32-36)	
8.	Chapter VII	Penalties and Procedures (Sections 41-50) Whosoever fails to comply with orders under the Act shall on conviction punishable with imprisonment upto 3 months or fine upto Rs. 5000/- or both. For continued contraventions additional fine of Rs. 1000/- per day may be imposed. Penalty for actions are glso outlined. Conviction under section 25, 26 may be upto six years with fine but not less than six months. Provisions for enhanced penalty has been made in Section 45-47.	Penalties and Procedures are outlined under (Sections 37-46) Whosoever fails to comply with provisions of section 21-22 or directions under 31 shall on conviction be punishable with imprisonment not less than one year and six months but may extend to six years with fine if failure continues penalties for certain acts and contravention of provisions of the act have been made under sections 38, 39. Offences by companies and Govt, departments have been mentioned under section 40, 41. Protection of Action in good faith	

S.No.	Title Chapter/ section	The Water (Prevention and Control) Act, 1974	The Air (Prevention and Control) Act, 1981	The Environment (Protection) Act, 1986
			(Section 42), cognizance of offences (Section 43), members officers and employees of Board to be public servants, reports and return and Bar of Jurisdiction are also provided.	
9.	Chapter VIII	Miscellaneous (Sections 51-60) Provisions under section 51 exist for the establishment of a Central Laboratory and under 52 a State Water Laboratory with functioning rules and other provisions. Section 53 deals with the appointment of persons as Govt, analysts, other provisions include about local authorities to assist (55), compulsory acquisition of land for the State Board (56), Returns and Reports (57), Bar of Jurisdiction (58), Protection of action taken in good faith (59), overriding effect (60), Powers to supercede Central and State Boards by Central and State Boards and make rules for the functioning (61-63).	Powers of State Govt, to supersede and special provisions of super session, dissolution of State Board Powers to amend the schedule and other powers of Central State Govts, are provided under Sections 47-54.	

Appendix 1.144: Schedule Giving List of Projects Requiring Prior Environmental Clearance Under EIA (Environmental Impact Assessment) Notification, 2006 (India)

S.N.	Project or Activity	Category with threshold limit		Conditions if any
		A	B	
1.	2.	3.	4.	5.
1.	**Mining, Extraction of Natural Resources, and Power Generation (for a specified production capacity)**			
1(a)	Mining of minerals	≥ 50 ha. of mining lease area Asbestos mining irrespective of mining area	< 50 ha and ≥ 5 ha. of mining lease area.	General Condition shall apply **Note** Mineral prospecting (not involving drilling) are exempted, provided the concession areas have got previous clearance for physical survey.
1(b)	Offshore and onshore oil and gas exploration, development and production	All projects	–	**Note** Exploration Surveys (not involving drilling) are exempted, provided the concession areas have got previous clearance for physical survey.
1(c)	River Valley projects	i. ≥ 50 MW hydroelectric power generation; (ii) ≥ 10, 000 ha. of culturable command area.	i. < 50 MW and ≥ 25 MW hydroelectric power generation; (ii) < 10, 000 ha. of culturable command area.	General Condition shall apply
1(d)	Thermal Power Plants	≥ 500 MW (coal/lignite/naptha and gas based); ≥ 50 MW (Peat coke, diesel and all other fuels)	< 500 MW (coal/lignite/ naptha and gas based); < 50 MW and ≥ 5 MW (Peat coke, diesel and all other fuels)	General Condition shall apply.
1(e)	Nuclear power projects and processing of nuclear fuel	All projects	–	–
2.	**Primary Processing**			
2(a)	Coal washeries	≥ 1 million ton/annum throughput of coal	< 1 million ton/annum throughput of coal	General Condition shall apply (If located within mining area, the proposal shall be appraised together with the mining proposal)
2(b)	Mineral beneficiation	≥ 0.1 million ton/annum mineral throughput	< 0.1 million ton/annum mineral throughput	General Condition shall apply (Mining proposal with Mineral beneficiation shall be appraised together for grant of clearance)

S.N.	Project or Activity	Category with threshold limit		Conditions if any
		A	B	
1.	2.	3.	4.	5.
3. Materials Production				
3(a)	Metallurgical industries (ferrous and non ferrous)	a. *Primary metallurgical industry* All projects	–	– General Condition shall apply for Sponge manufacturing
		b. *Sponge iron manufacturing* ≥ 200 TPD	Sponge iron manufacturing < 200 TPD	
		c. *Secondary metallurgical processing* industry All toxic and heavy metal producing units ≥ 20, 000 tonnes/ annum	Secondary metallurgical processing industry (i) All toxic and heavy metal producing units < 20, 000 tonnes/annum (ii) All other nontoxic secondary metallurgical processing industries > 5000 tonnes/annum	
3(b)	Cement plants	≥ 1.0 million tonnes/annum of production capacity	<1.0 million tonnes/ annum of production capacity. All Stand alone grinding units	General Condition shall apply
4. Materials Processing				
4(a)	Petroleum industry	All projects	–	–
4(b)	Coke oven plants	≥ 2, 50, 000 tonnes/annum	< 2, 50, 000 and ≥ 25, 000 tonnes/annum	
4(c)	Asbestos milling and asbestos based products	All projects	–	–
4(d)	Chlor-alkali industry	≥ 300 TPD production capacity ; or a unit located outside the notified industrial area/ estate	< 300 TPD production capacity ; or a unit located within a notified industrial area/estate	Specific Condition shall apply No new Mercury Cell based plants will be permitted and existing units converting to membrane cell technology are exempted from this Notification
4(e)	Soda ash Industry	All projects	–	–
4 (f)	Leather/skin/ hide processing industry	New projects outside the industrial area or expansion of existing units outside the industrial area	All new or expansion of projects located within a notified industrial area/ estate	Specific condition shall apply
5 Manufacturing/Fabrication				
5(a)	Chemical fertilizers	All projects	–	–
5(b)	Pesticides industry and pesticide specific intermediates (excluding formulations)	All units producing technical grade pesticides	–	–

S.N.	Project or Activity	Category with threshold limit		Conditions if any
		A	B	
1.	2.	3.	4.	5.
5(c)	Petro-chemical complexes (industries based on processing of petroleum fractions and natural gas and/ or reforming to aromatics)	All projects	–	–
5(d)	Man-made fibres manufacturing	Rayon	Others	General Condition shall apply
5(e)	Petrochemical based processing (processes other than cracking and reformation and not covered under the complexes)	Located outside the notified industrial area/ estate	Located in a notified industrial area/estate	Specific Condition shall apply
5(f)	Synthetic organic chemicals industry (dyes and dye intermediates; bulk drugs and intermediates excluding drug formulations; synthetic rubbers; basic organic chemicals, other synthetic organic chemicals and chemical intermediates)	Located outside the notified industrial area/estate	Located in a notified industrial area/estate	Specific Condition shall apply
5(g)	Distilleries	i. All Molasses based distilleries	–	General Condition shall apply
		(ii) All Cane juice/non-molasses based distilleries ≥ 30 KLD	All Cane juice/ non-molasses based distilleries < 30 KLD	
5(h)	Integrated paint industry	–	All projects	General Condition shall apply
5(i)	Pulp and paper industry, excluding manufacturing of paper from waste paper and manufacture of paper from ready pulp with out bleaching	Pulp manufacturing and Pulp and Paper manufacturing industry	Paper manufacturing industry without pulp manufacturing	General Condition shall apply
5(j)	Sugar Industry	–	≥ 5000 tcd cane crushing capacity	General Condition shall apply
5(k)	Induction/arc furnaces/ cupola furnaces 5 TPH or more	–	All projects	General Condition shall apply

S.N.	Project or Activity	Category with threshold limit		Conditions if any
		A	B	
1.	2.	3.	4.	5.
6 Service Sectors				
6(a)	Oil and gas transportation pipeline (crude and refinery/ petrochemical products), passing through national parks/ sanctuaries/ coral reefs/ ecologically sensitive areas including LNG Terminal.	All projects	–	–
6(b)	Isolated storage and handling of hazardous chemicals (As per threshold planning quantity indicated in column 3 of schedule 2 and 3 of MSIHC Rules 1989 amended 2000)	–	All projects	General Condition shall apply
7	Physical Infrastructure including Environmental Services.			
7(a)	Air ports	All projects	–	–
7(5)	All ship breaking yards including ship breaking units	All projects	–	–
7(c)	Industrial estates/parks/ complexes/areas, export processing Zones (EPZs), Special Economic Zones (SEZs), Biotech Parks, Leather Complexes	If at least one industry in the proposed inust-rial estate falls under the Category A, entire industrial area shall be treated as Category A, irrespecitve of . the area.	Industrial estates housing at least one Category B industry and area < 500 ha.	Special condition shall apply Note: Industrial Estate of area below 500 ha and not housing any industry of category A or B does not require clearance.
		Industrial estates with area greater than 500 ha. and housing at least one Category B Industry.	Industrial estates of area > 500 ha. and not housing any industry belonging to Category A or B.	
7(d)	Common hazardous waste treatment, storage and disposal facilities (TSDFs)	All integrated facilities having incineration and landfill or incineration alone	All facilities having landfill only	General Condition shall apply
7(e)	Ports, Harbours	≥ 5 million TPA of cargo handling capacity (excluding fishing harbours)	< 5 million TPA of cargo handling capacity and/ or ports/ harbours ≥ 10. 000 TPA of fish handling capacity	General Condition shall apply
7(f)	Highways	i. New National High-ways; and (ii) Expansion of National Highways greater than 30 km, involving additional right of way greater than 20 m involving land acquisition and passing through more than one State.	i. New State High-ways; and (ii) Expansion of National/State Highways greater than 30 km involving additional right of way greater than 20 m involving land acquisition.	General Condition shall apply

S.N.	Project or Activity	Category with threshold limit		Conditions if any
		A	B	
1.	2.	3.	4.	5.
7(g)	Aerial ropeways	–	All projects	General Condition shall apply
7(h)	Common Effluent Treatment Plants (CETPs)	–	All projects	General Condition shall apply
7(i)	Common Municipal Solid Waste Management Facility (CMSWMF)	–	All projects	General Condition shall apply
8. Building/ Construction projects/Area Development Projects and Townships.				
8(a)	Building and Construction projects	–	≥20, 000 sq. mtrs. of land area; and ≥ 1, 50, 000 sqm of built up area	# (built up area for covered construction; in the case of facilities. open to the sky, it will be the activity area).
8(b)	Townships and Area Development projects.	–	Covering an area ≥ 50 ha and or built up area ≥ 1, 50, 000 sq. mtrs.	++All projects under Item 8(b) shall be appraised as Category B1.

Note:

I. General Condition (GC)

Any project or activity specified in Category 'B' will be treated as Category A, if located in whole or in part within 10 km from the boundary of; (i) Protected Areas notified under the Wild Life (Protection) Act, 1972; (ii) Critically Polluted areas as notified by the Central Pollution Control Board from time to time; (iii) Notified Eco-sensitive area; (iv) inter-State boundaries and international boundaries.

Note:

II. Special Condition (SC)

If any Industrial Estate/Complex/Export processing Zones/Special Economic Zones/Biotech Parks/Leather Complex with homogeneous type of industries such as Items 4(d), 4(/), 5(e), 5(f), or those Industrial estates with pre-defined set of activities (not necessarily homogeneous, obtains prior environmental clearance, individual industries including proposed industrial housing within such estates/complexes will not be required to take prior environmental clearance, so long as the Terms and Conditions for the industrial estate/ complex are complied with (Such estates/complexes must have a clearly identified management with the legal responsibility of ensuring adherence to the Terms and Conditions of prior environmental clearance, who may be held responsible for violation of the same throughout the life of the complex/ estate).

Appendix 1.145: Environmental Legislation Enacted in India in Chronological Order

Year	Environmental legislation introduced
1974	The Water (Prevention and Control of Pollution) Act
1975	The Water (Prevention and Control of Pollution) Rules

1977	The Water (Prevention and Control of Pollution) Cess Act
1978	The Water (Prevention and Control of Pollution) Cess Rules
1981	The Air (Prevention and Control of Pollution) Act
1982/1983	The Air (Prevention and Control of Pollution) Rules
1986	The Environment (Protection) Act
1986	The Environment (Protection) Rules
1989	The Hazardous Waste (Management and Handling) Rules
1992/1993	Environmental (Protection) Rules-"Environmental Statement"
1993	Environmental (Protection) Rules-"Environmental Statement"
1994	Environmental (Protection) Rules-"Environmental Statement"
1995	The National Environment Tribunal Act

Appendix 1.146: ISO – 14000 (Environment Management System)

ISO 14000 refers to a family of voluntary standards developed by the International Organization for Standardization (ISO). The standards provide a framework for a strategic approach to an organization's environmental policies, plans and actions Using the framework, a company develops an environmental management system (EMS). The EMS is evaluated by a "certification body" to determine whether the EMS conforms to ISO 14000. If so, the organization is said to have "ISO 14000 certification."

ISO 14000 is quite different from most other ISO standards. It is known as a generic management system standard" Generic" means that the same standards can be applied:

- to any organization, large or small, whatever its product.
- Including whether its "product" is actually a service,
- In any sector of activity, and
- Whether it is a business enterprise, a public administration, or a government Department.

Management system refers to what the organization does to manage its processes or activities. ISO 14000 is concerned with the way an organization goes about its work, and not directly with the results of this work. The focus is on processes, not products.

ISO 14000 grew out of ISO's commitment to support sustainable development as discussed at the United Nations Conference on Environment and Development in Rio de Janeiro in 1992. Conversations among 20 countries, 11 international organizations, and more than 100 environmental experts actually began in 1991 to define the basic requirements of a new approach to environment-related standards. The first standards ISO 14000 and ISO 14001 – were published in September and October 1996, respectively. The "ISO 14000 family" consists of standards and guidelines relating to environmental management systems and supporting standards on terminology and specific tools, such as auditing. Essentially, the standards are concerned with what an organization does to minimize harmful effects on the environment caused by its activities: either during production or disposal, either by pollution or by depleting natural resources.

It is a management tool enabling an organization of any size or type to control the impact of its activities, products or services on the environment. It defines a structured approach for 1) setting environmental objectives and targets. 2) achieving these goals, and 3) demonstrating that they have been achieved. Because

ISO 14000 does not set levels of environmental performance, the standards can be implemented by a wide variety of organizations, whatever their current level of environmental maturity. However a commitment to compliance with applicable environmental legislation and regulations is required, along with a commitment to continuous improvement.

The Objectives of ISO 14000

ISO 14000 is likely to have the following effects:

- Lead to the harmonization of national rules, labels, and methods;
- Minimize trade barriers and related complications;
- Promote predictability and consistency;
- Lead to improvement of environmental performance;
- Lead to effective maintenance of regulatory compliance;
- Assist companies/organizations to demonstrate commitment and enhance public pasture
- Lend credibility to performance reporting;
- Support a worldwide focus on environment management;
- Sensitize the internal culture in organizations to environmental matters;
- Promote a voluntary consensus standards approach to environmental improvement

ISO has developed a three-pronged approach to meeting the needs of business, industry, governments and consumers in the field of the environment. If offers a wide-ranging portfolio of standardized sampling, testing and analytical methods to deal with specific environmental challenges. It has developed more than 350 International Standards (out of a total of some 14000) for the monitoring of such aspects as the quality of air, water and soil. These standards are a means of providing business and government with scientifically valid data on the environmental effects of economic activity. They also serve in a number of countries as the technical basis for environmental regulations.

The steps to ISO 14000 certification

1. Understand ISO 140001's need for controls, procedures and records.
2. Determine environmental impacts and laws relevant to organization.
3. Identify existing procedures to use in ISO, 140001 and new ones needed.
4. Write a Policy, Environmental Manual Management Procedures and Records.
5. Measure and record your environmental performance against objectives.
6. Audit, correct and improve your environmental system and procedures.
7. Undergo independent assessment and registration to ISO 14001.

ISO does not carry out assessments to check that standards are being implemented by users in conformity with the requirements of ISO standard. Rather, private sector suppliers or regulatory bodies that have been approved by a national accreditation body will carry out certifications.

Benefits of ISO 14000 certification

The ISO 14000 approach forces managers to take a critical look at all areas where their businesses have an environmental impact. This systematic approach can lead to following benefits: -

- Reduced cost of waste management.
- Savings in consumption of energy and materials.
- Lower distribution costs,
- Improved corporate image among regulators, customers and the public, and a
- Framework for continuous improvement of environmental performance.

Environmental Management System Requirements (sectionwise)

General requirements

The organization shall establish, document, implement, maintain and continually improve an environmental management system in accordance with the requirements of this international standard and determine how it will fulfill these requirements.

The organization shall define and document the scope of its environmental management system.

Environmental Policy

Top management shall define the organization's environmental policy and ensure that, within the defined scope of its environmental management system, it

a. Is appropriate to the nature, scale and environmental impacts of its activities, products and services.
b. Includes a commitment to continual improvement and prevention of pollution.
c. Includes a commitment to comply with applicable legal requirements and with other requirements to which the organization subscribes which relate to its environmental aspects.
d. Provides the framework for setting and reviewing environmental objectives and targets.
e. Is documented, implemented and maintained.
f. Is communicated to all persons working for or on behalf of the organization, and
g. Is available to the public

Planning

Environmental Aspects

The organization shall establish, implement and maintain a procedure(s)

a. To identify the environmental aspects of its activities, products and services within the defined scope of the environmental management system that it can control and those that it can influence taking into account planned or new developments, or new or modified activities, products and services, and
b. To determine those aspects that have or can have significant impact(s) on the environment (i.e. significant environmental aspects.)

The organization shall document this information and keep it up to date.

The organization shall ensure that the significant environmental aspects are taken into account in establishing, implementing and maintaining its environmental management system.

Legal and Other Requirements

The organization shall establish, implement and maintain a procedure(s) ensure that within the detained scope

a. To identify and have access to the applicable legal requirements and other requirements to which the organization subscribes related to its environmental aspects, and

b. To determine how these requirements apply to its environmental aspects.

The organization shall ensure that these applicable legal requirements and other requirements to which the organization subscribes are taken into account in establishing, implementing and maintaining its environmental management system.

Objectives, Targets and Programmes

The organization shall establish, implement and maintain documented environmental objectives and targets, at relevant functions and levels within the organization.

The objectives and targets shall be measurable, where practicable, and consistent with the environmental policy, including the commitments to prevention of pollution, to compliance with applicable legal requirements and with other requirements to which the organizations subscribes, and to continual improvement.

When establishing and reviewing its objectives and targets, an organization shall take into account the legal requirements and other requirements to which the organization subscribes, and its significant environmental aspects. It shall also consider its technological options, its financial, operational and business requirements, and the views of interested parties.

The organization shall establish, implement and maintain a programmed for achieving its objectives and targets. Programmed shall include.

a. Designation of responsibility for achieving objectives and targets at relevant functions and levels of the organization, and

b. The means and time-frame by which they are to be achieved.

Implementation and Operation

Resources, Roles, Responsibility and Authority

Management shall, ensure the availability of resources essential to establish, implement, maintain and improve the environmental management system. Resources include human resources and specialized skills. Organizational infrastructure, technology and financial resources.

Roles, responsibilities and authorities shall be defined, documented and communicated in order to facilitate effective environmental management.

The organization's top management shall appoint a specific management representative(s) who, irrespective of other responsibilities, shall have defined roles, responsibilities and authority for

a. Ensuring that an environmental management system is established, implemented and maintained in accordance with the requirements of this international Standard.

b. Reporting to top management on the performance of the environmental management system, for review, including recommendations for improvement.

Competence, Training and Awareness

The organization shall ensure that any person(s) performing tasks for it or on its behalf that have the potential to cause a significant environmental impact(s) identified by the organization is (are) competent on the basis of appropriate education, training or experience, and shall retain associated records.

The organization shall identify training needs associated with its environmental aspects and its environmental management system. It shall provide training or take other action to meet these needs, and shall retain associated records.

The organization shall establish, implement and maintain a procedure(s) to make persons working for it or on its behalf aware of

a. The importance of conformity with the environmental policy and procedures and with the requirements of the environmental management system.

b. The significant environmental aspects and related actual or potential impacts associated with their work, and

c. The environmental benefits of improved personal performance.

their roles and responsibilities in achieving conformity with the requirements of the environmental management system, and

a. the potential consequences of departure from specified procedures.

Communication

With regard to its environmental aspects and environmental management system, the organization shall establish, implement and maintain a procedures(s) for

a. internal communication among the various levels and functions of the organization,

b. receiving, documenting and responding to relevant communication from external interested parties.

The organization shall decide whether to communicate externally about its significant environmental aspects and shall document its decision. If the decision is to communicate, the organization shall establish and implement a method(S) for this external communication.

Documentation

The environmental management system documentation shall include:

a. the environmental policy, objectives and targets.

b. Description of the scope of the environmental management system.

c. Description of the main elements of the environmental management system and their interaction and reference to related documents.

d. Documents, including records, required by this international Standard and

e. Documents, including record, determined by the organization t.o- be necessary to ensure the effective planning, operation and control of processes that relate to its significant environmental aspects.

Control of documents

Documents required by the environmental management system and by this International Standard shall be controlled. Records are a special type of document and shall be controlled in accordance with the requirements given in 4.5.4.

The organization shall establish, implement and maintain a procedure(s) to

a. Approve documents for adequacy prior to issue,

b. Review and update as necessary and re-approve documents,

c. Ensure that changes and the current revision status of documents are identified,

d. Ensure that relevant versions of applicable documents are available at point of use,

e. Ensure that documents remain legible and readily identifiable,

f. Ensure that documents of external origin determined by the organization to be necessary for the planning and operation of the environmental management system are identified and their distribution controlled, and

g. Prevent the unintended use of obsolete documents and apply suitable identification to them if they are retained for any purpose.

Operational Control

The organization shall identify and plan those operations that are associated with the identified significant environmental aspects consistent with its environmental policy, objectives and targets, in order to ensure that they are carried out under specified conditions, by

a. Establishing, implementing and maintaining a documented procedure(s) to control situations where their absence could lead to deviation from the environmental policy, objectives and targets, and

b. Stipulating the operating criteria in the procedure(s) and

c. Establishing, implementing and maintaining procedures related to the identified significant environmental aspects of goods and services used by the organization and communicating applicable procedures and requirements to suppliers, including contractors.

Emergency Preparedness and Response

The organization shall establish, implement and maintain a procedure(s) to identify potential emergency situations and potential accidents that can have an impact(s) on the environmental and how it will respond to them.

The organization shall respond to actual emergency situations and accidents and prevent or miti@e associated adverse environmental impacts.

The organization shall periodically review and, where necessary, revise its emergency preparedness and response procedures, in particular, after the occurrence of accidents or emergency situations.

The organization shall also periodically test such procedures where practicable.

Checking

Monitoring and Measurement

The organization shall establish, implement and maintain a procedure(s) to monitor and measure, on a regular basis, the key characteristics of its operations that can have a significant environmental impact. The procedure(s) shall include the documenting of information to monitor performance, applicable operational controls and conformity with the organization's environmental objectives awd targets.

The organization shall ensure that calibrated or verified monitoring and measurement equipment is used and maintained and shall retain associated records.

maintained and shall retain associated records.

Evaluation of compliance

Consistent with its commitment to compliance, the organization shall establish, implement and maintain a procedure(s) for periodically evaluating compliance with applicable legal requirements.

The organization shall keep records of the results of the periodic, evaluations,

The organization shall evaluate compliance with other requirements to which it subscribes. The organization may wish to combine this evaluation with the evaluation of legal compliance referred to in or to establish a separate procedure(s)

The organization shall keep records of the results of the periodic evaluations.

Nonconformity, Corrective Action snd Preventive Action

The organization shall establish, implement and maintain a procedure(s) for dealing with actual and potential nonconformity (ies) and for taking corrective action and preventive action. The procedure(s) shall define requirements for

a. Identifying and correcting nonconformity (ies) and taking action(s) to mitigate their environmental impacts,
b. Investigating nonconformity (ies), determining their cause(s) and taking actions in order to avoid their recurrence.
c. Evaluating the need for action(s) to prevent nonconformity (ies) and implementing appropriate actions designed to avoid their occurrence.
d. Recording the results of corrective action(s) and preventive action(s) taken and
e. Reviewing the effectiveness of corrective action(s) and preventive action(s) taken

Actions taken shall be appropriate to the magnitude of the problems and the environmental impacts encountered.

The organization shall ensure that any necessary changes are made to environmental management system documentation.

Control of Records

The organization shall establish and maintain records as necessary to demonstrate conformity to the requirements of its environmental management system and of this international Standard, and the results achieved.

The organization shall establish, implement and maintain a procedure(s) for the identification, storage, protection retrieval retention and disposal of records.

Records shall be and remain legible, identifiable and traceable.

Internal Audit

The organization shall ensure that internal audits of the environmental management system are conducted at planned intervals to

a. Determine whether the environmental management system
   1. Conforms to planned arrangements for environmental management including the requirements of this International Standard, and
   2. Has been properly implemented and is maintained, and
b. Provide information on the results of audits to management.

   Audit programmed shall be planned, established implemented and maintained that address
   - The responsibilities and requirements for planning and conducting audits, reporting results and retaining associated records.
   - The determination of audit criteria, scope, frequency and methods.

Selection of auditors and conduct of audits shall ensure objectivity and the impartiality of the audit process.

Management Review

Top management shall review the organization's environmental management system, at planned intervals, to ensure its continuing suitability, adequacy and effectiveness. Reviews shall include assessing opportunities for improvements and the need for changes to the environmental management system, including the environmental policy and environmental objectives and targets. Records of the management reviews shall be retained, input to management reviews shall include:

a. Results of internal audits and evaluations of compliance with legal requirements and with other requirements to which the organization subscribes,
b. Communication(s) from external interested parties, including complaints,
c. The environmental performance of the organization,
d. The extent to which objectives and targets have been met,
e. Status of corrective and preventive actions,
f. Follow-up actions from previous management reviews,
g. Changing circumstances, including developments in legal and other requirements related to its environmental aspects, and
h. Recommendations for improvement.

The outputs from management reviews shall include any decisions and actions related to possible changes to environmental policy, objectives, targets and other elements of the environments management system,

Consistent with the commitment to continual improvement.

Summary of ISO 14000 Series Various Standards

ISO 14001 (2002): Environmental Management Systems: Specification with Guidance for use – This sets out the requirements for establishing and maintaining an environmental management system that assures a company of conformance with the environmental objectives that it sets for itself and also with any relevant legislation.

If desired, this can be audited and certified by national accrediting bodies.

ISO 14004 (1996): Environmental Management Systems: General Guidelines on Principles, systems and Supporting Techniques – This gives general guidance and practical advice on implementing or improving an EMS, and is supported by a number of examples.

ISO 14010 (1996): Guidelines for Environmental Auditing: General Principles – This sets out the general principles governing the audit process of an EMS, and sets out the necessary basic conditions that have to be satisfied before an audit can be carried out, such as the availability of adequate documentation, the provision of sufficient resources to support the audit, and the willing cooperation and assistance of the company whose EMS is being audited. Guidance on the structure of the report giving the audit findings is also provided. This standard is soon to be replaced by ISO 19011.

ISO 14011 (1996): Guidelines for Environmental Auditing: Audit Procedures: Auditing of Environmental Management Systems – This explains the usual objectives of an EMS audit to confirm that the EMS has been properly implemented and is meeting its objectives, is managed satisfactorily and is reviewed regularly to try to achieve continual improvement in its performance. The duties of lead and other auditors are set down, and also those of the company whose EMS is being audited. A summary of the necessary steps in executing the audit is also presented, and recommendations are made about the structure, content and distribution of the audit report. This standard is soon to be replaced by ISO 19011.

ISO 14012 (1996): Guidelines for Environmental Auditing: Qualification Criteria for Environmental auditors – This standard sets out the educational qualifications required, personal skills needed and the level of training and experience necessary for auditors who are appointed by Standards Bodies to carry out audits and certify an EMS. This standard is soon to be replaced by ISO 19011.

ISO 14015 (2001): Environmental Management Systems: Environmental Assessment of Sites and Organisations – This standard gives guidance on the conduct of an environmental assessment of a site or organisation. The usual circumstances in which such an assessment is carried out is when a company is considering purchasing another company, to ensure that the purchase would not involve it in large costs in rectifying nonconformance of the site with good environmental management principles, or with present or likely future environmental legislation. Like ISO 14001, it is applicable to a wide range of companies of all types, in all industries and of all sizes. The assessment will cover all present and proposed future operations, but will often have to consider past operations that may have caused ongoing environmental damage, such as contaminated land.

ISO 14020 (2001): Environmental Labels and Declarations: General Principles – This gives guidelines on the use of labels to indicate the environmental impact of products.

It requires that labels should be accurate and verifiable. Such labeling is provided for the benefit of environmentally conscious customers, who can decide whether or not to purchase a product, according to its environmental performance.

ISO 14021 (2001): Environmental Labels and Declarations: Self-declared Environmental Claims (Type II Environmental Labelling) – This sets common standard for the use of particular environmental labels and symbols. It specifies the requirements that must be satisfied for the application of various labels such as 'degradable' and 'recyclable'

ISO 14024 (2001): Environmental Labels and Declarations: Type I Environmental Labels: Principles and Procedures – This extends ISO 14020 by providing a framework for issuing licenses to permit the use of particular environmental labels on products, and to assess the continued compliance of the products at regular intervals.

ISO 14031 (2000): Environmental Management: Environmental Performance Evaluation Guidelines – This is provided as a management tool to help companies to assess whether its environmental performance is meeting the targets set. The document is particularly useful to companies that have not implemented a formal EMS, since it includes guidance on identifying the environmental impact of operations and setting targets for reduction in environmental damage.

ISO 14032 (2000): Environmental Management: Examples of Environmental Performance Evaluation – The document provides 17 case studies across a wide range of industries, showing how ISO 14031 has been implemented.

ISO 14040 (1997): Environmental Management: Life Cycle Assessment: Principles and Framework – This standard defines a suitable framework for conducting life-cycle assessments for products in terms of their environmental impact. Recommendations are made about setting environmental targets and defining relevant indicators of environmental performance for all stages in a product's life cycle, including raw material acquisition, product manufacture, product use and product disposal, at the end of its useful life. ISO 14040 defines the four phases in life-cycle assessment as defining goals, inventory analysis, impact assessment and interpretation. Further guidance on these four phases are provided in ISO 14041 – ISO 14043.

ISO 14041 (1998): Environmental Management: Life Cycle Assessment: Goal and Scope Definition and Inventory Analysis – This extends ISO 14040 by providing further guidance on defining the goals of life-cycle assessment and collecting data to assess environmental performance against the indicators set.

ISO 14042 (2000): Environmental Management: Life Cycle Assessment: Life Cycle Impact Assessment – This provides guidance on how the performance indicator data defined according to ISO 14041 should be used to assess the environmental impact of a product and identify opportunities to improve the product to reduce its environmental impact.

ISO 14043 (2000): Environmental Management: Life Cycle Assessment: Life Cycle Interpretation – This provides guidance on how the impact assessment made according to ISO 14042 should be summarized interpreted and discussed with respect to the environmental goals set according to ISO 14041

ISO 14048 (2002): Environmental Management: Life Cycle Assessment: Data Documentation Format – This provides a standard framework for the documentation of the performance indicators as specified in ISO 14041 and provides appropriate standards for data collection, quality and analysis

ISO 14049 (2000): Environmental Management: Life Cycle Assessment: Examples of Application of ISO 14041 to Goal and Scope Definition and Inventory Analysis – This provides a number of examples indicating how the various requirements of ISO 14041 might be implemented.

ISO 14050 (2002): Environmental Management Vocabulary – This defines all concepts, words and phrases used in ISO 14000 standards, to ensure that a common international understanding is achieved.

ISO 14061 (1998): Information to Assist Forestry Organisations in the Use of Environmental Management System Standards ISO 14001 and ISO 14004 – This provides information to assist forestry organisations in the use of ISO 14001 and ISO 14004. As well as guidelines specific to the forestry industry, appropriate reference material is provided and a number of case studies are included:

ISO 19011 (2002): Guidelines for Quality and/or Environmental Management Systems Auditing – This mainly copies the guidance on system auditing contained in ISO 14010-14012, but does so without specific reference to environmental systems, so that it is applicable either to quality assurance systems (ISO 9000) or to environmental management systems. This single document supersedes ISO 14010, ISO 14011 and ISO 14012 (and also corresponding documents in the ISO 9000 series

Appendix 1.147: All about Environmental Audit in Industries

Industrial pollution is creating a high risk environment. Various legislations viz. the water (Prevention and Control of pollution) Act. 1974 the Air (Prevention and Control of Pollution) Act. 1981 and the Environment (Protection) Act. 1986 have come into force. Industry and environment should go hand-in-hand so as to achieve sustainable development. Considerations of environmental factors at par with production helps in minimizing material losses and also in reduction of liabilities in the long run. Environmental auditing helps to achieve and maintain compliance with environmental laws and regulations as well as to help identify and correct unregulated environmental hazards.

The growing environmental pollution and the complexity of this problem with increasing risks from the regulatory controls needs an effective management too so as to prevent pollution.

Environmental audit is a technique which is part of industry's internal procedures in meeting their responsibilities towards better environment. A notification under the Environment (protection) Rules. 1986 requires industries to submit an environmental statement for the financial year concerned state Pollution Control Boards every year beginning. The submission of an environmental statement is applicable to the following.

1. Those who require consent under the Water (Prevention and Control of Pollution) Act, 1974:
2. These who require consent under the Air (Prevention and Control of Pollution) Act, 1981, and
3. Those who require authorization under Hazardous wastes (management and Handling) Rules, 1989.

Generally, audits may be categorized as

Internal-audits: using auditors from within the department, unit or facility being audited.

Cross-audits: using auditors from within the company but from different plants, facilities or regions

External-audits: using auditors who are wholly independent of the facility to be audited. They would be contractors or consultants i.e. not part of the company being audited.

There are advantages and disadvantages to each different type of audit, which is tabulated below:

Audit Type	**Advantages**	**Disadvantages**
Internal	Low cost. Low organizational disruption. Operational familiarity Good opportunities for cross Transfer of information	Least independent Least audit expertise
Cross-audit	More independent Opportunities for cross transfer More audit expertise	Higher cost More disruptive Less operational familiarity
External	Most independent Most audit expertise	Highest cost Most disruptive Low opportunity for information transfer. Little operational familiarity

There are many facets of an operation which may be subject to an environmental audit. Broadly, they can be split into the following types: management, technical and due diligence.

- **Management audits** cover, policy implementation, effectiveness of procedures, administrative systems, communications, staff motivation and training.
- **Technical audits** cover, legislative compliance, process/mass balance, waste reduction, energy conservation and product life cycle; and

Benefits of Environmental Audit: It ensures legislative compliance, management control, financial performance enhancement, public relations and staff education and motivation.

- It determines how well the process systems and pollution control systems are performing and identifies the operations of poor performance.
- It is a means for ensuring the continual improvement in environmental management;
- It is a good method of monitoring;
- It can assist efforts for sustainable development;
- It can help involve the people in environmental management;
- It may reduce the need for government inspections;
- It may help identify cost recovery through recycling and sale of by-products, etc.; and

- It may generate valuable data for regional or national environment people.
- It identifies potential cost savings which can be accrued through reduction in raw material consumption by way of waste minimization and adoption of recycle/recovery/reduction in pollution load:
- It increases awareness of environmental requirements. Policies and responsibilities.
- It helps in understanding the technical capabilities and attitude of the environmental organisation in a Company:
- It provides up-to-date environmental data base for use in plant modification emergencies etc.:
- Unravels surprises and hidden liabilities due to which regulatory risk and exposure to litigation can be reduced:
- It ensures independent verification. Identifies matters needing attention. And provides timely warning to management on potential future problems: and
- It helps to safeguard environment, and assists in complying with local, regional and national laws and regulations. With the Company's policy and with the environmental standards.

The benefits of environmental audit outlined above should be considered in association with the costs of implementation. Often, there is resistance to auditing in many organisation, which is seen as and cost without benefit. It should be stressed that short term cost will lead to long term benefit.

Objectives of Environmental Audit

The environmental audit helps in pollution control improved production. Safety and earth and conservation of natural resources. The objectives of environmental audit in an industry are:

1. To determine the mass balance of various materials used and the performance of various process equipment so as to identify usage of materials in excess than required, to review the conversion efficiencies of process equipment and accordingly fix up norms for equipment/operation performance and minimization of the wastes.
2. (a) to identify the areas of water usage and wastewater generation and determine the characteristics of wastewater:

   (b) to determine the emissions: their sources, quantities and characteristics: and

   (c) to determine the solid wastes and hazardous wastes generated their sources, quantities and characteristics.
3. To identify the possibilities of waste minimization and recovery and recycling of wastes.
4. To determine the performance of the existing waste treatment / control systems so as to modify or install additional or alternative control equipment accordingly.
5. To determine the impact on the surrounding environment (groundwater, stream, residential area, agricultural area, sensitive zone. etc.) due to the disposal of wastewater, emissions and solid wastes from the industry and accordingly identify suitable preventive measures if necessary.
6. To verify compliance with the standards and conditions prescribed by the regulatory bodies.
7. To check the effectiveness of (a) organizational set-up of the industry for decision-making and environmental management with special reference to their 'technical' view point. 'attitudinal' view point and training, and (b) environmental policy of the Company.

Environmental audit is a management tool comprising a systematic, documented periodic and objective evaluation of an organization performing functions of production within the mandated regulatory requirements.

The environmental audit studies comprise the following three basic activities:

i. Compilation of the complete information on the operation of the industrial facility and its potential sources of pollution through technical inspection, (pre-audit activities).

ii. Evaluation of the conditions surrounding the industrial facility in order to estimate possible impacts which may be caused and suggested measures for such situation. (Activities at industrial sites).

iii. Preparation and implementation of action plans for better control of the environment and environmentally related industrial activities, including further developmental activities of the areas (Post audit activities).

The advantages of environmental auditing is to help safeguard the environment and to substantiate compliance with local, regional and national laws and regulations, and with the company policy and standards.

General Approach of Environmental Audit

The general approach cover three main phases:

a. Collection of information
b. Evaluation of information collected
c. Formulation of conclusions including identification of aspects needing improvement.

These phases cover, pre audit preparation, a site visit normally involving interviews with persons and inspection of facilities and post visit activities.

i. **Pre audit Activities:**
   - Selection and review of site by audit team
   - Development of an audit plan, which defines the technical, geographic and time scope.
   - Arranging for sample collection, preservation transport and analysis.
   - Obtaining background information on the plant as well as the Criteria to used for evaluation.
   - Preparation of questionnaires

ii. **Activities at site:**
   - Identification and assessment of the management control system.
   - Date collection, collation, compilation and evaluation
   - Reporting audit findings

iii. **Post Audit Activities:**
   - Preparation of the final report
   - Development and follow up for implementation of a corrective action programme.

Problems Encountered During Audit

Problems which the audit team may face are variable with respect to facility. However what is expected and needs attention are:

i. Prior history of the site
ii. Age of relevant equipment
iii. Lack of records related to the relevant equipment
iv. Attitude of concerned personnel on site towards audit studies
v. Responses of the concerned management for implementation the corrective measures.

Pre-audit Activities

(a) Preliminary information collection:

Pre-audit activities include various preparatory works. Having chosen the industry to be audited preliminary information on the industry are to be obtained through a questionnaire The information include location of the industry with surrounding landuses, climatic conditions, products manufactured raw materials used details on water utilization. Wastewater generation and disposal gaseous emissions, solid waste/hazardous waste and organizational set up and policies of the Company for environmental management

The preliminary information received on the industry should be reviewed to identify main areas of concern. Thereafter it is required to prepare and organize audit team and resources, and allocate specific tasks to team members. Resources such as the sampling and monitoring equipment and laboratory facilities for analysis should be checked if available at site or else arrangements should be made for their availability loan from other industries etc. The visit programme should then be intimated to the industry. The prior intimation to the industry helps them convince the senior management and staff as various levels of the purpose of audit and the cooperation they have to extend to the audit team. The staff should not feel that the audit would lead to surfacing problems and hence they would be subject to criticism by the management. They should be made clear about the purpose and objectives of the audit and how beneficial it would be for the industry. This would also increase employees awareness towards waste reduction and promote input and support for the audit.

(b) **Audit team formation:**

Audit team should include employees from production quality control/laboratory R and D pollution control operations, technical staff for monitoring and analysis of waste samples and environment and an environment specialist. The number of people may vary from 4 to 8 depending on the size and complexity of the facility being audited.

The team should be sufficiently detached to provide an independent view. The members should be so chosen that they would not hesitate bringing out even criticism, owing to obligations with Supervisor.

It is important to have well-defined and systematic procedures which are known are understood by all concerned. The audit may take 3-10 days depending on the industry.

Effectiveness of audit is a direct result of the qualification, confidence, training and proficiency of the personnel who conduct audits. The team should understand regulatory requirements, relevant waste control technologies and their operations and process. They should have capability to examine, question, sample and analyse waste and interpret data. The management should be provided with a realistic assessment of environmental performance.

(c) **Activities at Site:**

The activities at site include deriving material balance, identifying waste flow lines, monitoring of characteristics, evaluating performance of pollution control equipment/system, assessing environmental quality, holding discussions with the management and finally preparing the draft report. Interviews should be carried out with various cross-sections of the staff engaged in production, laboratory/quality control R and D environment management etc. so as to understand different operational mechanisms. Having a fair idea on the manufacturing process, reconnaissance surveys should be made to be familiar with layout of the plant and process operations, and to understand possible impact on the surrounding environment, discussed.

Post Audit Activities: It includes evaluation of performance and adequacy of pollution control measures, evaluation of technical and attitudinal capabilities of staff, formulation of recommendation for the best practicable environment management systems, final report preparation with recommendation and follow up for implementation of recommendations

Basic steps involved in an environmental audit is shown below in the flowsheet

Basic Steps of an Environmental Audit

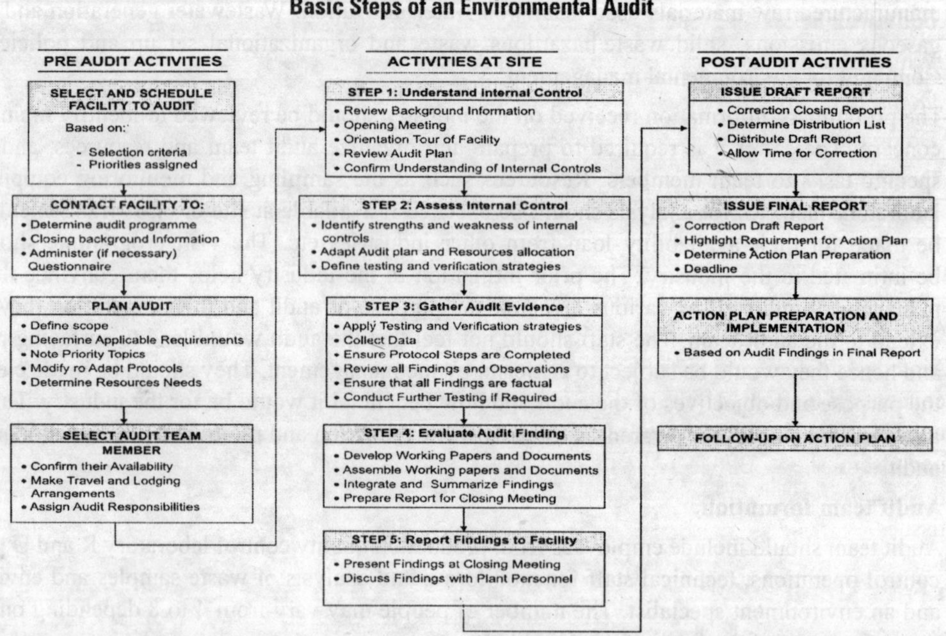

Post review activities involved in environmental audit is shown below in flowsheet.

Post Review Activities Involved in Environmental Audit

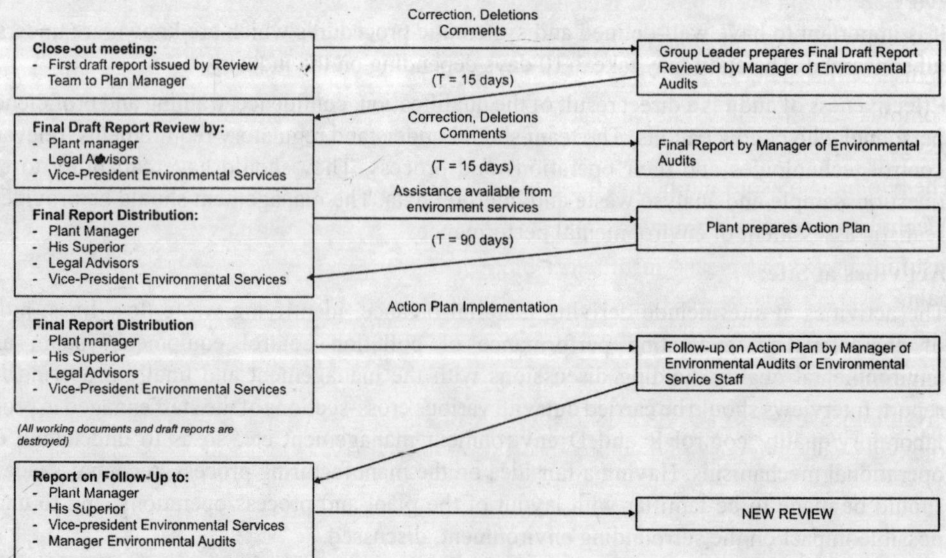

Appendix 1.148: Oxygen Requirement of Human Beings for Various Degree of Work

S.No.	Degree of work performed by an average man of 70kg body weight	Oxygen requirement, litres/ minute
1.	While taking rest	0.3
2.	Doing light work	0.3-1
3.	Doing moderate work	1-1.5
4.	Doing heavy work	1.5-2
5.	Doing very heavy work	2-6

Note:

a. Above figure is oxygen requirement; not air requirement

b. Air contains 21% by volume of oxygen. A fit human body tolerates exercise well at oxygen concentration from 21 to 18% in air. Between 18 to 17% oxygen, human body will probably not be affected adversely unless heavy work is undertaken when oxygen deficiency may cause unconsciousness. Between 17 to 16% oxygen, heavy work is not possible. Light activity under this condition will increase pulse rate and respiration rate in order to improve oxygen supply to the tissues.

Oxygen Fact Sheet

- The average individual takes in approximately 6 liters of air per minute in a resting state.
- If the body does not get a sufficient supply of oxygen, the body will not be able to utilize nutrients from food efficiently.
- Every action and every process the body goes through requires energy; energy comes from oxygen.
- Oxygen plays a pivotal role in the proper functioning of our immune system, especially in regards to disease resistance.
- Complex carbohydrates like vegetables and whole grains are considered oxygen-rich foods. Low oxygen = poor metabolism.
- The two main reasons our bodies are low in oxygen are poor diet and a lack of aerobic exercise.
- Alcohol, cigarettes, air pollutants, and stress also deplete our bodies of oxygen.
- Insufficient oxygen means insufficient biological energy, which can result in muscle fatigue and pain.
- The average person uses only about 50% of their lung capacity.
- Remember life starts with a breath and ends with a breath.
- Physiological effects of oxygen deficiency in working atmosphere is tabulated below (Asphyxia due to oxygen deficiency is rapid with no prior warning to the victim)

Oxygen content at atmospheric pressure, volume %	Physiological effects and symptoms
>20	No effect
11-14	Diminition of physical and intellectual performance without affected person's knowledge
8-11	Possibility of fainting after a short period without warning
6-8	Fainting within a few minates; resuscitation possible, if carried out immediately
0-6	Fainting and almost immediate death ensues; brain damge even if rescued.

Typical Examples of various degrees of work are tabulated below:

S.No.	Degree of work	Examples
1.	Rest	Sitting (quietly or with moderate arm movement)
2.	Light Work	Sitting or standing to control machines, performing light hand or arm work (eg. Using a table saw); occasionally walking; driving
3.	Moderate Work	Walking about with moderate lifting and pushing or pulling; walking at moderate pace; eg. Scrubbing in a standing position.
4.	Heavy Work	Pick and shovel work; digging; carrying; pushing/pulling heavy loads; walking at fast pace; eg. Carpenter sawing by hand.
5.	Very Heavy Work	Very intense activity at fast to maximum pace; eg. Shoveling wet sand

Appendix 1.149: Ozone Depletion Potential (ODP) of various CFC's relative to CFC - 11

S. No.	CFC (Chlorofluorocarbon) compounds	Relative ODP (Ozone depletion potential)
1.	CFC - 11	1.0
2.	CFC – 12	0.87
3.	CFC – 113	0.76
4.	CFC – 114	0.56
5.	CFC – 115	0.27
6.	HCFC - 22	0.043
7.	HCFC – 123	0.016
8.	HCFC – 124	0.017
9.	HCFC – 125	0
10.	HCFC – 134a	0
11:	CCl_4	1.1
12.	CH_3CCl_3	0.12

NOTES

The ozone layer is a layer of ozone gas about 20 km above the earth's surface, High up in our atmosphere, it performs a vital role in protecting humans and living things on earth from harmful UVB and UVC rays from the sun. The growing production and use of chemicals such as chlorofluorocarbons (CFCs) in aerosol sprays, refrigerators, insulation and airconditioning was- build-up of ozone-depleting chemicals in the atmosphere.

A reduced ozone layer causes a number of serious health risks for humans, including increased rates of skin cancer and eye cataract, with children being particularly vulnerable. There are also serious impacts for biodiversity. The increased concentration of UVB rays reduces the amount of plankton in the sea which, in turn, causes fish stocks to go down. These rays also have adverse effects on plant growth, thus reducing agricultural productivity. Another negative effect of a thinning ozone layer is the reduced lifespan of certain materials– such as certain plastics.

It is important to distinguish between ozone in the stratosphere – commonly referred to as the 'ozone layer' – and ground-level ozone. Ozone-depleting substances act on the stratospheric ozone layer. While an abundance of ozone in the ozone layer protects humans by shielding them from harmful UV radiation, excess amounts of ozone at ground level are bad for health because its strong oxidant properties are toxic for humans. Conversely, a greater intensity of UVB rays increases ground-level ozone.

Ozone-Depleting Substances (ODS)

The following bromine and chlorine-based chemicals have been found to destroy ozone and are regulated by international legislations. Chlorofluorocarbons (CFCs). CFCs are halo- alkanes containing both chlorine and fluorine. Because they were widely used -as refrigerants, propellants and cleaning solvents, they are the focus of early action under the Montreal Protocol. They are no longer used in most applications. CFCs also contribute to global warming, with a global warming potential (GWP) of 6 000-9 800 times that of carbon dioxide (over a period of 100 years).

- Methyl bromide (CH_3Br) an organic halogen with powerful ozone-depleting potential and highly toxic to humans. Before being regulated by the Montreal Protocol, it was widely used as a fumigant to control pests in soil, particularly for high-value crops such as strawberries and tomatoes. Although its use was banned in industrialised countries from 2005, a number of 'essential' or 'critical use' exemptions and also exemptions for quarantine and pre-shipment remain.
- Halons (halon-1211, halon-1301, halon- 2402). Halons are haloalkanes used primarily in fire-extinguishing systems with powerful ozone-depleting potential. Some applications are exempted, such as military applications where it has been difficult to find replacements with the required combination of properties.
- Carbon tetrachloride (CCl_4). In the early 20th century, carbon tetrachloride was widely used as a dry-cleaning solvent, as a refrigerant, and as a component in fire extinguishers. When severe adverse health effects became evident, safer alternatives were found for these applications. Large quantities of this substance were used to produce the freon refrigerants R-ll and R-12 (CFCs), which were later found to destroy ozone. Their use has been phased out under the Montreal Protocol, although carbon tetrachloride is still used as a feedstock to manufacture less destructive refrigerants.
- Hydrochlorofluorocarbons (HCFCs) are a class of haloalkanes in which not all hydrogen has been completely replaced by chlorine or fluorine. They are used primarily as CFC substitutes since their ozone-depleting effect is only about 10% that of CFCs. They are also currently being phased out.

An important feature of the Montreal Protocol is the ability to adapt it to meet new challenges. In the early days of the protocol the emphasis was on identifying ozone- depleting substances and agreeing on control measures. This has now been achieved. Attention in recent years has focused on implementation. Key challenges ahead include the following:

- **Critical/essential uses**. The overall quantity of methyl bromide used globally in soil fumigation has gone down substantially. In crops such as strawberry fruit cultivation, the .use of methyl bromide has been completely phased out in the EU. However, sectors where no technically or economically feasible alternatives are available are still exempt from the rules.
- **A number of medical applications:**, for example for treating asthma and other bronchial diseases, are still exempt from the rules on CFCs. Alternatives have now been developed in some countries

Potential (ODP)

Ozone Depletion & Global Warming Potential (GWP)

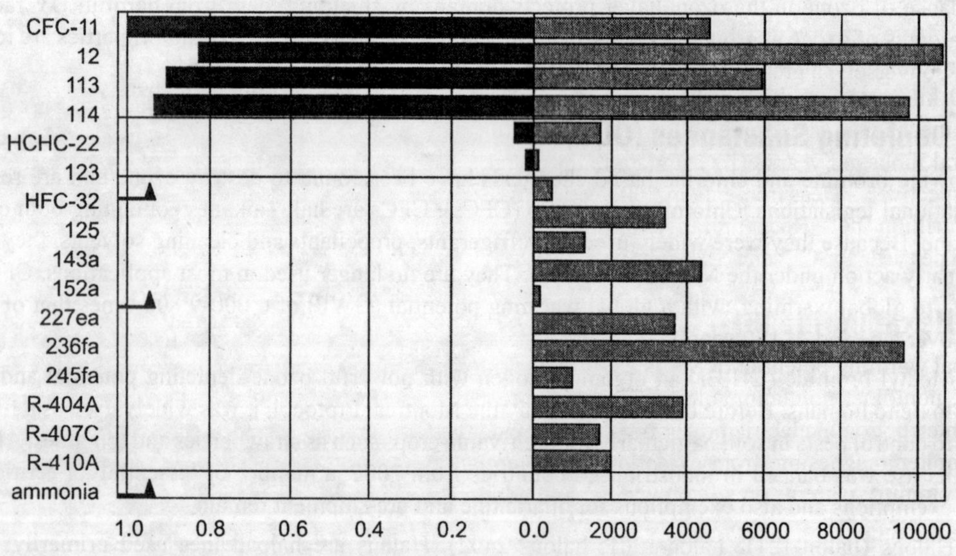

Potential (ODP) Ozone Depletion and Global Warming Potential (GWP)

Ozone Depletion Potential (ODP) is a simple measure of a compounds ability to destroy the stratospheric ozone. Global Warming Potential (GWP) is a measure of how much a given mass of greenhouse gas is estimated to contribute to global warming.

Hydrofluorocarbons (HFCs)

HFCs arc man-made chemicals, many of which have been developed as alternatives to ozone-depleting substances (ODS) for industrial, commercial, and consumer products. The global warming potentials of HFCs range from 140 (HCF-152a) to 11, 700 (HFC-23). The atmospheric lifetime for HFCs varies from just over a year for HFC-152a to 260 years for HFC-23. Most of the commercially used HFCs have atmospheric lifetimes less than 15 years; e.g., HFC-134a, which is used in automobile air conditioning and refrigeration, has an atmospheric life of 14 years.

The HFCs with the largest measured atmospheric abundances are (in order), HFC-23 (CHF_3), HFC-134a (CF_3CH_2F), and HFC-152a (CH_3CHF_2). The only significant emissions of HFCs before 1990 were of the chemical HFC-23, which is generated as a byproduct of the production of HCFC-22. HFCs are primarily used as a substitute for ozone-depleting chemicals. Between 1978 and 1995, HFC-23 concentrations have increased from 3 to 10 parts per trillion (ppt), and continue to rise. HFCs are primarily used as a substitute for ozone-depleting chemicals. Between 1978 and 1995, HFC-25 concentrations have increased from 3 to 10 parts per trillion (ppt), and continue to rise. Since 1990, when it was almost undetectable, global average concentrations of HFC - 134a have risen significantly to HFC-152 has increse steadily to about 0.3 ppt in 2000, however its relatively short life time years; has kept its atmospheric concentration below 1 ppt.

Perfluorocarbons (PFCs)

Primary aluminum production and semiconductor manufacture are the largest known man-made sources of two perfluorocarbons - CF_4 (tetrafluoromethane) and C_2F_6 (hexafluoroethane). The GWP of CF_4 and C_2F_6 emissions is equivalent to approximately 6, 500 and 9, 200 respectively. PFCs are also relatively minor substitutes for ozone-depleting substances (ODSs).

PFCs have extremely stable molecular structures and are largely immune to the chemical processes in the lower atmosphere that break down most atmospheric pollutants. Not until the PFCs reach the mesosphere, about 60 kilometers above Earth, do very high-energy ultraviolet rays from the sun destroy them. This removal mechanism is extremely slow and as a result PFCs accumulate in the atmosphere and remain there for several thousand years. The estimated atmospheric lifetimes for CF_4 and C_2F_6 are 50, 000 and 10, 000 years respectively. Measurements in 2000 estimate CF4 global concentrations in the stratosphere at over 70 parts per trillion (ppt). Recent relative rates of increase in concentrations for two of the most important PFCs are 1.3% per year for CF_4 and 3.2% per year for C_2F_6,

Sulfur Hexafluoride (SF_6)

The global warming potential of SF_6 is 23, 900, making it the most potent greenhouse gas. SF_6 is a colorless, odorless, nontoxic, nonflammable gas with excellent dielectric properties. SF_6 is used for insulation and current interruption in electric power transmission and distribution equipment, in the magnesium industry to protect molten magnesium from oxidation and potentially violent burning, in semiconductor manufacturing to create circuitry patterns on silicon wafers, and as a tracer gas for leak detection.

Like the other high GWP gases, there are very few sinks for SF_6, so all man-made sources contribute directly to its accumulation in the atmosphere. Measurements of SF_6 show that its global average concentration has increased by about 7% per year during the 1980s and 1990s, from less than ppt in 1980 to almost 4 ppt in the late 1990's.

Appendix 1.150: Concentration (ppm) of Trace Metals in Fossil Fuels (Coal & oil)

S.No.	Elements	Concentration (ppm) of trace metals in	
		Coal	Oil
1.	Mercury	0.012	10
2.	Selenium	3	0.17
3.	Cadmium	-	0.01

4.	Arsenic	5	0.01
5.	Molybednum	5	10
6.	Copper	15	0.14
7.	Nickel	15	10
8.	Lead	25	0.3
9.	Vanadium	25	50
10.	Manganese	50	0.1
11.	Zinc	50	0.25

Appendix 1.151: Composition of Fly Ash Generated by Coal Combustion

S.No.	Constituents of fly ash	% by weight
1.	Silica (SiO_2)	17.3 – 63.6
2.	Alumina (Al_2O_3)	9.8 – 58.A
3.	Potassium oxide (K_2O)	2.8 – 3.0
4.	Iron oxide (Fe_2O_3)	2.0 – 26.8
5.	Carbon (C)	0.4 – 36.2
6.	Sodium oxide (Na_2O)	0.2 – 0.9
7.	Sulfur (SO_2)	0.12 – 24.33
8	Calcium oxide (CaO)	0.12 – 14.3
9.	Phosphorous (P_2O_5)	0.07 – 47.2
10.	Magnesium (MgO)	0.06 – 4.77
11.	Carbonate (CO_3)	0.0 – 2.6
12.	Titanium (TiO_2)	0.0 – 2.8

Fly ash is a product of burning finely ground coal in a boiler to produce electricity. It is removed from the plant exhaust gases primarily by electrostatic precipitators, or baghouses and secondarily by scrubber systems. Physically, fly ash is a very fine, powdery material, composed mostly of silica. Nearly all particles are spherical in shape. Fly ash is generally light tan in color and consists mostly of silt-sized and clay-sized glassy spheres. This gives fly ash a consistency somewhat like talcum powder.

Fly ash is a pozzolan, a siliceous material which in the presence of water will react with calcium hydroxide at ordinary temperatures to produce cementitious compounds. Because of its spherical shape and pozzolanic properties, fly ash is useful in cement and concrete applications. The spherical shape and particle size distribution of fly ash also make it a good mineral filler in hot mix asphalt applications and improve the fluidity of flowable fill and grout.

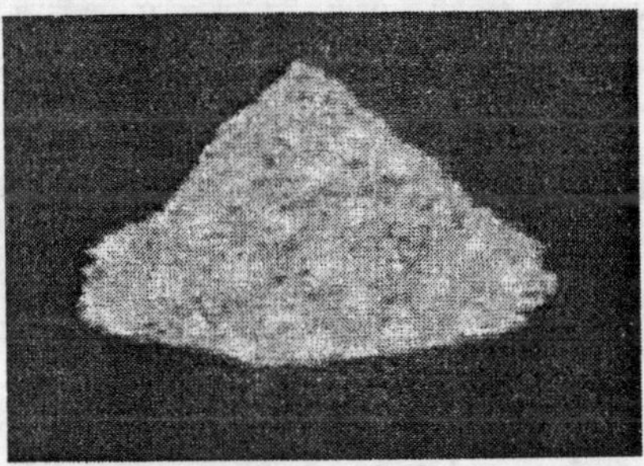

Fly ash applications include its use as a:

- Raw material in concrete products and grout
- Feed stock in the production of cement
- Fill material for structural applications and embankments
- Ingredient in waste stabilization and/or solidification
- Ingredient in soil modification and/or stabilization
- Component of flowable fill
- Component in road bases, sub-bases, and pavement « Mineral filler in asphalt
- Brick making row material

India produces about 70 million tons of coal ash per year from burning about 200 million tons of coal per year for electric power generation. Coal-ash management poses a serious environmental problem for India and requires a mission-mode approach.

At present about 10% ash is utilized in ash dyke construction and land filling (a technology developed and pioneered at IIT Kanpur) and only about 3% of ash is utilized in other construction industries. This is very much in contrast with 80% or more fly ash used in developed countries for the manufacture of bricks, cellular concrete blocks, road construction, land fill application, ceramics, agriculture, insulating bricks, recovery of **metals and cenospheres and dam constructions. Currently, about one acre per MW of land is needed for ash disposal.**

Benefits of Using Fly Ash

Concrete in its hardened state – with fly ash – shows improved performance with:

- **Greater strength**. Fly ash increases in strength over time, continuing to combine with free lime.
- **Decreased** permeability. Increased density and long-term pozzolanic action of fly ash, which ties up free lime, results in fewer bleed channels and decreases permeability.
- **Increased durability**. The lower permeability of concrete with fly ash also helps keep aggressive compounds on the surface, where destructive action is lessened. Fly ash concrete is also more resistant to attack by sulfate, mild acid, and soft (lime hungry) water.
- **Reduced** alkali silica reactivity. Fly ash combines with alkalis from cement that might otherwise combine with silica from aggregates, thereby preventing destructive expansion.
- **Reduced heat** of hydration. The pozzolanic reaction between fly ash and lime generates less heat, resulting in reduced thermal cracking when fly ash is used to replace a percentage of Portland Cement.

- **Reduced efflorescence**. Fly ash chemically binds free lime and salts that can create efflorescence. The lower permeability of concrete with fly ash can help to hold efflorescence-producing compounds inside the concrete.

The ball-bearing effect of fly ash in concrete creates a lubricating action when concrete is in its plastic state. This means:

- **Increased workability**. Concrete is easier to place with less effort, responding better to vibration to fill forms more completely.
- **Increased ease of pumping**. Pumping requires less energy; longer pumping distances are possible.
- **Improved finishing. Sharp**, clear architectural definition is easier to achieve, with less worry about in-place integrity.
- **Reduced bleeding**. Fewer bleed channels decreases porosity and chemical attack. Bleed streaking is reduced for architectural finishes. Improved paste to aggregate contact results in enhanced bond strengths.
- **Reduced segregation**. Improved cohesiveness of fly ash concrete reduces segregation that can lead to rock pockets.
- **Reduced slump loss**. More dependable concrete allows for greater working time, especially in hot weather.

MOEF Notification on Fly Ash

Ministry of Environment and Forests (MOEF), Govt, of India vide its notification (amendment) dated 3rd Nov 2009 has made it mandatory:

- **Within 100Km radius of a Thermal Power Plant**
  1. To use Fly Ash based Building products such as cement or concrete, fly ash bricks, blocks, tiles etc. in all construction projects
  2. To use Fly Ash in Road or Flyover Embankment construction
  3. To use Fly Ash in Reclamation of low lying areas
- **Within 50Km of a Thermal Power Plant (By Road)**
  1. To use Fly Ash in back filling of underground and open cast mines
- **Financial institutions to include a clause in their loan documents for compliance of this notification**

Appendix 1.152: Physical Properties of Water (SI units)

Tempreature, °C	Specific weight, y, kN/m³	Density, p, kg/m³	Modulus of elasticity, E/10⁶ kN/m²	Dynamic viscosity μ × 10³ N s/m²	Kinematic viscosity v ×10⁶, m²/s	Surface tension, † σ, N/m	Vapor pressure, Pv, kN/m²
0	9.805	999.8	1.98	1.781	1.785	0.0765	0.61
5	9.807	1000.0	2.05	1.518	1.519	0.0749	0.87

10	9.804	999.7	2.10	1.307	1.306	0.0742	1.23
15	9.798	999.1	2.15	1.139	1.139	0.0735	1.70
20	9.789	998.2	2.17	1.002	1.003	0.0728	2.34
25	9.777	997.0	2.22	0.890	0.893	0.0720	3.17
30	9.764	995.7	2.25	0.798	0.800	0.0712	4.24
40	9.730	992.2	2.28	0.653	0.658	0.0696	7.38
50	9.689	988.0	2.29	0.547	0.553	0.0679	12.33
60	9.642	983.2	2.28	0.466	0.474	0.0662	19.92
70	9.589	977.8	2.25	0.404	0.413	0.0644	31.16
80	9.530	971.8	2.20	0.354	0.364	0.0626	47.34
90	9.466	965.3	2.14	0.315	0.326	0.0608	70.10
100	9.399	958.4	2.07	0.282	0.294	0.0589	101.33

*At atmospheric pressure. †In contact with air.

Note:

The following points should be noted about the physical properties of water:

- **Water has a high specific heat.** Specific heat is the amount of energy required to change the temperature of a substance. Since water has a high specific heat, it can absorb large amounts of heat energy before it begins to get hot. It also means that water releases heat energy slowly when situations cause it to cool. Water's high specific heat allows for the moderation of the Earth's climate and helps organisms regulate their body temperature more effectively.
- **Water in a pure state has a neutral pH.** Pure water is neither acidic nor basic. Water changes its pH when substances are dissolved in it. Rain has a naturally acidic pH of about 5.6 because it contains natural derived carbon dioxide and sulfur dioxide.
- **Water conducts heat more easily than any liquid except mercury.** This characteristic features causes large bodies of liquid water like lakes and oceans to have essentially a uniform vertical temperature profile.
- **Water molecules exist in liquid form over an important range of temperature from 0 - 100° C** This range allows water molecules to exist as a liquid in most places on our planet.
- **Water is a universal solvent.** It is able to dissolve a large number of different chemical compounds. This feature also enables water to carry solvent nutrients in runoff, infiltration, groundwater flow, and also in living organisms.
- **Water has a high surface tension.** Water is adhesive and elastic, and tends to aggregate in drops rather than spread out over a surface as a thin film. This phenomenon also causes water to stick to the sides of vertical structures despite gravity's downward pull. Water's high surface tension allows for the formation of water droplets and waves, allows plants to move water (and dissolved nutrients) from their roots to their leaves, and the movement of blood through tiny vessels in the bodies of some animals.
- **Water molecules are the only substance on Earth that exists in all three physical states of matter: solid, liquid, and gas.** Incorporated in the changes of state are massive amounts of heat exchange. This feature plays an important role in the redistribution of heat energy in the Earth's

atmosphere. In terms of heat being transferred into the atmosphere, approximately 75 percent of this process is accomplished by the evaporation and condensation of water.

- **The freezing of water molecules causes their mass to occupy a larger volume.** When water freezes it expands rapidly adding about 9% by volume. Fresh water has a maximum density at around 4° Celsius. Water is the only substance on this planet where the maximum density of its mass does not occur when it becomes solidified.

Total volume of water in various water bodies/reservoir is tabulated below:

Water **Reservoir**	**Volume (Cubic km x 1, 000, 000)**	**Percent of Total**
Oceans	1370	97.25
Ice Caps and Glaciers	29	2.05
Groundwater	9.5	0.68
Lakes	0.125	0.01
Soil Moisture	0.065	0.005
Atmosphere	0.013	0.001
Streams and Rivers	0.0017	0.0001
Biosphere	0.0006	0.00004

Average residence time of water at the Earth's surface in various reservoirs is tabulated below:

Water **Reservoir**	**Average Residence Time**
Glaciers	20 to 100 years
Seasonal Snow Cover	2 to 6 months
Soil Moisture	1 to 2 months
Groundwater: Shallow	100 to 200 years
Groundwater: Deep	10, 000 years
Lakes	50 to 100 years
Rivers	2 to 6 months

Many processes work together to keep Earth's water moving in a cycle. There are five processes at work in the hydrologic cycle: condensation, precipitation, infiltration, runoff, and evapotranspiration. These occur simultaneously and, except for precipitation, continuously. Water vapour condenses to form clouds, which result in precipitation when the conditions are suitable. Precipitation falls to the surface and infiltrates the soil or flows to the ocean as runoff. Surface water (e.g., lakes, streams, oceans, etc.), evaporates, returning moisture to the atmosphere, while plants return water to the atmosphere by transpiration.

The water cycle - technically known as the hydrologic cycle - is the circulation of water within the earth's hydrosphere, involving changes in the physical state of water between liquid, solid, and gas phases. The hydrologic cycle refers to the continuous exchange of water between atmosphere, land, surface and subsurface waters, and organisms. In addition to storage in various compartments (the ocean is one such compartment); the multiple cycles that make up the earth's water cycle involve five main physical actions: ***evaporation, precipitation, infiltration, runoff, and subsurface flow***

Appendix 1.153: Annual Mean Concentration Range (mg/m³) for various Designated Pollution Level (in India)

Pollution level	Industrial		Residential	
	SO_2 and NO_2	SPM	SO_2 and NO_2	SPM
Low pollution level (L) (below 0.5 times the standard	0-40	0-180	0-30	0-70
Moderate pollution level (M) (between 0.5 to 1 times the standard)	40-80	180-360	30-60	70-140
High pollution level (H) (between 1 and 1.5 times the standard)	80-120	360-540	60-90	140-210
Critical pollution level (C) > (1.5 times the standard)	>120	>540	>90	>210
National standard (in Mg/m³) for SO_2 and NO_2 = 80, for CO = 2000, and for Respirable SPM (PM_{10}) (also called RPM) = 100				

Appendix 1.154: Guidelines for minimum stack height (Hmin) in India

a. For all plants, except thermal power plants, minimum stack height, H_{min} = 30 metres

b. For thermal power plants, H_{min} is as follows:

Boiler size, MW	Particulate emission standards, mg/Nm³			Minimum stack height for SO_2 dispersion, H_{min} (metres) Q=SO_2 emission rate in kg/hr
	Old plants (Before 1979)	New plants (After 1979)	Protected area	
<200	600	350	150	$H=14(Q)^{0.3}$
200-500	–	150	150	220
>, 500	–	–	–	275

c. $H_{min}=74Q^{0.27}$, where, Q= particulate emission rate in tons/hr

d. If stack height arrived at by using above criterias are different, then the higher value should be taken as the minimum stack height. But in no case stack height should be less than 30 metres.

Appendix 1.155: List of Pollution Control and Monitoring Equipments for Custom Duty Exemption (in India)

S.No.	Equipments exempted from custom duty
1.	Monitoring instruments for various ranges suitable for ambient air and stack monitoring for toxic/hazardous chemicals
2.	Gas detectors/alarms for measuring concentration of toxic/hazardous gases
3.	Personal samplers with rechargeable battery units
4.	Toxic gas monitoring kits containing personal samplers, sampler with filter head, filter holders, membrane filter of different pore sizes, small cyclones for respirable dust monitoring and activate carbon tubes for collection of specified chemical pollutants
5.	Direct reading calorimetric indicator tubes with accessories for monitoring toxic gases/vapours
6.	Long term direct reading detector tubes with peristaltic pump for monitoring toxic gases/chemicals
7.	Direct reading instruments for monitoring and evolution of air borne particles
8.	Direct reading instruments for air borne gases and vapours incorporating the principles of ultraviolet for infra red absorption, chromatography with flame ionisation or photo-ionisation detectors or micro-processor control solid state circuitry
9.	Platinum based catalyst and systems to incinerate formaldehyde in gaseous emissions
10.	Special incineration system for various toxic gases/solids
11.	Condensing heat exchangers for reducing thermal pollution through stacks
12.	Self-contained and compressed air/blower based breathing apparatus
13.	Full body protection suit for dealing with emergencies created due to leakage/spillage of highly toxic chemicals
14.	Hypalon suits for emergencies for hydrofluoric acid
15.	Ultrasonic or irradiation type thickness testers
16.	Instruments for liquid effluent control, namely (a) Ion chromatograph for anions b. Atomic absorption spectroscope for metal analysis c. Differential pulse polargraph for metal analysis
17.	Plate type compact clarifiers
18.	Sulzer or equivalent high efficiency backing for distillation columns for pollution control
19.	Special plastic media for fixed film technology used in anaerobic treatment of high strength organic effluents waste
20.	Analysers for carbon monoxide and hydrocarbons
21.	Auto analysers for phosphate and nitrogen
22.	Thermohydrograph
23.	Air sampling bags
24.	Resistivity measuring kits to analyse characteristics of flyash at different elevated temperatures

S.No.	Equipments exempted from custom duty
25.	Total organic carbon analyser
26.	Fluorimeter
27.	Dissolved oxygen meter
28.	Chemical oxygen demand (COD) meter
29.	Niskin water sampler
30.	Sediment sampler
31.	Ultracentrifuge micro-pure filtration units
32.	Ion selective electrodes for fluorides, cynamides, NO_3 and NH_3
33.	Chasis and engine dynamometer
34.	Carburettor flow bench

Appendix 1.156: Dictionary of Environmental Engineering Terms

Abatement: Reducing the degree or intensity of, or eliminating, pollution.

Absorption: The passage of one substance into or through another; e.g., an operation in which one or more soluble components of a gas mixture are dissolved in a liquid.

Acid deposition: A complex chemical and atmospheric phenomenon that occurs when emissions of sulfur and nitrogen compounds and other substances are transformed by chemical processes in the atmosphere, often far from the original sources, and then deposited on earth in either a wet or a dry form. The wet forms, popularly called "acid rain, " can fall as rain, snow, or fog. The dry forms are acidic gases or particulates.

Acid rain: Rain mixed with mainly nitric and sulfuric acid.

Activated carbon: A highly adsorbent form of carbon used to remove odors and toxic substances from liquid or gaseous emissions. In waste treatment, it is used to remove dissolved organic matter from wastewater. It is also used in motor vehicle evaporative control systems.

Activated sludge: Residue that results when primary effluent is mixed with bacteria laden sludge and then agitated and aerated to promote biological treatment. This speeds breakdown of organic matter in raw sewage undergoing secondary wastewater treatment.

Active ingredient: In any pesticide product, the component that kills, or otherwise controls, target pests. Pesticides are regulated primarily on the basis of active ingredients.

Acute exposure: A single exposure to a toxic substance that results in severe biological harm or death. Acute exposures are usually characterized as lasting less than 24 hours.

Acute toxicity: The ability of a substance to cause poisonous effects resulting in severe biological harm or death soon after a single exposure or dose; also, any severe poisonous effect resulting from a single short-term exposure to a toxic substance.

Adsorption: Adhesion of molecules of gas, liquid, or dissolved solids to a surface. An advanced method of treating wastes in which activated carbon removes organic matter from wastewater.

Advanced wastewater treatment: Any treatment of sewage that goes beyond the secondary or biological water treatment stage and includes the removal of nutrients such as phosphorus and nitrogen and a high percentage of suspended solids. Also called tertiary treatment of waste water.

Aeration: A process that promotes biological degradation of organic water. The process may be passive (as when waste is exposed to air), or active (as when a mixing or bubbling device introduces the air).

Aerobic: Life or processes that require, or are not destroyed by the presence of oxygen.

Aerobic Composting: It is a method of composting organic wastes using bacteria that need oxygen. This requires wastes to be exposed to air by turning or by forcing air through pipes that pass through the waste material.

Aerobic treatment: Process by which microbes decompose complex organic compounds in the presence of oxygen and use the liberated energy for reproduction and growth. Types of aerobic processes include extended aeration, trickling filtration, and rotating biological contactors.

Aerosol: A suspension of liquid or solid particles in a gas, that remain suspended from a few minutes to many months depending on particle size and weight.

Agricultural pollution: The liquid and solid wastes from farming, including runoff and leaching of pesticides and fertilizers; erosion and dust from ploughing; animal manure and carcasses; and crop residues and debris.

Airborne particulates: Total suspended particulate matter found in the atmosphere as solid particles or liquid droplets. The chemical composition of particulates varies widely depending on location and time of year. Airborne particulates include windblown dust, emissions from industrial processes, smoke from the burning of wood and coal, and the exhaust of motor vehicles.

Air mass: A widespread body of air that gains certain meteorological or polluted characteristics– for example, a heat inversion or smokiness– while standing in one location. The characteristics can change as the air mass moves away.

Air monitoring: It is the sampling for/and measuring of pollutants present in the atmosphere.

Air pollutant: Any substance in air that could, in high enough concentration, harm human beings, animals, vegetation, or material. Pollutants may include almost any natural or artificial composition of matter capable of being airborne. They may be in the form of solid particles, liquid droplets, gases, or combinations of these states. Generally, they fall into two main groups: (a) those emitted directly from identifiable sources and (b) those produced in the air by interaction between two or more primary pollutants or by reaction with normal atmospheric constituents, with or without photo activation. Exclusive of pollen, fog, and dust, which are of natural origin, about 100 contaminants have been identified. They fall into the following categories: solids, sulfur compounds, volatile organic chemicals (VOC's), nitrogen compounds, oxygen compounds, halogen compounds, radioactive compounds, and odors.

Air pollution: The presence of contaminant or pollutant substances in the air that do not disperse properly and interfere with human health or welfare or produce other harmful environmental effects, to humans, animals, vegetation or materials.

Air pollution episode: A period of abnormally high concentration of air pollutants, often due to low winds and temperature inversion, that can cause illness and death of human beings.

Algae: Simple rootless plants that grow in sunlit waters in relative proportion to the amounts of nutrients available. They can affect water quality adversely by lowering the dissolved oxygen in the water. Algae are food for fish and small aquatic animals.

Algal blooms: Sudden spurts of algal growth that can affect water quality adversely and that indicate potentially hazardous changes in local water chemistry.

Ambient air: Any unconfined portion of the atmosphere: open air, surrounding air.

Anaerobic: A life or process that occurs in, or is not destroyed by the absence of oxygen.

Aquifer: An underground geological formation, or group of formations, containing usable amounts of groundwater that can supply wells and springs.

Assimilation: The ability of a body of water to purify itself of pollutants.

Atmosphere: It is the gaseous mass or envelope of air surrounding the earth. The atmosphere from ground level up is further sub-divided into the troposphere, stratosphere, mesosphere and the thermosphere.

Attenuation: The process by which a compound is reduced in concentration over time, through adsorption, degradation, dilution, or transformation.

Background level: In air pollution control, the concentration of air pollutants in a definite area during a fixed period of time prior to the starting up or on the stoppage of a source of emission under control. In toxic substances monitoring, the average presence in the environment, originally referring to naturally occurring phenomena.

Bacteria (singular: bacterium): Microscopic living organisms that can aid in pollution control by consuming or breaking down organic matter in sewage or by similarly acting on oil spills or other water pollutants. Bacteria in soil, water, or air can cause human, animal, and plant health problems.

Baghouse filter: Large fabric bag, usually made of glass fibers, used to eliminate intermediate and large (greater than 20 microns in diameter) particles. This device operates in a way similar to the bag of an electric vacuum cleaner, passing the air and smaller particulate matter while entrapping the larger particulates.

Bar screen: In wastewater treatment, a device used to remove large solids.

Benthic organism (benthos): A form of aquatic plant or animal life found on or near the bottom of a stream, lake, or ocean.

Biochemical oxygen demand (BOD): A measure of the amount of oxygen consumed in the biological processes that break down organic matter in water. The greater the BOD, the greater the degree of pollution. BOD is generally understood to be BOD_5, the amount of dissolved oxygen so consumed in five days at 25 °C.

Biodegradable: The ability to break down or decompose rapidly under natural conditions and processes. Most of the organic wastes such as food, paper etc. are biodegradable materials.

Biological control: In pest control, the use of animals and organisms that eat or otherwise kill or outcompete pests.

Biological magnification: Refers to the process whereby certain substances such as pesticides or heavy metals move up the food chain, work their way into a river or lake, and are eaten by aquatic organisms such as fish, which in turn are eaten by large birds, animals, or humans. The sub stances become concentrated in tissues or internal organs as they move up the chain.

Biological oxidation: The way bacteria and microorganisms feed on and decompose complex organic materials; used in self-purification of water bodies and in activated sludge wastewater treatment.

Biological treatment: A treatment technology that uses bacteria to consume waste and thus break down organic materials.

Biomass: All the living material (wood/vegetation etc) in a given area; often refers to vegetation. Also called *biota*. Grown and produced expressly for use as fuel.

Biomonitoring: The use of living organisms to test the suitability of effluents for discharge into receiving waters and to test the quality of such waters downstream from the discharge. Analysis of blood, urine, tissues, etc., to measure chemical exposure in humans.

Biome: It is a climate region characterised by its dominant vegetation.

Biota: It is the flora and fauna of an area.

Biotic: Resources which are considered biotic and therefore renewable. The rainforests and tigers are examples of biotic resources.

Biotechnology: Techniques that use living organisms or parts of organisms to produce a variety of products–from medicines to industrial enzymes-to improve plants or animals or to develop microorganisms for specific uses such as removing toxics from bodies of water or for pesticides.

BOD_5: The amount of dissolved oxygen consumed in five days by biological processes breaking down organic matter.

Brackish water: A mixture of fresh and salt water.

Cadmium (Cd): A heavy metal element that accumulates in the environment.

Carbon adsorber: An add-on control device that uses activated carbon to absorb volatile organic compounds (VOCs) from a gas stream. The VOCs are later recovered from the carbon.

Carbon dioxide (CO_2): A colorless, odorless, non-poisonous gas that results from fossil fuel combustion and is normally a part of the ambient air. It is a green house gas causing global warming.

Carbon monoxide (CO): A colorless, odorless, poisonous gas produced by incomplete fossil fuel combustion. CO interferes with blood's ability to carry oxygen to the body's tissues and causes adverse health effects.

Carbon sink: It is a pool/reservoir that takes up released carbon from another part of the carbon cycle eg, if the net exchange between the biosphere and the atmosphere is towards the atmosphere, the biosphere is the source and the atmosphere is the sink.

Carcinogen: Any substance that can cause or contribute to the production of cancer.

Catalytic converter: An air pollution abatement device that removes pollutants from motor vehicle exhaust, either by oxidizing them into carbon dioxide and water or by reducing them to nitrogen and oxygen. It is a must for all new motor vehicles sold in India and other countries for automobile pollution control.

Catalytic incinerator: A control device that oxidizes volatile organic compounds (VOCs) by using a catalyst to promote the combustion process. Catalytic incinerators require lower temperatures than conventional thermal incinerators, yielding fuel and cost savings.

Cells: In solid waste disposal, holes in which waste is dumped, compacted, and covered with layers of dirt on a daily basis. The smallest structural part of living matter capable of functioning as an independent unit.

Chemical oxygen demand (COD): A measure of the oxygen required to oxidize all compounds in wastewater, both organic and inorganic.

Chemical treatment: Any one of a variety of technologies that use chemicals or a variety of chemical processes to treat waste, water.

Chlorinated hydrocarbons: A category which includes a class of persistent, broad-spectrum insecticides that linger in the environment and accumulate in the food chain. Among them are DDT, aldrin, dieldrin, heptaclor, chlordane, lindane, endrin, mirex, hexachloride, and toxaphene. Trichloroethylene (TCE), used as an industrial solvent, is also a chlorinated hydrocarbon.

Chlorinated solvent: An organic solvent containing chlorine atoms, e.g., methylene chloride and 1, 1, 1-trichloromethane, which is used in aerosol spray containers and in roadway paint.

Chlorination: The application of chlorine to drinking water, sewage, or industrial wastewater to disinfect or to oxidize undesirable compounds.

Chlorofluorocarbons (CFCs): A family of inert, nontoxic, and easily liquefied chemicals used in refrigeration, air conditioning, packaging, and insulation or as solvents and aerosol propellants. Because CFCs are not destroyed in the lower atmosphere, they drift into the upper atmosphere, where their chlorine components destroy ozone. Their use is currently restricted under Montreal Protocol.

Chronic toxicity: The capacity of a substance to cause long-term poisonous human health effects occuring over a long period of time (eg, months or years).

Cleanup: Actions taken to deal with a release or threat of release of a hazardous substance that could affect humans, the environment, or both. The term is sometimes used interchangeably with the terms *remedial action, removal action, response action, or corrective action.*

Coagulation: A clumping of particles in waste-water to settle out impurities; often induced by chemicals such as lime, alum, and iron salts.

Coliform index: A rating of the purity of water based on a count of fecal bacteria.

Coliform organism: Microorganisms found in the intestinal tracts of humans and animals. Their presence in water indicates fecal pollution and potentially dangerous bacterial contamination by disease-causing microorganisms.

Combined sewers: A sewer system that carries both sewage and stormwater runoff. Normally, its entire flow goes to a waste treatment plant, but during a heavy storm, the stormwater volume may be so great as to cause overflows. When this happens, untreated mixtures of stormwater and sewage may flow into receiving waters. Stormwater runoff may also carry toxic chemicals from industrial areas or streets into the sewer system.

Comminution: Mechanical shredding or pulverizing of waste; used in both solid waste management and wastewater treatment.

Compaction: Reduction of the bulk of solid waste by rolling and tamping.

Composting: The natural biological decomposition of organic material in the presence of air to form a humus-like material. Controlled methods of composting include mechanical mixing and aerating, ventilating the materials by dropping them through a vertical series of aerated chambers, or placing the compost in piles in the open air and mixing or turning it periodically.

Contaminant: Any physical, chemical, biological, or radiological substance or matter that has an adverse affect on air, water, or soil.

Cooling tower: A structure that helps remove heat from water used as a coolant, eg., in electric power generating plants. It controls thermal pollution of water streams/bodies.

Cover material: Soil used to cover compacted solid waste in a sanitary landfill.

Curie: A quantitative measure of radioactivity equal to 3.7×10^{10} disintegrations per second.

Cyclone collector: A device that uses centrifugal force to pull large particles from polluted air.

Decomposition: The breakdown of matter by bacteria and fungi; changes the chemical makeup and physical appearance of materials.

Deforestation: It is the practice or process that result in the land-use to non-forest users. Deforestation causes enhanced green house effect due to two reaons:

a. the burning/decomposition of wood releases CO_2.
b. trees that once removed CO_2 from the atmosphere in the process of photosynthesis are lost.

Denitrification: The anaerobic biological reduction of nitrate nitrogen to nitrogen gas.

Desertification: It is the progressive destruction/degradation of existing vegetative cover to form desert. This occurs due to overgrazing, deforestation, drought and the burning of extensive areas. It causes increased albedo, reduced atmospheric humidity, greater atmospheric dust loading which can cause wind erosion and/or atmospheric pollution.

Desulfurization: Removal of sulfur from fossil fuels to reduce pollution.

Detergent: Synthetic washing agent that helps to remove dirt and oil. Some detergents contain compounds that kill useful bacteria and encourage algal growth when they are discharged in wastewater that reaches receiving waters.

Digester: In wastewater treatment, a closed tank; in solid waste conversion, a unit in which bacterial action is induced and accelerated to break down organic matter and establish the proper carbon-to-nitrogen ratio.

Dioxin: Any of a family of compounds known chemically as dibenzo-p-dioxins It causes potential toxicity and contamination in commercial products. It is one of the most toxic man-made chemicals known.

Disinfectant: A chemical or physical process that kills pathogenic organisms in water. Chlorine is often used to disinfect sewage treatment effluent, water supplies, wells, and swimming pools.

Dispersant: A chemical agent used to break up concentrations of organic material such as spilled oil.

Disposal: Final placement or destruction of toxic, radioactive, or other wastes; surplus or banned pesticides or other chemicals; polluted soils; and drums containing hazardous materials from removal actions or accidental releases. Disposal may be accomplished through use of approved secure landfills, surface impoundments, land farming, deep well injection, ocean dumping, or incineration.

Dissolved oxygen (DO): The oxygen freely available in water; vital to fish and other aquatic life and for the prevention of odors. The level of dissolved oxygen is an indicator of the ability of a water body to support aquatic life. Secondary wastewater treatment and advanced wastewater treatment are generally designed to protect DO in waste-receiving waters.

Dissolved solids: Disintegrated organic and inorganic material contained in water. Excessive amounts make water unfit for drinking or for use in industrial processes.

Distillation: The act of purifying liquids through boiling so that the steam condenses to a pure liquid and the pollutants remain in a concentrated residue.

Dump: A site used to dispose of solid wastes without environmental controls.

Ecology: The relationship of living things to one another and their environment, or the study of such relationships.

Ecosystem: The interacting system of a biological community and its nonliving environmental surroundings.

Effluent: Wastewater–treated or untreated– that flows out of a treatment plant, sewer, or industrial outfall; generally refers to wastes discharged into surface waters.

Effluent limitation: Restrictions established by an environmental agency on quantities, rates, and concentrations in wastewater discharges.

Electrostatic precipitator (ESP): An air pollution control device that removes particles from a gas stream (smoke) after combustion occurs. The ESP imparts an electrical charge to the particles, causing them to adhere to metal plates inside the precipitator. Rapping on the plates causes the particles to fall into a hopper for disposal.

Emission: Pollution discharged into the atmosphere from smokestacks, other vents, and surface areas of commercial or industrial facilities, from residential chimneys; and from motor vehicle, locomotive, or aircraft exhausts.

Emission factor: The relationship between the amount of pollution produced and the amount of raw material processed. For example, an emission factor for a blast furnace making iron would be the number of kg of particulates per ton of iron produced.

Emission standard: The maximum amount of air-polluting discharge legally allowed from a single source (mobile or stationary).

Endangered species: They are the plant and animal species in danger of extinction.

Enrichment: The addition of nutrients (e.g., nitrogen, phosphorus, or carbon compounds) from sewage effluent or agricultural runoff to surface water. This process greatly increases the growth potential of algae and aquatic plants.

Environment: The sum of all external conditions affecting the life, development, and survival of an organism.

Environmental Impact assessment (ETA): A process whose breadth, depth, and type of analysis depend on the proposed project. EA evaluates a project's potential environmental risks and impacts in its area of influence and identifies ways of improving project design and implementation by preventing, minimizing, mitigating, or compensating for adverse environmental impacts and by enhancing positive impacts.

Environmental audit: An independent assessment of the current status of a party's compliance with applicable environmental requirements. An independent evaluation of a party's environmental compliance policies, practices, and controls.

Estuary: It is the region where fresh water from a river mixes with salt water from the sea.

Eutrophication: The slow aging process during which a lake, estuary, or bay evolves into a bog or marsh and eventually disappears. During the later stages of eutrophication the water-body is choked by abundant plant life as the result of increased amounts of nutritive compounds such as nitrogen and phosphorus. Human activities can accelerate the process, of eutrophication.

Evaporate emissions: They are the emissions from evaporating gasoline, which may occur during vehicle refueling, vehicle operation and even when the vehicle is parked. It can account for about two-thirds of hydrocarbon emissions from petrol driven vehicles on hot summer days.

Evapotranspiration: The loss of water from the soil both by evaporation and by transpiration from the plants growing in the soil.

Exposure: A potential health threat to the living organisms in the environment due to the amount of radiation or pollutant present in the environment.

Fabric filter: A cloth device that catches dust particles from industrial emissions.

Fauna: It is the total animal life in an area.

Fecal coliform bacteria: Bacteria found in the intestinal tracts of mammals. Their presence in water or sludge is an indicator of pollution and possible contamination by pathogens.

Fertilizer: Materials such as nitrogen and phosphorus that provide nutrients for plants. Commercially sold fertilizers may contain other chemicals or may be in the form of processed sewage sludge.

Filtration: A treatment process for removing solid (particulate) matter from water by passing the water through porous media such as sand or a manmade filter. The process is used to remove particles that contain pathogenic organisms.

Flocculation: The process by which clumps of solids in water or sewage are made to increase in size by biological or chemical action so that they can be separated from the water.

Flora: It is the total plant life in an area.

Flue gas: Vented air coming out of a chimney after combustion in the burner; can include nitrogen oxides, carbon oxides, water vapor, sulfur oxides, particles, and many chemical pollutants.

Flue gas desulfurization: A technology that uses a sorbent, usually lime or limestone, to remove sulfur dioxide from the gases produced by burning fossil fuels. Flue gas desulfurization is currently in use by major sulfur dioxide emitters such as thermal power plants.

Fluorides: Gaseous, solid, or dissolved compounds containing fluorine that result from industrial processes; excessive amounts in food can lead to fluorosis.

Fluorocarbon (FCs): Organic compounds analogous to hydrocarbons in which one or more hydrogen atoms are replaced by fluorine used as a propellant in aerosols, they are now primarily used in coolants and some industrial processes. FCs containing chlorine are called chlorofluorocarbons (CFCs). They modify the ozone layer in the stratosphere, thereby allowing more harmful solar radiation to reach the Earth's surface.

Fly ash: Noncombustible residual particles from the combustion process carried by flue gas.

Fossil fuels: They are the fuels formed from decaying plants and animals such as coal, oil and natural gas.

Fugitive emissions: Emissions not caught by a capture system.

Fumes: They are solid particles under 1 micron in diameter formed as vapor condenses on chemical reactions take place.

Garbage: It is the waste that is generated in the household, commercial areas, industries, etc.

Geige Muller counter: An electrical device that detects the presence of certain types of radioactivity.

Generator: A facility or mobile source that emits pollutants into the air or releases hazardous wastes into water or soil.

Granular activated carbon (GAC) treatment: A filtering system often used in small water systems and individual homes to remove organics. GAC can be highly effective in removing elevated levels of radon from water.

Global warming: It is the increase in the temperature of the earth as a result of emission of green house gases (GHG) such as CO_2.

Greenhouse effect: The warming of the Earth's atmosphere caused by a buildup of carbon dioxide or other trace gases; many scientists believe that this buildup allows light from the sun's rays to heat the Earth but prevents a counterbalancing loss of heat.

Green house gases (GHG): Gases such as CO_2, CH_4, H_2O vapor, chloroflorocarbons (CFC's), NO_2 etc which contribute to global warming.

Groundwater: The supply of fresh water found beneath the Earth's surface (usually in aquifers), which is often used, for supplying wells and springs. Because groundwater is a major source of drinking water, there is growing concern about areas where leaching agricultural or industrial pollutants or substances from leaking underground storage tanks are contaminating it.

Habitat: The place where a population (e.g., human, animal, plant, or microorganism) lives, and its surroundings, both living and nonliving.

Half-life: 1. The time required for a pollutant to lose half its effect on the environment. For example, the half-life of DDT in the environment is 15 years and that of radium is 1, 580 years. 2. The time required for half of the atoms of a radioactive element to undergo decay. 3. The time required for the elimination of half of a total dose from the body.

Hazardous wastes: By-products that can pose a substantial or potential hazard to human health or the environment when improperly managed. Substances classified as hazardous wastes possess at least one of four characteristics – ignitability, corrosivity, reactivity, or toxicity – or appear on special lists.

Hazy: A phenomenon that causes reduced visibility due to scattering of light.

Heavy metals: Metallic elements with atomic number greater than 20, such as mercury and lead. They can damage living things at low concentrations and tend to accumulate in the food chain.

Herbicide: A chemical pesticide designed to control or destroy plants, weeds, or grasses.

Herbivore: Animals that eat plants or part of plants.

Holding pond: A pond or reservoir, usually made of earth, built to store polluted runoff.

Hydrocarbons (HC): Chemical compounds that consist entirely of carbon-and hydrogen.

Hydrogen sulfide (H_2S): Gas emitted during organic decomposition and as a by-product of oil refining and burning. It smells like rotten eggs and, in heavy concentration, can cause illness.

Hydrology: The science dealing with the properties, distribution, and circulation of water.

Impoundment. A body of water or sludge confined by a dam, dike, floodgate, or other barrier.

Incineration: Burning of certain types of solid, liquid, or gaseous materials. A treatment technology involving destruction of waste by controlled burning at high temperatures, e.g., burning sludge to remove the water and reduce the remaining residues to a safe, non-burnable ash that can be disposed of safely on land, in some waters, or in underground locations.

Incinerator: A furnace for burning wastes under controlled conditions.

Indirect discharge: Introduction of pollutants from a nondomestic source into a publicly owned waste treatment system. Indirect dischargers can be commercial or industrial facilities whose wastes go into the local sewers.

Indoor air pollution: It occurs in buildings or other enclosed spaces as opposed to those occurring in outdoor or ambient air. Examples of indoor air pollutants include radon gas, Nox, smoke carbon monoxide, formaldehyde, asbestos, etc.

Infiltration: 1. The penetration of water through the ground surface into subsurface soil or the penetration of water from the soil into sewer or other pipes through defective joints, connections, or manhole walls. 2. A land application technique whereby large volumes of wastewater are applied to land and allowed to penetrate the surface and percolate through the underlying soil. See also Percolation

Inorganic chemicals: Chemical substances of mineral origin, not of basically carbon structure.

Inorganic waste: A waste consisting of materials other than plant or animal matter such as sand, glass etc.

Insecticide: A pesticide compound specifically used to kill or control the growth of insects.

Insolation: It is solar radiant energy received by the earth.

Instream use: Water use taking place within a stream channel, e.g., hydroelectric power generation, navigation, water quality improvement, fish propagation, or recreation.

Inversion: An atmospheric condition that occurs when a layer of warm air prevents the rise of cooling air trapped beneath it. This in turn prevents the rise of pollutants that might otherwise be dispersed and can cause an air pollution episode.

Ion exchange treatment: A water-softening method often found on a large scale at water purification plants; the treatment removes some organics and radium by adding calcium oxide or calcium hydroxide to increase the pH to a level at which the metals will precipitate out.

Lagoon, if: A shallow pond in which sunlight, bacterial action, and oxygen work to purify wastewater; also used for storage of wastewaters or spent nuclear fuel rods. A shallow body of water, often separated from the sea by coral reefs or sandbars.

Landfills: Sanitary landfills are land disposal sites for nonhazardous solid wastes at which wastes are spread in layers, compacted to the smallest practical volume, and covered at the end of each operating day. *Secure chemical landfills* are disposal sites for hazardous wastes that are selected and designed to minimize the chance of release of hazardous substances into the environment.

Leachate: A liquid that results when water collects contaminants as it trickles through wastes, agricultural pesticides, or fertilizers.

Leaching: The process by which soluble constituents are dissolved and carried down through the soil by a percolating fluid. Leaching may occur in farming areas and landfills and may result in hazardous substances entering surface water, groundwater, or soil.

Lead: It is a grey-white soft metal. Emitted in air from industrial operations (eg. lead smelter). Expose to lead causes brain and kidney damage and learning disabilities.

Liner: A relatively impermeable barrier designed to prevent leachate from leaking from a landfill. Liner materials include plastic and dense clay. An insert or sleeve for sewer pipes to pre-vent leakage or infiltration.

Mechanical aeration: Use of mechanical energy to inject air into water, causing a waste stream to absorb oxygen.

Methane: A colorless, nonpoisonous, flammable gas created by anaerobic decomposition of organic compounds eg, decomposing solid wastes at landfills, paddy fields etc.

Microbes: Microscopic organisms such as algae, viruses, bacteria, fungi, and protozoa, some of which cause disease.

Mitigation: Measures taken to reduce adverse impacts on the environment.

Mixed liquor: A mixture of activated sludge and water containing organic matter undergoing activated sludge treatment in an aeration tank.

Mobile source: A moving producer of air pollution, mainly forms of transport such as cars, trucks, motorcycles, and airplanes.

Modeling: An investigative technique using a mathematical or physical representation of a system or theory that accounts for all or some of its known properties. Models are used to test the effect of changes in system components on the overall performance of the system.

Monitoring: Periodic or continuous surveillance or testing to determine the level of compliance with statutory requirements or pollutant levels in various media or in humans, animals, and other living things.

Monitoring wells: Wells drilled at a site to collect groundwater samples for the purpose of physical, chemical, or biological analysis to determine the amounts, types, and distribution of contaminants in the groundwater beneath the site.

Mutagen: Any substance that can cause a change in genetic material.

Natural pollutants: Pollutants such as forest fire, volcanic eruptions, wind-blown dusts, sea salt spray etc.

Neutralization: Decreasing the acidity or alkalinity of a substance by adding to it alkaline or acidic materials, respectively.

Nitrate: A compound containing nitrogen that can exist in the atmosphere or as a dissolved gas in water and can have harmful effects on humans and animals. Nitrates in water can cause severe illness in infants and cows.

Nitric oxide (NO): A gas formed by combustion under high temperature and high pressure in an internal combustion engine. It changes to nitrogen dioxide in the ambient air and contributes to photochemical smog.

Nitrification: The process whereby ammonia in wastewater is oxidized to nitrite and then to nitrate by bacterial or chemical reactions.

Nitrogen dioxide (NO_2): The result of nitric oxide combining with oxygen in the atmosphere; a major component of photochemical smog.

Nitrogenous wastes: Animal or vegetable residues that contain significant amounts of nitrogen.

Nitrogen oxides (NO_x): Products of combustion from transport and stationary sources and major contributors to acid deposition and the formation of ground-level ozone in the troposphere.

Nonpoint sources: Pollution sources that are diffuse and do not have a single point of origin or are not introduced into a receiving stream from a specific outlet. The pollutants are generally carried off the land by storm-water runoff. The commonly used categories for nonpoint sources are agriculture, forestry, urban, mining, construction, dams and channels, land disposal, and saltwater intrusion.

Nutrient: Any substance assimilated by living things that promotes growth. The term is generally applied to nitrogen and phosphorus in waste-water but is also applied to other essential and trace elements.

Organic: Referring to or derived from living organisms. In chemistry, any compound containing carbon.

Organic chemicals/compounds: Animal- or plant-produced substances containing mainly carbon, hydrogen, and oxygen.

Organophosphates: Pesticide chemicals that contain phosphorus; used to control insects. They are short-lived, but some can be toxic when first applied.

Outfall: The place where an effluent is discharged into receiving waters.

Overburden: The rock and soil cleared away before mining.

Overland flow: Aland application technique that cleanses waste by allowing it to flow over a sloped surface. As the water flows over the surface, the contaminants are removed. The water is collected at the bottom of the slope for reuse.

Oxidation: The addition of oxygen, which breaks down organic waste or chemicals such as cyanides, phenols, and organic sulfur compounds in sewage by bacterial and chemical means. Oxygen combining with other elements. The process in chemistry whereby electrons are removed from a molecule.

Oxidation pond: A man-made lake or body of water in which liquid waste is consumed by bacteria. It is used most frequently with other water-treatment processes. An oxidation pond is basically the same as a sewage lagoon.

Ozone (O_3): Found in two layers of the atmosphere, the troposphere and the stratosphere. In the troposphere (the layer extending upto 11 kms from the Earth's surface), ozone is a chemical oxidant and major component of photochemical smog. In the stratosphere (the atmospheric layer beginning 11 to 50 kms above the Earth's surface), ozone is a form of oxygen found naturally that provides a protective layer shielding the Earth from the harmful health effects of ultraviolet radiation on humans and the environment.

Ozone depletion: Destruction of the stratospheric ozone layer that shields the Earth from ultraviolet radiation harmful to biological life. This destruction of ozone is caused by the breakdown of certain chlorine or bromine-containing compounds (chlorofluorocarbons or halons) that break down when they reach the stratosphere and catalytically destroy ozone molecules.

Particulates: Fine liquid or solid particles, such as dust, smoke, mist, fumes, or smog, found in air or emissions.

Pathogenic: Capable of causing disease.

Pathogens: Microorganisms that can cause disease in other organisms or in humans, other animals, and plants. They may be bacteria, viruses, or parasites and are found in sewage, in runoff from animal farms or rural areas populated with domestic or wild animals, and in water used for swimming. Fish and shellfish contaminated by pathogens, or the contaminated water itself, can cause serious illness.

Percolation: The movement of water downward and radially through the subsurface soil layers, usually continuing downward to the groundwater.

Permeability: The rate at which liquids pass through soil or other materials in a specified direction.

Persistence: Refers to the length of time a compound, once introduced into the environment, stays there. A compound may persist for less than a second or indefinitely.

Pesticide: Substance or mixture of substances intended for preventing, destroying, repelling, or mitigating any pest. Also, any substance or mixture of substances intended for use as a plant regulator, defoliant, or desiccant. Pesticides can accumulate in the food chain or contaminate the environment, if misused.

pH: A measure of the acidity or alkalinity of a liquid or solid material.

Phenols: Organic compounds that are byproducts of petroleum refining, tanning, and textile, dye, and resin manufacturing. Low concentrations cause taste and odor problems in water; higher concentrations can kill aquatic life and humans.

Phosphates: Certain chemical compounds containing phosphorus.

Phosphorus: An essential chemical food element that can contribute to the eutrophication of lakes and other water bodies. Increased phosphorus levels result from discharge of phosphorus-containing materials into surface waters.

Photochemical oxidants: Air pollutants formed by the action of sunlight on oxides of nitrogen and hydrocarbons.

Photosynthesis: The manufacture of carbohydrates and oxygen by plants from carbon dioxide and water in the presence of chlorophyll, using sunlight as an energy source.

Physical and chemical treatment: Processes generally used in large-scale wastewater treatment facilities. Physical processes may involve air stripping or filtration. Chemical treatment includes coagulation, chlorination, or ozone addition. The term can also refer to treatment of toxic materials in surface waters and groundwater, oil spills, and some methods of dealing with hazardous materials on or in the ground.

Phytoplankton: That portion of the plankton community comprised of tiny plants, e.g., algae.

Phytotoxic: Something that harms plants.

Plume: Visible or measurable discharge of a contaminant from a given point of origin; can be visible or thermal in water or visible in the air as, for example, a plume of smoke. The area of measurable and potentially harmful radiation leaking from a damaged reactor. The distance from a toxic release considered dangerous for those exposed to the leaking fumes.

Point source: A stationary location or fixed facility from which pollutants are discharged or emitted; any single identifiable source of pollution, e.g., a pipe, ditch, ship, ore pit, or factory chimney.'

Pollutant: Generally, the presence of matter or energy whose nature, location, or quantity produces undesired environmental effects. The term is also defined as the man-made or man-induced alteration of the physical, biological, and radiological integrity of environment including water.

Polychlorinated biphenyls (PCBs): a group of toxic, persistent chemicals used in transformers and capacitators for insulating purposes and in gas pipeline systems as a lubricant.

Polyelectrolytes: Synthetic chemicals that help solids to clump during sewage treatment.

Polyvinyl chloride (PVC): A tough, environmentally indestructible plastic that releases hydrochloric acid when burned.

Potable water. Water that is safe for drinking and cooking.

ppm/ppb: Parts per million/parts per billion, a way of expressing tiny concentrations of pollutants in air, water, soil, human tissue, food and/or other products.

Precipitation: Removal of solids from liquid waste so that the hazardous solid portion can be disposed of safely; removal of particles from airborne emissions.

Precipitators: Air pollution control devices that collect particles from an emission.

Precursor: In photochemical terminology, a compound such as a volatile organic compound (VOC) that "precedes" an oxidant. Precursors react in sunlight to form ozone or other photochemical oxidants.

Pretreatment: Processes used to reduce, eliminate, or alter the nature of wastewater pollutants from nondomestic sources before they are discharged into publicly treatment plants.

Primary wastewater treatment: First steps in wastewater treatment; screens and sedimentation tanks are used to remove most materials that float or will settle. Primary treatment results in the removal of about 30% of carbonaceous biochemical oxygen demand (BOD) from domestic sewage.

Putrescible: Able to rot quickly enough to cause odors and attract flies.

Pyrolysis: Decomposition of a chemical by extreme heat.

Radiobiology: The study of the effects of radiation on living things.

Radiation: Any form of energy propagated as rays, waves, or streams of energetic particles. The term is frequently used in relation to the emission of rays from the nucleus of an atom.

Raw sewage: Untreated wastewater.

Receiving waters: A river, lake, ocean, stream, or other watercourse into which wastewater or treated effluent is discharged.

Recycle/reuse: The process of minimizing the generation of waste by recovering usable products that might otherwise become wastes. Examples are the recycling of aluminum cans, waste paper, and plastic bottles.

Residual: Amount of a pollutant remaining in the environment after a natural or technological process has taken place, e.g., the sludge remaining after initial wastewater treatment, or particulates remaining in air after the air passes through a scrubbing or other pollutant removal process.

Reverse osmosis: A water treatment process used in small water systems by adding pressure to force water through a semipermeable membrane, Reverse osmosis removes most drinking water contaminants. It is also used in wastewater treatment. Large-scale reverse osmosis plants are under development.

Risk assessment: The qualitative and quantitative evaluation performed in an effort to define the risk posed to human health or the environment by the presence or potential presence and use of specific pollutants.

Rubbish: Solid waste, excluding wood waste and ashes, from homes, institutions, and workplaces.

Runoff: That part of precipitation, snowmelt, or irrigation water that runs off the land into streams or other surface water; can carry pollutants from the air and land into the receiving waters.

Salinity: The degree of salt in water.

Salts: Minerals that water picks up as it passes through the air and over and under the .ground and as it is used by households and industry.

Sand filters: Devices that remove some suspended solids from sewage. Air and bacteria decompose additional wastes filtering through the sand so that cleaner water drains from the bed.

Sanitary sewers: Underground pipes that carry off only domestic or industrial waste, not stormwater.

Sanitation: Control of physical factors in the environment that could harm human development, health, or survival.

Screening: Use of screens to remove course floating and suspended solids from sewage.

Scrubber: An air pollution device that uses a spray of water or reactant or a dry process to trap pollutants in emissions.

Secondary wastewater treatment: The second step in most water treatment systems, in which bacteria consume the organic parts of the waste. It is accomplished by bringing together waste, bacteria, and oxygen in trickling filters or in the activated sludge process. This treatment removes floating and settleable solids and about 90% of the oxygen-demanding substances and suspended solids. Disinfection is the final stage of secondary treatment.

Sedimentation: Letting solids settle out of waste-water by gravity during wastewater treatment.

Sedimentation tanks: Holding areas for waste-water in which floating wastes are skimmed off and settled solids are removed for disposal.

Sediments: Soil, sand, and minerals washed from land into water, usually after rain. Sediments pile up in reservoirs, rivers, and. harbors, destroying fish-nesting areas and holes of water animals and clouding the water so that needed sunlight may not reach aquatic plants. Careless farming, mining, and building activities will expose sediment materials, allowing them to be washed off the land after rainfalls.

Septic tank: An underground storage tank for wastes from homes having no sewer line to a treatment plant. The wastes go directly from the home to the tank, where the organic waste is decomposed by bacteria and the sludge settles to the bottom. The effluent flows out of the tank into the ground through drains; the sludge is pumped out periodically.

Settleable solids: Material heavy enough to sink to the bottom of a wastewater treatment tank.

Settling tank: A holding area for wastewater in which heavier particles sink to the bottom for removal and disposal.

Sewage: The waste and wastewater produced by residential and commercial establishments and discharged into sewers.

Sewage sludge: Sludge produced at a municipal water treatment works.

Sewer: A channel or conduit that carries waste-water and stormwater runoff from the source to a treatment plant or receiving stream. Sanitary sewers carry household, industrial, and commercial wastes. Storm sewers carry runoff from rain or snow. Combined sewers are used for both purposes.

Silt: Fine particles of sand or rock that can be picked up by the air or water and deposited as sediment.

Skimming: Using a machine to remove oil or scum from the surface of the water.

Slow sand filtration: Treatment process involving passage of raw water through a bed of sand at low velocity that results in the substantial removal of chemical and biological contaminants.

Sludge: A semisolid residue from any of a number of air or water treatment processes. Sludge can be a hazardous waste.

Slurry: A watery mixture of insoluble matter that results from some pollution control techniques.

Smelter: A facility that melts or fuses ore, often with an accompanying chemical change, to separate the metal. Emissions from smelters are known to cause pollution.

Smog: Fog made heavier and darker by smoke. Air pollution associated with oxidants.

Smoke: Particles suspended in air after incomplete combustion of materials.

Solid wastes: Nonliquid, nonsoluble materials, ranging from municipal garbage to industrial wastes that contain complex, and sometimes hazardous, substances. Solid wastes include sewage sludge, agricultural refuse, demolition wastes, and mining residues.

Solid waste disposal: The final placement of refuse that is not salvaged or recycled.

Solid waste management: Supervised handling of waste materials from their source through recovery processes to disposal.

Solidification and stabilization: Removal of wastewater from a waste or changing it chemically to make the waste less permeable and less susceptible to transport by water.

Stabilization: Conversion of the active organic matter in sludge into inert, harmless material.

Stable air: A mass of air that is not moving normally, so that it holds rather than disperses pollutants.

Stack: A chimney or smokestack; a vertical pipe that discharges flue gases etc.

Stack effect: Flue gas as in a chimney, that moves upward because it is warmer than the surrounding atmosphere.

Sterilization: In pest control, the use of radiation and chemicals to damage body cells needed for reproduction. The destruction of all living organisms in water or on the surface of various materials. In contrast, disinfection is the destruction of most living organisms in water or on surfaces.

Sulfur dioxide (SO_2): A heavy, pungent, colorless, gaseous air pollutant formed primarily by fossil fuel combustion.

Sump: A pit or tank that catches liquid runoff for drainage or disposal.

Surface water: All water naturally open to the atmosphere (rivers, lakes, reservoirs, streams, impoundments, seas, estuaries, etc.); also refers to springs, wells, or other collectors that are directly influenced by surface water.

Suspended solids: Small particles of solid pollutants that float on the surface of or are suspended in sewage or other liquids. They resist removal by conventional means.

Tailings: Residue of raw materials or waste separated out during the processing of crops or mineral ores.

Tertiary wastewater treatment: Advanced cleaning of wastewater that goes beyond the secondary or biological stage to remove nutrients such as phosphorus and nitrogen and most biochemical oxygen demand (BOD) and suspended solids.

Thermal pollution: Discharge of heated water from industrial processes that can affect the life processes of aquatic organisms.

Total suspended solids (TSS): A measure of the suspended solids in wastewater, effluent, or water bodies.

Toxic pollutants: Materials contaminating the environment that cause death, disease, or birth defects in organisms that ingest or absorb them. The quantities and length of exposure necessary to cause these effects can vary widely.

Toxic substance: A chemical or mixture that may present an unreasonable risk of injury to health or the environment.

Toxicity: The degree of danger posed by a substance to animal or plant life.

Trichloroethylene (TCE): A stable, low-boiling-point colorless liquid, toxic by inhalation. TCE is used as a solvent, as a metal degreasing agent, and in other industrial applications.

Trickling filter: A coarse biological treatment system in which wastewater trickles over a bed of stones or other material covered with bacterial growth. The bacteria break down the organic waste in the sewage and produce clean water.

Turbidity: Haziness in air caused by the presence of particles and pollutants. A similar cloudy condition in water due to suspended silt or organic matter.

Urban runoff: Stormwater from city streets and adjacent domestic or commercial properties that may carry pollutants of various kinds into sewer systems or receiving waters.

Vapor: The gaseous phase of substances that are liquid or solid at atmospheric temperature and pressure, e.g., steam.

Vapor capture system: Any combination of hoods and ventilation system that captures or contains organic vapors so that they may be directed to an abatement or recovery device.

Vinyl chloride: A chemical compound, used in producing some plastics. It is believed to be carcinogenic.

Volatile: Description of any substance that evaporates readily.

Volatile organic compound (VOC): Any organic compound that participates in atmospheric photochemical reactions; generally have a boiling point of less than 145°C Celsius.

Wastes: Unwanted materials left over from a manufacturing process. Refuse from places of human or animal habitation.

Wastewater treatment plant: A facility containing a series of tanks, screens, filters, and other processes by which pollutants are removed from water.

Wastewater: Spent or used water from individual homes, communities, farms, or industries that contains dissolved or suspended matter.

Water pollution: The presence in water of enough harmful or objectionable material to damage water quality.

Water quality criteria: Specific levels of water quality that, if reached, are expected to render a body of water suitable for its designated use. The criteria are based on specific levels of pollutants that would make the water harmful if used for drinking, swimming, farming, fish production, or industrial processes.

Wetlands: An area that is regularly saturated by surface water or groundwater and is subsequently characterized by a prevalence of vegetation adapted for life in saturated soil conditions. Examples include swamps, bogs, marshes, and estuaries.

CHAPTER 2

Terminologies, Definitions, Data, Units and their Conversions

Introductory concept of certain preliminary terms concerning the subject of environmental pollution control is required for better understanding of the subject mater. A first hand information is given below regarding them.

Terminologies Related to Air Pollution

AAQS-Ambient air quality standards

Abatement: Reduction in degree or intensity of pollution.

Abiotic component: Non-living component of the environment.

Aerosols: It means finely divided solid or liquid particles suspended in a gas (size ranger is 50 micron to sub-micron) and includes haze, smoke, dust, must, fume & fog.

Acid Rain: A condition in which natural precipitation becomes acidic after chemically reaching with pollutions in air.

Air Monitoring: It comprised of continuous sampling & measurement of air pollutions quantity present in outdoor atmosphere.

Air Pollution: Presence of abnormal quantity of obnoxious matter in the atmospheric air, which are detrimental to the health and/or welfare of the man.

Ambient air: The surrounding local air.

Anoxia: Deficiency of oxygen in blood or thereby in body tissues.

Antartica: The continent and nearby seas surrounding the south pole.

Bioconversion: A biological process to transform living organism into usable material.

Biota: All living organism including both plant and animal ie flora and fauna of an area.

Canopy: A leafy portion of a tree or shrub.

Chlorosis: The loss of chlorophyll in a plant, which is responsible for sickening & yellowing in plants & vegetation. IT occurs when chlorophyll formation is prevented.

Coalescence: Merging two drop of liquid into a single larger drop.

Coarse solid particles: solid particles of size ≥ 50 microns & also solid particles contained in or on liquid particles.

Colloids: Finely divided solids (5-200 millimicron size), which will not settle and lie between true solution & true suspension.

Contaminant: A substance not occurring in nature, but is introduced into atmosphere by human activity affecting its composition.

Cost/benefit ratio: An economic evaluation index relating control costs to benefits.

Crust: It is the outer shell of the earth. The crust rests on mantle and varies in thickness from an average of 6 Km in the ocean areal to 35-70 Kms in continental regions.

Cyclone separators: A mechanical used for separation of aerosols (liquid/wet solid/dry solid) from gas imposing a rapid whirling motion to the entering gas stream, by employing centrifugal force.

Digestion: Biochemical decomposition of organic matter. Digestion of sewage sludge occurs in tank, where it breaks dari into gas, liquid and mineral matter.

Dispersion: Particulate matter suspended in air or other gases. It is a method by which a pollutant spreads from its point of emission and becomes disputed in the atmosphere. It includes transport by winds & simultaneous spread by turbulence.

Dispersant: A chemical agent used to break up concentration of organic material such as spilled oil.

Dump: A site used to dispose of solid wastes without environment controls.

Dust: Finely divided irregular shaped particles of size 1-75 µm.

Effective chimney height: It is the sum of actual chimney height and the plume rise. Plume (smoke) rise is the rise of the effluent due to buogancy and efflux velocity. It separates fine particles into size fraction by their different rate of settling down under gravity in a stream of fluid.

Electrostatic Precipitators: An electrical equipment for removal of aerosols from a gas stream by electrically charging them and collecting on a plate/tube imparted with an opposite charge.

Ellutriation: A process of separating higher particles from heavier particles by washing solid waste with a slowly moving upward stream of fluid that carried higher particles with it.

Emission: The material which is being discharged into the outdoor atmosphere.

Environmental Impact Assessment (EIA): A detailed study report that discloses the environmental consequences of a course of action as an aid to decision making.

Environment: It refers to the man's total surrounding often specified as biological environment, air environment, water environment, marine environment etc.

Fauna and Flora: All the animal species (fauna) and plant species (flora) found in a country, area or region.

Floc: Small gelatinous mass formed in a liquid by aggregation of a number of fine suspended particles.

Flocculation: Bonding together of coagulated particles to from settelable or filterable solids.

Flora: Plant population of a region.

Fly ash: Finely divided particles of ash (may contain incompletely burned fuel) entrained in flue gases produced by the combustion of fossil fuels.

Fossil fuels: Fuels found inside the crust of the earth eg. Coal, crude oil, natural gas.

Fumes: Usually spherical solid particles (0.001 to 1 micron size) produced by condensation of metals from gaseous state.

Grain: Unit of mass, 1 grain= 0.0648 gram

Green House Gas: Those gases which trap infrared radiation given by the sum to produce green house effect leading to heating up of environment eg, CO_2, O_3, H_2O vapor, CH_4, N_2O etc.

Grit: Solid particles (of size 75 to 200 μn in micrometer) in air or flue gas, which fall out or deposit under their own weight.

Incineration: Process of burning solid/semi-solid/gaseous combustible waste to an inoffensive gas and a sterile residue (containing little or no combustible material).

Isokinetic: A condition of sampling in which the flow of gas into the sampling device (at the opening or face of the inlet) has the same flow rate and direction as the ambient atmosphere being sampled.

Leaching: The process by which nutrient chemicals or contaminants are dissolved and carried away by water or are moved into a lower layer of soil.

Lithosphere: The solid portion of the earth ie, the crust and the mantle.

Micron: 1 micron $(\mu) = 10^{-6}\,meter = 10^{-4}\,cm = 10^{-3}\,mm$

MINAS: Minimum national standards prescribed by central board for prevention and control of wastes for effluent discharges and stack emissions.

Mist: Liquid (water) droplet smaller than 10 micron in diameter suspended in the atmosphere reducing visibility between 1 and 2kms.

Natural atmospheric disperiods: It includes (with particle size range) the following:

$$haze \rightarrow 1\mu, Mist \rightarrow 1-5\mu, cloud\,/\,fog \rightarrow 5-200\mu$$

$$drizzle: 200-500\mu, Rain-500-8000\mu$$

Necrosis: Collapse and death of tissues.

Oxidants: Substances like peroxyacetyl or peroxyacyl nitrate (PAN-CH_3COONO_2), ozone (O_3)& nitrogen dioxide (NO_2) present in air, which are capable of oxidizing other chemicals/elements in oxidation-reduction type reactions.

PAN: Peroxyacetyl nitrate (a phytotoxicant) produced in the atmosphere by photochemical reaction (under the influence of sunlight) between hydrocarbons & oxides of nitrogen (NO_x).

Particulate matter: Any solid/liquid matter (0.0002 μ to 500μ in diameter) present in the atmosphere.

Parts per million (ppm): Number of parts of a gas (by volume) in a million (10^6) total parts. For ideal gases, pressure ratio = mole ratio = volume ratio

Phon: A unit of loudness.

Photochemical reaction: Chemical reaction occurring in the presence of sunlight in the atmosphere. Photochemical smog is produced by this reactor deriving energy from the ultraviolet rays of the sun.

Phytotoxic: Poisonous to plants.

Plankton: Organisms of minute size which floats/swims in waters of seas, rivers, ponds and lakes.

Plume: Flow/path of smoke from the chimney flue gas or vent. Plume of smoke may be visible but for some pollutions, it may not be visible.

Pollutant: Substances causing contamination of any part of the environment.

Polychlorinated biphenyl(PCB): It is a chlorinated hydrocarbon liquid used as insulator in high voltage transformers. Its use is being reduced due to evidence of its persistence and toxicity in the environment.

Ppb: Parts Per Billion (10^9).

Primary pollutant: Pollutants emitted directly from the sources like dust, smoke, ash, fame, SO_2, NO_2, H_2S, CO, NO, hydrocarbons, radioactive substances etc.

Receptor: Any thing that is affected by the pollutants eg, human beings are receptor of photochemical smog causing breathing trouble & irritation in eyes.

Respirator: A device used for breathing filtered air while working in an area having air contamination.

Ringelmann's chart: A chart wed to evaluate the degree of blackness of smoke from a chimney.

Saprophyte: A plant living on dead organic matter.

Scrubbers: Device used for removing dust, aerosols & gaseous pollutants from air by absorption or, reaction or contact with a liquid spray in a hollow or packed tower.

Secondary pollutant: Pollutants that are formed in the atmosphere by the chemical reaction among primary pollutants and normal atmospheric constituents eg, SO_3, H_2SO_4, O_3, No_2etc.

Sink: The material or medium, which consumer or interacts with a long lived pollutant eg, ocean are the sinks for atmospheric CO_2, ground water & subsoil water acts as a sink for the pesticides, marble acts as a sink for atmospheric sulphuric acid vapor etc.

Smog: It is a mixture of smoke, dust particles & small drops of fog. It characterizes certain type of atmospheric pollution by aerosols.

Smoke: Small gas bone liquid/solid particles (such as carbons ash -0.001 to 1 micron size) formed by incomplete combustion of fuels.

Sone: A unit of loudness.

Soot: A conglomeration of particles impregnated with tar formed by incomplete combustion of carbonaceous materials.

Sorption: A process consisting of either absorption or adsorption.

Stable atmosphere: State of atmosphere in which vertical air movements are restricted. This happens when there is temperature inversion.

Synergistic effect: This term is used to characterize the combined effect of air pollutants, when that effect is greater than or different from what would be expected, if the known effects of each of pollutants were simply added together.

Threshold limit value(TLV): It is the maximum allowable concentrations (MAC) of air bone substances/ pollutants, which do not cause adverse effect on human beings, when exposed for 8 hours a day for 5 continuous day in a week.

Tidal Volume: Amount of goses inhaled/exhaled in each breath maximum breathing capacity = 125 liters of gas breathed per minute.

Toxicology: Study of the harmful effects of toxic substances on living organisms in any ecosystem.

Toxic pollutants: Those pollutants which cave on inhalation/exposure, disease, physiological malfunction (eg, malfunction in reproduction), behaviourial abnormability, physical deformation in organisms/offsprings and finally death depending on the extent & duration of exposure.

Wash out: A process of removal of particulate and soluble matter from air by water drops during rain precipitation.

Terminology related to Water Pollution

Activated sludge process: removers organic matter from sewage by saturating it with air & biologically activé sludge.

Adsorption: Waste treatment technique in which carbon removes organic matter not responsive to clarification or biological treatment.

Aeration tank: Serves or the chamber for injecting air into water.

Algae: are plants which grow in sunlit water and are food for fish/small aquatic animals and like all plants put oxygen into water.

Bacteria: Smallest lining organism which eat the organic past of sewage.

BOD: Biological oxygen demand, is the amount of oxygen necessary in the water for bacteria which consume the organic sewage. It is used as a measure in telling how well a sewage treatment plant is working.

Chlorinator: Is a device for adding chlorine gas to sewage to kill infectious germs.

Coagulation: It is the clumping together of solids to make them settle out of sewage faster. Line, alum or polyelectrolytes are used for coagulation of solids.

Combined sewer: Carries both sewage and storm water runoff.

Comminutor: Is a device for catching &shredding of heavy solid matter in the primary stage of sewage treatment.

Diffused air: is a technique by which air under pressure is forced into sewage in an aeration tank.

Digestion: of sludge tanks place in a heated tank, where the material can decompose naturally & the odors can be controlled.

Effluent: is the liquid that comes out of a treatment plant after the completion of the treatment process.

Electrodialysis: is a process by which electricity attracts the mineral salts from sewage.

Incineration: Consists of burning the sludge to remove the water and reduce the remaining residues to a safe non-burnable ash which can be safely disposed on land, in water or underground to cations (caves).

Lateral sewers: Pipes that run under the street of a city and into which the sewers from homes/industries empty.

Lagoons: Ponds in which sunlight, algae & oxygen interact to restore water to a quality equal to effhurt from a secondary treatment plant.

Mechanical aeration: Forcing the sewage up through a pipe in a tank & then spraying over surface of the tank cavering waste stream to absorb oxygen from the atmosphere.

Microbes: Minute living plants/animals in the sewage. Acts as germs causing disease.

Mixed liquor: Effluent coming from aeration tank after sewage has been mixed with activated sludge & air.

Organic matter: Waste from home/industry which is of plant/animal origin.

Oxidation: Consuming or breaking down of organic water or chemicals in sewage by bacteria & chemical oxidants.

Oxidation Ponds: Man made lake or body of water in which wastes are consumed by bacteria. It is some as a sewage lagoon.

Sand filter: removes organic waste from sewage. Waste water is trickled over the bed of sand air & bacteria decompose the waste filtering through the sand.

Sanitary sewer: Pipes in a city that carry only domestic waste water. The storm water run off is taken care of by a separate system of pipes.

Secondary treatment: Second step in waste treatment system in which bacteria consumes the wastes. It is done by bringing the sewage and bacteria together in trickling filters or activated sludge process.

Sedimentation tanks: removes solid from sewage. Water is pumped to the tank where solids settle at the bottom or float on top as scum. Scum is skimmed off the top and solids on the bottom are pumped out to sludge digestion tanks.

Septic Tanks: Used to treat domestic wastes. Underground tanks receives the waste water directly from home. Bacteria in the sewage decomposes the organic wastes and the sludge settles on the bottom. Effluent flows out of the tank into ground through drains sludge is pumped out of the tanks periodically.

Sinking agents: Those materials which are applied to oils and hazardous substance spills to sink floating pollutants below the water surface.

Sludge: Solid matter that settles to the bottom of sedimentation tank and is disposed off by digestion or other methods.

SPM: Suspended particulate matter.

Storm sewers: Separate system of pipes that carry only sun offs from building & land during a storm.

TSS: Total suspended solids

Trickling filter: Bed of rocks/stones. Sewage is trickled over the beds so that the bacteria can break down the organic wastes. Bacteria collect on the rocks/stones through repeated use of the filters.

Thermal pollution: Pollution resulting from the discharge of heated waste water at temperatures that can be detrimental to aquatic life.

Terminology related to Waste Water Treatment

Activated carbon: Carbon particles obtained by carbonization of cellulosic material in the absence of air and possessing a high adsorptive capacity.

Activated sludge process: A biological waste-water treatment process in which a mixture of wastewater and activated sludge is agitated and aerated. The activated sludge is subsequently separated from the treated wastewater (mixed liquor) by sedimentation, and wasted or returned to the process as needed.

Adsorption: The adherence of a gas, liquid, or dissolved material on the surface of a solid.

Advanced treatment: Tertiary treatment of waste water.

Aeration: The bringing about of intimate contact between air and a liquid by one or more methods, such as spraying the liquid in the air, bubbling air through the liquid, or agitation of the liquid to promote surface absorption of the air.

Aeration period: The theoretical time, usually expressed in hours, during which mixed liquor is subjected to aeration in an aeration tank while undergoing activated sludge treatment it is equal to the volume of the tank divided by the volumetric rate of flow of the wastewater and return sludge. The theoretical time that water is subjected to aeration.

Agglomeration: The coalescence of dispersed suspended matter into larger flocs or particles that settle rapidly.

Agitator: Mechanical apparatus for mixing and/or aerating; a device for creating trubulence.

Alkaline: The condition of wastewater containing surfficient amounts of substances to raise the pH above 7.0.

Alkalinity: The capacity of water to neutralize acids. A term used to represent the content of carbonates, bicarbonates, hydroxides, and occasionally borates, silicates, and phospates in water. It is expressed in milligrams per liter of calcium carbonate, ($CaCO_3$) that would neutralize an equal amount of acid.

Alum: A common name for commercial grade aluminum sulfate, $[Al_2(SO_4)_3 \times H_2O]$

Ammonification: Bacterial decomposition of organic nitrogen to ammonia.

Anabolism: In biochemical process, when the external materials are necessary for the growth, maintenance and other metabolic activities for the cell of organisms, is known as anabolism.

Anaerobic: Requiring, or not destroyed by the absence of air or free elemental oxygen.

Bacteria: A group of universally distributed, rigid, essentially unicellular microscopic organisms lacking chlorophyll. Bacteria usually appear as spheroid, rod-like, or curved entities, but occasionally appear as sheets chains, or branched filaments. Bacteria are usually regarded as plants.

Bacteria, aerobic: Bacteria that require free elemental oxygen for their growth.

Bacteria, coliform group: A group of bacteria, predominantly inhabitants of the intestine of man or animals but also found on vegetatin including all aerobic and facultative anaerobic gram-negative, nonspore-forming bacilli that ferment lectors with gas formation.

Bacteria count: A measure of the concentration of bacteria.

Bacteria, facultative: Bacteria that can adapt themselves to growth in the presence, as well as in the absence of oxygen. May be referred to as facultative bacteria.

Bacteria, parasitic, : Bacteria that thrive on other living organisms.

Bacteria, pathogenic: Bacteria that may cause disease in the host organism by their parasitic growth.

Bacteria Plate: Number of colonies of bacteria grown on selected solid media at a given temperature and incubation period, usually expressed as the number of bacteria per milliliter of sample.

Bacteria, saprophytic: Bacteria that thrive on dead organic matter.

Biochemical: Pertaining to chemical change resulting from biological action. Measured by or expressed in terms of the ensuing chemical change.

Biochemical action: Chemical changes resulting from the metabolism of living organisms.

Biochemical oxygen demand (BOD): The quantity of oxygen used in the biochemical oxidation of organic matter in a specific time, at a specified temperature, and under specific conditions.

Biochemical oxygen demand, standard: Biochemical oxygen demand as determined under standard laboratory procedure for 5 days at 20°C expressed in milligrams per liter.

Biological Oxidation: The process whereby living organisms in the presence of oxygen convert the organic matter contained in wastewater into a more stable or mineral from.

Biological process: The process by which the life activities of bacteria and other micro organisms in search for food break down complex organic materials into simple, more stable substances. Self purification of polluted streams, sludge digestion, and all secondary wastewater treatments result from this process.

Blinding: Clogging of the filtering medium of a waste water filter.

Burner, waste gas: A gas device used for burning the waste gas from an anaerobic sludge digestion tank.

Carbonation: The diffusion of carbon dioxide gas through a liquid to render the liquid stable with respect to precipitation or dissolution of alkalineconstituents. Can also be used to lower the pH.

Catabolism: The biochemical process with provides the energy for the synthesis for new cells, as well as for the maintenance of other cell functions, is called catabolism.

Centrifuge: A mechanical device in which centrifugal force is used to separate solids from liquids and/or to separate liquids of different densities.

Chamber: A space enclosed by walls and a floor or a compartment, often prefixed by a descriptive word, such as "grit chamber", "screen chamber, " or "flushing chamber, " indicating its function.

Chamber, detritus: A settling tank of short detention period designed primarily to remove heavy settleable solids, usually without interrupting the flow of waste. Also called detritus detributs tank.

Chamber, flowthrough: The upper compartment of a two-storey sedimentation tank.

Chemical oxygen demand (COD): A measure of the oxygen-consuming capacity of inorganic and organic matter present in waste-water. It is expressed as the amount of oxygen consumed from a chemical oxidation in a specific test.

Chloramines: Compounds of organic amines or inorganic nitrogen with-chlorine.

Chlorinated lime: A combination of slaked lime and chlorine (also termed bleaching powder, chloride of lime, and hypochlorite of lime) in which calcium oxychloride is the active ingredient. When dissolved in water, it serves as a source of chlorine. It is relatively unstable.

Chlorination: The application of chlorine to wastewater, generally for the purpose of disinfection.

Chlorination, breakpoint: The addition of chlorine to wastewater until the chlorine demand has been satisfied and further additions result in a residual that is directly proportioned to the amount added beyond the breakpoint.

Chlorination pre: the application of chlorine to wastewater before waste any water treatment step.

Chlorine: An element ordinarily existing as a greenish yellow gas approximately 2.5 times as heavy as air. At atmospheric pressure and temperature of -34.5°C, the gas becomes liquid about 1.5 times as heavy as water.

Chlorine, available: A measure of the total oxidizing power of chlorinated lime and hypochlorites.

Chlorine contact chamber: A detention basin provided primarily to secure the diffusion of chlorine through the liquid. Also called chlorination chamber.

Chlorine demand: The difference between the amount of chlorine added to wastewater and the amount of residual chlorine remaining at the end of a specified contact time. The chlorine demand for given water varies with the amount of chlorine applied, time of contact, temperature, PH, and nature and the amount of impurities in the water.

Chlorine, dose: The amount of chlorine applied to a liquid, usually expressed in milligrams per liter.

Chlorine liquid: Elemental chlorine placed in a liquid state by a combination of compression and refrigeration of dry, purified chlorine gas. Liquid chlorine is shipped under pressure in steel containers.

Chlorine, residual: Chlorine (combined and free chlorine) remaining in wastewater at the end of a specific contact period.

Clarifier: It means sedimentation tank.

Coagulation: The destablization and initial aggregation of colloidal and finely divided addition, of a flocforming chemical or by biological processes.

Colloids: Finely divided solids <200 microns that will not settle but may be removed by coagulation biochemical action, or membrane filtration.

Combined available residual chlorine: That portion of the total residual chlorine remaining in Wastewater at the end of a specified contact period that will react chemically and biologically as chloramines or organic chloramines.

Combined sewer: A sewer intended to receive both wastewater and storm or surface water.

Combined wastewater: A mixture or surface runoff and other wastewater such as domestic or industrial wastewater.

Comminution: The process of cutting and screening solids contained in wastewater flow before they enter the flow pumps or other units in a wastewater treatment plant.

Comminutor: A grinder or shredder that converts bulky solid wastes into small particles.

Composting: A technique for processing solid wastes and/or wastewater sludge into a soil conditioner called compost.

Contamination: Any introduction into water of microorganisms, chemicals, wastes, or wastewater in a concentration that makes the water unfit for its intended use.

Decomposition: The breakdown of complex material into simpler substances by chemical or biological means.

Degreasing: The process of removing greases and oils from waste, wastewater, sludge, or solid wastes.

Denitrifiction: The removal of nitrogen or nitrogen compounds from wastewater.

Dexygenation: The depletion of DO dissolved oxygen in a liquid either under natural conditions associated with the biochemical oxidation of organic matter present or by the addition of chemical reducing agents.

Deposition: The act or process of settling solid material from a fluid suspension.

Detention period: The theoretical time required to displace the contents of a tank or treatment unit at a given rate of discharge (volume divided by rate of discharge). Also called detention time.

Detritus: The sand, grit, and other coarse material carried by wastewater.

Diffuser: A porous plate, tube, or other device through which air is forced and divided into minute bubbles for diffusion in liquids. Commonly made of porous ceramic, metal, or plastic, materials.

Digester: A tank in which sludge is placed to permit digestion to occur. Also called sludge digestion tank.

Digestion: The biological decomposition of organic matter in sludge, resulting in partial gasification, liquifaction, and mineralization.

Disinfection: The art of killing larger portion of microorganisms in or on a substance with the probability that all the pathogenic bacteria are killed by the agent used.

Diurnal: Having a daily cycle; recurring every day.

Dolomite lime: Lime containing 30 to 50 percent magnesia and-40 to 50 percent magnesia calcium oxide (CaO).

Dual media filtration: A filtration process that use a bed composed of two substances with different specific gravities such as anthracite coal and sand.

Educator: A device for maxing two fluids, such as air with water.

Effluent: Wastewater or other liquid, partially or completely treated, or in its natural state, flowing out of a reservoir, basin, treatment plant, or industrial treatment plant.

Ejector, pneumatic: A device for raising wastewater, sludge, or other liquid by entraining it in a high velocity stream of air or water.

Electrolyte:A substance that dissociates into two or more ions when it is dissolved in water.

Elutriation: A process of sludge conditioning whereby the sludge is washed with either fresh water or plant effluent to reduce the demand for conditioning chemicals and to improve settling or filtering characteristics of the solids. alkalinity is reduced in this process.

Eutrophication: The process of enrichment of water in nutrients (mainly by nitrogen and phosphorous) and so supporting the dense plant population, the decomposition of with kills animal life by depriving of oxygen and causing slow death to the water bodies, is known as eutrophication.

Facultative: Having the ability to live under both aerobic and anaerobic conditions.

Filter: A device or structure for removing solid or colloidal material, usually of a type that cannot be removed by sedimentation. The liquid is passed through a filtering medium that may be natural or artificial granular material, finely woven cloth, unglazed porcelain, specially prepared paper, or synthetic media.

Filter, high rate: A biological filter operated at a high average daily dosing rate usually between 10 to 40 mgd/acre including any recirculation of effluent.

Filter, intermittent: A natural or artificial bed of sand or other fine-grained material to the surface of which wastewater is intermittently applied in flooding doses and through which the liquid portion passes. Solids are retained in the upper layers of the bed under aerobic conditions.

Filter, low-rate: A biological filter designed to receive a small load of BOD per unit volume of filtering material and to have a low dosage rate per unit of surface area, usually 1 to 4 mgd/acre, and generally without recirculation.

Filter medium: (1) Any material through which wastewater or sludge is passed for the purpose of purification, treatment, or conditioning.

Filter press: A device operated mechanically for partially separating water from solid material.

Filter, sand: A filter in which sand is used as a filtering medium.

Filter, trickling: A filter consisting of a bed of coarse material, such as broken stone, clinkers, slate, slats, brush, or plastic materials, over which wastewater is distributed or applied in drops, films, or spray from troughs, drippers, moving distributors, or fixed nozzles and through which it trickles to the underdrains, giving opportunity for the formation of slimes on the media surfaces to clarify and oxidise the wastewater.

Filter, vacuum: A continuous filter consisting of a cylindrical drum mounted on a horizontal axis, covered with a filter cloth, revolving with a partial submergence in liquid. A vacuum is maintained under the cloth for the larger part of a revolution to extract moisture. The cake is then scrapped off.

Floc: Small gelatinous masses formed in a liquid by agglomeration of smaller particles.

Flocculation: In water and wastewater treatment, the agglomeration of colloidal and finely divided suspended matter after coagulation by gentle stirring by either mechanical or hydraulic means.

Floatation: The raising of suspended matter to the surface of the liquid in a tank as scum by aeration, evolution of gas, chemicals, electrolysis, heat, or bacterial decomposition and the subsequent removal of the scum by skimming.

Gate: A movable, watertight barrier for the control of liquid in a waterway.

Gate, sluice: A gate used for controlling the flow of water. It is constructed to slide vertically, and fastened into or against the walls of dams, tanks, conduits, or other structures.

Grease: A group of substances including fats, waxes, free fatty acids calcium and magnesium soaps, mineral oils, and certain other nonfatty materials.

Grease skimmer: A device for removing floating grease or scum from the surface of wastewater.

Grit: The heavy suspended mineral matter in wastewater, such as sand, gravel, and cinders.

Grit collector: A device placed in a grit chamber to convey deposited grit to a point of collection.

Heavy metals: Metals that can be precipitated by hydrogen sulfide in acid solution, for example, lead, silver, gold, mercury, bismuth, and copper.

Humus: The dark-colored, carboniferous residue in the soil resulting form the, decomposition of vegetable tissues of plants originally growing therein. Residues similar in appearance and behavior are found in well-digested sludges and in activated sludge.

Hydrated lime: Limestone that has been calcined and treated with water until the calcium oxide portion has been converted to calcium hydroxide.

Hydraulic loading: The flow (volume per unit time) applied to a unit process.

Hypochlorite: Calcium, sodium, or lithium hypochlorite.

Imhoff cone: A conically shaped one-liter graduated glass vessel used to measure the approximate volume of settleable solids in various liquids of wastewater origin during various settling periods.

Impeller: A set of vanes designed to rotate and to cause rotation of a mass of fluid.

Incilferation: The combustion (by burning) of organic matter in wastewater sludge solids after water evaporation from the solids.

Industrial wastes: The solid and liquid wastes from industrial processes, as distinct from domestic or sanitary wastes.

Industrial wastewater: Wastewater in which industrial wastes predominate.

Inflow: usually the rainfall that enters a sewer system through direct connections such as roof leaders and catch basins.

Influent: Wastewater or other liquid, raw or partially treated, flowing into a reservoir, basin, treatment process, or treatment plant.

Interstice: A pore or open space in rock or granular material not occupied by solid matter. It may be occupied by air, water, or other gaseous or liquid material. Also called void or void space.

Lagoon: A shallow body of water, as a pond or lake, that usually has a shallow, restricted inlet from the sea. A pond containing raw or partially treated wastewater in which aerobic or anaerobic stabilization occurs.

Lagoon, sludge: A basin used for the storage, digestion, or dewatering of sludge.

Liquor, mixed: A mixture of activated sludge and wastewater into which oxygen or air is added in the activated sludge process.

Liquor, supernatant: The liquor overlying deposited solids. The liquid in a sludge digestion tank that lies between the sludge at the bottom and the floating scum at the top.

Matter, inrganic: Chemical substances of mineral origin.

Matter, organic: Chemical substances of mineral or vegetable origin, or more correctly, of basically carbon structure. They include most carbon compounds; most are combustible and many are volatile.

Matter, suspended: Solids in suspension in wastewater or effluent; also called suspended solids (SS).

Metabolism: The total biochemical reactions (changes) by which nutritional activities of an organism are maintained involving assimilation and use of food, subsistence, growth and reproduction, is known as metabolism.

Microorganism: Minute organism, either plant or animal, invisible or barely visible to the naked eye.

Mixed liquor suspended solids (MLSS): The concentration of SS (suspended solids) carried in the aeration tank of an activated sludge process.

Most probable number (MPN): That number of organisms per unit volume that, would be more likely than any other number to yield the observed test result or that would yield the observed test result or that would yield the observed test result with the greatest frequency. Expressed as density or organisms per 100 milliliters. Results are computed from the number of positive findings of coliform group organisms resulting from multiple-portion decimal-dilution plantings.

Muriatic acid: Hydrochloric acid (HCl).

Nitrification: The conversion of nitrogenous matter into nitrates by bacteria.

Nutrients: Inorganic substances (K, N, P) involved in energy transfer and necessary for the growth and development of plants and animals.

Odor: The property or quality of a thing that stimulates or is perceived by the sense of smell.

Odor control: The prevention or reduction of objectionable odors by chlorination, aeration, or other processes.

Organic nitrogen: Nitrogen combined in organic molecules such as proteins, amines, and amino acids.

Outfall. The point, location, or structure where wastewater or drainage discharges from a sewer, drain, or conduit into the receiving waters.

Overflow rate: One of the criteria for the design of sedimentation tanks in treatment plants; the rate of flow per surface area of settling tank.

Oxidation, direct: The direct combination of substances with oxidants accomplished without benefit of living organisms. For example, oxidation of substances in wastewater by the, direct application of air or oxidizing agents such as chlorine.

Oxidation pond: Stabilization lagoon, exemplifies on oxidation pond.

Oxidation-reduction potential (ORP): The potential required to transfer electrons from the oxidant to the reductant. Used as a qualitative measure of the state of oxidation in wastewater treatment systems.

Oxygen balance: The relation between the BOD of a wastewater or treatment plant effluent and the oxygen available in the diluting water.

Oxygen deficiency: The additional quantity of oxygen required to satisfy the oxygen required to satisfy the oxygen requirement in a given liquid. Usually expressed in milligrams per liter. Any atmosphere containing less than 13 per cent oxygen is dangerous to human beings and is called an atmosphere with oxygen deficiency.

Oxygen, dissolved (DO): The oxygen dissolved in a liquid, usually expressed either in milligrams per liter or percentage of saturation.

Oxygen, residual: The DO content of a stream after deoxygenation has begun.

Oxygen, uptake rate: The amount of oxygen being used by an activated sludge system during a specific period time.

Parts per million (ppm): The number of weight or volume units of a minor constituent present with each 1 million units of the major constituent of a solution or mixture. Formerly used to express the results of most water and wastewater analyses, but more recently replaced by the ratio milligrams per liter (to which it is essentially equivalent).

Percolation: The flow or trickling of a liquid through a contact or filtering medium.

pH: A symbol denoting the negative logarithm of the hydrogen ion concentration in a solution. pH values run from 1 to 14. The number 7 indicates neutrality. > 7- alkaline, <7- Acidic.

Phosphorous: A highly reactive nonmetallic element, found in wastewater in three; principal forms orthophosphate ion, polyphosphates, and organic phosphorous compounds.

Plankton: The aggregate of passively floating, drifting, or weakly motile organisms in a body of water. The organisms are mostly microscopic.

Pollutional load: The quantity of material in a waste stream that requires treatment or exerts an adverse effect on the receiving system. The quantity of material carried in body of water that exerts a detrimental effect on so e subsequent use of that water.

Polyelectrolytes: Chemicals that consist of high-molecular weight molecules with many reactive groups situated along the length of the chain. Polyelectrolytes react with the fine particles in the waste and assist in bringing them together into larger and heavier masses for settling.

Potable water: Water that does not contain objectionable pollution, contamination, minerals, or infective agents and is considered satisfactory for human consumption.

Preaeration: A preparatory treatment of wastewater consisting of aeration to remove gases, add oxygen, or promote flotation of grease and aid coagulation.

Precipitation, chemical: Precipitation induced by the addition of chemicals.

Putrefaction: Biological decomposition of organic matter with the production of foul smelling products associated with anaerobic conditions.

Putrescibility: The relative tendency of organic matter to undergo decomposition in the absence of oxygen. The susceptibility of wastewater, effluent, or sludge to putrefaction. Term used to define stability of a polluted water or raw or partially treated wastewater.

Quicklime: A calcined material, the major part of which is calcium oxide or calcium oxide in natural association with a lesser amount of magnesium oxide. It is capable of slaking with water.

Rack, bar: A screen composed of parallel bars, either vertical or inclined, placed in a water-way to catch floating and large suspended debris and from which screenings may be raked.

Rate, filtration: The rate of application of wastewater to a filter, usually expressed in million gallons per acre per day or gallons per minute per square foot.

Rate, recirculation: The rate of return of part of the effluent from a treatment process to the incoming flow.

Reaeration: The absorption of oxygen into water where the water is in a condition of oxygen deficiency.

Reaeration, sludge. The continuous aeration of sludge after its initial aeration for the purpose of improving or maintaining its condition.

Recarbonation: The process of introducing carbon dioxide as a final stage in the lime-soda ash softening process to convert carbonates to bicarbonates and thereby stablize the solution against precipitation of carbonates.

Rotating biological reactor: A rotating bed of synthetic media that moves through a fixed trough into which wastewater is fed. The media is partially submerged in the wastewater, given opportunity for the formation of zoogleal slimes on the media surfaces to clarify and oxidize the wastewater.

Sanitary sewer system: A sewer intended to carry wastewater from homes, businesses, and industries.

Screen: A device with openings, generally of uniform size, used to retain or move suspended or floating solids in flowing water or wastewater and to prevent them from entering an intake or passing a given point in a conduit. The device may consist of parallel bars, rods, wires, grating, wire mesh, or perforated plate, and the opening may be of any shape, although they are generally circular or rectangular.

Screening: The removal of relatively coarse suspended & floating solids by straining through racks or screens.

Screenings grinder: A device that reduces the size of screened material by cutting, shredding, or macerating material removed from wastewater by screens.

Scum: The layer or film of extraneous or foreign matter that rises to the surface of a liquid and is formed there. A residue deposited on a container or channel at the water surface. A mass of matter that floats on the surface of wastewater.

Sedimentation: The process of deposition by gravity of suspended matter.

Self-purification: The natural processes occurring in a stream or other body of water that result in the reduction of bacteria, satisfaction of the BOD, stabilization of organic constituents, replacement of depleted DO, and the recovery of the stream biota.

Sewage: Wastewater is also called sewage.

Sewer: A pipe or conduit that carries wastewater or drainage water.

Sewer Gas: Gas evolved in sewers that results from the decomposition of the organic matter in wastewater. Any gas present in the wastewater system, even though it is from such sources as gas mains, gasoline, and cleaning fluid.

Shredder: A device that chops up screenings.

Skimming: The process of removing floating grease or scum from the surface of wastewater in a tank.

Slake: To mix with water so that a true chemical combination takes place, as in the slaking of lime.

Slimes: Substances of viscous organic nature, usually formed from microbiological growth.

Sloughings: Biological filter slimes that have been washed off the filter media.

Sludge: The accumulated solids separated from wastewater during processing, or deposits on bottoms of streams or other bodies of water. The precipitate resulting from chemical treatment, coagulation, or sedimentation of wastewater.

Sludge, activated: Biological floc produced in raw or settled wastewater by the growth of zoogleal bacteria and other organisms in the presence of DO (dissolved Oxygen) and accumulated in sufficient concentration by returning floc previously formed.

Sludge, age: In the activated sludge process, a measure of the length of time a particle of SS is retained in the treatment process, expressed in days.

Sludge bulking: A phenomenon that occurs in activated sludge plants by which the sludge occupies excessive volumes and volumes and will not concentrate readily.

Sludge conditioning: Treatment of liquid sludge to facilitate dewatering and enhance drainability.

Sludge dewatering: The process by which organic or volatile matter in sludge is gasified, liquefied, mineralized, or converted into more stable organic matter, through the activities of either anaerobic or aerobic organisms.

Sludge, raw: Settled sludge removed from sedimentation tanks, Frequently referred to as undigested sludge.

Sludge seeding: In biological treatment of wastewater and associated sludges, the inoculation of the unit process with biologically active sludge, resulting in acceleration of the initial stage of the process.

Sludge thickener. A tank or other equipment designed to concentrate wastewater sludge.

Sludge volume index (SVI): A numerical expression of the settling characteristics of activated sludge. The ratio of the volume in milliliters of sludge settled from a 1, 000-ml sample in 30 min: the concentration of mixed liquor in milligrams per liter multi lied by 1, 000. Reciprocal of SVI is called SDI (Sludge Density Index).

Solids, colloidal: Finely divided solids intermediate between dissolved and suspended particles.

Solids, nonsettleable: Wastewater matter that will stay in suspension for an extended period of time. Such period may be arbitrarily taken for testing purposes as 1 hour.

Solids, settleable: That matter in wastewater that will not stay in suspension during a preselected settling Period, such as 1 hour, but either settles to the bottom or floats to the top. In the Imhoff cone test, the volume of matter that settles to the bottom of the cone in 1 hour.

Solids; total (TS): The sum of dissolved and undissolved constituents in wastewater, usually stated in milligrams per liter.

Solids, volatile (VS): The quantity of solids in wastewater lost on ignition of the dry solids at 550°C.

Sparger: An air diffuser designed to give large bubbles, used singly or in combination with mechanical aeration devices.

Stability: The ability of any substance, such as wastewater, chemicals, or digested sludge, to resist chemical or biological change.

Stabilization lagoon: Any pond, natural or artificial, receiving raw or partially treated wastewater in which stabilization occurs because of sunlight, air, and microorganism.

Sump: A tankor pit that receives drainage and stores it temporarily and from which the drainage is pumped or rejected. A tank or pit that receives liquids.

Supcrnatant: Thc liquid standing above a sediment or precipitate.

Tertiary Treatment: Those processes that treat effluent from secondary treatment to remove or reduce nutrients, residual organics, and residual solids.

Total organic carbon (TOC): A measure of the amount of carbon in a sample originating from organic matter only. The test is run, by burning the sample and measuring the carbon dioxide produced.

Treatment, preliminary: The conditioning of a waste at its source before discharge to remove or neutralize substances injurious to sewers and treatment processes or to effect a partial reduction in load on the treatment process. In the treatment process, unit operations that prepare the liquor for subsequent major operations.

Treatment, primary: The first major (sometimes the only) treatment in a wastewater treatment works. The removal of a substantial amount of suspended matter but little or no colloidal and dissolved matter.

Treatment, secondary: The processes that remove or reduce SS. DS, and fine & colloidal solids and cause the reduction of organic material.

Turbidity: Optical measure of suspended matter in liquids.

Waste sludge: That portion of settled solids from the final sedimentation unit that is removed from the wastewater treatment processes to the solids handling facilities for ultimate disposal.

Wastewater: The spent water of a community. It may be a combination of the liquid and water-carried wastes from residences, commercial buildings, industrial plants, and institutions, together with any groundwater, surface water, and storm water that may be present.

Wastewater, domestic: Wastewater derived principally from dwellings, office buildings, institutions, and the like. It may or may not contain groundwater, surface water, or storm water.

Wastewater, sanitary: Domestic wastewater with storm and surface water excluded. Wastewater discharging from the sanitary fixtures of dwellings (including apartment houses and hotels), office buildings, industrial plants, or institutions. The water supply of a community after it has been used and discharged into a sewer.

Wastewater, storm: That portion of liquid, resulting from precipitation runoff, flowing in combined sewers during or after a period of rainfall.

Zooglea: A jelly-like matrix developed by bacteria. A major part of activated sludge floc and of trickling filter slimes.

Terminology related to Noise Pollution

Absorption co-efficient: Ratio of sound energy absorbed by a surface to the sound energy incident upon the surface. It depends upon the frequency and angle of incidence of sound waves.

Accoustical: Used in conjunction with apparatus/design connected with sound control.

Accoustic impotence: Product of density of a material and the velocity of sound in the material. It indicate the rate at which a unit mass of material accepts acoustic energy.

Attentuation of sound: Reduction in the intersity of sound.

Decibel (dB): A dimensionless unit defines as ten times the logarittum to the base ten of the ration between two quantites of power. Since sound power is related to the square of the sound pressure, a scale for sound (noise) measurement is defined as sound pressure level (SPL).

$$SPL = 10\log_{10}(1^2/10^2) = 20\log_{10}(p/p_0)dB$$

where,

p = sound pressure being measured

and p_0 = a reference sound preseaure (normally taken as 20 μN/m² ie, 0.0002 μ bar)

The use of decibel scale reduces the scale of sound pressures of 1:10⁶ to sound pressure level (SPL) of 0 to 120 dB; 0 dB indicating the reference level and 120 dB indicating the maximum level.

Frequency: The number of times a sound source vibrates in unit internal of time indicate as hertz (Hz) or cycles per second.

Frequency-Audio: Frequency detectable by a normal human ear (15 to 20000 Hz).

Frequency-Infresonic: Frequency below the audio frequency range.

Frequency-ultrasonic: Frequency above the audio frequency range.

Noise: Unwanted sound.

Noise reduction co-efficient: the average of the absorption co-efficient of a material at frequencies 250, 500, 1000 & 2000 Hz.

Obscuration: Concealing from right, in meteorology, it relates to restriction of visibility.

Octave band: The band of frequencies defined between a lower limiting frequency (f_1) and upper limiting frequency (f_2) If, $f_2 = 2f_1$, then f_1 f_2 from an octave band of frequencies. The centre frequency of an octave band is the geomantic mean of f_1 &f_2. Centre frequencies used intemationally in the noise measurement are 31.5, 63, 125, 250, 500, 1000, 2000, 4000, & 8000 Hz.

Olfactory Threshold: Pollutant concentration below the point at which none of the receptors detect moledour.

Resonance: The natural vibration of a volume of air or a panel of material at a particular frequency resulting from excitation by a sound of the same frequency.

Sound intensity: The average rate of sound energy transmitted in a specified direction through a unit area normal to this direction at the point of interest.

Sound power: The total sound energy radiated by a source per unit time.

Sound pressure: the instantaneous pressure created over the local atmospheric pressure, when a sound wave is propagated.

Speech interference level (SIL): A measure of the degree to which noise interferes with speech.

Threshold of hearing: Minimum effective sound pressure at which a sound become audible in the abcence of all other sounds. It corresponds to 20 N/m^2 or 0.0002 μbar or, 0 dB.

Threshold of feeling: Minimum effective sound pressure level (SPL) that will stimulate the ear to a feeling of discomfort or pain. It corresponds to 20 N/m^2 or 200 μbar, or 120 dB. SPL beyond 140 dB will cause immediate deafness.

Threshod odour: Lowest concentration of odour bearing gas at which only half of a panel of sniffers can detect is presence.

Transmission loss: The deorease in sound pressure during transmission from one point to another (eg, through a panel or wall etc.). It is also termed as sound insulation.

Zone of silence: A zone of relatively low audability eg, hospits, educational institution, court, etc.

Terminology related to Environmental Pollution Control

Activated Sludge: Sludge that has been aerated and subjected to bacterial action, used to speed breakdown of organic matter in row sewage during secondary waste treatment.

Actionograph: An instrument for recording incoming solar radiation.

Aerated lagoon: Abasin provided with mechanical or diffused oxygenation equipmentsin which organic waste are stablised.

Aerated pond: A natural or artificial waste water treatment pond in which mechanical or diffused air aeration used to supplement oxygen supply.

Aeration: To circulate oxygen through a substance, as in wastewater treatment where it aids in purification.

After burner: An exhaust gas incinerator used to control emissions of particulate matter. This device fitted after the main point of fuel combustion in a stational plant or vehicle to produce nearly complete combustion of the exhaust gases.

Air borne pathogen: A disease causing micro organism which travels in the air or on particles in air.

Air monitoring: Continuous sampling for and measuring of the quantity of air pollutants present in the outdoor atmosphere.

Air pollution index: An arbitrarily derived mathematical combination of air pollutants which gives a single number trying to describe the ambient air quality.

Back fill: The material used to refill a ditch or other excavation or the process of doing so.

Benthic region: The bottom layer of a body of water.

BOD: The dissolved oxygen required to decompose organic matter present in water.

Biological oxidation: The process by which micro-organisms decompose complex organic materials. The process is used in activated sludge waste water treatment and in self purification of bodies of water. It is also called biochemical oxidation.

Biological wastes: Wastes derived from living organisms.

Bloom: A proliferation of algae and/or higher aquatic plants in a body of water, often related to pollution.

Boom: A floating device used to contain oil on a body of water.

Breathing zone: That location in the atmosphere at which persons breathe. Generally, a layer 0-1.5m above ground level is considered as breathing zone.

Brackish water: A mixture of fresh and salty water.

Carcinogen: A substance/pollutant which causes carcinoma or cancer.

COD: A measure of the oxygen required to oxidize all compounds in water ie, both organic & inorganic compounds.

Coning: A cone shaped plume occurring in a near neutral atmosphere (adiabatic condition) when wind velocity is >32 Km/hr.

Chemical waste: Waste generated by chemical, photochemical, plastic, pharmaceutical, biochemical or microbiological manufacturing process.

Chlorosis: Loss of chlorophyll in a plant which produces sickening & yellowing of vegetation.

Chrome tan: The process of converling hides into leather using a form of chromium.

Chronic: It refers to injurious effects of air pollution, particularly effects on health. It implies lasting or long term morbid effects as against acute effects.

Co-efficient of haze: A measurement of visibility interference in the atmosphere. This is a unit for measuring smoke stains which depends on the reduction of light transmission through the filter after it has collected the smoke sample. Its objective is to measure haze over a distance of about 300 meters.

Coffin: A thick walled container (usually of lead) used for transporting radioactive materials.

Cold front: the sharp boundary between two extensive air masses where cold air is replacing warm air.

Combined sewers: A system that carries both sewage and storm water runoff.

Commercial solid waste: All types of solid wastes generated by stores, office, restaurants, warehouses and other non-manufacturing activities excluding residential and industrial wastes.

Compaction: The process of reducing the bulk of solid waste by rolling and tamping or compressing and crushing.

Cosmic radiation: Radiation of mostly atomic nuclei (protons) with every high energies, originating outside the earth's atmosphere. It is a part of natural background radiation. Some cosmic radiations are more energetic than any man made forms of radiation.

Curie: Measure of number of atoms disintegrating per second ie, strength of radio activity. Curie equals 37 billions disintegration per second approximately equal to the radio activity of 1 gm of radium.

Decontamination/detoxification: a) Process which will convert pesticides into non-toxic compounds.
b) Selective removal of radioactive materials from a surface or from within.

Demulsibility: The resistance of an oil to emulisification, or the ability of an oil to separate from any water with which it is mixed. Better the demulsibility rating, the more quickly the oil separates from water.

Dermal Toxicity: The ability of a patricide or toxic chemical to poison people or animals by touching the skin.

Digestor: A closed tank heated to 95°F (in waste water treatment) where sludge is subjected to intensified bacterial action.

Dissolved oxygen: A measure of amount of oxygen available for biochemical activity in a given amount of water. It is needed to support aquatic life. Low dissolved oxygen concentration results from inadequate waste treatment.

Ecosphere: The layer of earth and troposphere inhabited by or suitable for the existence of living organisms.

Ecosystem: The interacting system of a biological community and its non-living surroundings.

Effluent: Waste material discharged into the environment (treated or untreated). It generally refers to water pollution.

Environmental impact assessment (EIA): The report which identifies and anylses the environmental impacts of the applicants proposed source and feasible alternatives.

Evaporation ponds: Areas where sewage sludge or other wastes are dumped and allowed to dry out.

Fanning: Type of plume emitted under extreme inversion condition where mostly lateral diffusion takes place with very little vertical mixing.

Flue fed incinerator: An incinerator that is discharged through a shaft that functions as a chute for charging wastes and has a true to carry the product of combustion.

Flume: A natural or man-made channel that diverts water.

Flyash: Finely divided particles of ash entrained in flue gases arising from the combustion of fossil fuels like coal it may contain incompletely burned coal.

Fugitive emission: Any air pollutants emitted to the atmosphere other than from a stack.

Fumigation: Phenomenon in which air pollutants are brought rapidly to ground level when the air destabilises.

Heat island: Horizontal pocket of relatively warm air surrounded by cooler air often found over industrial complexes or city centre.

Hydrology: The science dealing with the properties, distribution and circulation of water.

Incineration: The controlled process in which solid, liquid or gaseous wastes are burned and changed into non-combustible gases.

Insolation: A rate at which direct solar radiation reaches the earth's surface.

Inversion: An inversion exists when air temperature increases with height resulting in damping of vertical movement of pollutants and attenuation of vertical spread of pollutants. During inversion, which may occur in isolated layer also, atmosphere is said to be stable.

Land fill: A disposal facility or part of a facility where hazardous waste is placed in or on land and which is not a land treatment facility, a surface impoundment or an injection well.

Lapse rate: Decrease of atmospheric temperature with height.

Lofting: Type of plane occurring when an inversion exists only below the plume and plume is inhibited from mixing downwards.

Looping: Plume having wavy character which occurs in a highly unstable atmosphere because of rapid mixing.

Municipal solid waste: Garbage, refuse, sludges, wastes and other discarded materials resulting from residential or non-residential operations/activities.

Mixing height: The height within which emitted pollutants are dispersed & mixed vertically, the vertical mixing being attenuated by presence of inversion layer above that height.

Organic refuse: Solid waste composed of carbon compounds and generally by-products of plants & animal life process eg, paper, wood, excreta, yard trimmings etc.

Phytotoxic: Poisonous to plants.

Plume: The flow or path of effluent from chimney or vent. while for smoke, the plume is visible, for may pollutants, it may not be so.

Pneumoconiosis: Disease of lung caused by inhalation of dust particles in occupation such as coal mining, quarrying & asbestos working; an excess of fibrous tissue forms in the lungs around the dust particles.

Poly electrolytes: Synthetic chemicals that help solids to clump during sewage treatment.

Quench tank: A water filled tank used to cool incinerator residues or hot material during industrial processor.

Reclamation: Therestoration of land, water or waste materials to usefulness through such methods as sanitary land fillings, waste water treatment & materials recovery.

Refuse: A term generally used for all solid waste materials.

Respirator: A device used by a person for breathing filtered air or artificial air from a cylinder while working in an area having air contamination.

Sanitation: Control of physical factors in the human environment that can harm health, development or survival.

Saporphytic: Organisms living on dead organic matter.

Sewage: Human body wastes and wastes from the toilet and other receptacles intended to receive or retain body wastes.

Sewage sludge: Any solid, semi-solid or liquid waste generated by a municipal waste water treatment plant, the ocean dumping of which may degrade or endanger human health, welfare or amenities or the marine environment & ecological system.

Sewer: A channel that carries waste water and storm water run off from the source to a treatment plant or receiving stream. Sanitary sewers carry households &commercial waste. Storm sewers carry run off from rain or snow. Combined sewer are used for both purposes.

Sewerage: Entire system of sewage collection, treatment & disposal. Also applied to effluents carried by the sewers.

Shredder: Amachine used to break up waste materials into smaller pieces by cutting or tearing.

Suspended particulates: Particles which are too fine to have an appreciable falling velocity and therefore tend to stay suspended in atmosphere for a considerable period.

Toxicity: The degree of danger posed by a substance to animal life or plant life.

Warm front: The sharp boundary between two extensive air masses where warm air is replacing cold air.

Water table: the level of ground water.

Water shed: The land area that drains into a stream.

Wind shear: Change in speed and direction of the wind usually with height.

Wind rose: Diagrammatic representation to show the distribution of wind direction experienced at a given location over a considerable period.

Zone of aeration: The zone between the land surface and the water table.

Zone of saturation: That part of earth's crust in which all voids are filled with water.

Zoning: A method of abating air pollution at a location by systematic planning of sites for dwelling and for factories/industries which discharge effluent taking into account the micrometeorological conditions, in and around the location.

Units and their Conversions

Length

[In-inch, ft-feet, m-metre, mm = millimeter, cm = centimeter, Å = Angstrom, km- kilometer, μm-micro-metre]

1 Å = 10^{-4} micron = 10^{-8} cm = 10^{-10}m, 1μm (micrometre) = 10^{-6}m, mm = 1000 microns

1 micron = 10^{-4} cm ; 200 mesh size = 74μm (micro-metre) = 0.074 mm = 74 microns

1 inch = 2.54 cms = 25.4 mm = 0.08333 ft = 0.0254m, 1μm = 1 micron = 10^{-6} m

1 ft = 12 inches = 30.48 cms = 0.3048 m = 0.3333 yard, 1 yard = 0.9144 m

1 m = 3.28 ft = 39.37 inches = 1.0936 yards = 100 cm = 1000 m = 10^{-3} km

1 mile =1.609 km (*i.e,* 5 miles = 8 km) = 5280 ft = 1760 yards = 1609 m

1 Nautical mile = 1852 m = 1.852 km = 1.15 mile ; 1 mm = 1000 microns

Area

1 barn = 10^{-28} m², 1 mm² = 10^{-6} m² = 0.00155 in² = 10.76 x 10^{-6} ft²

1 in² = 6.45 cm², 1 ft² = 0.0929 m² = 929.03 cm² = 144 in² = 0.093 m²

1 hectare = 10000 m² = 2.47 acres, 1 acre = 100 m², 1 cm² = 10^{-4} m²

1 acre = 4840 yard² = 0.4047 hectare = 4047 m² = 43560 ft² = 4840 yard²

1 m² = 10.76 ft² = 1550 in² = 1.196 yard² = 0.000247 hectare = 10^6 mm²

1 km² = 10^{12} mm² = 10.76 x 10^6 ft² = 10^6 m² = 1.196 x 10^6 yard² = 247.1 acre = 0.3861 mile²

1 mile² = 27.88 x 10^6 ft² = 3.097 x 10^6 yard² = 2.59 x 10^6 m² = 640 acre 2.59 km²

Volume

[ml-millilitre, KL-kilolitre, cc-cubic centimeter (cm³), dm-decimeter]

1 mm³ = 10^{-9} m³, 1 in³ = 16.39 cm², 1ft³ = 28.32 litres = 0.02832 m³

1 litre = 1000 ml = 1000 cc = 10^{-3} m³ = 0.22 U.K. gallon = 0.264 U.S. gallon = 1 dm³

1 imperial gallon = 1 U.K. gallon = 4.546 litres (dm^3) = 1.201 U.S. gallon = 4.546×10^{-3} m^3

1 U.S. gallon = 3.785 litres = 0.833 U.K. gallon

1 U.S. barrel (bb1) = 42 U.S. gallon = 34.973 U.K. gallon = 0.159 m^3 = 0.159 KL = 5.615 ft^3

1 ml = 10^{-6} m^3, 1 pint = 0.5682 litre = 1/8 U.K.gallon

1 m^3 = 1000 litre = 1 KL = 35.3148 ft^3 = 219.97 U.K. gallon = 264.17 U.S. gallon

1.308 $yard^3$ = 10^3 dm^3 = 61024 $inch^3$ = 10^9 mm^3 = 6.29 U.S. barrel

1 ft^3 = 0.028317 m^3 = 6.023 U.K. gallon = 7.481 U.S. gallon = 1728 in^3 = 28.32 dm^3

Specific Volume

1 litre/kg = 1 c.c/gm = 10^{-3} m^3/kg, 1 litre/gm = 1 m^3/kg

1 ft^3/kg = 2.8317×10^{-2} m^3/kg, 1 ft^3/1b = 6.2428×10^{-2} m^3/kg

Volumetric Flow Rate

1 ft^3/hr = 7865 mm^3/sec = 0.0166 ft^3/min = 7.865×10^{-6} m^3/sec, 1 ft^3/s = 1 cusec = 2.8317×10^{-2} m^3/s

1 ft^3/min = 60 ft^3/hr = 471.9 x 10-6 m^3/sec = 1 cumin = 4.7195×10^{-4} m^3/s

1 m^3/sec = 127.1×10^3 ft^3/hr = 2119 ft^3/min = 10^9 mm^3/sec

Velocity

1 in/sec = 60 in/min = 5 ft/min = 0.08333 ft/sec = 0.0254 m/sec = 24.4×10^{-6} km/sec

1 ft/sec = 720 in/min = 60 ft/min = 0.3048 m/sec = 304.8×10^{-6} km/sec

1 m/sec = 2368 in/min = 196.9 ft/min = 39.37 in/sec = 3.281 ft/sec = 10^{-3} km/sec

1 km/sec = 2368×10^3 in/min = 196.9×10^3 ft/min = 39.37×10^3 in/sec = 3281 ft/sec = 10^3 m/sec

1 km/hr = 0.6214 mile/hr, 1 mile/hr = 1.609 km/hr

Frequency

1 cycle/second (cps) = 10000 Hertz (Hz)

Mass

[lb-pound, oz-ounce]

1 grain = 0.0648 gm ; 1 mg = 10^{-6} kg ; 1 gm = 15.432 grains = 0.0353 (ounce) oz = 0.002205 lb ; 1 oz = 28.35 gm = 984.2×10^{-9} ton (U.K./long)

1 lb = 453.6 gm = 7000 grains = 0.454 kg = 16 oz = 446.4×10^{-6} ton (U.K./long)

1 long ton = 1 U.K. ton = 2240 lb = 1.12 short ton = 1.12 U.S. ton = 1.016 metric ton = 1.016×10^6 gm = 1016 kg,

1 U.S. ton (short) = 2000 lb = 0.8929 U.K. ton (long) = 0.9072 tonne = 907.2 kg

1 tonne = 1 metric ton = 1000 kg = 2204.6 lb = 0984 U.K. ton (long) = 1.1023 U.S. ton (short)

1 kg = 1000 gm = 2.205 lb = 984.2×10^{-6} ton (U.K./long)

1 slug = 14.59 kg, 1 oz = 28.35 gm, 1 tonne = 10^6 gm

Mass Flow Rate

1 lb/hr = 1.26×10^{-4} kg/s, 1 kg/hr = 2.7778×10^{-4}kg/s ; 1 tonne/day = 1.1574×10^{-2} kg/s, 1 tonne/hr = 2.7778×10^1 kg/s

1 tone (short)/day = 1.05 x 10^{-2} kg/s, 1ton(short)/hr = 2.52 x 10^{-1} kg/s ; 1 ton (long)/day = 1.176 x 10^{-2} kg/s, 1 ton (long)/hr = 2.8224 x 10^{-1} kg/s

Density and Concentration

[ppm-parts per million]

1 ppm = 1 ml/m^3 = 1 mg/kg ; 1% = 10000 ppm (for water, density = 1 gm/cc)

1 gm/ml = 1 gm/cc = 62.4 lb/ft^3 = 10.022 lb/gallon = 10^{-3} gm/m^3

1 lb/ft^3 = 16.02 kg/m^3 = 0.4536 kg/ft^3 = 578.7 x 10^{-6} lb/in^3;

1 kg/m^3 = 0.625 lb/ft^3 = 9.8 N/m^3 = 0.02832 kg/ft^3, 1kg/litre = 10 kg/m^3

1 kg sec^2/cm4 = 9.81 kg/m^3, 1 mg/ml = 1 kg/m^3, 1 mg/litre = 10^{-3} kg/m^3

1 m^3/kg = 16.019 lb/ft^3; 1 ft^3/lb = 0.062428 m^3/kg

1 kg/ft^3 = 35.31 kg/m^3 = 2.205 lb/ft^3 = 1.278 x 10^{-3} lb/in^3 = 35.31 x 10^{-6} gm/mm^3

1 lb/in^3 = 27.68 x 10^3 kg/m^3 = 1728 lb/ft^3 = 783.8 kg/ft^3 = 27.68 x 10^{-3} gm/mm^3

1 gm/mm^3 = 106 kg/m^3 = 62.43 x 10^3 lb/ft^3 = 28.32 x 10^6 kg/ft^3 = 36.05 lb/in^3

Absolute (or Dynamic) Viscosity

[cp-centipoise, Pa-pascond]

1 pose = 100 cp = 1 gm/ cm sec = 0.01 kgf sec/m^2 = 1 dyne sec/cm^2 = 10^{-1} Pa s = 0.1 N sec/m^2= 2.0885 x 10^{-3} lbf sec/ft^2

1 Pa.s = 10 poise = 0.02089 lbfs/ft^2

1 kgf sec/ cm^2 = 9.81 N/m^2 ; 1 kgf sec/m^2 = 98 poise = 0.205 lbf sec/ft^2

1 lb sec/ft^2 = 478.8 poise = 4.88 kgf sec/m^2 = 47.88 Pa.s

1 kg/m.hr = 2.7778 x 10^{-4} N.S/m^2, 1 lb/ft.hr = 4.1338 x 10^{-4} N.S/m^2 •

1 mN.S/m^2 = 1 centipoise = 10^{-3} N.S/m^2, 1 kg/M.S = 1 N.S/m^2

1 lb/ft.S = 1.4882 N.S/m^2

Kinematic Viscosity and Thermal Diffusivity

[cst—centistoke, rpm—rotation per minute]

1 stoke = 100 cst = 1 cm^2/sec = 0.36 m^2/hr = 3.875 ft^2/hr = 10^{-4} m^2/sec

1 m^2/s = 104 stokes = 10.76 ft^2/s

1 m^2/hr = 2.78 storkes = 10.76 ft^2, hr = 2.78 x 10^{-4} m^2/s

1 ft^2/hr = 0.26 storkes = 0.09 m^2/hr = 2.5806 x 10^{-5} m^2/s

1 ft^2/s = 0.09290 m^2/s

1 cst = 10^{-6} m^2/sec = 1 mm^2/s ; 1 rpm = 1.047 x 10^{-1} radian/sec

Force

[pdl—poundal, kN—kilo Newton]

1 pdl = 0.1383 N = 0.03108 lbf = 0.01410 kgf

1 N = 7.233 pdl = 0.2248 lbf = 0.1020 kgf

1 kN = 7233 pdl = 10^3 N = 224.8 lb = 102 kgf

1 lbf = 31.17 pdl = 4.448 N = 0.4536 kgf

1 kgf = 70.90 pdl = 9.803 N = 2.205 lbf

1 dyne = 10^{-5} N

Pressure

[mmwg—millimeter water gause, inwg—inch water gauge, mmHg—millimeter mercury gauge, psi—lbf/in^2, mB—mega bar, atm—atmospheric pressure, mmwc—millimeter water column, N—Newton, M—mega, Pa—pascal, kPa—kilo pascal]

1 atm = 760 mmHg = 29.92 in Hg = 14.7 psi = 1.013 bar = 1.013 kgf/cm^2 = 34 ft water = 10033 mmwc = 10033 kgf/m^2 = 1.013 x 105 N/m^2 = 1013.2 mB = 2116 lbf/ft^2 = 1.013 x 10^5 Pa

1 in wg = 1.87 mm Hg = 0.07351 in wg = 0.036 psi = 1.87 torr = 1.87 mm Hg = 249.1 N/m^2 = 249.1 Pa = 2.49 mB = 24.4 kg/m^2 = 25.4 mm wg = 5.202 lbf/ft^2

1 Pa = 4.015 x 10^{-3} in water = 295.3 x 10^{-6} in Hg = 145 x 10^{-6} psi = 64.75 x 10^{-9} tors/in^2

1 kPa = 20.89 lb/ft^2 = 0.145 psi = 10^{-2} bar = 64.75 x 10^{-6} ton/in^2 = 1 kN/m^2 = 102 mmwc

1 in Hg = 13.6 in wg = 3386 mB = 25.4 torr= 25.4 mm Hg = 0.0334 atm = 0.4912 psi = 0.03386 bar = 70.73 lb/ft^2 = 3386 Pa = 3386 N/m^2

1 ft water = 304.8 kgf/m^2 = 2986.1 N/m^2 ; 1 mB = 100 N/m^2

1 mm Hg = 1 torr = 133.3 Pa = 133.3 N/m^2 ; 1 μ bar = 1 millibar = 100 pa

1 Pascal = 1 N/m^2, 1 bar = 1 kgf/cm^2 = 10^5 N/m^2 = 105 Pa = 10^6 dynes/cm^2 = 401.5 in H_2O = 14.5 psi = 29.53 in Hg

1 psi = 0.07031 kgf/cm^2 = 68.95 mB = 6.895 x 103 N/m^2 = 6895 P = 0.06895 bar = 2.036 in Hg = 27.68 in water = 144 lb/ft^2

1 kgf/cm^2 = 14.7 psi = 10000 mm wc = 0.1 MN/m^2 = 9.81 N/cm^2 = 98kPa

1 kgf = 9.8 N, 1 lbf = 4.448 N, 1 MN/m^2 = 10 kgf/cm^2, 1 kgf/m^2 = 1 mmwc = 9.8067 N/m^2

1 lbf/ft^2 = 47.88 Pa = 47.88 N/m^2 = 0.1922 in water= 0.0141 in Hg = 6.944 x 10^{-3} psi = 478.8 x 10^{-6} bar

1 ata = 1 kgf/cm^2 = 9.8067 x 10^3 N/m^2, 1 N/mm^2 = 106 N/m^2

1 ton_f/inch^2 = 1.5477 x 10^5 N/m^2, 1 Poundal/ft^2 = 1.4881 N/m^2

Heat, Work, Power and Energy

[BTU—British Thermal Unit, CHU—Centigrade Heat Unit, J—Joule, kJ—kilo Joule, Cal—Calorie, Kcal—kilo calorie, kWh—kilowatt hour, W—watt, hp—horse power, min—minute, TNT—Tri-nitro-toulene, G—Giga = 10^9]

Heat

1 J = 07376 ft lbf = 0.2388 cal = 947.8 x 10^{-6} BTU = 277.7 x 10^{-9} kWh = 9.478 x 10^{-9} therm = 1 N.m

1 BTU = 252 calorie = 0.252 kcal = 778 ft lbf = 0.5556 CHU = 25030 pdl = 1055 J = 2.931 x 10^{-4} kWh

1 Therm = 2503 x 10^6 pdl = 105.5 x 10^6 J = 77.82 x 10^6 ft lbf = 25200 kcal = 10^5 BTU = 29.31 kWh

1 calorie = 0.003968 BTU = 4.186 J = 4.186 watt. sec. = 3.088 ft lbf = 1.163 x 10^{-6} kWh

1 kcal = 1000 calorie = 3.968 BTU = 0.00116 kWh = 3087 ft lb = 4186 J = 4.186 kJ = 2.205 CHU = 427 kg m = 0.001560 hp

1 ft.lbf = 03238 calorie = 0.1383 kg.m = 1.356 J = 1.356 N.m

1 kgf.m = 9.8067 J

1 hp.hr (metric) = 2.6477 x 10^6 J

1 hp.hr (British) = 2.6845 x 10^6 J

12.85 x 10^{-9} therm = 1.285 x 10^{-3} BTU = 3.776 x 10^{-7} kWh

1 ft lb/min = 0.0226 W, 1 ft lb/sec = 0.00136 kW

1 ton of TNT = 4.2 x 10^9 J = 4.2 10^6 kJ = 4.2 x 10^3 MJ = 4.2 GJ

Work and Heat Flow Rate

[sec—second, MW—mega watt = 10^6 W, kW—kilowatt, = 10^3 watt, hr—hour, N-Newton]

1 BTU/s = 1055.1 J/S, 1 BTU/min = 0.02356 hp, 1 CHU = 1.8 BTU = 0.453 kcal

1 BTU/hr = 0.2931 W = 0.2162 ft lbf/s = 0.00393 hp = 0.2931 J/S

1 Therm = 100, 000 BTU = 25200 kcal = 29.31 kWh = 2503 x 10^6 pdl = 105.5 x 10^6 J 77.82 x 10^6 ft lbf ; 1 m.Kgf/S = 9.8065 J/S

1 ton = 1 W. sec. = 107 ergs = 9.5 x 10^{-4} BTU = 02388 calorie = 1 N.m

1 ft lbf/s = 4.628 BTU/hr = 1.356 W = 0.01819 ph = 1.356 J/S

1 kcal/hr = 1.163 J/s, 1 MW = 10^6 W = 1000 kW, 1 N = 10^5 dyne, 1 therm/hr = 2.9308 x 104 J/S

1 MW day = 8.64 x 10^{10} J

1 W/m^2 = 0.3170 BTU/ft^2 hr = 1 J/S.m^2 ; 1 BTU/ft^2 hr = 3.155 W/m^2 = 30155 J/S.m^2

1 kcal/hr.m^2 = 1.163 J/S.m^2, 1 KW/m^2 = 1000 J/S.m^2, 1 kcal/hr.ft^2 = 12.518 J/S.m^2

Power and Energy

[W—watt, kW—kilowatt, MJ—mega joule]

1 hp = 0.746 kW = 746 W = 550 lb ft/sec. = 632.34 kcal/hr = 2545 BTU/hr

1 W = 3.412 BTU/hr = 44.24 ft.lb/min = 0.2389 cal/sec = 0.7376 ft.lbf/s = 0.01341hp

1 kW = 13.41 hp = 1000 W = 102 kg m/sec = 3412 BTU/hr = 737.6 ft.lbf/s ;

1 BTU/min = 17.58 W

1 kWh = 3.6 x 10^6 J= 2.655 x 10^6 ft lbf = 859.8 kcal = 3412 BTU = 3.6 MJ = 0.03412 therm Joule's equivalent = 778.3 ft.lb

Specific Heat

1 cal/gm . C° = 1 kJ/kg . °*C* = 1 BTU/lb . °*F* = 1 BTU/lb . °*R* = 4.1868 kJ/kg . °*K*

Enthalpy

1 cal/gm = 1 kcal/kg = 4.1868 kJ/kg, 1 BTU/lb = 2.326 kJ/kg

Calorific Value

[N–at N.T.P. *i.e.* °*C* and 760 mm Hg, S = at S.T.P.*i.e.*, *15°C* and 760 mm Hg]

1 kcal/kg = 1.8 BTU/lb = 0.0403 therm/ton = 4.187 kJ/kg

1 BTU/lb = 0.555 kcal/kg = 0.0224 therm/ton = 2.320 kJ/kg

1 BTU/Nft^3 = 8.9 kcal/Nm^3 = 0.038 J/N cm^3 = 0.038 MJ/Nm^3

1 kcal/Nm³ = 0.1124 BTU/N ft³, 1 therm/U.K. gallon = 5543.3 kcal/litre

10000 kcal/litre = 1.804 therm/U.K. gallon

1 cal/m³ = 0.11237 BTU/ft³

1 J/kg = 4.29 × 10⁻³ BTU/lb; 1 MJ/kg = 238 kcal/kg

1 J/cm³ = 27 BTU/ft³; 1 MJ/Nm³ = 238 kcal/Nm³

1 MJ/sm³ (dry) = 26.81 BTU/sft³ (dry); 1 MJ/sm³ (saturated) = 26.79 BTU/sft³ (saturated)

1 BTU/stf³ (dry) = 0.0373 MJ/sm³ (dry);

1 BTU/sft³ (saturated) = 0.03732 MJ/sm³ (saturated)

1 kg coal equivalent = 7000 kcal/kg = 2.9308 × 10⁴ kJ

1 kg oil equivalaent = 10180 kcal/kg = 4.2622 × 10⁴ kJ

Temperature

Conversion of degree Farenhite (°*F*) to degree centigrade/celcius (°*C*) and *vice versa* is done by the following formula:

$$\frac{C}{5} = \frac{F-32}{9}$$

From where, it follows that:

$$°C = (°F-32)\times\frac{5}{9} = (°F-32)\times 0.555$$

$$°F = °C\times\frac{9}{5}+32 = (°C+17.778)\times 1.8$$

$$°A \ (degree\ absolute) \ = \ °K \ (kelvin) = 273+°C$$

$$°F \ absolute = °R \ (Rankine) = 460+°F$$

Temperature difference :

$$1\ °K = 1°C = \frac{9}{5}°F = \frac{9}{5}°R$$

Heat Transfer Rate

1 BTU/ft².hr = 2.712 kcal/m²hr, 1 kcal/hr = 1.163 W

1 kcal/m².hr = 0.3687 BTU/ft² hr = 1.163 W/m²

1 BTU/in².hr = 390.6 kcal/m².hr ; 1 kcal/m³ hr = 1.163 W/m³

Heat Transfer Co-efficient

1 kcal/m² hr °C = 0.2048 BTU/ft2 hr °F = 1.163 W/m² °K = 1.163 J/S.m².°K

1 BTU/ft² hr °F = 4.88 kcal/m² hr °C = 20.44 kJ/m² hr °C = 5.678 J/S.m². °K = 5.678 W/m²°C = 135.6 x 10⁻⁶ cal/cm² sec °C

1 W/m²°C = 0.1761 BTU/ft².hr. °F = 23.88 x 10⁻⁶ cal/cm².sec. °C = 1 J/S.m². °K

1 kcal/hr. ft²°C = 1.2518 x 10 J/S.m².°K

1 kcal/hr. ft²°C = 4.1868 x 10⁴ J/S.m².°K

Radiation Co-efficient

1 kcal/m².hr.°K = 1.163 W/m². °K⁴

1 cal/cm².sec.°C = 41868 W/m².°C = 7373 BTU/ft².hr.°F

Thermal Conductivity

1 BTU/ft^2.hr°F/in thickness = 0.124 kcal/m^2.hr.°C/m = 0.344 cal/cm sec °C = 6.93 W/min.°K. = 1.4423 x 10^{-1} J/S.m.°K

1 BTU in/st^2 S °F = 519.2 W/m °C = 300 BTU/ft.hr.°F

1 BTU/ft^2 hr °F/ft = 1.488 kcal/m^{2}hr °C = 1.7308 J/S.m.°K

1 kcal/m hr °C = 0.672 BTU/ft hr °F = 8.063 BTU/ft^2 hr °F/inch = 1.163 W/m °K = 1.163 J/S.m. °K

1 cal/cm sec °C = 360 kcal/m hr °C = 242 BTU/ft hr °F = 2902.9 BTU in/ft^2 hr °F = 4.1868 x 10^2 J/S.m.°K

1 W/m °C = 0.5778 BTU/ft hr °F = 1.928 BTU in/ft^2 s °F

1 W/cm °C = 85.985 kcal/m.hr.°C = 57.779 BTU/ft.hr. °F

1 BTU/ft hr °F = 1.731 W/m°C = 3.333 x 10^{-3} BTU in/ft^{2}s. °F

1 kcal/hr.ft.°C = 3.8156 J/S.m. °K

SI (International System) Unit

SI units have been introduced to standardize the units throughout the world. India has also adopted this unit but still M.K.S. Unit is in vogue here. The base unit in SI unit system is also metre, kilogram and second.

Some of the important *SI derived units* having special names and sysmbols are tabulated in below table

Names and Symbols of SI Derived Units

S. No.	Quantity	Name	Symbol	Expression in terms of other SI units	*Expression in terms of base units*
1.	Force	Newton	N	–	kg m/sec^2
2.	Pressure	Pascal	Pa	N/m^2	kg m/sec^2
3.	Energy	Joule	J	N.m	m^2 kg/sec^2
4.	Absolute viscosity	Pascal second	PaS	–	kg/m.sec
5.	Heat capacity	Joule/Kelvin	J/K	–	m^{2}kg/sec^2.°K
6.	Thermal conductivity	Watt per metre Kelvin	W/m.°K	–	m kg/sec^3.°K
7.	Molar heat capacity	Joule per mole Kelvin	J/mol.°K	–	m^2 kg/sec^2 mol °K
8.	Moment of force	Newton-metre	N.m	–	m^{2}kg/s^2
9.	Heat flux density	Watt per square metre	W/m^2	–	kg/s^3
10.	Surface tension	Newton per metre	N/m	–	kg/s^2
11.	Specific heat capacity	Joule per kilogram kelvin	J/kg.°K	–	m^2/s^2.K
12.	Energy density	Joule per cubic metre	J/m^3	–	kg/m.s^2
13.	Specific energy	Joule per kilogram	J/kg	–	m^2/s^2
14.	Currents density	Ampere per square metre	A/m^2	–	A/m^2
15.	Electric field strength	Volt per metre	V/m	–	m kg/s^3.A

16.	Magnetic field strength	Ampere per metre	A/m	–	A/m
17.	Molar energy	Joule/mole	J/mol	–	$m^2kg/s^2.mol$
18.	Molar heat capacity	Joule per mole Kelvin	J/mol.K	–	m^2kg/s^2 mol.k
19.	Angular velocity	radian per second	rad/s	–	rad/s
20.	Angular acceleration	radian per second square	rad/s^2	–	rad/s^2

Abbreviations used for prefixes in SI units are tabulated below

Abbreviations used for Prefixes in SI units

Prefix	Symobl	Factor	Prefix	Symbol	Factor
deca	da	10^1	deci	**d**	10^{-1}
hecto	h	10^2	centi	**c**	10^{-2}
kilo	k	10^3	milli	**m**	10^{-3}
mega	M	10^6	micro	**μ**	10^{-6}
giga	G	10^9	nano	**n**	10^{-9}
tera	T	10^{12}	pico	**p**	10^{-12}
femto	f	10^{-15}	atto	**a**	10^{-18}

Appendices

Contents			Page No.

Appendix 2.1: Conversion Table for Gases/Vapors from mg/litre to ppm and vice versa (at STP ie, 25°C, 760 mm Hg)

Molecular weight	1 mg/litre = --------------- ppm	1 ppm = --------------- mg/litre
1	24450	0.0000409
2	12230	0.0000818
3	8150	0.0001227
4	6113	0.0001636
5	4890	0.0002045
6	4075	0.0002454
7	3493	0.0002863
8	6056	0.000327
9	2717	0.000368
10	2445	0.000409
11	223	0.00045
12	2038	0.000491
13	1881	0.000532
14	1746	0.000573
15	1630	0.000614
16	1528	0.000654
17	1438	0.000695
18	1358	0.000763
19	1287	0.000777
20	1223	0.000818
21	1164	0.000859
22	1111	0.0009
23	1063	0.000941
24	1019	0.000982
25	978	0.001022
26	940	0.001063
27	906	0.001104
28	873	0.001145
29	843	0.001186
30	815	0.001227
31	789	0.001268
32	764	0.001309
33	741	0.00135

Molecular weight	1 mg/litre = --------------- ppm	1 ppm = --------------- mg/litre
34	719	0.001391
35	699	0.001432
36	679	0.001472
37	661	0.001513
38	643	0.001554
39	627	0.001595
40	611	0.001636
41	596	0.001677
42	582	0.001718
43	569	0.001759
44	556	0.0018
45	543	0.00184
46	532	0.001881
47	520	0.001922
48	509	0.001963
49	499	0.002004
50	489	0.002045
51	479	0.002086
52	470	0.002127
53	461	0.002168
54	453	0.002209
55	445	0.00225
56	437	0.00229
57	429	0.002331
58	422	0.002372
59	414	0.002413
60	408	0.002454
61	401	0.002495
62	394	0.00254
63	388	0.00258
64	382	0.00262
65	376	0.00266
66	370	0.0027
67	365	0.00274
68	360	0.00278
69	354	0.00282

Molecular weight	1 mg/litre = --------------- ppm	1 ppm = --------------- mg/litre
70	349	0.00286
71	344	0.0029
72	340	0.00294
73	335	0.00299
74	330	0.00303
75	326	0.00307
76	322	0.00311
77	318	0.00315
78	313	0.00319
79	309	0.00323
80	306	0.00327
81	302	0.00331
82	298	0.00335
83	295	0.00339
84	291	0.00344
85	288	0.00348
86	284	0.00352
87	281	0.00356
88	278	0.0036
89	275	0.00364
90	272	0.00368
91	269	0.00372
92	266	0.00376
93	263	0.0038
94	260	0.00384
95	257	0.00389
96	255	0.00393
97	252	0.00397
98	249.5	0.00401
99	247	0.00405
100	244.5	0.00409
101	242.1	0.00413
102	239.7	0.00417
103	237.4	0.00421
104	235.1	0.00425
105	232.9	0.00429

Molecular weight	1 mg/litre = -------------- ppm	1 ppm = -------------- mg/litre
106	230.7	0.00434
107	228.5	0.00438
108	226.4	0.00442
109	224.3	0.00446
110	222.3	0.0045
111	220.3	0.00454
112	218.3	0.00458
113	216.4	0.00462
114	214.5	0.00466
115	212.6	0.0047
116	210.8	0.00474
117	209	0.00479
118	207.2	0.00483
119	205.5	0.00487
120	203.8	0.00491
121	202.1	0.00495
122	200.4	0.00499
123	198.8	0.00503
124	197.2	0.00507
125	195.6	0.00511
126	194	0.00515
127	192.5	0.00519
128	191	0.00524
129	189.5	0.00528
130	188.1	0.00532
131	186.6	0.00536
132	185.2	0.0054
133	183.8	0.00544
134	182.5	0.00548
135	181.1	0.00552
136	179.8	0.00556
137	178.5	0.0056
138	177.2	0.00564
139	175.9	0.00569
140	174.6	0.00573
141	173.4	0.00577

Molecular weight	1 mg/litre = --------------- ppm	1 ppm = --------------- mg/litre
142	172.2	0.00581
143	171	0.00585
144	169.8	0.00589
145	168.6	0.00593
146	167.5	0.00597
147	166.3	0.00601
148	165.2	0.00605
149	164.1	0.00609
150	163	0.00613
151	161.9	0.00618
152	160.9	0.00622
153	159.8	0.00626
154	158.8	0.0063
155	157.7	0.00634
156	156.7	0.00638
157	155.7	0.00642
158	154.7	0.00646
159	153.7	0.0065
160	152.8	0.00654
161	151.9	0.00658
162	150.9	0.00663
163	150	0.00667
164	149.1	0.00671
165	148.2	0.00675
166	147.3	0.00679
167	146.4	0.00683
168	145.5	0.00687
169	144.7	0.00691
170	143.8	0.00695
171	143	0.00699
172	142.2	0.00703
173	141.3	0.00708
174	140.5	0.00712
175	139.7	0.00716
176	138.9	0.0072
177	138.1	0.00724

Molecular weight	1 mg/litre = --------------- ppm	1 ppm = --------------- mg/litre
178	137.4	0.00728
179	136.6	0.00732
180	135.8	0.00736
181	135.1	0.0074
182	134.3	0.00744
183	133.6	0.00748
184	132.9	0.00753
185	132.2	0.00757
186	131.5	0.00761
187	130.7	0.00765
188	130.1	0.00769
189	129.4	0.00773
190	128.7	0.00777
191	128	0.00781
192	127.3	0.00785
193	126.7	0.00789
194	126	0.00793
195	125.4	0.00798
196	124.7	0.00802
197	124.1	0.00806
198	123.5	0.0081
199	122.9	0.00814
200	122.3	0.00818
201	121.6	0.00822
202	121	0.00826
203	120.4	0.0083
204	119.9	0.00834
205	119.3	0.00838
206	118.7	0.00843
207	118.1	0.00847
208	117.5	0.00851
209	117	0.00855
210	116.4	0.00859
211	115.9	0.00863
212	115.3	0.00867
213	114.8	0.00871

Molecular weight	1 mg/litre = -------------- ppm	1 ppm = -------------- mg/litre
214	114.3	0.00875
215	113.7	0.00879
216	113.2	0.00883
217	112.7	0.00888
218	112.1	0.00892
219	111.6	0.00896
220	111.1	0.009
221	110.6	0.00904
222	110.1	0.00908
223	109.6	0.00912
224	109.2	0.00916
225	108.7	0.0092
226	108.2	0.00924
227	107.7	0.00928
228	107.2	0.00933
229	106.8	0.00937
230	106.3	0.00941
231	105.8	0.00945
232	105.4	0.00949
233	104.9	0.00953
234	104.5	0.00957
235	104	0.00961
236	103.6	0.00965
237	103.2	0.00969
238	102.7	0.00973
239	102.3	0.00977
240	101.9	0.00981
241	101.5	0.00985
242	101	0.00989
243	100.6	0.00993
244	100.2	0.00997
245	99.8	0.01001
246	99.4	0.01005
247	99	0.01009
248	98.6	0.01013
249	98.2	0.01017

Molecular weight	1 mg/litre = --------------- ppm	1 ppm = --------------- mg/litre
250	97.8	0.01021
251	97.4	0.01027
252	97	0.01031
253	96.6	0.01035
254	96.3	0.01039
255	95.9	0.01043
256	95.5	0.01047
257	95.1	0.01051
258	94.7	0.01055
259	94.3	0.01059
260	93.9	0.01063
261	93.7	0.01067
262	93.3	0.01072
263	93	0.01076
264	92.6	0.0108
265	92.3	0.01084
266	91.9	0.01088
267	91.6	0.01092
268	91.3	0.01096
269	90.9	0.011
270	90.6	0.01104
271	90.2	0.01108
272	89.9	0.01112
273	89.6	0.01117
274	89.2	0.01121
275	88.9	0.01125
276	88.6	0.01129
277	883	0.01133
278	87.9	0.01137
279	87.6	0.01141
280	87.3	0.01145
281	87	0.01149
282	86.7	0.01154
283	86.4	0.01157
284	86.1	0.01162
285	85.8	0.01166

Molecular weight	1 mg/litre = --------------- ppm	1 ppm = --------------- mg/litre
286	85.5	0.0117
287	85.2	0.01174
288	84.9	0.01178
289	84.6	0.01182
290	84.3	0.01186
291	84	0.0119
292	83.7	0.01194
293	83.4	0.01198
294	83.2	0.01202
295	82.9	0.01207
296	82.6	0.01211
297	82.3	0.01215
298	82	0.01219
299	81.8	0.01223
300	81.5	0.01227

Appendix 2.2: Air Pollutant Concentrations Conversion under Various Conditions

Air pollutant concentrations, as measured or as calculated by air pollution dispersion modeling, must often be converted or corrected to be expressed as required by the regulations issued by various governmental agencies.

Such regulations involve a number of different expression of concentration. Some express the pollutants concentrations as ppmv (parts per million by volume) and some express the concentrations as mg/m^3 (milligrams per cubic meter), while others require adjusting or correcting the concentrations to reference conditions of moisture content, oxygen content or carbon dioxide content. Methods for converting concentrations from ppmv to mg/m^3 (and vice versa) and for correcting the concentrations to the required reference conditions are discussed below.

All of the concentrations and concentration corrections shown below apply only to air and other gases. They are not applicable for liquids.

(A) Converting Air Pollutant Concentrations

The conversion equations depend on the temperature at which the conversion is wanted (usually 25°C). At an ambient sea level atmospheric pressure of 1 atm (101.325 KPa or 1.01325 bar), the general equation is:

$$ppmv = mg / m^3 . \frac{(0.08205.T)}{M}$$

and for the reverse conversion ie, from mg/m^3 to ppmv, the equation is:

$$mg / m^3 = ppmv. \frac{M}{(0.08205.T)}$$

where,

mg/m^3 = milligrams of pollutant per cubic meter of air at sea level atmospheric pressure and temperature

$ppmv$ = air pollutant concentration, in parts per million by volume

T = ambient temperature in K = 273.15 + °C

0.08205 = Universal gas constant in atm·m³ (kmol· K)

M = Molecular weight of the air pollutant

NOTES

- 1 mol = absolute pressure of 101.325 KPa or 1.01325 bar
- 1 mol = 1 gram mole and 1 kmol = 1000 gram moles
- Although ppmv and mg/m³ have been used for the examples in all of the following sections, concentrations such as ppbv (i.e., parts per billion by volume), volume percent, mole percent and many others may also be used for gaseous pollutants.
- Particulate matter (PM) in the atmospheric air or in any other gas cannot be expressed in terms of ppmv, ppbv, volume percent or mole percent. PM is most usually (but not always) expressed as mg/m³ of air or other gas at a specified temperature and pressure.
- For gases. volume percent = mole percent
- 1 Volume percent = 10,000 ppmv (i.e., parts per million by volume) with a million being defined as 10^6.

(B) Correcting Air Pollutants Concentrations for Altitude

Air pollutant concentrations expressed as mass per unit volume of atmospheric air (e.g., mg/m³, µg/m³, etc.).at sea level will decrease with increasing altitude. The concentration decrease is directly proportional to the pressure decrease with increasing altitude. Some governmental regulatory jurisdictions require industrial sources of air pollution to comply with sea level standards corrected for altitude.

The change of atmospheric pressure with altitude can be obtained from this equation:

$$P_h = P.\left(\frac{288-6.5h}{288}\right)^{5.2558}$$

Given an air pollutant concentration at sea-level atmospheric pressure, the concentration at higher altitudes can be obtained from this equation:

$$C_h = C.\left(\frac{288-6.5h}{288}\right)^{5.2558}$$

where,

h = altitude, in km

P = atmospheric pressure at sea level

P_h = atmospheric pressure at altitude **h**

C = Air pollutant concentration, in mass per unit volume at sea level atmospheric pressure and specified temperature T

C_h = Concentration, in mass per unit volume at altitude **h** and at specified temperature T

As an example, given an air pollutant concentration of 260 mg/m³ at sea level, calculate the equivalent pollutant concentration at an altitude of 2800 meters:

$$C_h = 260\times[\{288-(6.5)(2.8)\}/288]^{5.2558} = 260\times 0.71 = 185 mg/m^3$$

The above equation for the decrease of air pollution concentrations with increasing altitude is applicable only for about the first 10 km of altitude in the troposphere (the lowest atmospheric layer) and is estimated to have a maximum error of about 3 percent. However, 10 km of altitude is sufficient for most purposes involving air pollutant concentrations.

(C) Correcting Air Pollutant Concentrations for Reference Conditions

Many environmental protection agencies have issued regulations that limit the concentration of pollutants in gaseous emissions and define the reference conditions applicable to those concentration limits. For example, such a regulation might limit the concentration of NOx to 55 ppmv in a dry combustion exhaust gas (at a specified reference temperature and pressure) corrected to 3 volume percent O_2 in the dry gas. As another's example, a regulation might limit the concentration of total particulate matter to 200 mg/m^3 of an emitted gas (at a specified reference temperature and pressure) corrected to a dry basis and further corrected to 12 volume percent CO_2 in the dry gas.

Environmental agencies often use the terms "dscm" or "scmd" to a "standard" cubic meter of gas. Since, there is no universally accepted set of "standard" temperature and pressure it is recommended that the reference temperature and pressure always be clearly specified when it is stating gas volumes or gas flow rates.

(C1) Correcting Air Pollutant Concentration

If a gaseous emission sample is analyzed and found to contain water vapor and a pollutant concentration of say 40ppmv, then 40 ppmv should be designated as the "wet basis" pollutant concentration. The following equation can be used to correct the measured "wet basis" concentration to a "dry basis" concentration:

$$C_{dry\ basis} = \frac{C_{wet\ basis}}{1-w}$$

where,

C = Concentration of the air pollutant in the emitted gas
w = fraction, by volume, of the emitted gas

As an example, a wet basis concentration of 40 ppmv in a gas having 10 volume percent water vapor would have a:

$$C_{dry\ basis} = 40 \div (1-0.10) = 44.4\,ppmv.$$

(C2) Correcting to a Reference Oxygen Content

The following equation can be used to correct a measured pollutant concentration in a dry emitted gas with a measured O_2 content to an equivalent pollutant concentration in a dry emitted gas with a specified reference amount of O_2.

$$C_r = C_m.\frac{(20.9 - reference\ volume\%O_2)}{(20.9 - measured\ volume\%O_2)}$$

where,

C_r = corrected concentration of a dry gas with a specified reference volume % O_2
C_m = measured concentration in a dry gas having a measured volume % O_2

As an example, a measured NOx concentration of 45 ppmv in a dry gas having 5 volume % O_2 is:

$$45 \times (20.9-3) \div (20.9-5) = 50.7\,ppmv\ of\ NOx$$

When corrected to a dry gas having a specified reference O_2 content of 3 volume %.

the measured gas concentration $\mathbf{C_m}$ must first be corrected to a dry basis before using the above equation.

(C3) Correcting Air Pollutant Concentration to a Reference Carbon Dioxide Content

The following equation can be used to correct a measure pollutant concentration in an emitted gas (containing a measured CO_2 content) to an equivalent pollutant concentration in an emitted gas containing a specified reference amount of CO_2.

$$C_r = C_m \cdot \frac{(20.9 - reference\ volume\%O_2)}{(20.9 - measured\ volume\%O_2)}$$

where,

C_r = corrected concentration of a dry gas with a specified reference volume % CO_2

C_m = measured concentration in a dry gas having a measured volume % CO_2

As an example, a measured particulates concentration of 200 mg/m³ in a dry gas that has a measure 8 volume % CO_2 is:

$$200 \times (12 \div 8) = 300\ mg/m^3$$

When corrected to a dry gas having a specified reference CO_2 content of 12 volume %.

Appendix 2.3: Conversion Factor between Volume and Mass Units of Air Pollutants Concentration (at 25°C, 760 mmHg) i.e., from ppm to μg/m³ and vice versa for some Gases

To convert from

S. No.	Air pollutants	Ppm (vol) to μg/m³, multiply by:	μg/m³ to ppm (vol), multipily by (× 10⁻³):
1	Ammonia (NH_3)	695	1.44
2	Carbon dioxide (CO_2)	1800	0.56
3	Carbon monoxide (CO)	1150	0.87
4	Chlorine (Cl_2)	2900	0.34
5	Ethylene (C_2H_4)	1150	0.57
6	Hydrogen chloride (HCl)	1490	0.37
7	Hydrogen fluoride (HF)	820	1.22
8	Hydrogen sulfide (H_2S)	1390	0.72
9	Methane (CH_4)	655	1.53
10	Nitrogen dioxide (NO_2)	1880	0.53
11	Nitric oxide (NO)	1230	0.81
12	Ozone (O_3)	1960	0.51
13	Peroxyacetylnitrate (PAN)	4950	0.20
14	Sulfur dioxide (SO_2)	2620	0.38

Appendix 2.4: Units of Measurement for Air Pollutants

Number concentrations:	**(# per volume of air, #/m³)**
• # of molecules per m^3	(highly reactive species, e.g., OH radical)
• # of particles per m^3	(particulate matter)
• # of cells or colony forming units per m^3	(biological)

Mass concentrations:	**(mass per volume of air)**
• ng/m^3	typical for metals and for SVOCs
• $\mu g/m^3$	typical for indoor VOCs and particulate matter
• mg/m^3	big sources, e.g., ETS, cooking, industrial hygiene

Molar concentrations: **(variations on *yi* = mol*i*/mol*air*)**

- Mole fraction (*yi*) = mol/mol
- % concentration = moles per 100 moles = $100 * y_i$
- Parts per million by volume (ppm_v) (or, just ppm)

$$1\,ppm = \frac{1\ mol\ of\ i}{10^6\ moles\ of\ air} = 10^{-6}\ \frac{moles\ of\ i}{moles\ of\ air} = 10^{-6} * y_i$$

- Parts per billion by volume (ppb_v) (or just ppb)

$$1\,ppb = \frac{1\ mol\ of\ i}{10^9\ moles\ of\ air} = 10^{-9}\ \frac{moles\ of\ i}{moles\ of\ air} = 10^{-9} * y_i$$

Appendix 2.5: Units of Measure and Conversion Table

meter (m): Unit of length in the metric system is based on meter. A softball bat is about a meter long. One meter is about 39 inches– slightly longer than a yard.

centimeter (cm): A centimeter is one one-humdredth (1/100) of a meter. One inch is about 2.5 cm; a penny is about 2 cm wide.

kilometer (km): A kilometer is equal to 1,000 meters. There are about one and a half kilometers in a mile.

millimeter (mm): One millimeter is equal to one one-thousandth (1/1,000) of a meter. A dime is slightly thicker than one mm. About 25 mm equal one inch.

Weight – *Units of weight in the metric system are based on the gram. For measuring things much smaller or larger than a gram, scientists use units that convert easily from grams – units that are 100, 1000, or 1,000,000 times smaller or larger than the gram.* ***Commonly*** *used one are included below.*

gram (g or gm): A paper clip weighs about 1 gram. An eighth of a teaspoon of sugar weighs about a gram. A penny is about 3 grams. There are about 28 grams in one ounce; 454 grams is about one pound.

kilogram (k or kg): A unit of mass (weight) in the metric system equal to 1,000 grams. A kilogram is about 2.2 pounds, which is about the same weight as 9 sticks of butter.

microgram (ug or mcg): A microgram is one one-millionth (1/1,000,000) of a gram, or about one half million times smaller than a pound. A regular aspiring tablet weighs about 400 milligrams.

milligram (mg): A milligram is one one-thousandth (1/1,000) of a gram, or about one half million times smaller than a pound. A regular aspirin tablet weighs about 400 milligrams.

nanogram (ng): A nanogram is one one- billionth (1/1,000,000,000) of a gram. It is 1,000 times smaller than a microgram.

picogram (pg): A picogram is one one-trillionth of a gram. One picogram is 1,000 times smaller than a nanogram.

Volume – *Units of volume in the metric system are primarily based on the liter for liquids, and the cubic meter for gases and solids.*

liter (l or L): This term is generally used when measuring a volume of liquid. A liter is slightly larger than a quart; about 3.8 liters are in a gallon.

milliliter (ml): A milliliter is one one-thousandth (1/1,000) of a litre. One teaspoon holds about 5 ml of water.

cubic meter (m^3 or cu.m.): This term is commonly used when measuring a volume of air. A cubic meter is the space inside of a box that is one meter wide, one meter high, and one meter deep. A softball bat is about a meter long (39 inches, or just over a yard). The cabinet under a kitchen sink has about one cubic meter of volume.

Concentration is measured using two units, because it is a measure of how much of one substance is contained in another. Typical units of concentration are described below.

micrograms per cubic meter (ug/m^3 or mcg/cu.m): A unit that expresses the concentration of a substance in air. A microgram per cubic meter is one one-millionth of a gram of substance per cubic meter or air.

micrograms per liter (ug/L or mcg/L): A unit that expresses the concentration of a substance in a solution. A microgram per liter is one one-millionth of a gram of substance per liter of solution. A ug/L of sugar in water is about the same as a teaspoon of sugar in 2,100,000 gallons of water. For low concentrations in water, a microgram per liter is about the same as a part per billion (ppb).

milligrams per cubic meter (mg/m3 or mg/cu.m): A unit that expresses the concentration of a substance in air. A milligram per cubic meter is one one-thousandth of a gram per cubic meter of air. The milligram (1,000 micrograms) is used with higher concentrations.

milligrams per kilogram (mg/kg): This unit expresses concentration by measuring the weight of a substance by the weight of the substance with the medium containing it. A milligram per kilogram is the same as one part per million. Mg/kg is also used when measuring a person or animal's exposure to a substance. For example, a 70 kg person (about 154 lbs) given one aspirin which has 325 mg of active ingredient would get a dose of 325 mg/70 kg, or 4.6 mg/kg.

milligrams per litre (mg/L): A unit that expresses the concentration of a substance in a solution. A milligram per liter is one one-thousandth of a gram of substance per liter of solution. One mg/l of sugar in water is about the same as a teaspoon of sugar in 2,100 gallons of water. For low concentrations in water, milligrams per litre (mg/l) is about the same as parts per million (ppm).

parts per billion (ppb): A way of expressing the concentration of a substance in air, water, soil or food. A microgram per kilogram is the same as one part per billion. One ppb means that there is one part of a substance for every billion parts of substance and the medium containing it. One ppb is about one drop of dye in 18,000 gallons of water or about 1 second in 32 years. One ppb is 1,000 times less than one ppm.

parts per million (ppm): The concentration of a substance in air, water or soil. A milligram per kilogram is the same as one part per million. One ppm means that there is one part of a substance for every million parts of substance and water or soil containing it. One ppm is about one drop of dye in 18 gallons of water, about one inch in 16 miles, or one penny in $10,000.

parts per trillion(ppt): The concentration of a substance in air, water or soil. A nanogram per kilogram is the same as one part per trillion. One ppt means that there is one part of a substance for every trillion parts of substance and air, water or soil containing it. One ppt is about one drop of dye in 18 million gallon of water or about 1 second in 32,000 years. One ppt is 1,000 times less than 1 ppb.

picocuries per liter (PCi/l): Unit of measure used to express the results of radioactivity tests in air and water. For radon gas, one picocurie per liter is the amount of radon in the air so that 2.2 atoms of radon decay during one minute in one litre of air.

Conversion Tables: Length

English		Metric
1 mile	=	1.61 kilometers
1 foot	=	30.5 centimeters
1 inch	=	2.54 centimeters
0.621 mile	=	1 kilometer
39.4 inches	=	1 meter

Metric Conversions (Length)

Unit		Number of meters
Kilometers	=	1,000
Meter	=	1
Centimeter	=	0.01
Millimeter	=	0.001
Micrometer (also micron)	=	0.000001

Mass and Weight

English		Metric
1 pound	=	0.454 kilograms
1 ounce	=	28.3 grams
2.20 pounds	=	1 kilogram
0.0353 ounce	=	1 gram

Metric Conversions (mass)

Unit		Number of grams
Kilogram (kg) also called metric tonne	=	1,000
Hectogram	=	100
Dekagram	=	10
Gram (g)	=	1
Milligram (mg)	=	0.001
Microgram (mcg)	=	0.000001
Nanogram	=	0.000000001
Picogram (pg)	=	0.000000000001

Expressing Concentrations (weight/weight)

1 part per million	=	1 milligram per kilogram
	=	1 microgram per gram
1 part per billion	=	1 microgram per kilogram
1 part per trillion	=	1 nanogram per kilogram

Volume (Liquids)

English		Metric
1 gallon	=	3.79 litres
1 quart	=	0.946 litres
1.06 quarts	=	1 litre
0.264 gallons	=	1 litre

Metric Conversions (Volume)

Unit		Number of grams
Kilolitre	=	1,000
Hectolitre	=	100
Dekalitre	=	10
Litre	=	1
Decilitre	=	0.1
Centilitre	=	0.01
Millilitre	=	0.001
Cubic Centimeter	=	0.001
Microlitre	=	0.000001

Expressing concentrations (weight/volume)*

1 part per million	=	1 milligram per liter*
	=	1 microgram per milliliter*
1 part per billion	=	1 microgram per liter*
1 part per trillion	=	1 nanogram per liter*

Volume (Gases)

English		Metric
1 cubic foot	=	0.0283 cubic meter
35.3 cubic feet	=	1 cubic meter

Metric Conversions (volume)

Unit		Number of cubic meter
1 cubic meter	=	1
1 liter	=	0.001
1 cubic centimeter	=	0.000001

Appendix 2.6: All about Particulate matter (PM) – Definition, Particle Size & Measurement, Size Ranges and Definition of Aerosols

Some Key Definitions for PM are:

- **Aerosol:** Mostly stable particles suspended in a gas
  - Refers to both gas and particles
- **Bioaerosol:** Aerosol of biological origin
  - Fungi, bacteria, viruses
  - Fungal spores, dust mite parts, pollen
- **Cloud:** Visible aerosol with defined boundaries
- **Dust:** Irregular, solid phase produced by mechanical processes
- **Fume:** Solid phase chains or clusters of small primary particles
  - Produced by condensation of vapors or gaseous combustion products
- **Haze:** Atmospheric aerosol that affects visibility
- **Mist and Fog:** Spherical liquid-particle aerosols
  - Formed by condensation and atomization
- **Smoke:** Results from incomplete combustion
- **Spray:** Liquid particle analog to dust

Particle sizes

- Usually referring to a characteristic dimension eg.,
  - Diameter for sphere
  - Diameter for fibers (e.g. asbestos)
  - Equivalent diameter for non-spherical particles
- Micrometer (μm)
  - 1 μm = 10^{-6}, = 1 microns (10^{-3} mm)
- Nanometer (nm)
  - 1 nm = 10^{-9} m

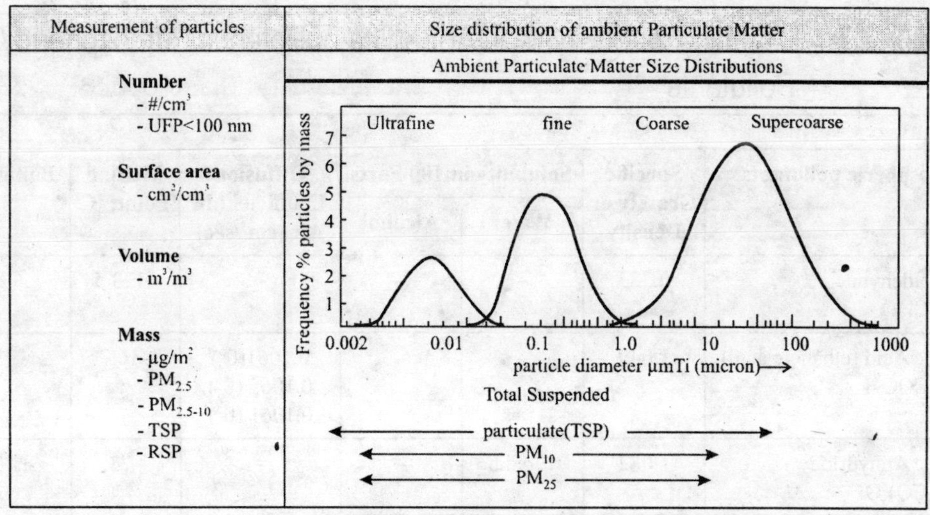

UFP- Ultrafine particles, TSP-Total suspended particulate
RSP-Respirable suspended particulate

Figure: (Ambient Particulate Matter Size Distributions)

Particle Diameter µm (microns)

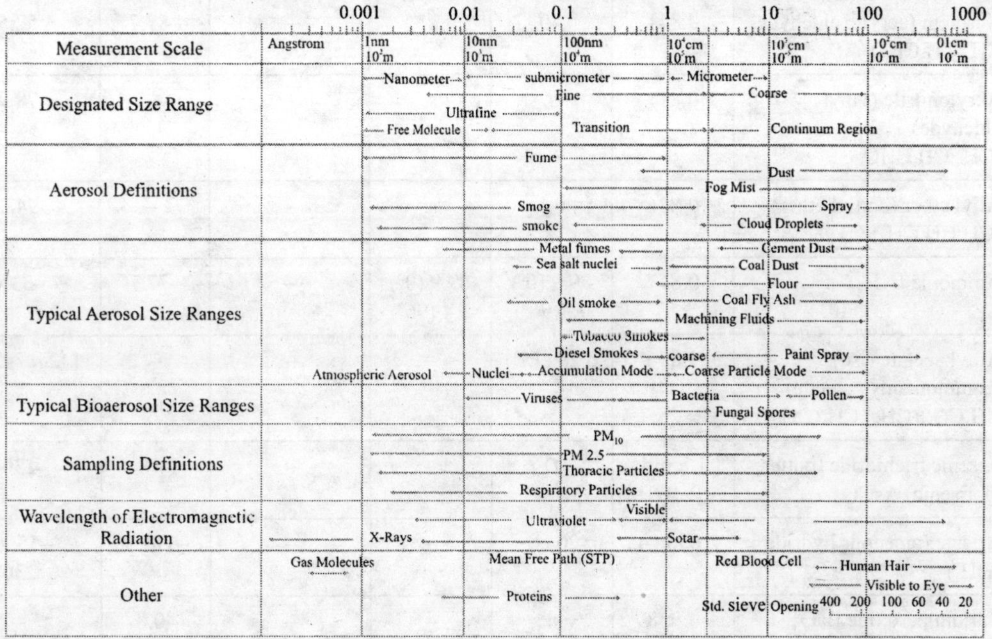

Particle size ranges and definitions for aerosols

UFP $PM_{2.5}$ PM_{10}

UFP- ultrafine particles, $PM_{2.5}$ - Particulate matter of diameter ≤ 2.5 Microns, PM_{10} - Particles of diameter ≤ 10 microns

Figure: [Particle Diameter, µm (microns)]

Appendix 2.7: Chemical and Physical Properties of Potential Atmospheric Air Pollutants

S. N.	Atmospheric pollutants	Specific Gravity or Density	Solubility in 100 Parts		Diffusion Coefficient in Air (cm^2/sec)	Melting Point, °C	Boiling Point, °C
			Water	Alcohol			
1	Acetaldehyde $CH_3.CHO$	0.783	–	–	–	–123.5	20.2
2	Acetic Acid (ethanoic acid) $CH_3.COOH$	1.049	–	–	0.1061(0°) 0.1065 (0°) 0.1061 (0°)	–16.7	118.1
3	Acetic Anhydride $(CH_3CO)_2O$	1.082	12c:d.h	–	–	–73	139.6
4	Acetone (propanone) $CH_3.CO.CH3$	0.792	–	–	–	–94.6	56.5
5	Acetylene terachloride (sym-tetra-chloroethane) $Cl_2CH.CHCl_2$	1.6000	i.	–	–	–36	146.3
6	Acrolein (acrylic aldehyde) CH_2:CH.CHO	0.841	40	s.	–	–87.7	52.5
7	Acrylonitrile (vinyl aldehyde) (CH_2:CH,CHO	0.811	s.	s.	–	–82	78.9
8	Allyl ether (daillyl ether) (CH:CH,$CH_2)_2$O)	0.826	0.3	s.	–	–	94.3
9	Ammonia (NH_3)	0.817	89.9 (0°) 7.4(96°)	89.9 (0°) 7.4 (96°)	–	–77.7	–33.4
10	Amyl acetate (iso) (common amyl acetate) $CH_3CO_2.(CH_2)_2CH:(CH_3)_2$	0.876	0.25 (15°)	–	–	–	142@742mm
11	Arsenic trichloride (butter of arsenic) $AsCl_3$	Liquid 2.163	D	–	–	–18	130
12	Arisine (arsenous hydride AsH_3)	2.695 (A)	20 cc.	–	–	–113.3	–55; d. 230
13	Barium peroxide BaO_2	4.958	v.sl.s.	–	–	–0.8	–
14	Benzene C_6H_6	0.879	0.07 (22°)	–	–	5.45	80.1
15	Benzene (mixture, gasoline)	Less than Water	–	–	–	–	–

S. N.	Atmospheric pollutants	Specific Gravity or Density	Solubility in 100 Parts		Diffusion Coefficient in Air (cm^2/sec)	Melting Point, °C	Boiling Point, °C
			Water	Alcohol			
16	Benzoyl chloride C_6H_5CO,Cl	1.212	d.	d.h.	–	–0.5	197.2
17	Benzyl amine (ω-amino-toluene) C_6H_5,CH_2,NH_2	0.982	–	–	–	–	184.5
18	Bromine Br_2	3.2	3.13 (30°)	s.	–	–7.2	58.78
19	Bromobenzene (phenyl bromide) C_6H_5Br	1.495	i.	s	–	–30.6	156.2
20	Bromoethane (vinyl bromide) C_2,H_5,Br	1.460	1.06(0°) 0.9(30°)	–	–	–118.9	38.4
21	Bromoethylene (vinyl bromide) CH_2,CH,Br	1.529	i.	–	–	–137.8	15.8
22	Bromomethane (methyl bromide) $CH_3.Br$	1.732	v. sl. s.	s	–	–93	45@758mm
23	Bromopropane (propyl bromide) CH_3,CH_2,CH_2Br	1.353	0.25 (20°)	–	0.085	–109.9	70.8
24	Bromopropane (allyl bromide) CH_3,CH,CH_2,Br	1.398	i.	–	–	–119.4	70.1@758mm
25	Ω-Bromotoluene (benzyl bromide) C_6H_5,CH_2,Br	1.443	i., sl. d.	–	–	–4	198.9
26	Bromoxylene (bromo-o-xylene[]) $Br,C_6H_3,(CH_3)_2$	1.365	i.	–	–	–	213.5
27	1,3-Butadiene (erythrene) $CH_2:CH,CH:CH_2$	0.621	i.	–	–	–108.9	–4.41
28	Butane (diethyl) CH_3,CH_2,CH_2,CH_3	liq. 0.60	15 cc. (17°/773)	1183 (17°/773)	–	–135	–0.6
29	n-Butanol C_2H_5,CH_2,CH_2OH	0.810	9 (15°)	–	0.0703 0.0681 (0°)	–79.9	117
30	2-Butanone (methylethyl ketone) CH_3,CO,C_2H_5	0.805	37	–	–	–85.9	79.6
31	Butene-1 (Butylene) $C_2H_5,CH:CH_2$	liq. 0.6	i.	v.s.	–	–130	–5
32	n-Butyl acetate $CH_3CO_2,(CH_2)_3,CH_3$	0.882	0.7	–	0.058	–76.3	125.1
33	n-Butyl cellosolve C_4H_4,O,CH_2CH_2OH	0.905	–	–	–	–	171.2

S. N.	Atmospheric pollutants	Specific Gravity or Density	Solubility in 100 Parts		Diffusion Coefficient in Air (cm^2/sec)	Melting Point, °C	Boiling Point, °C
			Water	Alcohol			
34	n-Butyl formate $HCO_2,(CH_2)_3,CH_3$	0.911	v.sl.s.	–	–	–	106.9
35	Butyl ether (n-dibutyl ether) $(C_2H_5,CH_2,CH_2)_2O$	0.769	< 0.05	–	–	–98	142.4
36	n-Butyl nitrite C_4H_9,O,NO	0.911	–	–	–	–	78
37	Cacodyl oxide $[(CH_3)_2As]_2O$	1.486	sl.s.	s.	–	–57	150
38	Cadmium Cd	8.65	i.	–	–	320.9	767
39	Carbon dioxide CO_2	Solid 1.56 (-79°)	179.7 cc (0°)	–	0.138	–56.5@ 5.2 atm.	subl. -78.5
40	Carbon disulfide CS_2	liq. 1.262 (22 °/20)	0.2 (0°)	–	0.0892 0.0883 (0°)	–108.6	46.3
41	Carbon monoxide CO	liq. 0.814 (-195 °/4)	0.0044 (0°)	–	–	–207	–192
42	Carbon terachloride CCl_4	1.595	0.097 (0°)	–	–	–22.6	76.8
43	Cellosolve (2-ethoxy-ethanol-l) $C_2H_5,O,(CH_2)_2OH$	0.931	–	–	–	–70	135.1
44	Chlorine Cl_2	liq. 1.56 (-33.6 °)	1.46 g (0°) 310 CC (10°)	–	–	–101.6	–34.6
45	Chlorine dioxide (chlorine peroxide) ClO_2	2.4 (11 °) A	2.000 cc (4°)	expl.	–	–76	9.9@730 mm expl. 100
46	Chlorine monoxide Cl_2O	2.9 A	2.000 cc (0°)	expl.	–	–20	3.8@766 mm expl. 100
47	Chloroacetone Cl,CH_2,CO,CH_3	1.162 (16 °)	–	–	–	–44.5	121
48	Chlorobenzene (phenyl chloride) C_6H_5,Cl	1.107 (20 °/4)	i.	–	–	–45.2	132.1
49	2-Chlorobutadiene (chloroprene) $CH_2:CCl,CH:CH_2$	0.958 (20 °/20)	sl. s.	–	–	–	59.4
50	o-Chloridiophenyl (chlorobiphenyl) Cl,C_6H_4,C_6H_5	–	i.	–	–	34	267

S. N.	Atmospheric pollutants	Specific Gravity or Density	Solubility in 100 Parts		Diffusion Coefficient in Air (cm^2/sec)	Melting Point, °C	Boiling Point, °C
			Water	Alcohol			
51	Chloroform (trichloromethane) $CHCl_3$	1.489	0.82 (20°)	–	–	–63.5	61.2
52	1 -Chloro-1-nitro-propane $C_2H_5,CH(Cl)NO_2$	1.209	< 0.8 (20°)	–	–	–	140
53	CHloropropane (n-propyl chloride) CH_3,CH_2,CH_2Cl	0.890	0.27 (20°)	–	–	–122.8	46.4
54	3-Chloropropene-1(allyl chloride) $CH_2:CH,CH_2Cl$	0.938	< 0.1	–	–	–136.4	44.6
55	Chloropicrin (nitrochloroform) NO_2,CCl_3	1.651	0.17 (18°)	37 cc 80 % al.	–	–64	112.3
56	ω-Chlorotoluene (benzyl chloride) C_6H_5,CH_2,Cl	1.100	i.	–	–	–39	179.4
57	Chloro-vinyldichloroarsine ClCH:CHAsCl_2	1.888	i.	–	–	–	190
58	Chloroxylene (o-xylychloride) $Cl,C_6H_3(CH_3)_2$	–	i.	–	–	< -20	189.5
59	Chromic acid (chromium trioxide)CrO_3	2.70	164.9 (0°)	s.	–	197 d.	–
60	o-Cresol CH_3,C_6H_4,OH	1.048	2.5	–	–	30.8	190.8
61	m-Cresol CH_3,C_6H_4,OH	1.034	0.5	–	–	10.9	202.8
62	p-Cresol CH_3,C_6H_4,OH	1.035	1.8	–	–	35	202
63	Cyanogen C_2N_2	liq. 0.866 (-172°)	450 CC (20°)	2,300 cc (20°)	–	–34.4	–20.5
64	Cyanogen bromide Br,CN	2.015	s.	s.	–	52	61.3 @750 mm
65	Cyanogen bromide Cl,CN	1.222 (0 °)	2.500 cc (20°)	10,000 cc (20°)	–	–6.5	12.5
66	Dichloroethyl ether (β, $β_1$-dichloro ether)	1.222	1.07 (20°)	s.	–	–	178.5
67	Dichlorodiethyl sulfide (mustard gas) $(Cl,CH_2,CH_2)_2S$	1.275	0.07 (25°)	s.	–	13.5	217 sl. d.
68	Dichloromethane (methylene chloride CH_2Cl_2)	1.336	2 (20°)	–	–	–96.7	40.5

S. N.	Atmospheric pollutants	Specific Gravity or Density	Solubility in 100 Parts		Diffusion Coefficient in Air (cm^2/sec)	Melting Point, °C	Boiling Point, °C
			Water	Alcohol			
69	Diethyl amine $(C_2H_5)_2NH$	0.712	v.s.	–	–	–38.9	55.5
70	Diethylene glycol (glycol ether) $(HO,CH_2,CH_2)_2O$	1.118	–	–	–	–	244.8
71	Diethyl amine $(CH_3)_2$ NH	0.680 (0 °/4)	v.s.	s.	–	–96	7.4
72	Dimethyl mercury (mercury dimethyl) $(CH_3)_2Hg$	2.954	v.sl.s.	v.s.	–	–	95
73	Diemthyl sulfate (methyl sulfate) $(CH_3)_2SO_2$	1.352 (0 °/4)	v.sl.s.	–	–	–26.8	188
74	Ethane (CH_3,CH_3)	0.546 (-88 °)	4.7 cc (20°)	150 cc	–	–172	–88.6
75	Ethyl acetate CH_3CO_2,C_2H_5	0.901	8.5 (15°)	–	0.0715	–82.4	77.1
76	Ethyl alcohol (ethanol) CH_3,CH_2OH	0.789	–	–	0.102	–112	78.4
77	Ethyl benzene (phenyl ethane) C_6H_5,C_2H_5	0.867	0.01 (15°)	–	0.0658	–94.4	136.2
78	n-Ethyl butyrate C_2H_5,CH_2,CO_2,C_2H_5	0.879	0.68 (25°)	–	0.579	–93.3	120
79	Ethyl ether (ether) $(C_2H_5)_2O$	0.708	7.5 (20°)	–	0.0778	–116.3	34.6
80	Ethyl formate $(HCOO,C_2H_5)$	0.923	11 (18°)	–	0.0840	–79	54
81	Hydrogen sulfide H_2S	1.1895 A	437 CC (0°) 186 cc (40°)	9.54 cc (15°)	–	–82.9	–59.6
82	Iodine l_2	4.93	0.0162 (0°)	s.	0.097	113.5	184.4
83	Mercury Hg	13.546	i.	–	–	–38.87	356.9
84	Methane CH_4	0.554	0.35 cc (20°)	47.1 cc (20°)	–	–182.6	–161.4
85	Methanol (methyl alcohol, wood alcohol) CH_3,OH	0.792	–	–	0.1325	–97.8	64.7

S. N.	Atmospheric pollutants	Specific Gravity or Density	Solubility in 100 Parts		Diffusion Coefficient in Air (cm^2/sec)	Melting Point, ºC	Boiling Point, ºC
			Water	Alcohol			
86	Methylacetate CH_3,CO_2,CH_3	0.924	33 (22º)	–	0.084	–98.7	57.1
87	Methyl butyrate C_2H_5,CH_2,CO_2,CH_3	0.898	1.7	–	0.0633	< -95	102.3
88	Methyl ethyl ketone (butanone) CH_3,CO,C_2H_5	0.805	37	–	–	–85.9	79.6
89	Isopropyl ether (iso-dipropyl ether) $[(CH_3)_2CH]_2O$	0.725	0.2	–	–	–60	68.7
90	Radon Rn	liq. 5.5	51 cc (0º)	–	–	–71	62
91	Styrene monomer (phenyl ethylene) $C_6H_5,CH:CH_2$	0.903	v. sl.s.	–	–	–31	145.5
92	Sulfur dioxide SO_2	2.264 A	22.8 (0º)	s.	–	–75.5	–10.0
93	Sulfuric acid H_2SO_4	1.834	–	d.	–	10.49	d. 340
94	Toluene $C_6H_5,CH:CH_2$	0.866	i.	–	–	–95	110.8
95	Trichloroethane (I.l.l) (methyl chloroform) CH_3,CCl_3	1.325	i.	–	–	–	74.1
96	Trichloroethylene (TCE) $ClCH:CCl_2$	1.466	0.1 (25º)	–	–	–73	87.2
97	Triethyl amine $ClCH:CCl_2$	0.729	–	–	–	–114.8	89.4
98	Turpentine $C_{10}H_{16}$	0.863	–	–	–	–	115.162
99	Vinyl chloride CH2:CH,Cl	0.908	sl. s.	s.	–	–160	–12
100	o-Xylene $C_6H_4(CH_3)_2$	0.881	i.	– lbs	–	–25	144

NOTES

i-insoluble, d-decomposes, v.sl.s-very slightly soluble, s-soluble, c-cold, d.h-decomposes hot, expl-explods, sl.s - slightly soluble, subl - sublimes, al- alcohol, v.s. - very soluble, abs. - absolute

Appendix 2.8: Flash Point, Explosive Limits, Auto Ignition Temperature and Specific Gravities of some Hazardous Chemicals

S. No.	Name of Chemicals	Flash point, °C	Auto ignition Temperature, °C	Explosive limit, % gas in gas/ air mixture		Vapor density/ (Air = 1)
				Lower Limit	Upper Limit	
1	Parathion	120 -160	–	–	–	–
2	2 - ethgl hexanol	–	–	–	–	4.49
3	50% methyl parathion	–	–	–	–	–
4	Acetaldehyde	–37.8	175	4	57	1.52
5	Acetone	–17.8	465	2.6	12.8	2.00
6	Acetone Cyanohydrin	73.9	687.8	2.2	12	2.93
7	Acetylene	–17.8	305	2.5	82	0.91
8	Acrylonitrile	–1.1	481.1	3.1	17	1.83
9	Ammonia (anhydrous)	Gas	651.1	16	25	0.6
10	Ammonium nitrate	27	–	–	–	–
10A	Amyl acetate - n	34	380	1.1	7.5	4.69
10B	Amyl alcohol - n	–	300	1.2	–	3.04
11	Arsenic pentoxide	–	–	–	–	–
12	Arsenic trioxide	–	–	–	–	–
13	Aviation turbine fuel (ATF)	–	–	–	–	–
14	Benzidine dihydrochloride	–	–	–	–	–
15	Benzene	–11.1	562.2	1.4	8	2.77
15A	Benzene	–17.5	286	1.4	5.9	–
16	Benzidine salts	–	–	–	–	–
17	Blast furnace gas	Gas	680	37	71	1.35
18	Bromine	–	–	–	–	5.5
19	BTXS raffinate	–	–	–	–	–
20	Butadiene	–76.1	420	2	11.5	1.87
20A	Butyl acetate-n	23	421	1.4	5.9	4.0
21	N-Butane	–60	405	1.9	8.5	2.046
22	Butanol - n	30	367	1.4	11.2	2.55

S. No.	Name of Chemicals	Flash point, °C	Auto ignition Temperature, °C	Explosive limit, % gas in gas/ air mixture		Vapor density/ (Air = 1)
				Lower Limit	Upper Limit	
23	Butylene	–	–	–	–	–
24	N - Buty raldehyde	–6.6	230	2.5	12.5	2.5
25	C-4. Raffinate	–	–	–	–	–
26	C_2/C_3 fraction (ethane + propane)	–	–	–	–	–
26A	Camphor	70	468	–	–	5.24
27	Carbofuran	–	–	–	–	7.9
28	Carbon- disulphide	–30	125	1.3	50	2.64
29	Carbon monoxide	Gas	608.9	12.5	74.2	–
29A	Carbon tetrachloride	–	–	–	–	–
30	Chlorine	Gas	–	–	–	2.49
31	Chlorine dioxide	–	–	–	–	–
32	Chloromethyl ether	< 23	–	–	–	–
32A	Chloroform	Gas	–	–	–	4.13
33	Cobalt metal, oxides as powder	–	–	–	–	–
33A	Coal tar oil/coal tar pitch	50/205	–	–	–	–
34	Coke oven gas	–	–	–	–	–
35	Crotonaldehyde	12.8	20.2	2.1	15.5	2.41
36	Crude benzel	–	–	–	–	–
37	Crude oil	–	–	–	–	–
38	Cyceohexane	< -6	310	2.8	–1.2	–
38A	Cyclhexanol	70	300	–	–	3.45
39	Cyciotrimethylene trinitramine	–	–	–	–	–
40	Cydo Hazamar	–	–	–	–	–
41	Defrosted crude Benzol	–	–	–	–	–
42	Demetion	45	463.0	1.0	5.3	> 1
42A	Denatured alcohol	16	400	–	–	1.6
43	Di-chioro Ethane	–	–	5.6	11.4	–

S. No.	Name of Chemicals	Flash point, °C	Auto ignition Temperature, °C	Explosive limit, % gas in gas/ air mixture		Vapor density/ (Air = 1)
				Lower Limit	Upper Limit	
44	Di-phiny/methane Di- isocyanate	–	–	–	–	–
45	Dichloropropane	21	–	3.1	–	–
46	Diesel	–	–	–	–	–
47	Dimethylamine (Anhydrous)	–17.8	400	2.8	14.4	1.55
48	EPN	–	–	–	–	–
49	Ethion	–	–	–	–	13.26
50	Ethyl acetate	–4.4	426.7	2.2	11	3.04
51	Ethyl alcohol	13.1	422.8	3.3	19	1.59
52	Ethyl benzene	15	433.2	1.2	6.8	3.66
53	Ethyl chloride	–50	518.9	3.8	15.4	2.22
54	Ethylene	–	490	2.7	36	0.98
55	Ethylene dichloride	12.8	412.8	6.2	15.6	3.4
55A	Ethylene glycol	110	407	3.2	–	2.14
56	Ethylene oxide	–20	428.9	3	100	1.52
56A	Ethyl-n propyl ether	–	–	1.9	24	–
57	Flammable gases	–	–	–	–	–
58	Flammable hydrocarbons	–	–	–	–	–
59	Flammable liquids	–	–	–	–	–
60	Formaldehyde	50-85	430	7	73	–
60A	Formaldehyde, 37% in water	55	423	–	–	1.03
61	Furnace oil	–	–	–	–	–
62	Gases mixture (CO)	–	–	–	–	–
63	Gasoline (From 50 To 100 Octane)	–45.6	280-456	1.3	6	3-4
64	N - heptane	–3.9	223	1.05	6.7	3.45
65	N- hexane	–23	225	1.2	7.5	2.97
66	High speed diesel	–	–	–	–	–
67	Hydrogen	–	400	4.1	74.2	0.069

S. No.	Name of Chemicals	Flash point, °C	Auto ignition Temperature, °C	Explosive limit, % gas in gas/ air mixture		Vapor density/ (Air = 1)
				Lower Limit	Upper Limit	
68	Hydrogen fluoride	–	–	–	–	0.7
69	Hydrogen sulfide	Gas	260	4	46	1.189
70	Iso- butanol	27.8	426.7	1.6	10.9	–
71	Iso- propanal	11.7	399	2.3	12.7	2.1
72	Iso- propyl alchol	–	455.6	2.5	12	2.07
73	Isobutylene	–	465	1.8	9.6	1.9
74	Isocyanate	–	–	–	–	–
74A	Jet fuel	45	230	0.6	5.6	–
75	Light diesel oil	–	–	–	–	–
76	Liquid oxygen	–	–	–	–	–
77	Low aromatic	–	–	–	–	–
78	LPG	60 to -104.4	405-466	2.2	9.5	1.5
78A	Maileic anhydride	105	475	–	–	3.38
79	Methand	12.2	470	6	36.5	1.11
79A	Methyl acetate	–10	501	3.1	16	2.56
80	Monomethyl amine	–	–	–	–	–
81	Mineral spirit	38	245	77	–	3.9
82	N – Butanol	–	–	–	–	–
83	N – Butyl acetate	22.2	425	1.4	7.5	–
84	Naphtha	39	226	0.8	5	–
84A	Naphthalene	80	526	0.9	5.9	4.42
85	Natural gas	–	–	–	–	–
86	Nickel metals oxide as powder	–	–	–	–	–
87	Nitroglycerene	–	270	–	–	4.84
88	Non – aromatic	–	–	–	–	–
88A	O – cresol	80	560	1.3	–	3.72
89	O- xylene	17	–	1	6	–
90	Octane	13.3	220	1	4.7	3.86
91	Oleum (SO_3)	–	–	–	–	2.76
92	P- xylene	25	530	1.1	7	3.66

S. No.	Name of Chemicals	Flash point, °C	Auto ignition Temperature, °C	Explosive limit, % gas in gas/ air mixture		Vapor density/ (Air = 1)
				Lower Limit	Upper Limit	
93	Parathion	120-171	–	–	–	–
94	Parathion – methyl	–	–	–	–	–
95	Pentaerythritol tetranitrate	–	–	–	–	–
96	Pentane	< -40	309	1.5	7.8	2.48
96A	Petroleum – crude	–6.5 to 32	–	–	–	–
97	Petroleum products – class A	–	–	–	–	–
97A	Petroleum ether	–18	–	1.4	5.9	2.50
98	Phorate	–42.8	–	–	–	–
98 A	Phenol	80	715	–	–	3.24
99	Phosogene	–	–	–	–	3.4
99A	Phathalic anhydride	165	585	1.7	10.5	5.1
100	Phosphamidon	–	–	–	–	–
101	Propane	–104.4	450	2.3	9.5	1.56
101A	Prophyl acetate –n	15	450	1.7	8.0	3.52
102	Propylene	–107.8	497.2	2	11	1.4
103	Propylene oxide	–32.2	465	2.8	37	–
104	Pyridine	20	482.2	1.8	12.4	2.73
105	RDX cyclonite	–	–	–	–	–
106	Re-gen.liq. Hydrocarbon gas (RLHG)	–	–	–	–	–
107	Rectified spirit	–	–	–	–	–
108	Sodium chlorate	–	–	–	–	–
109	Styrene	31	490	1.1	6.1	3.6
110	Styrene manomel	–	–	–	–	–
111	Sulphur dichloride	–	–	–	–	3.55
112	Sulphur dioxide	–	–	–	–	2.264
113	Sulphur trioxide	–	–	–	–	2.76
114	Superior kerosene oil	–	–	–	–	–
115	Tetraethyl lead	93.3	–	–	–	–

S. No.	Name of Chemicals	Flash point, °C	Auto ignition Temperature, °C	Explosive limit, % gas in gas/ air mixture		Vapor density/ (Air = 1)
				Lower Limit	Upper Limit	
116	Toluene	4.4	480	1.27	7	3.14
117	Toluene di-isocynate	132.2	–	0.9	9.5	6.0
117A	Trichloroethylene	Weakly flammable	–	10 (O_2)	65 (O_2)	4.53
118	Trichloromethane sulphenyl chloride	–	–	–	–	–
118 A	Turpentine	35	–	0.8	–	4.84
119	Vinyl acetate monomer	–7.8	426.7	2.6	13.4	3
120	Vinyl chloride monomer	–78.9	472.2	4	26	2.2
121	Xylene	17.3-28.9	–	1-1.11	7	–
122	P- xylene	25	530	1.1	7	3.66

Appendix 2.9: Immediately Dangerous to Life or Health (IDLH) Concentrations of various substances/chemicals (as per NIOSH as on 03/01/1995)

S. No.	Substances/chemicals	IDLH Value
1	Acetaldehyde	2,000 ppm
2	Acetic acid	50 ppm
3	Acetic anhydride	200 ppm
4	Acetone	2,500 ppm [LEL]
5	Acetonitrile	500 ppm
6	Acetylene terabromide	8 ppm
7	Acrolein	2 ppm
8	Acrylamide	60 mg/m^3
9	Acrylonitrile	85 ppm
10	Aldrin	25 mg/m^3
11	Allyl alcohol	20 ppm
12	Allyl chloride	250 ppm
13	Allyl glycidyl ether	50 ppm
14	2 Aminopyridine	5 ppm
15	Ammonia	300 ppm

S. No.	Substances/chemicals	IDLH Value
16	Ammonium sulfamate	1,500 mg/m^3
17	n-Amyl acetate	1,000 ppm
18	Sec-Amyl acetate	1,000 ppm
19	Aniline	100 ppm
20	o-Anisidine	50 mg/m^3
21	p-Anisidine	50 mg/m^3
22	Antimony compounds (as Sb)	50 Sb/m^3
23	Arsenic (inorganic compounds, as As)	5 mg As/m^3
24	Arsine	3 ppm
25	Azinphosmethyl	10 mg/m^3
26	Barium (soluble compounds, as Ba)	50 mg Ba/m^3
27	Benzene	500 ppm
28	Benzoyl peroxide	1,500 mg/m^3
29	Benzyl chloride	10 ppm
30	Beryllium compounds (as Be)	4 mg Be/m^3
31	Boron oxide	2,000 mg/m^3
32	Boron trifluoride	25 ppm
33	Bromine	3 ppm
34	Bromoform	850 ppm
35	1,3–Butadiene	2,000 ppm [LEL]
36	2-Butanone	3,000 ppm
37	2-Butoxyethanol	700 ppm
38	n-Butyl acetate	1,700 ppm
39	Sec-Butyl acetate	1,700 ppm
40	Tert-Butyl acetate	1,500 ppm
41	n-Butyl alcohol	1,400 ppm
42	Sec-Butyl alcohol	2,000 ppm
43	Tert-Butyl alcohol	1,600 pm
44	n-Butylamine	300 ppm
45	Tert-Butyl chromate	15 mg Cr(VI)/m^3
46	n-Butyl glycidyl ether	250 ppm
47	n-Butyl mercaptan	500 ppm
48	p-tert-Butyltoluene	100 ppm
49	Cadmium dust (as Cd)	9 mg Cd/m^3
50	Cadmium fume (as Cd)	9 mg Cd/m^3

S. No.	Substances/chemicals	IDLH Value
51	Calcium arsenate (as As)	5 mg As/m^3
52	Calcium oxide	25 mg/m^3
53	Camphor (synthetic)	200 mg/m^3
54	Carbaryl	100 mg/m^3
55	Carbon black	1,750 mg/m^3
56	Carbon dioxide	40,000 ppm
57	Carbon disulfide	500 ppm
58	Carbon monoxide	1,200 ppm
59	Carbon tetrachloride	200 ppm
60	Chlordane	100mg/m^3
61	Chlorinated camphene	200 mg/m^3
62	Chlorinated diphenyl oxide	5 mg/m^3
63	Chlorine	10 ppm
64	Chlorine dioxide	5 ppm
65	Chlorine trifluoride	20 ppm
66	Chloroacetatldehyde	45 ppm
67	Alpha-Chloroacetophenone	15 mg/m^3
68	Chlorobenzene	1,000 ppm
69	o-Chlorobenzylidene malononitrile	2 mg/m^3
70	Chlorobromomethane	2,000 ppm
71	Chlorodiphenyl (42% chlorine)	5 mg/m^3
72	Chlorodiphenyl (54% chlorine)	5 mg/m^3
73	Chloroform	500 ppm
74	1-Chloro-1-nitropropane	100 ppm
75	Chloropicrine	2 ppm
76	Bet-Chloroprene	300 ppm
77	Chromic acid and chromates	15 mg Cr(VI) /m^3
78	Chromium (II) compounds [as Cr(II)]	250 mg Cr(II) /m^3
79	Chromium (III) compounds [as Cr(III)]	25 mg Cr(III) /m^3
80	Chromium metal (as Cr)	250 mg Cr/m^3
81	Coal tar pitchvolatiles	80 mg/m^3
82	Cobalt metal, dust and fume (as Co)	20 mg CO//m^3
83	Copper (dusts and mists, as Cu)	100 mg Cu/m^3
84	Copper fume (as Cu)	100 mg Cu/m^3
85	Cotton dust (raw)	100 mg/m^3

S. No.	Substances/chemicals	IDLH Value
86	Crag (r) herbicide	500 mg/m^3
87	Cresol (o, m, p isomers)	250 ppm
88	Crotonaldehyde	50 ppm
89	Cumene	900 ppm [LEL]
90	Cyanides (as CN)	25 mg/m^3 (as CN)
91	Cyclohexane	1,300 ppm [LEL]
92	Cyclohexanol	400 ppm
93	Cyclohexanone	700 ppm
94	Cyclohexene	2,000 ppm
95	Cyclopentadiene	750 ppm
96	2,4-D	100 mg/m^3
97	DDT	500 mg/m^3
98	Decaborane	15 mg/m^3
99	Demeton	10 mg/m^3
100	Diacetone alcohol	1,800 ppm [LEL]
101	Diazomethane	2 ppm
102	Diborane	15 ppm
103	Dibutyl phosphate	30 ppm
104	Dibutyl phthalate	4,000 mg/m^3
105	o-Dichlorobenzene	200 ppm
106	p-Dichlorobenzene	150 ppm
107	Dichlorodifluoromethane	15,000 ppm
108	1,3-Dichloro 5,5-dimethylhydantoin	5 mg/m^3
109	1,1-Dichloroethane	3,000 ppm
110	1,2-Dichloroethylene	1,000 ppm
111	Dichloroethyl ether	100 ppm
112	Dichloromonofluoromethane	5,000 ppm
113	1,1-Dichloro 1-nitroethane	25 ppm
114	Dichlorotetrafluoroethane	15,000 ppm
115	Dichlorvos	100 mg/m^3
116	Dieldrin	50 mg/m^3
117	Diethylamine	200 ppm
118	2-Diethylaminoethanol	100 ppm
119	Difluorodibromomethane	2,000 ppm
120	Diglycidyl ether	10 ppm

S. No.	Substances/chemicals	IDLH Value
121	Diisobutyl ketone	500 ppm
122	Diisopropylamine	200 ppm
123	Dimethyl acetamide	300 ppm
124	Dimethylamine	500 ppm
125	N,N-Dimethylaniline	100 ppm
126	Dimethyl 1,2-dibromo 2,2-dichlorethyl phosphate	200 mg/m³
127	Dimethylformamide	500 ppm
128	1,1-Dimethylhydrazine	15 ppm
129	Dimethylphthalate	2,000 mg/m³
130	Dimethyl sulfate	7 ppm
131	Dinitrobenzene (o, m, p isomers)	50 mg/m³
132	Dinitroocresol	5 mg/m³
133	Dinitrotoluene	50 mg/m³
134	Di sec-octyl phthalate	5,000 mg/m³
135	Dioxane	500 ppm
136	Diphenyl	100 mg/m³
137	Dipropylene glycol methyl ether	600 ppm
138	Dndrin	2 mg/m³
139	Epichlorohydrin	75 ppm
140	EPN	5 mg/m³
141	Ethanolamine	30 ppm
142	2-Ethoxyethanol	500 ppm
143	2-Ethoxyethyl acetate	500 ppm
144	Ethyl acetate	2,000 ppm [LEL]
145	Ethyl acrylate	300 ppm
146	Ethyl alcohol	3,300 ppm [LEL]
147	Ethylamine	600 ppm
148	Ethyl benzene	800 ppm [LFL]
149	Ethyl bromide	2,000 ppm
150	Ethyl butyl ketone	1,000 ppm
151	Ethyl chloride	3,800 ppm [LEL]
152	Ethylene chlorohydrins	7 ppm
153	Ethylenediamine	1,000 ppm
154	Ethylene dibromide	100 ppm
155	Ethylene dichloride	50 ppm

S. No.	Substances/chemicals	IDLH Value
156	Ethylene glycol dinitrate	75 mg/m^3
157	Ethyleneimine	100 ppm
158	Ethylene oxide	800 ppm
159	Ethyl ether	1,900 ppm [LEL]
160	Ethyl formate	1,500 ppm
161	Ethyl mercaptan	500 ppm
162	N-Ethylmorpholine	100 ppm
163	Ethyl silicate	700 ppm
164	Ferbam	800 mg/m^3
165	Ferrovanadium dust	500 mg/m^3
166	Fluorides (as F)	250 mg F/m^3
167	Fluorine	25 ppm
168	Fluorotrichloromethane	2,000 ppm
169	Formaldehyde	20 ppm
170	Formic acid	30 ppm
171	Furfural	100 ppm
172	Furfuryl alcohol	75 ppm
173	Gycidol	150 ppm
174	Graphite (natural)	1,250 mg/m^3
175	Hafnium compounds (as Hf)	50 mg Hf//m^3
176	Heptachlor	35 mg/m^3
177	n-Heptane	750 ppm
178	Hexachloroethane	300 ppm
179	Hexachloronaphthalene	2 mg/m^3
180	n-Hexane	1,100 ppm [LEL]
181	2-Hexanone	1,600 ppm
182	Hexone	500 ppm
183	sec Hexyl acetate	500 ppm
184	Hydrazine	50 ppm
185	Hydrogen bromide	30 ppm
186	Hydrogen chloride	50 ppm
187	Hydrogen cyanide	50 ppm
188	Hydrogen fluoride (as F)	30 ppm
189	Hydrogen peroxide	75 ppm
190	Hydrogen selenide (as Se)	1 ppm

S. No.	Substances/chemicals	IDLH Value
191	Hydrogen sulfide	100 ppm
192	Hydroquinone	50 mg/m^3
193	Iodine	2 ppm
194	Iron oxide dust and fume (as Fe)	2,500 mg Fe/m^3
195	Isoamyl acetate	1,000 ppm
196	Isoamyl alcohol (primary and secondary)	500 ppm
197	Isobutyl acetate	1,300 ppm [LEL]
198	Isobutyl alcohol	1,600 ppm
199	Isophorone	200 ppm
200	Isopropyl acetate	1,800 ppm
201	Isopropyl alcohol	2,000 ppm [LEL]
202	Isopropylamine	750 ppm
203	Isopropyl ether	1,400 ppm [LEL]
204	Isopropyl glycidyl ether	400 ppm
205	Ketene	5 ppm
206	Lead compounds (as Pb)	100 mg Pb/m^3
207	Lindane	50 mg/m^3
208	Lithium hydride	0.5 mg/m^3
209	L.P.G.	2,000 ppm [LEL]
210	Magnesium oxide fume	750 mg/m^3
211	Malathion	250 mg/m^3
212	Maleic anhydride	10 mg/m^3
213	Mangnese compounds (as Mn)	500 mg Mn/m^3
214	Mercury compounds [except (organo) alkyls, as Hg]	10 mg Hg/m^3
215	Mercury (organo) alkyl compounds (as Hg)	2 mg Hg/m^3
216	Mesityl oxide	1,400 ppm [LEL]
217	Methoxychlor	5,000 mg/m^3
218	Methyl acetate	3,100 ppm [LEL]
219	Methyl acetylene	1,700 ppm [LEL]
220	Methyl acetylenepropadiene mixture	3,400 ppm [LEL]
221	Methyl acrylate	250 ppm
222	Methylal	2,200 ppm [LEL]
223	Methyl alcohol	6,000 ppm
224	Methylamine	100 ppm
225	Methyl (namyl) ketone	800 ppm

S. No.	Substances/chemicals	IDLH Value
226	Methyl bromide	250 ppm
227	Methyl Cellosolve (r) acetate	200 ppm
228	Methyl chloride	200 ppm
229	Methyl chloroform	2,000 ppm
230	Methylcyclohexane	700 ppm
231	Methylcyclonexanol	1,200 ppm [LEL]
232	o-Methylcyclohexanone	500 ppm
233	Methylene bisphenyl isocyanate	600 ppm
234	Methylene chloride	75 mg/m^3
235	Methyl formate	2,300 ppm
236	5-Methyl 3-heptanone	4,500 ppm
237	Methyl hydrazine	100 ppm
238	Methyl hydrazine	20 ppm
239	Methyl iodide	100 ppm
240	Methyl isobutyl carbinol	400 ppm
241	Methyl isocyanate	3 ppm
242	Metyl mercatan	150 ppm
243	Methyl methacrylate	1,000 ppm
244	Methyl styrene	700 ppm
245	Mica	1,500 mg/m^3
246	Molybdenum (insoluble compounds, as Mo)	5,000 mg Mo/m^3
247	Molybdenum (soluble compounds, as Mo)	1,000 mg Mo/m^3
248	Monomethyl aniline	100 ppm
249	Morpholine	1,400 ppm [LEL]
250	Naphtha (coal tar)	1,000 ppm [LEL]
251	Naphthalene	250 ppm
252	Nickel carbonyl (as Ni)	2 ppm
253	Nickel metal and other compounds (as Ni)	10 mg Ni/m^3
254	Nicotine	5 mg/m^3
255	Nitric acid	25 ppm
256	Nitric oxide	100 ppm
257	p-Nitroaniline	300 mg/m^3
258	Nitrobenzene	200 ppm
259	p-Nitrochlorobenzene	100 mg/m^3
260	Nitroethane	1,000 ppm

S. No.	Substances/chemicals	IDLH Value
261	Nitrogen dioxide	20 ppm
262	Nitrogen trifluoride	1,000 ppm
263	Nitroglycerine	75 mg/m^3
264	Nitromethane	750 ppm
265	1-Nitropropane	1,000 ppm
266	2-Nitropropane	100 ppm
267	Nitrotoluene (o, m, p isomers)	200 ppm
268	Octachloronapthalene	Unknown
269	Octane	1,000 ppm [LEL]
270	Oil mist (mineral)	2,500 mg/m^3
271	Osmium tetroxide (as Os)	1 mg Os/m^3
272	Oxalic acid	500 mg/m^3
273	Oxygen difluoride	0.5 ppm
274	Ozone	5 ppm
275	Paraquat	1 mg/m^3
278	Parathion	10 mg/m^3
279	Pentachlorophenol	1 ppm
280	n-Pentane	Unknown
281	2Pentanone	2.5 mg/m^3
282	Perchloromethyl mercaptan	1,500 ppm [LEL]
283	Perchloryl fluoride	1,500 ppm
284	Petroleum distillates (naphtha)	10 ppm
285	Phenol	100 ppm
286	p-Phenylene diamine	1,100 ppm [LEL]
287	Phenyl ether (vapor)	250 ppm
288	Phenyl etherbiphenyl mixture (vapor)	25 mg/m^3
289	Phenyl glycidyl ether	100 ppm
290	Phenylhydrazine	15 ppm
291	Phosdrin	4 ppm
292	Phosgene	2 ppm
293	Phosphine	50 ppm
294	Phosphoric acid	1,000 mg/m^3
295	Phosphorus (yellow)	5 mg/m^3
296	Phosphrous pentachloride	70 mg/m^3
297	Phosphorus pentasulfide	250 mg/m^3

S. No.	Substances/chemicals	IDLH Value
298	Phosphorus trichloride	25 ppm
299	Phthalic anhydride	60 mg/m^3
300	Picric acid	75 mg/m^3
301	Pindone	100 mg/m^3
302	Platinum (soluble salts, as Pt)	4 mg Pt/m^3
303	Portland cement	5,000 mg/m^3
304	Propane	2,100 ppm [LEL]
305	n-Propyl acetate	1,700 ppm
306	n-Propyl alcohol	800 ppm
307	Propylene dichloride	400 ppm
308	Propylene imine	100 ppm
309	Propylene oxide	400 ppm
310	n-Propyl nitrate	500 ppm
311	Pyrethrum	5,000 mg/m^3
312	Pyridine	1,000 ppm
313	Quinone	100 mg/m^3
314	Rhodium (metal fume and insoluble compounds, as Rh)	100 mg Rh/m^3
315	Rhodium (soluble compounds, as Rh)	2 mg Rh/m^3
316	Ronnel	300 mg/m^3
317	Rotenone	2,500 mg/m^3
318	Selenium compounds (as Se)	1 mg Se/m^3
319	Selenium hexafluoride	2 ppm
320	Silica, amorphous	3,000 mg/m^3
321	Silica, carystalline (respirable dust)	–
322	Cristobalite/tridymite:	25 mg/m^3
323	Quartz/tripoli:	50 mg/m^3
324	Silver (metal dust and soluble compounds, as Ag)	10 mg Ag/m^3
325	Soapstone	3,000 mg/m^3
326	Sodium fluoroacetate	2.5 mg/m^3
327	Sodium hydroxide	10 mg/m^3
328	Stibine	5 ppm
329	Stoddard solvent	20,000 mg/m^3
330	Strychnine	3 mg/m^3
331	Styréne	700 ppm
332	Sulfur dioxide	100 ppm

S. No.	Substances/chemicals	IDLH Value
333	Sulfuric acid	15 mg/m^3
334	Sulfur monochloride	5 ppm
335	Sulfur pentafluoride	1 ppm
336	Sulfuryl fluoride	200 ppm
337	2,4,5-T	250 mg/m^3
338	Talc	1,000 mg/m^3
339	Tantalum (metal and oxide dust, as Ta)	2,500 mg Ta/m^3
340	TEDP	10 mg/m^3
341	Tellurium compounds (as Te)	25 mg Te/m^3
342	Tellurium hexafluoride	1 ppm
343	TEPP	5 mg/m^3
344	Terphynyl (o, m, p isomers)	500 mg/m^3
345	1,1,1,2-Terachloro 2,2-difluoroethane	2,000 ppm
346	1,1,2,2,-Tetrachloro 1,2-difluoroethane	2,000 ppm
347	1,1,2,2-Tetrachloroethane	100 ppm
348	Tetrachloroethylene	150 ppm
349	Tetrachloronaphthalene	Unknown
350	Tetraethyl lead (as Pb)	40 mg Pb/m^3
351	Tetrahydrofuran	2,000 ppm [LEL]
352	Tetramethyl lead (as Pb)	40 mg Pb/m^3
353	Tetramethyl succinonitrile	5 ppm
354	Tetranitromethane	4 ppm
355	Tetryl	750 mg/m^3
356	Thallium (soluble compounds, as Tl)	15 mg Tl/m^3
357	Thiram	100 mg/m^3
358	Tin (inorganic compounds, as Sn)	100 mg Sn/m^3
359	Tin (Organic compounds, as Sn)	25 mg Sn/m^3
360	Titanium dioxide	5,000 mg/m^3
361	Toluene	500 ppm
362	Toleuene 2,4-diisocyanate	2.5 ppm
363	o-Toluidine	50 ppm
364	Tributyl phosphate	30 ppm
365	1,1,2-Trichloroethane	100 ppm
366	Trichloroethylene	1,000 ppm
367	Trichloronaphthalene	Unknown

S. No.	Substances/chemicals	IDLH Value
368	1,2,3-Trichloropropane	100 ppm
369	1,1,2-Trichloro 1,2,2-trifluoroethane	2,000 ppm
370	Triethylamine	200 ppm
371	Trifluorobromomethane	40,000 ppm
372	2,4,6-Trinitrotoluene	500 mg/m^3
373	Triorthocresyl phosphate	40 mg/m^3
374	Triphenyl phosphate	1,000 mg/m^3
375	Turpentine	800 ppm
376	Uranium (insoluble compouds, as U)	10 mg U/m^3
377	Uranium (soluble compounds, as U)	10 mg U/m^3
378	Vanadium dustVanadium fume	35 mg V/m^3
379	Vinyl toluene	35 mg V/m^3
380	Warfarin	400 ppm
381	Xylene (o, m, p isomers)	100 mg/m^3
382	Xylidine	900 ppm
383	Xylidine	50 ppm
384	Yttrium compounds (as Y)	500 mg Y/m^3
385	Zinc chloride fume	50 mg/m^3
386	Zinc oxide	500 mg/m^3
387	Zirconium compounds (as Zr)	25 mg Zr/m^3

NOTES

The current NIOSH definition for an IDLH condition is one that poses a threat of exposure to airborne contaminants when that exposure is likely to cause death or immediate or delayed permanent adverse health effects or prevent escape from such an environment. The purpose of establishing an IDLH value is **(1)** to ensure that the worker can escape from a given contaminated environment in the event of failure of the respiratory protection equipment and **(2)** is considered a maximum level above which only a highly reliable breathing apparatus providing maximum worker protection is permitted. In establishing the IDLH value, the following conditions must be assured:

A. The ability to escape without loss of life or immediate or delayed irreversible health effects. (Thirty minutes is considered the maximum time for escape so as to provide some margin of safety in calculating an IDLH value.)

B. The prevention of severe eye or respiratory irritation or other reactions that would hinder escape.

"Highly reliable" respirators (i.e., the most protective respirators) would be selected for emergency situations, fire fighting, exposure to carcinogens, entry into oxygen-deficient atmospheres, entry into atmospheres that contain a substance at a concentration greater an 2,000 times the NIOSH REL or OSHA PEL, and for entry into immediately dangerous to life or health conditions. These "highly reliable" respirators include either a self-contained breathing apparatus that has a full facepiece and is operated in a pressure-demand or other positive-pressure mode, or supplied-air respirator that has a full facepiece and is operated in a pressure-demand or other positive-pressure mode in combination with an auxiliary SCBA operated in a pressure-demand or other positive pressure mode.

This list includes IDLH values for 85 substances (e.g., benzene and methylene chloride) determined by NIOSH to meet the OSHA definition of "potential occupational carcinogen", For all of these substances, except ethylene oxide and crystalline silica, NIOSH recommends that the "most protective" respirators be worn by workers exposed at concentrations above the NIOSH REL, or at any detectable concentration when there is no REL. For ethylene oxide and crystalline silica, TIOSH recommends that the "most protective" respirators be worn in concentrations exceeding 5 ppm and 25 mg/rng, respectively.

The criteria utilized to determine the adequacy of existing IDLH values were a combination ofthose used during the SCP and a newer methodology developed by NIOSH. These criteria form a tiered approach with acute human toxicity data being used preferentially, followed next by acute animal inhalation toxicity data, and then finally by acute animal oral toxicity data to determine an updated IDLH value. When relevant acute toxicity data were insufficient or unavailable, then the use of chronic toxicity data or an analogy to a chemical with similar toxic effects was considered. In order to facilitate the revision process, secondary toxicological data were primarily used. Once a preliminary IDLH value was developed, it was compared to the existing IDLH value and to several other factors (e.g., existing short-term exposure guidelines and lower explosive limits).

The following "hierarchy" was followed to develop a "preliminary" value for the revised IDLH value:

A. Human acute toxicity data were used if sufficient to determine a concentration that for up to 30 minutes does not cause death, serious or irreversible health effects, or does not impair or impede the ability to escape.

B. Animal acute lethal concentration (LC) data were considered next. The only animal lethal concentration Clata used involved mammals; the vast majority of the data was from studies of rats, mice, guinea pigs, and hamsters. It was decided to generally use the lowest reliable LC data, with LC_{50} data preferred. If acute LC data determined during a 30-minute period were not available, then the data, based on a study. were "adjusted" to an equivalent 30-minute value using the following relationship:

$$\text{Adjusted } LC_{50} \text{ (30 minutes)} = LC_{50}(t) * (t/0.5)^{**} (l/n)$$

where: $LC_{50}(t) = LC_{50}$ determined over t hours

n = constant*

NOTES

*Ten Berge determined the relationship shown above based on experimental data. The constant "n" was determined by ten Berge to be less than 3.0 for 18 of the 20 substances studied. Although the individual "n" values determined by ten Berge were utilized when applicable during the review and revision of the original IDLHs, as a conservative estimate, an "n" = 3.0 was assumed when "adjusting" the LC data to 30 minutes for all other substances.

This equation with an "n" = 3.0 results in the following correction factors:

t (hours)	Correction factor
0.5	1.0
1	1.25
2	1.6
3	1.8
4	2.0
5	2.15
6	2.3
7	2.4
8	2.5

The LC values (after "adjusting" if necessary to 30 minutes) were divided by a safety factor of 10 to determine a "preliminary" IDLH for comparison purposes.

C. Animal lethal dose (LD) data were considered next. As was the case with the lethal concentration data, the only animal lethal dose data used involved mammals; the vast majority of the data were from studies of rats, mice, guinea pigs, and hamsters. It was decided to generally use the lowest LD data with oral LD_{50} data preferred. The LD data was used to determine the equivalent total dose to a 70-kg worker and, the air concentration containing this dose was determined by dividing by 10 cubic meters. [A worker breathing at a rate of 50 liters per minute for 30 minutes would inhale 1.5 cubic meters of air.] A "preliminary" IDLH for comparison purposes was determined by dividing these air concentrations by a safety factor of 10.

D. Chronic toxicity data were considered.if no relevant acute toxicity data existed. However, the fact that chronic exposures may have limited relevance to acute effects was taken into consideration.

E. When relevant toxicity data applying specifically to the chemicals in question were lacking, and if it was determined to be justified, then analogies to substances with similar acute toxic effects were considered.

F. All "preliminary" IDLHs derived during this update were checked against the following factors prior to establishing the final "revised" IDLH:

   1. Lower explosive limit (LEL):It was decided to restrict the "routine" entry into a possible explosive atmosphere to concentrations no greater than 10% of the LEL. However, OSHA considers concentrations in excess of 10% of the LEL to be a hazardous atmosphere in confined spaces
   2. RD_{50} data: An RDso is defined as the 10-minute exposure concentration producing a 50% respiratory rate decrease in mice or rats and can be used to estimate severe respiratory irritation. Prolonged exposure to an RDso concentration has been shown to produce respiratory tract lesions consistent with irritation.
   3. Other short-term exposure guidelines such as the American Industrial Hygiene Association's emergency response planning guidelines (ERPGs) and the National Research Council's emergency exposure guidance levels (EEGLs) and short-term public emergency guidance levels (SPEGLs), and occupational exposure standards or recommendations such as OSHA PELs, NIOSH RELs, or the American Conference of Governmental Industrial Hygienists (ACGIH) TLVs.
   4. Based on the NIOSH respirator decision logic, the revised IDLH values could not be greater than 2,000 times the NIOSH REL (or OSHA PEL).

Abbreviations used for IDLH Documentation

LC	: Lethal concentration
LC_{50}	: Concentration causing death in 50% of test animals
LD_{50}	: Dosage causing death in 50%
LEL	: Lower explosive limit
mg/kg	: Milligrams per kilogram of body weight
mg/m	: Milligrams per cubic meter of air
mppcf	: Millions of particles per cubic foot of air
NIOSH	: National Institute for Occupational Safety and Health
OSHA	: Occuptional Safety and Health Administration

PEL : Permissible exposure limit (OSHA)
Ppm : Parts per million parts of air
RD_{50} : Concentration producing a 50% decrease in respiratory rate following a 10-minute exposure
REL : Recommended exposure limit (NIOSH)
STEL : Short-term exposure limit
TLV : Threshold limit value (ACGIH)
TWA : Time-weighted average

Appendix 2.10: Terminology Related to Air Pollution

Absorption: The process in which incident radiant energy is retained by a substance. A further process always results from absorption: that is, the irreversible conversion of the absorbed radiation into some other form of energy within and according to the nature of the absorbing medium. The absorbing medium itself may emit radiation, but only after an energy conversion has occurred.

Absorptivity: (Also called absorption factor; infrequently called absorptive power.) A measure of the amount of radiant energy absorbed by a given substance of definite dimensions; defined as the ratio of the amount of radiant energy absorbed to the total amount incident upon that substance. The absorption may occur either at the surface of (as with an opaque body) or in transit through the substance.

Actinic: Pertaining to electromagnetic radiation capable of initiating photochemical reactions, as in photography or the fading of pigments. Because of the particularly strong action of ultraviolet radiation on photochemical processes, the term has come to be almost synonymous with ultraviolet, as in "actinic rays. "

Adiabatic Process: A thermodynamic change of state of a system in which there is no transfer of heat or mass across the boundaries of the system. In an adiabatic process, compression always results in warming, expansion in cooling.

Adsorption: The adhesion of a thin film of liquid or gas to a solid substance.

Advection: The process of transport of an atmospheric property solely by the mass motion (velocity field) of the atmosphere; also, the rate of change of the value of the advected property at a given point.

Often, particularly in synoptic meteorology, advection refers only to the horizontal or isobaric components of motion, that is, the wind field as shown on a synoptic chart.

Regarding the general distinction (in meteorology) between advection and convection, the former describes the predominantly horizontal, large-scale motions of the atmosphere while convection describes the predominantly vertical, locally induced motions.

Agglomeration: A colloidal system in which the dispersed phase is composed of either solid or liquid particles, and in which the dispersion medium is some gas, usually air.

There is no clear -cut upper limit to the size of particles comprising the dispersed phase in an aerosol, but as in all other colloidal systems, it is rather commonly set at 1 micron. Haze, most smokes, and some fogs and clouds may thus be regarded as aerosols. However, it is not good usage to apply the term to ordinary clouds whose drops are so large as to rule out the usual concept of colloidal stability. It is also poor usage to apply the term to the dispersed particles alone; an aerosol is a system of dispersed phase and dispersing medium taken together.

Agglomeration: The growth of a particle by collision with and assimilation of other particles. Ordinarily, the connotation is that all particles are solid.

Agglutination: The adhesion of solid particles together or to a solid surface.

Air: The mixture of gases comprising the earth's atmosphere. Since the composition of the atmosphere is slightly variable with respect to certain components, the term "pure air" has no precise meaning, but it is commonly used to imply freedom from non-gaseous suspensoids (dust, hydrometeors) and also freedom from such gaseous contaminants as industrial effluents.

By far the most important gas found in air, from the meteorological view-point, is water vapor.

The per cent by volume of those gases found in relatively constant amount in dry air is very nearly as follows:

nitrogen (N_2)	78.084 %
oxygen (O_2)	20.946%
argon (A)	0.934%
carbon dioxide (CO_2)	0.033%
neon (Ne)	0.0018%
helium (He)	0.000524%
methane (CH_4)	0.0002%
krypton (Kr)	0.000114%
hydrogen (H_2)	0.00005%
nitrous oxide (N_2O)	0.00005%
xenon (Xe)	0.0000087%

In addition to the above constituents there are many variable constituents. Chief of these is water vapor,which may vary from zero to volume percentages close to four per cent. Ozone, sulfur dioxide, ammonia, carbon monoxide, and other trace gases occur in small and varying amounts.

Air Conditioning: (Sometimes called climatization). The artificial control of humidity, temperature, "purity," and motion of the air within buildings and other enclosed spaces; also the operation of equipment for such controls. The Objective may be to secure either maximum human comfort or the best environment for a given industrial operation.

Air Current: Very generally, any moving stream of air; it has no particular technical connotation.

Air Pollution: The general term alluding to the undesirable addition to the atmosphere of substances (gases and liquids and solid particles) either that are foreign to the "natural" atmosphere or in quantities exceeding their natural concentrations.

Air Quality: The composition of air with respect to quantities of pollutants therein; used most frequently in connection with "standards" of maximum acceptable pollutant concentrations.

Ambient Air: Literally, the air moving around us; the air of the surrounding environment. It is used, generally, in opposition to closed, controlled environments of any kind.

Ammonia: (Chemical formula: NH3) A colorless gas with a sharp, irritating odor, having a density about six-tenths that of air at the same temperature and pressure.

Trace quantities of ammonia enter the atmosphere as a result of natural decay processes and from industrial pollution. Its high solubility in water should decrease its chances of remaining in the atmosphere under rainy conditions.

Anemometer: The general name for instruments designed to measure the speed (or force) of the wind.

Angstrom: (Abbreviated A, formerly Å.) A unit of length used in the measurement of the wavelength of light, x-rays, and other electromagnetic radiation and in the measurement of molecular and atomic diameters. One angstrom equals 10^{-8} cm or 10^{-4} microns.

The wavelength of visible light ranges from about 4000 to 7000 angstroms; whereas x-ray wavelengths and atomic diameters are of the order of a few angstroms. The unit JS named in honor of the nineteenth century Swedish spectroscopist, A. J. Angstrom.

Anticyclone: A large-scale atmospheric circulation system in which (in the northern hemisphere) the winds rotate clockwise. Because this is almost invariably accompanied by relatively high atmospheric pressure, the terms "anticyclone" and "high" are commonly used interchangeably.

Three characteristics of anticyclones are critical to air pollution. (1) Winds near an anticyclone center are very light. (2) The general outflow of air from an anticyclone causes the air aloft to subside and produce a widespread subsidence inversion, which traps pollutants at low levels. (3) An anticyclone is relatively cloud free, and the resulting sunshine induces photochemical smog.

Ash Fall: The gravitational settling of the fly ash through the atmosphere; also, the ash that has thus settled.

Background Level: In air pollution studies, the concentration of a pollutant that would exist in the absence of the particular source under study; a "standard" against which the contribution of the particular source can be compared.

Basin: With respect to air pollution, a geographical area where surrounding elevated terrain can, at least under some atmospheric conditions, impede the flow of air into and out of the area.

Biosphere: That transition zone between earth and atmosphere within which most forms of terrestrial life are commonly found; the outer portion of the geosphere and inner or lower portion of the atmosphere.

Bishop's Ring: A faint, broad, reddish-brown corona occasionally seen in dust clouds, especially those which result from violent volcanic eruptions. The angular radius of this ring's inner edge is about 20° and its angular width is about 10°, the exact dimensions being controlled by the particle -size of the dust layer.

It is named after Rev. S. Bishop of Honolulu, who first described the phenomenon.

Bivane: A bi-directional wind vane; a sensitive wind vane used in turbulence studies to obtain a record of the horizontal and vertical components of the wind. The instrument consists of two lightweight airfoil sections mounted orthogonally on the end of a counter-balanced rod which is free to rotate in the horizontal and vertical planes. The positions of the rod may be recorded by electrical techniques.

Black Body: A hypothetical "body" which absorbs all of the electromagnetic radiation striking it; that is, one which neither reflects nor transmits any of the incident radiation.

Black-Body Radiation: The electromagnetic radiation emitted by an ideal black body; it is the theoretical maximum amount of radiant energy of all wavelengths which can be emitted by a body at a given temperature.

Boundary Layer: In meteorology, a general term for the layer of air adjacent to the earth's surface. Several kinds of boundary layers (or the upper limits thereof) can be defined according to different physical or dynamical characteristics.

Unless otherwise specified in air pollution contexts, the concept of the planetary boundary layer should be assumed.

Buoyancy: (1) That property of an object that enables it to float on the surface of a liquid, or ascend through and remain freely suspended in a compressible fluid such as the atmosphere. Quantitatively, it may be expressed as the ratio of the specific weight of the fluid to the specific weight of the object; or, in another manner, by the weight of the fluid displaced minus the weight of the object. (2) (Or buoyant force; also called Archimedean buoyant force.) The up¬ward force exerted upon a parcel of fluid (or an object within the fluid) in a gravitational field by virtue of the density difference between the parcel (or object) and that of the surrounding fluid.

Burn Off: With reference to fog or low stratus cloud layers, to dissipate by daytime heating from the sun.

Calm: The absence of apparent motion of the air. In weather observing practice, the wind is reported as "calm" if its speed is less than one mile per hour (or one knot).

Carbon Dioxide: A heavy, colorless gas that enters the atmosphere as the result of efficient natural and artificial combustion processes; the fourth most abundant constituent of dry air.

There is substantial evidence that the amount of CO_2 in the atmosphere has increased by 10 percent, from about 300 ppm to 330 ppm, in the last half century, and that this increase has come about by the increased burning of carbonaceous fuels. Because of its strong absorption of infrared radiation, CO_2 has important influence on the atmosphere's radiation balance and, therefore, on the static stability of the atmosphere. The effects of increasing amounts of atmospheric CO_2 are, as yet, unknown; but it is worth noting that many present-day efforts to control pollution have the net effect of increasing the CO_2 output while decreasing the emission of other pollutants.

Carbon Monoxide: A colorless, odorless, and very toxic gas, found in trace quantities in the natural atmosphere, but also produced by the incomplete combustion of carbonaceous fuels.

Cascade Impactor: A low-speed impaction device for use in sampling both solid and liquid atmospheric suspensoids. The simple impactor consists of four pairs of jets and sampling plates working in series and designed in such a manner that each sampling plate collects particles of predominantly one size range.

Circulation: The flow or motion of a fluid in or through a given area or volume. This is the most general term applied to atmospheric motions on all scales.

Coagulation: Generally, any process that converts numerous' small particles into a smaller number of larger particles.

Coalescence: The merging of two liquid particles into a single particle.

Collection Efficiency: In general, the fraction of all particles initially moving on a collision course with a given impactor which actually do collide with and remain adhered to that. impactor.

Colloidal System: (Also called colloidal dispersion, colloidal suspension.) An intimate mixture of two substances one of which, called the dispersed phase (or colloid) is uniformly distributed in a finely divided state through the second substance, called the dispersion medium (or dispersing medium). The dispersion medium may be a gas, a liquid or a solid and the dispersed phase may also be any of these, with the exception that one does not speak of a colloidal system of one gas in another.

A system of liquid or solid particles colloidally dispersed in a gas is called an aerosol.

Combustion Nucleus: A condensation nucleus formed as a result of industrial or natural combustion processes.

The chemical nature of such nuclei may vary almost as much as can the nature of reactants in industrial combustion processes, but because of the prevalence of sulfur impurities in many coals, sulfur dioxide (SO_2) and sulfur trioxide (SO_3) are perhaps the most important of the substances acting as combustion nuclei. Both of these oxides are hygroscopic and hence can serve to nucleate atmospheric condensation processes. The role of the ammonium ion (NH_4^+) which is formed in some combustion processes is probably of almost equal importance to that of the sulfite (SO_3^{-2}) and sulfate (SO_4^{-3}) ions.

Concentration: General term for the amount of a substance contained in a unit area or volume, or relative to the amount of medium within which the substance is contained.

In air pollution, the concentration of particulates is expressed either as number of particles per unit volume of air or as weight of particles per unit volume of air. Gaseous pollutant concentrations are usually expressed as the ratio of the weight of the pollutant to the total weight of air- pollutant mixture.

Condensation: The physical process by which a vapor becomes a liquid or solid; the opposite of evaporation. In meteorological usage, this term is applied only to the transformation from vapor to liquid: any process in which a solid forms directly from its vapor is termed sublimation, as is the reverse process.

Condensation Nucleus: A particle, either liquid or solid, upon which condensation of water vapor begins in the atmosphere.

Condensation Trail: (Or contrail; also called vapor trail.) A cloud-like streamer frequently observed to form behind aircraft flying in clear, cold, humid air.

Convection: 1. In general, mass motions within a fluid resulting in transport and mixing of the properties of that fluid. Convection, along with conduction and radiation, is a principal means of energy transfer. Distinction is made between: free convection (or gravitational convection), motion caused only by density differences within the fluid; and forced convection, motion induced by mechanical forces such as deflection by a large-scale surface irregularity, turbulent flow caused by friction at the boundary of a fluid, or motion caused by any applied external force. 2. As specialized in meteorology, atmospheric motions that are predominantly vertical, resulting in vertical transport and mixing of atmospheric properties; distinguished from advection. Because the most striking meteorological results of convective motion occur in conjunction with the rising current of air (strong updrafts, convective clouds, etc.), convection often is used to imply only upward vertical motion, and in this sense it is opposed to subsidence.

Corrasion: (Or wind corrosion.) The abrasive action of wind-borne material, especially sand, dust, and ice crystals; a form of weathering.

Corrosion: The gradual deterioration of material by chemical processes, such as oxidation or attack by acids; if caused by an atmospheric effect, a form of weathering.

Of great Significance is the corrosion due to the combined effects of atmospheric temperature, humidity, and suspended impurities; for example, the rusting of iron, the direct effects upon a surface wetted by acid rain water, or, indirectly, the rotting of wood caused by the action of fungi or fungi of bacteria in the soil and in enclosed spaces.

Cyclone: 1. A large-scale atmospheric circulation system in which (in the northern hemisphere) the winds rotate counter clockwise. Because this is almost invariably accompanied by relatively low atmospheric pressure, the terms "cyclone" and "low" are commonly used interchangeably. Cyclones typically are associated with stormy weather and unstable air, and therefore serve to ameliorate air-pollution. 2. A device for removing particulates from a moving stream of gas by centrifugal force.

Deposition: The settling to earth of particulates previously suspended in the atmosphere.

Dew Point: (Or dew-point temperature.) The temperature to which a given parcel of air must be cooled at constant pressure and constant water-vapor content in order for saturation to occur. When this temperature is below 0ºC, it is sometimes called the frost point.

Diffuse Sky Radiation: (Or skylight; also called diffuse skylight, sky radiation.) Radiation reaching the earth's surface after having been scattered from the direct solar beam by molecules or suspensoids in the atmosphere. Of the total light removed from the direct solar beam by scattering in the atmosphere (approximately 25 percent of the incident radiation, about two thirds ultimately reaches the earth as diffuse sky radiation.

Diffusion: In meteorology, the exchange of fluid parcels (and hence the transport of conservative properties) between regions in space, in the apparently random motions of a scale too small to be treated by the equations of motion.

In meteorology the diffusion of momentum (viscosity), vorticity, water vapor, heat (conduction) .and gaseous components of the atmospheric mixture, are studied extensively. The atmospheric motions diffusing these properties may in many cases be of much larger scale than the molecular, the exchanging parcels being called eddies, and the diffusion equation extended by analogy to turbulent diffusion. It is this latter approach that is used in diffusion studies relating to air pollution.

Diffusnity: A measure of the rate of diffusion of a fluid property.

Dispersion: In air pollution terminology, loosely applied to the removal (by whatever means) of pollutants from the atmosphere over a given area; or, the distribution of a given quantity of pollutant throughout an increasing volume of atmosphere.

Downwash: The transport of pollutants downward to the earth's surface by a descending current of air.

Dust: Solid materials suspended in the atmosphere in the form of small irregular particules, many of which are microscopic in size. It imparts -a tannish or greyish hue to distant objects. The sun's disk is pale or colorless or has a yellowish tinge at all periods of the day,

Dust cannot be a stable component of the atmosphere because it must eventually fall back to the earth's surface when winds and turbulence become too weak to bear it aloft. Dust is due to many natural and artificial sources: volcanic eruptions, salt spray from the seas, blowing solid particles, plant pollen and bacteria, smoke and ashes of forest fires and industrial combustion processes, etc.

Dust Counter: (Also called Kern counter, nucleus counter.) General term for an instrument which measures the size and number of dust particles in the atmosphere.

Dust Fall: The gravitational settling of dust through the atmosphere; also, the dust that has thus settled.

Dust Horizon: The top of a dust layer which is confined by a low-level temperature inversion and has the appearance of the horizon when viewed from above against the sky. In such instances the true horizon is usually obscured by the dust layer.

Similarly defined are fog horizon, haze horizon, smog horizon, and smoke horizon.

Dust Loading: (Or particulate loading.) Engineering term for the concentration of dust (or other particulates) in the atmosphere.

Duststorm: (Or dust storm; also called duster, black blizzard.) An unusual, frequently severe weather condition characterized by strong winds and dust filled air over an extensive area. Prerequisite to a dust storm is a period of drought over an area of normally arable land, thus providing the very fine particles of dust which distinguish it from the more common sandstorm of desert regions.

Effluent: The mixture of substances, gases and liquids and suspended matter, discharged into the atmosphere (or ground, river, ocean) as the result of a given process. The presumption has been that nature provides for adequate dispersal of the effluent.

Electromagnetic Radiation: (Or electromagnetic energy; often called, simply radiation.) Energy propagated through space or through material media in the form of an advancing disturbance in electric and magnetic fields existing in space or in the media. The term radiation, alone, is used commonly for this type of energy, although it actually has a broader meaning.

Emission: With respect to air pollution, the release of pollutants into the atmosphere.

Entrainment: In meteorology, the mixing of environmental air into a preexisting organized air current so that the environmental air becomes part of the current.

Evaporation: (Also called vaporization.) The physical process by which a liquid or solid is transformed to the gaseous state; the opposite of condensation. In meteorology evaporation usually is restricted in use to the change of water from liquid to gas, while sublimation is used for the change from solid to gas as well as from gas to solid.

Eye Irriation: One effect of photochemical smog that is sufficiently general and qualitatively measurable that it is used as an index of smog severity.

Fallout: 1. The descent to the ground of dust and other air bourne debris, usually referring to debris resulting from an explosion, and especially applied to radioactive debris from an atomic or thermonuclear explosion. 2. In air pollution terminology, sometimes used for the settling out of very fine particles as opposed to larger particles of ash.

Flyash: Ash particles, frequently in the form of small, irregular flakes, that escape into the atmosphere and quite rapidly settle to the ground.

Fog: A hydrometeor consisting of a visible aggregate of minute water droplets suspended in the atmosphere near the earth's surface. According to international definition, fog reduces visibility below one kilometer. Fog differs from cloud only in that the base of fog is at the earth's surface while clouds are above the surface. When composed of ice crystals, it is termed ice fog.

Fog is easily distinguished from haze by its appreciable dampness and grey color. Mist may be considered as intermediate between for and haze; its particles are microscopic in size, it is less damp than fog, and does not restrict visibility to the same extent. There is no distinct line, however, between any of these categories.

Free Atmosphere: (Sometimes called free air.) That portion of the earth's atmosphere, above the planetary boundary layer, in which the effect of the earth's surface friction on the air motion is negligible, and in which the air is usually treated (dynamically) as an ideal fluid. The base of the free atmosphere is usually taken as the geostrophic wind level.

Front: In meteorology, generally, the interface or transition zone between two air masses of different density. Since the temperature distribution is the most important regulator of atmospheric density, a front almost invariably separates air masses of different temperature.

Fume: In air pollution terminology, very fine particulates, predominantly less than one micron in diameter.

Fumigation: The phenomenon in which pollutants, that had remained aloft in stratified, stable air, are brough rapidly to ground level when the air destabilizes.

Glare: Any hindrance to vision caused by scattering or reflection of light into an observer's line of sight.

Glare due to scattering of light by dust particles deposited on the windshield of an automobile, for example, is noticeable only when the driver is looking into the sun, since particles as large as most dusts scatter predominantly in the forward direction. Smaller atmospheric suspensoids, however, produce the more azimuthally uniform glare known as air-light which has an effect on visibility that is almost independent of the relative positions of sun and object viewed.

Greenhouse Effect: The heating effect exerted by the atmosphere upon the earth by virtue of the fact that the atmosphere (mainly its water vapor, carbon dioxide, ozone, and suspended particles) absorbs and re-emits radiation that otherwise would escape to space.

Half -Life: The time required for a system decaying at an exponential rate (such as an element in radioactive disintegration) to be reduced to one -half its initial size or intensity.

Haze: Fine dust or salt particles dispersed through a portion of the atmosphere; a type of lithometeor. The particles are so small that they cannot be felt or individually seen with the naked eye, but they diminish horizontal visibility and give the atmosphere a characteristic opalescent appearance that subdues all colors.

Many haze formations are caused by the presence of an abundance of condensation nuclei which may grow in size, due to a variety of causes, and become mist, fog, or cloud. Distinction is sometimes drawn between dry haze and damp haze, largely on the basis of differences in optical effects produced by the smaller particles (dry haze) and the larger particles (damp haze) which develop from slow condensation upon the hygroscopic haze particles. Dry haze particles, with diameters of the order of 0.1 micron, are small enough to scatter shorter wavelengths of light preferentially, though not according to the inverse fourth-power law of Rayleigh scattering. Such haze particles produce a bluish color when the haze is viewed against a dark background, for dispersion allows only the slightly bluish scattered light to reach the eye. The same type of haze, when viewed against a light background, appears as a yellowish veil, for here the principal effect is the removal of the bluer components from the light originating in the distant light-colored background. Haze may be distinguished by this same effect from mist, which yields only a gray obscuration, since in mist the particle sizes are too large to yield appreciable differential scattering of various wavelengths.

Hydrocarbons: A group of compounds that consist solely of the elements carbon and hydrogen. A variety of hydrocarbons exist in gasoline vapors, natural gas, and exhaust products. In photochemical smog, they react with atomic oxygen and ozone to form new reactive compounds.

Hydrogen: A colorless and odorless, highiy combustible gaseous element; the lightest and apparently most abundant chemical element in the universe. As a pure gas, however, it is found only in trace quantities in our atmosphere.

Hydrogen Sulfide: A colorless; flammable, dangerously poisonous gas having an offensive rotten egg) odor, deriving as an air pollutant mainly from inadequately controlled petroleum refineries.

Hydrometeor: Any product of condensation or sublimation of atmospheric water vapor, whether formed in the free atmosphere or at the earth's surface; also, any water particles blown by the wind from the earth's surface.

Hygroscopic: Pertaining to a marked ability to accelerate the condensation of water vapor.

In meteorology, this term is applied principally to those condensation nuclei composed of salts which yield aqueous solutions of a very low equilibrium vapor pressure compared with that of pure water at the same temperature. Condensation on hygroscopic nuclei may begin at a relative humidity much lower than 100 per cent (about 75 per cent for sodium chloride); while on so-called non-hygroscopic nuclei, which merely furnish sufficiently large (by molecular standards) wetrable surfaces, relative humidities of nearly 100 per cent are required. "Damp haze" is formed of hygroscopic particles in the process of slow growth in relatively dry air.

Impactor: (Or impactometer.) A general term for instruments which sample atmospheric suspensoids by impaction. In general, the instrument consists of a housing which constrains the air to flow past a sensitized sampling plate.

Infrared Radiation: (Abbreviated IR; also called long-wave radiation.) Electromagnetic radiation lying in the wavelength interval from about 0.8 microns to an indefinite upper boundary sometimes arbitrarily set at 1000 microns (0.01 cm). At the lower limit of this interval, the infrared radiation spectrum is bounded by visible radiation, while on its upper limit it is bounded by microwave radiation of the type important in radar technology.

Whereas visible radiation is generated primarily by intra-atomic processes, infrared radiation is generated almost wholly by larger-scale intra molecular processes, chiefly molecular rotations and internal vibrations of many types. Electrically symmetric molecules, such as the nitrogen and oxygen molecules which comprise most of the earth's atmosphere, are not capable of absorbing or emitting infrared radiation, but several of the tri-atomic gases, such as water vapor, carbon dioxide, and ozone are infrared active and play important roles in the propagation of infrared radiation in the atmosphere.

Insolation: In general, solar radiation received at the earth's surface.

Inversion: In meteorology, a departure from the usual decrease or increase with altitude of the value of an atmospheric property; also, the layer through which this departure occurs (the "inversion layer"), or the lowest altitude at which the departure is found (the "base of the inversion"). This term almost always means a temperature inversion.

Inversion Base: In standard practice, the lowest height in the atmosphere at which the temperature ceases to decrease with altitude; the bottom or base of an inversion layer.

Inversion Layer: A layer in the atmosphere through which the temperature remains constant or increases with altitude.

Konimeter: (Also spelled conimeter.) An instrument for determining the dust content of a sample of air. One form of the instrument consists of a tapered metal tube through which a sample of air is drawn and allowed to impinge upon glass slide covered with a viscous substance. The particles caught are counted and measured with the aid of a microscope.

Land Bree Ze: A coastal breeze blowing from land to sea, caused by the temperature difference when the sea surface is warmer than the adjacent land. Therefore, it usually blows by night and alternates with a sea breeze, which blows in the opposite direction by day.

Lapse Rate: The decrease of an atmospheric variable with height, the variable being temperature, unless otherwise specified.

Lithometeor: The general term for dry atmospheric suspensoids, including dust, haze, smoke and sand.

Mac: Abbreviation for "maximum acceptable concentration", with reference to air quality standards.

Meso-Scale: In meteorology: having characteristic spatial dimensions somewhere between 1 and 100 miles, usually implying between 5 and 50 miles. Mesoscale phenomena thus include the bulk of air pollution phenomena in addition to land and sea breezes, mountain winds, squalls and thunderstorms, etc.

Meteorological Range: (Also called standard visibility, standard visual range.) An empirically consistent measure of the visual range of a target; a concept developed to eliminate from consideration the threshold contrast and adaptation luminance, both of which vary from observer to observer. It is a function only of the extinction coefficient of the atmosphere at the time and place in question.

Meteorology: The study dealing with the phenomena of the atmosphere. This includes not only the physics, chemistry, and dynamics of the atmosphere, but is extended to include many of the direct effects of the atmosphere upon the earth's surface, the oceans, and life in general. The goals often ascribed to meteorology are the complete understanding, accurate prediction, and artificial control of atmospheric phenomena.

Micrometeorology: The study of micro-scale atmospheric phenomena.

Micron: A unit of length equal to one-millionth of a meter or one-thousandth of a millimeter. The micron is a convenient length unit for measuring wavelengths of infrared radiation, diameters of atmospheric particles, etc.

Micro-Scale: In meteorology, having characteristic spatial dimensions of about one mile and less. Typical micro-scale phenomena include the diffusion of effluents, the effects of structures on air flow, the retention of moisture by crops, etc.

Mist: According to international definition: a hydrometeor consisting of an aggregate of microscopic and more-or-less hygroscopic water droplets suspended in the atmosphere. It produces, generally, a thin, greyish veil over the landscape. It reduces visibility to a lesser extent than fog.

Mist is intermediate in all respects between haze (particularly "damp haze") and fog.

Mixed Layer: In air pollution terminology, the layer of air usually a sub-inversion layer, within which pollutants are mixed by turbulence and diffusion.

Mixing Height: (Or mixing depth.) The thickness of the atmospheric layer, measured from the surface upward, through which pollutants are presumed to mix by virtue of convection caused by daytime heating at the surface. The designation of a mixing height implies the existence of an inversion, in which case the mixing height is also the height of the inversion base.

Nephelometer: General name for instruments which measure, at more than one angle, the scattering function of particles suspended in a medium. Information obtained from such instruments may be used to determine the size of the suspended particles and the visual range through the medium.

Nitric Oxide: (NO) colorless gas that enters the atmosphere as a pollutant mainly via the exhaust of automobile engines. Its principal effect in smog is to combine with oxygen to form nitrogen dioxide (NO_2). NO_2 is then photo dissociated to produce NO and atomic oxygen, the first main step in the complex production of photochemical smog.

Nitrogen: A colorless, tasteless, odorless gaseous element. It is the most abundant constituent of the atmosphere, amounting to 78.09 per cent by volume of dry air.

Nitrogen enters the atmosphere from volcanoes and from the decay of organic matter. It is removed from the atmosphere by certain natural nitrogen-fixing bacteria for use in plant life processes; all a part of a complex nitrogen cycle.

Free nitrogen is very inactive, but once combined (by bacterial fixation, high pressure, high temperature, etc.) it takes part in many transformations.

Nitrogen Dioxide: (NO_2.) A reddish-brown gas that exists in photochemical smog as a secondary pollutant, having formed from the combination of nitric oxide (NO) and oxygen (O_2). It also photodissociates into the latter two components, the first step in the highly complex chain of reactions that constitutes photochemical smog.

The rate of production of NO_2 is greatly enhanced by the presence of hydrocarbons.

Nucleation: Any process by which the phase change of a substance to a more condensed state (condensation, sublimation, freezing) is initiated at certain loci within the less condensed state.

Nucleus: In physical meteorology, a particle of any nature upon which, or the locus at which, molecules of water or ice accumulate as a result of a phase change to a more condensed state; an agent of nucleation.

Obscuration: (Also called obscured sky cover.) In United States weather observing practice, the designation for the sky cover when the sky is completely hidden by surface -based obscuring phenomena.

Obscuring Phenomenon: (Also called obscuration.) In United States weather observing practice, any atmospheric phenomenon (this does not include clouds) which restricts the vertical visibility or slant visibility, that is, which obscures a portion of the sky from the point of observation. Potentially, all hydrometeors and lithometeors may be obscuring phenomena.

Oxidant: Generally, any oxidizing agent, that is, a substance that acquires electrons in a chemical reaction. Ozone (03) and atomic oxygen(O) are extremely effective oxidants.

As a measured constituent of polluted air, the oxidant level is a function of the reagent used for the measurement. Maximum measured values are on the order of one part per million. The most common reagent used is potassium iodide (Kl), which practically excludes the detection of 02 as a oxidant.

Oxygen: (O_2)In its free form, a colorless, tasteless, and odorless gaseous element; the second most abundant gas in the earth's atmosphere and a prerequisite of virtually all forms of animal life.

Except in local areas of photochemical smog, oxygen occurs primarily in molecular form (O_2) up to an altitude of about 20 kilometers.

Ozone: (O_3) An unstable, blueish colored gas with. a pungent odor. It is a much more effective oxidizing agent than 02 (ordinary molecular oxygen).

Quite apart from its natural production in the upper atmosphere, Ozone is an important component of photochemical smog where the required oxygen atoms (to combine with 02) are made available through the photodissociation of nitrogen dioxide.

Ozone Layer: Generally, any layer in the atmosphere in which there is a maximum of ozone concentration.

Ozonosphere - (Also called ozone layer.) The general stratum of the upper atmosphere in which there is an appreciable ozone concentration and in which ozone plays an important part in the radiative balance of the atmosphere. This region lies roughly between 10 and 50 krn, with maximum ozone concentration at about 20 to 25 km. The ozonosphere is produced and maintained by pro-cesses that are entirely independent of air pollution.

Partial Pressure: The pressure exerted by a single component of a gaseous mixture.

Particulates: Any liquid or solid particles suspended in or falling through the atmosphere.

As pollutants, particulates are particularly obnoxious in that they reduce visibility and create filth. They are produced primarily as the result of incomplete combustion (ash, soot). Some combustion products act as effective condensation nuclei, and droplets of various solutions are formed as a secondary process, some of which, notably sulfuric acid, are highly corrosive.

Photochemical Reaction: A chemical reaction which involves either the absorption or emission of radiation. It most commonly refers to reactions promoted by the absorption of sunlight.

In photochemical smog, reactions of this type produce the main secondary contaminants and are responsible for much of the observed diurnal variation in contaminant levels and effects.

Photochemical Smog: The type of smog typical of the Los Angeles basin but becoming more and more prevalent elsewhere. In general, it is any smog wherein secondary pollutants are produced by photochemical reactions.

The common photochemical smog of today has the following typical characteristics. Essential ingredients are the ultraviolet portion of sunlight, nitrogen dioxide (NO_2) and certain carbon compounds (notably hydrocarbons). NO_2 is the primary absorber of the sunlight, releasing highly energetic oxygen atoms. The carbon compounds react with the oxygen atoms (and with the ozone produced by the latter), forming new reactive compounds which, in turn, react with molecular oxygen and nitrogen oxides to create a great variety of additional compounds. The entire process is so complex and variable, and occurs at such small concentrations, that a majority of the secondary and tertiary reactions remain a matter for conjecture.

Photodissociation: The dissociation (splitting) of a molecule by the absorption of a photon. The resulting components may be ionized in the process (photo ionization).

Photooxidation: An oxidation reaction promoted by the absorption of radiation.

Planetary Boundary Layer: (Also called friction layer, atmospheric boundary layer.) That layer of the atmosphere from the earth's surface to the geostrophic wind level including, therefore, the surface boundary layer and the Ekman layer. Above this layer lies the free atmosphere.

Plume: The volume of air space containing any of the substance emitted from a point source. For practical purposes, the limits of a plume have to be arbitrarily defined according to some minimum concentration of the substance.

Pollutant: (Or contaminant.) With respect to the atmosphere, any substance within it that is foreign to the "natural" atmosphere or that exceeds its "natural" concentration in the atmosphere. The universal connotation is that a pollutant is potentially deleterious.

Air pollutants are categorized in several ways: according to their phase (gaseous or particulate, liquid or solid); their source (industrial, automotive, agricultural, etc.): and whether they enter directly from a source (primary) or form through chemical reactions in the atmosphere (secondary).

Precipitation: 1. Any of all of the forms of water particles, whether liquid or Solid, that fall from the atmosphere and reach the ground. It is a major class of hydrometeor; but is distinguished from cloud, fog, dew, rime, frost, etc. in that it must "fall"; and is distinguished from cloud and virga in that it must reach the ground. Precipitation includes drizzle, rain, snow, snow pellets, snow grains, ice crystals, ice pellets, and hail. 2. In general, the process of removing one or more components of a compound, mixture, or colloidal system by causing it to fallout in the form of a more condensed phase than the initial compound or mixture, or, in the case of a colloidal system or suspension, by inducing the suspended particles to separate from the medium.

Pressure Gradient: (Also, in meteorology, called barometric gradient.) The rate of decrease (gradient) of pressure in space at a fixed time. The term is sometimes loosely used to denote simply the magnitude of the gradient of the pressure field.

Pressure Pattern: (Sometimes called baric topography.) In meteorology, the general geometric characteristics of atmospheric pressure distribution as revealed by isobars on a constant-height chart; usually applied to cyclonic-scale features of a surface chart.

Pressure System: An individual cyclonic-scale feature of atmospheric circulation, commonly used to denote either a cyclone or anticyclone.

Primary Pollutant: An air pollutant having the same form as when it was emitted at its source.

Prevailing Visibility: In United States weather observing practice, the greatest horizontal visibility which is equalled or surpassed throughout half of the horizon Circle; it need not be a continuous half. In the case of rapidly varying conditions, it is the average of the prevailing visibility while the observation is being taken.

Prevailing Wind Direction: (Or prevailing wind.) The wind direction most frequently observed during a given period. The periods most frequently used are the observational day, month, season, and year.

Radiational Cooling: In meteorology, the cooling of the earth's surface and adjacent air, accomplished (mainly at night) whenever the earth's surface suffers a net loss of heat due to terrestrial radiation.

Radiation Fog: A major type of fog, produced over a land area when radiational cooling reduces the air temperature to or below its dew point. Thus, a strict radiation for is a nighttime occurrence, although it may begin to form by evening twilight and often does not dissipate until after sunrise.

Factors favoring the formation of radiation fog are (a) a shallow surface layer of relatively moist air beneath a dry layer and clear skies, and. (b) light surface winds.

Radioactive Fallout: The eventual descent to the earth's surface of radioactive matter placed in the atmosphere by an atomic or thermonuclear explosion.

Radiosonde: A balloon-borne instrument for the simultaneous measurement and transmission of meteorological data. The instrument consists of transducers for the measurement of pressure, temperature, and humidity; a modulator for the Conversion of the output of the transducers to a quantity which controls a property of the radio frequency signal; a selector switch which determines the sequence in which the parameters are to be transmitted; and a transmitter which generators the radio-frequency carrier.

Relative Humidity: (Popularly called humidity.) The (dimensionless) ratio of the actual vapor pressure of the air to the saturation vapor pressure. The corresponding ratios of specific humidity or of mixing ratio give approximations of sufficient accuracy for many purposes in meteorology. The relative humidity is usually expressed in per cent, and can be computed from psychrometric data.

Ridge: (Sometimes called wedge.) In meteorology, an elongated area of relatively high atmospheric pressure, almost always associated with and most clearly identified as an area of maximum anticyclonic curvature of wind flow. The locus of this maximum curvature is called the ridge line.

Salt Haze: A haze created by the presence of finely divided particles of sea salt in the air, usually derived from the evaporation of sea spray.

Scattering: (Also called scatter.) The process by which small particles suspended in a medium of a different index of refraction diffuse a portion of the incident radiation in all directions, In scattering no energy transformation results, only a change in the spatial distribution of the radiation. Along with absorption, scattering is a major cause of the attenuation of radiation by the atmosphere.

Scavenging: 1. Same as washout. 2. Chemically, the addition of a chemical to a mixture in order to remove an unwanted constituent or convert it to an acceptable form.

Sea Breeze: A coastal local wind that blows from sea to land, caused by the temperature difference when the sea surface is colder than the adjacent land. Therefore, it usually blows on relatively calm, sunny, summer days; and alternates with the Oppositely directed, usually weaker, nighttime land breeze. As the sea breeze regime progresses, the wind develops a component parallel to the coast, owing to the coriolis deflection.

Secondary Pollutant: An air pollutant that has undergone a chemical change since have been first emitted into the atmosphere; or a pollutant created by reactions involving primary pollutants.

Smog: As originally coined in 1905 by Des Voeux: a natural fog contaminated by industrial pollutants, a misture of smoke and fog. Today, it is the common term applied to problematical, largely urban, air pollution, with or without the "natural" fog; however, some visible manisfestation is almost always implied.

Smogs are consituted in great variety, but a major dichotomy exists between the "photochemical smogs" of nitrogen Oxides and hydrocarbons emitted mainly by automobile engines and, on the other hand, the sulfur-laden smogs produced by the large-scale combustion of fuel oil and coal. Both types contain carbon monoxida, carbon dioxide, and a variety of particulates.

Smoke: Foreign particulate matter in the atmosphere resulting from combustion process; a type of lithometeor.

When smoke is present, the disk of the sun at sunrise and sunset appears very red, andduringthe daytime has a reddish tinge. Smoke that has come a great distance from its source, such as from forest fires, usually has a light grayish or bluish color and is evenly distributed in the upper air.

Solar Radiation: The total electromagnetic radiation emitted by the sun.

To a first approximation; the sun radiates as a black body at a temperature of about 5700°K; hence about 99.9 per cent of its energy output falls within the wavelength interval from 0.15 microns to 4.0 microns, with peak intensity near 0. 47 microns. About one -half of the total energy in the solar beam is contained within the visible spectrum from 0. 4 to 0. 7 microns, and most of the other half lies in the near infrared, a small additional portion lying in the ultraviolet.

Soot: Fine particles of carbon produced by the incomplete combustion of carbonaceous fuels.

Source: A point, line, area, or volume at which mass or energy is added to a system, either instantaneously or continuously. Conversely, at a "sink," mass or energy is removed.

Examples of sources in the context of air pollution are as follows: a smokestack is a "point source"; a freeway or aircraft trajectory is a "line source"; an entire city is a "plane source. "

Spectrophotometer: A photometer which measures the intensity of radiation as a function of the frequency (or wavelength) of the radiation.

Stable: Pertaining to the atmosphere or an atmospheric layer in the condition of static stability, that is, an atmosphere whose temperature lapse rate and moisture distribution is such as to suppress the vertical exchange of air.

Stagnation: With respect to air pollution, the persistence of a given volume of stable air over a region, permitting an abnormal buildup of pollutants from sources within the region.

Static Stability: (Also called hydrostatic stability, vertical stability, convective or convectional stability.) The state of the atmosphere when it is stable relative to vertical displacements. Such an atmosphere tends to remain stratified, in that any air that is displaced vertically is subjected to a buoyant force that tends to restore it to its original level.

Static stability is determined primarily by the temperature lapse rate; an inversion layer is an extreme example of a statically stable layer.

Stratified Fluid: A fluid having density variation along the axis of gravity, usually implying upward decrease of density, i. e., a stratification characterized by static stability.

Sub-Inversion Layer: Any layer of the atmosphere lying beneath a temperature inversion, usually referring to the layer from the ground upward to the base of a low-level inversion.

Subsidence: A descending motion of air in the atmosphere, usually with the implication that the condition extends over a rather broad area.

Subsidence Inversion: A temperature inversion produced by the adiabatic warming of a layer of subsiding air. This inversion is enchanced by vertical mixing in the air layer below the inversion.

Sulfuric Acid: (H_2SO_4.) A strongly corrosive, dense oily liquid. Some sulfuric acid is emitted directly into the atmosphere by burning of sulfur-containing fuels, but it may also form by the reaction of sulfur oxides with atmospheric water.

Sulfur Dioxide: (SO_2) A colorless, extremely irritating gas or liquid; forms sulfurous acid (H_2SO_3) in combination with water. It enters the atmosphere as a pollutant mainly as a result of burning high-sulfur-content fuel oils and coal.

Irradiated SO_2 is a very effective producer of aerosols, a quality which is greatly enhanced by the presence of atmospheric water vapor.

Sunshine: Direct radiation from the sun, as opposed to the shading of a location by clouds or by other obstructions.

Because of variations in atmospheric turbidity (alone and in conjunction with the optical air mass), and because of the variable optical thickness of clouds, the distinction between sunshine and lack of sunshine is an arbitrary one that is largely dependent upon the type of sunshine recorder in use or upon the quality of subjective estimates.

Superior Air: An exceptionally dry mass of air formed by subsidence and usually found aloft but occasionally reaching the earth's surface during extreme subsidence processes.

It is often found above tropical maritime air, bounded by the trade-wind inversion.

Surface Boundary Layer: (Also called surface layer, friction layer, atmospheric boundary layer, ground layer.) That thin layer of air adjacent to earth's surface, extending up to the so-called anemometer level (the base of the Ekman layer). Within this layer the wind distribution is determined largely by the vertical temperature gradient and the nature and contours of the underlying surface; and shearing stresses are approximately constant.

Surface Inversion: (Or ground inversion.) A temperature inversion based at the earth's surface; that is, an increase of temperature with height beginning at the ground level.

This condition is due primarily to greater radiative loss of heat at and near the surface than at levels above. Thus, surface inversions are common over land prior to sunrise and in winter over high-latitude continental interiors.

Suspension: In physical chemistry, a system composed of one substance (suspended phase, suspensoid) dispersed throughout another substance (suspending phase) in a moderately finely divided state, but not so finely divided as to acquire the stability of a colloidal system. Given sufficient time, a suspension will, by definition, separate itself by gravitational action into two visibly distinct portions, whereas a colloidal system, by definition, is stable.

Dust in the atmosphere is an example of a suspension of a solid in a gas.

Telephotometer: A photometer that measures the received intensity of a distant light source. When specifically used to measure the transmissivity of the intervening atmosphere (or other medium), it is usually termed a transmissometer.

Tempera Ture Inversion: A layer in which temperature increases with altitude.

The principal characteristic of an inversion layer is its marked ~ stability, so that very little turbulent exchange can occur within it. Strong wind shears often occur across inversion layers, and abrupt changes in concentrations of atmospheric particulates and atmospheric water vapor may be encountered on ascending through the inversion.

When, in meteorological literature and discussion, an "inversion" is mentioned, a temperature inversion is usually meant.

Temperature Lapse Rate: The rate at which air temperature decreases with height in the atmosphere.

Terminal Fall Velocity: (Or terminal velocity.) The particular falling speed, for any given object moving through a fluid medium of specified physical properties, at which the drag forces and buoyant forces exerted by the fluid on the object just equal the gravitational force acting on the object; after which it falls at constant speed, unless it moves into air layers of different physical properties. In the atmosphere, the latter effect is so gradual that objects such as raindrops, which attain terminal velocity at great heights above the surface, may be regarded as continuously adjusting their speeds to remain at all times essentially in the terminal fall condition.

Thermal Island: (Also called heat island.) An area, generally around the center of urban buildup, where the average temperature is higher than that of the surroundings.

Thermal Pollution: The emission of high-temperature waste products into the environment.

To date, this is an important problem only in some contexts of water pollution where very large quantities of industrial coolant are injected into streams, estuaries and coastal waters.

Thermal Radiation: The electromagnetic radiation emitted by any substance as the result of the thermal excitation of its molecules.

Tracer: An easily detectable substance injected into the atmosphere for the purpose of subsequent measurement and reconstruction of its history (trajectory, diffusion, etc.)

Trajectory: (or path.) A curve in space tracing the points successively occupied by a particle in motion. At any given instant the velocity vector of the particle is tangent to the trajectory. In steady-state flow, the trajectories and streamlines of the fluid parcels are identical.

Transmissmty: (Also called transmittance, transmittancy.) A measure of the amount of radiation propagated through a given medium; defined as the ratio of transmitted radiation to the total radiation incident upon the medium.

Transmissometer: (Or telephotometer; also called transmittance meter, hazemeter.) An instrument for measuring the extinction coefficient of the atmosphere and for the determination of visual range.

Trough: In meteorology, an elongated area of relatively low atmospheric pressure; the opposite of a ridge. The axis of a trough is the trough line.

Turbidity: In meteorology, any condition of the atmosphere which reduces its transparency to radiation, especially to visible radiation.

Ordinarily, this is applied to a cloud-free portion of the atmosphere that owes its turbidity to air molecules and suspensoids such as smoke, dust, and haze, and to scintillation effects.

Ultraviolet Radiation: (Abbreviated UV.) Electromagnetic radiation of shorter wavelength than visible radiation but longer than x-rays; roughly, radiation in the wavelength interval from 10 to 4000 angstroms.

Such radiation has marked actinic and bacteriacidal action, and produces fluorescence in a number of substances. Ultraviolet radiation from the sun is responsible for many complex photochemical reactions characteristic of the upper atmosphere; e. g., the formation of the ozone layer through ultraviolet dissociation of oxygen molecules followed by recombination to form ozone. This portion of the solar spectrum accounts, however, for only about 4 or 5 per cent of the total energy transmitted in the solar beam; and almost all of this small portion is absorbed or reflected by the upper atmosphere, chiefly by oxygen molecules.

Vapor: Any substance existing in the gaseous state at a temperature lower than that of its critical point; that is, a gas cool enough to be liquefied if sufficient pressure were applied to it.

If any vapor is cooled sufficiently, say at constant pressure, it ultimately reaches a state of saturation such that further removal of heat is accompanied by condensation to the liquid phase. Except for states quite close to that of saturation, vapors exhibit the general properties of all gases. Quantitatively, however, vapors exhibit measurable departures from perfect-gas laws even in states well removed from that of saturation.

Vapor Pressure: (Also called vapor tension.) The pressure exerted by the molecules of a given vapor. For a pure, confined vapor, it is that vapor's pressure on the walls of its containing vessel; and for a vapor mixed with other vapors or gases, it is that vapor's contribution to the total pressure (i.e., its partial pressure).

In meteorology, vapor pressure is used almost exclusively to denote the partial pressure of water vapor in the atmosphere. Care must be exercised in interpreting the term's meaning as used in other branches of science .

Ventilation: With respect to air pollution, any process that serves to dilute pollutants through the atmosphere by thorough mixing or to move polluted air out of areas of potentially high concentration.

Visibility: The greatest distance in a given direction at which it is just possible to see and identify with thc unaided eye (a) in the daytime, a prominent dark object against the sky at the horizon, and (b) at night, a known, preferably unfocused, moderately intense light source. After. visibilities have been determined around the entire horizon circle, they are resolved into a single value of prevailing visibility for reporting purposes.

Visibility Meter: The general term for instruments used to make direct measurements of visual range in the atmosphere or of the physical characteristic of the atmosphere which determine the visual range. Visibility meters may be classified according to the quantities that they measure. Telephotometers and transmissometers measure the transmissivity or, alternatively the extinction coefficient of the atmosphere. Nephelometers measure the scattering function of the atmospheric suspensoids. A third category of visibility meters makes use of an artificial "haze" of variable density which is used to obscure a marker at a fixed distance from the meter.

Visible Radiation: Electromagnetic radiation lying within the wavelength interval to which the human eye is sensitive, the spectral interval from approximately 0.4 to 0.7 microns (4000 to 7000 angstroms). This portion of the electromagnetic spectrum is bounded on the Short-wavelength end by ultraviolet radiation, and on the long-wavelength end by infrared radiation.

Visual Range: (Or daytime visual range.) The distance, under daylight conditions, at which the apparent contrast between a specified type of target and its background becomes just equal to the threshold contrast of an observer; to be distinguished from the night visual range. The visual range is a function of the atmospheric extinction coefficient, the albedo and visual angle of the target, and the observer's threshold contrast at the moment of observation. Only in the so-called meteorological range does one have a visibility figure dependent only upon the extinction coefficient.

Washout: The sweeping out of airborne particulates by rain or snow.

Watcr Vapor: (Also called aqueous vapor, moisture.) Water substance in vapor form; one of the most important of all constituents of the atmosphere.

Its amount varies widely in space and time due to the great variety of both "sources" of evaporation and "sinks" of condensation that provide active motivation to the hydrologic cycle. Approximately half of all of the atmospheric water vapor is found below 2 km altitude, and only a minute fraction of the total occurs above the tropopause. Water vapor is important not only as the raw material for cloud and rain, but also as a vehicle for the transport of energy (latent heat) and as a regulator of planetary temperatures via the greenhouse effect. The amount of water vapor present in a given air sample may be measured in a number of different ways, involving, such concepts as the absolute humidity, mixing ratio, dew point, relative humidity, specific humidity, and vapor pressure.

Weathering: The mechanical, chemical, or biological action of the atmosphere, hydrometeors, and suspended impurities on the form, color, or constitution of exposed material; to be distinguished from erosion.

Mechanical weathering results from the disintegrating action of high temperature, large changes of temperature, frost, or the impact of windborne sand or water (e. g. frost action, corrosion, etc.). Chemical weathering is due to the chemical action of atmospheric constituents, especially acid impurities, in a moist atmosphere or in rain water (e. g. corrosion, oxidation, etc.). Biological agencies are mainly fungi which attack organic material (e. g. rotting, mildew, etc.).

Wind: Air in motion relative to the surface of the earth. Since vertical components of atmospheric motion are relatively small especially near the surface of the earth, meteorologists use the term to denote almost exclusively the horizontal component. Vertical winds are always identified as such.

Wind Vane: An instrument used to indicate wind direction.

Appendix 2.11: Terminologies Related to Solid Waste Pollution

Aerobic Composting: a method of composting organic wastes using bacteria that need oxygen. This requires that the waste be exposed to air, either via turning or by forcing air through pipes that pass through the material.

Anaerobic Digestion: a method of composition that does not require oxygen. This composting method produces methane. Also known as aerobic composting.

Ash: the noncombustible solid by-products of incineration or other burning process.

Autoclaving: sterilization via a pressurized, high –temperature steam process.

Biodegradable Material: any organic material that can be broken down by microorganisms into simpler, more stable compounds. Most organic wastes (e.g., food, paper) are biodegradable.

Bottom Ash: relatively coarse, noncombustible, generally toxic residue of incineration that accumulates on the grate of furnace.

Bulky Waste: large wastes such as appliances, furniture, and trees and branches, that cannot be handled by normal MSW processing methods.(MSW-Municipal Soild Waste)

Cell: the basic unit by which a landfill is developed. It is the general area where incoming waste is tipped, spread, compacted, and covered.

Chemocar: a special vehicle for the toxic and hazardous wastes from residences, shops, and institutions.

Cleaner Production: processes designed to reduce the wastes generated by production.

Co-disposal : the disposal of different types of waste in one area of a landfill or dump. For instance, sewage sludges may be disposed of with regular solid wastes.

Collection: the process of picking up wastes from residences, business, or a collection point, loading them into a vehicle, and transporting them to a processing, transfer, or disposal site.

Combustibles: burnable materials in the waste stream, including paper, plastics, wood and food and garden wastes.

Commingled: mixed recyclables that are collected after having been separated from mixed MSW.

Compactor Vehicle: a collection vehicle using high-power mechanical or hydraulic equipment to reduce the volume of solid waste.

Composite Liner: a liner system for a land-fill consisting of an engineered soil layer and a synthetic sheet of material.

Compost: the material resulting from composting. Compost, also called humus, is a soil conditioner and in some instances is used as a fertilizer.

Composting: biological decomposition of solid organic materials by bacteria, fungi and other organisms into a soil-like product.

Construction and Demolition Debris: waste generated by construction and demolition of building, such as bricks, concrete, drywall, limber, miscellaneous metal parts and sheets, packaging materials, etc.

Controlled Dump: a planned landfill that incorporates to some extent some of the features of a sanitary landfill: sitting with respect to hydrogeological suitability, grading, compaction in some cases, leachate control, partial gas management, regular (not usually daily) cover, access control, basic record-keeping, and controlled waste picking.

Curing: allowing partially composted materials to sit in a pile for a specified period of time as part of the maturing process in composting.

Disposal: the final handling of solid waste, following collection, processing, or incineration. Disposal most often means placement of wastes in a dump or a landfill.

Diversion Rate: the proportion of waste material diverted for recycling, composting, or reuse and away from landfill or incineration.

Drop-off Center: an area or facility for receiving compostable or recyclables that are dropped off by waste generators.

Energy recovery: the process of extracting useful energy from waste, typically from the heat produced by incineration or via methane gas from landfills.

Environment Impact Assessment (EIA): an evaluation designed to identify and predict the impact of an action or a project on the environment and human health and well-being. Can include risk assessment as a component along with economic and land use assessment.

Environment Risk Assessment: an evaluation of the interactions of agents, humans, and ecological resources. Comprised of human health risk assessment and ecological risk assessment, typically evaluating the probabilities and magnitude of harm that could come from environment contaminants.

Flaring the burning of methane emitted from collection pipes at a landfill.

Fluidized-bed Incinerator: a type of incinerator in which the stroker grate is replaced by a bed of limestone or sand that can withstand high temperature. The heating of the bed and the high air velocities used cause the bed to bubble, which give rise to the term fluidized.

Fly ash: the highly toxic particulate matter captured from the flue gas of an incinerator by the air pollution control system.

Garbage: in everyday usage, refuse in general.

Groundwater: water beneath the earth's surface that fills underground pockets (known as aquifers), supplying wells and springs.

Hazardous Waste: waste that is reactive, toxic, corrosive, or otherwise dangerous to living things and/or the environment. Many industrial by-products are hazardous.

Heavy Metals: metals of high atomic weight and density, such as mercury, lead and cadmium, that are toxic to living organisms and/ or the environment.

Humus: the end product of composting, also called compost.

Incineration: the process of burning solid waste under controlled conditions to reduce its weight and volume, and often to produce energy.

Inorganic Waste: waste composed of material other than plant or animal matter, such as sand, dust, glass and many synthetics.

Integrated Solid Waste Management: coordinated use of a set of waste management methods, each of which can play a role in an overall waste management.

In-vessel Composting: composting in an enclosed vessel or drum with a controlled internal environment, mechanical mixing and aeration.

Landfill Gases: gases arising from the decomposition of organic wastes; principally methane, carbon dioxide, and hydrogen sulfide. Such gases may cause explosions at landfills.

Landfilling: the final disposal of solid waste by placing it in a controlled fashion in a place intended to be permanent controlled dumps & sanitary landfills.

Leachate: liquid (which may be partly produced by deromosition of organic matter) that has seeps through a landfill or a compost pile and has accumulated bacteria and other possibly harmful dissolved or suspended materials. If uncontrolled, leachate can be contaminate both groundwater and surface water.

Leachate Pond: a pond or tank constructed at a landfill to receive the leachate from the area. Usually the pond is designed to provide some treatment of the leachate, by allowing settlement of solids or by aeration to promote biological processes.

Lift: the completed layer of compacted waste in a cell at a landfill.

Liner: a protective layer, made of soil and/or synthetic materials, installed along the bottom and sides of a landfill to prevent or reduce the flow of leachate into the environment .

Manual Landfill: a landfill in which most operations are carried out without the use of mechanized equipment.

Market Waste: primarily organic waste, such as leaves, skins, and unsold food, discarded at or near food markets.

Mass-burn Incinerator: a type o incinerator in which solid waste is burned without prior sorting or processing.

Materials Recovery: obtaining materials that can be reused or recycled.

Materials Recovery Facility: a facility for separating commingled recyclables by manual or mechanical means. Some facilities are designed to separate recyclables from mixed MSW, then bale and markets the recovered materials.

Methane : an odorless, colorless, flammable, explosive gas, CH, produced by anaerobically decomposing MSW at landfills.

Mixed Waste: unsorted materials that have been discarded into the waste stream.

Modular Incinerator: a relativelty small type of prefabricated soild waste combustion unit.

Monofill : a landfill intended for one type of waste only.

Municipal Solid Waste: all solids waste generated in an area except industrial and agricultural wastes. Sometimes includes construction and demolition debris and other special wastes that may enter the municipal waste stream. Generally excludes hazardous wastes except to the extent that they enter the municipal waste stream. Sometimes defined to mean all solid waste that a city authority accepts responsibility for managing in some way.

Municipal Solid Waste Management: planning and implementation of systems to handle MSW.

NGO: nongovernmental organization. May be used to refer to a range of organization from small community groups, through national organizations, to international ones. Frequently these are not-for-profit organizations.

Night Soil: human excreta.

NIMBY: "Not in My Back Yard ". An expression of resident opposition to the sitting of a solid waste facility based on the particular location proposed.

Open Dump: an unplanned "landfill" that incorporates few if any of the characteristics of a controlled landfill. There is typically no leachate control, no access control, no cover, no management, and many waste pickers.

Organic Waste: technically, waste containing carbon, including paper, plastics, wood, food wastes, and yard wastes. In practice in MSWM, the term is often used in a more restricted sense to mean material that is more directly derived from plant or animal sources, and which can generally be decomposed by microorganisms.

Pathogen: an organism capable of causing disease.

Pollution: the contamination of soil, water, or the atmosphere by the discharge of waste or other offensive materials.

Post-consumer Materials: materials that a consumer has finished using, which the consumer may sell, give away, or discard as wastes.

Primary Material: a commercial material produced from virgin materials used for manufacturing basic products. Example include wood pulp, iron ore, and silica sand.

Processing: preparing MSW materials for subsequent use or management, using processes such as baling, magnetic separation, crushing, and shredding. The term is also sometimes used to mean separation of recyclables from mixed MSW.

Putrescible: subject to decomposition or decay. Usually used in reference to food wastes and other organic wastes that decay quickly.

Pyrolysis: chemical decomposition of a substance by heat in the absence of oxygen, resulting in various hydrocarbon gases and carbon-like residue.

Recyclables: items that can be reprocessed into feedstock for new products. Common examples are paper, glass, aluminum, corrugated cardboard and plastic containers.

Recycling: the process of transforming materials into raw materials for manufacturing new products, which may or may not be similar to the original product.

Refuse: a term often used interchangeably with solid waste.

Refuse-derived Fuel (RDF): fuel produced from MSW that has undergone processing. Processing can include separation of recyclables and noncombustible materials, shredding, size reduction,and pelletizing.

Resource Recovery: the extraction and utilization of materials and energy from wastes.

Reuse: the use of a product more than once in its original form, for the same or a new purpose.

Rubbish : a general term for solid waste. Sometimes used to exclude food wastes and ashes.

Sanitary Landfill: an engineered method of disposing of solid waste on land, in a manner that meets most of the standard specifications, including sound sitting, extensive preparation, proper leachate and gas management and monitoring, compaction, daily and final cover, complete access control, and record-keeping.

Scrubber: emission control device in a incinerator, used primarily to control acid gases, but also to remove some heavy metal.

Secondary Material: a material recovered from post-consumer wastes for use in place of a primary material in manufacturing a product.

Secure Landfill: a disposal facility designed to permanently isolate wastes from the environment. This entails burial of the wastes in a landfill that includes clay and/ or synthetic liners, leachate collection, gas collection (in cases where gas is generated),and an impermeable cover.

Septage : sludge removed from a septic tank(a chamber that holds human excreta).

Set Out Container: a box or bucket used for residential waste that is placed outside for collection .

Sewage Sludge: a semi-liquid residue that settles to the bottom of canals and pipes carrying sewage or industrial waste waters, or in the bottom of tanks used in treating wastewaters.

Site Remediation: treatment of a contaminated site by removing contaminated solids or liquid or treating them on-site.

Source Reduction: the design, manufacture, acquisition, and reuse of materials so as to minimize the quantity and/ or toxicity of waste produced.

Source Separation: setting aside of compostable and recyclable materials from the waste stream before they are collected with other MSW, to facilitate reuse, recycling, and composing.

Special Wastes: wastes that are ideally considered to be outside of the MSW stream, but which sometimes enter it and must often be dealt with by municipal authorities. These include household hazardous waste, medical waste, construction and demolition debris, war and earthquake debris, tires, oils, wet batteries, sewage sludge, human excreta, slaughterhouse waste, and industrial waste.

Tipping Fee: a fee for unloading or dumping waste at a landfill, transfer station, incinerator, or recycling facility.

Tipping Floor: unloading area for vehicles that are delivering MSW to incinerator.

Transfer : the act of moving waste from a collection vehicle to a larger transport vehicles.

Transfer Point : a designated point, often at the edge of a neighborhood, where sma collection vehicle transfer waste to larger vehicles for transport to disposal sites.

Transfer Station: a major facility at which MSW from collection vehicles is considered loads that are transported by large trucks or other means to more distant final disposal facilities, typically landfill.

Vectors : organisms that carry disease causing pathogens. At landfill rodents, files, and birds are the main vectors that spread pathogens beyond the landfill site.

Virgin Matrials: any basic material processes that has not previously been used, for example wood-pulp trees, iron ore, crude oil, bauxite .

Waste characterization study: an analysis of samples from a waste stream to determine its composition.

Waste Collector: a person employed by a local authority or a private firm to collect waste from residences, business and community bins.

Waste Dealer: a middleman who buys recyclable materials from waste generators and itinerant buyers and sells them, after sorting and some processing, to wholesale brokers or recycling industries.

Waste Management Hierarchy: a ranking o waste management operation according to their environment or energy benefits. The purpose of the waste management hierarchy is to make waste management as environmentally sound as possible.

Waste Picker: a person who picks out recyclables from mixed waste wherever it may be temporarily accessible or disposed of.

Waste Reduction: all means of reducing the amount of waste that is produced initially and that must be collected by solid waste authorities. This ranges from legislation and product design to local programs designed to keep recyclables and composition out of the final waste stream.

Waste Stream: the total flow of waste from community, region, or facility.

Waste-to-energy (WTE): a facility that uses solid waste materials(processed or raw) to produce energy. WTE plants include incinerators that produce steam for district heating or industrial use, or that generate electricity; they also include facilities that convert landfill gas to electricity .

Water Table : level below the earth's surface at which the ground becomes saturated with water.

Wetland : an area that is regularly wet or flooded and has a water table that stands at or above the land surface for at least part of the year.

Windrow: an elongated pile of aerobically composting materials that are turned are periodically to expose the materials to oxygen and to control the temperature to promote biodegradation.

Working Face: the length and width of the row in which waste is being deposited at a landfill. Also known as the tipping face.

Worm Castings: the material produced from the digestive tracts of worms as they live in earth or compost piles. The castings are rich in nitrates, potassium, phosphorous, calcium, and magnesium.

Worm culture: a relatively cool, aerobic composting process that uses worms and microorganisms. Also known as venniculture.

Yard waste: leaves, grass clippings, prunings, and other natural organic matter discarded from yards and gardens.

Appendix 2.12: Noise Pollution Related Terminologies and Units

S. No.	Noise control terms	Description
1	Decibel, dB	A unit of sound pressure equal to 20 times the logarithm (base 10) of the magnitude of a sound pressure p to the reference sound pressure p_1. The reference pressures in air and water and 0.00002 and 0.1 N/m^2, respectively. The unit of dBA is the measured A-scale sound level. Since most sound-measuring instruments measure sound pressure, decibel usually refers to sound pressure level.

2	Frequency	The number of periods (cycles) repeated over a unit of time and is expressed as cycles per sec (cps) or Hertz (Hz).
3	Intensity	The average rate of sound energy transmitted through a unit of area in the direction specified and is expressed in watts per square meter, W/m^2.
4	Loudness	A subjective attribute and is expressed qualitatively, such as quiet or loud. Quantitative systems for loudness levels use phons or sones.
5	Noise	An undesired or unwanted audible sensation, sound, or an intermittent oscillation or vibration.
6	Phons	The average sound pressure level, dB, relative to a pure 1000 Hz, and is expressed on an arithmetic scale.
7	Sones	The loudness of a 1000-cps, or 1000-Hz, sound at 10-dB intensity above a listener's threshold; is expressed on a logarithmic scale.

Appendix 2.13: Terminologies Related to Radioactive/Nuclear Pollution

Absorbed dose: the energy imparted by ionizing radiation per unit mass of irradiated material. The units of absorbed dose are the rad and the gray (Gy).

Activity: the rate of disintegration (transformation) or decay of radioactive material. The units of activity are the curie (Ci) and the Becquerel (Bq).

Annual limit on intake (ALI): the derived limit for the amount of radioactive material taken into the body of an adult worker by inhalation or ingestion in a year.

Background radiation: Radiation from cosmic sources; naturally occurring radioactive material, including radon (except as decay product of source or special nuclear material); and global fallouts as it exists in the environment from the testing of nuclear explosive devices or from past nuclear accidents.

Becquerel (Bq): one Becquerel = 1 distinegration per second

Collective dose: the sum of the individual doses received in a given period of time by a specified population from exposure to a specified source of radiation.

Committed does equivalent (HT, 50): the dose equivalent to organs or tissues of reference (T) that will be received from an intake of radioactive material by an individual during the 50-year period following the intake.

Committed effective dose equivalent (HE, 50): the sum of the products of the weighting factors applicable to each of the body organs or tissues that are irradiated and the committed does equivalent to these organs or tissues.

Curie: one curie = 3.7×10^{10} disintegrations per second = 3.7×10^{10} becquerels = 2.22×10^{12} distintegrations per minute.

Deep-dose equivalent (Hd): which applies to external whole-body exposure, is the dose equivalent at a tissue depth of 1 cm (1000 mg/cm^2).

Derived air concentration (DAC): the concentration of a given radionuclide in air which, if breathed by the reference man for a working year of 2, 000 hours under conditions of light work (inhalation rate 1.2 cubic meters of air per hour), results in an intake of one ALI.

Dose or radiation dose: a generic term that means absorbed dose, dose equivalent, effective dose equivalent, committed does equivalent, committed effective dose equivalent, or total effective dose equivalent.

Dose equivalent (HT): the product of the absorbed dose in tissue, quality factor, and all other necessary modifying factors at the location of interest. The units of dose equivalent are the rem and sievert (Sv).

Gray (Gy): one gray is equal to an absorbed dose of 1 J/kg (100 rads).

High-radiation area: an area, accessible to individuals, in which radiation levels from radiation sources external to the body could result in an individual receiving a dose equivalent in excess of 0.1 rem (1 mSv) in 1 hour at 30 cm from the radiation source or 30 cm from any surface that the radiation penetrates.

Occupational dose: the dose received by an individual in the course of employment in which the individual's assigned duties involve exposure to radiation or to radioactive material. Occupational dose does not include dose received from background radiation or from any medical administration the individual has received.

Public dose: the dose received by a member of the public from exposure to radiation or the release of radioactive materials. Public dose does not include occupational dose or doses received from background radiation, or from any medical administration the individual has received.

Quality Factor (Q): a measure of potential damage caused to a tissue from various type and energies of radiation.

Rad: is the special unit of absorbed dose. One rad is equal to an absorbed dose of 100 ergs/gram or 0.01 J/kg (0.01 Gy).

Radiation area- means: an area, accessible to individuals, in which radiation levels could result in an individual receiving a dose equivalent in excess of 0.005 rem (0.05 mSv) in 1 hour at 30 cm from the radiation source or from any surface that the radiation penetrates.

Rem (acronym for "Roentgen equivalent mammal"): the dose equivalent in rems is equal to the absorbed dose in rads multiplied by the quality factor (1 rem = 0.01 sievert).

Sievert: the dose equivalent in sieverts is equal to the absorbed dose in grays multiplied by the quality factor (1 V = 100 rems).

Very high radiation area: an area, accessible to individuals, in which radiation levels from radiation sources external to the body could result in an individual receiving an absorbed dose in excess of 500 rads (5 Gy) in 1 hour at 1 m from a radiation source or 1 m from any surface that the radiation penetrates.

Appendix 2.14: Terminologies related to environmental science

Abiotic- non-living chemical and physical factors of the environment.

Absorpotion pit (soakaway) – a hole dug in permeable ground and filled with broken stones or granular material and usually covered with earth allowing collected water to soak into the ground.

Acclimation – the process of an organism adjusting to chronic change in its environment.

Acid mine drainage – the outflow of acidic water from metal mines or coal mines.

Acid rain – rain or other forms of precipitation that is unusually acidic.

Adsorption – one substance taking up another at its surface.

Aerobic – requiring air or oxygen; used in reference to decomposition processes that occur in the presence of oxygen.

Aerosols – solid or liquid particles suspended within the atmosphere.

Afforestation – planting new forests on lands that have not been recently forested.

Agroforestry – (sustainability) an ecologically based farming system, that, through the integration of trees in farms, increases social, environmental and economic benefits to land users.

Air pollution – the modification of the natural characteristics of the atmosphere by a chemical, particulate matter, or biological agent.

Albedo – reflectance; the ratio of light from the Sun that is reflected by the Earth's surface, to the light received by it. Unreflected light is converted to infrared radiation (heat), which causes atmospheric warming. Thus, surfaces with a high albedo, like snow and ice, generally contribute to cooling, whereas surfaces with a low albedo, like forests, generally contribute to warming. Changes in land use that significantly alter the characteristics of land surfaces can alter the albedo.

Algal bloom – the rapid and excessive growth of algae; generally caused by high nutrient levels combined with other favourable conditions. Bloom can deoxygenate the water leading to the loss of wildlife.

Alternative fuels – fuels like ethanol and compressed natural gas CNG that produce less pollutants emissions than the traditional fossil fuels.

Anaerobic – not requiring air or oxygen; used in reference to decomposition processes that occur in the absence of oxygen.

Anoxic – with abnormally low levels of oxygen.

Anthropogenic – man-made, not natural.

Application efficiency – (sustainability) the efficiency of watering after losses due to runoff, leaching, evaporation, wind etc.

Aquaculture – the cultivation of aquatic organisms under controlled conditions.

Aquifer – a bed or layer yielding water for wells and springs etc.; an underground geological formation capable of receiving, storing and transmitting large quantities of water. Aquifer types include; confined (sealed and possibly containing "fossil" water); unconfined(capable of receiving inflow); and Artesian (an aquifer in which the hydraulic pressure will cause the water to rise above the upper confining layer).

Arable land – land that can be used for growing crops.

atmosphere – general name for the layer of gases around a material body; the Earth's atmosphere consists, from the ground up, of the troposphere (which includes the planetary boundary layer or peplosphere, the lowest layer),stratosphere, mesosphere, ionosphere (or thermosphere), exosphere and magnetosphere.

autotroph – an organism that produces complex organic compounds from simple inorganic molecules using energy from light or inorganic chemical reactions.

Available water capacity – that proportion of soil water that can be readily absorbed by plant roots.

Avoidance – (sustainability) the first step in the waste hierarchy where waste generation is prevented (avoided).

backflow – movement of water back to source e.g. contaminated water in a plumbing system.

baffle - an obstruction to trap debris in drainage water.

Bioaccumulation – the accumulation of a substance, such as a toxic chemical, in the tissues of a living organism.

Biocoenosis (alternatively, biocoenose or biocenose) – all the interacting organisms living together in a specific habitat (or biotope).

Biodegeadable – capable of being decomposed through the action of organisms, especially bacteria.

Biodiversity – the variety of life in all its forms, levels and combinations; includes ecosystem diversity, species diversity, and genetic diversity.

Bioelement – an element required by a living organism.

Bioenergy – used in different senses; in its most narrow sense it is a synonym for biofuel, (fuel derived from biological sources). It encompasses also biomass, the biological material used as a biofuel,

Biofuel – the fuel produced by the chemical and/or biological processing of biomass. Biofuel will either be a solid (e.g. charcoal), liquid (e.g. ethanol) or gas (e.g. methane).

Biogas – landfill gas and sewage gas, also called biomass gas.

Biogeochemical cycle – a circuit or pathway by which a chemical element or molecule moves through both biotic ("bio-") and abiotic ("geo-") parts of an ecosystem.

Biogeochemical cycles – the movement of chemical elements between organisms and non-living components of the atmosphere, aquatic systems and soils.

Biological oxygen demand (BOD) – a chemical procedure for determining how fast biological organism use up oxygen in a body of water.

Biological pest control a method of controlling pests (including insects, mites, weeds and plant diseases) that relies on predation, parasitism, herbivory, or other natural mechanisms.

Biological productivity – (bioproductivity) the capacity of a given area to produce biomass; different ecosystems (i.e. pasture, forest, etc.) will have different levels of bioproductivity. Biological productivity is determined by dividing the total biological production (how much is grown and living) by the total area available.

Biomass – the materials derived from photosynthesis (fossilized materials may or may not be included) such as forest, agricultural crops, wood and wood wastes, animal wastes, livestock operation residues, aquatic plants, and municipal and industrial wastes; the quantity of organic material present in unit area at a particular time mostly expressed as tons of dry matter per unit area; organic matter that can be used as fuel.

Biome – a climatic and geographically defined area of ecologically similar communities of plants, animals and soil organisms, often referred to as ecosystems.

Biophysical – the living and non-living components and processes of the ecosphere. Biophysical measurements of nature quantify the ecosphere in physical units such as cubic metres, kilograms or joules.

Bioregion – (ecoregion) an area comprising a natural ecological community and bounded by natural borders.

Bioremediation – a process using organisms to remove or neutralise contaminants (e.g. petrol), mostly in soil or water.

Biosolids – nutrient-rich organic materials derived from wastewater solids (sewage sludge) that have been stabilized through processing.

Biosphere – the part of the Earth, including air, land, surface rocks, and water, within which life occurs, and which biotic processes in turn alter or transform.

Biosphere – the zone of air, land and water at the surface of the earth that is occupied by living organisms; the combination of all ecosystems on Earth and maintained by the energy of the Sun; the interface between the hydrosphere, geosphere and atmosphere.

Biotic pontential – the maximum reproductive capacity of a population under optimum environmental conditions.

Biotic – relating to, produced by, or caused by living organisms.

Blackwater – household wastewater that contains solid waste i.e. toilet discharge.

Bluewater – collectible water from rainfall; the water that falls on roofs and hard surfaces usually flowing into rivers and the sea and recharging the ground water. In nature the global average proportion of total rainfall that is blue water is about 40%. Blue water productivity in the garden can be increased by improving irrigation techniques, soil water storage, moderating the climate, using garden design and water-conserving plantings; also safe use of grey water.

Brundtland Commission Report – a UN report, Our Common Future, published in 1987 and dealing with sustainable development and the policies required to achieve it, which the report characterizes as " development that meets the needs of the present without compromising the ability of future generations to meet their own needs."

C3 & C4 plants – C4 plants comprise about 5% of all plants, are most abundant in hot and arid conditions, and include crops like sugar cane and soybeans. During photosynthesis they form molecules with 4-carbon atoms and saturate at the given level of CO2. C3 plants, the other 95%, photosynthesis to form 3 carbon molecules and increase photosynthesis with as CO2 levels increase.

Cancer – a group of diseases in which cells are aggressive (grow and divide without respect to normal limits), invasive (invade and destroy adjacent tissues), and sometimes metastatic (spread to other locations in the body).

Carbon budget – a measure of carbon inputs and outputs for a particular activity.

Carbon credit - a market-driven way of reducing the impact of greenhouse gas emissions; it allows an agent to benefit financially from an emission reduction. There are two forms of carbon credit, those that are part of national and international trade and those that are purchased by individuals. Internationally, to achieve Kyoto Protocol objectives, 'caps' (limits) on participating country's emissions are established. To meet these limits countries, in turn, set 'caps' (allowances or credits; I convertible and transferable credit = 1 metric tone of CO2 (emissions) for operators. Operators that meet the agreed 'caps' can then sell unused credits to operators who exceed 'caps'. Operators can then choose the most cost-effective way of reducing emissions. Individual carbon credits would operate in a similar way cf. carbon offset.

Carbon cycle – the biogeochemical cycle by which carbon is exchanged between the biosphere, geosphere, hydrosphere, and atmosphere of the Earth.

Carbon Dioxide Equivalent (CO_2e) – the unit used to measure the impacts of releasing (or avoiding the release of) the seven different greenhouse gases; it is obtained by multiplying the mass of the greenhouse gas by its global warming potential. For example, this would be 21 for methane and 310 for nitrous oxide.

Carbon dioxide – a gas with the chemical formula CO_2; the most abundant greenhouse gas emitted from fossil fuels.

Carbon equivalent (C-e) – obtained by multiplying the CO_2-e by the factor 12/44.

Carbon footprint – a measure of the carbon emissions that are emitted over the full life cycle of a product or service and usually expressed as grams of CO_2-e.

Carbon labeling – use of product labels that display greenhouse emissions associated with goods.

Carbon neutral – activities where net carbon inputs and outputs are the same. For example, assuming a constant amount of vegetation on the planet, burning wood will add carbon to the atmosphere in the short term but this carbon will cycle back into new plant growth.

Carbon pool – a storage reservoir of carbon.

Carbon sink – any carbon storage system that cause a net removal of greenhouse gases from the atmosphere.

Carbon source – opposite of carbon sink; a net source of carbon for the atmosphere.

Carbon stocks – the quantity of carbon held within a carbon pool at a specified time.

Carbon taxes – a surcharge on fossil fuels that aims to reduce carbon dioxide emissions.

Carcinogen – a substance, radionuclide or radiation that is an agent directly involved in the promotion of cancer or in the facilitation of its propagation.

Carrying capacity – the maximum population that an ecosystem can sustain (biocapacity).

Catchment area - the area that is the source of water for a water supply whether a dam or rain water tank.

Cell – (biology) the structural and functional unit of all known living organisms and is the smallest unit of an organism that is classified as living.

CFC(chlorofluorocarbon) – CFCs are potent greenhouse gases which are not regulated by the Kyoto Protocol since they are covered by the Montreal Protocol.

Chlorofluorocarbons – one of the more widely known family of haloalkanes.

Circular metabolism – a system in which wastes, especially water and materials, are reused and recycled cf. linear metabolism.

Cleaner production – the continual effort to prevent pollution, reduce the use of energy, water and material resources and minimize waste – all without reducing production capacity.

Climate change – a change in weather over time and/or region; usually relating to changes in temperature, wind patterns and rainfall; although may be natural or anthropogenic, common discourse carries that climate change is anthropogenic.

Cogeneration – the simultaneous production of electricity and useful heat from the combustion of the same fuel source.

Cohousing – clusters of houses having shared dining halls and other spaces, encouraging stronger social ties while reducing the material and energy needs of the community.

Commercial and industrial waste – solid waste generates by the business sector as well as that created by government, schools and tertiary institutions. Does not include that from the construction and demolition industry.

Commingled materials – materials mixed together, such as plastic bottles, glass, and metal containers. Commingled recyclable materials require sorting after collection before they can be recycled.

Compensation point – the point where the amount of energy produced by photosynthesis equals the amount of energy released by respiration.

Compost – the aerobically decomposed remnants of organic residues that may be used as a soil assistive.

Confined aquifer – aquifers that have the water table above their upper boundary and are typically found below unconfined aquifers.

Construction and demolition waste – includes waste from residential, civil,& commercial construction and demolition activities, such as fill material (e.g. soil), asphalt, bricks and timber. It waste excludes construction waste which is included in the municipal waste stream. It does not generally include waste from the commercial and industrial waste stream.

Consumption (ecology) – the use of resources by a living system, the inflow and degradation of energy that is used for system activity.

Contour ploughing – the farming practice of plowing across a slope following its contours. The rows formed have the effect of slowing water run-off during rainstorms so that the soil is not washed away and allows the water to percolate into the soil.

Corporate Social Responsibility (CSR) – Integration of social and environmental policies into day-to-day corporate business.

Crop coefficient – A variable used to calculate the evapotranspiration of a plant crop based on that of a reference crop.

Crop evapotranspiration – It is the crop water use – the daily water withdrawal.

Crop rotation – the practice of growing a series of dissimilar types of crops in the same space in sequential seasons for various benefits such as to avoid the buildup of pathogens and pests that often occurs when one species is continuously cropped.

Cullet – crushed glass that is suitable for recycling by glass manufacturers.

Cultural eutrophication – the process that speeds up natural eutrophication because of human activity.

Culvert – drain that passes under a road or pathway, may be a pipe or other conduit.

Cyanobacteria (Cyanophyta or blue-green algae) – a phylum of bacteria that obtain their energy through photosynthesis.

Cyclone – intense low pressure weather systems; mid-latitude cyclones are atmospheric circulations that rotate clockwise in the Southern Hemisphere and anti-clockwise in the Northern Hemisphere and are generally associated with stronger winds, unsettled conditions, cloudiness and rainfall. Tropical cyclones (which are called hurricanes in the Northern Hemisphere) cause storm surges in coastal areas.

DDT – a chlorinated hydrocarbon used as a pesticide that is a persistent organic pollutant (pop).

Decomposers – consumers, mostly microbial, that change dead organic matter into minerals and heat.

Deforestation – the conversion of forested areas to non-forest land for agriculture, urban use, development, or wasteland.

Desalination – producing potable or recyclable water by removing salts from salty or brackish water. This is done by three methods; distillation/freezing; reverse osmosis using membranes and electrodialysis; ion exchange. All these method are energy intensive.

Desert – an area that receives an average annual precipitation of less than 250 mm or an area in which more water is lost than falls as precipitation.

Desertification – the degradation of land in arid, semi arid and dry sub-humid areas resulting from various climatic variations, but primarily from human activities.

Detritivore – animals and plants that consume detritus (decomposing organic material), and in doing so contribute to decomposition and the recycling of nutrients.

Detritus - non-living particulate organic material (as opposed to dissolved organic material).

Dieback – (arboriculture) a condition in trees or woody plants in which peripheral parts are killed, either by parasites or due to conditions such as acid rain.

Dioxin – any one of a number of chemical compounds that are persistent organic pollutants and are carcinogenic.

Diversion rate – (waste disposal) the proportion of a potentially recyclable material that has been diverted out of the waste disposal stream and therefore not directed to landfill.

Divertible resource – The proportion of water runoff and recharge that can be accessed for human use.

Downcycling – (waste management) recycling in which the quality of an item is diminished with each recycling.

Downstream – those processes occurring after a particular activity e.g. the transport of a manufactured product from a factory to the wholesale or retail outlet.

Drainage – that part of irrigation or rainfall that runs off an area or is lost to deep percolation.

Drawdown – drop in water level, generally applied to wells or bores.

Dredging – the repositioning of soil from an aquatic environment, using specialized equipment, in order to initiate infrastructural and/or ecological improvements.

Drinking water – (Potable water) – water fit for human consumption in accordance with World Health Organisation (who) guidelines.

Drip irrigation – A drip hose placed near the plant roots so minimizing deep percolation and evaporation.

Driver – (ecology) any natural or human-induced factor that directly or indirectly causes a change in an ecosystem. A direct driver is one that unequivocally influences ecosystem processes and that can be measured.

Drop off centre – (waste management) a location where discarded materials can be left for recycling.

Drought – an acute water shortage relative to availability, supply and demand in a particular region. An extended period of months or years when a region notes a deficiency in its water supply. Generally, this occurs when a region receives consistently below average precipitation.

Dryland salinity – accumulation of salts in soils, soil water and ground water; may be natural or induced by land clearing.

Eco - a prefix now added to many words indicating a general consideration for the environment e.g. ecohousing, ecolabel, ecomaterial.

Eco-asset – a biological asset that provides financial value to private land owners when they are maintained in or restored to their natural state.

Ecolabel – seal or logo indicating a product has met a certain environmental or social standards.

Ecological Footprint (Eco-footprint, Footprint) – a measure of the area of biologically productive land and water needed to produce the resources and absorb the wastes of a population using the prevailing technology and resource management schemes; a measure of the consumption of renewable natural resources by a human population, be it that of a country, a region or the whole world given as the total area of productive land or sea required to produce all the crops, meat, seafood, wood and fibre it consumes to sustain its energy consumption and to give space for its infrastructure.

Ecological niche – the habitat of a species or population within its ecosystem.

Ecological succession – the more-or-less predicatable and orderly changes in the composition or structure of an ecological community with time.

Ecological sustainability – the capacity of ecosystems to maintain their essential processes and function and to retain their biological diversity without impoverishment.

Ecologically sustainable development – using, conserving and enhancing the human community's resources so that ecological processes, on which all life depends, can be maintained and enriched into the future.

Ecology – the scientific study of living organism and their relationship to one another and their environment; the scientific study of the processes regulating the distribution and abundance of organisms; the study of the design of ecosystem structure and function.

Ecosystem boundary – the spatial delimitation of an ecosystem usually bases on discontinuities of organisms and the physical environment.

Ecosystem services – the role played by organisms, without charge, in creating a healthy environment for human beings, from production of oxygen to soil formation, maintenance of water quality and much more. These services are now generally divided into four groups, supporting, provisioning, regulating and cultural.

Ecosystem - a dynamic complex of plant, animal and microorganism communities and their non-living environment all interacting as a functional unit.

E-cycling – recycling electronic waste.

Effective rainfall – the volume of rainfall passing into the soil; that part of rainfall available for plant use after runoff, leaching, evaporation and foliage interception.

Effluent – a discharge or emission of liquid, gas or other waste product.

El Nino – a warm water current which periodically flows southwards along the coast of Ecuador and Peru in South America, replacing the usually cold northwards flowing current; occurs once every five to seven years, usually during the Christmas season (the name refers to the Christ child); the opposite phase of an El Nino is called a La Nina.

Emission standard – a level of emissions that, under law, may not be exceeded.

Emissions intensity – emissions expressed as quantity per monetary unit.

Emissions – substances such as gases or particles discharged into the atmosphere as a result of natural processes of human activities, including those from chimneys, elevated point sources, and tailpipes of motor vehicles.

Endangered species – a species which is at risk of becoming extinct because it is either few is number, or threatened by changing environment or predation parameters.

Energetics – the study of how energy flows within an ecosystem; the routes it takes, rates of flow, where it is stored and how it is used.

Energy – a property of all systems which can be turned into heat and measured in heat units.

Available energy – energy with the potential to do work (exergy);

Delivered energy – energy delivered to and used by a household, usually gas and electricity;

Direct energy – the energy being currently used, used mostly at home (delivered energy) and for fuels used mainly for transport;

Embodied energy – the energy expended over the entire life cycle of a good or service OR the energy involved in the extraction of basic materials, processing/manufacture, transport and disposal of a product OR the energy required to provide a good or service;

Enhanced greenhouse effect - the increase in the natural greenhouse effect resulting from increases in atmospheric concentrations of greenhouse gases due to emissions from human activities.

Environment – the external conditions, resources, stimuli etc. with which an organism interacts.

Environmental indicator – physical, chemical, biological or socio-economic measure that can be used to assess natural resources and environmental quality.

Environmental movement (environmentalism) – both the conservation and green movements; a diverse scientific, social, and political movement. In general terms, environmentalists advocate the sustainable management of resources and stewardship of the natural environment through changes in public policy and individual behaviour. In its recognition of humanity as a participant in ecosystems, the movement is centered around ecology, health, and human rights.

Environmental science – the study of interactions among physical, chemical, and biological components of the environment.

Epidemiology – the study of factors affecting the health and illness of populations, and serves as the foundation and logic of interventions made in the interest of public health and preventive medicine.

Erosion – displacement of solids (sediment, soil, rock and other particles) usually by the agents of currents such as, wind, water, or ice by downward or down-slope movement in response to gravity or by living organisms.

Escherichia coli (E. coli) – a bacterium used as an indicator of faecal contamination and potential disease organisms in water.

Estuary – a semi- enclosed coastal body of water with one or more rivers or streams flowing into it, and with a free connection to the open sea.

Eutrophication – the enrichment of waterbodies with nutrients, primarily nitrogen and phosphorus, which stimulates the growth of aquatic organisms.

Eutrophication – an increase in chemical nutrients, typically compounds containing nitrogen or phosphorus, in an ecosystem.

Evapotranspiration – the water evaporating from the soil and transpired by plants.

E-waste – electronic waste, especially mobile phones, televisions and personal computers.

Externality – (environmental economics) by-products of activities that affect the well-being of people or damage the environment, where those impacts are not reflected in market prices. The costs (or benefits) associated with externalities do not enter standard cost accounting schemes. The environment is often cited as a negatively affected externality of the economy.

Extinction event – a sharp decrease in the number of species in a relatively short period of time.

Extinction – the cessation of existence of a species reducing biodiversity.

Fertigate – apply fertilizer through an irrigation system.

Fertilizers – compounds given to plants to promote growth; they are usually applied either through the soil, for uptake by plant roots, or by foliar feeding, for uptake through leaves.

Food chain (food webs, food networks and/or trophic networks) – the feeding relationships between species within an ecosystem.

Food security – food produced in sufficient quantity to meet the full requirements of all people i.e. total global food supply equals the total global demand. For households it is the ability to purchase or produce the food they need for a healthy and active life. For national food security, the focus is on sufficient food for all people in nation and it entails a combination of national production, imports and exports. Food security always has components of production, access and utilization.

Footprint – (Ecological Footprint) in a very general environmental sense a "footprint" is a measure of environmental impact. However, this is usually expressed as an area of productive land needed to counteract the impact.

Forage – the plant material (mainly plant leaves) eaten by grazing animals.

Forest – land with a canopy cover greater than 30%

Fossil fuel – any hydrocarbon deposit that can burned for heat or power, such as coal, oil and natural gas (produces carbon dioxide when burnt); fuels formed from once-living organisms that have become fossilized over geological time.

Fossil water – groundwater that has remained in an aquifer for thousands or millions of years; when geologic changes seal the aquifer preventing further replenishment, the water becomes trapped inside and is then referred to as fossil water. Fossil water is a limited resource and can only be used once.

Freon – DuPont's trade name for its odourless, colorless, nonflammable, and noncorrosive chlorofluorocarbon and hydrochlorofluorocarbon refrigerants, which are used in air conditioning and refrigeration systems.

Fresh water – water containing no significant amount of salt; potable water suitable for all normal uses.

Front – (weather) the boundary between warm (high pressure) and cold (low pressure) air masses

Fugitive emissions – These are gases emitted from fuel production itself including, processing, transmission, storage and distribution processes, and including emissions from oil and natural gas exploration, venting, and flaring, as well as the mining of coal.

Gene pool – the complete set of unique alleles in a species or population,

Generalist species – those able to thrive in a wide variety of environmental conditions and can make use of a variety of different resources.

Gene – a locatable region of genomic sequence, corresponding to a unit of inheritance, which is associated with regulatory regions, transcribed regions and/or other functional sequence regions.

Genetic diversity – one of the three levels of biodiversity that refers to the total number of genetic characteristics.

Greenhouse effect – the process in which the emission of infrared radiation by the atmosphere warms a planet's surface.

Greenhouse gas – components of the atmosphere that contribute to the greenhouse effect.

Green manure – a type of cover crop grown primarily to add nutrients and organic matter to the soil.

Groundwater – water located beneath the ground surface in soil pore spaces and in the fractures of lithologic formation.

Geosphere – the solid part of planet Earth, the main divisions being the crust, mantle, and liquid core. The lithosphere is the part of the geosphere that consists of the crust and upper mantle.

Geothermal – energy – energy derived from the natural heat of the earth contained in hot rocks, hot water, hot brine or steam.

Global dimming – a reduction in the amount of direct solar radiation reaching the surface of the earth due to light diffusion as a result of air pollution and increasing levels of cloud. A phenomenon of the last 30-50 years.

Global warming potential (GWP) a system of multipliers devised to enable warming effects of different gases to be compared.

Global warming – the observable increase in global temperatures considered mainly caused by the human induced enhanced greenhouse effect trapping the Sun's heat in the Earth's atmosphere.

Globalisation – the expansion of interactions to a global or worldwide scale; the increasing interdependence, integration and interaction among people and organisations from around the world. A mix of economic, social, technological, cultural, and political interrelationships.

Green architecture – building design that moves towards self-sufficiency sustainability by adopting circular metabolism.

Green design – environmentally sustainable design.

Green power – Electricity generated from clean, renewable energy sources (such as solar, wind, biomass and hydro power) and supplied through the grid.

Green products – and services – products or services that have a lesser or reduced effect on human health and the environment when compared with competing products or services that serve the same purpose. Green products or services may include, but are not limited to, those which contain recycled content, reduce waste, conserve energy or water, use less packaging, and reduce the amount of toxics disposed or consumed.

Green purchasing – purchasing goods and services that minimize impacts on the environment and that are socially just.

Green waste (green organic material or green organics, sometimes referred to as "green wealth") – plant material discarded as non-putrescible waste – includes tree and shrub cutting and pruning's, grass clipping, leaves, natural (untreated) timber waste and weeds (noxious or otherwise).

Green – (sustainability) like 'eco' – a word frequently used to indicate consideration for the environment e.g. green plumbers, green purchasing etc., sometimes used as a noun e.g the Greens.

Greenhouse effect – the insulating effect of atmospheric greenhouse gases (e.g., water vapor, carbon dioxide, methane, etc.) that keeps the Earth's temperature about 16 ^{0}C warmer than it would be otherwise (enhanced greenhouse effect).

Greenhouse gases – any gas that contributes to the greenhouse effect; gaseous constituents of the atmosphere, both natural and from human activity, that absorb and re-limit infrared radiation. Water vapor (H_2O) is the most abundant greenhouse gas. Greenhouse gases are a natural part of the atmosphere and include carbon dioxide (CO_2), methane (CH_4, persisting 9-15 yrs with a greenhouse warming potential (GWP) 21 times that of CO_2), nitrous oxide (N_2O) persists 120 years and has a GWP of 310), ozone (O_3), hydrofluorocarbons, perfluorocarbons and sulfur hexafluoride. (SF_6)

Greenlash – dramatic changes in the structure and dynamic behaviour of ecosystems.

Greenwashing – companies that portray themselves as environmentally friendly when their business practices do not back this up. Generally applies to excessive use of green marketing and packaging when this does not take account of the total ecological footprint.

Greenwater – water replenishing soil moisture, evaporating from soil, plant and other surfaces, and transpired by plants. In nature the global average amount of rainfall becoming green water is about 60%. Of the green water about 55% falls on forests, 25% on grasslands and about 20% on crops. We can increase green water productivity by rainwater harvesting, increases infiltration and runoff collection. Green water cannot be piped or drunk (cannot be sold) and is therefore generally ignored by water management authorities but it is crucial to plants in both nature and agriculture and needs careful management as an important part of the global water cycle.

Greywater – household waste water that has not come into contact with toilet waste; includes water from baths, showers, bathrooms washing machines, laundry and kitchen sinks.

Groundwater – water found below the surface – usually in porous rocks, or soil, or in underground aquifers.

Habitat – an ecological or environmental area that is inhabited by a particular species.

Hard waste – household garbage which is not normally accepted into rubbish bins by local councils, e.g. old stoves mattresses.

Heat – energy derived from the motion of molecules; a form of energy into which all other forms of energy may be degraded.

Herbicide – a chemical the kills or inhibits growth of a plant.

Herbivory – predation in which an organism known as an herbivore, consumes principally autotrophs such as plants, algae and photosynthesizing bacteria.

Heterotroph (chemoorganotrophy) – an organism that requires organic substrates to obtain its carbon for growth and development.

Homoclime – a region with the same climate as the one under investigation.

Homeostatis – the property of either an open system or a closed system, especially a living organism, that regulates its internal environment so as to maintain a stable, constant condition.

Horton overland flow – the tendency of water to flow horizontally across land surfaces when rainfall has exceeded infiltration capacity and depression storage capacity.

Household metabolism – the passage of food, energy, water, goods, and waste through the household unit in a similar way to the metabolic activity of an organism (Industrial metabolism).

Humus – organic material in soil lending it a bark brown or black colouration.

Humus – semi- persistent organic matter in the soil that can no longer be recognised as tissue.

Hydrocarbons – chemicals made up of carbon and hydrogen that are found in raw materials such as petroleum, coal and natural gas, and derived products such as plastics.

Hydroelectric power – the electrical power generated using the power of falling water.

Hydrological cycle – (water cycle) – natural cycle of water from evaporation, transpiration in the atmosphere, condensation (rain and snow), and flows back to the ocean (e.g. rivers).

Hydrosphere – all the Earth's water; this would include waster found in the sea, streams, lakes and other waterbodies, the soil, groundwater, and in the air.

Incineration – combustion (by chemical oxidation) of waste material to treat or dispose of that waste material.

Indicator species – any biological species that defines a trait or characteristic of the environment.

Industrial agriculture – a form of modern farming that involves industrialized production of livestock, poultry, fish, and crops.

Infiltration – movement of waster below topsoil to the plant roots and below.

Infiltration – the process by which water on the ground surface enter the soil.

Indicators – quantitative markers for monitoring progress towards desired goals.

Insecticide – a pesticide used to control insects in all development forms.

Integrated Pest Management - a pest control strategy that uses an array of complementary methods: natural predators and parasites, pest-resistant varieties, cultural practices, biological controls, various physical techniques, and the strategic use of pesticides.

Intercropping - the agricultural practice of cultivating two or more crops in the same space at the same time.

In-stream – the use of freshwater where it occurs, usually within a river or stream: it includes hydroelectricity, recreation, tourism, scientific and cultural uses, ecosystem maintenance, and dilution of waste,

Integrated pest management – pest management that attempts to minimize chemical use by using several pest control options in combination. The goal is not a climate all pest but to reduce pest populations to acceptable levels; an ecologically based pest control strategy that relies heavily on natural mortality factors and seeks out control tactics that disrupt these factors as little as possible.

Integrated product life –cycle management – management of all phases of goods and services to be environmentally friendly and sustainable.

Intergovernmental Panel on Climate Change (IPCC) – the IPCC was established in 1988 by the World Meteorological Organization and the UN Environment Programme to provide the scientific and technical foundation for the United Nations Framework Convention on Climate Change (UNFCCC), primarily through the publication of periodic assessment reports.

Internal water footprint – the water embodied in goods produced within a country (although these may be subsequently exported) ef. External waster footprint.

Irrigation scheduling – watering plants according to their needs.

Irrigation – watering of plants, no matter what system is used.

ISO 14001 – The international standard for companies seeking to certify their environmental management system International Organisation for Stardardisation (ISO) 14001 standard was first published in 1996 specifying the requirements for an environmental management system in organization (companies and institutions) with the goal of minimizing harmful effects on the environment and the goal of continual improvement of environmental performance.

Kerbside collection – collection of household recyclable materials (separated or comingled) that are left at the kerbside for collection by local council services.

Keystone species – a species that has a disproportionate effect on its environment relative to its abundance, affecting many other organisms in an ecosystem and help in determine the types and numbers of various others species in a community,

Kyoto Protocol – an international agreement adopted in December 1997 in Kyoto, Japan. The Protocol sets binding emission targets for developed countries that would reduce their emissions on average 5.2 percent below 1990 levels.

Land use, Land-use change and forestry – land uses and land-use changes can act either as sinks or as emission sources. It is estimated that approximately one-fifth of global emissions result from its activities. The Kyoto Protocol allows parties to receive emissions credit for certain activities that reduce net emissions.

Landfill – solid waste disposal in which refuse is buried between layers of soil, a method often used to reclaim low-lying ground; the word is sometimes used as a noun to refer to the waste itself.

Landfill gas – the gas emissions from biodegrading waste in landfill, including CO_2, CH_4, and small amounts of nitrogen, oxygen with traces of toluene, benzene and vinyl chloride.

Landfill prohibition – The banning of a certain material or product type from disposal to landfills. Occurs occasionally, for example, where a preferable waste management option is available.

Landfill – a site for the disposal of waste materials by burial and is the oldest form of waste treatment.

Leaching – the movement of chemical in the upper layers of soil into lower layers or into groundwater by being dissolved in water.

Lithosphere – the solid outermost shell of a rocky planet.

Leachate (waste) – the mixture of water dissolved solids (possibly toxic) that accumulates as water passes through waste and collects at the bottom of a landfill site.

Life cycle (of a product) – All stages of a product's development, from raw materials, manufacturing through to consumption and ultimate disposal.

Life Cycle Analysis (LCA) – an objective process to evaluate the environmental impacts associated with a product, process, or activity. A means of identifying resource use and waste released to the environment, and to assess management options.

Life support systems – The biophysical processes "that sustain the productivity, adaptability and capacity for renewal of lands, waters, and / or the biosphere as a whole.

Lilacwater – recycled water that is unsuitable for drinking.

Linear metabolism – direct conversion of resources into wastes that are often sent directly to landfill

Loam – a soil composed of sand, silt, and clay in relatively even concentration (about 40-40-20% concentration respectively), *locally existing capacity – the total ecological production that is found within a country's territories. It is usually expressed in hectares based on world average productivity.

Magma – molten rock that forms beneath the surface of the Earth (or any other terrestrial planet) that often collects in a magma chamber and is ejected by volcano's

Manure – organic matter used as fertilizer in agriculture.

Mauna Loa record – the record of measurement of atmospheric CO_2 concentrations taken at Mauna Loa Observatory, Mauna Loa, Hawaii, since March 1958. This record shows the continuing increase in average annual atmospheric CO_2 concentrations.

Maximum soil water deficit – amount of water stored in the soil that is readily available to plants.

Microorganism – an organism visible only through a microscope.

Middle East – 15 countries, Bahrain, Islamic Rep. Iran, Iraq, Israel, Jordan, Kuwait, Lebanon, Oman, Qatar, Saudi Arabia, Syria, United Arab Emirates, Yemen.

Mobile garbage bin – A wheeled kerbside container for the collection of garbage or other materials.

Monoculture – the practice of producing or growing one single crop over a wide area.

Mortality rate – generally understood as the total number of deaths per 1000 people of a given age group.

Mulch – any composted or non-composted organic material, excluding plastic, that is suitable for placing on **soil surfaces to restrict moisture loss from the soil and to provide a source of nutrients to the soil.**

Municipal waste – solid waste generated from domestic premises (garbage and hard wise) and council activities such as street sweeping, litter and street tree lopping. Also includes waste dropped at transfer stations and construction waste from owner/occupier renovations.

Natural – the existing air, water land, and energy resources from which all resources derive. Main functions include resource production (such as fish, timber or cereals), waste assimilation (such as CO2 absorption, sewage decomposition), and life support services (UV protection, biodiversity, waster cleansing, climate stability). The environmental services that must be maintained so that human development can be sustainable.

Natural resources – naturally occurring substances that are considered valuable in their relatively unmodified (natural form.

Natural selection – the process by which favorable heritable traits become more common in successive generations of a population of reproducing organisms, and unfavorable heritable traits become less common.

Neighbourhood environment improvement plan – plans developed by a local community including residents, special interest groups, local government, local industry and government agencies.

Nematocide – a chemical that kills nematodes.

Nickel cadmium – batteries – batteries typically used in appliances such as power tools and mobile phones. Cadmium is a heavy metal that poses risk to human and ecosystem health.

Noise pollution (environmental noise) – displeasing human or machine created sound that disrupts the activity or happiness of human or animal life.

Nonpoint source pollution – water pollution affecting a water body from diffuse sources, rather than a point source which discharges to a water body at a single location.

Appendix 2.15: Terminologies Related to Ecology

Absorption: The taking in of water and dissolved minerals and nutrients across cell membranes.

Aerobic: Pertaining to the presence of free oxygen. Aerobic organisms require oxygen for their life processes.

Anaerobic: Pertaining to the absence of free oxygen. Anaerobic organisms do not require oxygen for their life processes, infact oxygen is toxic to many of them. Most anaerobic organisms are **bacteria** or archaeans.

Autotroph: Any organism that is able to manufacture its own food. Most **plants** are autotrophs, as are many **protists** and **bacteria**. Autotrophs may be **Photoautotrophic**, using light energy to manufacture food, or **chemoautotrophic**, using chemical energy.

Benthic: Organisms that live on the bottom of the ocean are called benthic organisms. They are not free-floating like pelagic organisms.

Biological/biotic factors: Living factors such as decomposers, scavengers and predators.

Canopy: Layer of vegetation elevated above the ground, usually of tree branches and epiphytes. In tropical forests, the canopy may be more than 30 metres above the ground.

Carnivore: Literally, an organism that eats meat. Most carnivores are animals, but a few fungi, plants, and protists are as well.

Decomposer: An organism that breaks down the tissue and/or structures of dead organisms.

Decomposition: The breakdown of dead organic material by detrivores or saprophytes.

Detritus: Accumulated organic debris from dead organisms, often an important source of nutrients in a food web.

Detrivore: Any organism which obtains most of its nutrients from the detritus in an ecosystem.

Disease: Organisms suffer from disease when their normal function is impaired by some genetic disorder, or more often from the activity of a parasite or other organism living within them. Many diseases are caused by viruses, Bacteria, or fungi.

Ecosystem: All the organisms in a particular region and the environment in which they live. The elements of an ecosystem interact with each other in some way, and so depend on each other either directly or indirectly.

Environment: The place in which an organism lives, and the circumstances under which it lives. Environment includes measures like moisture and temperature, as much as it refers to the actual physical place where an organism is found.

Food chain/food web: All the interactions of predator and prey, included along with the exchange of nutrients into and out of the soil. These interactions connect the various members of an ecosystem, and describe how energy passes from one organism to another.

Generalist: Organism which can survive under a wide variety of conditions, and does not specialize to live under any particular set of circumstances.

Grassland: Region in which the climate is dry for long periods fo the summer, and freezes in the winter. Grasslands are characterized by grasses and other erect herbs, usually without trees or shrubs. Grassland occur in the dry temperate interiors of continents.

Groundwater: Water found underground as a result of rainfall, ice and snow melt, submerged rivers, lakes, and springs. This water often carries minerals. These minerals can accumulate in the remains of buried organisms and eventually cause fossilization.

Habitat: The place and conditions in which an organism lives.

Halophile: Organism which lives in areas of high salt concentration. These organisms must have special adaptations to permit them to survive under these conditions.

Herbivore: Literally, an organism that eats plants or other autotrophic organisms. The term is used primarily to describe animals.

Host: Organism which serves as the habitat for a parasite, or possibly for a symbiont. A host may provide nutrition to the parasite or symbiont, or simply a place in which to live.

Ingestion: The intake of water or food particles by "swallowing" them, taking them into the body cavity or into a vacuole.

Inorganic: Not containing carbon. Not from living things. Ex., minerals, water, oxygen, etc.

Intertidal: The coastal zone measuring from the lowest to the highest tide mark. The intertidal zone is subject to alternating periods of flooding and drying.

Limnology: the study of river system ecology and life.

Litter: Leaf litter, or forest litter, is the detritus of fallen leaves and bark which accumulate in forests.

Macroscopic: Objects or organisms that are large enough to be seen with the naked eye.

Marine: Refers to the ocean.

Microscopic: Objects or organisms that are too small to be seen with the naked eye.

Monsoonal: Describes a climate pattern with a wind system that changes direction with the seasons; this pattern is dominant over the Arabian Sea and Southeast Asia.

Morphology: The form and structure of anything, usually applied to the shapes, parts, and arrangement of features in living and fossil organisms.

Niche: The portion of the environment which a species occupies, defined in terms of the conditions under which an organism can survive, and may be affected by the presence of other competing organisms.

Nitrogen fixation: the conversion of gaseous nitrogen into a form usable by plants. Usually by bacteria.

Nocturnal: Active only at night.

Nutrient: Any element or simple compound necessary for the health and survival of an organism. This includes air and water, as well as food.

Nutrient cycling: All the processes by which nutrients are transferred from one organism to another. For instance, the carbon cycle includes uptake of carbon dioxide by plants, ingestion by animals, and respiration and decay of the animal. Animals who do not restrict their diet to just plants or other animals.

Organic: Pertaining to compounds containing carbon. Also refers to living things or the materials made by living things.

Parasite: An organism that lives on or within a host (another organism); it obtains nutrients from the host without benefiting or killing (although it may damage) the host; **parasitism**- a type of symbiotic relationship in which one organism benefits and the other does not.

Pathogenic: Organism which causes a disease within another organism.

Pelagic: Pelagic organisms swim through the ocean, and may rise to the surface, or sink to the bottom. They are not confined to live on the bottom as benthic organisms do.

Periphyton: Dense strands of algal growth that cover the water surface between the emergent aquatic plants. *Spirogyra* is commonly responsible for this growth.

Photic zone: Region of the ocean through which light penetrates; and the place where photosynthetic marine organisms live.

Phytoplankton: Tiny, free-floating, photosynthetic organisms in aquatic systems.

Plankton: Very small, free-floating organisms of the ocean or other aquatic systems, including phytoplankton, which produce their own nutrients through photosynthesis, or zooplankton, which get their nutrients from organisms.

Pollinator: Animal which carries pollen from one seed plant to another, unwittingly aiding the plant in its reproduction. Common pollinators include insects, especially bees, butterflies, and moths, birds, and bats.

Predator: Organism which hunts and eats other organisms. This includes both carnivores, which eat animals, and herbivores, which eat plants.

Prey: Organism hunted and eaten by a predator.

Producer: Any organism which brings energy into an ecosystem from inorganic sources. Most plants and many protists are producers.

Rain shadow: the dry region on the leeward side of a mountain range, where rainfall is noticeably less than on the windward side.

Riparian: Having to do with the edges of streams or rivers.

Salinity: A measure of the salt concentration of water. Higher salinity means more dissolved salts.

Saprophyte: Organism which feeds on dead and decaying organisms, allowing the nutrients to be recycled into the ecosystem. Fungi and bacteria are two groups with many important saprophytes.

Scavenger: An organism that feeds upon dead and dying organisms.

Seaweed: Any large photosynthetic protist, including rhodophytes and kelps. Seaweeds are not true plants, but like plants they can make their own food.

Substrate: "Supporting surface" on which an organism grown. The substrate may simply provide structural support, or may provide water and nutrients. A substrate may be inorganic, such as rock or soil, or it may be organic, such as wood.

Symbiosis: A relationship between two organisms that live in intimate contact with each other; includes mutualism (both organisms benefit, they rely on each other for survival), parasitism (one organism benefits at its host's expense) and commensalism (one partner benefits and the other is neither benefitted nor harmed);

Terrestrial: Living on land, as opposed to marine or aquatic.

Tropical: Region in which the climate undergoes little seasonal change in either temperature or rainfall. Tropical regions of the earth lie primarily between 30 degrees north and south of the equator.

Upwelling: the raising of benthic nutrients to the surface waters. This occurs in regions where the flow of water brings currents of differing temperatures together, and increases productivity of the ecosystem.

Zooplankton: Tiny, free-floating organisms in aquatic systems. Unlike phytoplankton, zooplankton cannot produce their own food.

Appendix 2.16: Terminologies related to Environmental Health

Absorption: The process of taking in, as when a sponge takes up water. Chemicals can be absorbed into the bloodstream after breathing or swallowing. Chemicals can also be absorbed through the skin to the bloodstream and the transported to other organs. Not all of the chemical breathed, swallowed, or touched is always absorbed.

Acute: Occurring over a short time, usually a few minutes or hours. An acute exposure can result in short term or long term health effects. An acute effect happens within a short time after exposure.

Ambient: Surrounding. Ambient air usually means outdoor air (as opposed to indoor air).

Aquifer: An underground source of water. This water may be contained in a layer of rock, sand or gravel.

Background level: A typical level of a chemical in the environment. Background often refers to naturally occurring or uncontaminated levels. Background level in one region of the state may be different than those in other areas.

Biological monitoring: Analyzing chemicals, hormone levels or other substance in biological materials (blood, urine, breath, etc.) as a measure of chemical exposure, health status, etc. in humans or animals. A blood test for lead is an example of biological monitoring.

Body burden: The total amount of a chemical in the body. Some chemicals build up in the body because they are stored in body organs like fat or bone or are eliminated very slowly.

Central Nervous System (CNS): The part of the nervous system that includes the brain and the spinal cord.

Chronic: Occurring over a long period of time, several weeks, months or years.

Cohort study: A study in which a group of people with a past exposure to chemicals or other risk factors are followed over time and their disease experience compared to that of a group of people without the exposure.

Composite sample: A sample which is made by combining sample from two or more locations. The sample can be of water, soil or another medium.

Concentration: The amount of one substance dissolved or contained in a given amount of another substance or medium. For example, sea water has a higher concentration of salt than fresh water does.

Dermal: Referring to the skin. Dermal absorption of pollutants means absorption through the skin.

Detection limit: The smallest amount of substance that a laboratory test can reliably measure in a sample of air, water, soil or other medium.

Dose: The amount of substance to which a person is exposed.

Epidemiology: The study of the occurrence and cause of health effects in human populations. An epidemiological study often compares two groups of people who are alike except for one factor such as exposure to a chemical or the presence of a health effect. The investigators try to determine if the factor is associated with the health effect.

Exposure: Contact with a chemical by swallowing, by breathing or by direct contact (such as through the skin or eyes). Exposure may be either short term (acute) or long term (chronic).

Exposure assessment: A process that estimates the amount of a chemical that enters or comes into contact with people or animals. An exposure assessment also describes how often and for how long an exposure occurred, and the nature and size of a population exposed to a chemical.

Feasibility Study: A study that compares different ways to clean up a contaminated site. The feasibility study recommends one or more actions to remediate the site.

Gradient: the change in a property over a certain distance. For example, lead can accumulate in surface soil near a road due to automobile exhaust. As you move away from the road, the amount of lead in the surface soil decreases. This change in the lead concentration with distance from the road is called a gradient.

Health assessment for contaminated sites: Determination of actual or possible health effects due to environmental contamination or exposure. It includes a health-based interpretation of all the information known about the situation. The information may come from site investigations (environmental sampling and studies), exposure assessments, risk assessments, biological effects etc.

Health registry: A record of people exposed to a specific substance (such as a heavy metal), or having a specific health condition (such as cancer or a communicable disease).

Ingestion: Swallowing (such as eating or drinking). Chemicals in or on food, drink, utensils, cigarettes, hands, etc. can be ingested. After ingestion, chemicals may be absorbed into the blood and distributed throughout the body.

Inhalation: Breathing. People can take in chemicals by breathing contaminated air.

Interim Remedial Measure: An action taken at a contaminated site to reduce the chances of human or environmental exposure to site contaminants. Interim remedial measures are planned and carried out before comprehensive remedial studies. They can prevent additional damage during the study phase, but don't interfere in any way with the need to develop a complete remedial program. An example of an interim remedial measure is removing drums of chemicals to a storage facility from a site that has drums sitting in an empty field.

Latency period: the period of time between exposure to something that causes a disease and the onset of the health effect. Cancer caused by chemical exposure may have a latency period of 5 to 40 years.

Leaching: As water moves through soils or landfills, chemicals in the soil may dissolve in the water thereby contaminating the groundwater. This is called leaching.

Maximum Contaminant Level (MCL): the highest (maximum) level of a contaminant allowed to go uncorrected by a public water system under federal or state regulations. Depending on the contaminant, allowable levels might be calculated as an average over time, or might be based on individual test results. Corrective steps are implemented if the MCL is exceeded.

Media: Elements of surrounding environment that can be sampled for contamination; usually soil, water, or air. Plants, as well as humans (when sampling blood, urine, etc.) and animals.

Metabolism: All the chemical reactions that enable the body to work. For example, food is metabolized (chemically changed) to supply the body with energy. Chemicals can be metabolized by the body and made either more or less harmful.

Morbidity: Illness or disease. A morbidity rate for a certain illness is the number of people with that illness divided by the number of people in the population from which the illnesses were counted.

Odor threshold: The lowest concentration of a chemical that can be smelled. Different chemicals have different odor thresholds. Also, some people can smell a chemical at lower concentration than others can.

Organic: Generally considered as originating from plants or animals, and made primarily of carbon and hydrogen. Scientists use the term organic to mean those chemical compounds which are based on carbon.

Permeability: the property of permitting liquids or gases to pass through. A highly permeable soil, such as sand, allows a liquid to pass through quickly. Clay has a low permeability.

Persistence: the quality of remaining for a long period of time (such as in the environment or the body). Persistent chemicals (such as DDT and PCBs) are not easily broken down.

Plume: An area of chemicals moving away from its source in a long band or column, A plume, for example, can be a column of smoke from a chimney or chemicals moving with groundwater.

Quality assurance and quality control: A system of procedures, checks and audits to judge and control the quality of measurements and reduce the uncertainty of data. Some quality control procedures include having more than one person review the findings and analyzing a sample at different times or laboratories to see if the findings are similar.

Remedial Investigation (RI): An in-depth study (including sampling of air, soil, water and waste) of a contaminated site needing remediation to determine the nature and extent of contamination.

Remediation: Correction or improvement of a problem, such as work that is done to clean up or stop the release of chemicals from a contaminated site. After investigation of a site, remedial work may include removing soil and/or drums, capping the site or collecting and treating the contaminated fluids.

Risk: Risk is the possibility of injury, disease or death. For example, for a person who has measles, the risk of death is one in one million.

Risk assessment: A process which estimates the likelihood that exposed people may have health effects. The four steps of a risk assessment are: hazard identification (Can this substance damage health?); dose-response assessment (What dose causes what effect?); exposure assessment (How and how much do people contact it?); and risk characterization (combining the other three steps to characterize risk and describe the limitations and uncertainties).

Risk management: the process of deciding how and to what extend to reduce or eliminate risk factors by considering the risk assessment, engineering factors (Can procedures or equipment do the job, for how long and how well?), social, economic and political concerns.

Route of exposure: The way in which a person may contact a chemical substance. For example, drinking (ingestion) and bathing (skin contact) are two different routes of exposure to contaminates that may be found in water.

Safe: Strictly, free from harm or risk. Exposure to a chemical usually has some risk associated with it, although the risk may be very small. However, many people use the word safe to mean something that has a very low risk or one that is acceptable to them.

Solubility: The largest amount of a substance that can be dissolved in a given amount of a liquid, usually water. For a highly water-soluble compound, such as table salt, a lot can dissolve in water. Motor oil is only slightly soluble in water.

Target organ: An organ (such as the liver or kidney) that is specifically affected by a toxic chemical.

Volatile: evaporating readily at normal temperatures and pressures. The air concentration of a highly volatile chemical can increase quickly in a closed room.

Volatile organic compound (VOC): An organic chemical that evaporates readily. Petroleum products such as kerosene, gasoline and mineral spirits contain VOCs. Chlorinated solvents such as those used by dry cleaners or contained in paint strippers are also VOCs.

Appendix 2.17: Definition and Meaning of Medical terms related to Effects of Environmental Pollutants

Medical Terms	Meaning/Definition
A	
Acromegaly	Enlargement of the extremities
Adeniod	Resembling a gland
Afebrile	Without fever
Anesthesia	Without feeling or sensation
Arteriosclerosis	Hardening of the arteries

Medical Terms	Meaning/Definition
B	
Blepharoptosis	Downward displacement of the eyelids
Bradycardia	Slow heartbeat
C	
Cardiovascular	Pertaining to the heart and vessels
Cerebroma	Tumor of the brain
Chondritis	Inflammation of the cartilage
Cyanosis	Condition of blueness
Cystitis	Inflammation of the bladder
D	
Dentalgia	Toothache/pain
Dermatophytosis	Condition of fungus of the skin
Dyspnea	Difficult breathing
E	
Enecphalitis	Inflammation of the brain
Endocarditis	Inflammation within the heart
Endometritis	Inflammation within the uterus
Erythrocyte	Red cell
G	
Gastroenteritis	Inflammation of the stomach and intestine
Glossitis	Inflammation of the tongue
Glucosuria	Sugar in the urine
H	
Hematemesis	Vomiting of blood
Hematoma	Tumor filled with blood
Hematuria	Blood in the urine
Hepatic	Pertaining to the liver
Hepatitis	Inflammation of the liver
Hepatosplenomegaly	Enlargement of the liver and spleen
Hyperemesis	Excessive vomiting
Hypertension	High blood pressure
Hypertrophy	Excessive growth
Hypotension	Low blood pressure
I	
Intercostal	Between the ribs
Intra-Abdominal	Inside the abdomen
K	
Keratitis	Inflammation of the cornea of eye
L	
Lipoma	Fatty tumor
M	

Medical Terms	Meaning/Definition
Monocyte	Single or one cell
Myalgia	Pain in the muscle
Myospasm	Involuntary contraction of a muscle
N	
Necrosis	Condition of dead tissue
Nephrolithiasis	Enlargement of the kidney
Nocturia	Night urine
O	
Oliguria	Scant urine
Osteoarthritis	Inflammation of the bones and joints
Otoscopic	Pertaining to examination of the ear
P	
Pericarditis	Inflammation around the heart
Plyneuritis	Inflammation of many nerves
Polyphagia	Excessive eating
Pyogenic	Producing pus
R	
Rhinitis	Inflammation of the nose
S	
Salpingitis	Inflammation of the tubes
T	
Tachycardia	Fast heartbeat
Thrombophlebitis	Inflammation of a vein with a clot

Appendix 2.18: Terminologies Related to Environmental Safety

Acute: Acute means sudden or brief. Acute can be used to describe either an exposure or a health effect. An acute exposure is a short-term exposure. Short-term means lasting for minutes, hours or day. An acute health effect is an effect that develops either immediately or a short time after an exposure. Acute health effects may appear minutes, hours or even days after an exposure.

Aerosol: An aerosol is a collection of very small particles suspended in air. The particles can be liquid (mist) or solid (dust or fume). The term aerosol is also commonly used for a pressurized container (aerosol can) which is designed to release a fine spray of a material such as paint.

Auto-ignition Temperature: The auto-ignition temperature is the lowest temperature at which a material begins to burn in air in the absence of a spark or flame. Many chemicals will decompose (break down) when heated. The auto-ignition temperature is the temperature at which the chemicals formed by decomposition begin to burn. Auto-ignition temperatures for a specific material can vary by one hundred degrees Celsius or more, depending on the test method used. To avoid the risk of fire or explosion, materials must be stored and handled at temperatures well below the auto-ignition temperature.

Bioaccumulation: Bioaccumulation describes the process by which a chemical accumulates in a living organism – either from the surrounding environment (air, water or soil) or from other sources, for example by consuming food containing the substance.

Bioconcentration: Bioconcentration describes the process by which a chemical can build up in a living organism to levels higher than those found in the surrounding environment. For example, a fish will have a higher concentration of a chemical in its body than is present in the water it swims in. Bioconcentration refers to the uptake of the chemical from water only – not all sources of exposure to the chemical.

Biodegradation: Biomagnification is the accumulation of higher concentrations of a chemical in a living organism at each successive level in the food chain. The chemical is transferred to organisms further up the food chain as they consume organisms lower in the food chain.

Biotransformation: This is the process by which living organisms transform chemicals in the environment.

Biohazardous Infectious Material: A biohazardous infectious material is a material that contains organisms which can cause disease in humans or animals. Included in this category are bacteria, viruses, fungi and parasites. Because these organisms can live in body tissues or fluids (blood, or urine), the tissues and fluids are also treated as toxic. For example, a person exposed to a blood sample from someone with hepatitis B may contract the disease.

BOD: BOD stands for biological oxygen demand or biochemical oxygen demand. It is used as a method of determining how much contamination has entered a water supply. It is used primarily for waters which receive pollution from sewage and industrial wastes.

BOD Test: A test that measures the amount of oxygen that is consumed as microbial life decomposes organic matter in water.

Boiling Point: The boiling point is the temperature at which the material changes from a liquid to a gas. Below the boiling point, the liquid can evaporate to form a vapour. As the material approaches the boiling point, the changes from liquid to vapour is rapid and vapour concentrations in the air can be extremely high. Airbone gases and vapours may pose fire, explosion and health hazards.

Sometimes, the boiling point of a mixture is given as a range of temperatures. This is because the different ingredients in a mixture can boil at different temperatures.

Carcinogen, Carcinogenic, Carcinogenicity: A carcinogen is a substance which can cause cancer. Carcinogenic means able to cause cancer. Carcinogenicity is the ability of a substance to cause cancer.

Chemical Family: The chemical family describes the general nature of the chemical. Chemicals belonging to the same family often share certain physical and chemical properties and toxic effects. However, there may also be important differences. For example, toluene and benzene both belong to the aromatic hydrocarbon family. However, benzene is a carcinogen, and toluene is not.

Chemical Formula: The chemical formula, sometimes called the molecular formula, tells which elements (carbon, hydrogen, oxygen, and so on) make up a chemical. It also gives the number of atoms of each element in one unit or molecule of the chemical. The chemical formula can be used to confirm the identity of ingredients or to indicate the presence of a potentially hazardous element.

Chemical Reactivity: Chemical reactivity is the ability of a material to undergo a chemical change. A chemical reaction may occur under conditions such as heating, burning, contact with other chemicals, or exposure to light. Undesirable effects such as pressure buildup, temperature increase or formation of other hazardous chemicals may result.

Chronic: Chronic means long-term or prolonged. It can describe either an exposure or a health effect. A chronic exposure is a long-term exposure. Long-term means lasting for months or years. A chronic health effect is an adverse health effect resulting from long-term exposure or a persistent adverse health effect resulting from a short-term exposure.

COD: COD is stands for chemical oxygen demand. COD is the amount of oxygen used to oxidize reactive chemicals in water. It is a measure of water quality.

Combustible: Combustible means able to burn. Broadly speaking, a material is combustible if it can catch fire and burn. The terms combustible and flammable both describe the ability of a material to burn. Commonly, combustible materials are less easily ignited than flammable materials.

Corrosive Material: A corrosive material can attack (corrode) metals or human tissues such as the skin or eyes. Corrosive materials may cause metal containers or structural materials to become weak and eventually to leak or collapse. Corrosive materials can burn or destroy human tissue on contact and can cause effects such as permanent scarring or blindness.

Density: The density of a material is its weight for a given volume. Density is usually given in units of grams per milliliter (g/mL) or grams per cubic centimeter (g/cc). Density is closely related to specific gravity (relative density). The volume of a material in a container can be calculated from its density and weight.

Embryo: An embryo is an organism in the early stages of its development prior to birth. In humans, the embryo is the developing child from conception to the end of the second month of pregnancy.

Embryotoxic, Embryotoxicity: Embryotoxic means harmful to the embryo. Embryotoxicity is the ability of a substance to cause harm to the embryo.

Engineering Controls: Engineering controls help reduce exposure to potential hazards either by isolating the hazard or by removing it from the work environment. Engineering controls include mechanical ventilation and process enclosure. They are important because they are built into the work process.

Engineering controls are usually preferred to other control measures such as the use of personal protective equipment. Substitution of a less hazardous material or industrial process is the best way to reduce a hazard and is often considered to be a type of engineering control.

Environmental Fate: The environmental fate of a chemical is a statement of what tends to happen to the chemical when it is released into the environment. Fate depends on properties of the chemical such as solubility, vapour pressure, partition coefficient, stability and reactivity. A chemical may preferentially end up in air, in water, or in soil, and may break down quickly or slowly.

Evaporation Rate: The evaporation rate is a measure of how quickly the material becomes a vapour at normal room temperature. Usually, the evaporation rate is given in comparison to certain chemicals, such as butyl acetate, which evaporate fairly quickly. For example, the rate might be given as "0.5 (butyl acetate =1)." This means that, under specific condition, 0.5 grams of the material evaporates during the same time that 1 gram of butyl acetate evaporates. Often, the evaporation rate is given only as greater or less than 1, which means the material evaporates faster or slower than the comparison chemical.

In general, a hazardous material with a higher evaporation rate presents a greater hazard than a similar compound with a lower evaporation rate.

Explosive Limits: Explosive limits specify the concentration range of a material in air which will burn or explode in the presence of an ignition source (spark or flame). Explosive limits may also be called flammable limits or explosion limits.

The lower explosive limit (LEL) is the lowest concentration of gas or vapour which will burn or explode if ignited. The uppoer explosive limit (UEL), is the highest concentration of gas or vapour which will burn or explode if ignited. From the LEL to the UEL, the mixture is explosive. Below the LEL, the mixture is too lean to burn. Above the UEL, the mixture is too rich to burn. However, concentrations above the UEL are still very dangerous because, if the concentration is lowered (for example, by introducing fresh air), it will enter the explosive range.

In reality, explosive limits for a material very since they depend on many factors such as air temperature. Therefore, the values given on a Material Safety Data Sheet are approximate.

The explosive limits are usually given as the percent by volume of the material in the air. One percent by volume is 10, 000 ppm. For example, gasoline has a LEL of 1.4% and UEL of 7.6%. This means that gasoline vapours at concentration of 1.4% to 7.6% (14, 000 to 76, 000 ppm) are flammable or explosive.

Exposure Limits: An exposure limit is the concentration of a chemical in the workplace air to which most people can be exposed without experiencing harmful effects. Exposure limits should not be taken as sharp dividing lines between safe and unsafe exposure. IT is possible for a chemical to cause health effects, in some people, at concentrations lower than the exposure limit.

Exposure limits have different names and different meanings depending on who developed them and whether or not they are legal limits. For example, Threshold Limit Values (TLVs) are exposure guidelines developed by the American Conference of Governmental Industrial Hygienists (ACGIH). Permissible Exposure Limits (PELs)are legal exposure limits in the U.S.

Exposure limits have not been set for many chemicals for many different reasons. For example, there may not be enough information available to set an exposure limit. Therefore, the absence of an exposure limit does not necessarily mean the material is not harmful.

There are three different types of exposure limits in common use:

1) **Time-weighted average (TWA) exposure limit** is the time-weighted average concentration of a chemical in air for a normal 8-hour work day and 40-hour work week to which nearly all workers may be exposed day after day without harmful effects. Time-weighted average means that the average concentration has been calculated using the duration of exposure to different concentrations of the chemical during a specific time period. In this way, higher and lower exposures are averaged over the day or week.
2) **Short-term exposure limit (STEL)** is the average concentration to which workers can be exposed for a short period (usually 15 minutes) without experiencing irritation, long-term or irreversible tissue damage, or reduced alertness. The number of times the concentration reaches the STEL and the amount of time between these occurrences can also be restricted.
3) **Ceiling (C) exposure limit** is the concentration which should not be exceeded at any time.

Extinguishing Media: Extinguishing media are agents which can put out fires involving the material. Common extinguishing agents are water, carbon dioxide, dry chemical, “alcohol” foam, and halogenated gases (Halons). It is important to know which extinguishers can be used so they can be made available at the worksite. It is also important to know which agents cannot be used since an incorrect extinguisher may to work or may create a more hazardous situation. IF several materials are involved in a fir, an extinguisher effective for all of the materials should be used.

Fetotoxic, Fetotoxicity: Fetotoxic means the substance is harmful to the fetus/foetus. Fetotoxicity describes the ability of a substance to harm the fetus.

Fetus/Foetus: A fetus is an organism in the later stages of development prior to birth. IN humans, it is the unborn child from the end of the second month of pregnancy to birth.

First Aid: First aid is emergency care given immediately to an injured person. The purpose of first aid is to minimize injury and future disability. In serious cases, first aid may be necessary to keep the victim alive.

Flammable/Flammability: Flammable means able to ignite and burn readily. Flammability is the ability of a material to ignite and burn readily.

Flammable Gas: A flammable gas is a gas which can ignite readily and burn rapidly or explosively. Flammable gases can be extremely hazardous in the workplace; for example:

- If the gas accumulates so that its lower explosive limit (LEL) is reached and if there is a source of ignition, an explosion may occur.
- If there is inadequate ventilation, flammable gases can travel a considerable distance to a source of ignition and flash back to the source of the gas.

Flammable Liquid: A flammable liquid gives off a vapour which can be readily ignited at normal working temperatures.

Flammable liquids can be extremely hazardous in the workplace; for example:

- If there is inadequate ventilation, vapours can travel considerable distances to a source of ignition and flash back to the flammable liquid.
- It may be difficult to extinguish a burning flammable liquid with water because water may not be able to cool the liquid below its flash point.

Flammable Solid: A flammable solid is a material which can ignite readily and burn vigorously and persistently. Flammable solids may be hazardous because heat from friction (for example, surfaces rubbing together) or heat from processing may cause a fire. Flammable solids in the form of a dust or powder may be particularly hazardous because they may explode if ignited.

Flash Back: Flash back occurs when a trail of flammable gas, vapour or aerosol is ignited by a distant spark, flame or other source of ignition. The flame then travels back along the trail of gas, vapour or aerosol to its source. A serious fire or explosion could result.

Flash Point: The flash point is the lowest temperature at which a liquid or solid gives off enough vapour to from a flammable air-vapour mixture near its surface. The lower the flash point, the greater the fire hazard. The flash point is an approximate value and should not be taken as a sharp dividing line between safe and hazardous condition s. The flash point is determined by a variety of test methods which give different results.

Fumes: Fumes are very small, airborne, solid particles formed by the cooling of a hot vapour. For example, a hot zinc vapour may form when zinc-coated steel is welded. The vapour then condenses to form fine zinc fume as soon as it contacts the cool surrounding air. Fumes are smaller than dusts and are more easily breathed into the lungs.

Gas: A Gas is a material without a specific shape or volume. Gases tend to occupy an entire space uniformly at normal room pressure and temperature. The terms vapour and fume are sometimes confused with gas.

General Ventilation: General ventilation, also known as dilution ventilation, is the removal of contaminated air from the general area and the brining in of clean air. This dilutes the amount of contaminant in the work environment. General ventilation is usually suggested for non-hazardous materials.

Hazard, Hazardous: Hazard is the potential for harmful effects. Hazardous means potentially harmful. The hazards of a material are evaluated by examining the properties of the material, such as toxicity, flammability and chemical reactivity, as well as how the material is used. How a material is used can very greatly from workplace to workplace and, therefore, so can the hazard.

Hazardous Combustion Products: Hazardous decomposition products are formed when a material decomposes (breaks down) because it is unstable, or reacts with common materials such as water or oxygen (in air). This information should be considered when planning storage and handling procedures.

Henry's Law Constant: Henry's Law Constant is a measure of the tendency of a chemical to evaporate from a solution in water. It indicates whether a chemical accumulates in water or in air at equilibrium.

Hepatotoxin: Hepatotoxins are agents that can cause toxic effects on the liver.

IDLH: IDLH stand for Immediately Dangerous to Life or Health. For the purposes of respirator selection, the National Institute for Occupational Safety and Health (NIOSH) defines the IDLH concentration as the airborne concentration that poses a threat of exposure to airborne contaminants when that exposure is likely to cause death or immediate or delayed permanent adverse health effects or prevent escape from such an environment. The purpose of establishing an IDLH exposure concentration is to ensure that the worker can escape from a given contaminated environment in the event of failure of the respiratory protection equipment.

Incompatible Materials: Incompatible materials can react with the product or with components of the product and may:

- destroy the structure of function of a product;
- cause a fire, explosion or violent reaction; or
- cause the release of hazardous chemicals.

Inert Ingredient: An inert ingredient is anything other than the active ingredient of a product. It may be a solvent, colorant, filler or dispersing agent. In some cases, inert ingredients may be hazardous.

Ingestion: Ingestion means taking a material into the body by mouth (swallowing).

Inhalation: Inhalation means taking a material into the body by breathing it in.

Irritancy/Irritation: Irritancy is the ability of a material to irate the skin, eyes, nose, throat or any other part of the body that it contacts. Signs and symptoms of irritation include tearing in the eyes and reddening, swelling, itching and pain of the affected part of the body.

Irritancy is often described as mild, moderate or severe, depending on the degree of irritation caused by a specific amount of the material. Irritancy may also be described by a number on a scale of 0 to 4, where 0 indicates no irritation and 4 means severe irritation. Irritancy is usually determined in animal experiments.

LC50: LC stands for lethal concentration. LC50 is the concentration of a material in air which causes the death of 50% (one half) of a group of test animals. The material is inhaled over a set period of time, usually 1 or 4 hours. The LC50 helps determine the short-term poisoning potential of a material.

LD50: LD stands for lethal dose. LD50 is the amount of a material, given all at once, which causes the death of 50% (one half) of a group of test animals. The LD50 can be determined for any route of entry, but dermal (applied to skin) and oral (given by mouth) LD50's are most common. The LD50 is one measure of the short-term poisoning potential of a material.

Local Exhaust Ventilation: Local exhaust ventilation is the removal of contaminated air directly at its source. This type of ventilation can help reduce worker exposure to airborne materials more effectively than general ventilation. This is because it does not allow the material to enter the work environment. It is usually recommended for hazardous airborne materials.

Mechanical Ventilation: Mechanical ventilation is movement of air by mechanical means (for example, a wall fan). There are two kinds of mechanical ventilation: general ventilation and local exhaust ventilation.

Melting Point: The melting point is the temperature at which a solid material becomes a liquid. The freezing point is the temperature at which a liquid material becomes a solid.

It is important to know the freezing or melting point for storage and handling purposes. For example, a frozen or melted material may burst a container. As well, a change of physical state could alter the hazards of the material.

mg/m³: The abbreviation mg/m³ stands for milligrams (mg) of a material per cubic metre (m³) of air. It is a unit metric measurement for concentration (weight/volume). The concentrations of any airborne chemical can be measured in mg/m³, whether it is a solid, liquid, gas or vapour.

Miscible: Miscible means able to be mixed. Two liquids are said to be miscible if they are partially or completely soluble in each other. Commonly, the term miscible is understood to mean that the two liquids are completely soluble in each other.

Mist: A mist is a collection of liquid droplets suspended in air. A mist can be formed when spraying or splashing a liquid. It can also be formed when a vapour condenses into liquid droplets in the air.

mL: mL stands for milliliters (mL).

mm Hg: The abbreviation mm Hg stands for millimeters (mm) of mercury (Hg). It is a common unit of measurement for the pressure exerted by gases such as air. Normal atmospheric pressure is 760 mm Hg.

Molecular Weight: The molecular weight of a chemical is a number showing how heavy one molecule (or unit) of the chemical is compared to the lightest element, hydrogen, which has a weight of 1. The molecular weight has various technical uses, such as calculating conversions from parts per million (ppm) to milligrams per cubic metre (mg/m³) in air.

MSDS: MSDS stands for Material Safety Data Sheet. The MSDS is a document that contains information on the potential health effects of exposure and how to work safely with the material. It is an essential starting point to a health and safety program. It contains hazard evaluations on the use, storage, handling, and emergency procedures all related to the material.

Mutagen/Mutagenic/Mutagenicity: A mutagen is a substance which can cause changes in the DNA of cells (mutations). Mutagenic means able to cause mutations. Mutagenicity is the ability of a substance to cause mutations.

DBA determines the characteristics that children inherit from their parents. DNA also determines how cells in the body divede or reproduce.

A number of mutagenicity tests are used to screen chemicals for possible carcinogenicity or reproductive effects. This is because there is some evidence that mutations may increase the risk of cancer and reproductive problems such as infertility or birth defects. However, mutagenicity test results are not very reliable predictors of these effects. One reason for this is that the human bdy can repair mutations while most mutagenicity tests cannot.

Mutagenicity is an early indicator of potential hazard, and often there is very little other evidence available on possible carcinogenic or reproductive effects.

Natural Ventilation: Natural ventilation is a type of general ventilation which depends on natural instead of mechanical means for air movement. Natural ventilation can depend on the wind or the wind or the difference in temperature from one area to another to move air through a building. Therefore, it is unpredictable and unreliable.

Nephrotoxins: Nephrotoxins are agents that can cause toxin effects on the kidney.

Neurotoxins: Neurotoxins are agents that can cause toxic effects on the nervous system.

NIOSH: NIOSH stands for National Institute for Occupational Safety and Health. NIOSH is a branch of the United States government which undertakes research and develops occupational health and safety standards.

Nuisance Dust/Nuisance Particulate: Nuisance particulate is a term used by ACGIH (American Conference of Governmental Industrial Hygienists) to describe airborne materials (solids and liquids) which have little harmful effect on the lungs and do not produce significant disease or harmful effects when exposures are kept under reasonable control. Nuisance particulates may also be called nuisance dusts. High levels of nuisance particulates in the air may reduce visibility and can get into the eyes, ear and nose. Removal of this material by washing or rubbing may cause irritation.

Odour Threshold: The odour threshold is the lowest concentration of a chemical in air that is detectable by smell. The odour threshold should only be regarded as an estimate. This is because odour thresholds are commonly determined under controlled laboratory conditions using people trained in odour recognition.

As well, in the workplace, the ability to detect the odour of a chemical varies from person to person and depends on conditions such as the presence of other odours materials.

Odour cannot be used as a warning of conditions since workers may become used to the smell (adaption), or the chemical may numb the sense of smell, a process called olfactory fatigue. However, if the odour threshold for a chemical is well below its exposure limit, odour can be used to warn of a problem with respirator.

OSHA: OSHA stands for Occupational Safety and Health Administration. It is the branch of the United States government which sets and enforces occupational health and safety regulation. For example, OSHA sets the legal exposure limits in the United States, which are called Permissible Exposure Limits (PELs).

Oxidizing Agent, Oxidizing Material: An oxidizing agent or material gives up oxygen easily or can readily oxidize other materials. Examples of oxidizing agents are oxygen, chlorine and peroxide compounds. These chemicals will support a fire and are highly reactive.

PEL: PEL stands for Permissible Exposure Limit. PELs are legal limits in the United States set by the Occupational Safety and Health Administration (OSHA).

Personal Protective Equipment (PPE): Personal protective equipment is clothing or devices worn to help isolate a person from direct exposure to a hazardous material or situation. Recommended personal protective equipment includes protective clothing, respiratory protection and eye protection.

The use of personal protective equipment is the least preferred method of protection from hazardous exposures. It can be unreliable and, if it fails, the person can be left completely unprotected. This is why engineering controls are preferred. Sometimes, personal protective equipment may be needed along with engineering controls. For example, a ventilation system (an engineering control) reduces the inhalation hazard of a chemical, while gloves and an apron (personal protective equipment) reduce skin contact. In addition, personal protective equipment can be an important means of protection when engineering controls are not practical: for example, during an emergency or other temporary conditions such as maintenance operations.

pH: The pH is a measure of the acidity or basicity (alkalinity) of a material when dissolved in water. It is expressed on a scale from 0 to 14. Roughly, pH can be divided into the following ranges:

- pH 0 – 2 strongly acidic
- pH 3 – 5 Weakly acidic

- pH 6 – 8 Neutral
- pH 9 – 11Weakly basic
- pH 12 – 14 Strongly basic

Under the Canadian Controlled Products Regulations, materials with pH values of 0-2 or 11.5-14 may be classified corrosive. Corrosive materials must be stored and handled with great care.

Polymer: A polymer is a natural or man-made material formed by combining units, called monomers, into long chains. The word polymer means many parts. Examples of polymers are starch (which has many sugar units), polyethylene (Which has many ethylene units) and polystyrene (Which has many styrene units).

Most man-mad polymers have low toxicity, low flammability and low chemical reactivity. In these ways, polymers tend to be less hazardous than the chemicals (monomers) from which they are made.

Polymerize/Polymerization: Polymerization is the process of forming a polymer by combining large numbers of chemical units or monomers into long chains. Polymerization can be used to make some useful materials. However, uncontrolled polymerization can be extremely hazardous. Some polymerization processes can release considerable heat, can generate enough pressure to burst a container or can be explosive. Some chemicals can polymerize on their own without warning. Others can polymerize upon contact with water, air or other common chemicals. Inhibitors are normally added to products to reduce or eliminate the possibility of uncontrolled polymerization.

ppb: ppb stand for parts per billion.

ppm: The abbreviation ppm stands for parts per million. It is a common unit of concentration of gases or vapour in air. For example, 1 ppm of a gas means that 1 unit of the gas is present for every 1 million units of air. One ppm is the same as 1 minute in 2 years or 1 inch in 16 miles.

Process Enclosure: Process enclosure means that the operation in which the material is used is completely enclosed. A physical barrier separates the worker from the potential health or fire hazard. Process enclosure is usually recommended if the material is very toxic or flammable.

Pyrophoric: Pyrophoric chemicals are defined as chemicals which will ignite spontaneously in air at a temperature of or below.

Reactive Flammable Material: A reactive flammable material is a material which is a dangerous fire risk because it can react readily with air or water. This category includes any material which:

- is spontaneously combustible, that is, a material which can react with air until enough heat builds up that it begins to burn;
- can react vigorously with air under normal conditions without actually catching fire;
- gives off dangerous quantities of flammable gas on reaction with water; or
- becomes spontaneously combustible when it contacts water or water vapour.

Reactive flammable materials must be kept dry and isolated from oxygen (in air) or other oxidizing agents. Therefore, they are often stored and handled in an atmosphere of unreactive gas, such as nitrogen or argon.

Reproductive Effects: Reproductive effects are problems in the reproductive process which may be caused by a substance. Possible reproductive effects include reduced fertility in the male or female, menstrual changes, miscarriage, embryotoxicity, fetotoxicity, teratogenicity, or harmful effects to the nursing infant from chemicals in breast milk.

Most chemicals can cause reproductive effects if there is an extremely high exposure. In these cases, the exposed person would experience other noticeable signs and symptoms caused by the exposure. The signs and symptoms act as a warning of toxicity. Chemicals which cause reproductive effects in the absence of other significant harmful effects are regarded as true reproductive hazards. Very few workplace chemicals are known to be true reproductive hazards.

Reproductive Toxicity: Materials which have reproductive toxicity is adverse effects on fertility.

Sensitization: Sensitization is the development, over time, of an allergic reaction to a chemical. The chemical may cause a mild response on the first few exposures but, as the allergy develops, the response becomes worse with subsequent exposures. Eventually, even short exposures to low concentrations can cause a very severe reaction.

There are two different types of occupational sensitization: skin and respiratory. Typical symptoms of skin sensitivity are swelling, redness, itching, pain, and blistering. Sensitization of the respiratory system may result in symptoms similar to a severe asthmatic attack. These symptoms include wheezing, difficulty in breathing, chest tightness, coughing and shortness of breath.

Solubility: Solubility is the ability of a material to dissolve in water or another liquid. Solubility may be expressed as a ratio or may be described using words such as insoluble, very soluble or miscible.

Solvent: A solvent is a material, usually a liquid, which is capable of is solving another chemical. Chemicals commonly called solvents can dissolve many different chemicals. Examples of common solvents are water, ethanol, acetone, hexane and toluene.

Specific Gravity: Specific gravity is the ratio of the density of a material to the density of water. The density of water is about 1 gram per cubic centimeter (g/cc). Materials which are lighter than water (specific gravity less than 1.0) will float. Most materials have specific gravities exceeding 1.0, which means they are heavier than water and so will sink. Knowing the specific gravity is important for planning spill clean-up and fire fighting procedures. For example, a light flammable liquid such as gasoline may spread and, if ignited, burn on top of a water surface.

Stability: Stability is the ability of a material to remain unchanged in the presence of heat, moisture or air. An unstable material may decompose, polymerize, burn or explode under normal environmental conditions. Any indication that the material is unstable gives warning that special handling and storage precautions may be necessary.

STEL: STEL stands for Short-Term Exposure Limit.

STP: STP stands for Standard Temperature and Pressure (O°C Celsius and one atmosphere pressure).

Synergistic/Synergism: Synergism means that exposure to more than one chemical can result in health effects greater than expected when the effects of exposure to each chemical are added together. Very simply, it is like saying 1 + 1 = 3. When chemicals are synergistic, the potential hazards of the chemicals should be re-evaluated, taking their synergistic properties into consideration.

Target Organ Effects: Chemicals are identified as having target organ effects if there is statistically significant evidence of an acute or chronic health effect determined in a scientifically valid study. The following agents would be included (the list is not all-inclusive): hepatotoxins, agents which damage the lungs (including irritants), agents which act on the hematopoietic system, neurotoxins, nephrotoxins, reproductive toxins (mutagens, embryotoxins, teratogens and reproductive toxins), cutaneous hazards (chemicals which affect the dermal layer of the skin) and eye hazards (chemicals which affect the eye or visual capacity). There are no maximum dose criteria for chronic toxicity studies.

Teratogen/Teratogenic/Teratogenicity: A teratogen is a substance which can cause birth defects. Teratogenic means able to cause birth defects. Teratogenicity is the ability of a chemical to cause birth defects. Teratogenicity results from a harmful effect to the embryo or the fetus/foetus.

Thermal Decomposition Products: Thermal decomposition products are chemicals which may be formed when the material is heated but does not burn. These chemicals may be toxic, flammable or have other hazards. The chemicals released and their amounts vary depending upon conditions such as the temperature. The thermal decomposition products may be quite different from the chemicals formed by brining the same material (hazardous combustion products). It is important to know which chemicals are formed by thermal decomposition because this information is used to plan ventilation requirements for processes where a material may be heated.

TLV: TLV stands for Threshold Limit Value. It is the occupational exposure limit established by the American Conference of Governmental Industrial Hygienists (ACGIH). TLVs are adopted by some governments as their legal limits.

TLV-C: TLV-C stands for the ACGIH (American Conference of Governmental Industrial Hygienists) Threshold Limit Value-Ceiling.

Toxic/Toxicity: Toxic means able to cause harmful health effects. Toxicity is the ability of a substance to cause harmful health effects. Descriptions of toxicity (e.g. low, moderate, severe, etc.) depend on the amount needed to cause an effect or the severity of the effect.

TWA: TWA stands for Time-Weighted Average.

uG: uG stands for microgram, a unit of mass.

Unstable (Reactive): A chemical is identified as unstable (reactive) if in the pure state, or as produced or transported, it will vigorously polymerize, decompose, condense, or will become self-reactive under conditions of shock, pressure or temperature.

Vapour: A vapour is the gaseous form of a material which is normally solid or liquid at room temperature and pressure. Evaporation is the process by which a liquid is changed into a vapour. Sublimation is the process by which a solid is changed directly into the vapour state.

Vapour Pressure: Vapour pressure is a measure of the tendency of a material to from a vapour. The higher the vapour pressure, the higher the potential vapour concentration. In general, a material with a high vapour pressure is more likely to be an inhalation or fire hazard than a similar material with a lower vapour pressure.

Ventilation: Ventilation is the movement of air. One of the main purposes of ventilation is to remove contaminated air from the workplace.

VOC: VOC stands for Volatile Organic Compound.

Volatile/Volatility: Volatile means a material can evaporate. Volatility is the ability of a material to evaporate. The term volatile is commonly understood to mean that a material evaporates easily. Volatility is commonly expressed as the "% volatile." The percent volatile can vary from 0% (none of the material will evaporate) to 100% (all of the material will evaporate if given enough time).

Appendix 2.19: Terminologies Related to Environmental Toxicology

Abiotic Degradation: Breakdown of a chemical by processes other then living organism, such as photo degradation and chemical reactions (*e.g.hydrolysis*)

Absorbed Dose: The amount of a substance that actually enters into the body, usually expressed as milligrams of substance per kilogram of body weight

Absorption: The process whereby a substance moves from outside the body into the body.

Acceptable Daily Intake(ADI): The amount of a chemical to which a person can be exposed each day over a long period of time*(usually lifetime)* without suffering harmful effects.

Acetylcholine: An improvement chemical in the body having physiological function, including the neurotransmission of electrical impulses across synapses of nerve endings.

Acetylcholinesterase: An enzyme present in nervous tissue, muscle, and red blood cells that catalyzes the hydrolysis of acetycholine to choline and acetic acid.

Active Transport: The movement of substance across a membrane requiring energy.

Acute Dose: The amount of a substance administered or received over a very short period of time (*minutes or hours*), usually withn 24 hours.

Acute Effect: An effect that occurs almost immediately (*hours/days*) after a single or brief exposure to a toxie agent. Generally, acute effects will be evident within 14 days.

Adsorption: The process of attracting and holding a substance to a surface. For example, a substance may adsorb onto a soil particle.

Aerosols: Aerosols are airborne particulates. They may be solids or liquid droplets.

Allergy: An immune hypersensitivity reaction of body tissues to allergens that can affect the sking (*urticaria*), respiratory tract (*asthma*), gastrointestinal tract (*vomiting and nausea*) or produce a systemic circulatory response (*anaphylactic response*).

Alveoli: The air sacs at the ends of the tracheo-bronchial tree in which gases are exchanged between inhaled air and the pulmonary capillary blood.

Ambient environment: The surrounding environment. This can refer to ambient air, ambient water, or ambient soil.

Anemia: A condition in which there is reduced or impaired red blood cells or hemoglobin resulting in an inadequate capacity of the blood to transport oxygen to body tissues.

Anoxia: An insufficient (*below normal*) supply of oxygen in the body tissues.

Antagonism: An interaction between two chemicals in which one decreases the expected toxic effect of the other.

Antibody: An antibody is a protein molecule (*immunoglobulin with a unique amino acid sequence*) that only interacts with a specific or closely related foreign substance (*antigen*). The antibody is induced (*a response of the immune system*) as a result of prior exposure to the antigen.

Anticholinergic Effects: Neurological effects resulting from the blockage of acetylcholine which transmits impulses across nerve junctions.

Antidote: A remedy for counteracting a poison.

Asphyxiant: A relatively non-toxic gas that in high concentrations in the air results in insufficient oxygen which can cause hypoxia.

Average Daily Intake: The amount of a chemical to which a person consumes over a period of a day. It is determined by multiplying typical concentration of the chemical in drinking water, air, and food by an average daily intake factor such as 2 liters of water per day.

Benign tumor: A tumor that grows only at the site of origin; and does not invade adjacent tissues or metastasize. It is generally treatable.

Bioaccumulation: The build-up of a substance inn a biological organism such that the level in the organism is greater than in environmental source of the substance. *same as* Bioconcentration.

Bioactivation: The metabolic process whereby a parent substance is chemically changed to a daughter substance with enhanced biological activity.

Bioassay: A laboratory study used to determine the ability of a substance to produce a particular biological effect.

Bioavailability: The physical and /or biological state of a substance rendering it capable of being absorbed into the body.

Biodegradation: Breakdown of a chemical into smaller less complex molecules by microorganisms in environmental media (*e.g., soil, water, sediment*).

Biological Half-Life: The time required to eliminate one-half the quantity of a substance from the body.

Biomarker: Indicators of events occurring in biologic systems due to a xenobiotic. The type of indicators are exposure, effects, or susceptibility.

Biotransformation: Conversion of a chemical from one form to another by a biological organism.

Body Burden: The concentration of a substance which has accumulated in the body.

Bone Marrow: The tissue within the internal open space of bones (*e.g.shaft of long bones*) in which the blood-forming elements exist.

Bronchioles: The very small branches of the tracheo-bronchial tree of the respiratory tract which terminate in the alveoil.

Cancer: An uncontrolled growth of abnormal cells, creating a tumor that can invade surrounding tissues and may spread (*metastasis*) to distant organs.

Cancer Slope Factor: A key risk assessment parameter. It is an estimate of the probability that an individual will develop cancer if exposed to a specified amount to chemical (*mg/kg*) every day for a lifetime.

Carcinogen: A compound which is capable of causing cancer.

Carcinogenic: The ability of a substance to cause caner.

Carcinogenicity: The complex process whereby normal body cells are transformed to cancer cells.

Case-Control Study: A type of epidemiology study to investigate the cause of the toxicity. It compares the exposure histories of humans who have a particular toxic effect with that of normal individuals.

Ceiling Level: The maximum allowable concentration of the chemical in the workplace for a specific period of time (*usually 15 minutes*).

Cell Transformation: The change of a cell from one form to another. The term is generally used to denote the change from normal to malignant.

Cholinergic Effect: Neurological effects resulting from the transmission by acetylcholine of impulses across synaptic junctions between nerves.

Cholestasis: A liver condition in which excretion of bile salts via the bile duct is inhibited resulting in bile salts backing up into liver cells.

Chromosome: One of a group of structures that from in the nucleus of a cell during cell division. Chromosomes, bearing the DNA, carry the genetic code for the organism.

Chromosome Aberration: Changes in chromosome structure.

Chronic Dose: Substance administered or received gradually over a long period of time (*months to years*).

Chronic Effect: An effect that either shows up a long time after an exposure (*the latency period*) or an effect that results from a long term (*chronic*) exposure.

Cirrhosis: A chronic condition of the liver in which liver cells are replaced by fibrous cells.

CNS: The central nervous system consisting of the brain and spinal cord.

Cohort Study: An epidemiology study in which a cohort (*group*) of individuals with exposure are followed over time to compare disease occurrence.

Conjugation: A metabolic process in which chemical groups are attached to foreign substances in the body, usually making the conjugated chemical more water soluble and easier to eliminate from the body.

Control Group: A group of animals or humans in a study that are treated the same as the exposed groups but without receiving specific exposure.

Cornea: The transparent front surface of the eye.

Corrosion: Direct chemical action that results in irreversible damage at the site of contact. It is manifested by ulceration, necrosis, and scar formation.

Cross-Sectional Study: A type of epidemiology study that tests for the prevalence of a disease or clinical parameter among one or more exposed groups (*e.g., the prevalence of respiratory conditions among furniture makers*).

Demyelination: The loss of the myelin sheath (*insulation*) around a nerve.

Dermal Sensitization Test: An assay for immune hypersensitivity of the skin.

Dermal Toxicity: Toxicity of the skin which can range from mild irritation to corrosivity, hypersensitivity, and skin cancer. It can result from direct contact or internal distribution of the xenobiotic to the skin.

Detoxification: A metabolic process whereby a parent substance is changed to a daughter product (*metabolite*) that has lessened toxicity.

Disposition: The term used to describe the kinetics of a substance in the body. It encompasses absorption, distribution, metabolism, and elimination of a chemical.

Distribution: Movement of a substance from the site of entry to other parts of the body.

DNA: Deoxyribonucleic acid. The very large molecules in the nucleus of the cell, that carry the genetic information.

Dominant Lethal Assay Test: A mutagenicity test that can detect heritable dominant lethal mutation present in the sperm as the result of exposure to a substance.

Dosage: The determination of quantity of a substance received that incorporates the size, frequency, and duration of doses (*e.g., 10 mg every 8 hours for 5 days*).

Dose: the amount of a substance received at one time. Dose is usually expressed as administered or absorbed dose (*e.g., milligrams material/kilogram of body weight*).

Dose-Response Assessment: The relation between dose levels and associated effects.

Dose-Response Curve: A graphical representation of the quantitative relationship between doses of a substance and specific biological effects.

Draize Test: The test for eye irritation in which the test substance is placed on the eyes of white rabbits and observed for 72 hours.

Drug Idiosyncrasy: An abnormal susceptibility or sensitivity to a drug.

Drug Tolerance: the unusual ability to endure, without effect, a quantity of a drug that would normally preduce biological activity.

Ecologic: A term referring to the general environment.

Ecotoxicity: The toxic effects on environmental organisms other than humans.

ED50: Effective dose 50%. The estimated dose that causes some specific effect (*usually desirable*) for 50% of the population.

ED99: Effective dose 99%. The estimated dose that causes some specific effect (*usually desirable*) for 99% of the population.

Edema: The retention of fluid in an organ or in the body.

Effluent: The discharge of waste from a plant or other source into the environment.

Embryo: An early stage of the development of the unborn offspring in which cell differentiation proceeds rapidly along with the formation of the major organs. In humans this stage occurs from about 3 weeks unit 8-9 weeks after conception.

Embryotoxic: The harmful effects of a substance on the developing embryo.

Environmental Fate: The fate of a substance following its release into the environment. It includes the movement and persistence of the substance.

Enzyme: A protein which serves as a catalyst for chemical reactions in cells.

Enzyme Activation: The increase in levels of an enzyme as the result of stimulation by another chemical substance.

Enzyme Inhibitor: A substance which causes a decrease in levels of an enzyme.

Epidemiology: The study of the relative characteristics of exposed and nonexposed human populations for the purpose of detecting harmful effects.

Epidermis: The outer layer of the skin.

Excretion: A process whereby substances (*or metabolites*) are eliminated from the body.

Exposure: Contact with a foreign substance, usually by inhalation, ingestion, or skin contact.

Exposure Assessment: The analysis or estimation of the intensity, frequency and duration of human exposures to an agent.

Exposure Dose: The amount of a substance in the environment to which a person is subjected.

Femtogram (fg) : An extremely minute quantity, 1×10^{-15} gram.

Fetus: The unborn offspring in the postembryonic period, after major structres have been outlined. In humans this occurs from 8-9 weeks after conception until birth.

Fibrosis: The formation of scar tissue in an organ, generally by replacement of functional organ cells by non-functional firbrous tissue.

Geno: The smallest subunit of a chromosome that contains a genetic message.

Gene Mutation: A change in the DNA sequence within a gene.

Genetic Toxicity: Toxic effects that result from damage to DNA and altered genetic expression.

Germ Cell: Reproductive cells which give rise to sperm or ova.

Glomerulus: The highly vascular structure in the kidney where much of the fluid portion of the blood (*serum*) is filtered and passes into the kidney tubules, carrying with it toxins and many other materials present in the serum.

Hazard: The inherit adverse effect of substance.

Hazard Identification: Characterization of the innate adverse toxic effects of an agent.

Hepatic Cancer: Cancer of liver.

Hepatic Necrosis: Death of liver cells (*hepatocytes*).

Hepatitis: Inflammation of the liver

Hepatotoxicity: Toxicity of the liver and associated bile duct and gall bladder.

Hepatotoxin: A systemic poison whose target organ is the liver.

Hypersensitivity: A state of altered immune reactivity in which the body reacts with an exaggerated response to a foreign agent.

Hypoxia: A partial reduction in the oxygen concentration supplied to cells or tissues.

Idiosyncrasy: An abnormal sensitivity to some drug or other substance which is perculiar to the individual.

IDLH: Immediately Dangerous to Life and Health. An estimate for the maximum level of exposure from which a person could exit in 30 minutes without escape-imparing symptoms or irreversible health effects.

Immunotoxicity: Toxicity of the immune system. It may take several forms hypersensitivity (*allergy and autoimmunity*) immunodeficiency, and uncontrolled proliferation (*leukemia and lymphoma*).

***In Vitro*:** Outside a living organism (*e.g., in a test tube*).

In Vivo: Occurring within a living organism.

Interactions: Refers to measures of effects of simultaneous exposure to two or more substance. The four types of interactions are additive, antagonistic, potentiation, or synergistic.

Irritation: Local tissue reaction without involvement of an immunologic mechanism. It is a reversible inflammation.

Latency Period: The period of time between an exposure and onset of toxicity.

LC0: Lethal Concentration 0%. The calculated concentration of a gas at which none of the population is expected to die.

LC10: Lethal Concentration 10%. The calculated concentration of a gas at which 10% of the population is expected to die.

LC10: Lethal Concentration 50%. The calculated concentration of a gas at which 50% of the population is expected to die.

LC90: Lethal Concentration 90%. The calculated concentration of a gas at which 90% of the population is expected to die.

LD0: Lethal Dose 0%. The estimated dose at which none of the population is expected to die.

LD10: Lethal Dose 10%. The estimated dose at which 10% of the population is expected to die.

LD50: Lethal Dose 50%. The estimated dose at which 50% of the population is expected to die.

LD90: Lethal Dose 90%. The estimated dose at which 90% of the population is expected to die.

Lipid Soluble: Capable of being dissolved in fat or in solvents that dissolve fat. Usually non-ionized compounds.

LOAEL: Lowest Observed Adverse Effect Level. The lowest dose in a study in which there was an observed toxic or adverse effect.

MAC: Maximum Allowable Concentration. A recommendation for the highest level for long-term exposure to a chemical which will not produce adverse effect.

Malignant Cell: A cancer cell that has the potential to invade surrounding tissues or spread to other areas of the body (*metastasis*).

Malignant Tumor: A tumor that can invade surrounding tissues or metastasize to distant sites resulting in life-threatening consequences.

Margin of Safety (MOS): The ratio of the dose that is just within the lethal range *(LD01)* to the dose that is 99% effective *(ED99)*, LD01/ED99. A ratio of greater than 1 gives comfort to the physician whereas an a ratio of less than 1 denotes caution.

Mechanism of Action: The specific manner by which a substance causes a particular effect.

Median Toxic Dose : The dose level at which 50% of the population will experience toxic effects.

Metabolism: The conversion of a chemical from one form to another. *same as* Biotransformation.

Metastasis: The movement of diseased cells, in particular cancer cells, from the site of orgin to another location in the body.

MG/kg: A commonly used dose that stands for mg of a substance per kg of body weight.

MG/kg/day: A commonly used dosage that stands for mg of a substance per kg of body weight on daily basis.

Mg/M³: An exposure unit used to express concentriations of particulates in the air, standing for milligrams of compound per cubic meter of air.

Microgram (µg): A commonly used unit of weight consisting of one millionth (1×10^{-6}) of a gram.

Micronucleus Test: A test for mutagenicity in which bone marrow or peripheral blood cells are examined for the presence of micronuclei *(broken pieces of chromosomes surrounded by a nuclear membrane)*.

Milligram (mg): The most commonly used unit of measure in medicine and toxicity consisting of one thousandth of a gram (1×10^{-3} g).

Minimal Risk Levels (MRL's): A risk level calculated by the ATSDR for noncancer end points. The MRL is an estimate of daily human exposure to a substance that is likely to be without an appreciable risk of adverse effects over a specified duration of exposure. MRL's are derived for acute *(14 days or less)*, intermediate *(15-364 days)*, and chronic *(365 days or more)* duration exposures for either inhalation or oral routes.

Mouse Specific Locus Test: The major test for gene mutation i whole animals. Exposed mice are bred and observed for induced hereditary changes.

MTD: Maximum Tolerated Dose. The highest dose used in an animal cancer test which can be tolerated without serious weight loss or othe toxic effects.

Multi Hit Model: The least conservative quantitative risk assessment model. It assumes that several interactins are needed befroe a cell can be killed, damaged, or transformed into a cancerous cell.

Mutagen: A substance that causes mutations (genetic damage).

Nanogram (ng): A unit of weight consisting of one billionth of a gram (1×10^{-9} g).

Necrosis: Death, usually referring to cell death.

NEL: No Effect Level. A quantity of a substance that is below the threshold on the dose-response curve.

Neonates: Newborn animals.

Neoplasm: An uncontrolled and progressive growth of cells. It may be benign or malignant. *same as* Tumor.

Nephrotoxin: A systemic poison whose target is the kidney.

Neurotoxicity: Toxicity to cells of the central nervous system *(brain and spinal cord)* and the peripheral nervous system *(nerves outside the CNS)*.

Neurotoxin: A systemic poison whose target organ is the nervous system.

NOAEL: No Observed Adverse Effect Level. The highest dose in a toxicity study at which there were no toxic or adverse effects observed.

Non-Clinical Laboratory Study: A study of a pharmaceutical performed with laboratory animals which provides the basis for human clinical investigations.

Octanol/Water Partition Coefficient: the ratio of the amount of a substance that will dissolve in octanol versus the amount that will dissolve in water. The higher the octanol/water partition coefficient the greater the tendency of substance to be stroed in fatty tissues.

Odds Ration (O/R): A statistical calculation in a case-control study involving the ratio of risk of an exposed group to that of an unexposed group. An O/R=2 means that the exposed group has twice the risk of the non-exposed group.

One Hit Model: This is the most conservative quantitative cancer assessment model. It assumes that a single molecular event can induce a cell transformation which leads to cancer.

Organogenesis: the period of fetal development when the embryo differentiates and specific organs develop.

Passive Transfer: the movement across a membrane by simple diffusion.

Pathology: The branch of medicine that involves the functional and structural changes in tissues and organs which are caused by disease.

PEL: Permissible Exposure Level. The standard stipulated for the highest safe level of exposure to a chemical in the workplace.

Percutaneous: Movement through the skin.

Pharmacokinetics: Quantitation of the time course of chemical absorption, distribution, metabolism, and elimination.

Pharmacology: the science that deals with the origin, nature, chemistry, effects and use the drugs.

Phototoxic: The enhanced toxicity of a substance in or on the skin due to exposure to light *(usually ultraviolet light)*.

Picogram (pg): A unit of weight consisting of one quadrillionth of a

Poison: A substance capable of causing toxicity when absorbed into the body in a relatively small quantity.

Polyploidy: An increase in the normal number of chromosomes.

Power of the Study: The statistical ability of a study to detect an effect.

PPB: Parts per billion. The number of units of a substance in a billion units. PPB is a common concentration unit for dilute samples of dissolved substance or airborne substance.

PPM: Parts per million - the number of units of a substance in a million units. PPM is a common concentration unit for dilute samples of dissolved substances or airborne substances.

Primary Dermal Irritaiton Test: A test with laboratory animals *(usually rabbits)* that determines dermal toxicity of a substance when applied to the skin. It is manifest mainly by erythema. edema, and eschars.

Reference Dose (RfD) : The estimate of a lifetime daily exposure level for humans that is likely to be without risk of harmful effects. RfD's are acceptable safety levels for chronic non-carcinogenic and developmental effects. The process used to derive an RfD is a modification of that used to derive an ADI.

Relative Risk: A statistical calculation of the ratio of disease in an exposed population to that of an unexposed population.

Reproductive Toxicity: Toxicity of the male or female reproductive system. Toxic effects may include damage to the reproductive organns or offspring.

Respiratory Toxicity: Toxicity of the upper *(nose, pharynx, larynx, and trachea)* or lower *(bronchi, bronchioles, and lung alveoli)* respiratory system.

Risk: The probability that a hazard or effect will occur at a specific level of exposure.

Risk Assessment: The process by which the probability that an adverse effect will occur at a defined exposure level is determined.

Risk Characterization: The final stage in the risk assessment process which involves the prediction of the frequency and severity of effects in exposed population.

Safety Factor: Factors used in the calculation of acceptable humans or environmental exposures. They are applied to data from laboratory experiments or epidemiology studies. Factors of 10 are normally used to account for such uncertainties in the data on which risk assessments are made. Similar to uncertainty factors.

Selective Toxicity: Selective toxicity refers to species differences in toxicity between two species simultaneously exposed.

Sensitization: An immune capability developed following an individual's exposure to a specific antigen. Subsequent exposure results in an immune reaction.

Sensitizer: A substance that causes an allergic immune response.

Slope of the Dose-Response Curve: Rate of buildup of toxic effects with increasing doses.

Solubility: Ability of a substance to be dissolved in a solvent. The solubility is expressed according to the solvent, e.g., solubility in acetone, etc.

Somatic Cell: A body cell other than germ cell.

Standard Deviation: The statistical calculation denoting the variability of responses to an exposure. One standard deviation incorporates 68% of the responses while two standard deviations incorporates 95% of the responses.

Steatosis: Lipid accumulation in hepatocytes.

Structure-Activity Relationship: The process by which the toxicity of a substance can be predicted based on its similarity in structure to that of other chemicals for which the toxicity is known.

Subchronic Toxicity: the adverse effects of a substance resulting from repeated exposure to a toxic agent over a period of several weeks or months.

Systemic toxin: A toxin that affects the entire body or many organs.

Target Organ: An organ in which a exnobiotic exerts a toxic effect.

TD0: Toxic Dose 0%. The estimated dose none of the population is expected to exhibit toxic effects.

TD50: Toxic Dose 50%. The estimated dose at which 50% of the population exhibit toxic effects.

TD90: Toxic Dose 90%. The estimated dose at which 90% of the population exhibit toxic effects.

Teratogen: A compound that causes birth defects in a developing fetus.

Teratogenesis: The process by which a substance causes the abnormal development of tissues or organs in a developing fetus.

Teratogenicity: The development of birth defects as the result of exposure to a teratogenic toxicant.

Therapeutic Index (TI): The ratio of the dose needed to produce the desired therapeutic response to the dose producing toxicity.

Threshold Dose: The dose at which a toxic effect is first encountered.

Threshold Limit Value (TLV) : A recommendation by the ACGIH for the highest level of exposure to a chemical that is safe.

Tolerance: The ability to endure unusually large doses of a substance without ill effect. Toxic effects are decreased with continued exposure to the substance.

Total Dose: The sum of all individual doses which may be received over a period of time.

Toxicant: An agent that produces adverse effects when absorbed into the body.

Toxicokinetics: The pharmacokinetics of a toxic chemical.

Toxicologist: A person who studies harmful effects of chemicals including the mechanisms by which the effects are produced and the probability that the effects will occur under specific exposure conditions.

Toxicology: The study of the harmful interactions of chemicals on living organisms and biological systems.

Toxin: A specific protein produced by certain plants, animals and microorganisms that is highly toxic to other organisms *(snake venom).*

CHAPTER 3

Classification of Environmental Pollutants

Environmental pollutants are classified in the following five ways as tabulated in table 3.1.

Table 3.1: Classification of Environmental Pollutants

Classification based on the	Types of pollutants
A) States of matter of the pollutants	i) Gaseous pollutant, and ii) Particulate matter
B) Origin of the pollutants	i) Primary pollutants, and ii) Secondary pollutants
C) Chemical composition of the pollutants	i) Organic gaseous pollutant, and ii) Inorganic gaseous pollutants
D) Biodegradability of the pollutants	i) Biodegradable pollutant, and ii) Non-biodegradable pollutant
E) Nature of polluting agents	i) Chemical agents ii) Physical agents iii) Biological agents

(A) Classification Based on the States of Matter

(i) Gaseous pollutants: They are produced by the combustion of fossil fuels (coal, oil, natural gas etc.) used for the production of power & energy for domestic and industrial usage besides emissions from vehicles driven by petrol/diesel/natural gas. Examples are sulphur dioxide, oxides of nitrogen (NO_x) & carbon monoxide etc.

Gaseous pollutants behave much as the air itself & they do not settle out.

Table 3.2 gives the major gaseous pollutants and their sources.

Table 3.2: Gaseous Pollutants and their Sources

S. No.	Gaseous Pollutants	Their Sources of generation
1	Carbon Dioxide (CO_2)	Fossil fuels (coal, oil, natural gas, wood etc.) combustion, chemical manufacturing process, vehicular emission including jets etc.
2	Carbon monoxide (CO)	Fossil fuels incomplete combustion, iron & steel making, vehicular transport sector, petroleum refining etc.
3	Sulphur Dioxide (SO_2)	Combustion of coal, coke, furnace oil, transport vehicles, sulphuric acid manufacture etc.
4	Hydrogen sulphide (H_2S)	Pulps & paper industries, fertilizer plants

S. No.	Gaseous Pollutants	Their Sources of generation
5	Nitrogen Oxides (NOx)	Combustion of coal, oil & natural gas, vehicular emission, petroleum refining, manufacturing process of chemicals, nitric acid manufacture etc.
6	Hydrocarbons (HC)	Petrol/diesel engines, combustion of fossil fuels, petroleum refining etc.
7	Halogenated hydrocarbons	By product of degreasing, dry cleaning and the use of solvents.
8	Chlorine gas (Cl_2) & hydrogen Chloride gas (HCl)	Paper & cotton bleaching plants, Petroleum refining etc.
9	Hydrogen fluoride (HF)	Phosphoric acid plants, superphosphate plants etc.
10	Ammonia (NH_3)	Nitric acid plants, Nitrogenous fertilizer plants.

ii) Particulate matter pollutants: A particulate consists of five solids (eg, dust, smoke & fume) or, liquid droplets suspended in air (eg, mist, fog & spray). Large particulates are grit, flyash, dust & soot and the smaller ones are smoke, mist &aerosols.

The larger particulates settle out quickly and they produce their effects near the source of emission. Moderate size particulates travel further but eventually settle out at some distance from the source depending on size, emission rate & meteorological factors. The smallest particulates (1 μm size) behave almost like a gas and remain suspended indefinitely & are readily transported by wind currents. During the time, the particulates remain suspended in air, it is known as aerosols. The nature & magnitude of air pollution problem caused by particulate matter depends upon four factors: a) concentration of particulates in air. b) size rage of particles. c) chemical composition. d) rate of settling or dust fall.

The size range and the characteristics of various particulate pollutants are tabulated in Table 3.3.

Table 3.3: Size Range and Characteristics of Various Particulate Pollutants

S. No.	Pollutants	Size range (Diameter)	Characteristics
1	Grit	> 500 μm	Solid particles suspended in air
2	Dust	0.25 - 500 μm	Solid particles of natural or industrial origin usually formed by disintegration process.
3	Smoke	< 2 μm	Gas borne solid formed by incomplete complete combustion of fuels consisting of tiny particles of ash, carbon, oil etc.
4	Fumes	< 1 μm	Solid suspended in air, which are generated by condensation of vapors by sublimation, distillation, calcinations or chemical reaction process.
5	Mist	< 2 μm	Liquid droplets suspended in air, which arise from vapor condensation, chemical reaction or atomization of a liquid.
6	Fog	–	It is a naturally occurring mist.
7	Smog	–	It represents the combination of smoke & fog.
8	Aerosols	< 1 μm	Solid or liquid particles suspended in air or some other gases.

NOTES

1 mm = 1000 microns = 1000 micrometer (μm)

(B) Classification Based on the Origin of the Pollutants

i) Primary pollutants: These are the pollutants, which are emitted directly from the sources as a result of natural events or human activities like ash, smoke, dust, SO_2, CO, CO_2, H_2S, NO_x etc.

Primary air pollutants include the following:

i. Finer particles (< 100 μm in diameter)
ii. Coarse particles (> 100 μm in diameter)
iii. Sulphur compounds (SO_x, SO_2, SO_3, H_2S, RSH)
iv. Oxides of nitrogen (NO_x, ie, NO, N_2O, NO_2)
v. Carbon monoxide
vi. Halogen compounds
vii. Organic compounds
viii. Radioactive compounds

Many of the commonly encountered pollutants like SO_2, NO_2 & hydrocarbons are primary as they are produced from the combustion of fuels and are found in the atmosphere in the same form as emitted from the source.

ii) Secondary pollutants: Those pollutants which are not directly emitted to the atmosphere, but are formed in the atmosphere by some chemical or photochemical reaction among primary pollutants are one or more of air constituents eg, acid mist, formaldehyde, sulphur trioxide, ozone, peroxyacetylnitrate(PAN) are photochemical smog etc. As for example, SO_2 (a primary pollutant) emitted from combustion of fuels reacts with oxygen of atmospheric air to produce SO_3 (a secondary pollutant).

The chemical reactions in the atmosphere which converts primary pollutants into secondary pollutants are oxidation, dissociation and dissolution. The five most common primary pollutants are CO, SO_x, NO_x, hydrocarbons and particles.

Chemical/photo chemical transformation of some primary air pollutants to secondary air pollutants is tabulated in table 3.4.

Table 3.4: Chemical/Photochemical Transformation of Primary Pollutants to Secondary Pollutants

S. No.	Primary pollutants emitted from source	Conversion mechanism/reaction	Secondary pollutants produced
1	Sulphur dioxide (SO_2)	Oxidation in atmosphere produces sulphur trioxide which further coming in contact with water vapor forms sulphuric acid: $SO_2 + \frac{1}{2}O_2 \rightarrow SO_3$ (Oxidation) $SO_3 + H_2O \rightarrow H_2SO_4$ (dissolution) $SO_2 + H_2O \rightarrow H_2SO_3$ (dissolution)	Sulphur, Trioxide (SO_3), sulphuric acid and sulphurous acid.
2	Hydrogen sulphide (H_2S)	$H_2S + O_2 \rightarrow SO_2 + H_2$ (Oxidation in atmosphere)	Sulphur dioxide

S. No.	Primary pollutants emitted from source	Conversion mechanism/reaction	Secondary pollutants produced
3	Nitrogen dioxide (NO_2)	Absorbs light and dissociates into NO & free oxygen $NO_2 \rightarrow NO + \frac{1}{2}O_2$ (Photochemical dissociation)	Nitric Oxide
4	Carbon dioxide (CO_2)	$CO_2 + H_2O \rightarrow H_2CO_3$ (With water it forms carbonic acid)	Carbonic acid
5	Hydrogen fluoride (HF)	With water, it forms hydrofluoric acid	Hydrofluoric acid
6	Hydrocarbons & nitric oxide	A cycle of reaction between hydrocarbons, NO, NO_2 and Oxygen produces photochemical smog.	Photochemical smog

Other air pollutants are metals like Be, Ba, Cd, Fe, Mn, Pb, Hg, Ni, Zn (from mining & metallurgy), non-metals like As, P, Se (from the combustion of fuels, fertilizers) and radioactive substance from nuclear power plants.

(C) Classification of Pollutants Based on Chemical Composition

i) Organic gaseous pollutants: Organic gaseous pollutants encountered include hydrocarbons, ketones, aldehydes armines, alcohol etc.

ii) Inorganic gaseous pollutants: Inorganic gaseous pollutants encountered include, carbon compounds (eg, CO), Sulphur compounds (eg, SO_2, SO_3, H_2SO_4 etc.), nitrogen compounds (NO, NO_2, NH_3 etc.).

Classification of pollutants based on chemical composition is tabulated in Table 3.5.

Table 3.5: Classification of Pollutants Based on Chemicals Composition

Nature of Gas	Sub-classes	Typical pollutants
Inorganic gases	i. Oxides of sulphur (SO_x) ii. Oxides of nitrogen (NO_x) iii. Oxide of carbon iv. Other inorganics	SO_2, SO_3 NO, NO_2 CO, CO_2, (CO_2 is not strictly considered a pollutant) H_2S, HF, NH_3, Cl_2
Organic gases	i. Hydrocarbons ii. Ketones & aldehyde iii. Other organics	Hexane, benzene, ethylene, methane, butane, butadiene, acetylene Acetone, formaldehyde Amines, alcohols, carboxylic acids, chlorinated hydrocarbon, benzo-(a)-pyrene

(D) Classification of Pollutants Based on Biodegradability

i) Biodegradable pollutants: These are the pollutants which are easily decomposed by the micro-organisms either by nature itself or by suitable treatment and thus are not harmful eg, cow dung, domestic sewage etc. However if those are present in excess in the environment, they do not undergo degradation completely and thus become pollutant.

ii) Non-biodegradable pollutants: These are the pollutant substance which do not undergo degradation completely or degrade very slowly, eg, DDT, mercury, aluminum etc. Their presence in the environment even in very small amounts is very harmful for the human beings as well as the plants. They may also react with the other compounds present in the environment and produce even more toxic compounds.

According to another system of classification, air pollutants are classified as given below in table 3.6.

Table 3.6: Classification of Air Pollutants

S. No.	Major Classes	Sub-classes	Typical members of sub-classes
1	Particulates	Solid Liquid	Dust, smoke, fumes, flyash, carbon, lead, heavy metals Mist, spray, oil, grease, acids
2	Gases-organic	Hydro carbons (alkenes, alkenes & aromatics) Aldehydes, ketones & alkynes other organics eg, alcohols, alkylhalide, amines	Hexane, benzene, ethylene, methane, butane, butadiene, ethane Formaldehyde, acetone, acetylene Alcohols, chlorinated hydrocarbons eg, PCB (poly chloro biphenyl), ethylene dichloride, methyl amine
3	Gases-inorganic	oxides of carbon (CO_X) oxides of sulphur (SO_x) oxides of nitrogen (No_x) Photochemical oxidants Other inorganics	CO, CO_2 SO_2, SO_3 NO, NO_2 O_3 H_2S, HF, NH_3, Cl_2, F
4	Other pollutants	–	Radioactive substances, pesticides, aeroallergens

(E) Stationary and Mobile sources of air pollution

Air pollution occurs from gaseous emissions from industries, thermal power plants, automobile exhausts domestic combustion etc. Depending upon whether the air pollutants is emitted from stationary or mobile sources they are classified as given below in table 3.7.

Table 3.7: Stationary and Mobile Sources of Air Pollution Stationary Sources

Single or point sources	Multiple or area sources
i. Industrial stack emission ie, chimneys of thermal power plants, steel plants, oil refineries, pulp & paper plants etc. ii. Municipal incinerators of solid wastes iii. Open burning of solid wastes iv. Industrial fossil fuel (coal, oil & gas) combustion	i. Domestic combustion of fuels in residential areas, offices, hotels, apartments, hospitals etc. ii. On site incinerator iii. Commercial & institutional heating by coal, oil & gas iv. Open burning of fuels/wastes v. Evaporation losses

Mobile Sources

Line Sources	Area sources
i. Highway automobile/vehicles ii. Trains/railroad locomotives iii. Channel vessels	i. Air - crafts (air ports) ii. Light, medium& heavy duty motor vehicles iii. Rail yards Locomotives iv. Port vessels

(E) Classification of Environmental Pollutants Based on the Type and Nature of Polluting Agents is Tabulated below in table 3.8.

Table 3.8: Environmental Pollutants Classified on the Basis of Nature of Polluting Agents

S. No.	Type of polluting agents	Examples of environmental pollutants
1	Physical agents	Heat/thermal pollution, odour pollution, noise pollution, radiation pollution
2	Chemical agents	Gases, particulates, solid & liquid wastes, pesticides, petroleum & its products, heavy metals
3	Biological agents	Micro-organisms, increasing population

Appendices

Contents			Page No.

Appendix 3.1: Classification of Pollutants

S. No.	Major Classes	Sub-classes	Typical Members of Sub-classes
1	Organic Gases (Hydrocarbons)	Alkanes	Ethane
		Alkenes	Ethylene
		Alkynes	Acetylene
		Alkyl Halides	Ethylenedichloride
		Aldehydes	Formaldehyde
		Ketones	Acetone
		Amines	Methyl Amine
		Alcohols	Ethanol
		Aromatics	Benzene
2	Inorganic Gases	Photochemical Oxidants	Ozone
		Oxides of Nitrogen	Nitrogen Dioxide, Nitric Oxide
		Oxides of Sulfur	Sulfur Dioxide, Sulfur Trioxide
		Oxides of Carbon	Carbon Monoxide, Carbon Dioxide
		Halides	Chlorine, Flourine
		Miscellaneous	Ammonia, Hydrogen Sulfides
3	Particulates	Solid Particulates	Dust, Smoke
		Liquid Particulates	Mist, Spray
			Heavy Metals
4	Other Pollutants Include:		
	• Radioactive Substances	—	—
	• Pesticides	—	—
	• Aeroallergens	—	—

Appendix 3.2: Some Examples of Stationary Sources of Pollution and the Pollutants Emitted

S. No.	Stationary sources of pollution	Pollutants emitted
1	Coal Fired Power Stations	SO_2, NO_x, PM, VOCs, ...
2	Sulphuric Acid Plants	SO_2, Sulphuric acid mist, SO_3
3	Boilers, Combustion Plants	SO_2, NO_x, PM, VOCs, ...
4	Nitric Acid Plants	NO_x
5	Fertiliser Plants	PM (ammonium nitrate, phosphate rock, etc)
6	Oil refineries	SO_2, NO_x, PM, VOCs, ...

7	Glass manufacture	SO_2, NO_x, PM
8	Landfills	CH_4, H_2S, odourous gases
9	Incinerators	Dioxins, SO_2, NOx, PM, VOCs, ...
10	Open burning of solid waste	Dioxins, PM, VOCs,
11	Mines and smelters	PM, NO_x, SO_2, ...
12	VOC storage facilities, paint shops, dry cleaners etc...	VOCs

VOC- Volatile Organic Compounds, PM-Particulate Matter

Appendix 3.3: Sources of Primary Air Pollutants (ie, CO, SO_x, NO_x, HC, SPM) from various Usage Sectors (in India)

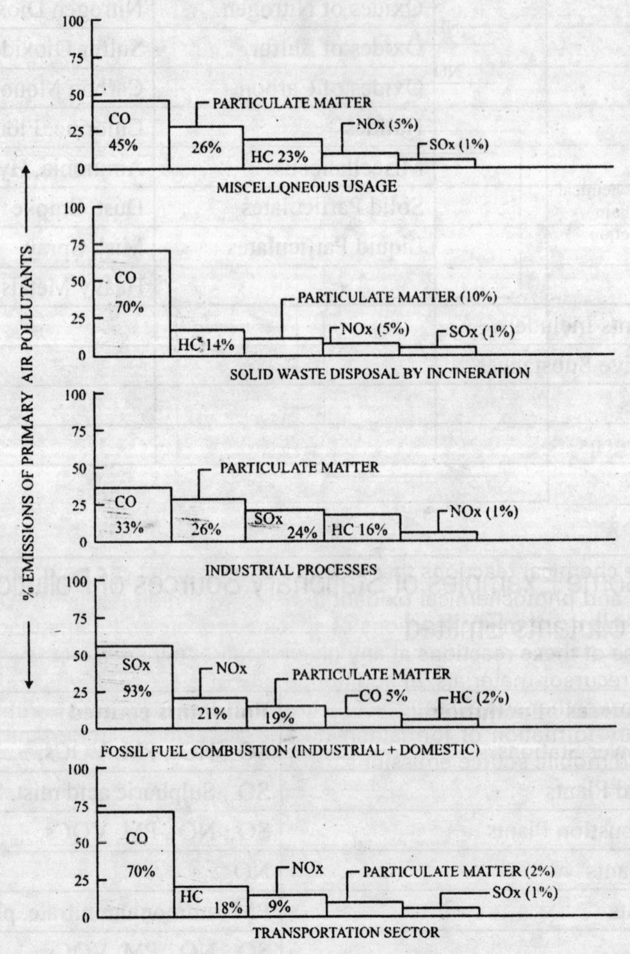

Appendix 3.4: Type of Reaction Involved in the Formation of Secondary Pollutants from Primary Pollutants

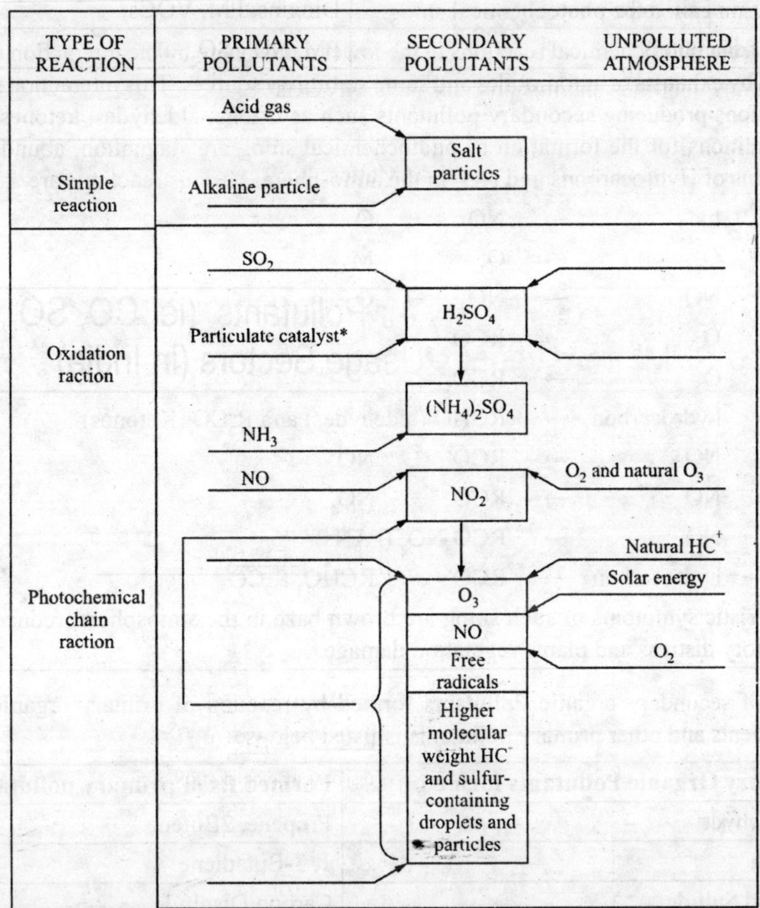

NOTES

Examples of large-scale chemical reactions that result in products that can be hazardous to health include the formation of ozone and photochemical oxidant in areas with high levels of NOx and volatile organic emissions. In addition, there are many reactions of specific organic pollutants that generate air toxics of concern. The importance of these reactions at any given location depends on the emissions and resulting concentrations of the precursor materials. In addition, many of these reactions are catalyzed directly and indirectly by sunlight, so weather and climatic factors are important in judging the importance of secondary reactions. The secondary formation of formaldehyde and acrolein are important in many regions with significant industrial and mobile source emissions.

Formation of Secondary Air Pollutants

Hydrocarbons are mainly involved in photochemical reactions in the atmosphere and result in the formation of a variety of toxic secondary pollutants. The presence of some secondary pollutants in the atmosphere results in a condition called the photochemical smog.

Smog arises from photochemical reactions in the lower atmosphere by the interaction of Hydrocarbons and NO_x released by exhausts of automobiles and some stationary sources. This interaction results in a series of complex reactions producing secondary pollutants such as Ozone, aldehydes, ketones and peroxyacyl nitrates. The conditions for the formation of photochemical smog are stagnation, abundant sunlight and high concentrations of Hydrocarbons and NO_x in the atmosphere. Various reactions are as follows:

$NO_2 + h\nu \rightarrow NO + O$

$O + O_2 \rightarrow O_3 + M$

$O_3 + NO \rightarrow NO_2 + O_2$

Hydrocarbon + O → $RCO^{\cdot}$

$RCO^{\cdot} + O_2 \rightarrow RCO_3^{\cdot}$

$RCO_3^{\cdot}$ + hydrocarbon → RCHO (aldehydes) and R_2CO (Ketones)

$RCO_3^{\cdot} + NO \rightarrow RCO_2^{\cdot} + NO_2$

$RCO_2^{\cdot} + NO \rightarrow RCO + NO_2$

$RCO_3^{\cdot} + NO_2 \rightarrow RCO_3NO_2$ (PANs)

O_3 + hydrocarbon → $RCO_2^{\cdot}$ + RCHO, R_2CO

The characteristic symptoms of such smog are brown haze in the atmosphere, reduced visibility, eye irritation, respiratory distress and plant /vegetation damage.

A few examples of secondary organic Pollutants formed by reaction of primary organic pollutants with normal air constituents and other primary pollutants is listed below:

S. No.	Secondary Organic Pollutants formed	Formed from primary pollutants
1	Acetaldehyde	Propene, 2Butene
2	Acrolein	1, 3-Butadiene
3	Carbonyl Sulfide	Carbon Disulfide
4	O-Cresol	Toluene
5	Formaldehyde	Ethene, Propene
6	Hydrogen Chloride	Nitric Acid, Chlorinated Organics
7	Methylethyl Ketone	Butane, Branched Alkenes
8	N-Nitroso-N-Methylurea	N-Methylurea
9	N-Nitrosodiethylamine	Dimethylamine
10	N-Nitrosomopholine	Morpholine
11	Phosgene	Chlorinated Solvents
12	Propionaldehyde	1-Butene

The overall lifetime of a chemical in the environment is equal to the sum of the atmospheric lifetime when considering only physical processes plus the lifetime when considering only chemical processes. This equation is the same as saying that the overall rate constant for pollutant removal/destruction is equal to the sum of the rate constant for physical removal plus the rate constant for chemical reaction. This relationship follows from the nature of first-order reaction kinetics, and is only an approximate description of actual physical processes. It is a useful approximation, however, that can be used to evaluate the importance of atmospheric processes for many pollutants. For any given air toxic, overall persistence in the atmosphere depends on particle-vapor partitioning behavior, particle size distribution (if the material is non-volatile), and susceptibility to various types of chemical reactions.

Appendix 3.5: Ambient Primary and Secondary Air Pollutants having Adverse Health Effects

S. No.	Primary air pollutants	S. No.	Secondary air pollutants
1	Dusts	1	Organic aerosols
2	Soots	2	Nitrate salts
3	Metal compounds	3	Sulfate salts
4	Suspended particulates	4	Ozone
5	Carbon monoxide	5	Nitrogen dioxide
6	Sulfur dioxide	6	Peroxyacetyl nitrate
7	Nitrogen Oxides	7	Nitric acid
8	Nitric Oxide	8	Nitrous acid
9	Nitrogen dioxide	9	Suspended particulates
10	Volatile organic compounds (VOC's)	10	Sulphuric acid

Appendix 3.6: Sources of Sulphur Dioxide (SO_2) Generation

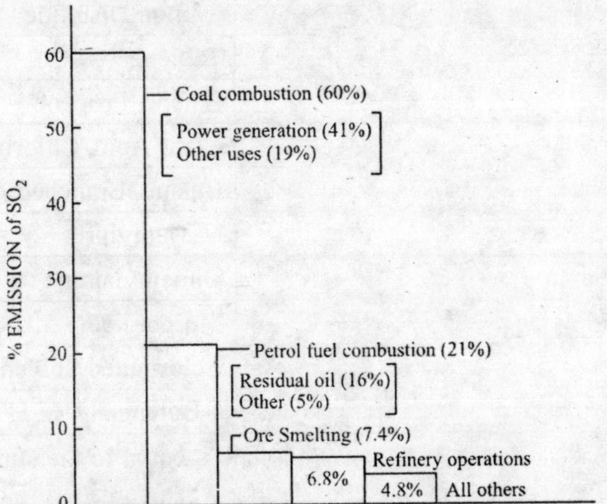

Appendix 3.7: Diameter Ranges of Air Pollutants in Micron

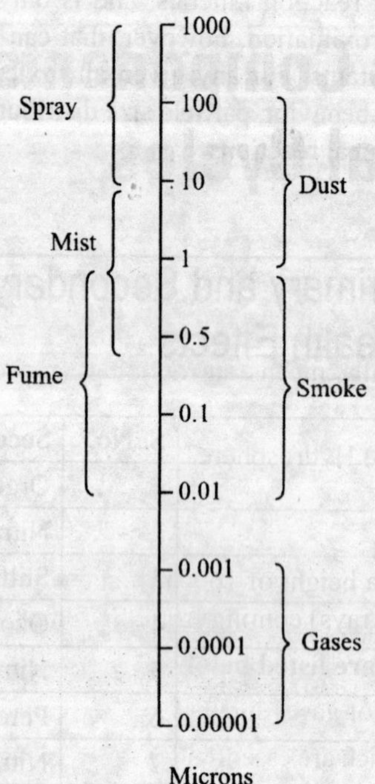

1 micron = 10^{-6} m = 1 μm (micro meter)

1 mm = 1000 microns = 1000 micrometers (μm)

CHAPTER 4

Environment: Its Components and Bio-Geochemical Cycles

The term environment of earth literally means surroundings and comprises of four major components, namely:

(A) Atmosphere (B) Hydrosphere (C) Lithosphere (D) Biosphere

(A) Atmosphere

It is a cover of gases extending up to a height of 1600 km above the earth's surface and protects the life on earth from harmful radiations (cosmic rays) coming from the sun or the outer space.

Salient Features of atmosphere are listed below:

- It refers to protective blanket of gases surrounding the earth.
- It contains all those gases, which are essential for the existence of life on the earth eg, oxygen present in it is required by human beings & animals for respiration, **carbon dioxide** present in it is needed for photosynthesis in plants, and the **Nitrogen** present in it is needed for plants indirectly, as it is first converted into water soluble compounds like ammonia & then added into the soil. Nitrogen supply is also used by nitrogen fixing bacteria & ammonia manufacturing plants to yield chemically bound nitrogen essential for life.
- Its major components are N_2 & O_2 and the minor components include Ar, CO_2& some trace gases. Bulk of these gases are present within 5 km above the earth surface.
- It is an important carrier of water vapor, which are needed for all forms of life.
- It is divided into five different regions of varying attitudes namely, Troposphere, Stratosphere, Mesosphere, Thermosphere (ionosphere) and Exosphere.
- Ozone present in it absorbs the harmful ultraviolet radiations ie, cosmic rays from outer space & major portion of electromagnetic radiation from the sun.
- It transmits only nearly ultraviolet visible, near infra-red radiation (300–2500 nm) & radio waves (0.10–40μ), while filtering out tissue damaging ultraviolet radiation (< 300 nm).
- It maintains the heat balance of the earth by absorbing infra red radiation emitted by the sun & remitted from the earth. For the some reason, it controls, weather & climatic conditions.

(B) Hydrosphere

It contains that part of the environment, which contains water in the forms of sea, ground water, ocean, glaciers, river, lakes, polar ice caps, ponds etc. About 75% of the earth surface is covered by hydrosphere.

Most of it is in the oceans containing salt, while fresh water is present in lakes/river/ponds which flows into them from rain or melting of snow etc.

Salient features of hydrosphere are listed below:

- It includes all water resources like oceans, sea, rivers, lakes, streams, reservoir, glaciers, ice caps & ground water.
- 97% of earth water supply lies in oceans (containing 3.5 % salt), 2% of water resource is locked in glaciers & polar ice caps and only 1% is found as fresh water (surface water, rivers, lake, streams & ground water).
- Usage of water includes thermal power plants (50%), irrigation (30%), industrial consumption (13%) & domestic use (7%).
- Contamination of surface water lakes place by pesticides &fertiliser in agricultural runoff water, human & animal waste in sewage and industrial waste.
- Water born diseases occur due to sewage.
- Hydrosphere maintains the availability of fresh water to the biosphere through the hydrological (water) cycle.
- Oceans (a major component of water cycle) absorb more CO_2 than atmosphere, besides being store house of water, salt, minerals & food. It is the largest sink (receptor of pollutants) of the planet.

(C) Lithosphere

It is the solid component of the earth consisting of soil, rocks, mountains etc. the upper most part of the earth's crust contains weathered rocks as well as organic matter and is called **soil.** This is the most important part of lithosphere, because plant is grown on this part. It is also a store house of minerals.

Salient Features of lithosphere is listed below:

- Solid components of earth is termed as the lithosphere.
- It includes outer parts of the **solid earth** and refers to the mineral concentrated in earth's crust and to the complex mixture of minerals, organic matter, water & air making up soil.
- Soil is the most important part of the lithosphere.
- Earth is a cold, spherical, solid planet of the solar system which spins on its axis & resolves around the sun at a certain constant distance.
- Lithosphere comprises of three main layers namely crust, mantle and outer & inner core.
- Core (composed of nicked & iron) refers to the central fluid or vaporised sphere (having diameter of 2500 km from the centre).
- Mantle (is in molten state) extends 29000 Km above the core.
- Mantle of earth accounts for 67% of its total mass & 90% of its total volume.
- Crust forms the outermost solid zone of the earth and is about 8 to 40 Km above the mantle.
- Crust surface is covered with soil where plants grow and decay. Organic matter in soil is decomposed by micro-organisms thus forming bio-mass which mixes with soil fauna.
- Soil on the earth surface in terms of total diameter of the earth is only a tissue thin layer.
- Major component of soil are air, water, minerals inorganic matters & organic matters (comprising of decaying plant biomass, bacteria, fungi, animals, earthworms etc,).
- Earth's crust (64 to 96 Km from surface) is rich in silica &aluminum.

(D) Biosphere

It is that part of the lithosphere, hydrosphere & atmosphere, where living organisms interact with these parts and thus live together. For example, green plants during photosynthesis gives out oxygen which is added into the atmosphere, animals inhale oxygen & give out carbon dioxide, which is used by plants for photosynthesis.

Salient feature of biosphere are listed below:

- It refers to the realm of living organisms & their interaction with the environment ie, atmosphere, hydrosphere and lithosphere.
- It includes living organisms & the surroundings in which they can exist.
- Biosphere and environment considerably influence each other.
- CO_2 & O_2 levels of atmosphere are entirely based on the plant kingdom.
- Green plants are responsible for a accumulation of oxygen in the atmosphere through photosynthesis and decay.
- Biological world is intimately related to energy flow in the environment and water chemistry.
- Biosphere is influenced by the chemistry of the environment and the turn influences the chemistry of lithosphere & hydrosphere especially.
- Life on earth occupies a thin skin extending few kilometers above and below its surface which is called biosphere.
- In the biosphere, the radiant energy is converted to chemical form (carbohydrates) through the process of photosynthesis. Only then the energy transfer takes place from chemical to mechanical and heat forms during cellular metabolism.

Environmental segment has been shown in figure 4.1

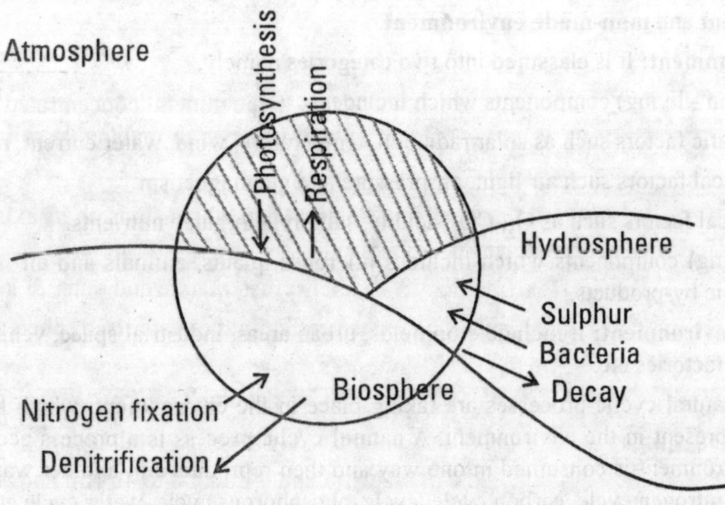

Figure 4.1: Environmental Segments

Cross section of the planet (earth) showing its different layers is shown below in figure 4.2.

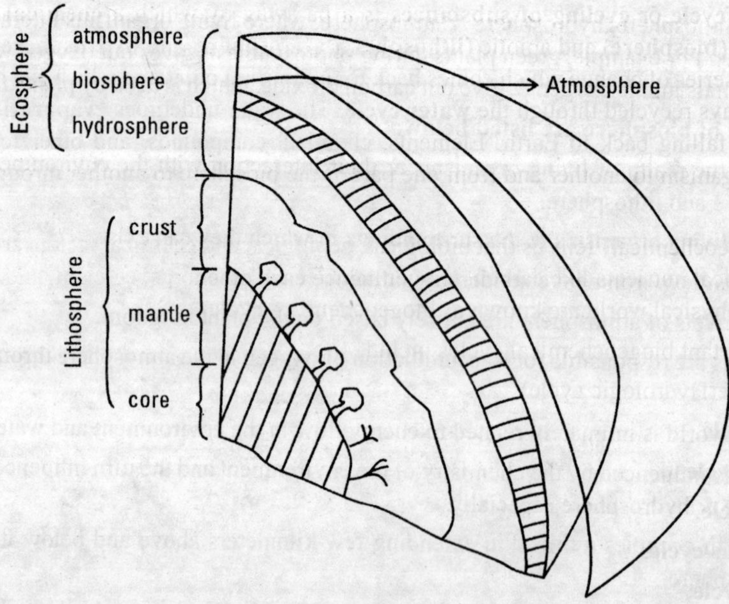

Figure 4.2: Cross-section of planet showing its different layers.

Environment encompasses everything that surround an organism. Earth is the only planet that supports life out of 9 planets, meteorites & satellites in our solar system. Our environment is broadly classified into **natural environment** and **man-made environment**.

Natural environment: It is classified into two categories namely:

a) Abiotic (non - living) components which includes:

   i. Climatic factors such as solar radiation, temperature, wind, water current, rainfall.

   ii. Physical factors such air light, air pressure and geomagnetism.

   iii. Clinical factors such as O_2, CO_2, acidity, salinity, inorganic nutrients.

b) Biotic (living) components which includes microbes, plants, animals and all living organisms & their organic by-products.

Man - made environment: It include crop fields, urban areas, industrial space, vehicles, power plants, telecommunication factories etc.

A number of natural cyclic processes are taking place in the environment, which keeps a balance of different materials present in the environment. A natural cyclic process is a process according to which a material of the environment is consumed in one way and then reproduced in another way, for example, we have oxygen cycle, nitrogen cycle, carbon cycle, cycle, phosphorous cycle, water cycle etc.

Environmental chemistry deals with the chemical phenomenon occurring in the environment.

Environmental science deals with the study of sources of pollution and the methods of monitoring & controlling the pollutants.

Biogeochemical Cycle

A **biogeochemical cycle** or **cycling of substances** is a pathway by which a **chemical substance** moves through both biotic (**biosphere**) and abiotic (**lithosphere**, **atmosphere**, and **hydrosphere**) compartments of **Earth**. A cycle is a series of change which comes back to the starting point and which can be repeated. Water, for example, is always recycled through the **water cycle**. The water undergoes **evaporation**, **condensation**, and **precipitation**, falling back to Earth. Elements, chemical compounds, and other forms of matter are passed from one organism to another and from one part of the biosphere to another through biogeochemical cycles.

The term "biogeochemical" tells us that biological, geological and chemical factors are all involved. The circulation of chemical nutrients like carbon, oxygen, nitrogen, phosphorus, calcium, and water etc. through the biological and physical world are known as biogeochemical cycles.

The most important biogeochemical cycles include the :

- **water cycle**(Hydrologic cycle)
- **carbon cycle**,
- **nitrogen cycle**,
- **oxygen cycle**,
- **phosphorus cycle**,
- **sulphur cycle**,
- and the **rock cycle**

The Importance of Biogeochemical Cycles

Biogeochemical cycles serve a variety of functions at ecosystem level and in ensuring survival of various organisms including humans. Below are some of the importances of biogeochemical cycles.

- Biogeochemical cycles enable the transformation of matter from one form to another. This transformation enables the utilization of matter in a form specific to particular organisms. For example humans utilize water in liquid form. Through the hydrological cycle, water vapour is condensed to liquid and ice converted to liquid water. Nitrogen, despite its abundance in the atmosphere it's often the most limiting nutrient for plant growth. This problem occurs because most plants can only take up nitrogen in two solid forms: ammonium ion (NH_4^+) and the ion nitrate (NO_3^-). Therefore, biogeochemical cycles enable the provision of elements to organisms in utilizable forms.
- Biogeochemical cycles enable the transfer of molecules from one locality to another. Some elements such as nitrogen a re highly concentrated in the atmosphere, but some of the atmospheric nitrogen is transferred to soil through the nitrogen cycle (which is a biogeochemical cycle).
- Biogeochemical cycles facilitate the storage of elements. Elements carried through the biogeochemical cycles are stored in their natural reservoirs, and are released to organisms in small consumable amounts. For example through the nitrogen cycle and with the help of the nitrogen fixing bacteria, green plants are able to utilize nitrogen in bits though it is abundant in the atmosphere.
- Biogeochemical cycles assists in functioning of ecosystems. An ecosystem is a system that properly functions in a state of equilibrium, and when ever any imbalances occur, the ecosystem through the biogeochemical cycles restores to the equilibrium state; this may take a few days or many years. The adjustment is such that the disturbing factor is eliminated.

- Biogeochemical cycles link living organisms with living organisms, living organisms with the non living organisms and nonliving organisms with non living organism. This is because all organisms depend on one another and most especially, the biotic (living component) and a biotic component of the ecosystem are linked by flow on nutrients engineered by the biogeochemical cycles.
- Biogeochemical cycles regulate the flow of substances. Since the biogeochemical cycles pass through different spheres, the flow of elements is regulated because each sphere has a particular medium and the rate at which elements flow is determined by the viscosity and density of the medium. Therefore elements in the biogeochemical cycles flow at differing rates with in the cycle and this regulates the flow of the elements in those cycles.

Lifespan and Rate of Biogeochemical Cycles

Lifespan of biogeochemical cycles is described in terms of the time a particular element or molecule of a substance being carried in the biogeochemical cycle takes to make one complete cycle. Each Biogeochemical cycle has its life span, ranging from several days to millions of years. For example a water droplet of average size may stay in the atmosphere for about ten days before precipitation, and carbon atoms may reside in the earth crust for the age of the Earth.

The speed of the cycles depends on the medium in which the molecule being cycled is and the surrounding conditions. For this reason, climatic conditions have a significant impact on the biogeochemical cycles .

Cycles that involve molecules or ions in a gaseous state are generally shorter than cycles that involve solid or liquid state transfer. Because of the slow rate at which molecules move through the lithosphere, the attainment of some elements of nutritional value by some living organisms in the hydrosphere, and counteract this deficiency, movement from terrestrial biosphère to the ocean (via stream flow) is usually compensated by movement either through the atmosphere (such as with nitrogen and carbon) or by weathering (such as with phosphorous or **calcium**). Therefore increased or decreased transport by stream flow severely disrupts the cycles of nutrients that lack a gaseous phase.

Reservoirs, Exchange Pools and Resident Time of Elements in the Biogeochemical Cycles

Each of the elements has its biogeochemical cycle and each cycle with a unique pathway characterized by reservoirs, exchange pools and resident times.

A reservoir is a place or region or location where a biogeochemical element is in its highest concentration. Elements being cycled are held and stored for some time in reservoirs, for example coal or fossil fuels are reservoirs for carbon. Some elements have several reservoirs while others have one.

When chemicals are held for only short periods of time, they are being held in **exchange pools**. Examples of exchange pools include plants and animals, which temporarily use elements in their systems and release them back into the air or surrounding medium. Generally, reservoirs are abiotic (nonliving) factors while exchange pools are biotic (living) factors. Carbon is held for a relatively short time in plants and animals when compared to fossil fuel deposits. The amount of time that a chemical is held in one place is called its **resident time**.

The term **Influx** is commonly used in describing biogeochemical cycles to refer to the difference between the amount of elements entering a reservoir and the amount leaving the reservoir.

Human Activities and their Influence on Biogeochemical Cycles and Climate Change

Life on earth is inextricably linked to climate through a variety of interacting cycles and feedback loops. In recent years there has been a growing awareness of the extent to which human activities, such as **deforestation** and fossil fuel burning, have directly or indirectly modified the biogeochemical and physical processes involved in determining the earth's climate. These changes in atmospheric processes can disturb a variety of {C}ecosystem services that humans depend upon. In addition to helping to maintain relative climate stability ecosystem services protect living organisms on earth from the sun's harmful ultraviolet rays, mediate runoff and **evapotranspiration** and regulate nutrient cycling.

Humans clearly disrupt many, if not all biogeochemical cycles and in the process threaten many ecosystems. In recent years human activities have directly or indirectly affected the biogeochemical cycles that determine climatic conditions of earth. It is imperative to mention that, managing and understanding environmental problems caused by climate change would require an understanding of the biogeochemical cycles. Biogeochemical cycles always involve equilibrium states: a balance in the cycling of the element between spheres. However, overall balance may involve elements distributed on a global scale and that is why a disruption in one cycle causes a disruption in all other cycles. Below is a summary of how human activities have contributed to disruption of biogeochemical cycles. For impacts on specific cycles, the reader should refer to the sites where these cycles are presented.

Use of phosphorus fertilizers: Human influences on the phosphorus cycle come mainly from the introduction and use of commercial synthetic fertilizers. Use of fertilizers mainly has affected the phosphorus and nitrogen cycles. Plants may not be able to utilize all of the phosphate fertilizer applied; as a consequence, much of it is lost from the land through the water run-off. The phosphate in the water is eventually precipitated as sediments at the bottom of the water body. In certain lakes and ponds this may be redissolved and recycled as a problem nutrient. Animal wastes or manure may also be applied to land as fertilizer. If misapplied on frozen ground during the winter, much of the fertilizer may be lost when ice melts and forms runoff. In certain areas very large feed lots of animals, may result in excessive run-off of phosphate and nitrate into streams. Other human sources of phosphate are in the out flows from municipal sewage treatment plants. Without an expensive tertiary treatment, the phosphate in sewage is not removed during various treatment operations. Again an extra amount of phosphate enters the water.

Mining of Fossil fuels: Humans have interfered with the carbon cycle where fossil fuels have been mined from the earth crust. Had fossils not been discovered prior to industrial revolution, they could have remained there until now. Carbon dioxide is number one green house gas contributing to global warming and climate change. Additionally, clearing of vegetation that serve as carbon sinks has increased the concentration of carbondioxide in the atmosphere.

Production of Sulphur dioxide: Human impact on the sulfur cycle is primarily in the production of **sulfur dioxide** (SO_2) from industry (e.g. burning coal) and the internal combustion engine. Sulfur dioxide can precipitate onto surfaces where it can be oxidized to sulphate in the soil (it is also toxic to some plants), reduced to sulphide in the atmosphere, or oxidized to sulphate in the atmosphere as sulphiric acid (a principal component of acid rain). Sulphur compounds play a big role in the climate system because they are important for the formation of clouds.

Additionally, a lot of sulphur is brought into the air by **volcanic eruptions**. A strong eruption can emit particles up to the stratosphere hence leading to cooling down of the planet.

Cultivation of legumes and use of nitrogen fertilizers:

As a result of extensive cultivation of legumes, creation of chemical fertilizers, and pollution emitted by vehicles and industrial plants, human beings have more than doubled the annual transfer of nitrogen into biologically available forms. Humans have significantly contributed to the transfer of nitrogen gases from Earth to the atmosphere, and from the land to aquatic systems through four main processes:

The application of nitrogen fertilizers to crops has caused increased rates of denitrification and leaching of nitrate into groundwater. The additional nitrogen entering the **groundwater** system eventually flows into streams, rivers, lakes, and estuaries. In these systems, the added nitrogen can lead to **eutrophication**.

- Increased deposition of nitrogen from atmospheric sources because of fossil fuel combustion and forest burning. Both of these processes release a variety of solid forms of nitrogen through combustion.
- Livestock ranching. Livestock release a large amounts of ammonia into the environment from their wastes. This nitrogen enters the soil system and then the hydrologic system through leaching, groundwater flow, and runoff.
- Sewage waste and septic tank leaching.

Water as a "Driver" of Biogeochemical Cycles

The water cycle is powered from solar energy, more than 85% of the global evaporation occurs from the oceans, reducing their temperature by cooling by evaporation. Without the cooling effect of evaporation the greenhouse effect would lead to a much higher surface temperature of about 67^0 C and a warmer planet.

While the water cycle is itself a biogeochemical cycle, flow of water over and beneath the Earth is a key component of the cycling of other biogeochemical cycles. Runoff is responsible for almost all of the transport of eroded sediment and phosphorus from land to water bodies. The salinity of the oceans is derived from erosion and transport of dissolved salts from the land. Eutrophication of lakes is primarily due to phosphorus, applied in excess to agricultural fields in fertilizers, and then transported overland and down rivers. Both runoff and groundwater flow play significant roles in transporting nitrogen from the land to water bodies. Runoff also plays a part in the carbon cycle, through the transport of eroded rock and soil.

The climatic conditions determine the speed and the life span of the biogeochemical cycles. Scientists are still investigating the pathways that the biogeochemical cycles take and how human activity has affected these cycles. There are still many questions that are unanswered.

Generalized Biogeochemical Cycle

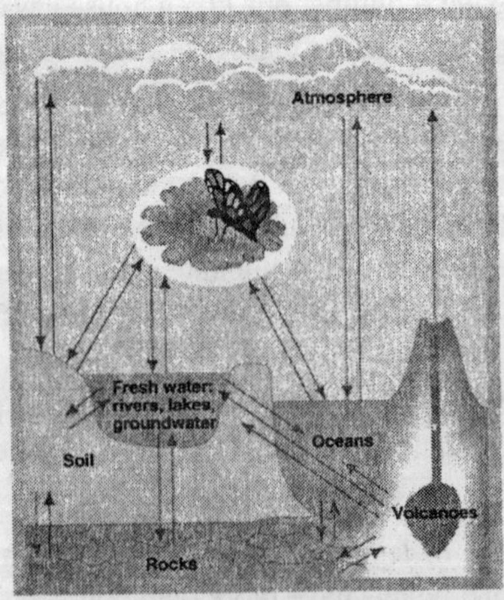

Generalized biogeochemical cycle. The major parts of the biosphere are connected by the flow of chemical elements and compounds. In many of these cycles, the biota plays an important role. Matter from the Earth's interior is released by volcanoes. The atmosphere exchanges some compounds and elements rapidly with the biota and oceans. Exchanges of materials between rocks, soils, and the oceans are generally slower by comparison.

Hydrologic(Water) Cycle

The hydrologic cycle begins with the **evaporation** of water from the surface of the ocean. As moist air is lifted, it cools and water vapor **condenses** to form clouds. Moisture is **transported** around the globe until it returns to the surface as **precipitation**. Once the water reaches the ground, one of two processes may occur; 1) some of the water may evaporate back into the atmosphere or 2) the water may penetrate the surface and become **groundwater**. Groundwater either seeps its way to into the oceans, rivers, and streams, or is released back into the atmosphere through **transpiration**. The balance of water that remains on the earth's surface is **runoff**, which empties into lakes, rivers and streams and is carried back to the oceans, where the cycle begins again.

Evaporation of warm surface water increases the amount of moisture in the colder, drier air flowing immediately above the lake surface. With continued evaporation, water vapor in the cold air **condenses** to form ice-crystal clouds, which are **transported** toward shore.

By the time these clouds reach the shoreline, they are filled with snowflakes too large to remain suspended in the air and consequently, they fall along the shoreline as **precipitation**. The intensity of lake effect snowfall can be enhanced by additional lifting due to the hills along the shoreline. Once the snow begins to melt, the water is either absorbed by the ground and becomes **groundwater**, or goes returns back to the lake gap-off.

Estimate of Global Water Distribution

Water source	Water volume, in cubic kilometers	Percent of freshwater	Percent of total water
Oceans, Seas, & Bays	1, 338, 000, 000	–	96.5
Ice caps, Glaciers, & Permanent Snow	24, 064, 000	68.7	1.74
Groundwater	23, 400, 000	–	1.69
Fresh	10, 530, 000	30.1	0.76
Saline	12, 870, 000	–	0.93
Soil Moisture	16, 500	0.05	0.001
Ground Ice & Permafrost	300, 000	0.86	0.022
Lakes	176, 400	–	0.013
Fresh	91, 000	0.26	0.007
Saline	85, 400	–	0.006
Atmosphere	12, 900	0.04	0.001
Swamp Water	11, 470	0.03	0.0008
Rivers	2, 120	0.006	0.0002
Biological Water	1, 120	0.003	0.0001

The hydrologic cycle, showing the transfer of water from the oceans to the atmosphere to the continents and back to the oceans again.

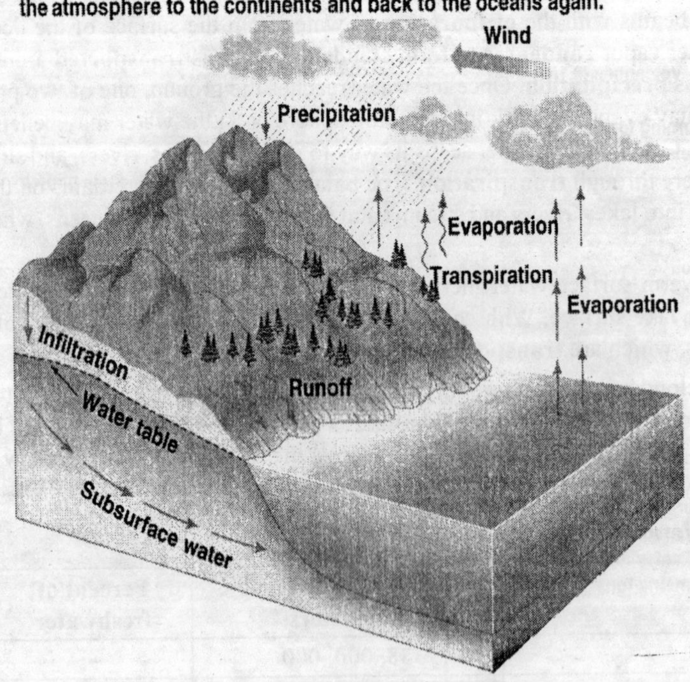

Carbon Cycle

The carbon cycle is one of the most important to humans because it is important to our existence:

- one of the primary elements forming human tissues.
- necessary to plants, the basis of human food

and because it is important to the climate system which sets the background for our environment.

- carbon dioxide (CO_2) and methane (CH_4) are greenhouse gases which help set global temperatures.

Global Carbon Storage

Location	Amount (gigatons carbon)
Atmosphere	750
Land plants	610
Soil and detritus	1, 500
Surface ocean	1, 020
Intermediate and deep ocean	37, 890
Sediments	78, 000, 000

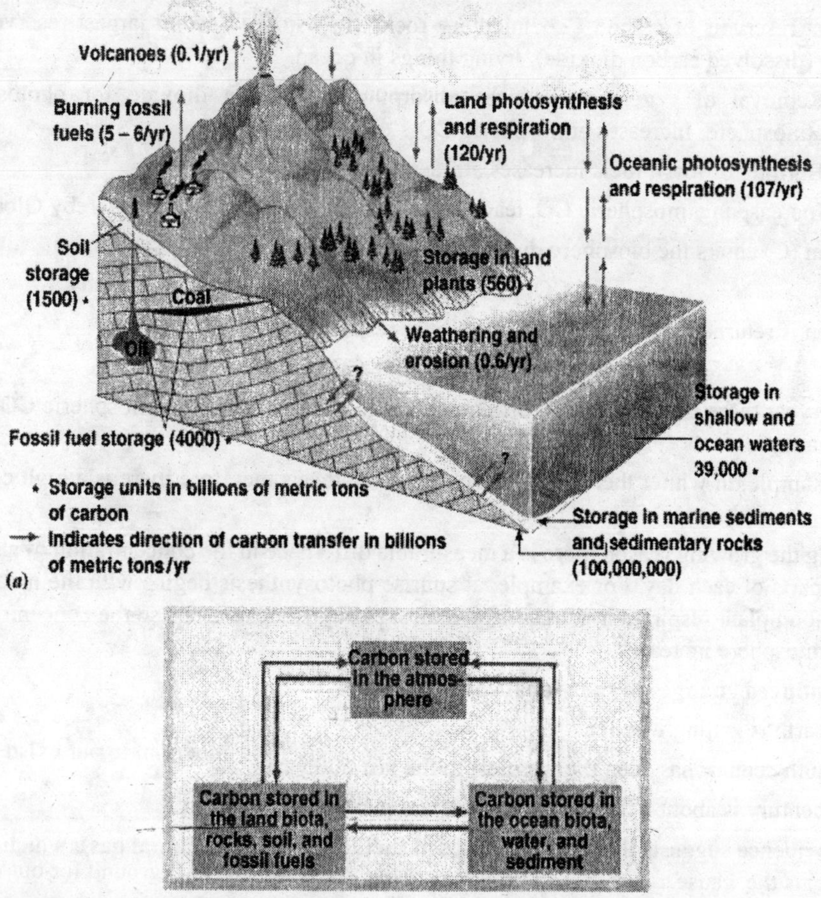

(a) Generalized global carbon cycle. (b) Parts of the carbon cycle simplified to illustrate the cyclic nature of the movement of carbon.

Carbon Cycle Factsheet

- Is an atmospheric cycle.
- Carbon is required for formation of organic compounds in living things.
- C in carbon dioxide in atmosphere and in water is moved to C in glucose by photosynthesis by producers.
- C in glucose is moved to C in carbon dioxide by cellular respiration.
- C in glucose is moved to C in organic molecules by synthesis reactions in living things.
- C in organic molecules is moved to C in carbon dioxide by combustion.
- C in organic molecules in organisms is moved to C in fossil fuels over millions of years by pressure, heat, and bacterial action.

- C in limestone ($CaCO_3$) is released slowly to C in carbon dioxide when exposed to oxygen and/or water.
- Largest reservoir of carbon - sedimentary rocks (limestone) Second largest reservoir of carbon - ocean (dissolved carbon dioxide), living things in ocean.
  - Removal of vegetation reduces absorption of carbon dioxide for photosynthesis from atmosphere. Increases atmospheric CO_2.
  - Burning of fossil fuels increases atmospheric CO_2.
  - Increase in atmospheric CO_2 leads to increased Greenhouse Effect thereby Global Warming.
- Carbon (C) enters the biosphere during photosynthesis:

$$CO_2 + H_2O \rightarrow C_6H_{12}O_6 + O_2 + H_2O$$

- Carbon is returned to the biosphere in cellular respiration:

$$O_2 + H_2O + C_6H_{12}O_6 \rightarrow CO_2 + H_2O + \text{energy}$$

- Every year there is a measurable difference in the concentration of atmospheric CO_2 in phase with the seasons.

  For example, in winter there is almost no photosynthesis therefore there is a high concentration of CO_2.
- During the growing season there is a measurable differnece in the concentration of atmospheric CO_2 over parts of each day. For example, at sunrise photosynthesis begins with the uptake of CO_2, by afternoon plant respiration increases, at sunset photosynthesis stops so the concentration of CO_2 in the atmosphere increases.

Human induced changes in the global carbon cycle:

- The Earth is getting warmer.
- The 20th century has been the warmest in the last 600 years.
- This century is about 1 degree Fahrenheit warmer than last century.
- The evidence suggests that burning of fossil fuel (eg. coal, oil, natural gas), which emits CO_2 as a waste, is the cause.
- CO_2 is a «Green House» gas - it traps heat at the Earth›s surface. (H_2O vapor and methane are also examples of green house gases)

Signs that the climate is warming:

- Plants start blooming 8 days earlier in spring than 11 years ago.
- Birds from the United Kingom lay eggs earlier.
- Buds on trees appear earlier and leaves fall later in the Northern Hemisphere.
- Alaska, North West Canada, and Siberia have warmed up as much as 5 degrees Fahrenheit in the last 30 years.

Nitrogen Cycle

Nitrogen cycle is important because:

- nitrogen is a necessary nutrient
- nitrogen is part of acid rain

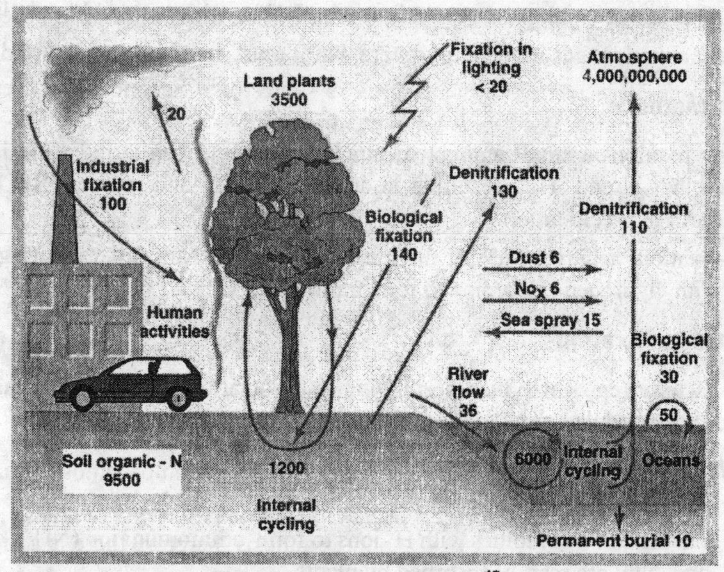

The global nitrogen cycle. Pools (□) and annual (↔) flux in $10^{12} gN_2$. Note that the industrial fixation of nitrogen is nearly equal to the global biological fixation.

Nitrogen Cycle

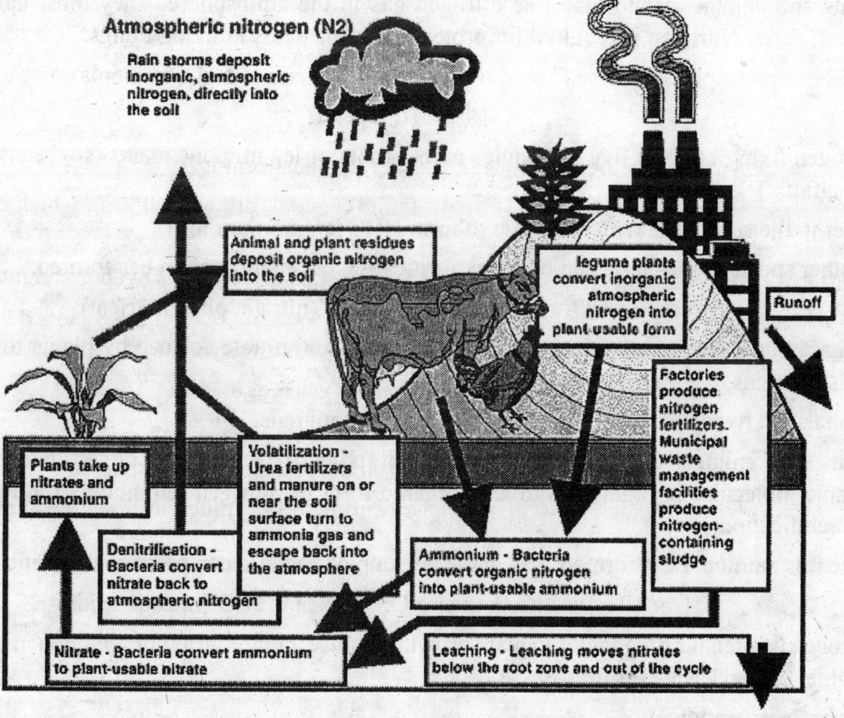

- Nitrogen (N) is an essential constituent of protein, DNA, RNA, and chlorophyll.
- N is the most abundant gas in the atmosphere, but it must be fixed or converted into a usable form.

Nitrogen Fixation Methods

1. High energy fixation- a small amount of atmospheric nitrogen is fixed by lightening. The high energy combines N and H_2O resulting in ammonia (NH_3) and nitrates (NO_3). These forms are carried to Earth in precipitation.
2. Biological fixation: achieves 90% of the nitrogen fixation. Atmospheric nitrogen (N_2) is split and combined with hydrogen (H) atoms to form ammonia (NH_3).

Who performs nitrogen fixation?

- symbiotic bacteria (eg. Rhizobium spp.)living in association with leguminous (plants in the pea family), and root-noduled non- leguminous plants.
  - free-living anaerobic bacteria
  - blue-green algae (cyanobacteria)

  Once NH_3 is in the soil it combines with H^+ ions to form ammonium ion (NH_4), or without it to form NO_3. NH_4^+ and NO_3 are readily absorbed by plants.

Nitrogen Cycle Fact Sheet

- Is an atmospheric cycle.
- Plants and animals cannot use free nitrogen gas in the atmosphere. They must have nitrogen in "fixed" form. Nitrogen is required for proteins, nucleic acids in living things.
- Free N_2 in atmosphere is "fixed" by nitrogen-fixing bacteria to NH_3 (ammonia):

$$N_2 + 3H_2 \rightarrow 2NH_3$$

- Nitrogen fixing bacteria live in nodules on the roots of leguminous plants (soybeans, peas, clover, and alfalfa.)
- Water in the soil reacts with ammonia to form NH_4^+ (ammonium ion)
- Another species of bacteria can perform nitrification once ammonium has formed:

$$NH_4^+ \rightarrow NO_2^- \text{(nitrite; toxic)} \rightarrow NO_3^- \text{(nitrate; plant nutrient)}$$

- Assimilation - absorption of ammonia, ammonium ion, nitrate for use by plants to make nucleic acids, proteins
- Animals get fixed nitrogen by eating plants or other animals.
- Plants and animals are broken down by still other bacteria that convert nitrogen-containing organic molecules in organisms to an inorganic form of nitrogen (ammonia or ammonium ion) = ammonification
- Once this ammonia has formed, still another group of bacteria can perform denitrification:

$$NH_3 \text{ or } NH_4^+ \rightarrow NO_2^- \text{ and/or } NO_3^- \rightarrow N_2 \text{ and } N_2O \text{ (nitrous oxide)}$$

- Nitrogen is often limiting factor in plant growth because ammonia, ammonium ion, nitrate are water-soluble: can be leached from soil.
- Human Intervention:

- In the atmosphere:

$$N_2 + O_2 \rightarrow 2NO \text{ (nitric oxide) produced when burning fuel or forests.}$$

(Heat combines N_2 and O_2 present in atmosphere)

$$NO + O_2 \rightarrow NO_2 \text{ (nitrogen dioxide gas)}$$

$$NO_2 + H_2O \rightarrow HNO_3 \text{ (nitric acid - dissolved in water causes acid deposition)}$$

- N_2O (nitrous oxide) released from decomposition of fertilisers and waste.
- Excess nitrogen added to aquatic systems by runoff of artificial fertilizer, farm waste, discharge of sewage. This stimulates growth of algae. Breakdown of algae by aerobic decomposers depletes water of oxygen.

Oxygen Cycle

Oxygen, like carbon and hydrogen, is a basic element of life. In addition, in the form of O_3 ozone, it provides protection of life by filtering out the sun's UV rays as they enter the stratosphere. In addition to constituting about 20% of the atmosphere, oxygen is ubiquitous. It also occurs in combination as oxides in the Earth's crust and mantle, and as water in the oceans.

Early in the evolution of the Earth, oxygen is believed to have been released from water vapor by UV radiation and accumulated in the atmosphere as the hydrogen escaped into the earth's gravity. Later, photosynthesis became a source of oxygen. Oxygen is also released as organic carbon in CHO, and gets buried in sediments.

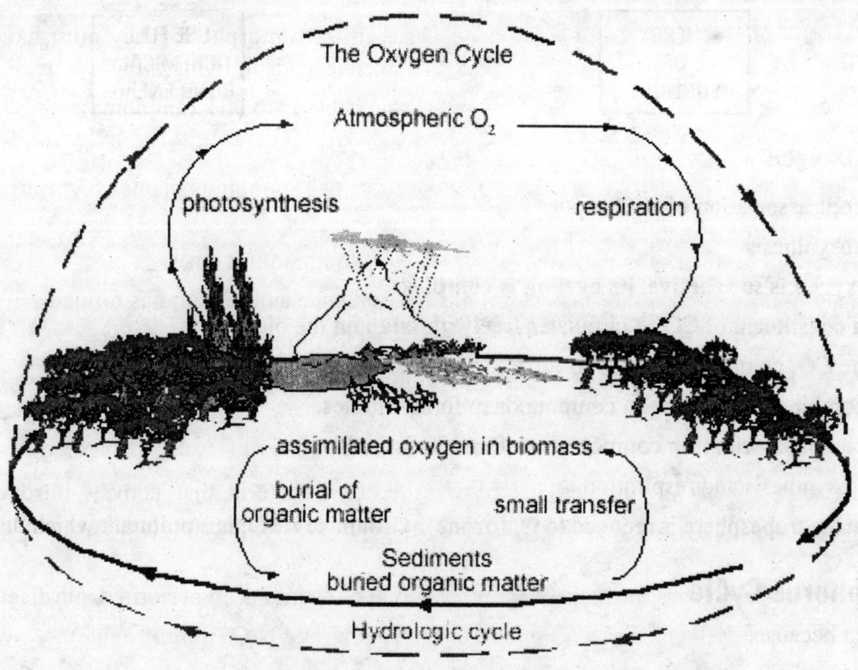

Figure: The Oxygen Cycle

Oxygen is highly reactive. A colorless, odorless gas at ordinary temperatures, it turns to a bluish liquid at -183° C. Figure above shows an overview of oxygen cycling in nature. The environment of oxygen in numerous reactions make it hard to present a complete picture.

Oxygen is vital to us in many ways (besides for breathing). Water can dissolve oxygen and it is this dissolved oxygen that supports aquatic life. Oxygen is also needed for the decomposition of organic waste. Wastes from living organisms are "biodegradable" because there are aerobic bacteria that convert organic waste materials into stable inorganic materials. If enough oxygen is not available for these bacteria, for example, because of enormous quantities of wastes in a body of water, they die and anaerobic bacteria that do not need oxygen take over. These bacteria change waste material into H_2S and other poisonous and foul-smelling substances. For this reason, the content of biodegradable substances in waste waters is expressed by a special index called "biological oxygen demand" (BOD), representing the amount of oxygen needed by aerobic bacteria to decompose the waste.

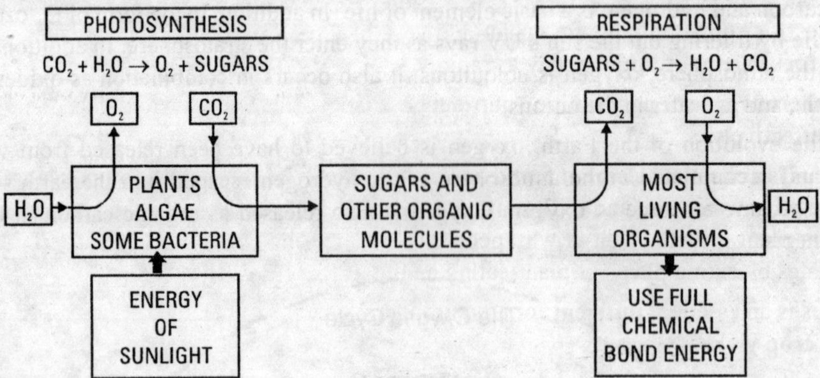

Sources of Oxygen

1. photodisassociation of H_2O vapor
2. photosynthesis

Since oxygen is so reactive, its cycling is complex:

1. As a constituent of CO_2 it circulates freely throughout the biosphere.
2. Some CO_2 combines with Ca to form carbonates.
3. O_2 combines with nitrogen compounds to form nitrates.
4. O_2 combines with iron compounds to form ferric oxides.
5. Photosynthesis and respiration
6. O_2 in the troposphere is reduced to O_3 (ozone). Ground level O_3 is a pollutant which damages lungs.

The Phosphorus Cycle

It is important because:

- Phosphorus is a necessary, *limiting* nutrient
- Phosphate runoff causes *eutrophication*

Reservoirs: (in millions of metric tons)

Earth's Crust: 20, 000, 000, 000 (recoverable : ~20, 000)

Ocean: 100, 000

Freshwater: ~100

Land Plants: ~3000

Soils: ~100, 000 (most of the phosphorus is in rocks that are unrecoverable)

- Phosphorous has no stable gas phase, so addition of P to land is slow (low rain P).
- Most P in plants cycles between living and dead plants... addition by weathering is small compared to cycling within plants.
- Humans have greatly accelerated P transfer from rocks to plants and soils (about 5x faster than weathering).
- Natural transfer of P from ocean to land is very small.
- Sources for human mining are guano and very old rocks formed in shallow seas which dried up . Such rocks are not forming today as rapidly.
- Phosphorous is a strongly limiting nutrient because it cannot be transferred from the ocean to plants very effectively.

Phosphorus is a chemical **element** found on Earth in numerous **compound** forms, such as the phosphate **ion**(PO_4^{3-}), located in water, soil and **sediments.** The quantities of phosphorus in soil are generally small, and this often limits plant growth. That is why people often apply phosphate fertilisers on farmland. Animals absorb phosphates by eating plants or plant-eating animals.

Phosphorus is an essential **nutrient** for animals and plants. Insufficient phosphorus in the soil can result in a decreased **crop** yield.

Phosphorus moves in a cycle through rocks, water, soil and sediments and organisms.

The key steps of the **phosphorus cycle** are :

- Over time, rain and **weathering** cause rocks to release phosphate ions and other minerals. This **inorganic** phosphate is then distributed in soils and water.
- Plants take up inorganic phosphate from the soil. The plants may then be consumed by animals. Once in the plant or animal, the phosphate is incorporated into **organic** molecules such as DNA. When the plant or animal dies, it decays, and the organic phosphate is returned to the soil.
- Within the soil, organic forms of phosphate can be made available to plants by **bacteria** that break down organic **matter** to inorganic forms of phosphorus. This process is known as **mineralisation.**
- Phosphorus in soil can end up in waterways and eventually oceans. Once there, it can be incorporated into sediments over time.

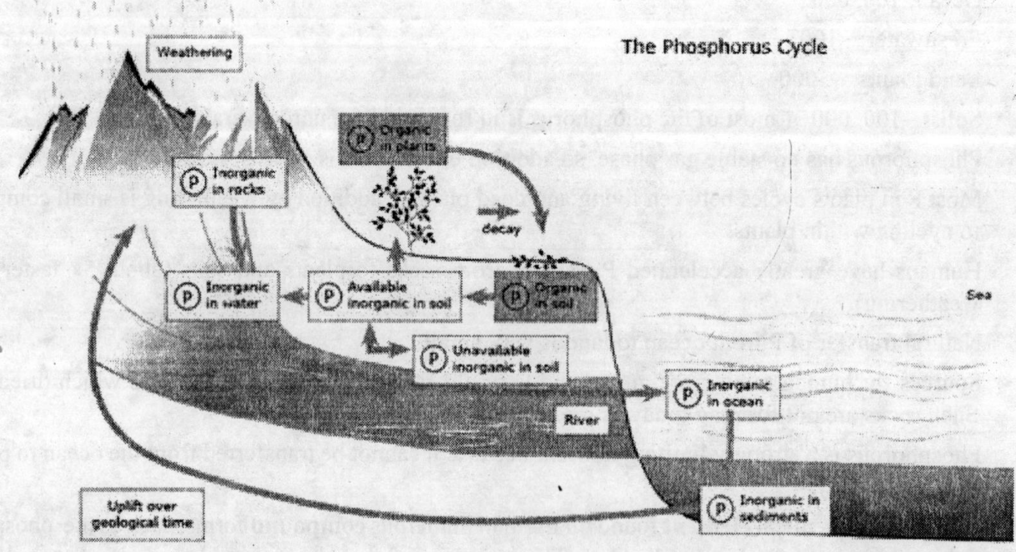

Figure: The phosphorus cycle

Since most of our phosphorus is locked up in sediments and rocks, it's not available for plants to use. A lot of the phosphorus in soils is also unavailable to plants. Soil phosphorus becomes unavailable to plants through several routes:

- **Bacteria**: Bacteria convert plant-available phosphate into organic forms that are then not available to plants. Although other bacteria make phosphate available by mineralisation, the contribution of this is small.
- **Adsorption**: Inorganic (and available) phosphorus can be chemically bound (adsorbed) to soil particles, making it unavailable to plants. **Desorption** is the release of adsorbed phosphorus from its bound state into soil solution.
- **pH**: Inorganic phosphorus compounds need to be soluble to be taken up by plants. This depends on the acidity (pH) of the soil. If soils are less than pH 4 or greater than pH 8, the phosphorus starts to become tied up with other compounds, making it less available to plants.

Many plant crops need more phosphorus than is dissolved in the soil to grow optimally. In addition, crops are usually harvested and removed – leaving no decaying **vegetation** to replace phosphorus. Therefore, farmers replenish the phosphorus 'pool' by adding fertilisers or **effluent** to replace the phosphorus taken up by plants.

Phosphorus availability and unavailability to plants

Up take

pH-dependent chemical reactions

microbet

Minerals containing phosphate

P

Immobilisation

Mineralisation

Chemically bound to soil

Adsorption

Figure: Phosphorus availability to plants

Farmers replenish phosphorus through the use of phosphatic fertilisers. Adjusting the pH of the soil for efficient plant uptake of phosphate should be done prior to **fertilisation**. For example, adding **lime** reduces soil acidity, which provides an environment where phosphate becomes more available to plant. When fields are overfertilised, phosphate not utilised by plants can be lost from the soil through **leaching** and water **run-off**. This phosphate ends up in waterways, lakes and **estuaries**. Excess phosphate causes excessive growth of plants in waterways, lakes and estuaries leading to **eutrophication**.

The Sulfur Cycle

Sulphur is one of the components that make up proteins and vitamins. Proteins consist of amino acids that contain sulphur atoms. Sulphur is important for the functioning of proteins and enzymes in plants, and in animals that depend upon plants for sulphur. Plants absorb sulphur when it is dissolved in water. Animals consume these plants, so that they take up enough sulphur to maintain their health.

Most of the earth's sulphur is tied up in rocks and salts or buried deep in the ocean in oceanic sediments. Sulphur can also be found in the atmosphere. It enters the atmosphere through both natural and human sources. Natural recourses can be for instance volcanic eruptions, bacterial processes, evaporation from water, or decaying organisms. When sulphur enters the atmosphere through human activity, this is mainly a consequence of industrial processes where sulphur dioxide (SO_2) and hydrogen sulphide (H_2S) gases are emitted on a wide scale.

When sulphur dioxide enters the atmosphere it w ill react with oxygen to produce sulphur trioxide gas (SO_3), or with other chemicals in the atmosphere, to produce sulphur salts. Sulphur dioxide may also react with water to produce sulphuric acid (H_2SO_4). Sulphuric acid may also be produced from demethylsulphide, which is emitted to the atmosphere by plankton species.

All these particles will settle back onto earth, or react with rain and fall back onto earth as **acid deposition**. The particles will than be absorbed by plants again and are released back into the atmosphere, so that the sulphur cycle will start over again.

Sulfur (S), is a brittle, yellow, tasteless, and odorless non-metallic element. It comprises many vitamins, proteins, and hormones that play critical roles in both climate and in the health of various ecosystems. The majority of the Earth's sulfur is stored underground in rocks and minerals, including as sulfate salts buried deep within ocean sediments.

The sulfur cycle contains both atmospheric and terrestrial processes. Within the terrestrial portion, the cycle begins with the **weathering** of rocks, releasing the stored sulfur. The sulfur then comes into contact with air where it is converted into sulfate (SO_4). The sulfate is taken up by plants and microorganisms and is converted into organic forms; animals then consume these organic forms through foods they eat, thereby moving the sulfur through the food chain. As organisms die and decompose, some of the sulfur is again released as a sulfate and some enters the tissues of microorganisms. Natural sources that emit sulfur directly into the atmosphere, include volcanic eruptions, breakdown of organic matter in swamps & tidal flats, and the evaporation of water.

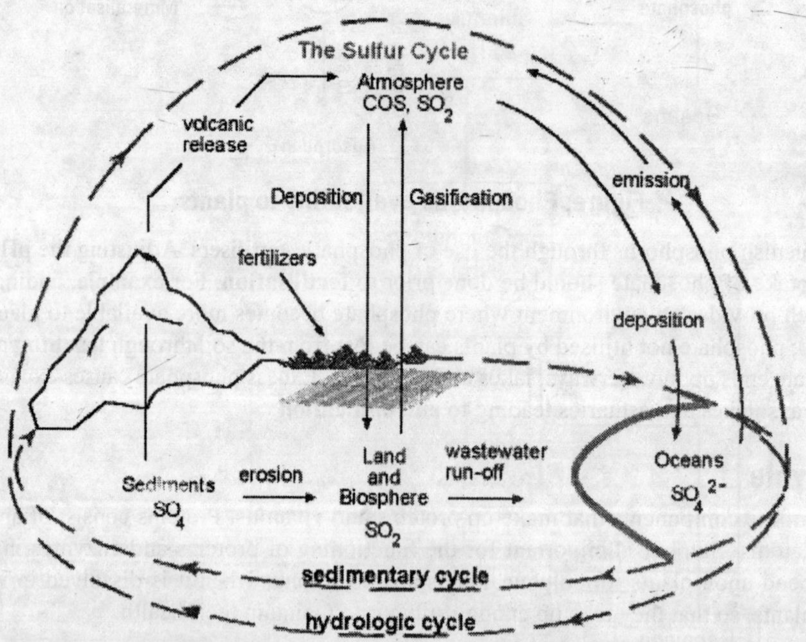

Sulfur eventually settles back into the Earth or comes down with rainfall. A continuous loss of sulfur from terrestrial ecosystem runoff occurs through drainage into lakes & streams, and eventually oceans. Sulfur also enters the ocean through fallout from the Earth's atmosphere. Within the ocean, some sulfur cycles through marine communities, moving through the food chain. A portion of this sulfur is emitted back into the atmosphere from sea spray. The remaining sulfur is lost to the ocean depths, combining with iron to form ferrous sulfide which is responsible for the black color of most marine sediments.

Since the Industrial Revolution, human activities have contributed to the amount of sulfur that enters the atmosphere, primarily through the burning of fossil fuels and the processing of metals. Emissions from these activities, along with nitrogen emissions, react with other chemicals in the atmosphere to produce tiny particles of sulfate salts which fall as **acid rain**, causing a variety of damage to both the natural & man-made environments, such as the chemical weathering of buildings. However, as particles and tiny airborne droplets, sulfur also acts as a regulator of global climate. Sulfur dioxide and sulfate aerosols absorb **ultraviolet radiation**, creating cloud cover that cools cities and may offset global warming caused by the **greenhouse effect.**

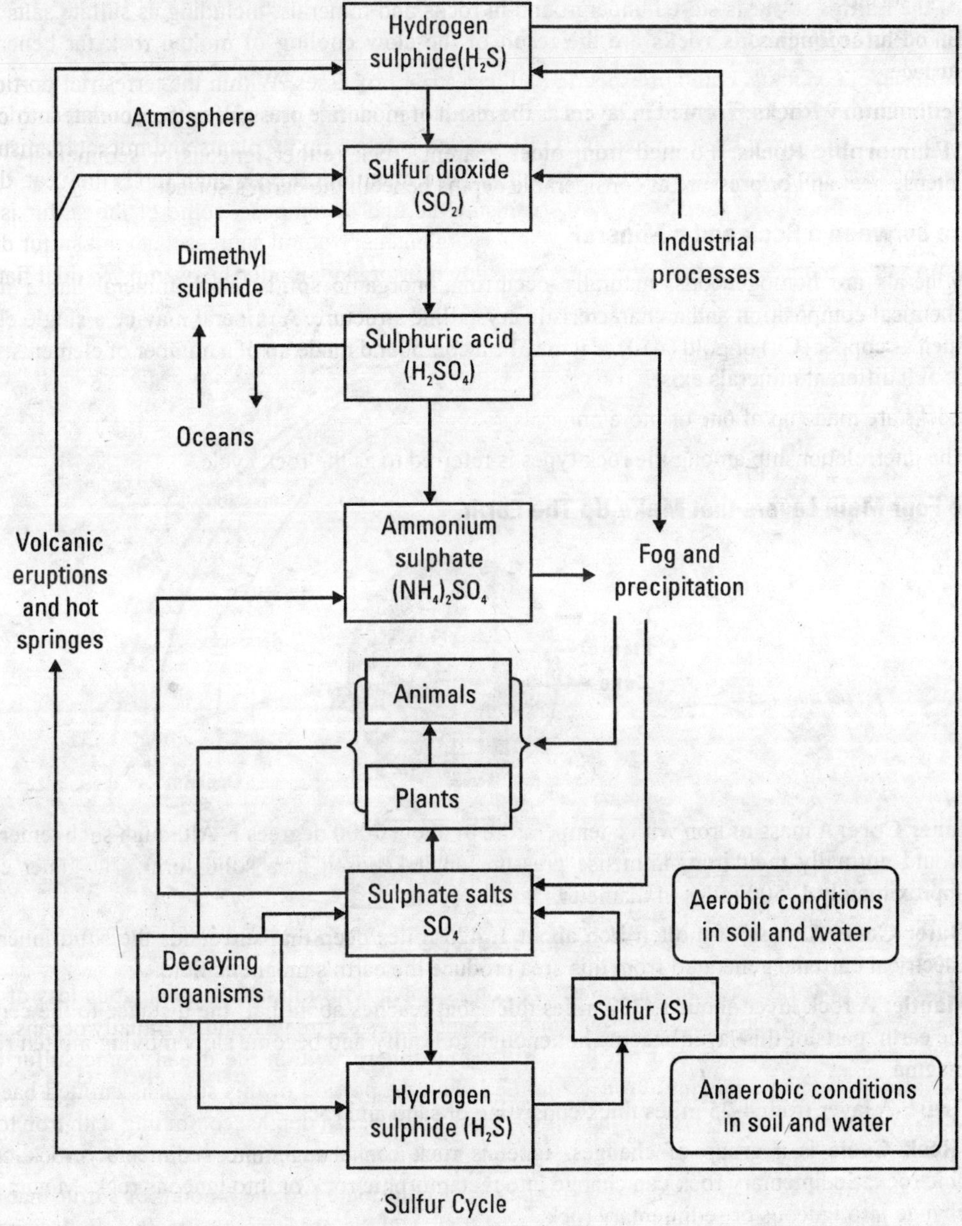

Figure: Sulfur Cycle

The Rock Cycle

There are three types of rocks. They are:

1. **Igneous Rocks:** Formed from the cooling of molten rock.

   A. **Volcanic igneous rocks** formed from molten rock that cooled quickly on or near the earth's surface.

   B. **Plutonic igneous rocks** are the result of the slow cooling of molten rock far beneath the surface.

2. **Sedimentary Rocks:** Formed in layers as the result of moderate pressure on accumulated sediments.
3. **Metamorphic Rocks:** Formed from older "parent" rock (either igneous or sedimentary) under intense heat and/or pressure at considerable depths beneath the earth's surface

Difference between a Rock and a Mineral

- Minerals are homogeneous, naturally occurring, inorganic solids. Each mineral has a definite chemical composition and a characteristic crystalline structure. A mineral may be a single element such as copper (Cu) or gold (Au), or it may be a compound made up of a number of elements. About 2, 500 different minerals exist.
- Rocks are made up of one or more minerals.
- The interrelationship among the rock types is referred to as the rock cycle.

There are Four Main Layers that Make Up The Earth

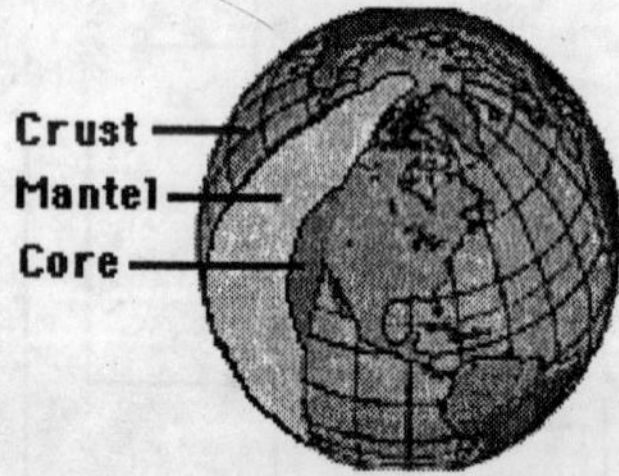

1. **Inner Core:** A mass of iron with a temperature of about 7000 degrees F. Although such temeratures would normally melt iron, immense pressure on it keeps it in a solid form. The inner core is approximately 1, 500 miles in diameter.
2. **Outer Core:** A mass of molten iron about 1, 425 miles deep that surrounds the solid inner core. Electrical currents generated from this area produce the earth's magnetic field.
3. **Mantle:** A rock layer about 1, 750 miles thick that reaches about half the distance to the center of the earth. parts of this layer become hot enough to liquify and become slow moving molten rock or magma.
4. **Crust:** A layer from 4-25 miles thick consisting of sand and rock.

The Rock Cycle is a group of changes. Igneous rock can change into sedimentary rock or into metamorphic rock. Sedimentary rock can change into metamorphic rock or into igneous rock. Metamorphic rock can change into igneous or sedimentary rock.

Igneous rock forms when magma cools and makes crystals. Magma is a hot liquid made of melted minerals. The minerals can form crystals when they cool. Igneous rock can form underground, where the magma cools slowly. Or, igneous rock can form above ground, where the magma cools quickly.

When it pours out on Earth's surface, magma is called lava, the same liquid rock matter that comes out of volcanoes.

On Earth's surface, wind and water can break rock into pieces. They can also carry rock pieces to another place. Usually, the rock pieces, called sediments, drop from the wind or water to make a layer. The layer can be buried under other layers of sediments. After a long time the sediments can be cemented together to make sedimentary rock. In this way, igneous rock can become sedimentary rock.

All rock can be heated. Inside Earth there is heat from pressure & friction. There is also heat from radioactive decay. Heat bakes the rock. Baked rock does not melt, but it does change. It forms crystals. If it has crystals already, it forms larger crystals. Because this rock changes, it is called metamorphic. Metamorphosis can occur in rock when they are heated to 300 to 700 O°C.

When Earth's tectonic plates move around, they produce heat. When they collide, they build mountains and metamorphose the rock.

The rock cycle continues. Mountains made of metamorphic rocks can be broken up and washed away by streams. New sediments from these mountains can make new sedimentary rock. The rock cycle never stops.

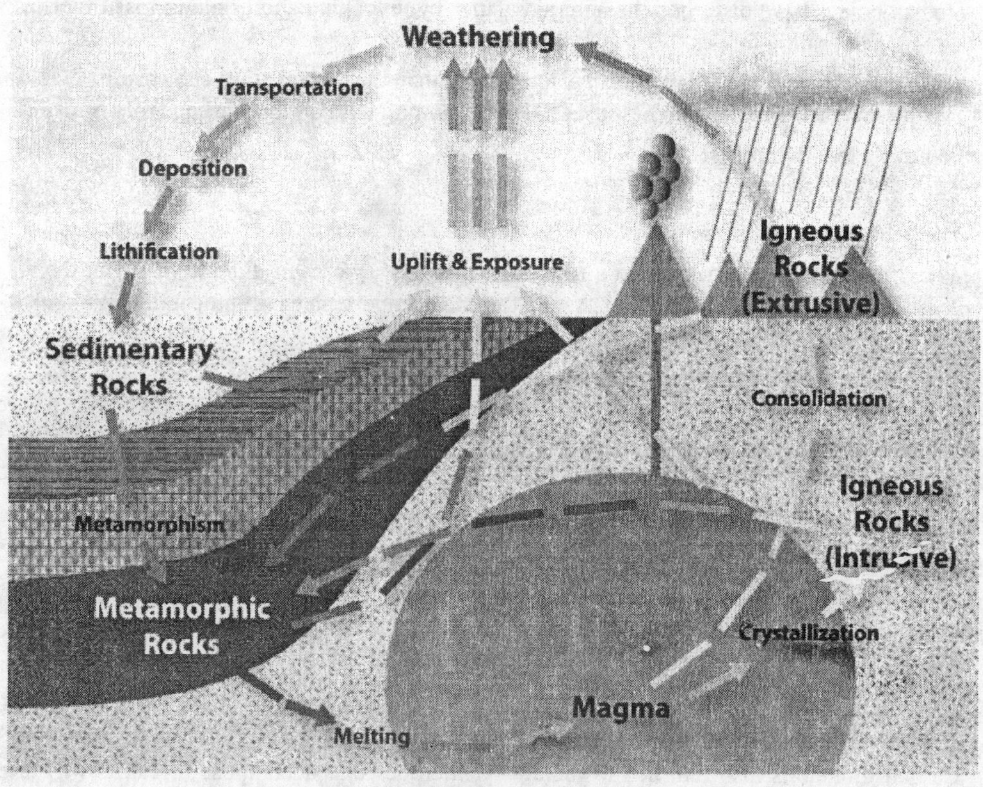

Figure: Rock Cycle

Environmental Cycles

Environmental cycle is a natural process in which elements are continuously cycled in various forms between different compartments of the environment (eg, air, water, soil, organisms). Examples include the water cycle, carbon cycle, oxygen cycle, nitrogen cycle, phosphorous cycle, sulphur cycle etc.

Carbon Cycle

The **carbon cycle** includes the uptake of carbon dioxide by plants through, its ingestion by animals and its release to the atmosphere through respiration and decay of organic materials. Human activities like the burning of fossil fuels contribute to the release of carbon dioxide in the atmosphere.

The Carbon Cycle fact sheet:

- Carbon dioxide is normally found in very small amounts in the atmosphere (0.03%).
- Burning fossil fuels produces carbon dioxide.
- The carbon cycle consists of a natural balance between photosynthesis and respiration in living things.
- Fossil fuels from when decay is prevented. Their formation happened slowly over millions of year, locking up huge amounts of carbon.
- Human beings have disturbed the natural balance by unlocking and bringing fossil fuels as well as cutting and brining forests.
- Since today more carbon dioxide is being released than the carbon sinks can absorb, CO2 levels are rapidly increasing to their highest ever. Global warming with all its consequences is resulting.

Carbon cycle has two parts:

- The biological carbon cycle
- The geological carbon cycle

Biological carbon cycle: It is the role of carbon as a gas ie, carbon dioxide (CO_2) in the air during photosynthesis whereby plants make food that is the most important environmentally. Without it, living things on earth could not survive.

Plants make the food (glucose) from carbon dioxide, water and sunlight. During this process called photosynthesis, carbon dioxide is removed from the atmosphere to make the glucose-and oxygen is released into it. Plants use glucose for energy and to make tissues such as cellulose-sad lignin.
The equation describing photosynthesis is:

Energy from sun: **Carbon dioxide + water → glucose + oxygen chlorophyll**

Since animals can't make their own food, they must eat plants or other animals that eat plants. In this way the carbon in the air is used to make the bodies of animals. The carbon dioxide that is removed from the air by photosynthesis is returned to it in two main ways:

- The carbon that makes up the food, glucose is used as energy by plants and animals, when it combines with, oxygen in a process called respiration. Respiration takes place in every cell. Once waste product of this is carbon dioxide which is carbon dioxide which is then returned to the atmosphere.
- Carbon dioxide is also returned to the air by microbes that decompose (break down) the bodies of living things after their death in the presence of oxygen. This can only happen when the dead remains are open to the atmosphere.

Glucose + oxygen → energy + CO_2 + water

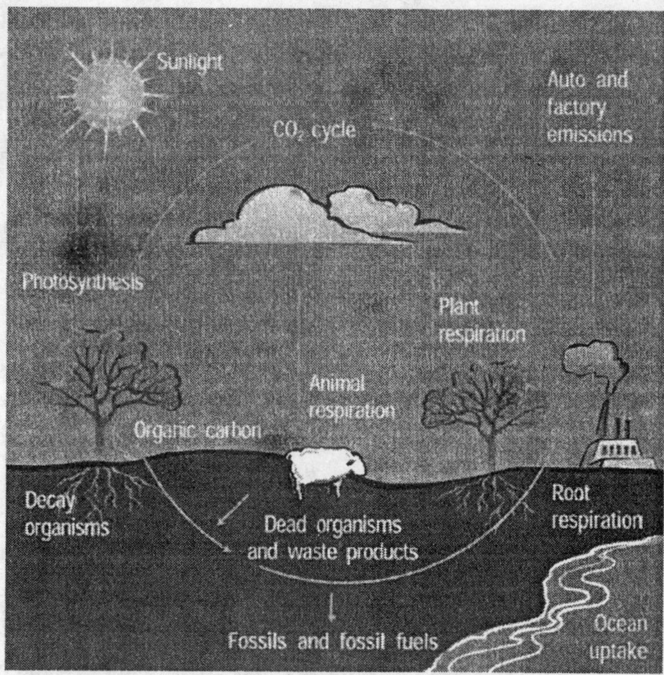

Figure: The carbon cycle

The carbon cycle has been shown in fig.

The carbon cycle has been in a natural balance for millions of years. Around the same amount of CO_2 is removed from the air by photosynthesis as is returned to it through respiration and the decay of dead plants and animals.

The geological carbon cycle: A small leak can occur if dead plants or animals get 'locked up' within rocks before they decay and release their carbon dioxide back to the' air. Over millions of years, this small leak has resulted in most of the carbon in the cycle being locked into rocks (97%) or fossil fuels (2.9%).

Fossil fuels such as coal, oil and natural gas are formed when decay cannot occur because oxygen formed is absent. Coal was formed about 250 million years ago when many parts of the world were covered by subtropical swamps. Some of the dead vegetation became covered by sand and mud and was compressed into coal. Oil and natural gas resulted about 150 million years ago from the remains of Dead Sea creatures falling to the sea bed and being covered by sand and mud that was compressed into rock. As a result of the temperature rising in these underground rocks and the action of anaerobic bacteria, the sea creatures were converted into oil and gas. This locking of carbon happened slowly over millions of years.

Some carbon dioxide is also removed from the air by dissolving in sea water. The world's oceans (just like the forests) are called 'carbon sinks' because they can take carbon dioxide out of the air. Cold water can dissolve mote carbon dioxide than warm water.

Most carbon is locked up in sedimentary rocks such as chalk and limestone that were made from the shells of tiny sea creatures. These sea creatures took the carbon dioxide out of the sea water to make their shells. When they dies they fell to the sea bed and were compressed by the weight of sand and mud deposited on top of them. Uplift has since raised these rocks so that many of them now form hills on land.

Throughout human history until now the carbon cycle has been in balance So that the proportions of carbon in each part of the cycle have remained the same. Human activities are now changing this balance, increasing the amount of carbon dioxide in the air with devastating effects on the environment including acid rain and global warming.

The spread of industrialisation, agriculture and the increased use of fossil fuels have all interfered with the natural carbon cycle. When coal, oil or natural gas are burnt carbon dioxide is released back into the atmosphere. For the last 200 years this has been happening at a much faster rate than it was originally removed when the fuels were formed.

Burning fossil fuels can also result in acid rain 'acid rain' this falls on limestone can cause a chemical reaction that releases even more carbon dioxide into the air.

Forest clearance also has an impact. Because research shows that in mature forests, photosynthesis, respiration and decay balance out, if trees are cut or clear-felled, then there is less change to the overall carbon balance. But most forests are cleared for agriculture and are burnt. Burning releases vast amounts of carbon dioxide very quickly.

Trees and oceans act as carbon sinks. This means that they can take carbon dioxide out of the air. Planting new forests is an effective way of removing carbon dioxide from the air since when a tree is growing it is needing a lot of carbon for energy and tissue growth.

Oceans can also remove carbon from the air but this amount has an upper limit. Eventually the seawater becomes saturated and can dissolve no more. Because cold water absorbs more carbon dioxide than warm water, as the oceans warm through global warming, the ocean becomes saturated and begins to release carbon dioxide back into the air. This in turn may cause air temperatures to rise and cause a 'runaway' temperature increase.

Today, levels of carbon dioxide in the air are at, their highest ever. Only about a half of new releases of carbon dioxide into the air are being 'mopped up' by the oceans. So the concentration of atmospheric carbon dioxide is increasing.

Water (Hydrological) Cycle

The water cycle is the process by which water travels in a sequence from the air (condensation) to the earth (precipitation) and returns to the atmosphere (evaporation). It is also referred to as the hydrologic cycle.

Human use of water can transform the water cycle through irrigation or the construction of dams, for example.

The water cycle fact sheet:

- The amount of accessible fresh water in the earth is limited. Global consumption is doubling every 20 years. Now sources are becoming expensive to develop and treat.
- Water supplies are not evenly distributed. Many countries already experience a water shortage and water costs are rising.
- Agriculture uses most water followed by industry and cities.
- Water supplies today are increasingly becoming polluted.
- The consequences of today's growing water use include a disruption of groundwater supplies, ecosystems wildlife and people's livelihoods; increased pollution and an increased likelihood of conflict between cities, states and nations over access to water.

Just as there is a constant cycling of carbon in a 'biological' and 'geological' cycle, the same is true for water. Water is essential for all living things. Without it there could be no life as it form an essential part of photosynthesis the process by which plants capture the sun's energy and use it for living and growth. While water is 'cycled' through living things, it also becomes trapped or locked into rocks. Aquifers hold 99% of the world's fresh water and while some are recharged by fresh water, many hold water that seeped into them millions of years ago.

In anyone year, heat from the sun is enough to evaporate about one a quarter metres of seawater. Two thirds of this vapour that escapes the sea falls back into it as rain, but the rest drops from clouds that move over land. This runs off as rivers, soaks into aquifers, is trapped in dams or is taken up by plants and then released into the air through evapo-transpiration. Most of it is returned to the oceans within ten days, which means that often the, same water is used over and over again.

For millions of years, people made little impact on this natural cycle as they made use of perennial rivers, natural springs and lakes. Gradually however with population growth, Industrialisation and the needs of urban dwellers for food and water, more and more water has been extracted from this natural cycle. Because global water consumption is doubling every 20 years and new sources are becoming scarcer, within 30 years there will not be enough water to go round. At the rate the world is developing and population is growing, water will become the most pressing environmental and developmental issue this century.

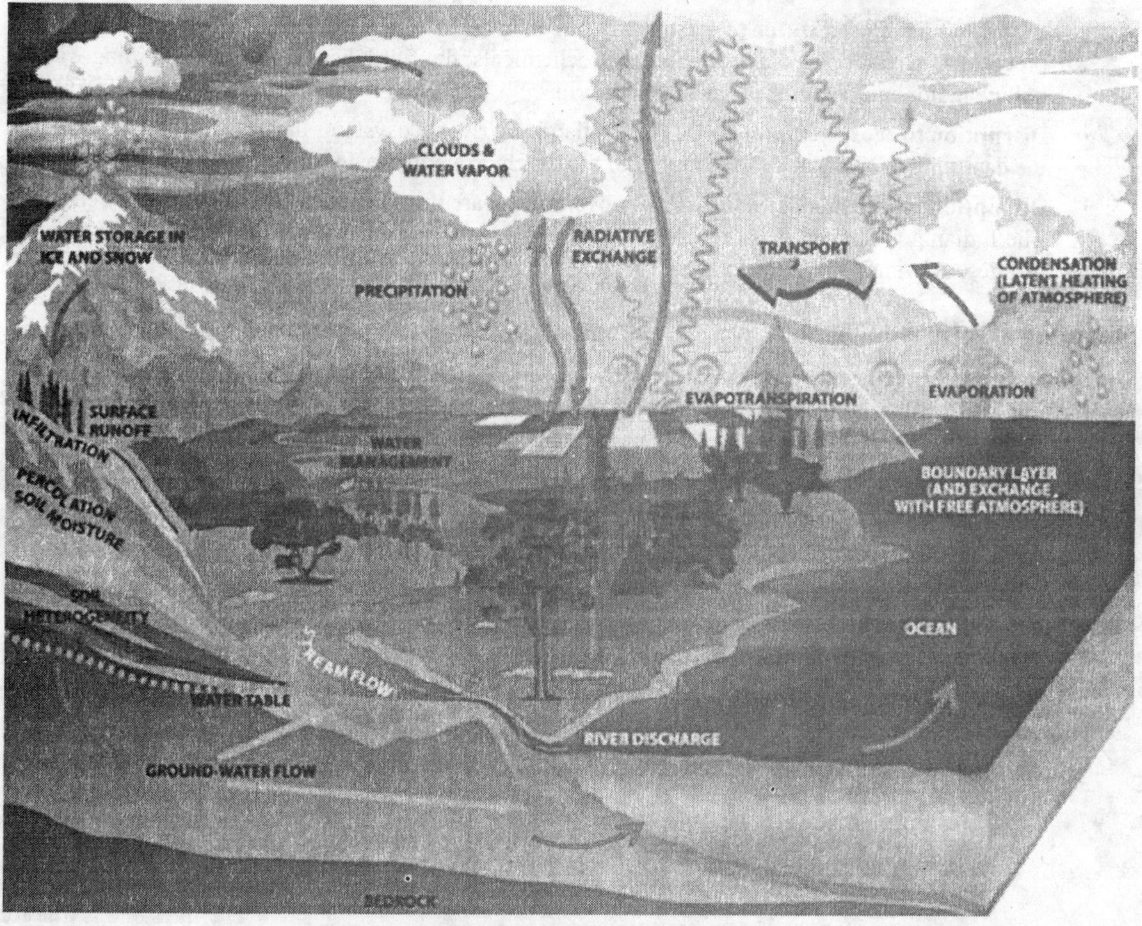

Figure: The Water/Hydrological Cycle

Water cycle has been shown in figure

Human activity and the water cycle:

- Agriculture uses 87% of all water globally with rice and cotton requiring the most water.
- Industry uses the next.
- Domestic consumption is the least.

Some environmental consequences of interrupting the cycle include:

- **Decrease in ground water supplies:** 99% of all the world's fresh water is found in aquifers. These are waterlogged seams of rock or gravel. The deeper they are, the more costly it is to exploit them while some aquifers are still recharged by rainfall, others contain water that became trapped in them thousands of years ago and cannot be replenished. As aquifers run dry, water has to be sourced further away from where it is required, increasing its cost.
- **Increased pollution:** Overpumping of groundwater has resulted in the little water that remains containing increased concentrations of pollutants. For example, up to 1 million wells in Bangla Desh were found to contain arsenic resulting in slow poisoning, skin lesions and cancer.
- **Industries have used water to wash away harmful chemicals:** Used as part of the manufacturing process. As a result pesticides, nitrates, petrochemicals, fluorides, heavy metals and mining wastes pollute major aquifers in the 'developed' world.
- **Disruption to ecosystems and wildlife:** Wetland and river ecosystems have been disrupted through the damming of rivers and the pumping of too much groundwater for irrigation.
- **Disruption in livelihoods:** Increasingly governments are thinking of bigger water supply schemes. The Indian government plans to divert to link 14 major rivers including the Brahmaputra and the Ganges, evicting 3 million people as 300 reservoirs are built.

Nitrogen Cycle

The **nitrogen cycle** involves the uptake of nitrogen form the atmosphere by a process called fixation which is carried out by microbes or industrial processes. Decomposition of biological waste by microbes can return nitrogen to the atmosphere. Nitrogen is mainly used by humans as a fertilizer in farmlands, but its excessive usage can lead to serious problems (such as eutrophication).

The Nitrogen cycle fact sheet:

- All life requires nitrogen in order to live and grow (make proteins and nucleic acids)
- Nitrogen in the air is 'fixed' in a cycle that involves four processes: fixation, decay, nitrification and denitrification. These processes change nitrogen gas to nitrates, decaying materials to ammonia and then finally back to nitrates or nitrogen.
- Human activities have disturbed this cycle and added more nitrogen to both land and air through the manufacture of fertilisers, the burning of fossil fuels and car emissions.
- An excess of nitrates in the ground has resulted in lake eutrophication and excess nitrates

contaminating water and food supplies. An excess of nitrogen oxide in the air has led to acid rain.

Nitrogen is essential to all life on earth. Nitrogengas makes up 78% of the air we breathe. In its various forms it passes through a series of natural processes from the air, to the soil, to plants and animals and then back to the soil and the air.

All life requires nitrogen compounds to make proteins and nucleic acids. But most organisms cannot use the nitrogen that is in the air. Plants must secure it in a 'fixed' or inorganic form such as nitrates (NO_3), ammonia (NH_3) or urea $(NH_2)_2CO_2$. Animals secure their nitrogen (and all other) compounds from plants or animals that have fed on plants.

Four processes participate in the cycling of nitrogen through the biosphere:

Nitrogen fixation: Nitrogen (N_2) is quite inert and must be broken apart before it combines with other atoms to form useful compounds. This can happen in three ways:

- **Atmospheric fixation:** Where the energy released by lightning breaks apart the nitrogen molecules and enables their atoms to combine with oxygen forming nitrogen oxides. These dissolve in rain forming nitrates that are carried to the ground. Atmospheric nitrogen contributes about 58% of all nitrogen fixed.
- **Industrial fixation:** through the Haber where nitrogen is combined with hydrogen, under great pressure and heat to make ammonia (NH_3) Ammonia can be used as a fertiliser but most of it is processed further to make urea and ammonium nitrate.
- **Biological fixation:** certain bacteria can fix nitrogen. Some live in a symbiotic relationship with plants of the legume family (soybeans alfalfa), while others live in a similar relationship with other plants. Others exist freely in the soil while cyanobacteria are essential to maintaining the fertility of semi aquatic environments such as rice paddies. In the process ammonia is formed, which is quickly incorporated into protein and other organic nitrogen compounds.
- **Decay:** The proteins made by plants are eaten by herbivores. Carnivores then prey on herbivores for their protein. Most of this protein is used to make muscle and other tissue but some is used for energy and the nitrogen is not needed. Mammals excrete waste nitrogen as urea in their urine. Animal dung doesn't contain much nitrogen based matter but the urine that is mixed with it is an important nitrogen source, though other material in the dung helps to conserve water and nutrients and adds to soil fertility. These materials are then used by two types of micro-organisms of decay:
  - **Nitrification:** Nitrifying bacteria turn them first into ammonia, then to nitrites and finally to nitrates a form that plants can take up and use
  - **De-nitrification:** In the absence of air, denitrifying bacteria break down the nitrogen compounds and release them back into the air as nitrogen molecules.

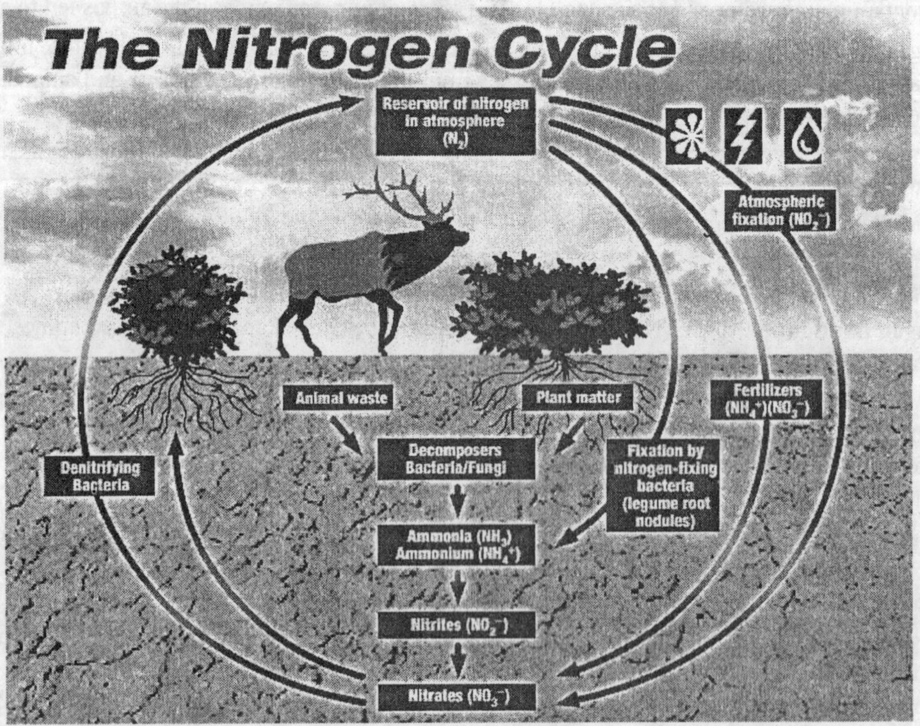

Figure: The Nitrogen Cycle

Human activities and Nitrogen Cycle:

The Nitrogen cycle is shown in figure above.

Today as a result of the 'green revolution' the use of artificial fertilisers is widespread. Agriculture is now responsible for around a half of all the nitrogen fixation taking place on earth through:

- The use of fertilizers
- The growing of legumes

While yields have increased, the misuse or overuse of fertilisers has had serious environmental consequences.

Eutrophication is the name given to the process that occurs in lakes, dams or bays when excess nitrates or phosphates get into the water. Although it is a natural process in the aging of lakes, human activities may greatly accelerate it. Agricultural run-off, urban run-off, the discharge from food factories, leaking septic and sewage systems and the discharge of sewage all increase the flow of these 'nutrients'. When this is washed into a lake it has an excessive 'fertilising' effect causing algae to grow quickly or 'bloom'. Algal blooms hurt water ecosystems in two ways:firstly they cloud the water and block sunlight underwater grasses to die. Because these grasses provide food and shelter for water organisms, the spawning and nursery habitat is destroyed and waterfowl have less to eat when grasses die off. Secondly when the algae die and decompose, oxygen is used up. Dissolved oxygen in the water is essential to most organisms living in water such as fish and crabs. In the absence of oxygen other bacteria multiply giving the water a bad smell and rendering it undrinkable. Limiting fertiliser use treating sewage properly compo sting and recycling and controlling run-off all help to prevent eutrophication.

Acid rain: Burning fossil fuels, forests and the increased use of road and air transport have all resulted in large quantities of nitrogen oxide (as well as other gases) being released into the air. As a result local concentrations of nitrogen oxide have risen so much that when dissolved in rain water it has been precipitated as nitric acid or 'acid rain'.

Health issues: We are exposed to nitrogen in the food we eat, the water we drink and the air we breathe. Drinking water sources are susceptible to watershed run-off. Ground water contamination from over-fertilisation of plant life, human and animal waste discharges can occur. Excessive amounts of nitrogen as nitrate in drinking water can cause health problems especially in babies and young infants. Removing the nitrates is expensive and adds to the cost of water.

Phosphorous Cycle

The **phosphorus cycle** involves the uptake of phosphorus by organisms. Phosphorus in the environment is mainly found in rocks, and natural weathering processes can make it available to biological systems. After decomposition of biological waste, it can accumulate in large amounts in soils and sediments. Phosphorus is used by humans as a fertilizer in farmlands and in detergents. Overuse of phosphorus can lead to eutrophication. Unlike carbon and nitrogen, phosphorus does not have a gaseous phase on the earth. This leads to major difference in its cycle. It is an, essential element, one of the "big six", required in large quantities by all forms of life. Phosphate is often a limiting element for plant growth.

Phosphate exists in the atmosphere as part of small particles. In addition, phosphorous tends to form compounds that are relatively insoluble in water and therefore are not readily eroded as part of the hydrological cycle. Phosphorous occurs generally in its oxidized state as phosphate, which combine with calcium, potassium, magnesium and iron to, form minerals found in soils and in water. As a result, the rate of transfer of phosphorous is slow in comparison to carbon and nitrogen.

Phosphorous enters the biotic system through uptake by plants and some bacteria. Plants can take up phosphorous in its oxidized form as phosphate (a common ion) Phosphate minerals are relatively insoluble in water and therefore, phosphorous is available slowly through weathering of rocks or rock particles in the soil. It is lost from the soil as suspended particles through run-off and is transported by rivers and streams to the ocean.

Phosphorous returns from the ocean to the land by way of ocean feeding birds. These birds feed on tiny ocean plankton. Plankton thrive when nutritionally essential chemical elements occur. These occurs in areas of oceanic upwellings, which are rising currents. Upwelling carry nutrients including phosphorous, from the depths to the surface. Thus upwellings are among the most fertile and biologically productive regions of oceans. Many major fisheries of the world occur where there are upwellings.

The fish-eating birds nest on off shore islands where they are protected from predators. Over time their nesting sites become covered with their phosphorous laden excrement.

In dry climates, the Guano hardens into a rock like mass that may be up to 40 m thick. These have provided source of the major sources of phosphorous for fertilizers.

Another source of phosphorus fertilizers is the mining of sedimentary rocks made up of marine animal's fossils.

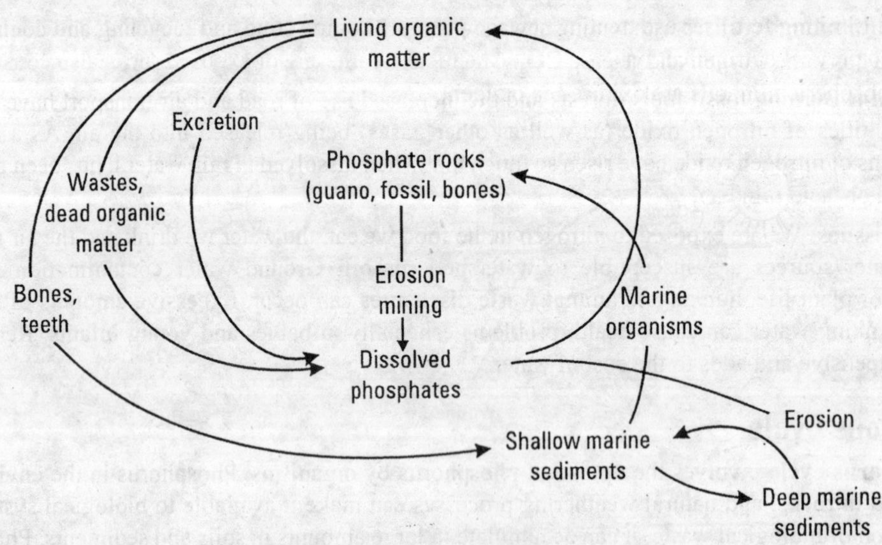

Figure: The phosphorus cycle

The phosphorous cycle has been shown in figure above.

Human influences in the phosphate cycle which is mainly from the introduction and use of commercial synthetic fertilizers. The phosphate is obtained through mining of deposits of calcium phosphate call *Apatite*. Huge quantities of sulphuric is used in the conversion of the phosphate rock into a fertilizer product called "super phosphate". Plants use limited amount of phosphate from the fertilizer and the balance is lost from the soil through the water run-off and eventually precipitates as sediments at the bottom of water bodies. In certain lakes and ponds this may be redissolved and recycled as nutrient for water plants.

Oxygen Cycle

It is the biogeochemical cycle that describes the movement of oxygen within and between its three main reservoirs: the atmosphere (air), the total content of biological matter within the biosphere (the global sum of all ecosystems), and the lithosphere (earth's crust). Failures in the oxygen cycle within the hydrosphere (the combined mass of water found on, under, and over the surface of a planet) can result in the development of hypoxic zones. The main driving factor of the oxygen cycle is photosynthesis, which is responsible for the modern earth's atmosphere and life.

The largest reservoir of earth's oxygen is within the silicate and oxide minerals of the earth's crust and mantle (99.5%). Only a small portion has been released as free oxygen to the biosphere (0.01%) produces sugars and oxygen from carbon dioxide and water:

Photosynthesizing organisms include the plant life of the land areas as well as the phytoplankton of the oceans. An additional source of atmospheric oxygen comes from photolysis, whereby high energy ultraviolet radiation breaks down atmospheric water and nitrous oxide into component atmos. The free H and N atoms escape into space leaving O_2 in the atmosphere.

Oxygen is mainly lost from the atmosphere by respiration and decay mechanisms in which animal life and bacteria consume oxygen and release CO_2. Chemical weathering of exposed rocks also consume oxygen because lithospheric minerals are oxidised in oxygen.

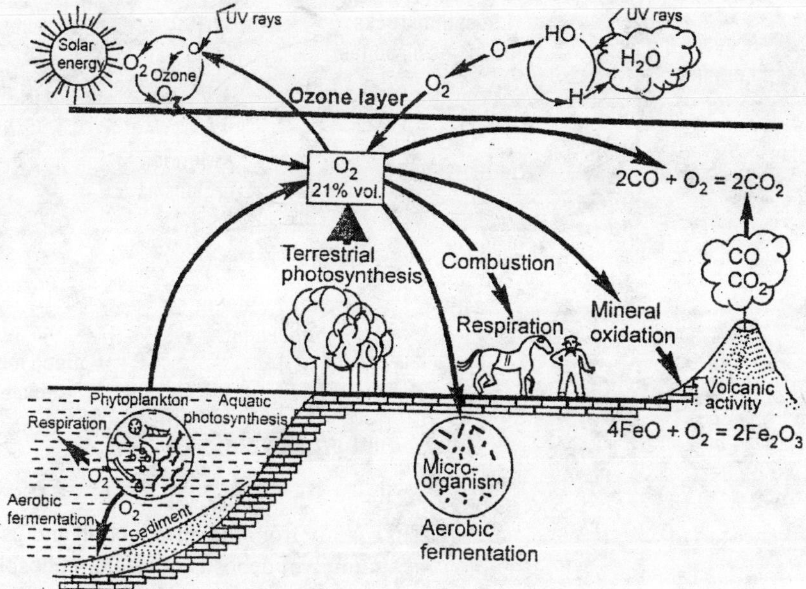

Figure: Oxygen cycle

Oxygen cycle is shown in figure above.

Oxygen is also cycled between biosphere and lithosphere. Marine organism in the biosphere create $CaCO_3$ shell materials that is rich in oxygen. When the organism dies, its shell is deposited on the shallow sea floor and buried over time to create the limestone rock of the lithosphere. Weathering processes initiated by organisms can also free oxygen from the lithosphere. Plants and animals extract nutrient minerals from rocks and release oxygen in the processes.

The presence of atmospheric oxygen has lead to the formation of Ozone and the ozone layer within the stratosphere. This ozone layer absorbs the harmful UVradiation.

[illegible]

[illegible]

Figure: Oxygen cycle

[illegible]

[illegible]

[illegible]

Appendices

Contents **Page No.**

Appendix 4.1: Biosphere Encompassing Atmosphere, Ecosphere, Hydrosphere & Lithosphere

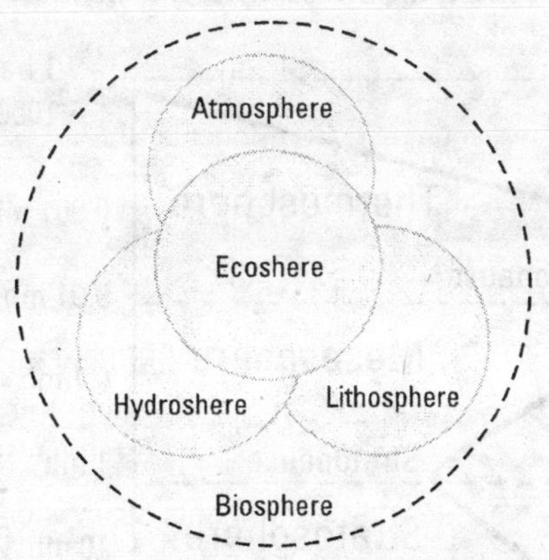

NOTES

Abiotic and biotic environmental components are as follows:

Abiotic environmental components	Hydrosphere (Ocean, lake, river, ground water)
	Lithosphere (solid earths, soil)
	Atmosphere
Biotic environmental components	Living organisms (animal, plant, fungi, bacteria, virus)
	Dead organic matters

Appendix 4.2: Various Layers of Lithosphere

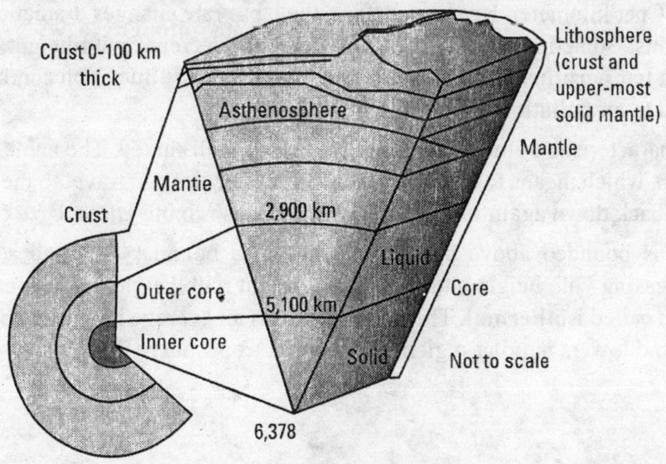

Appendix 4.3: Atmospheric Layers and their Features

Each layer of the atmosphere has distinct characteristics. There are 5 main layers within the atmosphere. They are the troposphere, the stratosphere, the mesosphere, the thermosphere, and the exosphere.

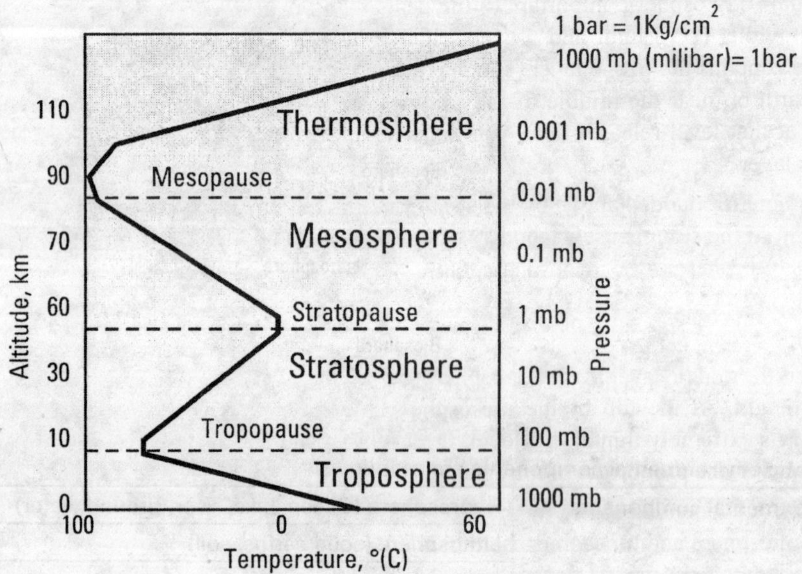

Figure: Layers of atmosphere.

Troposphere

The troposphere is the lowest layer of the atmosphere. This is the layer where most weather takes place. Most thunderstorms don't go much above the top of the troposphere (about 11 km). In this layer, pressure and density rapidly decrease with height, and temperature generally decreases with height at a constant rate. The change of temperature with height, is known as the **lapse rate**. The standard lapse rate for the troposphere is a decrease of about 6.5°Cper kilometer. Near the surface, the lapse rate changes dramatically from hour to hour on clear days and nights. Sometimes the temperature does not decrease with height, but increases. Such a situation is known as a **temperature inversion**. Persistent temperature inversion conditions, which represent a stable layer, can lead to air pollution episodes.

The other main characteristic of the troposphere is that it is well-mixed. The name troposphere is derived from the Greek *tropein*, which means to turn or change. Air molecules can travel to the top of the troposphere (about 11 km up) and back down again in a just a few days. This mixing encourages changing weather.

The troposphere is bounded above by the tropopause, a boundary marked as the point where the temperature stop decreasing with height and becomes constant with height. Any layer where temperature is constant with height is called **isothermal**. The tropopause has an average height of about 11 km (it is higher in equatorial regions and lowers in polar regions).Above the troposphere is the stratosphere.

Stratosphere

The stratosphere is the layer above the troposphere, characterized primarily as a **stable, stratified** layer (hence, stratosphere) with a large temperature inversion throughout. The main impact the stratosphere has on weather is that its stable air prevents large storms from extending much beyond the tropopause.

The other main stratospheric important to life deals with **ozone**. Ozone is the triatomic from of oxygen that absorbs ultraviolet (UV) light and prevents it from reaching the earth's surface at dangerous levels. The stratosphere contains the ozone layer. The maximum concentrations of ozone are at about 25 km above the surface of earth or, near the middle of the stratosphere. The interaction between UV light, ozone, and the atmosphere at that level releases heat, warming the atmosphere and helping to create the temperature inversion in this layer.

The stratosphere is bounded above by the stratopause, where the atmosphere again becomes isothermal the average height of the stratopause is about 50 km. This is at about the 1 mb (0.1 kPa) pressure level. The layer above the stratosphere is the mesosphere.

Mesosphere

The mesosphere is the middle layer in the atmosphere (hence, mesosphere). The temperature in the mesosphere decreases with height. At the top of the mesosphere, air temperature reaches its coldest value, around −90 O°C. The air is extremely thin at this level. Over 99.9 percent of the atmosphere's mass lies below the mesosphere. However, the proportion of nitrogen and oxygen at these levels is about the same as at sea level.

The mesosphere is bounded above by the mesopause. The average height of the mesopause is about 85 km, where the atmosphere again becomes isothermal. This is at around the 0.005 mb(0.0005 kPa) pressure level.Above the mesosphere is the thermosphere.

Thermosphere

The thermosphere is a warm layer above the mesosphere. In this layer, there is a significant temperature inversion. The few molecules that are present in the thermosphere receive extraordinary amounts of energy from the sun, causing the layer to warm. Though the measured temperature is very hot, if you exposed your skin to the thermosphere, the perceived temperature would be very cold. Because there are so few molecules present, there would not be enough molecules bombarding your body to transfer heat to your skin.

Above the thermosphere is the exosphere. There is no well defined boundary between the thermosphere and the exosphere (i.e., there is no boundary layer called the thermopause).

Exosphere

The exosphere is the region where molecules from the atmosphere can overcome the pull of gravity and escape into outer space. The atmosphere slowly diffuses into the viod of space. The exosphere usually begins about 500 km up but there is no definable boundary to mark as the end of the thermosphere and the beginning of the exosphere. Even at heights of 800 km, the atmosphere is still measurable. However, molecule concentrations are very small and considered negligible.

Appendix 4.4: Factsheet of Environmental Components

Environment means surroundings. It includes the gaseous envelope surrounding earth, (the **Atmosphere**), the mass of water above &below the earth, (the **Hydrosphere),** the land masses that support life, (the **Lithosphere),** and the whole plethora of microbes, plants, & animals, collectively referred as, **Biosphere**. The living world has been sustained for over 200 million years by other components. "Environment is the sum of all substances and forces external to an organism which determines its existence and regulate its processes". Earth is a unique planet of Solar System which has free water, oxygen & life. The pyramid of life is supported by a complex set of interactions and exchanges between water, air, land and life.

The study of environment in totality is an arduous task. The age of earth is nearly 6000×10^6 years which shows that for a very long time no life existed on earth.

Atmosphere

The atmosphere constitutes a gaseous envelope that surrounds the earth and held in place by the forces of gravity. It is essential to life in terms of supplying oxygen for breathing and circulating carbon-dioxide. The composition of atmosphere is generally uniform upto an altitude of 80km. This portion of atmosphere is called **Homosphere** and is made up of 78.084% Nitrogen and 20.946% oxygen (by volume). The rest is made up of mainly inert gases (expressed as parts per million) as shown below:

Argon (Ar)	9340 ppm, (inert gas)
Neon (Ne)	182 ppm, (inert gas)
Helium (He)	53 ppm, (inert gas)
Krypton (Kr)	1.2 ppm, (inert gas)
Xenon (Xe)	1.0 ppm, (inert gas)
Radon (Rn)	Traces (Radio active/gas)
Carbon dioxide (CO_2)	(CO_2) 390 ppm, essential for life
Ozone (O_3)	Traces (Toxic)

Other components which are highly variable are, Sulfur-dioxide (SO_2) Carbon-mono-oxide (CO), Methane (CH_4), Nitrous oxide (N_2O) etc. The concentration of water vapour is highly variable near the earth's surface. The atmosphere also contains particulate matter (dust) in varying amounts. Other emissions may also add, hydrocarbons (HC), oxides of Nitrogen ($NO + NO_2$), Chlorofluorocarbons, spores & pollens (from moulds & plants).

The layer beyond homosphere extends upto 10, 000 km (80 km – 10, 000 km) is called **Heterosphere.** The composition of this layer is different at different levels as shown below:

Molecular Nitrogen layer (N_2)	80 km – 200 km
Atomic Oxygen layer (O)	200 km – 1100 km
Helium layer (He)	1100 km – 3500 km
Hydrogen layer (H_2)	3500 km – 10, 000 km

The limit of each layer is arbitrary. Beyond 10, 000 km, the space is interplanetary space, which has Hydrogen in approximately the same concentration as in Hydrogen layer.

The atmospheric pressure decreases with altitude almost exponentially. This determines the characteristics of upper atmosphere. There pressure at sea level is one atmosphere. At high altitudes the pressure is low as gases expand and normal reactive species persists for long periods of time.

The **thermal structure of atmosphere** is distinct and well defined. It consists of three layers in Homosphere and one in Heterosphere.

The atmosphere is stratified on the basis of temperature and densities. The lowest layer of atmosphere is **Troposhere.** It extends from ground to 10-16 km (10 km at poles and 16 km at equator) and is characterized by a gradual reduction in temperature, (termed as **lapse rate).** The cooling is because a parcel of air looses heat when it rises due to reduction in pressure. A dry parcel of air (if neither gains nor looses temperature from surroundings) cools at the rate of 0.96°C/100 m and is known as **Dry Adiabatic Lapse Rate (DALR).** Such conditions of atmosphere are rare. Troposhere has highly variable water vapour. It contains most of water and clouds. Therefore the cooling could be higher or lower than DALR. The troposphere ends at **tropopause.** The temperature of tropopause at poles and equator are different, and is much smaller at poles, due to equatorial bulge.

Temperature of atmosphere at pole if is 0°C than at tropopause it will be -96°C, where as if the temperature of atmosphere at equator is 40°C the temperature at tropopause will be -113°C.

The intense cold conditions at tropopause serves as a barrier that causes water vapour to condense to ice and the so the water vapour does not reach the next layer.

At tropopause (10 – 16 km), there is a gradual change of temperature reduction. The temperatures beyond rise to as high as ground temperatures. This layer is known as **Stratosphere.** The rise in temperature is due to the presence of Ozone, which absorbs UV radiations. At an altitude of 50 km and more, ions are more active and the region is also called ionosphere. Ultraviolet radiations are responsible for this phenomenon.

The stratosphere gradually through **Stratapause** changes over to **Mesosphere.** It extends upto 85 km and is very windy & turbulent. The mesosphere is characterized by decrease in temperature. The absence of high levels of radiation absorbing species in mesosphere results in decrease in temperature to about – 83° to -92°C. The upper regions of Mesosphere defines a region called Exosphere from which molecules and ions do not escape. This is the last layer of Homosphere. Mesosphere is followed by a region with rarified gases which absorb high energy radiations (= 100 nm) and reach temperatures as high as 1100°C – 1650°C. This is known as **Thermosphere.** Troposphere, Stratosphere and Mesosphere lie in **Homosphere** and Thermosphere in **Heterosphere**.

The air pressure exerted by atmosphere on objects at the surface of earth is equal to the weight of overlying air. At sea level, it is 10.3 metric tons/m^2 and is defined as 1 atmosphere. It is measured as 1013.25 millibar or 1000 mb. Since air pressure acts in all directions it is not noticed as a force. The density of air increase with increase in pressure. (If the temperature of dry air is kept constant, the density of air varies in direct proportion to pressure). The density of air at an altitude of 12 km where most of, subsonic jets fly is about 1/4th the density at sea level.

At any location, the variation in pressure is small, but these variations can abruptly changes the weather. During high pressure the skies are clear but low pressures indicate strong weather. The density ad pressure drop rapidly with increasing altitude. Pressure may become negligible but does not end.

Dynamism of atmosphere (energy) is maintained by a ceaseless flow of energy. The sun is the source of all energy and it drives the atmosphere. The energy in the form of electromagnetic radiations reaching the top of earth's atmosphere is distributed over a wide range of wave lengths as shown below:

Type of Radiation	Wavelength, micronsymbol	Fate of Radiations
Infra red (50%)	0.71 – 100	Absorbed by $CO_2$and H_2O mostly below 10 km
Visible (41%)	0.40 – 0.71	Transmitted to earth undiminished
Ultraviolet (9%) (UV)	0.10 – 0.40 < 0.12	All absorbed by O_2 and N_2 above 100 km
	0.12 – 0.18	All absorbed by O_2 above 50 km
	0.18 – 0.30	All absorbed by O_3 between 25 – 50 km
	0.30 – 0.34	Part absorbed by O_3
	0.34 – 0.40	Transmitted to earth undiminished

The total energy emanating from Sun is 4×10^{27} cal/min. A significant amount of energy is prevented from reaching the earth's surface by gaseous constituents. The energy in the radiation is absorbed by gas molecule, warms the atmospheric layer. As shown above, uv-solar radiations (wave length < 0.3 symbol) are completely absorbed high in the atmosphere (50 km) and infrared are substantially depleted through absorption by CO_2 and water vapour at low altitudes (< 10.0 km). The visible solar radiations are able to pass through atmospheric gases. The atmospheric O_3 is able to absorb UV radiations in the wave length of 0.18 – 0.30 symbol in stratosphere.

The fate of incoming solar radiations which are not absorbed by atmospheric gases is shown below figure.

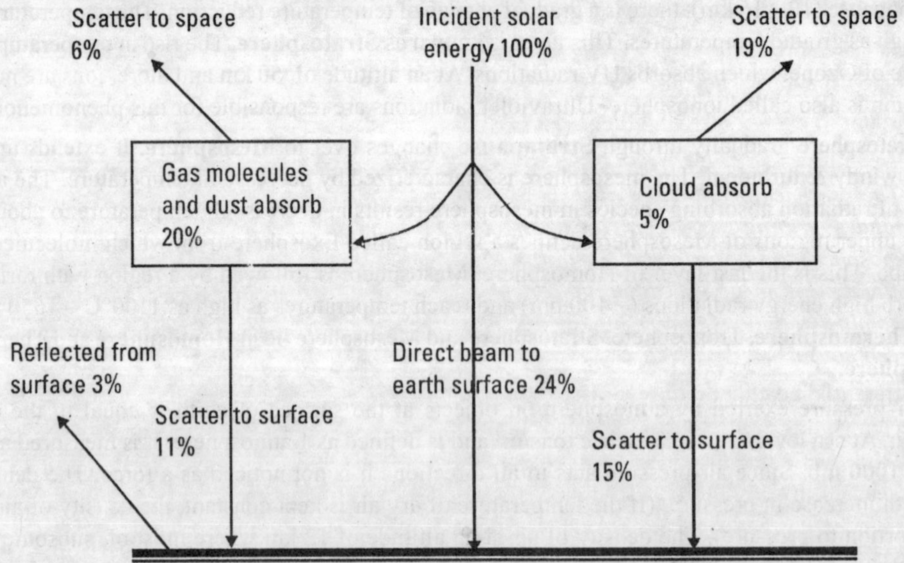

Figure: Fate of Incoming Solar Radiations

Scattering and absorption is done by aerosols (aerosols are particles dispersed in gas) as well as gases. This depends on the size, composition, altitude and humidity.

The reflectivity of a surface is called Albedo. The albedo is equal to reflected energy divided by incoming radiation. The earth's albedo not only depends on characteristics of the surface but also the angle of incidence (of solar radiations). The albedo of various surfaces are as follows:

Surface	Albedo
Snow	0.5 – 0.9
Water	0.03 – 0.8
Sand	0.2 – 0.3
Grass	0.2 – 0.25
Soil	0.15 - 0.25
Forest	0.05 – 0.25

The energy absorbed by earth is used up in driving the atmosphere in terms of wind, waves, ocean currents, hydrological cycle and photosynthesis and the rest is sent back to space as terrestrial radiations.

The peak intensity of these radiations occur at a wave length of 10 symbol(infra red). In contrast the peak intensity of solar radiations is 0.45 symbol wave length.

Since solar radiations and terrestrial radiations represent different portions of electromagnetic spectrum, their interaction with atmosphere differ. Nineteen percent of solar radiations are absorbed by atmosphere whereas, bulk of terrestrial radiations are absorbed by CO_2, O_3 and water vapor. The atmosphere, therefore, through the properties of cloud, water vapour and CO_2 act as a thermal blanket. This thermal blanket is termed as green house effect.

Green House Effect

Other gases that absorbs infra-red radiations are methane (CH_4), hydrocarbons (HC), nitrogen oxides (NO + NO_2) and many free ions. They also contribute to Green House effect.

The amount of energy reaching the surface of earth after absorption, and reradiation is approximately 2 cal/cm²/min.

* *The presence of infra red absorbing gases and water vapour in the atmosphere contributes to **Global Warming** the **Green House effect** by allowing incoming. Solar radiations to penetrate the atmosphere to the surface of earth (peak intensity (green), while reabsorbing infra red radiations emanating from it. The differential behaviour of the atmosphere towards out going radiations / incoming radiations play a Similar role like that of the glass roof in small horticulture system i.e. a green houses.*

The differential heating of air gives rise to horizontal pressure gradients leading to horizontal movement of air. The temperature differences between atmosphere at poles and at equator and between continents and over the oceans are responsible for large air movements. The earth's rotation modifies the horizontal movement. The air flow normally is from high pressure to low pressure (from cold area to hot areas).

Movement of the pressure system, diurnal heating and cooling of earth produces patterns of wind movement. This varies both temporally and spatially.

Hydrosphere

The sum total of water bodies is termed **Hydrosphere.** Water is an essential element for all living beings. The amount of water in living individuals vary between 70 - 80% by weight. Nature has maintained the some ratio globally also. 70% of globe is covered by oceans which constitutes nearly 97.2% of total water. Two percent water is locked in the icy sheets of Arctic and Antarctic regions. The remaining = 1.0% water represent the inland waters. On a global scale, total abundance of water is not a problem; the problem is availability at the right place, at the right time and in the right form. Water is a heterogeneous resource that can be found in either a liquid, solid or a gaseous form at number of locations on/or below the surface of earth. Depending on specific location of water, the residence time may vary from a few days to thousands of years. Further, more than 99% of earth's water is unavailable for beneficial human use (because of salinity in sea water or location, icy sheets). The water budget is depicted below:

Location	**Surface area, km^2**	**Water volume, km^3**	**% of total water**	**Estimated residence time**
Oceans	361×10^6	1230×10^6 km^3	97.2	Thousands of years
Atmosphere	510×10^6	12700	0.001	9 days
Rivers and streams	-	1200	0.001	2 weeks
Ground Water upto the depth of 0.8 km	130×10^6	4×10^6	0.31	Hundreds of years
Lakes (Fresh Water)	855×10^3	123, 000	0.009	Tens of years
Icy caps	282×10^5	286×10^5	2.15	Upto 10, 000 years

The oceans contain 1230×10^6 km^3 of water near the surface of earth upto varying depths (1.6 – 2.4 km). It contains saline water of nearly uniform composition. The approximate concentration of total salt is 34.45 gm/1000 gm water (as shown below):

Sodium chloride (NaCl)	23gm/1000 gm
Magnesium chloride ($MgCl_2$)	5 gm/1000 gm
Sodium sulfate (Na_2SO_4)	4 gm/1000 gm
Calcium chloride ($CaCl_2$)	1 gm/1000 gm
Potassium chloride (KCl)	0.7 gm/1000 gm
Others	0.75 gm/1000 gm
	Total = 34.45 gm/1000 gm

The Sea Water contains many other elements. Its composition has remained unchanged for thousands of year.

Glaciers are icy sheets formed on land when accumulated snow is compressed and hardened into ice by the weight of overlying layers. Sea ice is formed from sea water attached to the land and floats on the sea surface. The glaciers occur at high mountain ranges and may be km's long and few hundred meters thick.

Bulk of the world's ice is tied up in sheets that cover, Green land, and Antarctica land masses. Formation is possible by intense cold climate. The icy sheets of Green land cover an area of 1.74×10^6 km^2 at an average thickness of 1600 m. The Antarctic sheets cover 13.0×10^6 km^2 with the thickness of 2300 m. The ice sheets like glaciers are in motion carrying to oceans the ice formed about 10 m/year.

Hydrologic Cycle

Nearly 99% water of the globe is present in oceans or locked in icy sheets of arctic and Antarctic regions. The remaining 1% resides in atmosphere, in river & lakes, in soil and subsurface layers. This fraction approximately 1000 km^3/yr is involved in cycling and maintaining flows in surface waters. This water determines the physical environment and terrestrial plant and animal life. The process that maintains the flow of water through the terrestrial and atmospheric components of hydrosphere is termed **Hydrological Cycle.**

The Hydrological Cycle includes:

- Evaporation of water from oceans, other water bodies and soil.
- Evapotranspiration of water from plants.
- Horizontal transport of atmospheric water from one place to another, either as vapour or as liquid droplets.
- Precipitation in which atmospheric water condenses and fall on oceans, land as rain, sleet, hail or snow.
- Runoff in which water that has fallen on continents as precipitation finds its way flowing on land and under the surface back to oceans.

The water quality in nature is highly variable and depends on its source*. Water is an excellent solvent. It picks up impurities during flow. Other impurities are thrust upon it during its manifold use. The inland waters have varying amounts of salts present. The relative abundance of ions (by weight) make them hard or soft.

More precisely water quality is related to its use. The uses may be in stream or off stream.

In Stream uses: Navigation, hydroelectric power generation, recreation, taking away of waste waters and preservation of wild life and fish

Off Stream uses: Drinking and other domestic uses, washing, industry, agriculture and other consumptive uses

There are well designed criteria of water quality for different uses. These criteria are based on physical, chemical and bacteriological parameters.

Physical parameters: Temperature; Conductance; Total, suspended, dissolved volatile and fixed residues.

Chemical parameters: pH; acidity; alkalinity; hardness; chlorides; sulfates; fluorides; phosphates; dissolved gases, oxygen and CO_2; organic material, BOD, COD; Nitrogen, saline, Kjeldahl, nitrates and metals.

Bacteriological parameters: Total bacterial count, MPN Coliform, MPN faecal Coliform, MPN Streptococcus.

**Fresh water: Water with Salt concentration of less than 0.01%*

Salt water: Water with salt concentration around 3.0%

Brackish water: Water in estuaries, a mixture of fresh and seawater where rivers enter sea.

Lithosphere

The outermost layer of Earth's surface is called crust. It ranges in thickness from about 6 km beneath the oceans to as much as 75 km below the mountain ranges. This layer floats over a dense material "mantle" beneath. The combination of crust and hard upper layer of mantle is called **Lithosphere.** In general, land systems which supports biota and human life is referred as Lithosphere. They comprise an ecosystem which is subjected to maximum onslaught by anthropogenic activities.

Lithosphere is composed of four subsystems viz.

- Minerals
- Sedimentary strata
- Clays and
- Soils

Each system differs from others in physical and chemical properties. They respond differently to pollution.

Minerals are commonly occurring inorganic solids with well defined crystalline structures. The crust consists of rocks (Rocks are conglomerate of one or more minerals). Most land systems consist of Silica, Oxygen (75% w/w), Aluminium (8%), Iron (5%) Calcium (4%), Sodium (3%), Potassium (3%) and Magnesium (2%) Together they constitute 99% of earth's crust.

The **sediments**consist of mixture of clays, silt, sand, minerals and organic matter. Land, sediments of lakes and rivers are made up of sedimentary rocks. The characteristics of the strata depends on its origin and transport. The matter is predominantly carried by water and some time by wind. The sediments at the bottom of a water body is determined by physical, physic-chemical and biological processes. Many minerals, organic matter, recalcitrant compounds exist in sediments. Rivers carry sedimentary material through erosion and transport through flows and transfers them to the bottom. These deposits are called "alluvium".

The sediments are important repositories for trace metals. They may exist in mobile (ionic) form or i the form of complexes or chelates of insoluble hydroxides.

Claysare silicate materials containing Aluminium and are defined by size and other inorganic constituents. In water they are distributed as colloidal material. They are largely hydrated aluminium and silicon oxides. The most common clay minerals are "illite", "chlorite" and "kaolinite"

$$[Al\ Mg\ Na\ (OH)_2\ Si_4O_{10}],\ /\ [Al_3\ k\ SiO_3.\ (OH)_4],\ /\ [Al_2.\ Si_2O_5\ (OH)_4]$$

Montomorillonitelillite kaolinite

Clays are made up of very fine grains having layered structure of sheets of SiO_2 alternating with sheets of Al_2O_3. Two or three sheets make up a unit layer. Some clays may absorb water and cause swelling of clays. Clays bind cations, such as Ca, Mg, Al, which protect them from leaching.

Soil constitutes the most important component of land system. It grades into the parent bed material. The soil is a finely divided rock derived material containing organic matter, capable of supporting vegetation. Each soil has a unique morphology resulting from interaction of climate, living plants, animals and parent rock material. The organic content consists of biomass in various stages of decay, varying population of bacteria / fungi. The pores in the soil are filled with oxygen, CO_2 whose concentration vary between 15 – 35% and 0.1 – 3% v/v respectively. Soil profile is shown below in figure.

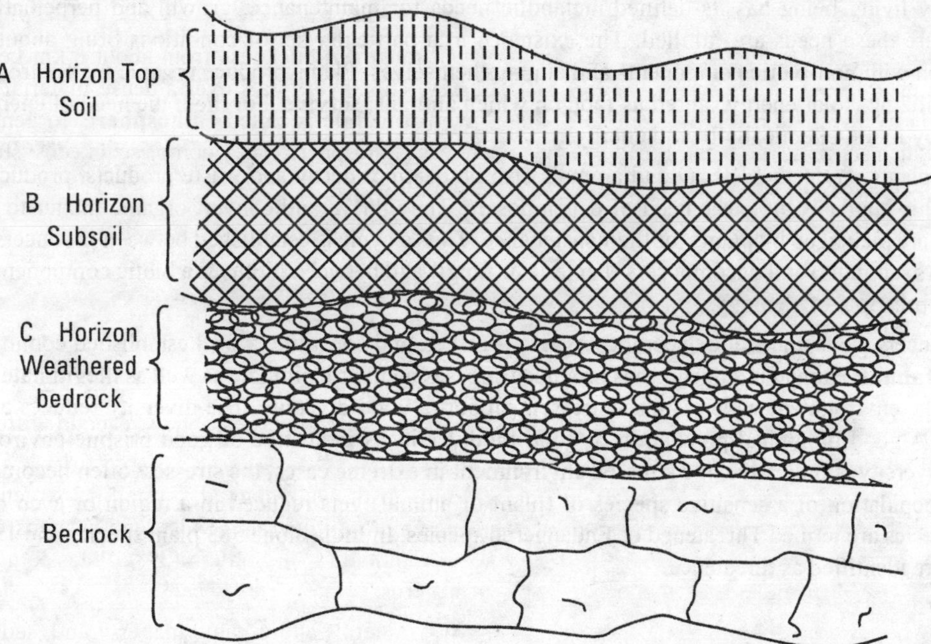

Soil Profile

Soil is arranged in Horizons (layers). The arrangement of horizons is termed soil profile. They are more or less parallel to surface and differ from another by texture, color, porosity and consistency. The top layer of the soil is A-Horizon. Maximum biological activity takes place in this layer. It contains most of the organic matter symbol5% of solid fraction. The next layer is B-Horizon or subsoil. This receives organic matter and salts from top soil by leaching. The C-Horizon is the weathered layer of parent rock material.

The mechanical properties of soil are particle size. The major categories are:

Gravel	2-64 mm
Sand	0.06 – mm
Silt	0.006 – 0.06 mm
Clay	<0.006 mm

The soil also provides nutrients for plants.

Biosphere

Biosphere comprises a very large component of environment and includes a vast array of living beings. The biological spectrum is:

Cell symbol, Tissue symbol, Organ symbol, Organ system, symbol Organism, symbolPopulatin, symbol Community

The living organization exist at cellular level (algale, bacteria, fungi and protozoa) or at multi cellular level (all other animals and plants). Entities which exist at sub cellular (molecular) level are phages, plant and animal viruses.

Every living being has its defined metabolic needs for maintenance, growth and perpetuation. They exist where these needs are fulfilled. The existence in a particular set of conditions bring about constant interaction with non-biotic components and among themselves. Plants produce organic matter from mineral matter utilizing, solar energy, stored as food on which animals, survive. For them the need of energy is also fulfilled by food (Chemically bound energy).

The plants on death & decay and animals also on death & decay and waste products, produce organic matter. This matter is the basic need of decomposers. They thrive and convert organic matter to minerals. Minerals are picked up by plants for their sustenance. Thus a cycle is established between producers (plants), consumers (animals) and decomposers (bacteria & fungi) and between biotic & a biotic components. This is the basic principie of **Ecology.**

Producers, consumers and decomposers live together, and constitute a well established community in a defined **Habitat.** Human indulgence has produced stresses by their legitimate as well as illegitimate activities of life. The environment which is least stressed support a good diversity. The diversity reduces as stresses increase. A good biological population or good biodiversity is indicative of good pristine environment. A low biodiversity is indicative of a stressed environment in extreme cases, the stresses/ often become so large that the population of a sensitive species of (plant or animal) gets reduced in a region or even disappear. These species are termed Threatened or Endangered species. In India alone 583 plant species and 158 animal species are identified as threatened.

Appendix 4.5: Some Basic Facts about the Earth

The earth is the third planet from the Sun at distance of about 150 million kilometers, which is called an "astronomic unit". The earth is 12760 kilometers in diameter. It is not an ideal globe. At the equator there are little bumps and at the poles it is flatter than it is at the rest of the world. The southern part has more bumps than the northern part. The Earth's circumference is 40070 kilometers. Some basic facts about our planet earth is detailed in this table.

Average Diameter	12760 kms (12756/12713 kmat equator/pole)
Average Circumference	40070 kms
Orbiting time	365.26 days
Surface area	510 Million sq.kms
Orbiting speed	29.8 km/second
Land area	149 Million sq.kms (29.2%)
Ocean area	361 Million sq.kms (70.8%)
Mass	5854 billion billion tons
Human population	Around 7 Billion
Distance from Sun	150 million km
Average surface temperature	14.3 °C
Age	4.6 billion years

The surface temperatures on our planet fluctuate between −88ºC (in Siberia) and +58°C (in Death Valley, California, USA). The temperature in the Earth's core is about 10000–12000 °C at a pressure of about 3 millions times our air pressure at sea level. About 70% of the Earth's surface consists of the ocean's water and hence Earth is called "the blue planet". The oceans contain about 97% of all the water on our planet. The oceans have high salt content. The Earth is the only planet in our solar system that has an atmosphere.
A cut of earth is shown below in figure:

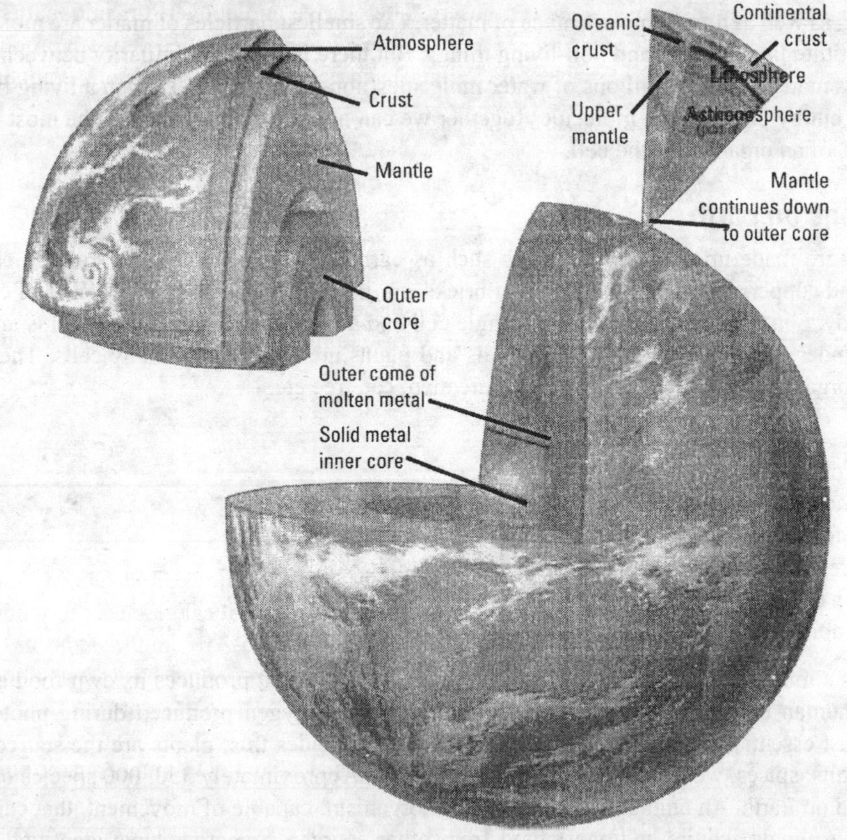

Earth

The Earth is made up of 3 major and distinctly different units: the core, the mantle-crust system and the atmosphere-hydrosphere system and its composition is dominated by: Fe (iron) ~35%, O (oxygen) ~30%, Si (silicon) ~15%, Mg (Magnesium) ~13%, Ni (nickel) ~2.5% and S (sulfur) ~2%. More than 80 chemical elements occur naturally in the Earth and its atmosphere. The Crust is made mostly from oxygen and silicon, with aluminium, iron, calcium, magnesium, sodium, potassium, titanium and traces of 64 other elements. The Upper Mantle is made up of iron and magnesium silicates; The Lower is silicon and magnesium sulphides and oxides. The Core is mostly iron, with little nickel and traces of sulphur, carbon, oxygen and potassium.

There is a vast variety of million of things present on the earth and around us. These are living, non living and dead things. Plants and animals are the living things. There are so many non-living things present on earth such as air, water, weather, climate, temperature stone iron etc.

Dead things are those which were once living but now they do not show the characteristics of living things such as wooden table, leather belt, leather shoe, basket of bamboo etc.

Non-living things can neither respond nor adapt to their environment. They can be altered by external forces in their environment, but only living things can change their habits or metabolisms to adapt to ambient changes. Non-living things cannot move under their own power, whereas all living things can move deliberately.

The living as well as non-living are made of matter. The smallest particles of matter are molecules. These particles constitute living things and non-living things. But there is some dissimilarity between these two. A cup of water is made of simply billions of water molecules(non-living thing).Though a living being is made of molecules, but by putting such molecules together we can not make an organism. The most basic unit, or structural unit, of an organism is the cell.

Living Things on Earth

Living things are made up of many elements such as carbon, oxygen, hydrogen, sodium, calcium, iron, phosphorus and copper. As a wall is made up of bricks, similarly living things are made up of cells. A cell is the unit of body. Some animals are of only a single cell and are called unicellular organisms and these cells can be seen under the microscope. Other animals and plants are made up of many cells. These are called multicellular organisms and their few examples are man, cat, tree etc.
Living things

- have definite shape and size.
- can move.
- need food.
- respire, excrete, reproduce and grow.
- have a definite life span. and
- have ability to sense environment.

A plant is a multi-cellular organism, incapable of movement, that produces its own food as well as for animals and human beings by the process of photosynthesis. Oxygen produced during photosynthesis is one of the most essential requirements of life on the earth. Besides this, plants are the sources of cereals, vegetables, fruits, spices, wood etc required for life support. Approximately 350, 000 species of plants have been identified on Earth. An animal is a multi-cellular organism, capable of movement, that cannot produce its own food. Animals acquire andingest food from other sources. Scientists have identified more than a million different animal species.

The living and non-living environment affects the survival of an organism. The non-living environment has a major influence on the lives of all other living and non-living things that form an ecosystem. There are many non-living (or abiotic) factors that influence where an organism can live. The survival of any organism will depend on how well it has adapted to these factors and how well it fits into a specific area. All organisms need to be in harmony with their environment in order to maintain stable conditions in their bodies.

Organism

All organisms are made of cells. A cell is a unit or building block with a definite structure and function. The number of cells in an organism varies from one to millions, depending on the size of the organism. Organisms may vary in size and shape, but all organisms are similar in that they can function independently and carry out life processes. A life process is a function that is essential for an organism to remain alive and produce new members of the species. When the life processes stop, the organism dies.

Unicellular organisms are too small in size and measured in microns(1 micron = 1/1000 mm) and also called micro-organism or microbes. Bacteria, fungi, virus and algae are important classes of micro-organisms. They are also called decomposers because they decompose or break down the dead bodies of plants and animals. They consume some of the decomposed products as food and release simpler compounds like C_2O, H_2O and N_2 into the environment They take part in environment. They take part in numerous natural processes and produce various type of products for our use, which are important to us in many ways. Few examples are

- large scale production of alcohol and vinegar;
- formation of curd from milk and bread from flour;
- They decomposes sewage in sewage treatment plant; and
- several medicines and vaccines are produced in the presence of micro-organism.

Habitat

The place where an organism (or a whole community of organisms) lives naturally is called its habitat. The natural habitat of the Zebra, for example, is dry grassland, while that of the whale is the ocean. Habitats on land are called terrestrial and that in water are called aquatic.

Species

A species is a category within the classification system. Living things of the same type belong to the same species, For example humans are one species and dogs are another species.

Individuals of the same species can reproduce to make more individuals of the same species. Two individuals belonging to different species cannot normally reproduce together. If they do, their offspring is usually infertile and unable to reproduce.

For example, animals called tigers are produced when a male lion and a female tiger reproduce. But a tiger cannot have offspring. This means that lions and tigers are different species.

Non-Living Things

The non-living things include air, water, soil, climate and weather humidity sunlight, temperature atmospheric pressure with many other factors also. Soilis often comprised of both living and non-living elements. Living things affect with the change to non-living substance. We can understand with the following few examples:

Soil

Soil is not only the covering of the Earth's surface and the substance that plants need but it also provides a home for many living things including such as human, micro-organisms, worms, mice, snakes, bacteria, rabbits, fungi etc.

Sunlight

Sunlight affects the living conditions of all organisms to a great extent. The existence of plants and all other organisms would cease without the sun. Light provides energy to the plants which they need to carry out photosynthesis. This production of food inside plant bodies initiates the chain of life and would not take place if there was no light.

Temperature and Humidity

Temperature and humidity greatly influence the lives of organisms. When living things get too hot or too cold, their bodies do not function properly. Processes such as digestion, respiration, excretion and reproduction take place at an optimum (most favorable or best) temperature range, That is why many desert creatures sleep during the extremely hot days and emerge in the cool of the night to feed and engage in courtship.

Water

Slightly less than three-quarters of the world is covered with water, Water is held in oceans, lakes, rivers, glaciers and as sub-surface water. It also exists in the bodies of all organisms and in the air. The characteristics of water can profoundly affect the lives of the organisms that depend on it.

Weather

Some organisms may also need to adapt to other weather conditions such as wind, air currents and storms. Plants that live in regions that are frequently hit by strong winds need to develop strong root systems that enable them to stay alive. Similarly, organisms that live in areas that experience frequent floods or wild storms need to adapt to the changing conditions.

An organism can usually survive the weather if it has adequate shelter and protection. Different plants and animals require different types of shelter or protection to survive. Rocks, hollow logs, trees and burrows can all provide shelter for different organisms.

Atmospheric Pressure

Because of the gravitational force of the earth the atmospheric gases are pulled towards the surface. This phenomenon leads to the air pressure being more near the surface of the earth and it decreases as the altitude or the height increases. All organisms can again survive only in particular range of atmospheric pressure and when the air pressure is low, especially in higher altitudes some may find it difficult to breathe because of the insufficient amount of oxygen present at a height. In the depth of the oceans the atmospheric pressure increases with the increasing depth and again this induces only certain kinds of animals and plants to survive in certain specific ocean regions.

Chemical Components

Carbon, Oxygen, Hydrogen and Nitrogen are four natural elements that constitute 97% of the molecules of life. Of the92 known natural elements in Earth, 25 elements are found in living beings. Organic and inorganic compounds derived from plants and other food sources ensure the smooth functioning of life process of organisms.

Appendix 4.6: Types of Environment and Environmental Exposure Route, Pathways, Types and Effects

1. Types of Environments

Environment can be defined in a number of ways

A. Inner versus outer environment

B. Personal versus ambient environment

C. Gaseous, liquid, and solid environments

D. Chemical, biological, physical, and socioeconomic environments

A. Inner Versus Outer Environment

This refers to the human body and consists of the environment within the body and the environment outside of the body. The human body has three protective barriers against outside environmental contaminants.

- The skin, which protects the body from contaminants outside the body;
- The gastrointestinal (GI) tract, which protects the inner body from contaminants (toxins) that have been ingested,
- The membranes within the lungs, which protect the inner body from contaminants that have been inhaled.

However, each of these barriers are liable to damage under certain conditions. Contaminants can penetrate to the inner body through the skin by dissolving the layer of wax that we have covering our oil (sebaceous) glands. The GI Tract is also very vulnerable to compounds that are soluble and can be easily absorbed and taken into the body cells. However, the body has ways of protecting the GI tract. For example, unwanted material can be vomited through the mouth, or rapidly excreted through the bowels (as in the case of diarrhea).

The lungs are the most important route for toxic substances, and they are also the most fragile. Airborne materials that are inhaled can be deposited in the lungs, and, if they are soluble, they can be absorbed. A number of mechanisms protect the lungs, such as, simple coughing, or cleansing by "macrophages" that engulf and promote the removal of anything foreign.

Unless a toxicological agent or environmental contaminant penetrates one of the three barriers that protect the body, it will not get into the inner environment, and even if it does get in, there are other ways to remove it. For example, materials entering the circulatory system (arteries, veins, etc.) can be detoxified in the liver or excreted through the kidneys.

B. Personal Versus Ambient Environment

Personal environment represents the environment that we can control. The ambient (working) environment represents the environment over which we have no control. It is thought that the working environment poses the greatest threat to health, but some health experts believe that the personal environment, influenced by a number of factors, is the most important for our well being. Factors which are important for the personal environment include hygiene, diet, sexual practices, exercise, use of tobacco, drug and alcohol use, and frequency of medical checkups.

C. Gaseous, Liquid, and Solid Environments

Our environment exists in one of three forms: gas, liquid, or solid, each of which can be polluted. Particulate (large particles) and gases are released into the air (gaseous), sewage and liquid wastes are discharged into water (liquid); and solid wastes, such as plastics and toxic chemicals are disposed of on land (solid). People interact with all of these environments.

D. Chemical, Biological, Physical, and Socioeconomic Environments

These types of environments could affect people's health.

- Chemical factors and contaminants include toxic wastes and pesticides in the environment; chemicals used at home (cleaning products) and by industry; and preservatives for food.

- Biological factors include different forms of disease organisms in food and water, and those that can be transmitted by insects and animals, and person-to-person contact.
- Physical factors include elements that may influence health and well-being, such as injuries and deaths from accidents, loud and excessive noise, extreme temperatures (heat and cold), and the effects of radiation.
- Socioeconomic factors - are hard to measure, but significantly affect the lives and the health of-people, Low socioeconomic status increases death and illness rates.

2. Routes of Exposure

It has been estimated that about 70, 000 chemicals are used worldwide. Because of this, we are exposed to a number, of chemicals in our home, at work, and in the general environment. Trace amounts of toxic chemicals are present in the food, the air, and the drinking water. Exposure to toxic substances occurs through the three major routes listed below.

a. The skin (dermal absorption)
b. The respiratory tract (inhalation)
c. The digestive tract (ingestion)
d. Other routes of exposure

A. Dermal Absorption

Contact with the skin is the most common path of toxic substance exposure. The skin is composed of three layers:

- **Epidermis (outer layer):** The outermost layer is the stratum *corneum (carnified layer)*. This is the structure that determines the rate of absorption of substances through the epidermis. For example, a pesticide such as Malathion, which easily penetrates the stratum corneum, moves quickly through the other layers of the skin and is rapidly absorbed into the bloodstream. DDT, another type of pesticide, does not easily penetrate the stratum corneum, so the rate of absorption is much slower.
- **Dermis (inner layer):** The inner layer of the skin is sometimes referred to as the true skin. In animal hides, this is the layer that turns to leather when chemically processed. The dermis is the source of oxygen and of nutrients for the idermis. The hair follicles, sweat glands, and sebaceous (oil) glands are found in this layer. These structures playa limited role in the absorption of substances across the skin.
- **Subcutaneous fatty tissue:** This layer provides a cushion for the underlying structures and allows the skin to move to some extent.

Factors affecting dermal absorption of toxic substances include

- The condition of the skin. An intact stratum corneum (epidermis) is an effective barrier to absorption of some toxic chemicals. However, physical damage to the protective barrier, such as a cut or abrasion, allows toxic substances to penetrate the epidermis and enter the dermis where they more readily enter the bloodstream and are carried to other parts of the body.
- The chemical makeup of the substance. Inorganic chemicals and substances are not easily absorbed through intact, healthy skin (such as cadmium, lead, mercury, and chromium. Organic chemicals dissolved in water do not easily penetrate the skin because the skin is impermeable to water. However, organic solvents, such as paint thinner or gasoline, are easily absorbed through the epidermis.
- Increasing the concentration of the toxic substance or the exposure time can increase the rate or amount of material absorbed.

B. Inhalation

Inhalation is the easiest and fastest means of exposure to toxic substances because toxic substances are readily absorbed in the respiratory tract. The lining of the respiratory tract is not effective preventing absorption of toxic substances into the body. The respiratory tract consists of the nasal passages, trachea (windpipe), larynx (voice box) and the lung. The following factors affect inhalation of toxic substances:

- Concentration of toxic substance in the air,
- Solubility of substance in the blood and tissue,
- Respiration rate,
- Length of exposure,
- Condition of respiratory tract, and
- Size of toxic particle.

C. Ingestion

Ingestion of toxic substances usually occurs accidentally or unknowingly. The digestive tract consists of the mouth, the esophagus (food canal), stomach, and intestine (large and small). The major function of the digestive tract is to digest and absorb the foods we eat. Physical and chemical factors affect the absorption of toxic substances. Absorption is affected because of the structure of the body and the length of time food containing the substance remains in the body. Once a chemical is absorbed, its effects depend on its concentration in the target organs, its chemical and physical form, what happens to it after it is absorbed, and how long it remains in the tissue or organ of choice. After being taken up in the blood, a chemical is be quickly distributed throughout the body; it may be moved from one organ or tissue to another (translocation), or changed into a new compound (biotransformation).

D. Other Routes of Exposure

The eye is a common point of contact for toxic substances. The primary point of contact for toxic substances is the cornea. Acidic and basic compounds are the most common types of exposures that do damage. Depending on the amount of damage, the cornea may be able to repair itself. The outer layer of the eye is made up of connective tissue called the *sclera*. In the front of the eye, this form into the cornea, which is transparent. The middle layer of the eye is called the *choroid*, which forms the *iris*, in the front of the eye. The innermost layer is the *retina*, which produces images. Two compartments within the eye contain a fluid-like substance called the *aqueous humor*, and a transparent jelly-like substance called the vitreous *humor*. All nutrients and oxygen must diffuse through the aqueous humor in the back to the cornea to repair the damaged tissue. Injections are another common route for exposure.

Injections are mainly used in laboratory studies on experimental animals. Following are the different types of injections:

- Intravenous injections (into a vein).
- Intramuscular injections (into the muscle).
- Intra peritoneal injections (into the abdominal cavity).
- Intradermal injections (into the skin).
- Subcutaneous injections (under the skin).

3. Pathway of Exposure

- An exposure pathway as the process by which an individual is exposed to contaminants that originate from some source of contamination. For exposure to occur, a completed exposure path way must exist. A completed exposure pathway exists when all of the following five elements are present:

- A source of contamination, for example a smoke stack on a factory;
- Media for the contaminant to travel, such as groundwater, surface soil, surface water, air, subsurface soil, sediment, and biota (animal and plant life);
- A point of exposure, or a place where people actually come into contact with the contaminated material;
- A route of exposure, or how contaminants enter or contact the body (i.e. ingestion, inhalation, dermal contact, and dermal absorption); and
- A receptor population, or those persons who are exposed or potentially exposed to the contaminants.

A potential exposure pathway exists when one or more of the elements is missing, but available information indicates that exposure is likely. An incomplete exposure pathway exists when one or more of the elements is missing and available information indicates that exposure is not expected to occur.

4. Types of Exposure

Toxic chemicals generally produce the greatest effect and the most rapid response when inserted directly into the bloodstream. Occupational exposure generally occurs from breathing contaminated air (inhalation) and/or direct or extended contact of the skin with the substance (dermal exposure).In contrast, accidental and suicidal poisoning occurs most frequently by oral ingestion. The types of exposures are

- Acute, which is exposure to a chemical for 24 hours or less.
- Chronic, which is exposure to a chemical for more than 3 months.
- Sub-acute, which is exposure to a chemical for 1 month or less.
- Sub-chronic, which is exposure to a chemical between 1 to 3 months.

5. Effects After Exposure

Local effects are seen at or near the body part or parts where exposure occurred. For example, inhaling particles can result in irritation of the respiratory tract, resulting in effects ranging from sneezing to chest pains and difficulty inbreathing. An ant bite leads to redness and swelling at the bite location.

Some substances are absorbed into the bloodstream and are then carried to other parts of the body, where their effect. These types of substances usually cause their effect in one or two target body organs, Whether or not these effects occur depends on the concentration of the chemical in the target organ. The concentration in the organ is dependent on the absorption, distribution, biotransformation, and excretion of the substance. Biotransformation occurs when a substance is changed from one form to another, which may also change the toxic properties of the substance. It usually occurs ill several steps, primarily in the liver, but it may also occur in other tissue like the kidneys, lungs, and digestive tract.

Some substances are absorbed from the bloodstream and stored in tissues where they may not cause an adverse effect. For example, lead can be stored primarily in the long bones of the body, but when released, has a toxic effect on the nervous system.

6. Excretion of Toxins

The rate (speed) at which a toxic substance is removed from the body determines whether it will have a toxic effect. The longer a chemical is in the body, the greater the likclihood of damage. The main way a chemical is excreted from the human body is through the urine, but the kidneys, the lungs and the liver are also important in removing certain chemicals from the body. The kidney eliminates the greatest number of toxins than any other tissue/organ. The lungs eliminate substances that are in the gaseous phase, like carbon dioxide. The liver removes substances like lead or DDT (pesticide) by excreting them into bile, which is made by the liver and travels to the small intestine. From there the substance can be absorbed in the feces and then eliminated through excretion. Neither the, sweat glands nor the GI tract are important routes for excreting of Toxic substance. Skin, hair and breast milk are pathways, although minor, for excretion.

Appendix 4.7: Typical Temperature Variation with Height in A Hot Tropical Region and Polar Region During Summer and Winter (T-Marks the Stropopause on each Curve; and S is the Stratopause)

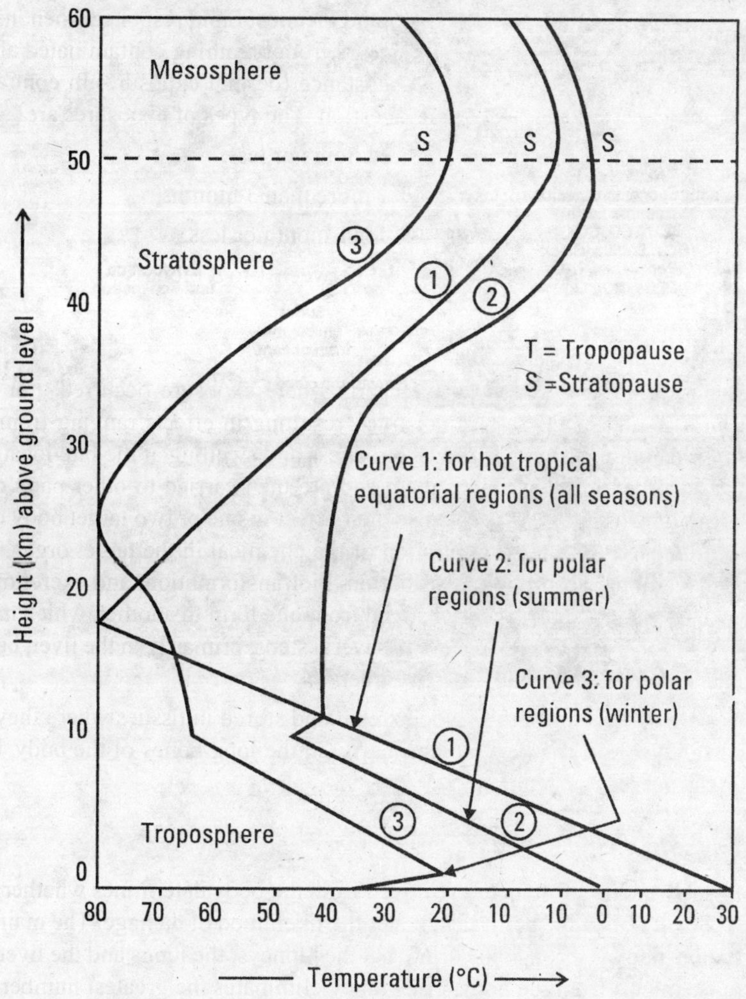

Appendix 4.8: Paths of Various Atmospheric Pollutants Dispersion in the Environment

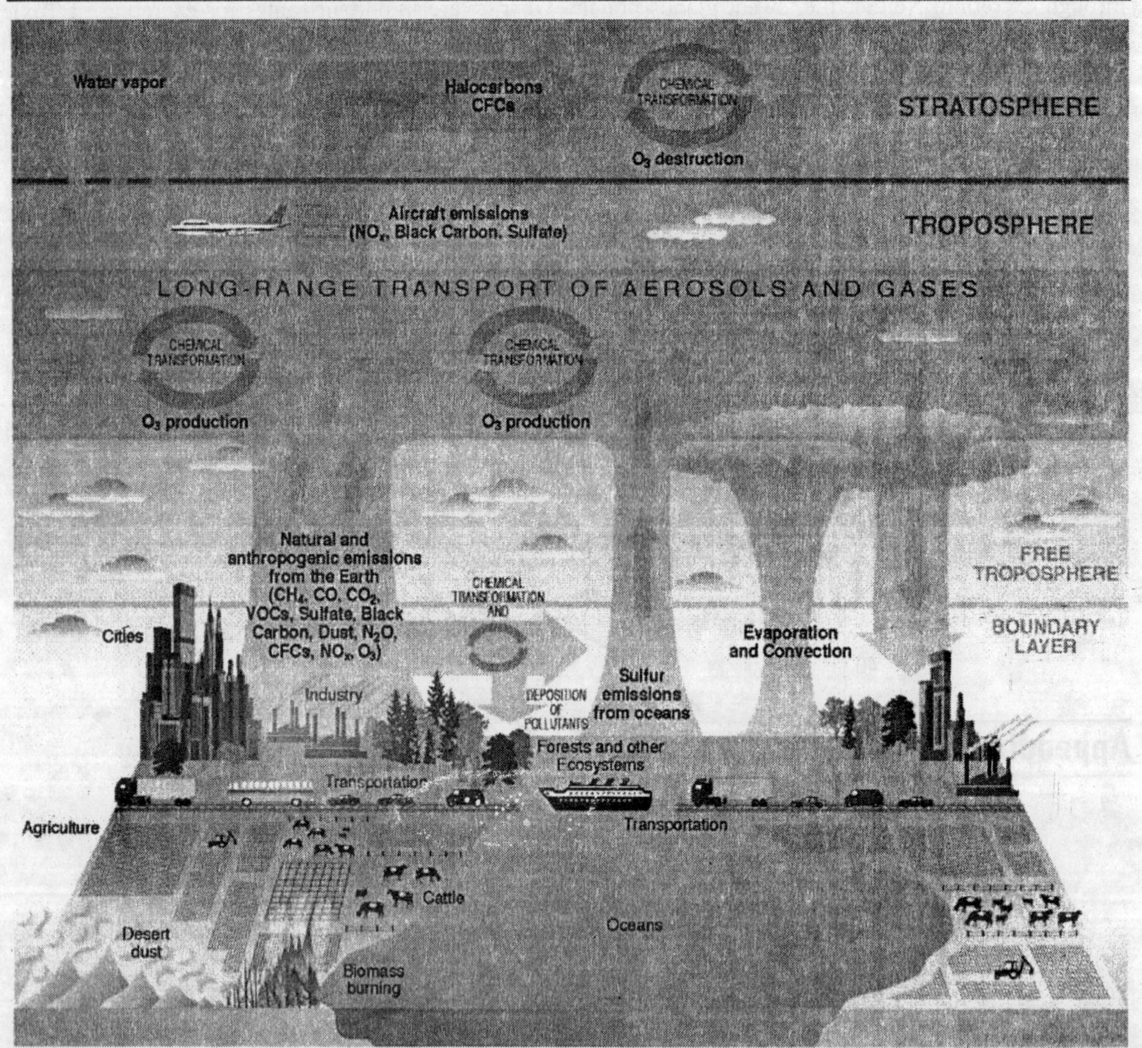

Appendix 4.9: Environmental Pollutants Sources Pathways

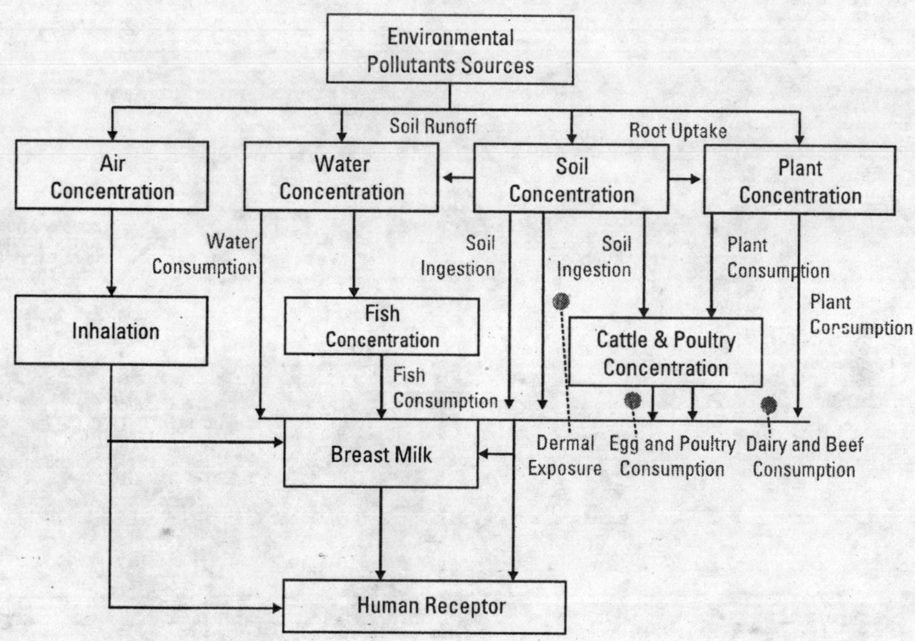

Appendix 4.10: Water/Hydrological Cycle

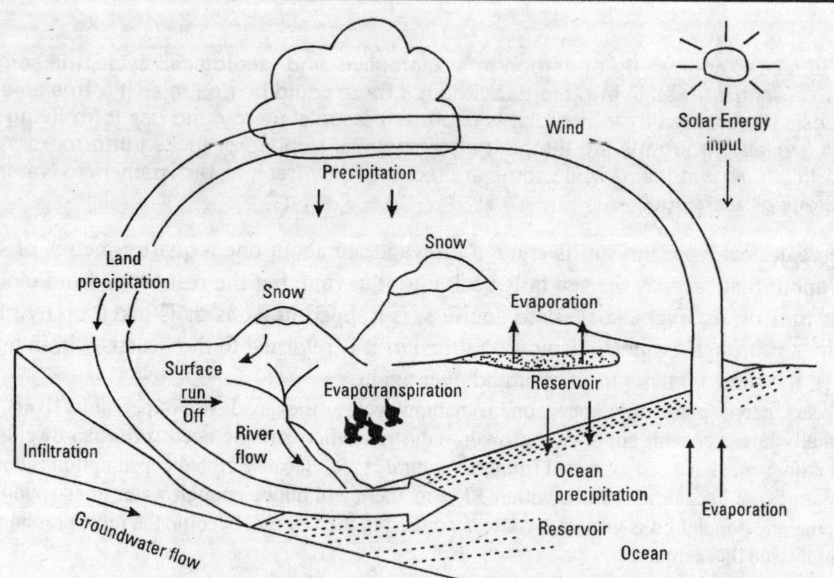

Appendix 4.11: Human Activity and the water cycle

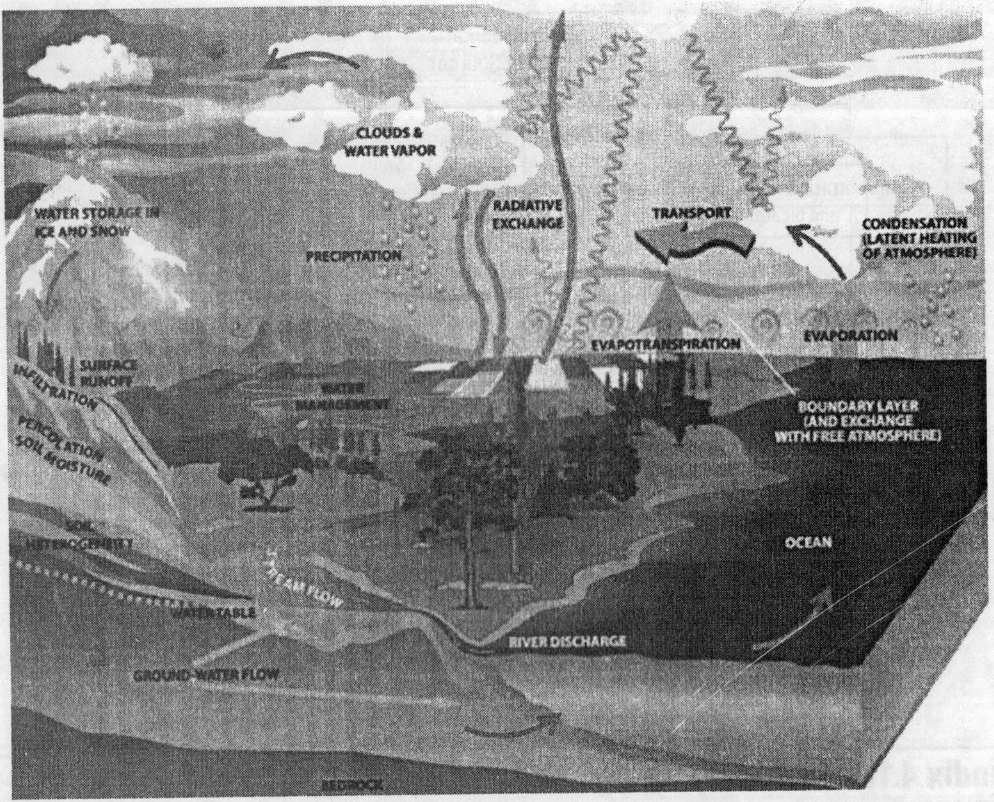

NOTES

Just as there is a constant cycling of carbon in a 'biological' and 'geological' cycle, the same is true for water. Water is essential for all living things. Without it there could be no life as it forms an essential part of photosynthesis the process by which plants capture the sun's energy and use it for living and growth. While water is 'cycled' through living things it also becomes trapped or locked into rocks. Aquifers hold 99% of the world's fresh water and while some are recharged by fresh water, many hold water that seeped into them millions of years ago.

In anyone year, heat from the sun is enough to evaporate about one a quarter metres of seawater. Two thirds of this vapour that escapes the sea falls back into it as rain, but the rest drops from clouds that move over land. This runs off as rivers, soaks into aquifers, is trapped in dams or is taken up by plants and then released into the air through evapo-transpiration. Most of it is returned to the oceans within ten days, which means that often the same water is used over and over again.

For millions of years, people made little impact on this natural cycle as they made use of perennial rivers, natural springs and lakes. Gradually however with population growth, industrialisation and the needs of urban dwellers for food and water, more and more water has been extracted from this natural cycle. Because global consumption is doubling every 20 years and new sources are becoming scarcer, within 30 years there will not be enough water to go round. At the rate the world is developing and population is growing, water according to the UN will become the most pressing environmental and developmental issue this century.

Appendix 4.12: The Aquatic Carbon Cycle

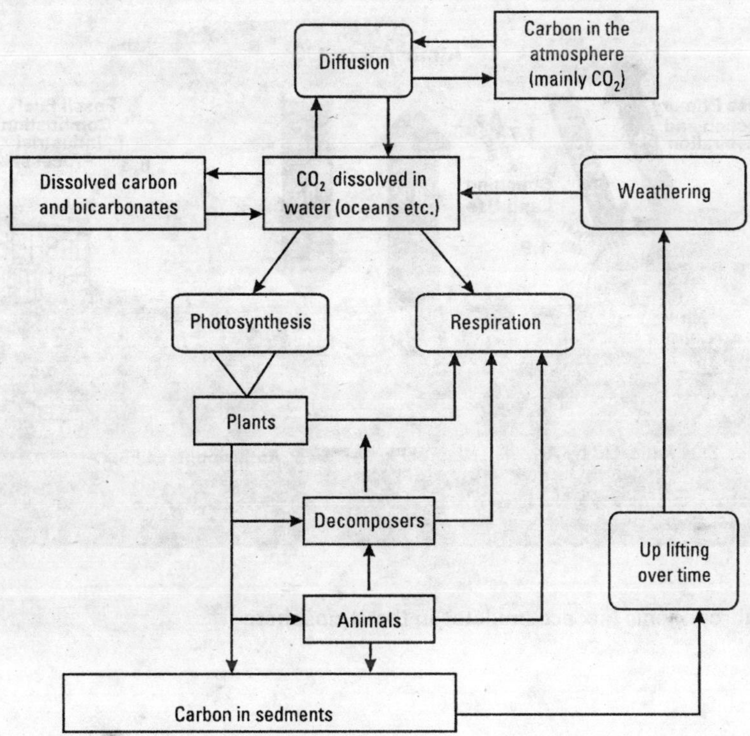

Appendix 4.13: The Terrestrial Carbon Cycle

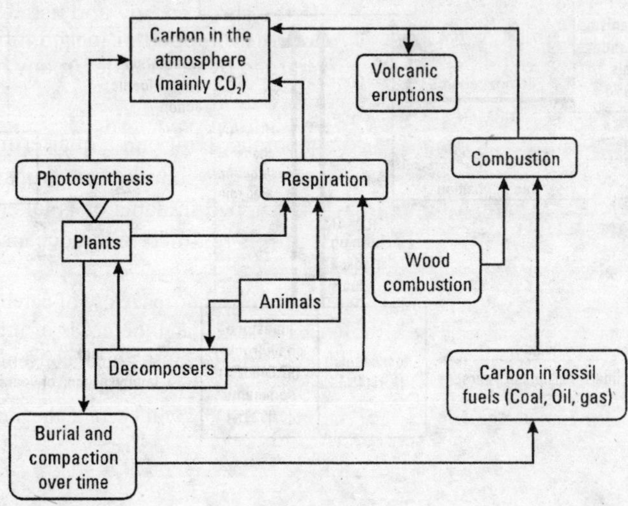

Appendix 4.14: Global Carbon Cycle (Billion metric tons (CO_2)

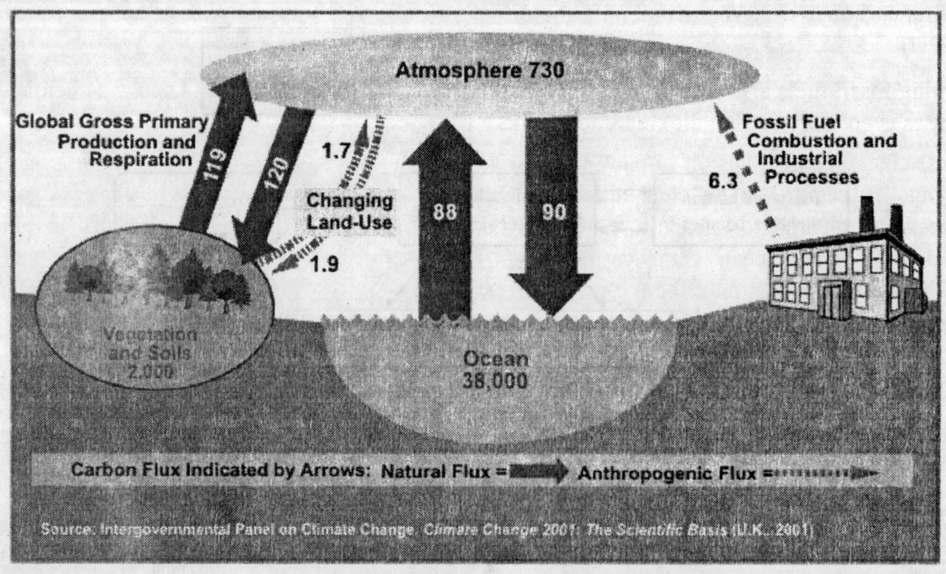

NOTES

About ½ of the anthropogenic flux accumulates in the atmosphere

Appendix 4.15: Human activity and the carbon cycle

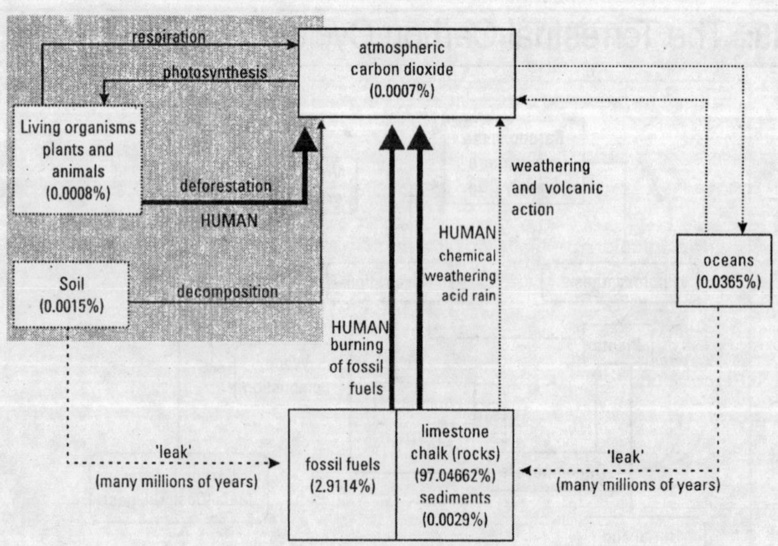

NOTES

The spread of industrialisation, agriculture & the increased use of fossil fuels have all interfered with the natural carbon cycle. When coal, oil or natural gas are burnt carbon dioxide is released back into the atmosphere.

Burning fossil fuels also results in acid rain and if this falls on limestone can cause a chemical reaction that releases even more carbon dioxide into the air.

Forest clearance also has an impact, because in mature forests, photosynthesis, respiration and decay balance out, if trees are cut or clear-felled, then there is less change to the overall carbon balance. But most forests are cleared for agriculture and are burnt. Burning releases vast amounts of carbon dioxide very quickly. Trees and oceans act as carbon sinks. This means that they can take carbon dioxide out of the air. Planting new forests is an effective way of removing carbon dioxide from the air since when a tree is growing, it need a lot of carbon for energy and tissue growth.

Oceans can also remove carbon from the air but this amount has an upper limit. Eventually the seawater becomes saturated and can dissolve no more. Because cold water absorbs more carbon dioxide than warm water, as the oceans warm through global warming, the ocean becomes saturated and begins to release carbon dioxide back into the air. This in turn may cause air temperatures to rise and cause a 'runaway' temperature increase.

Appendix 4.16: Components of Environment and Environmental Cycle

Abiotic Components	Biotic Components
Sunlight	Primary producers
Temperature	Herbivores
Precipitation	Carnivores
Water or moisture	Omnivores
Soil or water chemistry (e.g., P, NH_4+) etc.	Detritivores etc.

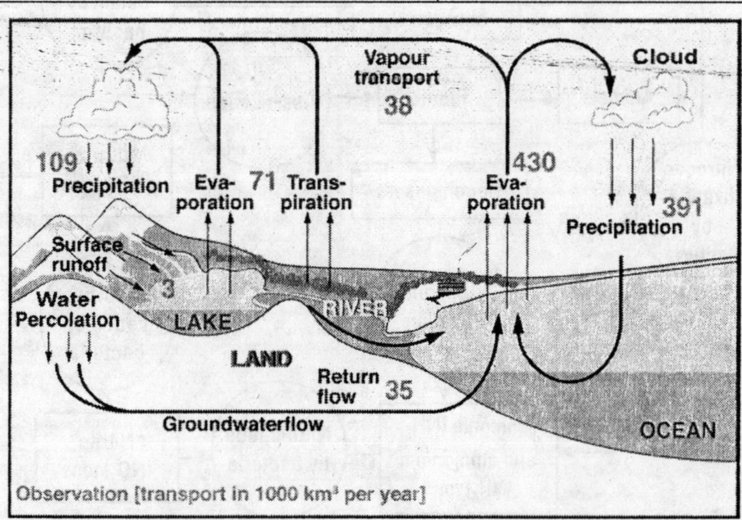

Appendix 4.17: Carbon dioxide (CO_2) Cycle

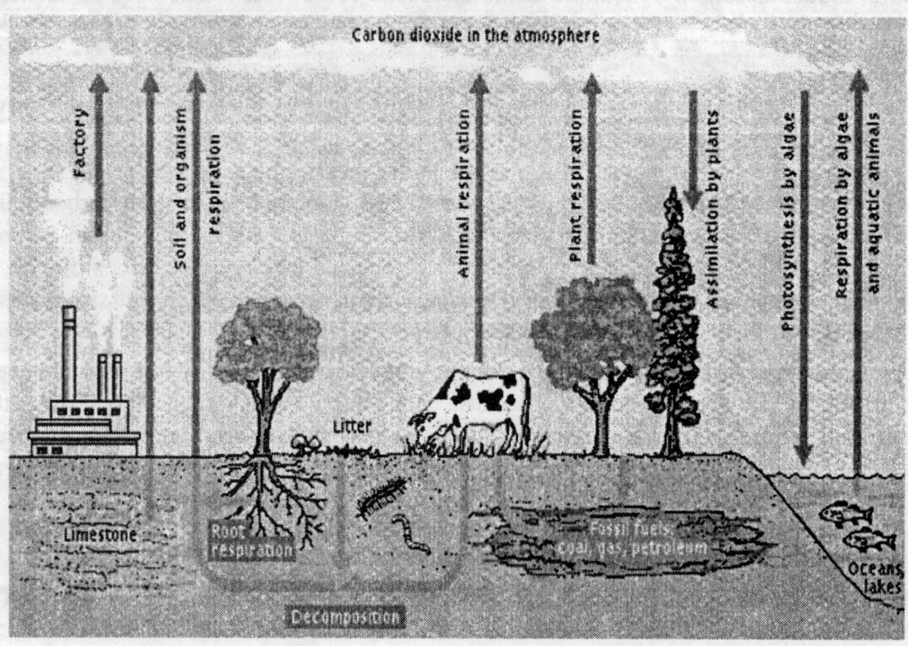

Appendix 4.18: Nitrogen Cycle

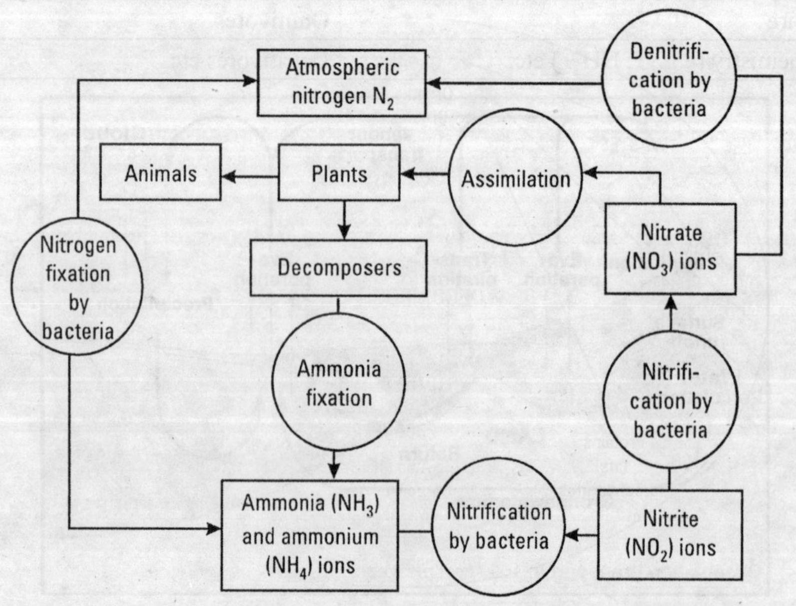

Appendix 4.19: Aerobic Carbon, Nitrogen, Phosphorus and Sulfur Cycle

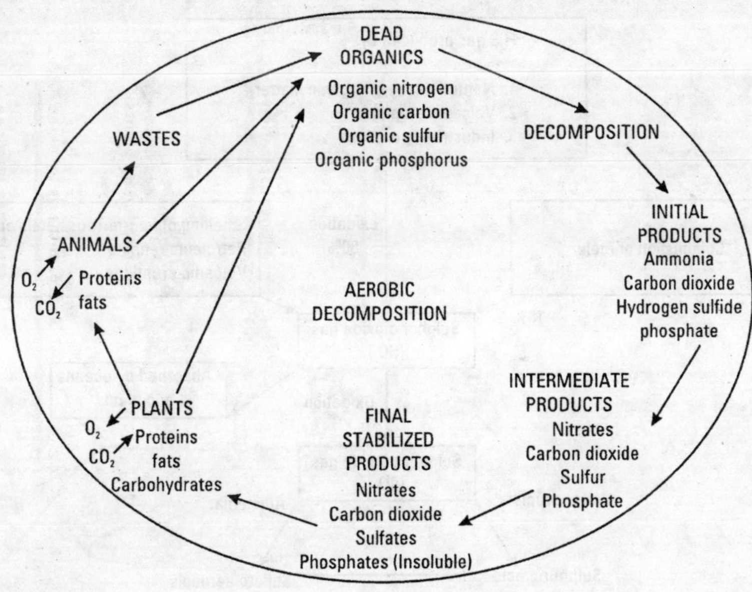

Appendix 4.20: Anaerobic Carbon, Nitrogen, Phosphorus and Sulfur Cycles

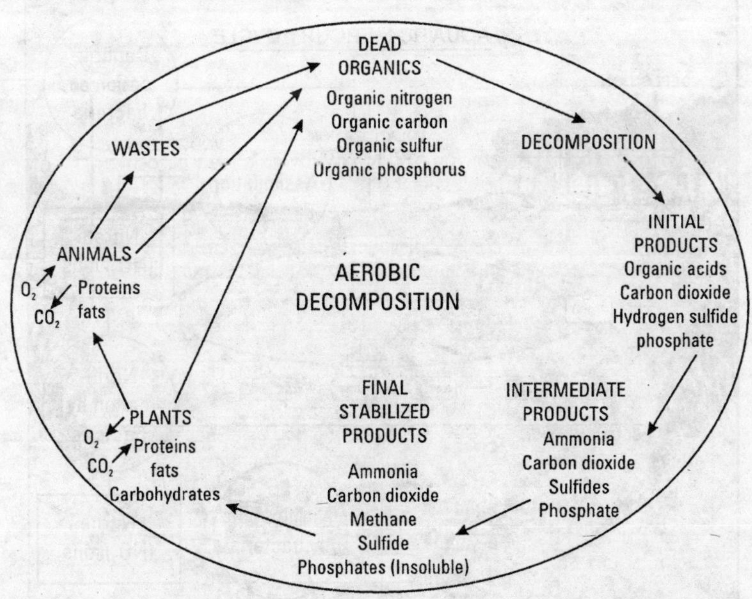

Appendix 4.21: Production and Disposal Cycle of Sulfur Dioxide

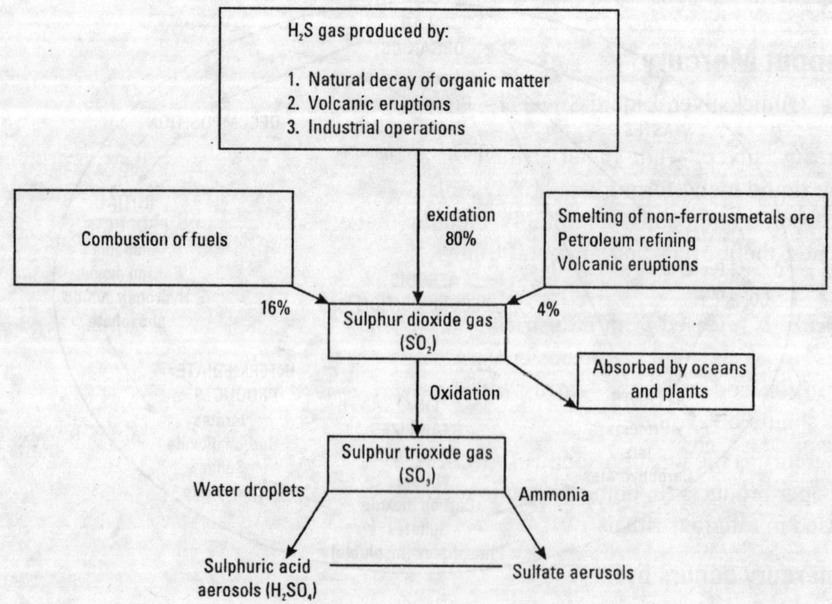

Appendix 4.22: Aquatic Mercury (Hg) Cycle

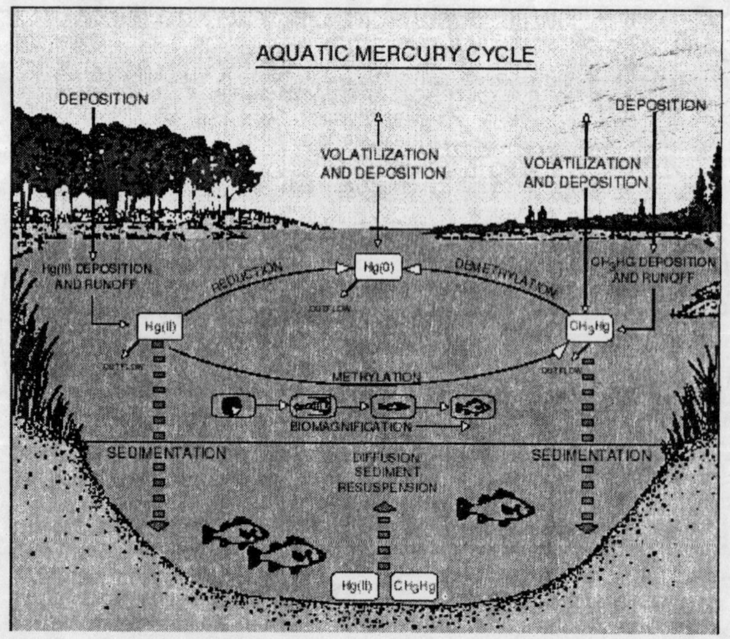

Note: Mercury cycling pathways in aquatic environments are very complex. The various forms of mercury can be converted from one form to another, the most important is conversion to methyl mercury (CH_3Hg^-) which is the most toxic form. Ultimately, mercury ends up in sediments, fish and wildlife or evades back to the atmosphere by volatalisation.

Factsheet about Mercury

(Also known as: Quuicksilver, Liquid Silver)

Mercury is a heavy, silvery-white metal element. It can exist as a liquid at room temperature or as a solid crystal salt. The liquid metal form gives off invisible, odorless, toxic vapors. Mercury can also be found in organic (with carbon) compounds. Commonly, metallic mercury can be found in thermometers, barometers, electrical switches, thermostats and in dental fillings. Mercury has also been found in some **skin products sold abroad**

When mercury is released from industries into the air, it can travel long distances and be deposited on soil and in lakes. In lakes, small organisms change the mercury to a form of organic mercury (methlmercuy) that builds up in the bodies of fish. **eating too much of certain types of fish** containing high levels of methylmercury should be avoided.

Organic mercury is the most poisonous form. It is used as a fungicide and preservative for seeds, wood products, and paper products. In homes, organic mercury can be found in latex paints, and metallic mercury is sometimes used in religious rituals

Exposure to mercury occurs by:

Breathing: People can be exposed by breathing mercury vapors. This type of exposure can happen in the workplace, or in homes where mercury is spilled. People who use exterior latex paints that contain mercury in unventilated areas may be exposed and could become seriously ill.

Drinking/Eating: People can be exposed to mercury by eating fish or shellfish caught in contaminated waters. Some dental fillings contain mercury and blood levels may be elevated for a short time after teeth are filled. Mercury can enter the body when contaminated water is used for drinking or for preparing food.

Touching: People who work with exterior latex-paints containing mercury can absorb mercury through their skin. If a water supply is contaminated, people can absorb mercury as they bathe or use the water for other purposes

Drinking water standards are set at 2 parts per billion (ppb) of mercury. Stop drinking water containing more than 2 ppb of mercury. If levels of mercury are very high in water, avoid washing, bathing, or using the water for other purposes

Harmful health effects of mercury:

Metallic liquid mercury generally does not absorb very well when it swallowed. Breathing its vapors is very dangerous. When metallic mercury is touched it can slowly pass through the skin.

The following health effects can happen immediately or shortly after exposure to high levels of mercury:

- Neurological effects, confusion, hand tremors
- Chills
- Chest tightness, bronchitis, pneumonia
- Abdominal pain, nausea, vomiting, and loss of appetite
- Bleeding gums

- Leg pains and burning sensation in feet
- Lung and Kidney damage
- Skin rashes

Children and infants can develop a specific allergic reaction to mercury.

The following health effects can occur after several years of exposure to mercury (more than 10 ppb in air):

Cancer: There is no evidence that mercury causes cancer.

Reproductive Effects: Symptoms can include menstrual problems, possible miscarriages and damage to unborn babies.

Organ Systems: People's nervous systems and kidneys are very sensitive to mercury and are easily damaged. Symptoms of damage include blood in urine, shaking, burning pain in legs and feet, sleep disturbance, personality changes, irritability and memory loss.

In general, chemicals affect the same organ systems in all people who are exposed. A person's reaction depends on several things, including individual health, previous exposure to chemicals, and personal habits such as smoking or drinking. It's also important to consider the length of exposure to the chemical; the amount of chemical exposure; and whether the chemical was inhaled, touched, or eaten.

Determination of exposure to mercury:

Blood, urine, hair and breast milk can all be tested for mercury. Normal levels of mercury in urine can vary, but are generally less than 15 micrograms per liter. Doctors do additional medical tests to check kidney and nervous system functions.

Appendix 4.23: Environmental Exposure Effects Of Toxic Heavy Metals, Benzene, PCBs and Pesticides

Toxic Heavy Metals

Heavy metals are everywhere in nature as components of the earth's crust. Plants can absorb and accumulate metals, which may be toxic. Industrial development has resulted in exposure to heavy metals in people because of increased production of products such as arsenic cadmium lead, mercury, chromium, and zinc.

Arsenic

- Arsenic, one of the most toxic metals, is produced at a very high rate. It is found primarily in a trivalent form or a pentavalent form. The pentavalent form is not as toxic as the trivalent form. The trivalent form of arsenic is the most toxic and it is deposited primarily in the respiratory tract Excretion is primarily via the urine, although excretion may also occur by the natural shedding of the skin and through sweat.

  Potable water usually has a very small arsenic concentration. It is also present in very small amounts in wine and seafood. This is why the diet should be taken into account when determining arsenic exposure. The primary routes of exposure to arsenic are ingestion and inhalation. Arsenic tends to collect in skin, hair, and nails. It is removed mainly through the urine, a few days after it is ingested. This can result in some upset stomach. Measuring arsenic in the urine is the main way to detect arsenic exposure.

Acute exposure to arsenic may result in death, fever, anorexia, and liver enlargement. Chronic exposure may result in poisoning of the nervous system, liver damage, and peripheral vascular disease, which could result in gangrene of the lower limbs. This condition is more commonly known as "black foot disease" and was a phenomenon that occurred in Taiwan from arsenic contamination of the drinking water supply. In addition, skin cancer and lung cancer may result from chronic arsenic exposure. Chronic arsenic exposure has also been associated with leukemia, kidney, and bladder cancers, dermatitis, hyper pigmentation, and keratosis (or arsenical keratosis).

Workers are occupationally exposed to arsenic fumes and dust such as in smelting industries. The major source of occupational exposures to arsenic is in industries involved in the manufacture of pesticides, herbicides, and other agricultural products.

Cadmium

- Cadmium is a by product of the mining and smelting of lead and zinc. It does not corrode and is primarily used for electroplating activities. Cadmium can gather and concentrate in plants. It has also contaminated irrigation waters and is found in fertilizers. Shellfish represent a major source of cadmium in the diet (100–1, 000 microgram per kilogram [μg/kg]), Cadmium is also found in tobacco, each cigarette has approximately 1–2 μg of cadmium. Approximately 1 μg of cadmium may be found in one liter of breast milk. Very little cadmium is absorbed through the ingestion route, and it is not easily absorbed.

  Acute toxicity from cadmium exposure occurs primarily through ingestion of contaminated beverages or food. This could result in nausea, vomiting, and abdominal pain. Acute toxicity through inhalation may result in chemical pneumonia and fluid in the lung. Irritation of the nose and throat, coughing, dizziness, weakness, chills, fever, chest pains, and labored breathing are also symptoms. Acute cadmium toxicity through inhalation may result in a condition known as metal fume fever. Chronic toxicity may result in chronic obstructive pulmonary disease, emphysema, and kidney disease. It may also result in adverse affects to the cardiovascular system and the skeleton.

Lead

- Lead is primarily used in the manufacture of batteries, plastics, china, ceramic glass, and paint products. It is the most widespread toxic metal on earth and is the second most hazardous substance found at sites. Scientific evidence links exposure to lead and adverse health effects due to acute and chronic exposure to lead.

  The primary route of lead exposure is ingestion. This could occur through eating lead-based paint flakes and chips (e.g., pica-like behavior), breathing contaminated dust, drinking contaminated water, and absorbing lead from lead-contaminated glaze in pottery. Children are particularly at risk. Blood lead levels are consistently higher for poor and minority children and for residents of largely urban areas. The fetal brain may be particularly sensitive to the toxic effects of lead because of the immaturity of the blood-brain barrier in the fetus.

  The toxic effects of lead include lead encephalopathy (brain disease) in children, resulting in lethargy, vomiting, irritability, loss of appetite, and dizziness. In adults, lead causes high blood pressure, and adverse reproductive effects (lowered sperm count and sperm motility). One classic symptom of lead toxicity are the Burton's lines, which are purple-blue discolorations of the gums. The primary measure for treating lead toxicity is through chelation therapy (binding of lead with other metals to remove it from the body).

Mercury

- Mercury is the third most toxic substance in the environment. Approximately half of all mercury is used to produce vapor lamps, fluorescent tubes, thermometers, and electrical products. It exists in a number of forms, which may affect different parts of the body. Organic mercury primarily affects the brain. Methyl mercury is the most toxicological form of the element and, by its accumulation in the central nervous system (CNS), may result in neurotoxic effects in adults and toxicity in the fetuses of mothers exposed to methyl mercury during pregnancy. Metallic mercury is slowly absorbed by the gastrointestinal system and is not as toxic as methyl mercury. Inorganic mercury (mercury salts) primarily affects the kidneys. Exposure to mercuric salts may lead to abdominal cramps and bloody diarrhea.

  Chronic mercury exposure may lead to tremor and personality disturbances and permanent CNS damage may result from methyl mercury exposure. Acute mercury exposure can be assessed by measuring the level of mercury in blood. Chronic exposure is best assessed by measuring the amount of mercury in urine. Chelation therapy is typically used in acute mercury poisoning.

Benzene

Benzene is a component and a by-product of gasoline. Benzene is widespread in the environment, and is one of the most prevalent solvents. In addition, it has been used as a solvent in rubber, inks, adhesives, and transformer fluids. Benzene dissolves easily in the bloodstream and quickly diffuses from the lungs to the blood and is metabolized in the liver to compounds that interact with cellular DNA. In addition, it is lipid-soluble and is easily absorbed by the skin and by cells lining the digestive tract.

Inhalation is the predominant route of exposure to the toxic effects of benzene, particularly in the workplace. Acute exposure to benzene may result in CNS effects, which may lead to unconsciousness and death. There is no antidote for acute benzene poisoning. Chronic benzene exposure may result in bone marrow damage, resulting in anemia. Symptoms of chronic benzene exposure may include fatigue and anorexia. Leukemia is a classic outcome from chronic low-level exposure to benzene with a latency period of about 15 years.

Polychlorinated Biphenyls (PCBs)

PCBs are another major contaminant of concern in communities. They were used in plasticizers, adhesives, and as dielectric fluids in capacitors. In humans, they accumulate in fat tissue and milk, which is a major route of excretion. PCBs cause liver cancer in rats and mice and is classified as a probable carcinogen in humans by EPA. PCBs are persistent in the environment and bioaccumulate upward in the food chain. A non-occupational source of PCBs is in fish from contaminated water. Cultures where subsistence fishing is a primary source of food would subsequently represent populations of concern regarding PCB-contaminated fish.

High-level exposure to PCBs may cause a classic dermatological condition called chloracne. Developmental and fetotoxic effects may also be observed in humans. Occupational exposures to PCBs primarily occurs in the production of electrical equipment. PCBs have been banned from further production as a material for the production of electrical equipment and chronic workplace exposure is now uncommon.

- Pesticides

Several classes of pesticides cause adverse human health effects in humans. These major classes of pesticides include insecticides, herbicides, fungicides, fumigants, and rodenticides. Seasonal agricultural workers have an increased risk of harmful human health effects due to pesticide exposure.

- Insecticides include the organochlorides, organophosphates, and the carbamates. Organochloride insecticides affect the peripheral nervous system (PNS) through dermal absorption, inhalation, and ingestion. Organochloride compounds alsodecrease antibody production, placing a person at risk for infection. DDT is aorganochloride insecticide that persists and bioaccumulates in the environment, which is why it is no longer manufactured as a pesticide. Organophosphorous exposure may result in headache, anxiety, chest tightness, seizures; loss of consciousness, abnormal heart beat, and liver dysfunction. In addition, Organophosphorous pesticides, like malathion, seem to enhance the immune response in some circumstances. Parathion has been known to decrease antibody production.
- Herbicides such as 2, 4, 5-T, 2, 4, -D, and the classic contaminant, 2, 3, 7, 8, - TCDD (dioxin) are toxic to both animals and humans. Liver problem and nerve damage may result from chronic herbicide exposure, while chloracne is a classic symptom of herbicide dermal exposure.
- Fungicides are used in the treatment of fruit trees and vegetables and have a relatively low toxicity. Skin irritation, headache, nausea, vomiting, lethargy, and dermatitis are classic symptoms of some fungicides, such as creosote and hexachlorobenzene.
- Fumigants are used to eradicate insects, bacteria, and rodents. Fumigants are typically used on fruits, vegetable, ships, and buildings. Methyl bromide is a classic fumigant that may result in dermatitis, pulmonary irritation, headache, nausea, vomiting, dizziness, and dementia from exposure via inhalation and dermal exposure.
- Rodenticides are used primarily to eradicate rats, mice, rabbits, and gophers. Warmfarin is rodenticide that causes severe adverse health effects.

CHAPTER 5

Atmosphere and its Constituents

The gaseous envelope surrounding the earth is known as **atmosphere**. Two major components of dry and clean atmospheric air are nitrogen (78.09%) & oxygen (20.94%). Minor components of air are Argon (0.94%), He, Ne, CH_4, carbon dioxide (0.034%) etc. Air can hold water vapor from 0.1 to 5% by volume & also contains traces of noble gases (neon, helium, krypton, xenon), hydrogen, methane, sulphur dioxide, ammonia, ozone etc.

As we go higher above the surface of earth, the density as well as pressure of the atmosphere decreases. Total mass of atmosphere is about 5×10^{15} tonnes, which is about one millionth of the total mass of the earth.

Different parts of the atmosphere absorb &reflect different amounts of heat of the solar radiation, here the temperature of the atmosphere varies from −100 to 1200 °C. There are large variation in the atmosphere with regards to composition, temperature, humidity & intensity of sunlight.

Many chemical reactions (due to oxygen in atmosphere) and photochemical reactions (due to absorption of solar radiation) occur in atmosphere, which play a significant role in governing the chemical species present in the atmosphere. Therefore, different types of reactions are observed under varying atmospheric conditions.

Composition of unpolluted dry atmospheric air is given in table 5.1.

Regions/Structure of the Atmosphere

The atmosphere is divided into mainly four regions above the earth's surface namely, **Troposphere, Stratosphere** (or, ozonosphere), **Mesosphere & Thermosphere** based on temperature variation with height, decrease in density and chemical composition. There is one region above the thermosphere lying in the range of 500–1600 km above earth and is called exosphere (containing mainly atomic & ionic oxygen, hydrogen and helium). Beyond **Exosphere** is the unbounded area called inter-stellar space.

Characteristics of various regions of atmosphere is given in table 5.2 which includes temperature variation with height above the earth's surface and gases/species present in various regions.

Table 5.1: Composition of Unpolluted Dry Atmospheric Air

S.No.	Type of constituents	Constituents	Volume fraction	Total mass, × 10^6 metric tons	Estimated residence time	Microgram/ Cu.m, µg/m³	Weight fraction
1	Major constituents	Nitrogen (N_2)	78.09%	3,850,000,000	continuous	8.95×10^8	75.51%
		Oxygen (O_2)	20.94 %	1,180,000,000	continuous	2.74×10^8	23.15%

S.No.	Type of constituents	Constituents	Volume fraction	Total mass, × 10^6 metric tons	Estimated residence time	Microgram/ Cu.m, μg/m³	Weight fraction
2	Minor constituents	Argon (Ar)	0.94%	65,000,000	Continous	1.52×10^7	1.28%
		Carbon dioxide (CO_2)	0.032%	2,500,000	2–4 years	5.67×10^5	0.046%
3	Trace constituents	Neon (Ne)	18 ppm	64,000	Continous	1.49×10^4	12.5 ppm
		Helium (He)	5.2 ppm	3,700	upto 2 million years	8.50×10^2	0.72 ppm
		Methane (CH_4)	1.3 ppm	3,700	4–7 years	7.87×10^2	1.2 ppm
		Krypton (Kr)	1 ppm	15,000	Continous	3.43×10^3	2.90 ppm
		Hydrogen (H_2)	0.5 ppm	180	Not known	4.13×10^1	0.03 ppm
		Nitrous oxide (N_2O)	0.25 ppm	1,900	4 years	9.0×10^2	1.5 ppm
		Carbon monoxide (CO)	0.10 ppm	500	6 months	–	–
		Xenon	0.08 ppm	–	–	4.29×10^1	–
		Ozone (O_3)	0.02 ppm	200	upto 2 months	–	–
		Sulphur dioxide (SO_2)	0.001 ppm	11	4 days	–	–
		Nitrogen dioxide (NO_2)	0.001 ppm	8	5 days	–	–
			Total	5×10^{15} tons			

Table 5.2: Characterics of Main Regions of Atmosphere

Regions	Height from earth's surface, Km	Temperature range, °C	Gases/species present	Remarks
Troposphere	0–11	Decreases from 15 to – 56°C with height	N_2, O_2, Ar, CO_2, H_2O Vapor	It is domian of all living organisms including plants & animals. Fall in temperature with height due to the fall of pressure & density of air
Stratosphere	11–50	Increases from −56 to -2°C	N_2, O_2,O_3,O atoms	Temperature increases in this region due to the absorption of ultraviolet radiation of the sun by ozone present in this region, hence, it is also called ozonosphere. From 11 to 20 km, temp is constant at −56 °C
Mesosphere	50–85	Decreases from –2 to –92 °C	N_2, O_2, O_2^+, NO^+	Characterised by rapid decrease in temperature. Contains gases in ionised form.
Thermosphere	85–500	Increases from –92 to 1200°C	O_2^+, O^+, NO^+, e^-	Characterised by rapid rise in temperature. Gases present in ionised from reflect back the radiowaves to the earth, which forms the basis of wireless communication.
Exosphere	500–1600	–	Atomic & ionic oxygen, hydrogen and helium	Highest region of the atmosphere
Iner-stellar space	>1600	–	–	–

Various regions of atmosphere is shown below in the **figure** 5.1

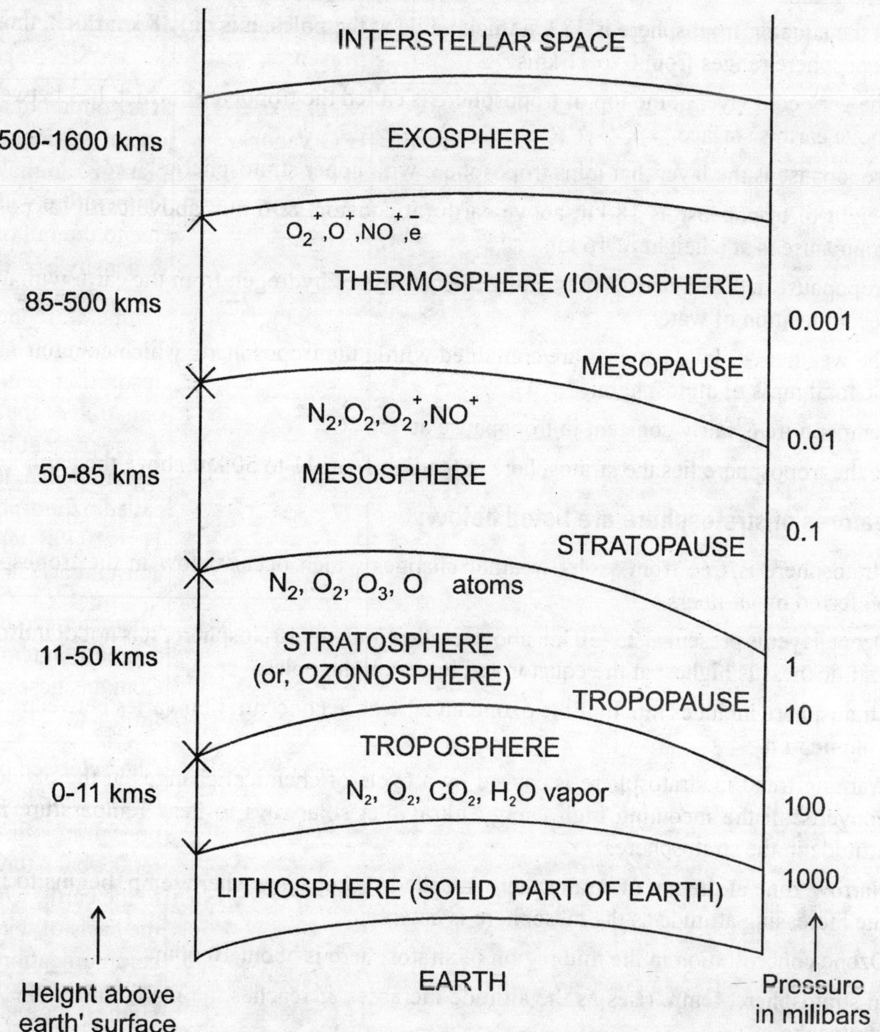

Figure 5.1: Regions of atmosphere with temperature & pressure distribution with height above earth's surface

Salient features of troposphere are listed below which is the lowest layer of atmosphere:

- About 70% of the mass of atmosphere is found in the troposphere which extends from 0 to 11 km above the earth's surface.
- Temperature drops with the increase in altitude. The saying that "higher we go, cooler it is" applies to this region.
- Temp. drops to −56 °C by the time an altitude of 11–20 km is reached, within which it is constant at −56 °C in troposphere.
- Rate of temp. drop is 6.5 °C/km in troposphere a **figure** which is called <u>normal environment lapse</u>

rate (ELR).

- At the equator, troposphere is 18 km thick while at the poles, it is only 8 km thick, though height of troposphere ranges from 0 to 11 kms.
- The very cold layer at the top of troposphere is called the tropopause which lies between 10–20 km above earth's surface.
- Tropopause is the layer that joins troposphere with upper stratospheric layer.
- Height of tropopause is 18 km above earth (at equator) & 6 mm above earth (at poles). In India, tropopause is at a height of 16 km.
- Tropopause acts as a barrier preventing the net loss of hydrogen from the earth with accompanying decomposition of water.
- The weather & cloud patterns are contained within the troposphere, which contains about 2/3 rd of the total mass of atmosphere.
- Temperature is fairly constant in tropopause at −56 °C.

Above the troposphere lies the stratosphere extending from 11 to 50 km above the earth's surface

Salient features of stratosphere are listed below:

- Stratosphere is free from violet weather changes which occur below in the troposphere. So it is preferred by jet liners.
- Ozone layer is present at 25–30 km above earth within the stratosphere. It is not of uniform thickness. Its thickness is highest at the equator and lowest at the poles.
- Stratosphere includes much of the ozone layer, whose concentration varies between 1 to 10 ppm by volume in it.
- Warning trend in stratosphere is caused by a cycle of chemical changes in the ozone layer, which converts all the incoming high energy ultraviolet solar rays to heat Temperature increases with attitude in the stratosphere.
- Narrow zone at the top of stratosphere is called stratopause, where temp. begins to fall again with the increasing attitude as the mesophere is entered.
- Ozone concentration in the mid region of stratosphere is about 10 ppm.
- In stratosphere, temp. rises as the altitude increases & reaches the maximum at the upper limit of stratosphere.
- Air planes travels in stratosphere.
- Global transport of pollution, particularly debris above ground, atomic bomb test & volcanic eruptions occur in the stratosphere.
- Absorption & scattering of solar energy also occurs in the stratosphere.
- Stratosphere is of interest to communication engineers because of radio/television communication.
- The absorption of ultraviolet radiation (all below 290 nm most between 290 & 320 nm) by ozone in the stratosphere prevents it from reaching earth, thereby protecting the biological life on the earth.

Salient features of next upper layer ie, mesosphere are listed below:

- Mesosphere is a relatively calm region, where few energy releasing reactions occur. Here, temperature decreases with altitude.

- Mesosphere, which is above stratosphere extends from 50 to 85 kms & its temperature drops to between −80 to −90 °C at the mesopause, a thin zone where temp. stabilises & soon starts to rise again as the thermosphere is ascended.
- Upper limit of mesosphere is called mesopause.
- Ionised gases present in mesosphere (N_2, O_2, O_2^+, NO^+) reflect the radio waves sent from the surface.
- Mesosphere & zones above it has no significance in the field of air pollution & its control.

Salient features of the next upper layer called thermosphere (ionosphere) are listed below:

- In this zone which extends from 85 to 500 kms the temperature becomes very high, but the air has such a low density that it can hold very little heat. All atoms in this region are ionised (hence called ionosphere also) ie, their electrons are stripped of by the sun. Free electrons flashing off in this area make thermosphere a good conductor.
- Any denser object in thermosphere will be very hot in the sunlight, but very cold at night.
- Temperature falls with increasing altitude in thermosphere due to decrease in the level of ozone which is a radiation absorbing species.
- Maximum temperature in thermosphere is 1430 °C over north pole (870 °C over the equator) due to absorption of highly energetic radiation of wavelength less than 200 nm by the species in this region.
- Atmospheric pressure decreases almost exponentially in the thermosphere. Pressure decreases by a factor of 10 for each 20 km rise in altitude.
- Upto 80 km from earth atmospheric components remains constant & hence the long zone upto 80 km is called homosphere.
- Mesosphere & thermosphere collectively is called ionosphere because they contains gases in the ionised form.

  Not much is known about the region above thermosphere which is called exosphere. It extends from 500 to 1600 Kms, which contains mainly atomic & ionic oxygen, hydrogen and helium. Beyond exosphere is the unbounded area called interstellar space.
- 75% of the whole atmosphere lies within 16 kms from the earth's surface & about 99% of the atmosphere is below an altitude of 30 kms.
- Life is present in the troposphere only. The observable universe has a diameter of about 25 billion light years and the study of air pollution is limited to only about 30 kms from earth's surface.
- Atmosphere is an insulating blanket protecting the earth. It softens the intense heat & light of the sun & stratospheric ozone present in it acts as a protecting umbrella that absorbs harmful ultraviolet radiation coming from the sun.
- Atmosphere is bound to earth by gravity unlike satellites like moon which does not hold an atmosphere due to its low gravitational power.
- Without the atmosphere, the temperature at the equator during day & night would have been 82°C & -140°C respectively.
- Limits of atmosphere is not sharply defined as it becomes progressively thin with increasing distance from the earth. There is no sharp boundary between the atmosphere and the outer space.
- Without atmosphere, there would be no sound, light and conventional long distance radio communication as it depends upon electrons in the upper atmosphere. There would be no lightenings,

no thunders, no clouds, no wind, no snow, no fire & no pollution in absence of atmosphere.

- Pollutants emitted into atmosphere from various sources are transported & dispersed. The air borne cycle is initiated with emission of pollutants followed by transport, diffusion or concentration in the atmosphere and completed by deposition on soil, vegetation, livestock, water surfaces & other objects and these are finally washed out of atmosphere by rain or precipitation.
- The lowest temperature in troposphere is about -60°C, in stratosphere -88°C, in mesosphere -100°C and in the thermosphere it continuously rises.
- The highest temperature in troposphere is 20°C, in stratosphere it is 80°C, in mesosphere it is 100°C and in thermosphere it is >500°C.
- An air-craft flies upto a maximum height of 20 kms and experience a minimum temperature of -60°C. The highest peak of Himalaya namely Gaurishankar is at a height of 6–8 km with the minimum temperature of -40°C.
- In the upper atmosphere, we have atomic oxygen & nitrogen (O&N). It is an atomic layer while the middle air contains diffusion of O_2 & N_2 and is called as homopause or diffusion layer, while the ground bottom layer contains a mixture of oxygen & nitrogen.

Appendices

Contents — **Page No.**

Appendix 5.1: Major Regions of the Atmosphere and their Characteristics

Region of the atmosphere	Altitude range, km	Temp, range, °C	Important Chemical components present	Pressure range, milibars
Troposphere	0–11	15 to −56	N_2, O_2, CO_2, H_2O	1000–100
Stratosphere	11–50	−56 to −2	O_3	100–1
Mesosphere	50–85	−2 to −92	O_2, NO^+	1–0.01
Thermosphere	85–500	−92 to 1200	O_2^+, O^+, NO^+	0.01–0.002

NOTES

The atmosphere is essential for life on Earth as a source of oxygen for organisms, carbon dioxide for plants, stabilization of surface temperatures (the good greenhouse effect), and protection from damaging ultraviolet radiation from ozone (O_3) in the stratosphere. Although the atmosphere extends far above Earth's surface, over 99% of it is within a few kilometers of the surface. In fact, if Earth were a classroom globe, virtually all of the atmosphere's air would be in a layer the thickness of the varnish on the globe. A distinctive feature of atmospheric chemistry is the ability of energetic photons of ultraviolet radiation to put large amounts of energy into a single molecule as shown here for the photodissociation of stratospheric O_2 leading to ozone formation. Major atmospheric characteristics are shown below diagrammatically.

Appendix 5.2: Temperature Profile of the Atmosphere

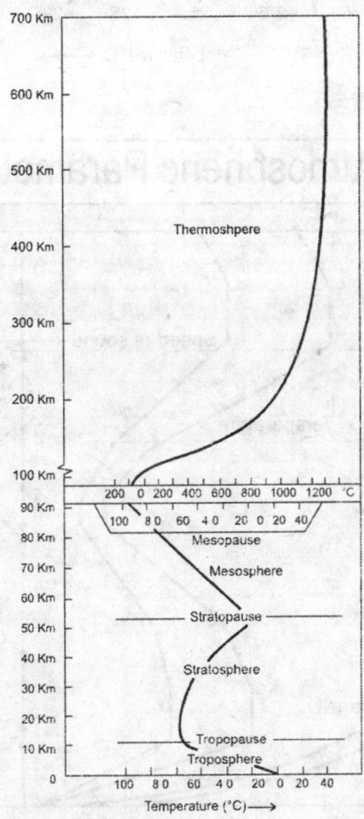

Appendix 5.3: Variation of Pressure and Temperature with Altitude in the Atmosphere

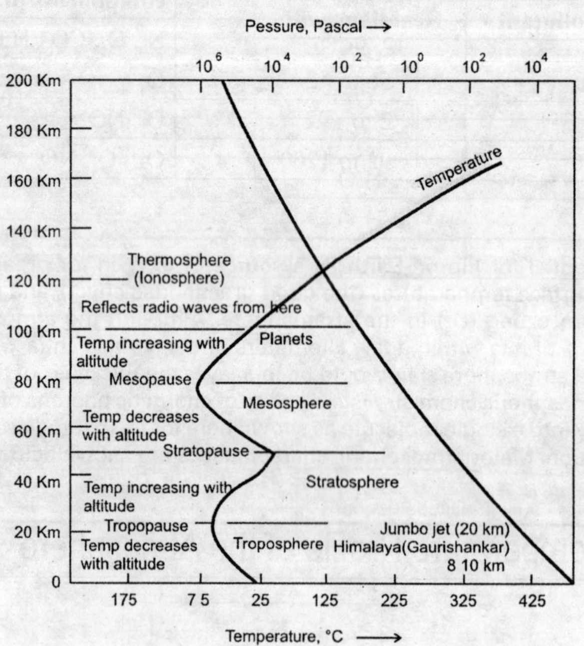

Appendix 5.4: Standard Atmospheric Parameters

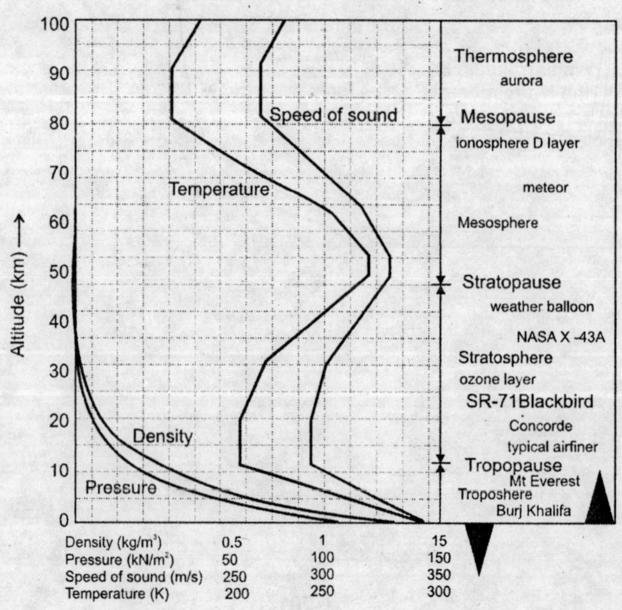

Appendix 5.5: Reaction of Primary Air Pollutants in the Atmosphere to Produce The Secondary Air Pollutants

S.N0.	Primary air pollutant	Reactions with	Secondary air pollutant produced
1	CO_2	H_2O	H_2CO_3
2	H_2S	O_2	H_2SO_4
3	SO_2	H_2O	H_2SO_4
4	NO_2	H_2O	HNO_3, HNO_2, N_2O
5	HF(g)	H_2O	HF
6	SiF_4	H_2O	H_2SiF_6

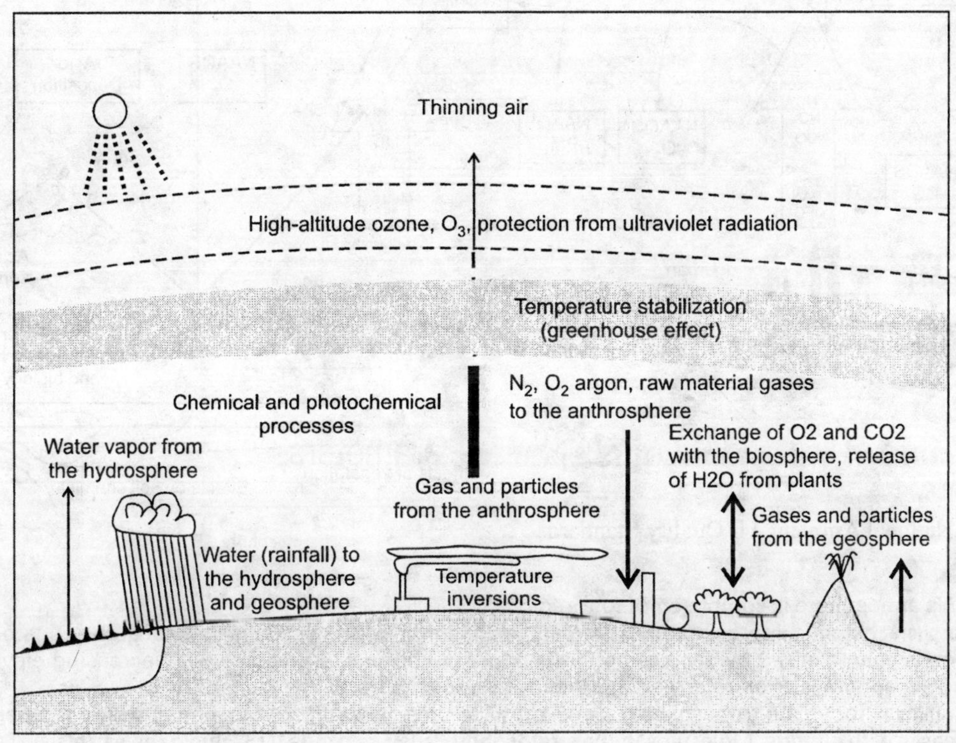

Appendix 5.6: (Interaction of trace gases in lower atmosphere) Composition and Chemistry of Atmosphere

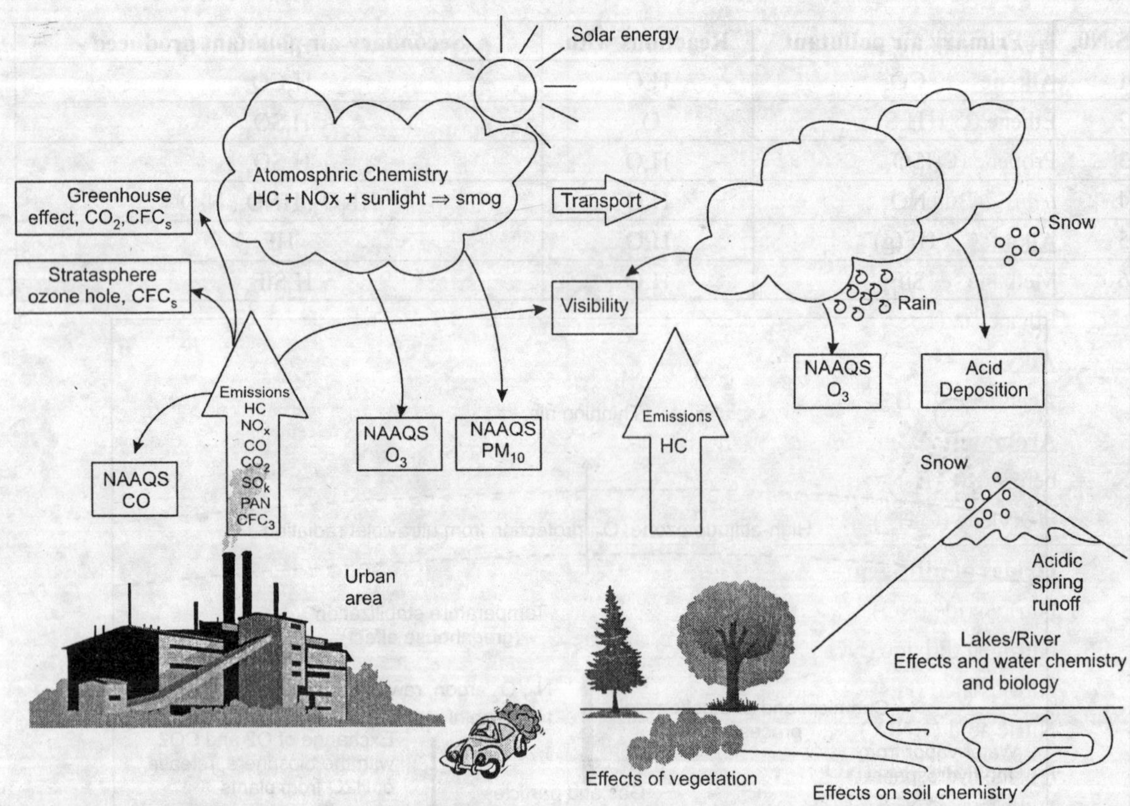

NAAQS-National Ambient Air Quality Standards

NOTES

The earth's atmosphere is comprised of four regions:

(1) **troposphere**, extends approximately 11 km above the earth's surface and is thickest at the equator. The region is characterized by decreasing temperature & barometric pressure with increasing elevation. The troposphere is often referred to as the lower atmosphere.

(2) **stratosphere**, above the troposphere is a 40 km wide area called the stratosphere, where temperature increases with altitude. Molecules in the stratosphere, such as ozone (O_3), shield the earth's surface by absorbing ultraviolet radiation.

(3) **mesosphere**, starts at about 50 km above the earth's surface and extends to an altitude of approximately 85 km. Temperatures in this region decrease as altitude increases and can reach as low as –140 °C. Radiant energy from the sun places molecules in this region into an excited state.

(4) **thermosphere**, is the thickest of the atmospheric layers, extending up to 600 km above the earth. Temperatures increase with altitude in this region, reaching up to 1700 °C. Because of the high temperatures, chemical reactions occur much faster. This region is also known as the upper atmosphere.

Appendix 5.7: Classes and Examples of Various Compounds Present in The Atmosphere

Hydrocarbons:– **Alkenes** Ethene (C_2H_4) Propene (C_3H_6) *trans*-2-Butene **Alkanes** Methane (CH_4) Ethane (C_2H_6) **Alkynes** Acetylene (C_2H_2) **Aromatics** benzene (C_6H_6) *m*-Xylene (C_6H_{10})	**Oxygenated hydrocarbons:–** Aldehydes Formaldehyde (HCHO) Acetylaldehyde (CH_3CHO) Other aldehydes (RCHO) **Acids** Formic acid (HCOOH) Acetic acid (CH_3COOH) **Alcohols** Methanol (CH_3OH)
Oxides of nitrogen:– Nitric oxide (NO) Nitrogen dioxide (NO_2) Nitrous acid (HNO_2) Nitric acid (HNO_3) Nitrogen trioxide (NO_3) Dinitrogen pentoxide (N_2O_5) Ammonium nitrate (NH_4NO_3)	**Oxides of sulfur (SO_X):–** Sulfur dioxide (SO_2) Sulfur trioxide (SO_3) Sulfuric acid (H_2SO_4) Ammonium bisulfate[$(NH_4)HSO_4$] Ammonium sulfate[$(NH_4)_2SO_4$] **Oxidants:–** PAN ($CH_3COO_2NO_2$) Ozone (O_3)
Free radicals:– Atomic oxygen (O) Atomic hydrogen (H) Hydroxyl (OH^-) Hydroperoxyl ($HO_2^=$) Acyl (RCO) Peroxyacyl ($RCOO_2^=$)	

Appendix 5.8: Energy Exchanges in the Atmosphere–Unit-Wm^{-2}

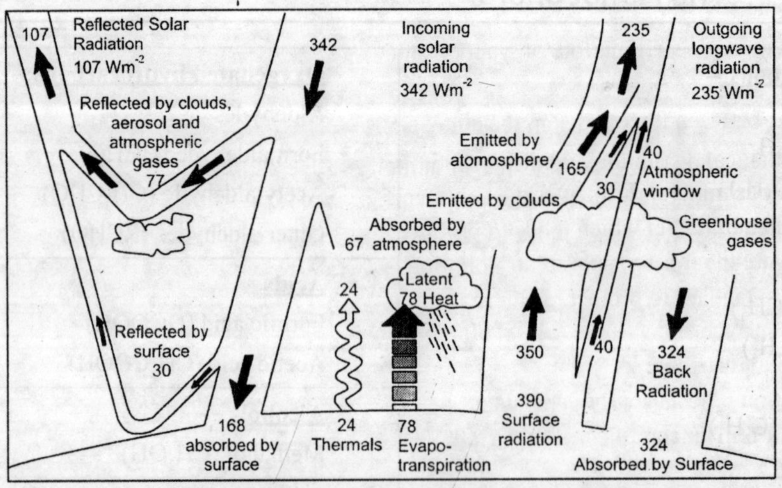

NOTES

Above figure is earth's annual and global mean energy balance. The amount of incoming solar radiation absorbed by earth and atmosphere is balanced by earth and atmosphere releasing the same amount of outgoing long wave radiation. About half of the incoming solar radiation is absorbed by earth's surface. This energy is transferred to the atmosphere by warming the air in contact with the surface (thermals), by evapotranspiration and long wave radiates radiation that is absorbed by clouds and greenhouse gases. The atmosphere in turn radiates long wave energy back to earth as well as out to space.

Appendix 5.9: Origin of the Atmospheric Aerosol

Aerosol: dispersed condensed matter suspended in a gas
Size range: 0.001 μm (molecular cluster) to 100 μm (small raindrop)

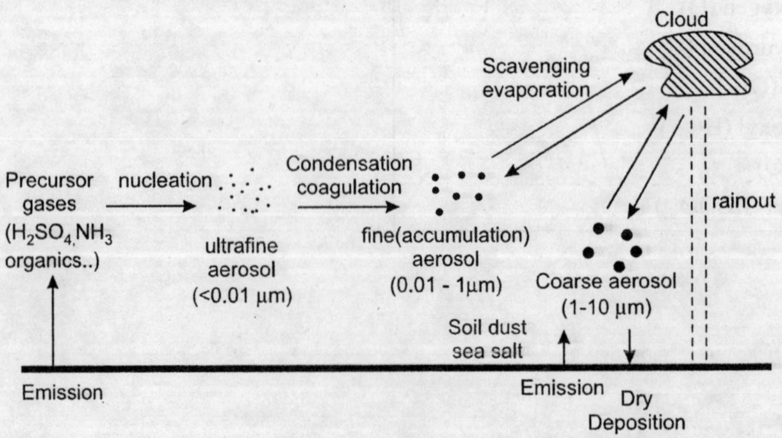

Environmental importance of aerosols include:

- Health (respiration), visibility, radiative balance, cloud formation, heterogeneous reactions, delivery of nutrients.
- Condensation & sublimation of vapors and the formation of smokes in natural & man-made combustion.
- Reactions between trace gases in the atmosphere through the action of heat, radiation, or humidity.
- The mechanical disruption and dispersal of matter at the earth's surface, either as sea spray over the oceans, or as mineral dusts over the continents.
- Coagulation of nuclei which tends to produce larger particles of mixed constitution
- Aerosols are the most visible of air pollution classes.
- The $PM_{2.5}$ standard to protect health is 15 μg/m^3.
- Number, surface and volume (mass) all play a role.
- Log-normal distributions.
- Composition reflects sources and sinks.
- Large and still uncertain impact on climate, aerosols inhibit gentle rains but exacerbate severe storms.

Appendix 5.10: Formation, Fate and Sizes of Atmospheric Aerosols

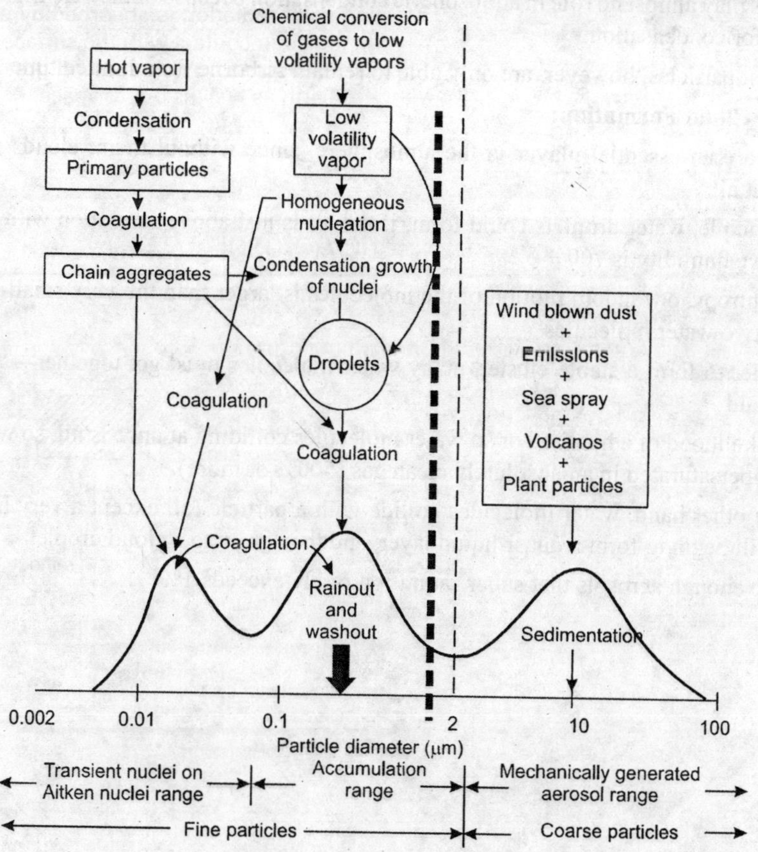

Aerosol Size Naming Convention

Usually divided into three size groups (D-diameter)

1. Aitken Nuclei → 2×10^{-3} μm < D ≤ 0.2 μm
2. Large Nuclei → 0.2 μm < D ≤ 2.5 μm
3. Giant Nuclei → D > 2.5 μm

Other Naming Convention of aerosols

- Nucleation mode → D ≤ 0.1 μm
- Accumulation or coagulation mode → 0.1 μm < D ~ 1 μm

Thought to be most important in natural cloud formation

- Coarse Particle Mode → D ~ 1 μm

Abundance and size of Aerosols in the Atmosphere:

- Aerosol concentration is highly variable in space and time. Concentrations are usually highest near the ground and near sources.
- A concentration of 10^5 cm^{-3} is typical of polluted air near the ground, but values may range from 2 orders of magnitude higher in very polluted regions to several lower in very clean air.
- Radii range from ~ 10^{-7} cm for the small ions to more than 10 μm (10^{-3} cm) for the largest salt and dust particles.
- Small ions play almost no role in atmospheric condensation because of the very high supersaturations required for condensation.
- The largest particles, however, are only able to remain airborne for a limited time

Aerosols and Cloud Formation:

- Aerosols are an essential player in the atmosphere, since without them, clouds would form only rarely if at all.

  → In principle, water droplets could form, if the air is just above saturation with water vapor, i.e., the relative humidity is 100 + δ %.
- But the entropy, or random motion of the molecules is larger than the very small energy that holds together two water molecules

  → In order to form a stable cluster, many water molecules must get together— enough to form a quasi-liquid

  → The likelihood of a hundred or so water molecules colliding at once is nil, so water vapor can be highly supersaturated in an absolutely clean gas (500% or more).
- If, on the other hand, water molecules collide with a particle (all except a very hydrophobic one), then it will begin to form a quasi-liquid layer, and then grow to a cloud droplet.
- There are enough aerosols that super saturation rarely exceeds 1%.

Appendix 5.11: Atmospheric solid particles by size

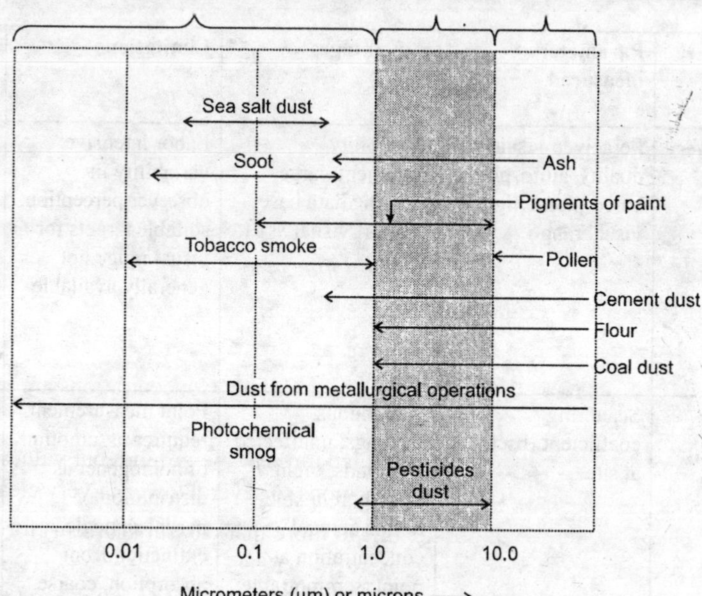

Appendix 5.12: Factors Determining Visibility in the Atmosphere

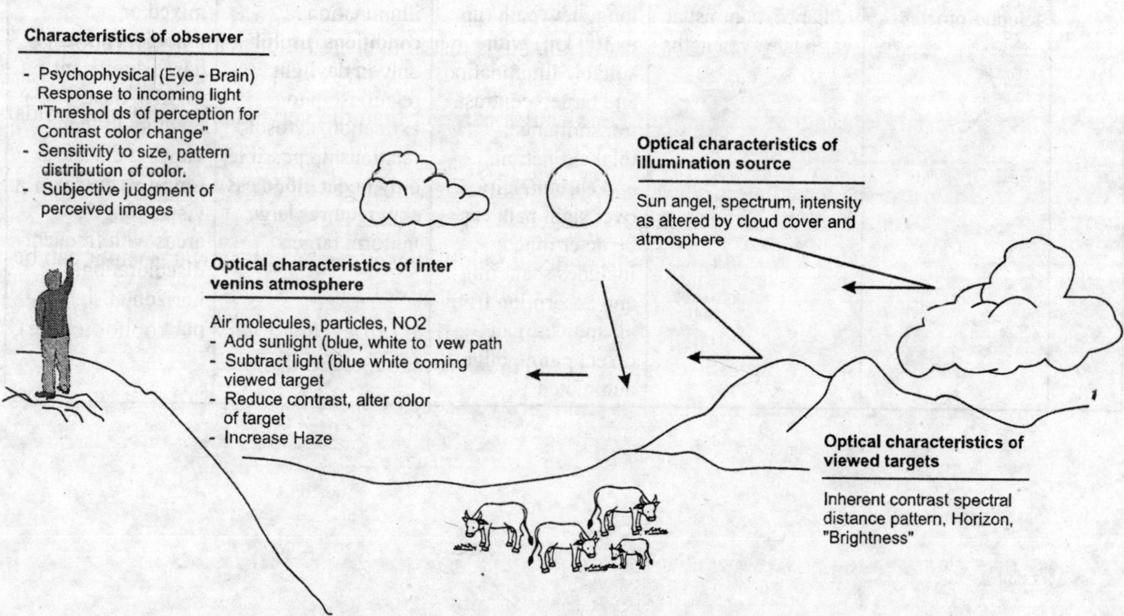

Appendix 5.13: Atmospheric Visibility Monitoring Methods

S.N0.	Visibility monitoring Methods	Parameters measured	Advantages	Limitations	Preferred use
1	Human observer	Perceived visual quality, atmospheric color, plume light, visual range	Flexibility, judgment; large existing data base (airport visual range)	Labor intensive; variability in observer perception; suitable targets for visual range not generally available	Complement to instrumental observations; areas with frequent plume blight, discoloration; visual ranges with available target distances
2	Integrating nephelometer	Scattering coefficient (bscat) at site	Continuous readings; unaffected by clouds, night; b_{scat} directly relatable to fine aerosol concentration at a point; semiportable; used in a number of previous studies; sensitive models available; automated	Point measurement, requires assumption of homogeneous distribution of particles; neglects extinction from absorption, coarse particles (>3–10 µm; must consider humidity effects at high relative humidity.	Areas experiencing periodic, well-mixed general haze; medium to short viewing distances; small absorption coefficient (b_{abs}); relating to point composition measurements
3	Multiwavelength telephotometer	Sky and/or target radiance, contrast at various wavelengths	Measurement over long view path (up to 100 km) with suitable illumination and target, contrast transmittance, total extinction, and chromaticity over sight path can be determined; includes scattering and absorption from all sources; can detect plume blight; automated	Sensitive to illumination conditions; useful only in daylight; relationship to extinction, aerosol relationship possible only under cloudless sky; requires large, uniform targets	Areas experiencing mixed or inhomogeneous haze, significant fugitive dust; medium to long viewing distances (one-fourth of visual range); areas with frequent discoloration; horizontal sight path

S.N0.	Visibility monitoring Methods	Parameters measured	Advantages	Limitations	Preferred use
4	Transmissometer	Long path extinction coefficient (b_{ext})	Measurement over medium view path (10–25 km); measures total extinction, scattering and absorption; unaffected by clouds, night	Calibration problems; single wavelength; equivalent to point measurement in areas with long view paths (50–100 km); limited applications to date still under development	Areas experiencing periodic mixed general haze, medium to short viewing distance areas with significant absorption (b_{abs})
5	Photography	Visual quality, plume blight, color, contrast (limited)	Related to perception of visual quality; documentation of vista conditions	Sensitive to lighting conditions; degradation in storage; contrast measurement from film subject to significant errors	Complement to human observation, instrumental methods; areas with frequent plume blight, discoloration
6	Particle samplers	Particles	Permit evaluation of causes of impairment	Not always relatable to visual air quality; point measurement	Complement to visibility measurements
7	Hi volume sampler	TSP	Large data base, amenable to chemical analysis; coarse particle analysis	Does not separate sizes; sampling artifacts for nitrate, sulfate; not automated	Not useful for visibility sites
8	Cascade impactor	Size-segregated particles (more than two stages)	Detailed chemical, size evaluation	Particle bounce, wall losses; labor intensive	Detailed studies of scattering by particles <2 µm
9	Dichotomous and fine particle samplers (several fundamentally different types)	Fine particles (<2.5 µm) coarse particles (2.5–15 µm) inhalable particles (0–15 µm)	Size cut enhances resolution, optically important aerosol analysis, low artifact potential, particle bounce; amenable to automated compositional analysis; automated versions available; large networks under development	Some large-particle penetration; 24 hr or longer sample required in clean areas for mass measurement; automated version relatively untested in remote locations	Complement to visibility measurement, source assessment for general haze, ground-level plumes

Appendix 5.14: Major Components, Sources and Effects of $PM_{2.5}$ (Particulate matter of dia ≤ 2.5 microns)

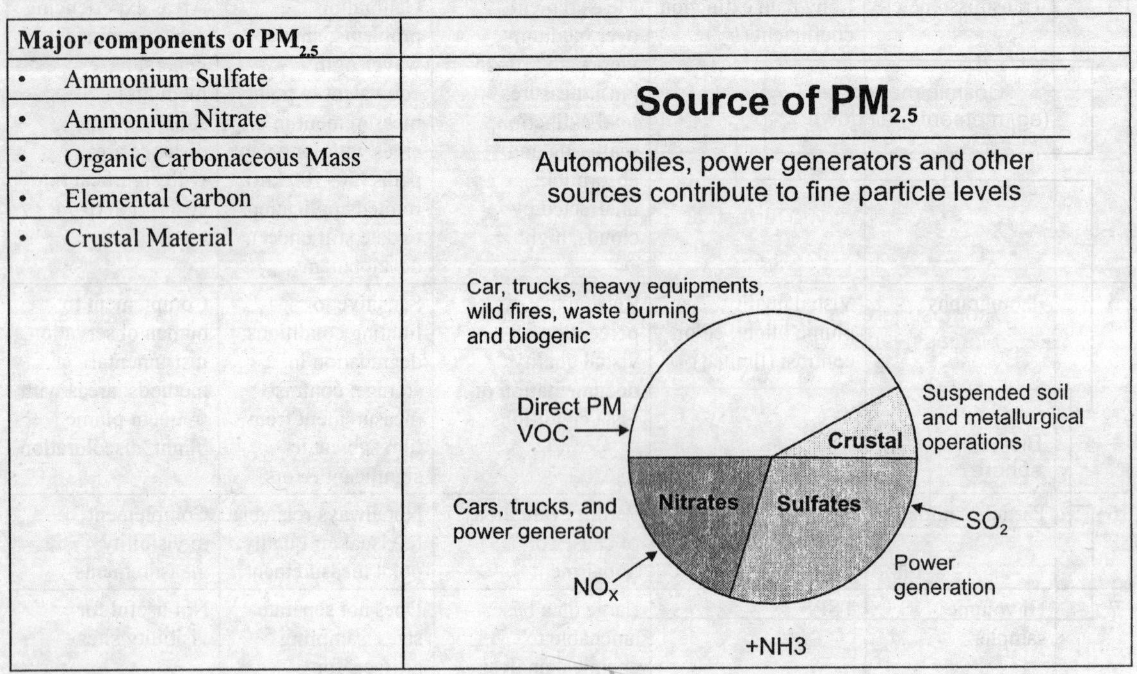

Major components of $PM_{2.5}$
• Ammonium Sulfate
• Ammonium Nitrate
• Organic Carbonaceous Mass
• Elemental Carbon
• Crustal Material

NOTES

1. The chemistry is complicated and particle formation is dependent on the other pollutants and atmospheric conditions. Ammonia from sources such as fertilizers and animal feed operations contribute to the formation of sulfates and nitrates that exits in the atmosphere as ammonium sulfates and ammonium nitrate.
2. All $PM_{2.5}$ constituents are not equally toxic.
3. Sulfate ion, iron, nickel, and zinc in $PM_{2.5}$ cause mortality.
4. Vanadium, elemental carbon, and nickel in $PM_{2.5}$ cause – Cardiovascular diseases and respiratory hospitalizations.
5. Elemental carbon, organic carbon, and nitrates in $PM_{2.5}$ cause – Cardiovascular deaths.
6. Elemental carbon in $PM_{2.5}$ cause –Cardiovascular hospital admissions.

Appendix 5.15: Interrelationship of Subsystems of the Climate Mechanism in Atmosphere

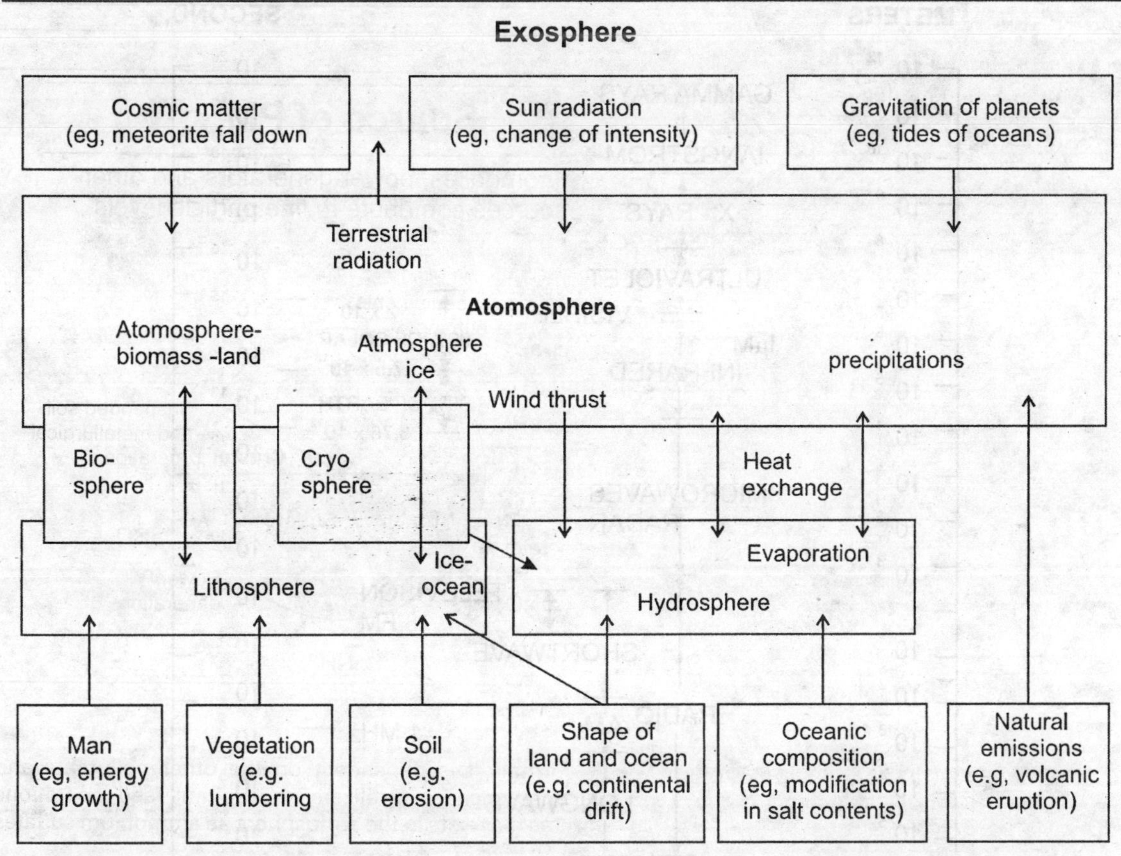

Appendix 5.16: Elecromagnetic Spectrum-Wave Length and Frequency

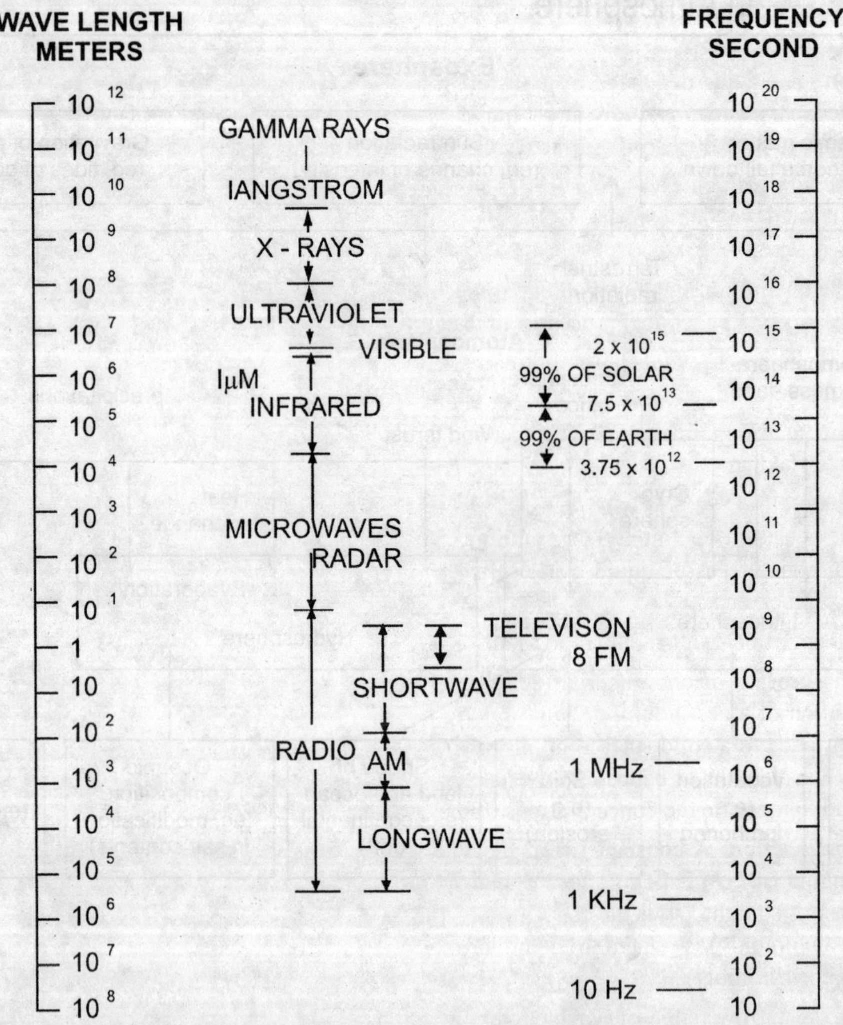

NOTES

The regions of solar and earth radiation.

Appendix 5.17: Terminologies Related to Atmospheric Science and Chemistry

Acid: A species which reacts in liquid water to generate hydrogen ions (conventionally represented as cations H^+ or hydronium, H_3O^+); anions (e.g., sulphate, SO_4^{2-}; nitrate, NO_3^-) which were associated with the H^+ in the acid are also released. Important acids in the atmosphere include sulfuric acid (H_2SO_4), nitric acid (HNO_3), and organic acids (e.g., formic acid, HCO_2H; acetic acid, CH_3CO_2H, etc.). A molecular entity capable of donating a proton to a base, (i.e. a "proton donor") or the corresponding chemical species. For example: H_2O, H_3O^+, CH_3CO_2H, H_2SO_4, HSO_4^-.

Acid deposition: The acids deposited by transfer from the atmosphere either by precipitation (rain, fog, snow, etc.), called *wet deposition,* or by direct transfer to the surface of the earth induced by the flow of acid-containing air masses over the earth's surface, called *dry deposition.*

Acid rain: Rain with pH values < about 5.6; commonly results from acids formed from pollutants. "Pure" rain water equilibrated with atmospheric CO_2 and naturally occurring acids in relatively clean air usually has a pH > 5.6

Actinic flux: The quantity of light available to molecules at a particular point in the atmosphere and which, on absorption, drives photochemical processes in the atmosphere.

Abatement: Action taken to reduce air pollution which involves the use of control equipment or some new process. This refers to a reduction or lessening as opposed to elimination of a type of discharge or pollutant.

Absorbance: The logarithm (must be specified as to base 10, lg, or base e, ln) of the reciprocal of transmittance [$\ln(I_o/I_t)$ or $\ln(I_o/I_t)$], where I_o and I_t are the monochromatic radiances (intensities) of light incident on and transmitted through, respectively.

Absorber: A device used commonly for sampling by absorption in which a gaseous or liquid material is removed from another gas or liquid by selective absorption; these include: scrubber, impinger, packed column, spray chamber, etc.

Absorptance, absorption-factor: The ratio of absorbed to incident radiant or luminous flux.

Absorption: The process of one material (absorbent) being retained by another (absorbate); this may be the physical solution of a gas, liquid, or solid in a liquid, attachment of molecules of a gas, vapour, liquid, or dissolved substance to a solid surface by physical forces, etc. In spectrophotometry, absorption of light at characteristic wavelengths or bands of wavelengths is used to identify the chemical nature of molecules, atoms or ions and to measure the concentrations of these species.

Absorption cross section: A constant (σ) in the following relation describing the absorption of light of a given wavelength: $\ln(I_o/I_t) = [A]l\sigma$; here I_o and I_t are the monochromatic radiances (intensities) of the incident and transmitted light, $[A]$ is the concentration (molecules cm^{-3}), and l is the path length (cm); σ (units of cm^2/molecule commonly) is related to the molar cross section κ by; $\sigma = \kappa/N_A$, where N_A is the Avogadro constant ($6.02213 \times 10^{23}\ mol^{-1}$).

Absorption line: A narrow range of wavelengths in which a substance absorbs light; a series of discrete absorption lines can be used as an unambiguous identification for many relatively simple chemical species.

Absorption spectrum: The wavelength dependence of the absorption cross section (or absorption coefficient); usually represented as a plot of absorption cross section versus wavelength λ (or $1/\lambda$) of the light.

Accretion: The process by which aerosols grow in size by external addition of various chemical species; a form of agglomeration.

Adiabatic lapse rate: The rate of decrease in temperature with increase in altitude of an air parcel which is expanding slowly to a lower atmospheric pressure without exchange of heat; for a descending parcel it is the rate of increase in temperature with decrease in altitude. Theory predicts that for dry air it is equal to the acceleration of gravity divided by the specific heat of dry air at constant pressure (approximately 9.8 °C/km). The moist adiabatic lapse rate is less than the dry adiabatic lapse rate and depends on the moisture content of the air mass.

Adsorbate: A molecular species of gas, dissolved substance or liquid which adheres to or is adsorbed in an extremely thin surface layer of a solid substance. The material accumulated on the surface of an adsorbent by the process of adsorption.

Adsorbent: An adsorbent is the solid material to which another species is adsorbed in a surface layer. A condensed phase at the surface of which adsorption may occur.

Adsorber: Solid material used to capture either a gas or liquid; often activated carbon is employed as the solid because of its high surface area per unit mass.

Adsorption: The process by which molecules of gas, dissolved substances, or liquids adhere to the surface of solids through either weak physical forces (physical adsorption) or stronger chemical forces (chemical adsorption).

Advection: The transport of air, its properties (such as heat), trace materials, fog, cold air, etc., solely by mass motion of the atmosphere, generally in a horizontal direction.

Aeration: The process by which a volume filled with a liquid becomes permeated with air or another gas; aeration is often accomplished by spraying the liquid into the air, bubbling air through a liquid, or agitating the liquid to promote surface absorption of air.

Aerometer: Instrument used to measure the density of gases.

Aerometric measurements: Measurements of the temperature, pressure, air motion (velocity vectors), or other physical measurements to characterize an air mass.

Aerosol: Mixtures of small particles (solid, liquid, or a mixed variety) and the carrier gas (usually air); owing to their size, these particles (usually less than 100 μm and greater than 0.01 μm in diameter) have a comparatively small settling velocity and hence exhibit some degree of stability in the earth's gravitational field. An aerosol may be chracterized by its chemical composition, its radioactivity, the particle size distribution, the electrical charge and the optical properties. Aerosols are dispersions in gases. In aerosols the particles often exceed the usual size limits for colloids. If the dispersed particles are solid, one speaks of "aerosols of solid particles", if they are liquid of "aerosols of liquid particles".

Agglomeration: The growth of particles in size due to an attachment of other particles following collisions. The term refers to aerosol particles, cloud droplets, etc.

Air, composition of pure: The composition of air is variable with respect to several of its components (e.g., CH_4, CO_2, H_2O) so "pure" air has no precise meaning; it is commonly considered to be air which is free of dust, aerosols, and reactive gaseous contaminants of anthropogenic origin. The composition of the major components in dry air is relatively constant (percent by volume given): nitrogen, 78.084; oxygen, 20.946; argon, 0934; carbon dioxide, 0.033; neon, 0.0018; helium, 0.000524; methane, 0.00016; krypton, 0.000114; hydrogen 0.00005; nitrous oxide, 0.00003; xenon, 0.0000087. The concentration of carbon dioxide, methane, nitrous oxide, the chlorofluorocarbons, and some other species of anthropogenic origin are increasing measurably with time. Relative clean air which is free of most reactive anthropogenic pollution (NO, NO_2, SO_2, non-methane hydrocarbons, etc.), often used as a reference sample in the calibration and operation of instruments.

Air monitoring station: A site at which monitoring of the concentration of one or more pollutants is carried out.

Air pollutant: A substance, gaseous material or aerosol which has been introduced into the air (either by human activity or by natural processes) in sufficient concentration to produce a measurable effect on humans, animals, vegetation, or materials (monuments, etc.): SO_2, NO_2, H_2S, CO, hydrocarbons, etc.

Air pollution: Usually the presence of substances in the atmosphere, resulting either from human activity or natural processes, present in sufficient concentration, for a sufficient time and under circumstances such as to interfere with comfort, health or welfare of persons or the environment.

Air pollution control: Measures taken to control the purity of the air to some degree specified by law.

Air pollution index or air quality index: A mathematical combination of the concentrations of air pollutants (weighted in some fashion to reflect the estimated health impact of the specific pollutant) which gives an approximate numerical measure of the quality of the air at a given time. These indices have little scientific basis but have been used to inform the public (in a qualitative fashion) of the degree of pollution present at a given time. It is recommended that the actual measured pollutant concentrations be used by all information services when possible with simultaneous reference given to the corresponding concentrations which are considered by health authorities to be hazardous to human health.

Analytical unit; analyser: An assembly of sub-units comprising: suitable apparatus permitting the introduction and removal of the gas, liquid or solid to be analysed and/or calibration materials; a measuring cell or other apparatus which, from the physical or chemical properties of the components of the material to be analysed, gives signals allowing their identification and/or measurement; signal processing devices (amplification, recording) or, if need be, data processing devices.

Aneroid barometer: An instrument for monitoring the atmospheric pressure in which no liquid is employed, but rather changes in pressure between the atmosphere and a closed vessel bend a diaphram which moves a pointer on a scale.

Anticyclone: A large system of winds that rotate about a centre of high atmospheric pressure, clockwise (viewed from above) in the northern hemisphere and counterclockwise in the southern hemisphere.

Ash: The solid residue which remains after the combustion of a fuel such as coal. Ash consists largely of heat treated mineral matter, but it may contain some products of the incomplete combustion of the fuel as well.

Atmosphere (of the earth): The entire mass of air surrounding the earth which is composed largely of nitrogen, oxygen, water vapour, clouds (liquid or solid water), carbon dioxide, together with trace gases and aerosols;

Atmosphere (unit of pressure measurement): A unit of pressure measurement which has been employed in describing the pressure of gases in storage tanks. It is roughly related to the average pressure of the atmosphere at sea level. One atmosphere = 1.01325×10^5 N m^{-2} (Pa) ≡ 1013.25 mbar ≡ 760 Torr. The use of the atm unit of pressure measurement is no longer recommended by international scientific bodies; it has been recommended that standard pressure be redefined as 10^5 Pa.

Arrester: Equipment designed to remove particles from a gaseous medium.

Aspirator: Any apparatus that produces a movement of a fluid by suction (e.g., a squeeze bulb, pump, Venturi, etc.)

Anthropogenic: Produced by human activities.

Background concentration (level): Synonymous with baseline concentration. The concentration of a given species in a pristine air mass in which anthropogenic impurities of a relatively short lifetime are not present. The background concentrations of relatively long-lived molecules, methane, carbon dioxide, halocarbons (CF_3C1, CF_2Cl_2, etc.), and some other species continue to rise due to anthropogenic input, so the composition of background air is undergoing continual change. Background concentration of a given species is sometimes considered to be the concentration of that impurity in a given air mass when the contribution from anthropogenic sources under study is absent.

Baffle chamber: A chamber used in incinerator design to promote the settling of fly ash and coarse particulate matter by changing the direction and/or reducing the velocity of the gases produced by the combustion of the refuse.

Bag filter: A large bag constructed of a suitable fabric which is often tubular in shape, into which a particle-containing air stream flows. Modern bags are constructed of a fabric which is capable of collecting all but very fine particles in the gas stream. The efficiency of the removal of particles of various size ranges changes with the amount of particles captured by the filter and the filtering time. The bag operates on the same principle as the one on a household vacuum cleaner.

Baghouse: An installation which contains many bag filters in parallel so that the resistance to air flow in a large installation is not seriously increased by the addition of these controls.

Base: A species which when dissolved in water generates hydroxide (OH^-) ions or is capable of reacting with an acid to form a salt. *Base* is a molecular entity capable of accepting a proton from an acid (i.e. a "proton acceptor") or the corresponding chemical species. For example: OH^-, H_2O, $CH_3CO_2^-$, HSO_4^-, SO_4^{2-}. A *Lewis base* is a molecular entity (and the corresponding chemical species) that is able to provide a pair of electrons and thus capable of co-ordination to a Lewis acid, thereby producing a Lewis adduct.

Beer-Lambert Law: An equation useful in the description of light absorption by a compound of concentration $[A]$ (often in mol L^{-1}) and with an optical path length l (often in cm): $\ln(I_o/I_t) = [A]l\varepsilon$; here I_o and I_t represent the radiance (intensity) of the incident and the transmitted light, respectively, and ε is the decadic molar absorption coefficient with common units of L mol^{-1} cm^{-1} coherent SI units mol^{-1} m^2. For a collimated beam having a unique pathlength through the absorbing medium, this law applies to irradiance as well.

Biosphere: That part of the globe that encompasses all forms of life on the earth. It extends from the ocean depths to a few thousand meters of altitude in the atmosphere, and includes life forms at the earth's surface, in soils, and constituents which exchange materials with atmospheres, oceans, or surfaces.

Bivane: A wind vane used in turbulence studies to obtain horizontal and vertical components of the wind vector.

Blowdown: Hydrocarbons purged during refinery shutdowns and startups which should be piped to storage systems for safe venting, flaring, or recovery. This term also applies to the purging of water in boiler operation, and serves in the control of dissolved solids in the boiler water.

Boundary layer: That well-mixed region of the lower atmosphere in which the turbulence is maintained largely by convective buoyancy induced by the upward heat flux originating from the solar-heated surface of the earth. During the afternoon this often extends from 1 to 5 km in height. *Surface boundary layer* is that region of the lower atmosphere where the shearing stress is constant. It is separated by the Ekman layer from the *free atmosphere*, where the behavior of the atmosphere approaches that of an ideal fluid in approximate geostrophic equilibrium (horizontal coriolis force balances the horizontal pressure force at all points in the field).

Breeze: A term sometimes used to describe very fine particles of coke. Also a meteorological term for a gentle wind.

Brownian motion: The movement of particles in colloidal system such as an aerosol caused by collision with the molecules in the fluid in which the particles are embedded.

Bubbler: An apparatus used to absorb certain water soluble components in a gas stream for later analysis. Usually it involves the use of a glass fritted tube which forces the air into small bubbles of high surface area during operation.

Calibration component: A component of a calibration mixture, present in the gaseous or vapour state, quantitatively and qualitatively defined, and used directly for testing and for calibration.

Calibration gas mixture: A gas mixture of known composition, generally comprising one or more calibration components and a complimentary gas.

Calibration mixture: A gaseous or liquid mixture of known composition, generally comprising one or more calibration components and an inert diluent, used directly for testing and calibration of analytical instruments.

Carbon black: Finely divided forms of carbon made by the incomplete combustion or thermal decomposition of natural gas, higher molecular weight hydrocarbons or other carbon-containing compounds.

Carbon dioxide (CO_2): Colorless, odorless gas which is the major carbon-containing product of combustion of carbon compounds. Its removal in the atmosphere involves the photosynthetic cycle in plants which convert CO_2 into carbohydrate and other organic molecules important in plant growth, as oxygen is released. It is also removed by the oceans where it is converted to carbonate deposits. Its concentration in the atmosphere varies with position on the earth, season, time of day, etc. However its average mixing ratio in air has increased from about 315 to 390 ppmv during the last 35 years, presumably as a result of increased burning fossil fuels.

Carbon monoxide (CO): Colorless, odorless gas which is highly poisonous even at the level of 0.2% in air. It is a product of the incomplete combustion of hydrocarbon fuels, and it is also formed in the clean atmosphere as a result of the oxidation of methane and its oxidation product formaldehyde (CH_2O); mixing ratios of CO in air masses are usually of the order of 100 ppbv near sea level.

Carrier gas: A gas introduced in order to transport a sample for analytical purposes. In gas chromatography it is the gas which is passed continuously through the column and whose passage promotes the elution of the components of the sample. The carrier gas together with the portions of the sample present in this phase constitutes the mobile phase.

Catalysis: The alteration of the rate of a chemical reaction (usually increase in the rate) by the addition of some substance (the catalyst) which does not undergo a net chemical change. Cases occur with certain reactants in which the addition of a substance reduces the rate of a particular reaction, for example, the addition of an inhibitor in a chain reaction or a poison in a catalytic reaction. The term "negative catalysis" has been used for these phenomena, terms such as *inhibition or poisoning* are preferred.

Ceramic filter: A component of a stack sampling system which is suitable for high temperature use; also known as a ceramic thimble.

Charcoal: A carbonaceous solid residue which is formed in the destructive distillation of wood. It has a very high surface area per unit weight which makes it especially useful as an absorber of various gases and vapours.

Chemisorption (or chemical adsorption): Adsorption which results from chemical bond formation (strong interaction) between the adsorbent and the adsorbate in a monolayer on the surface. Adsorption in which the forces involved are valence forces of the same kind as those operating in the formation of chemical compounds. If the molecule is adsorbed without fragmentation, the adsorption process is non-dissociative. Adsorption of carbon monoxide is frequently of this type.

Chemiluminescent methods of detection (analysis): Methods which utilize gas or liquid phase reactions between two molecules which produces a third molecule in an excited state. The wavelength distribution of the light emission from the excited molecule is characteristic of the species; in some circumstances this may be used to identify and to determine the concentration of one of the reacting species.

Chemiluminescence analyser: Instrument consisting of a reaction chamber with separate inlets for the sample and reagent gas, an optical filter, a photomultiplier and a signal processing device. The reactive gas is introduced in excess. The quantity of light produced is proportional to the sample flow rate and the concentration of the measured substance in the sample under specified temperature and pressure conditions. The filter limits the wavelength to the region of interest and helps to eliminated interferences.

Chimney (stack): A vertical flue for conducting cooled combustion products of a process to the atmosphere. This term also includes structures and opening of any kind from or through which smoke, grit, dust or fumes may be emitted.

Chimney effect: A vertical movement of a localized mass of air or other gases which occurs due to local temperature differences.

Chimney height (stack height), effective: A height used for the purposes of calculating the dispersion of emitted gases from a chimney, and which differ from the real chimney height by an amount which depends on such factors as the exit velocity, buoyancy effects and wind speed; it may also be affected by the local topography. It denotes the maximum height of the centre of a plume path above the level of the ground. The effective height may be above or below the actual chimney height, although the former case is most common.

Chromatographic analysis (chromatography): A method of separation of different chemical species based upon selective adsorption. Gas chromatography involves the separation of components (gases or vapours) in a gaseous stream (carrier gas) on a column of some suitable sorbent (stationary phase). Nonvolatile components are often separated by liquid chromatography where a solution of the components is separated by partition between a liquid mobile phase and a "solid" stationary phase. In both cases, the stationary phase may either be a liquid (often chemically bonded to the support) in which the components selectively dissolve, or a solid on which they are selectively adsorbed. Ion chromatography and ion exchange chromatography are useful for the separation and analysis of the ions present in rain-water or cloud-water. The stationary phase may be packed in a column, spread as a layer, or distributed as a film, etc.,; in these definitions chromatographic bed is used as a general term to denote any of the different forms in which the stationary phase may be use. The mobile phase may be gaseous or liquid.

Cloud: An aerosol of the atmosphere which is dense enough to be perceptible to the eye. Usually refers to an assembly of water-droplet aerosols suspended in the atmosphere, although dust clouds are also recognized.

Coagulation (flocculation): Process of converting a finely divided or colloidally dispersed suspension of one substance (usually a solid) in a liquid into larger-size particles which do settle or precipitate under the influence of gravity. Aluminum sulfate or other salt which contains a trivalent or divalent metal ion is sometimes used for this purpose. For atmospheric aerosol, coagulation denotes the process by which particles grow due to the attachment of smaller, more mobile particles subject to Brownian motions. Aitken particles coagulate efficiently with larger particles. When a sol is colloidally unstable (i.e. the rate of aggregation is not negligible) the formation of aggregates is called coagulation or flocculation.

COH, coefficient of haze: One technique of measurement of the amount of filterable particulate matter suspended in air which has been used in the past depends upon drawing a measured sample of air (usually 1000 linear feet) through a paper or membrane filter. A measurement is made of the intensity of light transmitted through the dust spot formed relative to that transmitted through an identical clean filter. The dirtiness of the air is reported in terms of the COH unit. This relates to the quantity of particulate material which produces an optical density, $\ln(I_0/I)$, of 0.01 when measured by light transmission at 400 nm and relative to the transmission of an identical dust-free filter taken as 100%. Thus a filter which transmitted 50% relative to the clean filter has an absorbance of 0.301 or 30.1 COH units. This is not a recommended measure of filterable particulate matter since the size, color and other properties of the aerosol and the air in which is suspended affect the results.

Collection efficiency: A term which characterizes an entire sampling and sample pretreatment procedure, usually represented as a percentage of the original amount of the analyte which is left for measurement (signal formation) after having passed through this procedure. This term also applies to the efficiency of collection of an air pollutant by an arrestment plant. For particulates, the collection efficiency is size related.

Collector: A device for removing and retaining contaminants from air or other gases. Term which is often applied to cleaning devices in exhaust systems. Also used to designate a device for removing and retaining samples from media in different environmental compartments to be investigated. A collector is sometimes used to describe a ***scavenger***. A solid substance added to or formed within a solution to collect a micro or macro-component.

Colorimeter: An instrument used for color measurement based on optical comparison with standard colors.

Combustion chamber: The region within which solids, vapours and gases from the primary chamber are burned and the some settling of fly ash takes place.

Combustion gases: Gases and vapours produced in furnaces, combustion chambers or in open burning.

Concentration, gas and liquid solutions: This represents the quantity of matter (mol, g, molecules, or other) per unit of volume (cm^3, L, m^3, or other). The IUPAC **GB** designation of this quantity is the amount of substance concentration (with the **SI** base unit of mol m^{-1}, but practical units of mol dm^{-3} or mol L^{-1}). For a gas phase species it is common to describe the concentration as the number of molecules of the particular species per cm^3 of air; this unit is most common in gas phase kinetics and in calculations involving chemical changes in the atmosphere. The number concentration or number density, often employed in physics and in physical chemistry, describes the number of molecules per cm^3; however in writing the units for this quantity, molecules are understood but omitted; only cm^{-3} is written. The most common method of describing concentration of a given solute in the liquid phase (such as cloud-water) is the amount of the compound or ion (in moles) present per litre of the solution. One mole of a given species is 6.02213×10^{23} (Avogadro constant) of these species. The *mixing ratio* is also employed as a measure of relative concentration. For a non-reactive gas, the mixing ratio does not change with altitude although the absolute concentration (molecules cm^{-3}) decreases. Thus one part-per-million by volume (1 ppmv) is a concentration equivalent to one molecule of a given species present in one million molecules (10^6) of air. Similarly one part-per-billion by volume (1 ppbv) represents a molecular ratio of one molecule for every 10^9 molecules of air; one part-per-trillion by volume (1 pptv) represents one molecule per 10^{12} molecules of air. Concentrations of solids suspended in air are sometimes given in units of mass of the species per unit volume (e.g., micrograms per cubic meter of air). At times the mass fraction is used in contrast to the use of the volume, molecule, or mole fraction (e.g., ppmv); thus one part-per-million by mass (ppmm) is one unit of mass of a given species per million mass units of the mixture.

Concentration, ground level: The concentration of a chemical species, normally a pollutant, in air, usually measured at a specific height above the ground.

Concentration, particle: Commonly expressed in several ways: mass concentration (usually as µg m^{-3}) or number concentration (number of particles cm^{-3}); modern instrumentation allows measurement of the number of particles as a function of size as well as the total number present in a given air volume. For atmospheric aerosols, this is a complex distribution for which diameters range from below 0.01 to above 100 µm; the particles making the highest contribution to the total number density are in the size range below 0.1 µm, those contributing most to the total surface area are in the 0.1 to 1.0 µm range, while those with the highest contribution to the volume or mass of the aerosol come from both the 0.1 to 1.0 and 1.0 to 100 µm ranges.

Condensation: The physical process of converting a material from a gaseous or vapour phase to a liquid or solid phase; this commonly results when the temperature is lowered and/or the vapour pressure of the material is increased. The tendency exists for condensation to occur when the partial pressure of a given component of a gaseous mixture at a given temperature exceeds the vapour pressure of the liquid or solid form of that component at the given temperature.

Condensation nuclei (sometimes symbolized as CN): A particle, either liquid or solid, or an ion upon which condensation of water vapour (or other substances) begins in the atmosphere. Condensation nuclei are usually very small hygroscopic aerosols (0.001 to 0.1 μm in diameter), but these are not as abundant as the smaller particles. The number of CN which are active (initiate condensation) in a given air mass may be a function of the relative humidity. Usually CN are counted as the active nuclei at about 300% relative humidity, while cloud condensation nuclei (CCN) are counted as the number of active nuclei at relative humidity less than or equal to 102%.

Contaminant, air: A substance, gaseous material or aerosol, which is present in an air mass at levels greater than in clean air. An air contaminant has been added commonly by anthropogenic activity.

Continuous air monitoring: A series of monitoring stations which are established near large cities for the continuous monitoring of gaseous pollutants, SO_2, NO, NO_2, CO, hydrocarbons, O_3 etc.

Convection (as applied to air motion): Vertical motion of the air induced by the expansion of the air heated by the earth's surface and its resulting buoyancy.

Cracking: The thermal or catalytic decomposition of a compound such as a hydrocarbon into chemical species of smaller molecular weight.

Critical temperature: That temperature, characteristic of each gas, above which it is not possible to liquefy a given gas.

Critical point: The temperature and pressure at which the liquid and vapour intensive properties (density, heat capacity, etc.) become equal. It is the highest temperature (critical temperature) and pressure (critical pressure) at which both a gaseous and a liquid phase of a given compound can coexist.

Cryogenic: Term used to describe low temperature processes, apparatus, etc.; usually applied to systems operated at the temperature of liquid nitrogen, helium or other condensed gas which boils at a very low temperature (at atmospheric pressure).

Cumulative sample: A sample collected over an extended period of time.

Cut off (as applied to aerosol sizes): The size of particles at which the retention efficiency of an instrument device drops below a specified value under defined conditions.

Cyclone (collector): A dust, grit, or droplet separator utilizing essentially the centrifugal force derived from the motion of the gas. The flow of gases containing suspended particles into the device is transformed into a confined vortex from which centrifugal forces tend to drive the suspended particles to the wall of the cyclone body. The agglomerated particles are subsequently removed from the cyclone by gravitational action.

Cyclone (meteorology): A large circulatory wind system around a region of low atmospheric pressure.

Dark reaction (darkness reaction): A chemical reaction that does not require or depend on the presence of light. Contrasts with a photochemical reaction which is initiated by light absorption by one or more of the reactants.

Deliquescence: The process that occurs when the vapour pressure of the saturated aqueous solution of a substance is less than the vapour pressure of water in the ambient air. When water vapour is collected by the pure solid compound, a mixture of the solid and liquid, or an aqueous solution of the compound forms until the substance is dissolved and is in equilibrium with its environment; at this time the vapour pressure of water over the aqueous solution will equal the partial pressure of water in the atmosphere in contact with it. A crystalline salt aerosol particle will deliquesce in the atmosphere when the relative humidity surpasses a characteristic value, the so-called *deliquescence point*.

Demister: Apparatus made of wire mesh or glass fibre which is used to help remove acid mist as in the manufacture of sulfuric acid. Demisters are also components of wet arrestment plants.

Density: The mass per unit volume of a gas, liquid, or solid under specified conditions (temperature, pressure, etc.).

Denuder system (tube or assembly): An apparatus used to separate gases and aerosols (over a given diameter) which is based upon the difference in diffusion velocity between gases and aerosol particles. Usually a tube containing a selective internal wall coating which removes the gaseous compounds at the wall.

Deodorizer: Equipment for the removal of noxious gases and odors, which may consist of combustion, absorption, or adsorption units.

Deposition: Deposition is normally considered to be one of two types: *dry deposition* is the process by which aerosols and gases in the air are deposited on the surface of the earth (soil, water, rock, plants, etc.); this is termed "dry" deposition even when the receptor surface is moist. *Wet deposition* is that process which involves the transport of chemicals to the surface of the earth by water droplets or snow crystals which scavenge pollutants as they form and fall through the atmosphere.

Deposition velocity: The ratio of flux density (often given in units of g cm^{-2} s^{-1}) of a substance at a sink surface to its Concentration in the atmosphere (corresponding units of g cm^{-3}). While the units of this ratio are clearly those of velocity (in this case cm s^{-1}), the ratio is not a flow velocity in the normal sense of the word.

Desorption: The removal of a substance from a surface on which it has become absorbed. Desorption is accelerated by heating and exposure of the surface to a low pressure (or vacuum). The converse of adsorption, i.e. the decrease in the amount of adsorbed substance.

Desulfurization: The process by which sulfur is removed from a material such as coal or oil. It may involve one of many techniques including elutriation, froth flotation, laundering, magnetic separation, chemical treatment, etc.

Detection: The process of establishing the presence of air pollutants by means of specific instrumentation and sampling.

Detection limit, lower: The minimum concentration of a compound in an air sample which can be determined by an analytical method with a given statistical probability. Usually the lower detection limit is defined as three-times the standard deviation of the noise of an analytical method under the assumption that its distribution is Gaussian. In this case a concentration can be detected with a probability of 99.7%. This is sometimes referred to as the absolute detection limit.

Detection limit (threshold): The minimum concentration of substance which produces an observable response.

Detector: An instrument or part of an instrument which indicates the presence of air quality characteristic by means of some specific spectroscopic or chemical property of the pollutant.

Dew point: That temperature at which water vapour present in the atmosphere is just sufficient to saturate it. When air is cooled below the dew point, the excess of water vapour appears as tiny droplets or crystals of ice depending on the temperature of the air mass. This term is also applicable to acid gases as in the problem of acid corrosion and acid smut formation.

Diameter, equivalent: The diameter of a spherical particle which will give identical geometric, optical, electrical or aerodynamic behavior to that of the particle (non-spherical) being examined; sometimes referred to as the Stokes diameter for particles in non-turbulent flows.

Diffraction: A modification which light (or electron, neutron beams, etc.) undergoes in passing by the edges of opaque bodies or through narrow slits or in being reflected from ruled surfaces (or crystalline materials). The light waves, owing to their wave-like nature appear to be deflected and produce fringes of parallel light and dark bands corresponding to regions of constructive reinforcement or destructive interference, respectively, of the waves.

Diffuser: A porous plate or tube, commonly made of carborundum, alundum, or silica sand, through which air is forced and divided into minute bubbles for diffusion in liquids.

Diffusion: The spreading or scattering of a gaseous or liquid material. Eddy diffusion in the atmosphere is the process of transport of gases due to turbulent mixing in the presence of a composition gradient. Molecular diffusion is the net transport of molecules which results from their molecular motions alone in the absence of turbulent mixing; it occurs when the concentration gradient of a particular gas in a mixture differs from its equilibrium value. Eddy diffusion is the most important mixing process in the lower atmosphere, while molecular diffusion becomes significant at the lower pressures of the upper atmosphere.

Diluent gas: A gas of known quality introduced for analytical purposes so that it quantitatively lowers the concentration of the components of a gaseous sample.

Dioxin: Colloquial (short) name of a toxic by-product (and sometimes contaminant) of herbicides; the full name of this species is: 2,3,7,8-tetrachlorodibenzo[*b,e*][1,4]dioxin.

Discomfort threshold: The lowest value (e.g., concentration of an impurity, etc.) at which a sensation of discomfort is perceived; a measure which varies from person to person.

Dispersion: The dilution of a pollutant by spreading in the atmosphere due to diffusion or turbulent action (eddy diffusion).

Diurnal variation: Indicates variations which follow a distinctive pattern which recurs with a daily cycle.

Dobson unit: Unit sometimes used in the description of the total ozone in a column of air overhead. It is given as the thickness (in units of 10^3 cm or 10^5 m) of that layer which would be formed by the total ozone in a column reduced to the pressure of 760 Torr and temperature of 0 °C.

Dosage: As applied to an air pollutant in an exposure chamber, dosage is commonly defined as the concentration of the pollutant times the duration of exposure.

Downwash: As applied to the action of chimney gases, it is the downward motion of the chimney gases brought on by eddies which form in the ice of a chimney when the wind is blowing. It may result in bringing the flue gases to the ground prematurely.

Droplet: A small liquid particle. The size of droplets encountered in the atmosphere extends over a wide range; e.g., liquid aerosol solutions which make up the fine particle fraction of continental tropospheric aerosol are usually <2 μ*m* in diameter. Cloud water droplets usually have diameters in the range of 5 to 70 μm, while rain droplets commonly have diameters ranging from 0.1 to 3 mm.

Dry bulb temperature: In psychrometry, the temperature of the gas measured by a dry bulb thermometer.

Drying agents: Substances which remove water (liquid or gaseous).

Dust: Small, dry, solid particles projected into the air by natural forces, such as wind, volcanic eruption, and by mechanical or manmade processes such as crushing, grinding, milling, drilling, demolition, shoveling, conveying, screening, bagging, and sweeping. Dust particles are usually in the size range from about 1 to 100 μm in diameter, and they settle slowly under the influence of gravity.

Dust collector: A device for monitoring dust emissions. Also the equipment used to remove and collect dust from process exhaust gases; this may employ simply sedimentation (dustfall jars, coated slides, papers, etc.), inertial separation (cyclones, impactors, impingers, etc.), precipitation (thermal and electrostatic) or filtration.

Dust fall: Solid particles in the air which fall to the ground under the influence of gravity.

Dynamic range of an analyser: The ratio between the maximum usable indication and the minimum usable indication (detection limit). A distinction may be made between the linear dynamic range, where the response is directly proportional to concentration, and the dynamic range where the response may be non-linear, especially at higher concentrations.

Eductor: A liquid pump for mixing air with water which operates under a jet principle using the liquid under pressure as the operating medium.

Eddy: In turbulent fluid motion, a blob of the fluid that has some definitive character and moves in some way differently from the main flow.

Eddy dispersion (diffusion): The process by which substances are mixed in the atmosphere or in any fluid system due to eddy motion.

Effluent: Any spent liquors or other waste material which are emitted by a source (waste from plating shops, pickling tanks, sewage treatment plants, chemical manufacturing plants, etc.).

Einstein, unit: 1 mole (6.02213×10^{23}) of quanta or photons of electromagnetic radiation; for monochromatic light of wavelength λ, a light beam of energy E (J s^{-1}) striking a surface, will have $E/(6.022 \times 10^{23} \times hc/\lambda)$ Einstein striking per sec; h is the Planck constant (6.626×10^{-34} J s) and c is the velocity of light (2.998×10^{8} ms^{-1}).

Electrochemical methods of detection (analysis): Methods in which either current or potential is measured during an electrochemical reaction. The gas or liquid containing the trace impurity to be analysed is sent through an electrochemical cell containing a liquid or solid electrolyte and in which an electrochemical reaction specific to the impurity takes place.

Electrochemical methods, amperometric detection: The current is proportional to the concentration, of the species generating the current.

Electrochemical methods, coulometric detection: The current is directly proportional to the flow rate of the substance involved in the electrochemical reaction, and the amount of charge which flows is proportional to the amount of substances taking part in the reaction.

Electrochemical methods, potentiometric detection: The potential of a cell is related to the concentration (activity) of a reactant which is a component of the cell fluid.

Electrostatic filter: Filters for which an electrostatic charge is applied to the filter element. A fibrous filter material is often pleated in between V-shaped supports consisting of electrostatically charged, metal rods which are insulated from the supporting frame.

Electrostatic precipitator: A device which separates particles from a gas stream by passing the carrier gas between pairs of electrodes across which a unidirectional, high-voltage potential is placed. The particles are charged before passing through the field and migrate to an oppositely charged electrode. These devices are very efficient collectors of small particles, and their use in removing particles from power plant plumes and in other industrial applications is widespread.

Eluate: The effluent from a chromatographic bed emerging when elution is carried out.

Elute: To remove sorbed materials from a sorbent by means of a fluid called the eluent. **GB**: To chromatograph by elution chromatography. This term is preferred to the term develop, which has been used in paper chromatography and in thin-layer chromatography. The process of elution may continue until the components have left the chromatographic bed.

Elutriation: The process of separating the lighter particles of a powder from the heavier ones by means of an upward directed stream of fluid (gas or liquid).

Emission: The total rate at which a solid, liquid, or gaseous pollutant is emitted into the atmosphere from a given source; usually expressed as mass per unit time. *Primary emissions* are those substances which are emitted directly to the atmosphere (e.g., NO, SO_2, etc.), while secondary emissions are formed from the primary emissions through thermal or photochemical reactions (e.g., ozone, aldehydes, ketones, sulfuric acid, nitric acid, etc.). The point or area from which the discharge takes place is called the *source*; the area in which the emission or its transformed products (e.g., in the case of aerosols, acidic deposition, etc.) may be deposited is called the receptor area or sink. *Emission* may be applied to noise, heat, etc., as well as pollutants.

Emission control equipment: Air pollution control equipment which either converts the pollutant chemically to a non-polluting substance or collects the pollutant by some means including gravity settling chambers, inertial separators, cyclonic separators, filters, electrical precipitators, scrubbers, (spray towers, jet scrubbers, Venturi scrubbers, inertial scrubbers, mechanical scrubbers and packed scrubbers). Certain gases and odoriferous compounds are controlled by combustion, absorption (spray chambers, mechanical contactors, bubble cap or sieve plate contactors, and packed towers), and adsorption units (packed beds or fluidized beds).

Emission flux: The emission per unit area of the appropriate surface of an emitting source.

Emission inventory: A systematic collection of information concerning the air pollution emissions in a given area. Usually the types of sources (power plants, refineries, etc.) and the height at which the discharge takes place as well as the source contribution (composition and rate of discharge) are summarized. The variation of these emissions with time of day, and month of the year are desired information in inventories as well.

Emission source: Several types of emission sources exist: Point sources of emission include: power plants, incinerators, refineries, steel mills, chemical plants, etc. Often the emissions for these sources are released from elevated stacks. Mobile sources of emission include: automobiles, trucks, buses, ships, airplanes, etc.

Enforcement: The act of preventing the discharge into the atmosphere from any source operating without a permit or in violation of the terms of a permit or rules, regulations, or orders of an pollution control agency.

Entrainment: The act of forming a mist or fog droplets of a liquid carried off by the vapours of a boiling liquid or from a liquid through which bubbles of gas or vapour are passing rapidly.

Evaporation: The physical process by which a liquid substance is converted to a gas or vapour. This may occur at or below the normal boiling point of the liquid (the temperature at which a liquid boils at 1 atmosphere pressure), and the process is endothermic.

Excitance, radiant, spectral: The derivative of the incident radiant power with respect to the projected area of a specified emitting surface (W cm^{-2}). With the adjective *spectral*, differentiation with respect to wavelength is also implied (W cm^{-2} nm^{-1}). Irradiance and excitance are similar but refer to energy *arriving* at versus *leaving* from a surface, respectively.

Explosivity limits (or explosion limits); flammable limits: The concentration limits, usually of a substance in air, between which combustion will be self-sustaining.

Exposure: Subjecting a person, animal, plant, or material to an environment containing a significant concentration of an air pollutant.

Exposure chamber: In environmental studies, this is a suitable room or chamber in which the atmosphere may be controlled (relative humidity, pollutant concentrations, temperature, light intensity, etc.) and in which exposure to animals, plants, or materials can be made. The exposure of humans to small concentrations of any toxic or seemingly non-toxic chemical in chambers is a very hazardous practice which should be avoided; errors in the controls of the contaminant concentration and allergic reactions, etc., can occur with risk to life.

Eye irritation: Experienced by many persons upon exposure to some compounds (often lachrymators) in polluted air: SO_2, SO_3, HC1, formic acid, peroxyacetyl nitrate (PAN), formaldehyde, acrylaldehyde, etc.

Fallout: A measurement of air contamination consisting of the mass rate at which solid particles deposit from the atmosphere.

Fanning: In blast furnace operation, the idling period between the blowing periods when the blast pressure is reduced to a minimum. This also applies to plume behaviour under inversion conditions.

Filter: A porous material on which solid particles present in air or other fluid which flows through it are largely caught and retained. Filters are made with a variety of materials: cellulose and derivatives, glass fibre, ceramic, synthetic plastics and fibres. Filters may be naturally porous or be made so by mechanical or other means. Membrane/ceramic filters are prepared with highly controlled pore size in a sheet of suitable material such as polyfluoroethylene, polycarbonate, or cellulose esters. Nylon mesh is sometimes used for reinforcement. The pores constitute 80-85% of the filter volume commonly, and several pore sizes are available for air sampling (0.45 to 0.8 μm are commonly employed).

Filtration: The process of segregation of phases; e.g., the separation of suspended solids from a liquid or gas, usually by forcing a carrier gas or liquid through a porous medium.

Flame photometry: The use of emission spectroscopy in the ultraviolet and visible regions to identify and estimate the amounts of various elements which are excited in a flame, an arc or high voltage spark.

Flare, solar: A bright eruption of energy from the sun's chromosphere.

Flash point: The lowest temperature at which a substance, e.g., fuel oil, will give off a vapour that will flash or burn momentarily when ignited.

Flocculation: A process of contact and adhesion whereby the particles of a dispersion form larger-size clusters. Synonymous with *agglomeration* and *coagulation*.

Floccule: A small loosely aggregated mass of material suspended in or precipitated from a liquid; a cluster of particles.

Flow rate: The volume or mass flow per unit time.

Flow regulator: A device used in sampling to maintain constant airflow for a given period of time.

Flue: A passage for the conduction of combustion gases in an incinerator; synonymous with chimney. The term also applies to structures and openings of any kind used to transport gases in operations other than incineration.

Flue gas: Waste gas from the combustion process.

Flue gas scrubber: Equipment for removing fly ash and other objectionable materials from the products of combustion by means of sprays or wet baffles.

Fluorimeter: An instrument used to measure the intensity and the wavelength distribution of the light emitted as fluorescence from a molecule excited at a specific wavelength or wavelengths within the absorption band of a particular compound. Characteristic fluorescence bands may be used to identify specific pollutants such as the polynuclear aromatic hydrocarbons. *Excitation spectra* of impurities can be observed by scanning the wavelength of the excitation light which is incident on the sample over a range of wavelengths and observing the relative intensity of the fluorescence emitted at a given wavelength. These spectra are also characteristic of the impurity.

Flux: The number or mass of particles or molecules which pass through a given unit area of surface per unit time. *Net vertical flux* of a species is the difference between the downward and upward flux of that species,

Flux density, photon: At a given point in space, the number of photons incident in a time interval on a suitably small sphere centred at that point, divided by the cross-sectional area of that sphere and by the time interval.

Fly ash: Finely divided particles of ash entrained in flue gases resulting from the combustion of fuel.

Fly ash collector: Equipment to remove fly ash from the products of combustion.

Fog: A general term applied to a suspension of droplets in a gas. In meteorology, it refers to a suspension of water droplets resulting in a visibility of less than 1 km. An aerosol of liquid particles, in particular a low cloud.

Fog horizon: The top of a fog layer which is confined by a low-level temperature inversion and gives the appearance of the horizon (which it actually obscures) when viewed from above, against the sky.

Fossil fuels: Fuels such as coal, oil, and natural gas which have formed over many years through the decomposition of deposited vegetation which was under extreme pressure of an overburden of earth.

Fragment ion in mass spectrometry: An ion obtained when a molecule in the gaseous mixture under analysis is split into fragments during the ionization process (impact with the ionizing electron beam). An ion produced by the loss of one or more fragments from a parent molecular ion.

Freezing out: Term used in combustion for the analysis of incinerator flue gas components (largely the organic fraction) in which a series of traps at progressively lower temperatures are employed. In more general use, the term implies the removal of a condensable gas or liquid by condensation in a trap at low temperatures.

Fumes: In popular usage, a term often taken to mean airborne effluents, unpleasant and malodorous, which might arise from chemical processes.

Fumigation: An atmospheric phenomenon in which pollution, retained by an inversion layer near its level of emission, is brought rapidly to ground level as the inversion breaks up. This term also applies to the exposure of material (e.g., grain) to chemicals to kill insects, etc.

Gas analysis installation: Assembly for the purpose of determining one or more components in a gaseous mixture. It generally comprises the following elements: sample probe, region of primary treatment of the gases, region of secondary treatment of gases, points for venting to the atmosphere and for calibration sample injection, transfer line, and the analytical unit.

Gas black: (also carbon black, channel black, furnace black). Finely divided carbon (graphite) produced by incomplete combustion or thermal decomposition of natural gas.

Gas chromatography: A method of separating the components of a gas mixture. The sample is passed, in a stream of carrier gas, through a column which has different retention properties relative to the components of interest. Different components pass through the column at different rates and are detected as they elute from the column. Comprises all chromatographic methods in which the moving phase is a gas. The word *chromatography* itself implies that a stationary phase is present in addition to the moving phase

Gas chromatography, detector in: Generally a device which provides an analogue output proportional to the instantaneous concentration of the substance or its instantaneous eluting rate in the carrier gas.

Gas chromatography, electrochemical detector: An electrochemical cell which responds to certain substances in the carrier gas eluting from the column. The electrochemical process may be an oxidation, reduction or a change in conductivity.

Gas chromatography, electron capture detector: A small radioactive source containing ^3H or ^{63}Ni ionizes the molecules of the carrier gas (nitrogen or argon-methane), and a potential difference creates a small current. This current is reduced when an electronegative substance (such as a halocarbon) is introduced. The reduction in current is a measure of the concentration of the electronegative substance. The detection limit (threshold) varies greatly according to the substances to be analysed and can reach a mixing ratio of 10^{-12}. The linear dynamic range may be 10^4 but the maximum measuring value generally lies below 1 ppmv.

Gas chromatography, flame ionization detector: The gases emerging from the column are fed into a hydrogen flame across which an electrical potential is placed. Certain molecules ionize easily in the flame and the current produced is proportional to the instantaneous flow rate of the eluted component. The detector is relatively insensitive to inorganic molecules and is most used for organic compounds. Concentrations below 1 ppmv are easily detected. The linear dynamic range is between 10^3 and 10^5

Gas chromatography, flame photometric detector: The eluent from the column is fed into a hydrogen-rich flame and produces light emission. Optical filters are used to select the wavelength range of the emission which is characteristic of specific atoms (usually sulfur or phosphorous). The detector is very specific, depending on the choice of optical filters. It can detect the S- and P-containing compounds down to 10^{-3} ppmv, but the detector is nonlinear.

Gas chromatography, helium ionization detector: A weak beta source and a high potential raise the helium atom of the carrier gas to a metastable state. All other substances having an ionization potential lower than 18 ev are ionized, and the current that results is used to measure the components. The detector is usually employed to measure inorganic compounds at concentrations between 0.1 and 10 ppmv. It has a linear range of about l^2 but is somewhat unstable and requires great care to ensure the helium purity and to eliminate all leaks in the system.

Gas chromatography, mass spectrometric detector: A mass spectrometer can be used as a detector and gives qualitative and quantitative data on the various eluted substances. The mass spectrum of the eluted compounds provide evidence beyond the elution time .as to the chemical nature of the species.

Gas chromatography, photoionization detector: Radiation from an ultraviolet lamp ionizes certain species in the carrier gas. A potential difference is applied and the resulting ionization current is detected. The detector is only useful for substances with ionization potentials below about 11 ev. This makes it quite useful for detecting one component of a combined eluent when the other component, for instance nitrogen, has a high ionization potential. The detector has a small linear dynamic range and is capable of detecting substances below 1 ppmv.

Gas chromatography, thermal conductivity detector: In general, two cells arranged in a bridge configuration detect the change in thermal conductivity of the gas at the output of the column. This detector is sensitive to any substance with thermal conductivity different from that of the carrier gas. The lowest detectable limit is between 0.5 and 100 ppmv. The linear dynamic range is of the order of 10^3. This type of detector is often used for measuring components at relatively high concentrations.

Gas chromatography, ultrasonic detector: Sound is generated in a reference cell and a measuring cell. An eluted component passing through the measuring cell changes the velocity of sound in the cell. This change is detected by a phase shift of acoustic signals between the two cells. This detector is usually employed for inorganic gases in the region where thermal conductivity detectors are not sufficiently sensitive. The detection threshold is of the order of 0.1 ppmv. The linear dynamic range is of the order of 10^3.

Gas-liquid chromatography; partition chromatography: Comprises all gas-chromatographic methods in which the stationary phase is a liquid distributed on a solid support. Separation is achieved by partition of the components of a sample between the phases. In this and in gas-solid chromatography the first word specifies the mobile phase and the second word the stationary phase. Liquid stationary phases are supported on solids, but the solid support affects the chromatographic process. For classification, the term relating to the predominant effect is chosen.

Gas plume: The stack effluent which may assume various geometrical forms and concentration distributions which are dependent on the degree of turbulence: **a) Looping plumes** are formed under conditions of high convective turbulence which may result from the intense solar heating of the earth's surface or from mechanical turbulence induced from nearby hills upwind of the stack; **b) Coning plumes** which are reasonably symmetrical, occur under more nearly neutral thermal conditions when both convective and mechanical turbulence are relatively low; **c) Fanning plumes** occur under stable conditions, often at night, when the vertical component of the mechanical turbulence is suppressed more than the horizontal component so that the horizontal plume width is greater than its vertical thickness; **d) Fumigating plume** is caused by unstable air below and a stable layer above which results in a downward dispersion of gases which is greater than that seen under stable conditions; often occurs in the morning following a night of marked stability as the sun heats the ground to create the instability from the ground upward; **e) Lofting plume** is caused by conditions inverse to those causing fumigation and results in a greater diffusion upward; the meteorology of late afternoon and early evening often favors this condition.

Gravimetric methods: A determination by weight; e.g., in the older method of Cl ion determination, a weighed amount of sample is dissolved in water, $AgNO_3$ solution added, AgCl is precipitated, dried and weighed. From the known mass fraction of silver in AgCl, the weights of the initial sample and that of the AgCl precipitated, the percentage of chlorine in the sample can be calculated readily.

Greenhouse effect: Heating effect produced by certain gases (e.g., CO_2, O_3 etc.), which by virtue of their characteristic infrared absorption, lower the earth to space of long wavelength radiation but allow transmission of shorter wavelength radiation inward from the sun to the earth.

Grit: Airborne solid particles in the atmosphere which are of natural or manmade origin and which remain in suspension for some time; the size of the grit particles is defined as greater than 75 μm in diameter (retained on a 200 mesh British Standard sieve).

Gustiness: Intensity of turbulence; the ratio of the root mean square of wind velocity fluctuations to the mean wind velocity.

Halocarbons/halogenated hydrocarbons: The substitution of one or more of the hydrogen atoms in an alkane or alkene by halogen atoms (chlorine, fluorine, bromine, or iodine) generates useful compounds for use in many fields. Those hydrocarbons in which all of the hydrogen atoms are replaced by halogens are designated as halocarbons. These have been used extensively in refrigeration, "aerosol" spray cans, manufacture of insulation, etc. Two common anthropogenic halocarbons are: trichlorofluoromethane (CCl_3F), designated by the Dupont Company as Freon-1 1; dichlorodifluoromethane (CCl_2F_2), designated by the DuPont Company as Freon-12. Several other halocarbons and alkyl halides are used commercially and also appear as trace gases in the atmosphere: tetrachloromethane (or carbon tetrachloride, CCl_4), 1,1,l-trichloroethane (or methyl chloroform, CCl_3CH_3), trichloroethene (or trichloroethylene, $CCl_2=CCHCl$), tetrachloroethene (or perchloroethylene, $CCl_2 = CCl_2$), dichloromethane (CH_2Cl_2). The lifetimes of many of the chlorofluorocarbons (e.g., $CClF_3$, CCl_2F_2, etc.) are very long in the troposphere, and the major destruction of these species occurs in the stratosphere where photodecomposition of the halocarbons produces C1-atoms which are involved in pathways to stratospheric ozone destruction. International agreements are anticipated to control the manufacture and release of the very long-lived halocarbons, and replacement of these by species which are destroyed within the troposphere is anticipated.

Haze: A state of reduced visibility (1–2 km) resulting from the increased light scatter due to the presence of fine dust or aerosol particles (H_2SO_4, NH_4HSO_4, products of the ozone-terpene reactions, etc.).

Haze horizon: The top of a haze layer which is confined by a low-level temperature inversion so that it gives the appearance of the horizon which it may obscure.

Humidity: A general term referring to the water content of a gas.

Hydrocarbons: Compounds which are composed only of carbon and hydrogen atoms. There are several types of hydrocarbons which are in common use as fuels or in the preparation of polymers, etc.; these differ in the type of bonding present in the molecules

Hydrocracking unit: Used in the thermal decomposition of heavy (high molecular weight) hydrocarbons to smaller (low molecular weight) hydrocarbons; high pressures of hydrogen and a special catalyst are employed. Sulfur compounds in the fuel are reduced to H_2S, and the final hydrocarbon product can be obtained relatively sulfur-free.

Hydrometeor: Any condensed water particle in the atmosphere of sufficient size to be potentially capable of undergoing precipitation (in fogs, clouds, some hazes, rain-drops, snow-flakes, etc.).

Hydrosphere: The gaseous, liquid, and solid water of the earth (oceans, icecaps, lakes, rivers, etc.) as distinguished from the *lithosphere* and the *atmosphere*.

Hygrometry and moisture analysis: The measurement or indication of the water content of the ambient air or of a sample of gas.

Hygrometer: Instrument used to measure the water vapour content of the atmosphere. The designs and operations of the various instruments now employed are very different. See the following types:

Hygrometer, capacitance: Hygrometer using the capacitance variations of a capacitor whose dielectric medium consists of the gas or of a material in contact with this gas.

Hygrometer, dew point (cooled surface condensation): Instrument in which the sample is passed over a cooled surface. The temperature at which dew forms on the cooled surface is a function of the water content of the gas passing over the surface.

Hygrometer, electrical: A hygrometer whose sensitive element has electrical properties which vary with the humidity of the gas which traverses the hygrometer.

Hygrometer, electrolytic: Hygrometer using a hygroscopic substance (for example, diphosphorus pentoxide, P_2O_5) which is transformed into an electrolyte (phosphoric acid, H_3PO_4) in contact with the moisture in the gas. The electrolyte (phosphoric acid) is electrolysed continuously and the electrolysis current is measured. At a constant flow of the gas to be analysed, the electrolysis current is a linear function of the water concentration.

Hygrometer, frost point: Instrument in which the sample is passed over a cooled surface. The temperature at which frost forms on it is a function of the water content of the gas passing over the surface.

Hygrometer, mechanical: An apparatus containing an element (hair, goldbeater's skin, carbon-film) whose dimension or mass varies as a function of its water content.

Hygrometer, psychrometric: Instrument by which the relative humidity of the atmosphere may be determined. It is generally composed of two temperature sensors, one of which measures the temperature of the air; the other sensor is moistened with water and senses the cooling due to evaporation of water. The temperature difference between the two sensors is a function of relative humidity; sometimes referred to as a "wet- and dry-bulb" hygrometer.

Illuminance: Photometric counterpart of irradiance (not of radiance or intensity). Common units are footcandle, lumens cm^{-2}, lumens m^{-2}. Not a radiometric quantity, but a photometric quantity.

Immission: The transfer of pollutants from the atmosphere to a "receptor"; for example, pollutants retained by the lungs. It does not have the same meaning as ground level concentration, but is the opposite in meaning to emission.

Immission dose: The integral of the immission flow into the receptor over the exposure period.

Immission flux: The immission rate divided by the unit surface area of the receptor.

Immission rate: The mass (or other physical quantity) of pollutant transferring per unit time into a receptor.

Impaction: A forcible contact of particles of matter with a surface.

Impactor, cascade: An instrument used for the classification of aerosols according to size and for possible subsequent chemical analysis. Air is drawn through a series of orifices of decreasing size; the air flow is normal to collecting surfaces on which aerosols are collected by inertial impaction. The particles, separated stepwise by their momentum differences into a number of size ranges, are collected simultaneously.

Impingement: Equivalent to impaction; often refers to impaction on a liquid surface.

Impinger: A sampling instrument employing impingement for the collection of particulate matter. Common types are: a) the midget impinger employing impingement in 1-10 cm^3 water, b) the standard impinger employing impingement in 75 cm^3 water, and c) dry impingers. Impingers are also suitable for sampling certain gases and vapours.

Incinerator: Equipment in which solid, liquid, or gaseous combustible wastes are ignited and burned. Types include flue-fed and multiple-chamber incinerators with several stages of combustion.

Individual perception threshold (IPT): A term used in odor testing which signifies the lowest concentration of a particular species at which a subject indicates both an initial positive and repeated response.

Inert gas: A non-reactive gas under particular conditions. For example nitrogen at ordinary temperatures, and the noble gases (helium, argon, krypton, xenon and radon) are unreactive toward most species.

Inertial separator: Any dry type collector which utilizes the relatively greater inertia of particles to effect their removal from a gas stream; e.g., cyclonic and impingement separators, gravity settling chambers, and high-velocity gas reversal chambers.

Infrared gas analyser: Instruments with various degrees of sophistication are employed to monitor certain species with characteristic infrared absorption bands. E.g., relatively simple systems are employed to detect carbon monoxide (CO) in air; these are built with matching cells, one containing a reference CO sample with the air to be tested in the other. Filtered infrared light which lies largely within the CO absorption region is passed through both cells. In one use of the instrument, the pressure difference which results from the preferential heating of the cell with the higher CO concentration is measured. The signal can be calibrated to yield the CO concentration. Sophisticated infrared systems involving Fourier transform spectrometers or infrared laser diodes are employed to detect particular molecular species of interest in atmospheric chemistry (NO_2, H_2O_2, CH_2O, CO, HNO_3, etc.) and to study their reactions in the real atmosphere or in simulated atmospheres in the laboratory. Long path lengths for the infrared beam are required to detect the small concentrations of most species.

Instantaneous (spot) sampling: Obtaining a sample of the atmosphere in a period which is short compared with the duration of the sampling exercise. Such samples are often called "grab" samples. These are useful for the analysis of hydrocarbons and other complicated mixtures of trace gases which are relatively stable in a stainless steel canister or tank and can be transported back to the laboratory for chromatographic or other analysis which cannot be done satisfactorily in the field. The use of plastic bags to collect instantaneous samples and to store them for analysis is not recommended. It is very difficult to eliminate pinholes, the reproducibility of heat seals used in constructing the bags is not easy to achieve, gases can permeate through the bag walls, and impurities can be introduced through off-gassing of the bag walls.

Intensity, radiant, spectral (I): Radiant power per unit solid angle (W sr^{-1}). With the adjective spectral, differentiation with respect to wavelength is also implied (W sr^{-1} nm^{-1}). Unfortunately, several other conflicting definitions of intensity exist, including the power per unit area, and the power per unit solid angle and projected area. The latter definition is common in atmospheric radiative transfer literature, and is synonymous with *radiance*. In spectroscopic literature and in Beer-Lambert law studies, the term *intensity* is used to describe the radiant power of collimated light entering a detector; this definition is synonymous with *irradiance*. These ambiguous uses of *intensity* are so common as to be unavoidable, but unless the definition given above is intended, it is best to use the word only in a generic sense to signify the "amount of light", and then to use the appropriate radiative quantity for numerical calculations. **GB:** The energy per unit time entering a small sphere in a time interval divided by the area of the great circle of the sphere and the time interval. For monodirectional radiation, this reduces to the energy flux density.

Inversion height: The height above ground level at which there is change in sign of the normal temperature "lapse rate", dT/dz (the rate of change of the temperature with height). Several temperature inversions may be present in the air over a given site at different altitudes as result of various meteorological factors.

Inversion, temperature: A departure from the normal decrease of temperature with increasing altitude. A temperature inversion may be produced, for example, by the movement of a warm air mass over a cool one. Intense surface inversions may form over the land during nights with clear skies and low winds due to the radiative loss of heat from the surface of the earth. The temperature increases as a function of height in this case. Poor mixing of the pollutants generally occurs below the inversion, since the normal convective process which drive the warmer and lighter air at ground level to higher altitudes is interrupted as the rising air parcels encounter the warmer air above. Temperature inversions near the surface are particularly effective in trapping ground level emissions.

Ion: An atomic or molecular particle having a net electric charge.

Ionization potential in mass spectrometry: The minimum electron energy necessary for the production of the molecule ion. The term *ionization energy* is preferred.

Isobar: Lines on a plot joining points of equal barometric pressure in the atmosphere.

Isokinetic line: A line in a given surface connecting points with equal wind speed; also called *isotach* or *isovel*.

Isokinetic sampling: A technique for collecting airborne particulate in which the sampling device has a collection efficiency of unity for all sizes of particles in sampled air, regardless of wind velocity and direction of the instrument. The air stream entering the collector has a velocity (speed and direction) equal to that of the air in the gas stream just ahead of the sampling port of the collector.

Isotherm: Lines joining points of equal temperature in the atmosphere.

Isotropic: A quantity which is independent of direction. *Anisotropic* and *nonisotropic* refer to quantities which are direction dependent.

Lachrymator: A substance which produces a flow of tears in a person or animal (e.g., acetyl chloride, acrylaldehyde, etc.).

Lapse rate: The variation of an atmospheric variable with height; unless otherwise stated the variable is temperature.

Lapse rate, temperature: The rate of change of temperature with altitude (d*T*/d*z*). The rate of temperature decrease with increase in altitude which is expected to occur in an unperturbed dry air mass is 9.8×10^{-3} °C m^{-1}. This is called the *dry adiabatic lapse rate*. The lapse rate is taken as positive when temperature decreases with increasing height. For air saturated with H_2O, the lapse rate is less because of the release of the latent heat of water as it condenses. The average tropospheric lapse rate is about 6.5×10^{-3} °C. The lapse rate has a negative value within an inversion layer.

LIDAR: An acronym for Light Detection And Ranging. It is a real time remote sensing technique for the determination of aerosol and trace gas concentrations in air by measurement of scattered laser radiation.

Light scattering: The redirection of a light beam due to interactions with molecules (Rayleigh and Raman scattering) and aerosols (Mie scattering). Scattered light received at any point in the atmosphere (sometimes called sky radiation) is a very important component of the total radiation received from the sun; it is dependent on the solar zenith angle, elevation, aerosol concentration, etc. **GB**: The deflection of a light beam by discrete variation in refractive index due to the presence of particles or by spatial refractive index fluctuations. For *elastic scattering* there is no wavelength shift. For *inelastic scattering* there are wavelength shifts due to molecular transitions (Raman effect, *fluorescence*) and for *quasielastic scattering* there are wavelength shifts and line broadening due to time dependent processes. In *single scattering* there is only one scattering encounter whereas in *multiple scattering* there is successive rescattering of the radiation within the scattering medium.

Lithometeor: A particle of dry substance in the atmosphere, as contrasted to a hydrometeor.

Lithosphere: The crust of the earth, usually thought of as discrete from and in contact with the hydrosphere and the atmosphere.

Luminance: Photometric counterpart of radiance, producing the visual sensation called brightness. Typical units are: candela m^{-2} (nit), candela cm^{-2} (stilb), foot lambert (2.426 nit). As with all photometric quantities, luminance does not refer to a specific wavelength, but applies to light emitted by a standard source (formerly a "standard international candle", now a blackbody radiator emitting at the temperature of solidifying platinum, 2042 K). Conversion from photometric units to radiometric units (e.g., J s^{-1}) requires convolution over wavelength of the relative spectral response of the human eye (photopic response tables).

Luminescence: The emission of light by a molecule or ion which is at temperatures below those required for incandescence. Luminescence can originate from electronically excited states of molecules or ions. If the excited state is short lived, as is the case for excited singlet to ground state singlet electronic transitions, the emission follows excitation within a short time period (about 10^{-9} to 10^{-6} s commonly) and is called *fluorescence*. The fluorescence from the polycyclic aromatic compounds provides a useful means of identification and quantification of these species. If the emitting state undergoes a spectroscopically forbidden transition, for example, an excited triplet to ground state singlet electronic transition, then the emission may be delayed (greater than about 10^{-3} s) following excitation; this emission is called *phosphorescence*. Some molecules (e.g., butanedione) give strong phosphorescence which is useful in identification. Chemiluminescence is the emission which results from an excited state formed in a chemical reaction which does not involve light absorption; for example, luminol oxidation by H_2O, generates chemiluminescence which can be used for analytical purposes.

Macrometeorology: Study of the largest-scale aspects of the atmosphere, e.g., general global circulation.

Manometer, U-tube: An instrument for measuring pressure differences. Usually a liquid such as mercury, water, or some low vapour pressure liquid of the desired density is placed in a U-tube. One end of the tube is evacuated and the other is open to the system in which the pressure is to be measured. Differences between the lengths of the liquid columns in the arms (coupled with a knowledge of the density of the liquid) are used to monitor the pressure changes.

Mass spectrometer: An instrument used in gas analysis which operates by ionising gaseous atoms or molecules, generally, with electrons of relatively low energy. The parent ions and ion fragments produced are accelerated electrically into a mass separator (involving electric and magnetic fields) which separates ions of specific mass to charge ratio. In a *quadrupole mass spectrometer* the mass separator involves four parallel rods between which a fixed direct current voltage is applied together with a superimposed radiofrequency voltage. *Chemical ionization mass spectrometry* involves the creation of reagent ions by transfer of electrons or protons from an initial ion to a reagent molecule. In *time-of-flight* mass spectrometry, ions produced in an ionizing pulse are accelerated by a given voltage toward a receiver electrode, and the ions of the different mass-to-charge ratios are resolved by their different arrival times at the detector.

Mass spectrometer, ion source: generally an assembly composed of: a) an ionization chamber in which a stream of electrons flows from a hot filament across a stream of gas to collector. The potential between filament and collector is usually between 50 and 70 v; b) a device for the acceleration of these ions.

Mass spectrometer, focusing system; deflection system: Assembly permitting the separation of ions according to their mass to charge ratio.

Mass spectrometer, ion collector: Device for the capture of selected ions such as a Faraday cup collector (with a d.c. amplifier) or an electron multiplier.

Mass spectrometry, mass peak in: Record of the ion current, at a specific mass to charge ratio, received by the collector.

Mass spectrometry, mass range in: Range of mass numbers which can be characterized by a mass spectrometer with sufficient resolution to differentiate adjacent peaks.

Mass spectrometry, mass spectrum in: The series of signals at mass numbers corresponding to the ions produced by the sample; each compound has a characteristic mass spectrum. This is sometimes referred to as the *cracking pattern.*

Maximum allowable concentration: The maximum concentration of a pollutant which is considered harmless to healthy adults during their working hours, assuming they breathe uncontaminated air at all other times.

Maximum emission concentration: Standards for maximum concentration of air pollutant emission from stationary or mobile sources.

Maximum storage life: Period during which there is no change in concentration in excess of the value of the uncertainty of the concentration.

Measured value (of air quality characteristic): Estimated value of the air quality characteristic derived from instrument readings; this usually involves calculations related to the calibration process and conversion to required quantities.

Measurement resolution (in atmospheric trace component analysis): The minimum value above which the difference of two values of air quality characteristic can be distinguished with 95% probability.

Measurement, upper limit of (in atmospheric trace component analysis): Highest value of the air quality characteristic which can be measured by an instrument; its variations, caused for example by instability, are expected to lie within specified limits. The difference between the lower detection limit and the upper limit of measurement constitutes the dynamic range of the instrument.

Mesopause: That region of the atmosphere between the mesosphere and the thermosphere at which the temperature is a minimum.

Mesoscale: In meteorology, the size or scale of phenomena smaller than ordinary cyclones or weather systems but larger than such microscale phenomena as the thickness of the boundary layer, the wakes of objects, etc. Thunderstorms involve mesoscale processes, and other meteorological events the size of cities are usually mesoscale processes.

Mesosphere: That region of the atmosphere which lies above the stratopause (about 47–52 km) and below the mesopause (about 80–90 km) and in which temperature decreases with increasing height; this is region in which the lowest temperatures of the atmosphere occur.

Meteorological range: The distance $L_v = 3.9/b_{scat}$; this is the distance over which an average observer could just see a large black object against the horizon sky during daytime, under isotropic conditions of b_{scat} and illumination.

Method (as employed in atmospheric trace component analysis): Procedure for sampling and analysing one or more air quality characteristics. The accuracy may be established using either a reference material or reference procedures. Two or more methods are considered *equivalent methods* if the values for their statistical and functional performance characteristics (for example bias, precision, sensitivity, etc.) and tolerances in the presence of specified interferant(s) and under specified operating conditions, fall within minimum specified limits.

Microclimatology: The science that deals with the climate of restricted areas and investigates their phenomena and causes.

Micrometeorology: The study of the meteorological processes on scales from a millimeter or less up to tens or hundreds of meters; e.g., meteorology of a local site that is usually small and often is confined to a shallow layer of air next to the ground.

Middle atmosphere: The combined stratosphere and mesosphere in the atmosphere.

Mist: A qualitative term applied to a suspension of droplets in a gas. In the atmosphere a mist produces a generally thin, grayish veil over the landscape. It reduces visibility to a lesser extent than fog but somewhat more than haze (visibility of less than 2 km but greater than 1 km).

Mixing height: The height to which significant mixing of added pollutants occurs within the atmosphere. In reference to stack gases, it is considered the height at which stack effluent begins mixing with the atmosphere as it leaves the stack.

Mixing ratio: In meteorology, the dimensionless ratio of the mass of a substance (such as water vapour) in an air parcel to the mass of the remaining substances in the air parcel. For trace substances, this is approximated by the ratio of the mass of the substance to the mass of air. However, in the case of water vapour the mass of dry air is used. In atmospheric chemistry, mixing ratios (molecular, molar, by volume, as well as by weight) are used to describe relative concentrations of atmospheric trace gases and impurities;

Mobile phase in chromatography: The carrier fluid with the chemical to be analysed passing through the chromatographic column.

Mobility: In aerosol physics, the velocity of a particle per unit applied force.

Monitoring: In a broad sense of the term, repeated measurements to follow changes over a period of time regular measurement of pollutant levels in relation to some standard or in order to assess the effectiveness of a system of regulation and control.

Monitoring systems: Automatic systems placed in a smokestack, a work place, or in the ambient atmosphere which measure and record the amounts of specified air pollutants which are present.

Monthly averages (with reference to atmospheric component analysis): For reporting integrated analyses of ambient air on a monthly rate, an average value is calculated; i.e., it is given on the basis of a 30-day month.

Nephelometry: Analytical methods which depend on the measurement of the intensity of scattered light emanating from an illuminated volume of an aerosol. The ratio of scattered intensity to illuminating intensity is compared with a standard of known properties.

Network, air sampling: A number of air sampling stations which are established in a given geographical region at which periodic measurements of both pollutant concentrations and meteorological quantities (wind speed, direction, rain fall, humidity, etc.) are made to determine the extent and the nature of the air pollution and to establish trends in the concentrations of the air pollutants with time.

Nitric acid (HNO_3): A major acid formed in the atmosphere from the reaction of HO-radicals with NO_2 ($HO + NO_2 + M(N_2/O_2) \rightarrow HNO_3 + M$), also through the reaction of N_2O_5 with water droplets ($N_2O_5 + H_2O$ (liquid) $\rightarrow 2HNO_3$), and by reaction of NO_3 with CH_2O or certain other organic species ($NO_2 + CH_2O \rightarrow HNO_3 + HCO$). This is a strong acid and one of the two major acidic components of acid deposition; H_2SO_4 is the other.

Nitric oxide (NO): The major oxide of nitrogen emitted from combustion sources such as power plants, the exhaust of mobile sources, incinerators, etc. It is an important source of NO_2 through its oxidation by hydroperoxyl (HO_2) and alkylperoxyl, acylperoxyl, and arylperoxyl radicals (RO_2) and ozone in the atmosphere.

Nitrogen dioxide (NO_2): A common impurity in polluted atmospheres which arises from the oxidation of NO released from combustion sources. Its photodecomposition in sunlight of wavelengths less than 420 nm is an important step in the formation of ozone in the polluted troposphere: $NO_2 + h\nu \rightarrow NO + O$; $O + O_2 + M(N_2/O_2) \rightarrow O_3 + M$. It is also a major source of nitric acid in the troposphere through its reaction with the hydroxyl radical (HO): $HO + NO_2 + M(N_2/O_2) \rightarrow HNO_3 + M$.

Nitrogen pentoxide (dinitrogen pentoxide, N_2O_5): An oxide of nitrogen which forms in the atmosphere through the oxidation of NO_2 by O_3: $NO_2 + O_3 \rightarrow NO_3 + O_2$; $NO_3 + NO_2 \rightarrow N_2O_5$. Its formation is favored in polluted air masses of high O_2, and NO_2 content and during the night. It is a precursor to nitric acid through its reaction with water (liquid) or aqueous aerosols.

Nitrogen trioxide (NO_3): A transient species formed in the atmosphere by reaction of NO_2 with O_3 ($NO_2 + O_3 \rightarrow NO_3 + O_2$). It is normally in equilibrium (or more likely approaching equilibrium) with N_2O_5 through the reactions, $NO_3 + NO_2$ À N_2O_5. NO_3 absorbs sunlight strongly and is photodissociated rapidly during the day. Its concentration rises at sundown, and it is most readily detected in the atmosphere at night through its strong absorption bands in the 620–660 nm region.

Nitrous acid (HONO): A weak acid formed in the troposphere. During the day HONO is formed by the reaction, $HO + NO \rightarrow HONO$, but it is rapidly photodecomposed in sunlight ($HONO + h\nu \rightarrow + HO + NO$) so that only a very small concentration of HONO is present at that time. In urban areas a buildup of HONO occurs during the night from unknown sources. Possible sources include homogeneous and heterogeneous reactions such as: $HO_2 + NO_2 \rightarrow HONO + O_2$; or $2N0_2 + H_2O \rightarrow HONO + HNO_3$.

Nitrous oxide (N_2O): This oxide of nitrogen is released into the atmosphere through biological activity and from other sources. It does not absorb light which is present within the troposphere, and it is unreactive toward transient species including the HO-radical. Its major loss mechanism involves transport to the stratosphere where it undergoes photodecomposition and reaction with $O(^1D)$ to form NO in part. It is present in air at about 0.3 ppmv.

Nuisance threshold: That concentration of an air pollutant that is considered objectionable. In the case of a substance with an objectionable odor, it is the smallest concentration of the substance which can be detected by a human being (nose).

Odor threshold: The concentration of a compound which produces an odor which is detectable by a human being (nose). For certain compounds this threshold is very low (e.g., 1 part in 10^9 for certain sulfides).

Open burning: The occasional burning of outdoor stores of lumber, sawdust, scrapped cars, tires, textiles, open dumps, etc.

Overfire: High-velocity air jets issuing from nozzles in a furnace enclosure to provide turbulence and oxygen to aid combustion, or to provide cooling air; *overfire* air is also known as *secondary* air.

Oxidant (with reference to atmospheric chemistry): A very qualitative term which includes any and all trace gases which have a greater oxidation potential than oxygen (for example, ozone, peroxyacetyl nitrate, hydrogen peroxide, organic peroxides, NO_3, etc.). It is recommended that alternative, more definitive terms be used which define the specific oxidant of interest whenever possible.

Oxidation: The chemical process by which the oxidation state of an atom in a element, ion, or compound is increased.

Oxygen, molecular (O_2): A reactive species which comprises about 21% by volume of the gases present in the earth's atmosphere. It is a necessary species for the existence of most forms of life on earth. It is a highly reactive gas which is capable of combining with most of the other elements.

Ozone (O_3): A highly reactive compound which occurs in the atmosphere. It is formed in the stratosphere through the photolysis of molecular oxygen at wavelengths less than 240 nm: ($O_2 + h\nu \rightarrow + 2O$; $O + O_2 + M(N_2/O_2) \rightarrow O_3 + M$). The stratospheric ozone acts as a protective shield which filters the wavelengths of sunlight which enter the lower atmosphere to those longer than about 290 nm. Ozone is a common secondary pollutant which is formed in a polluted air mass through the action of sunlight on NO_2 ($NO_2 + h\nu \rightarrow O + NO$; $O + O_2 + M(O_2/N_2) \rightarrow O_2 + M$). This is the most abundant of the oxidants formed in photochemical smog.

Ozone hole: A region of the stratosphere over Antarctica in which a marked decrease in the concentration of ozone has been observed in the Antarctic spring in recent years. The origin of this phenomenon is not yet established, but several theories based on both physical (transport related) and chemical processes (involvement of the halocarbons and their products of oxidation) have been suggested.

PAH: Policyclic aromatic hydrocarbons;

Particle (atmospheric): A small discrete mass of solid or liquid matter.

Particle counter: A device for measuring the number of suspended particles (in a certain size range) per volume unit of a gaseous or liquid medium.

Particle size: To describe the size of liquid or solid particles (aerosol) the average or equivalent diameter is used. For nonspherical particles collected in an impactor, for example, the aerodynamic diameter of a particle of arbitrary shape and density refers to the size of a spherical particle of unit density that would , deposit on a given impactor surface.

Particle size distribution: The size of the liquid or solid particles in the atmosphere usually extends from >0.01 to <100 μm in diameter. In the earth's atmosphere the distribution function which describes the number of particles as a function of diameter, mass, or surface area of the aerosol can be determined reasonably well with modern instrumentation.

Particulate matter: A general -term used to describe airborne solid or liquid particles of all sizes. The term aerosol is recommended for general use in describing airborne particulate matter.

Passive sampler: A device for preconcentration of trace substances from gaseous media based on molecular diffusion without controlled conveyance of the gas to be investigated (e.g., work place air).

PCB: Polychlorinated biphenyls; toxic compounds which have been employed as insulating fluids in some electrical transformers.

Peak, chromatographic: The portion of a differential chromatogram recording the detector response or eluate concentration while a single compound emerges from the column. If separation is incomplete, two or more components may appear as one unresolved peak.

Peak concentration (trace atmospheric component): The highest concentration of a given trace component which was measured with a continuous analyser during a specified sampling period.

Permeation tube: A device used for dynamic preparation of test gas mixtures by means of controlled permeation of a gaseous analyte out of a container through polymer material into a carrier gas stream. These devices containing certain condensable gases (e.g., NO_2, SO_2, etc.) when operated at closely controlled temperatures can be used as primary standards calibrated in terms of the weight loss per unit time.

Photochemical reaction: A reaction involving an electronically excited state of a molecule or atom which can occur following light absorption of sufficient energy per quantum (usually ultraviolet or visible radiation). In electronically excited molecules one of several types of photochemical reaction may occur; photodecomposition into molecular fragments or radicals, photodecomposition into stable, smaller products, photoisomerization, intersystem crossing (a singlet to triplet state, etc.), internal energy conversion (e.g., excited singlet to vibrationally excited, singlet ground state), fluorescence, phosphorescence, electronic or vibrational energy quenching, etc.

Photochemical Smog: The resulting mix of ozone, peroxyacetyl nitrate, aldehydes and other oxidation products of the hydrocarbons, oxides of nitrogen, aerosols, etc., which, form in highly polluted urban atmospheres under conditions of stagnation and high sunlight intensity in cities such as Los Angeles, California. Most large cities in the world show some degree of photochemical smog generation today. Photochemical smog is a highly oxidizing atmosphere in contrast with the SO_2-rich pollution and fogs which characterized many cities such as London, England, many years ago. The characteristics of photochemical smog also include eye-irritation, visibility reduction, damage to certain sensitive plants and trees, and possible pulmonary and other health-related problems in humans and animals.

Photochemistry: The chemistry of excited electronic states of atoms and molecules formed by the absorption and tight by an atom or molecule. The primary processes in photochemistry of a given compound are the various reaction channels of the electronically excited species. These can be molecular decomposition into stable molecules or radicals, molecular isomerization, fluorescence, phosphorescence, electronic quenching by collision, eta. The reactions of the molecular fragments which are formed photochemically, are correctly referred to as thermal reactions, in that they are analogous to the reactions induced following thermal decomposition at high temperatures. The term photochemistry is sometimes used incorrectly by atmospheric scientists to refer both the photochemical reactions of the electronically excited species as well as the purely thermal reactions of the atoms, free radicals and molecules which are present in the atmosphere.

Photolysis: Although it means literally the cleavage of a bond by light, it has been used incorrectly to indicate the act of irradiating a substance with light. Specific terms such as photodecomposition, photoisomerization, photodissociation, which describe the physical consequences of the irradiation are more descriptive.

Photometry: The measurement of light over wavelengths which produce visual sensation (approximately 380-780 nm). Thus photometric quantities do not refer to a specific wavelength but to the light emitted by a standard source (formerly a "standard international candle", now a blackbody radiator emitting at the temperature of solidifying platinum, 2042 K). Luminance and illuminance are the photometric analogues of the radiometric quantities radiance and irradiance, respectively, but conversion from photometric units (e.g., lm cm^{-2}) to radiometric units (e.g., J s^{-1} cm^{-2}) requires convolution over wavelength of the spectral radiation with the relative spectral response of the human eye. The standard response is called the "spectral luminous efficiency of radiation", and is tabulated for daylight adapted vision in photopic response tables.

Photophoresis: In aerosol physics, the motion of particles due to the influence of light. In many cases, this amounts to a special form of thermophoresis due to the heating of the particles by the light.

Photosynthesis: A metobolic process involving plants and some types of bacteria (e.g., Chromataceae, Rhodospirillaceae, Chlorobiaceae) in which light energy absorbed by chlorophyll and other photosynthetic pigments results in the reduction of CO_2 followed by the formation of organic compounds. In plants the overall process involves the conversion of CO_2 and H_2O to carbohydrates (and other plant material) and the release of O_2.

Phytotoxicant: An agent which produces a toxic effect in vegetation.

Plant damage: Damage to living plants by certain air pollutants (SO_2, certain metal fluorides, O_3, peroxyacetyl nitrate, nitrogen oxides, etc.) often is evident as leaf tissue collapse, chlorosis, and growth alteration.

Plume: The gaseous and aerosol effluents which are emitted from a chimney and the volume of space which they occupy (often visible)! The shape of the plume and the concentrations of pollutants in it at various points along the path of the plume are sensitive functions of the meteorology, local topography, and the chemistry which occurs in the plume. Urban plumes are observed over many urban areas and downwind of these areas in which a combination of man sources of pollution are concentrated.

Polycyclic compounds: Compounds of a structure containing more than one ring (cyclic series of covalent bonds); see; also polycylic aromatic hydrocarbons. Atoms of elements other than carbon and hydrogen may be present in these ring structures (nitrogen, sulfur, oxygen, etc.).

Potential temperature: The temperature that a dry air parcel would have if lowered or raised adiabatically to level of 1000 mbar pressure (or other arbitrary standard pressure).

Precipitation, in chemistry: The sedimentation of a solid material from a liquid solution in which the material is present in amounts greater than its solubility in the liquid.

Precipitation, electrostatic: Separation of particles or droplets suspended in a gas or air. A large potential differenceis required between the spaced electrodes in the precipitator. The charged particles are attracted to an electrode of opposite charge and collected.

Precipitation, in meteorology: Rain, snowfall, hail, etc.

Precursor: A chemical compound which is released into the atmosphere, undergoes chemical change, and leads to a new (secondary) pollutant, is called a precursor of that species.

Pressure, static: The pressure of a fluid at rest, or in motion exerted perpendicularly to the direction of flow.

Primary mixture: A mixture obtained directly from two or more components intended for the preparation of more dilute calibration mixtures (called secondary or tertiary mixtures).

Primary pollutant: A pollutant emitted directly into the air from identifiable -sources (e.g., SO_2, NO, hydrocarbons, etc.). Secondary pollutants, such as ozone, are generated within the atmosphere through chemical changes which occur in primary pollutants.

Probe (in stack gas sampling): A device, commonly in the form of a tube, used for sampling or measurement inside a duct, stack, volcano, vent, etc.

Psychrometry: The use of a wet-and-dry bulb thermometer for measurement of atmospheric humidity.

Quantum of light: A photon or "particle" of light; this has an energy given by $E = hc/\lambda$ where h is the Planck constant, c the velocity of light, and λ the wavelength of light.

Radiance: The radiance, $L(\lambda, \theta, \phi)$ is defined as $L(\lambda, \theta, \phi) = dQ(\cos\theta \, ds \, d\omega \, dt \, d\lambda)$, where dQ is the radiant (light) energy in the wavelength interval λ to $\lambda + d\lambda$ which is transported across a particular surface element of area ds, in time dt, originating from a solid angle dω which is at an angle θ to the surface normal Thus, the radiance specifies the energy emanating from an extended source, measured per unit time, unit solid angle, unit wavelength, and unit area of a surface perpendicular to the direction of incidence. In contrast to the actinic flux and the irradiance, the radiance has complete information about the direction of incidence of the light. The adjective "spectral" is often used with "radiance" to emphasize its wavelength dependence. The term "intensity" is often used instead of "radiance".

Radical (or free radical): A reactive atom or a fragment of a stable molecule which contains an odd number of bonding electrons. These reactive species have very short lifetimes in the lower atmosphere and low concentrations, although they are responsible for much of the chemical transformations in the atmosphere. Typical radical species encountered in atmospheric chemistry are: HO, HO_2, Cl, ClO, NO_3, CH_3O_2, etc. The concentrations show strong diurnal variations. NO and NO_2 are technically radicals as well (since they have an odd-number of bonding electrons), but they do not exhibit the high reactivity normally associated with typical radicals. Some chemists (from Great Britain and elsewhere) follow a practice of coupling with the formula of a radical, a dot placed near the atom in the radical which is judged to have the highest probability for the location of the odd electron. Thus the H-propyl and isopropyl radicals may be written as $CH_3CH_2CH_2^{\cdot}$ and $(CH_3)_2CH^{\cdot}$, respectively. Although this practice is not followed today by many of the atmospheric scientists, its use has some value in that it allows a rapid recognition of a radical species and of its reactive site. ***Biradicals*** contain two unpaired electrons, e.g., $O(^3P)$ CH_2(triplet), CH_2OO (Criegee intermediate), O_2(ground state) etc. *Radical ions* are charged radicals (most common in solution phase reactions). At least in the context of physical organic chemistry, it seems desirable to cease using the adjective "free" in the general name of this type of *chemical species* and molecular entity, so that the term free radical may in future be restricted to those radicals which do not form parts of ***radical*** *pairs.*

Radiometry: The measurement of quantities associated with radiant energy. The quantities may also describe the variation of the energy with respect to other variables such as wavelength, time, position, direction (solid angle), area normal to the light, or projected area of emitting or receiving surfaces. For example, see *intensity, radiance) irradiance.* If the light is monochromatic, it is sometimes convenient to replace the radiant energy by the corresponding number of photons (or quanta) which is obtained by dividing the energy by hc/λ where h is the Planck constant, c the velocity of light, and λ the wavelength of the light.

Radiosonde: A miniature radio transmitter with instruments in a package that is carried aloft (e.g., by an unmanned balloon) for broadcasting every few seconds by means of precise tone signals or other suitable method, the humidity, temperature, pressure, or other parameter.

Rain out: The mechanism by which small particles in the clouds are removed by the formation of rain-drops; this is a different mechanism from *wash out* which is a process which occurs below cloud level. Both the terms *rain out* and *wash out* have not always been used in accordance with these definitions. For clarity they should be replaced by the term, ***in-cloud*** *scavenging* and ***below-cloud*** *scavenging,* respectively.

Rayleigh scattering: The scattering of light by particles and molecules which are much smaller than the wavelength of the light. In the ideal case, the process is one of pure dipole and induced dipole interactions with the electric field of the light wave. The scattering cross section for light of wavelength λ is proportional to λ^{-4}.

Reflectance, spectral, directional: For a surface receiving radiation, ratio *of* the reflected radiant exitance to the incident irradiance. It may depend on wavelength and on direction of incidence, as indicated by the prefixes.

Relative density: The ratio of the density of a given substance to the density of a reference material at specified conditions of temperature and pressure; this is called *specific gravity.*

Relative humidity: The ratio, often expressed as a percentage, of the partial pressure of water in the atmosphere at some observed temperature, to the saturation vapour pressure of pure water at this temperature.

Remote sensing, *(in atmospheric sciences):* The determination of substances in the atmosphere, or in emissions, or of meteorological parameters in the atmosphere, by means of instruments not in immediate physical contact with the sample being examined.

Residence time: a) Instrumentation: The time required for air or reagent parcel to pass from the entrance to the exit of an instrument. Often this is approximated as the ratio of the interior volume of the device to the flow rate, b) **Atmospheric**: The average time a molecule or aerosol spends in the atmosphere after it is released or generated there. For compounds with well defined sources and emission rates, this is estimated by the ratio of the average global concentration of a substance to its production rate on a global scale. It is, a function of not only the emission rates but the loss rates by chemical and physical removal processes.

Response time: Time taken for an instrument to respond to a rapid change in value of the air quality characteristic. It consists of two major parts: a) Lag time, the time taken to reach 10% of the final change in instrument reading; b) rise time (fall time), the time taken to pass from 10% to 90% of the final change in instrument reading. For instruments where transient oscillations occur in the approach to the final instrument reading, the rise time is considered to be the time taken for the oscillations to fall to less than 10% of the final change in instrument reading.

Retention efficiency (in particle separation): The ratio of the quantity of particles retained by a separator to the quantity entering it (generally expressed as a percentage).

Retention time in gas chromatography: The time between the beginning of the injection of a sample and the moment where there is emergence of the peak maximum of a stated component. This time, is quite reproducible with a given pure component when the flows, temperature, etc., are carefully controlled.

Ringelmann chart: A chart which has been used in air pollution evaluation for assigning the degree of blackness of smoke emanating from a source. The observer compares the shades of grey (white to black) with a series of shade diagrams formed by horizontal and vertical black grid lines on a white background. A corresponding number, the Ringelmann number, is then assigned to the describe the best match; numbers range from 0 (white) to 5 (black). This method of pollution evaluation, although relavent to the inforcement of legislation still in force in many countries, is not recommended for use today since more quantitative indicators of the efficiency of the combustion are now available.

Sample, air: A quantity of air (sometimes of known volume), which is assumed to be representative of the air mass under investigation, and which is examined for air quality characteristics.

Sample fine. Line provided to remove a representative sample of the gas to be analysed and to conduct it to the sample point. It may include devices such as filters, dryers, or condensers (primary and secondary treatments of gas) which are necessary to prepare the sample for analysis.

Sample point: Point in the gas analysis installation beyond which it is assumed that no alteration of the sample occurs and the sample can be considered representative.

Sample probe: Device inserted into the gas to be sampled and to which is connected a sample line or a container for collecting the sample.

Sampler, dichotomous: Device for dividing a polydispersed aerosol particle population into two size fractions during sampling. The fractionation is based on the momentum differences of the particles which allow the .larger particles to pass through a zone of stagnant gas.

Sampling, continuous: Sampling, without interruptions, throughout an operation or for a predetermined time.

Sampling, cryogenic: The collection of trace compounds from gaseous media by co-condensation with a major constituent (e.g., water vapour, CO_2, N_2, Ar) of the matrix.

Sampling, discontinuous: Sampling in which the sample flow is periodically interrupted but not necessarily at equal time intervals.

Sampling grab: The taking of a sample (often in an evacuated bulb) in a very short time; preferred terms are instantaneous sampling or spot sampling.

Sampling, sequential: Repetitive sampling from the same gas to be analysed or from several gases to be analysed successively.

Sampling train: The complete assemblage of equipment necessary to sample atmospheres.

Sanitary land fill: An engineered burial of refuse. The refuse is dumped into trenches and compacted by bulldozer, where, it is hoped, aerobic metabolism by microorganisms decomposes the organic matter to stable compounds (H_2O, CO_2 etc.). Moisture is essential for the biological degradation and groundwater assists the process except when it fills air voids and prevents the transport of oxygen to the refuse. Land fills of unsatisfactory design can be major sources of air, water and soil pollution.

Saturation vapour pressure: The pressure exerted by a pure substance (at a given temperature) in a system containing only the vapour and condensed phase (liquid or solid) of the substance.

Scale: In meteorology, the size of the system, Microscale processes are the subject of micrometeorology, synoptic scale processes are studied in synoptic meteorology, etc.

Scanning in mass spectrometry: Variation of one or more operating parameters (electric, magnetic, etc.) of a spectrometer which makes it possible to cause the appearance, in the ion collector, of a succession of ions with increasing or decreasing mass to charge ratio.

Scattering (light): An interaction of light with an object that causes the light to be redirected in its path.

Scattering angle: The angle between the direction of propagation of the scattered and incident (or transmitted) light.

Scavenging: The removal of pollutants from the atmosphere by natural processes, including scavenging by cloud water, rainout, and washout. This type of removal process is termed precipitation scavenging. Scavenging of airborne pollutants at surfaces of plant, soil, etc., is termed dry deposition

Scrubbing: A process used in gas sampling or gas cleaning in which components in the gas stream are removed by contact with a liquid surface or a wetted packing, on spray drops, droplets, or in a bubbler, etc.

Secondary pollution (emissions): The products of the primary pollutants which form through photochemical and thermal reactions in the atmosphere (O_3, peroxyacetyl nitrate, etc.).

Sedimentation: In the atmospheric sciences, the process of removal of an air borne particle from the atmosphere due to the effect of gravity.

Settling chamber: Chamber designed to reduce the velocity of gases in order to permit the settling out of fly ash. It may be either part of, adjacent to, or external to an incinerator.

Settling velocity: The terminal rate of fall of a particle through a fluid as induced by gravity or other external force.

Sink: the sink is the receptor for material which is removed from the atmosphere. Because of long range transport of many pollutants such as SO_2, sulfuric acid and its salts, the sink region can be many hundreds of kilometers from the source region of the pollutants.

Smog: The term originated in Great Britain as a popular derivation of "smoke-fog" and appears to have been in common use before World War I. It originally referred to the heavy pollution derived largely from coal burning (largely smoke filled air, rich in sulfur dioxide), and it probably was largely a reducing atmosphere. More common today in cities is an oxidising atmosphere which contains ozone and other oxidants;

Smog chamber: A large confined volume in which sunlight or simulated sunlight is allowed to irradiate air mixtures of atmospheric trace gases (hydrocarbons, nitrogen oxides, sulfur dioxide, etc.) which undergo oxidation. In theory these chambers allow the controlled study of complex reactions which occur in the atmosphere. However, ill-defined wall reactions which generate some molecular and radical species (e.g., HONO, CH_2O, HO-radicals, etc.) and remove certain products (H_2O_2, HNO_3, etc.), the use of reactant concentrations well above those in the atmosphere, ill-defined light intensities and wavelength distribution within the chamber, and other factors peculiar to chamber experiments require that caution be exercised in the extrapolation of results obtained from them to atmospheric systems.

Smoke: A visible aerosol made up of small gas-borne particles resulting from the incomplete combustion of organic matter. It consists of carbon, carbon-rich products and all other dispersible product particles from the incomplete combustion. This does not include steam (condensed water vapour).

Smoke abatement: Legal measures which a control agency of a community may take to enforce laws and regulations concerning smoke emission.

Smoke alarm: An instrument which can provide an objective method of continuous measurement of the smoke density. Such devices are often depend upon the attenuation of light, and they can actuate an alarm when the attenuation exceeds some predetermined amount.

Smokestack: A chimney which carries the products of combustion away from a combustion chamber.

Smuts: Agglomerates of soot that become detached from the wall of a chimney and are swept out by the flue gas and then fall in the vicinity of the stack.

Soiling: Visible damage to materials by deposition of air pollutants.

Solar radiation: The electromagnetic radiation emitted by the sun. The total range of wavelengths of light emitted by the sun (99.9% in the range from 150 to 4000 nm) is filtered on entering the earth's atmosphere, largely through the absorption by oxygen, ozone, water vapour, and carbon dioxide. Near sea level only light of wavelengths longer than about 290 nm is present. The light from 290–400 nm is effective in inducing important photochemical processes since absorption by the important trace gases, ozone, nitrogen dioxide, aldehydes, ketones, etc., is significant in this region.

Soot: Aggregations of black carbonaceous particles formed during incomplete combustion and which are deposited before being emitted from a chimney.

Sorption: Process by which a substance (sorbate) is sorbed (adsorbed or absorbed) on or in another substance (sorbent). It is sometimes difficult or impossible to discriminate experimentally between adsorption and absorption: in such cases it is convenient to use the non-committal term sorption (together with its derived terms, sorbent, sorbate, sorptive).

Source: In atmospheric chemistry, the place, places, or group of sites or areas where a pollutant is released into the atmosphere. Point sources, elevated sources, area sources, multiple sources are often identified.

Spectrometry, optical (in trace component analysis of air): Methods of identification of substances and determination of their concentration based upon the measurement of light transmittance at selected wavelengths or bands of wavelength

Spectrophotometry: Methods using a photometer to measure light intensity as a function of wavelength.

Stability, static: In meteorology, the stability of the atmosphere in the vertical direction, to vertical displacements. If r is the dry adiabatic lapse rate, dry air is stable if the lapse rate $(-dT/dz) < \Gamma$, unstable if $(-dT/dz) > \Gamma$, and neutral if $-(dT/dz) = \Gamma$;

Stack gas: Gaseous waste products discharged to the atmosphere through a stack.

Stack height selection: The degree of dispersion of the stack effluent increases with the height of the stack. The maximum ground level concentration of pollutants and the point from the stack at which this occurs are complex functions of the wind turbulence and other meteorological factors as well as stack height. There are physical and practical limitations which control the choice of stack height: cost, air traffic hazards, aesthetic considerations, aerodynamic factors, vibration, materials, etc.

Stack sampling: Collection of representative gaseous and particulate samples of matter flowing through a duct or stack. Acceptable performance should indicate a collection efficiency of 95 ± 5 percent. Samples should be taken under isokinetic conditions to obtain an accurate representation of the particle size distribution in the effluent.

Stationary phase in gas chromatography: Solid or liquid which has reversible sorption properties, for the components to be analysed. The non-mobile phase in the chromatographic bed, on which the separation depends. For example, in gas-solid chromatography and liquid-solid chromatography the active solid, and in gas-liquid and liquid-liquid chromatography the liquid, but not the solid support.

Stratopause: That region of the atmosphere which lies between the stratosphere and the mesosphere and in which a maximum in the temperature occurs.

Steady state: A term used in reaction kinetics to describe the near constant (steady) concentration of reactive intermediate species in a reacting system when it is formed at a rate which is approximately equal to its rate of removal. In the atmosphere very reactive transient species such as $O(^1D)$ and HO have very short lifetimes and achieve a steady state in a very short time. Longer lived transient radicals (e.g., organylperoxyl radicals and NO_3) and reactive molecules (e.g., N_2O_5) achieve a steady state much more slowly (or never achieve one), since the rates of their formation and/or destruction may vary significantly during the lifetime of the species. The steady state concentration of a highly reactive species does adjust quickly to changes in the conditions such as the reactant concentrations, and, if the reactions forming or destroying the intermediate have a photochemical component, the actinic flux. In this case when the nearly constant concentration of the reactive transient is achieved, the system is said to be in a ***photostationary state***.

Stochastic sampling: Random sampling.

Stratosphere: The atmospheric shell lying just above the troposphere which is characterized by an increasing temperature with altitude. The stratosphere begins at the tropopause (about 10–15 km height) and extends to a height of about 50 km, where the lapse rate changes sign at the stratopause and the beginning of the mesosphere.

Sulfur dioxide (SO_2): A major oxidation product of sulfur-containing compounds in fuels and the major sulfur species emitted from power plants, furnaces, and other combustion sources. It is oxidized eventually to sulfuric acid and its salts in the atmosphere by reaction with HO-radicals in the gas phase, and more rapidly in cloud-water, rain, and in ground water, where it exists largely as HSO_3 ion, by reaction with dissolved H_2O_2, O_3, and other oxidizing agents.

Suspended matter: All particulate material which persists in the atmosphere or in a flue gas stream for lengthy periods because the particles are too small in size to have an appreciable falling velocity.

Synoptic scale: In meteorology, the size or scale of ordinary weather systems or cyclones; typically 1000 km horizontally.

Thermosphere: Atmospheric shell extending from the top of the mesosphere to outer space. It is a region of more or less steadily increasing temperature with height, starting at 70 or 80 km. It includes the exosphere and most or all of the ionosphere

Toxicity: The quality of being poisonous. It may be expressed as a fraction indicating the ratio of the smallest mass of the material which has some specified probability (often 50%) of causing death of some specific animal to the mass of the animal.

Tracer: A foreign substance mixed with or attached to a given substance to enable the distribution or location of the latter to be determined subsequently. There are several types of tracers which are used: **a)** A physical tracer is one that is attached by physical means to the object being traced; **b)** A chemical tracer is a chemical with properties similar to those of the substance being traced with which it is mixed homogeneously; **c)** An isotopic tracer is a unique isotope, either radioactive or an enriched, uncommon stable isotope, of the element to be traced; **d) A radioactive tracer** is a physical or chemical tracer having radioactivity as Its distinctive property which allows detection at small concentrations and hence after large transport distances. The composition of aerosols in the troposphere has been used as a qualitative tracer of air masses. The elemental analyses (determined by neutron activation, X-ray fluorescence, etc.) of the aerosols transported from various sources or source regions sometimes have characteristic patterns which are used to define qualitatively the origin of tropospheric aerosols collected in other geographical regions.

Transfer line: Line provided to carry the sample to be analysed from the sample point to the analytical unit without altering the composition of the sample.

Transmittance: The ratio of the transmitted spectral radiance (intensity), I_t of monochromatic light (single wavelength or small band of wavelengths) to that of the incident spectral radiance (intensity), I_o: I_t/I_o.

Troposphere: The lowest layer of the atmosphere, ranging from the ground to the base of the stratosphere (tropopause) at 10–15 km of altitude depending on the latitude and meteorological conditions. About 70% of the mass of the atmosphere is in the troposphere. This is where most of the weather features occur and where the chemistry of the reactive anthropogenic species released into the atmosphere takes place.

Tropopause: The region of the atmosphere which joins the troposphere and stratosphere, and where the decreasing temperature with altitude, characteristic of the troposphere ceases, and the temperature increase with height which is characteristic of the stratosphere begins.

Turbulence: Atmospheric eddies, mechanical or thermal, which are produced primarily by shearing stress or by convection. Atmospheric turbulence is primarily responsible for the dilution and mixing of atmospheric pollutants. Turbulent fluctuation in stable air are mainly of high frequency with typical periods of seconds (mechanical turbulence). The fluctuation often constitute major deformations of flow and are capable of transporting momentum, energy, and suspended matter at rates far in excess of the rate of transport by molecular diffusion and conduction in a nonturbulent or laminar flow.

Undetermined components: Components of the sample, the concentrations of which are not measured during the analysis,

Visibility: Defined as the greatest distance at which a black object of suitable dimensions can be seen and recognized against the horizon sky, or, in the case of night observations, could be seen and recognized if the general illumination were raised to the normal daylight level. The criterion of recognizing the object, not just seeing the object without recognition, is used. Transmissometers, telephotometers, and sun photometers are devices used to measure the degree of transmission of light. Nephelometers (integrating) are used to measure visibility by way of the light scattering from aerosols in the air mass.

Volume measuring device, gases: Several different devices exist with which gas volumes are measured: **a) Dry test meter:** A device which consists of bellows and valves. Gas fills the bellows to a predetermined volume; this volume of gas is then released, and the bellows return to their original position. The cycle is repeated and the cyclic motion activates an indicator, **b) Wet test meter:** A device consisting of a cylindrical chamber divided into compartments. The cylinder is partially submerged in a fluid, usually water, and the gas to be measured enters below the water line. As the gas flows into a compartment the fluid is displaced and the cylinder turns. When one compartment has been filled the next one fills, and a continuous rotary motion of the cylinder results. The rotary motion is indicated on a dial.

Wash out: The removal from the atmosphere of gases and sometimes particles by their solution in or attachment to rain-drops as they fall.

Wet bulb temperature: In psychrometry, the temperature of the sensor or the bulb of a thermometer in which a constantly renewed film of water is evaporating. The temperature of the water used to renew the film must be at the temperature of the gas.

Wind rose: A diagram designed to show the distribution of wind direction experienced at a given location over a considerable period of time. Usually shown in polar coordinates (distance from the origin being proportional to the probability of the wind direction being at the given angle usually measured from the north). Similar diagrams are sometimes used to summarize the average concentrations of a given pollutant seen over a considerable period of time as a function of direction from a given site (sometimes called a **Pollution rose**).

Zero gas or zero sample; Substance or mixture of substances resembling, as closely as possible, the matrix of the actual air sample to be measured, but characterized by a value of the air quality characteristic which is not detectable by the method used.

CHAPTER 6

Introduction to Air Pollution

Definition of Air Pollution

"Presence of one or more contaminants in the outdoor atmosphere in such quantities and of such duration, which tend to be injurious to human, plant, animal life & property, or which unreasonably interfere with the comfortable enjoyment of life or property or the conduct of business."

Air pollution is largely confined to the **troposphere**, which is the lowest region of atmosphere extending upto **11 kms from the earth's surface**. A healthy man inhales about 30-60 kg air per day.

Air pollutants are categorized as **primary pollutants** (which enter the atmosphere in the same form as emitted from source e.g., dust, SO_2, NOx, hydrocarbons etc.) and **secondary pollutants** (reaction products of primary pollutants & constituents of atmosphere, e.g., photochemical smog, peroxy acyl nitrate-PAN, ozone etc.)

Nature has long contaminated the air with **sand & dust storms, forest fires and volcanic eruptions**, that throw tons of particles & gases in the atmosphere. However, pollutants released into atmosphere by **industry and automobiles** has surpassed nature's contribution thousand fold. The magnitude of air pollution has increased immensely with industrialization and urbanization added by population explosion.

Global air pollution, if unchecked will adversely affect the **ecological balance, man, materials, animals & vegetation**. In India, air pollution problem is presently not very serious. However, continuous expansion & new addition of industries, development of new technologies products, increasing number of automobile vehicles coupled with population growth are introducing a great number & quantity of pollutants in air particularly in large urban areas.

Environment pollution is a side effect of rapid industrialization and technological revolution and it includes air pollution, water pollution & land pollution. Chemical substances discharged into the atmosphere from man made sources or released from natural sources (called primary pollutants) are often converted into secondary pollutants in the biosphere (which comprises of earth's crust, its atmosphere & all living beings and is divided into smaller ecosystems) in presence of water, oxygen and ultraviolet solar energy.

Identification of air pollution

There are three methods of identifying air pollution:

A. **Sensory perception/recognition by individual** by:

i. Strong odours
ii. Reduction in visibility
iii. Feel of grit under foot
iv. Acid taste in the mouth
v. Eye irritation

B. **Physical measurement of pollutants** by standard methods of sampling & analysis

C. **Effects on buildings, plants, materials & animals**:

i. Retarded growth of plant
ii. Deleterious effect on buildings & other materials
iii. Ill effect on the health of animals & human beings

Units of measurement of air pollutants: Following units are normally used

ppm (parts per million) – volume (at S.T.P) of pollutant contained in 1 million (10^6) volumes of air.
1% = 10000 ppm
S.T.P = 25°C & 760 mmHg column
N.T.P = 0°C & 760 mmHg column

i. Microgram/m³ (mass unit) – $\mu g/m^3$, $1 \mu g = 10^{-6}$ gm
ii. $mg/m^3 = 10^{-3}\ gm/m^3$
iii. 1 Milligram = 1000 microgram (μg) i.e., $10\ mg/m^3 = 10000\ \mu g/m^3$
iv. $1\ ml/m^3 = 1\ c.c/m^3 = 10^{-6}\ m^3/m^3 = 1$ ppm
v. $1°A = 10^{-4}$ micron $= 10^{-8}$ cm $= 10^{-10}$ m
vi. 1 μm (micro-metre) $= 10^{-6}$ metre = 1 micron $= 10^{-3}$ mm
vii. 1 mm = 1000 microns

Air pollutants are normally expressed in ppm (parts per million) or micrograms per cubic meter (μg/m³)

- For water, 1 ppm = 1 part of pollutant in 10^6 (1 million) parts of water $= 1\ mg/10^6\ mg = 1$ mg/litre
- But for air, the same formula does not hold good, as its volume depends on its pressure & temperature
- For all gases, 1 gm mole occupies 22.4 litres (22400 c.c) or, 1 kg mole occupies 22.4 Nm³ or, 1 gm mole occupies 22.4×10^{-3} Nm³
- At N.T.P (i.e., 0°C & 760 mm Hg) volume of gas is expressed in Nm³ [normal (for NTP) cubic metre] and at S.T.P (i.e., 25°C & 760 mm Hg), volume is expressed as Sm³
- It follows that 1 gm mole of SO_2 (i.e., 64 gm of SO_2) or, 1 gm mole of NO (i.e., 30 gm of NO) or 1 gm mole of NO_2 (i.e., 46 gm of NO_2) or, 1 gm mole of O_3 (i.e., 48 gms of ozone)...all these gases occupy a volume of 22.4 litres at N.T.P (0°C & 760 mm Hg) and 24.5 litres at S.T.P (25°C & 760 mm Hg)
- Conversion of ppm into μg/m³ (microgram per cubic metre) for gaseous pollutants:

$$1\ \text{ppm} = \frac{1\ \text{part of gaseous pollutant}}{10^6\ \text{parts of air}} = \frac{1\ \text{m}^3\ \text{of gas}}{10^6\ \text{m}^3\ \text{of air}}$$

If 'M' is the molecular weight of the gas, then 'M' gm of it occupies 22.4×10^{-3} m³ at N.T.P & 24.5×10^{-3} m³ at S.T.P

$$\therefore 1\ \text{ppm} = \frac{M/(22.4 \times 10^{-3})\ \text{gm}}{10^6\ \text{m}^3} = \frac{M}{22.4} \times 1000\ \mu g/m^3 \text{ at N.T.P} = \frac{M}{24.5} \times 1000\ \mu g/m3 \text{ at S.T.P}$$

$$\text{Since, } \frac{P_1V_1}{T_1} = \frac{P_2V_2}{T_2} \text{ and, } 0\ ^0K \text{ (absolute temperature)} = 273 + °C$$

General form of conversion factor for gaseous pollutant from ppm to μg/m³ is as follows:

$$1 \text{ ppm} = 44.64 \text{ M} \times \frac{273}{273 + t} \times \frac{P_2}{P_1} \ \mu g/m^3$$

where,

M = molecular weight of the gaseous pollutant

t = Gas temperature, °C

P_1 = Gas pressure at STP (25°C & 760 mmHg)

P_2 = Air pressure

Based on the above formula, the conversion factor for some common **gaseous** pollutants from ppm to µg/m³ has been calculated at NTP (0°C & 760 mmHg) & STP (25°C & 760 mmHg) and tabulated below:

Conversion factor for some gaseous pollutants from ppm to µg/m³

S.No.	Gaseous air pollutants	Chemical formula	Molecular weight	ppm to µg/m³ conversion factor	
				At NTP (0°C, 760 mmHg)	At STP (25°C, 760 mmHg)
1.	Sulphur dioxide	SO_2	64	2860	2620
2.	Carbon monoxide/ Ethylene	CO/C_2H_4	28	1250	1145
3.	Nitric oxide	NO	30	1343	1230
4.	Nitrogen dioxide	NO_2	46	2054	1880
5.	Ozone	O_3	48	2142	1960
6.	Carbon dioxide/Ethylene oxide/Acetaldehyde	$CO_2/C_2H_4O/CH_3CHO$	44	1964	1796
7.	Hydro carbons (as methane)	CH_4	16	715	655
8.	Peroxyacyl nitrate (PAN)	$CH_3\ (CO)O_2\ NO_2$	121	5348	4945
9.	Ammonia	NH_3	17	759	694
10.	Benzene	C_6H_6	78	3482	3184
11.	Acetone	CH_3COCH_3	58	2589	2367
12.	Naphthalene	$C_{10}H_8$	128	5714	5224
13.	Phosgene	$COCl_2$	99	4420	4040
14.	Phosphine	PH_3	34	1517	1388
15.	Carbon disulphide	CS_2	76	3393	3102
16.	Nitrobenzene	$C_6H_5NO_2$	123	5491	5020
17.	Phenol	C_6H_5OH	94	4196	3837

S.No.	Gaseous air pollutants	Chemical formula	Molecular weight	ppm to µg/m³ conversion factor	
				At NTP (0°C, 760 mmHg)	At STP (25°C, 760 mmHg)
18.	Aniline	$C_6H_5NH_2$	93	4152	3796
19.	Carbon tetra chloride	CCl_4	154	6875	6286
20.	n-hexane	$CH_3(CH_2)_4CH_3$	86	3839	3510

Types of air pollution: It is of following three types:

A. **Grit (>75µm) & dust (1-75µm)** – Examples are – sea salt, sand, road dust, pollen, mineral dust from quarries, stone crushers, coal dust from chimney etc.

B. **Smoke** – It is carbonaceous particles (of size < 1 µm), soot, fire dust, droplets of tar/oil or smoke. Produced by incomplete combustion of fuels. Smoke mixed with fog produces **smog**.

C. **Gaseous pollutants** - SOx (oxides of sulphur), CO, NO_x (oxide of nitrogen), hydrocarbons, H_2S, ground level ozone, fumes of acid etc.

Magnitude of pollution

i. A thermal power plant burning 4000 tons coal/day (containing 1% sulphur) emits 120 tons equivalent of H_2SO_4/day

ii. Similarly, a thermal power plant burning 3000 tons of furnace oil/day (containing 3.3% sulphur) emits 300 tons equivalent of H_2SO_4/day.

Classifications of Air Pollutants

Air pollution can be classified as follows:

A. **Natural contaminants, eg:**
   i. Natural fog
   ii. Pollen grains
   iii. Bacteria
   iv. Products of volcanic eruptions

B. **Aerosols (Particulates), eg:**
   i. Dust
   ii. Smoke
   iii. Mist
   iv. Fog
   v. Fumes

C. **Gases & vapors, eg:**
   i. SO_2, SO_3, H_2S, mercaptans (RSH)
   ii. NO, NO_2, NH_3
   iii. CO, CO_2, O_3 (ground level)

iv. HCl,HF

v. Hydrocarbons, aldehydes

Pollens grains:

- These are small grains from the feathers of flowers.
- Sources are weeds, grasses & trees.
- Its size is 10 to 50 μ.
- Thousands of pollen grains are liberated due to wind pollination
- Its effects include asthma, allergic response to sensitive people.
- Some people get hay fever and this system disappears at the end of the pollen season.
- Other disorders are bronchitis, bronchial asthma & dermatitis

Grit: Solid particle suspended in air with diameter > 500 mm (microns). Heavier dust particles & aerosols in the atmosphere include industrial, domestic or natural sources.

Aerosols (Particulates):

- It is a colloidal system in which a gas is the dispersion medium and the dispersed phase is solid or liquid.
- Size of aerosols is (< 1 μm).

Forms of Aerosols:

a. **Dust:**
   - Temporary suspension of dust in air
   - They do not flocculate unless subjected to the electrostatic forces.
   - They settle under the action of gravity.
   - Crushing & grinding are the dust sources.
   - Size of dust is 0.25 – 500 μm.
   - Fly ash size emitted from chimney is 3 – 80μ.
   - Other sources of dust generation include combustion of fuel, material handling processing, earth moving, house cleaning, sand blasting, engine exhaust, poultry feeding, crop spraying, and rubber tyre abrasion.

b. **Smoke:**
   - Finely divided particles, both combustible & incombustible
   - Size of smoke particle is < 2 mm (microns)
   - Smoke is a measure of fuel combustion efficiency.
   - Smoke is produced due to following reasons:

     i. Supply of insufficient combustion air
     ii. Insufficient time provided for combustion
     iii. Maintaining low temperature in combustion chamber.
     iv. Improper mixing of fuel and air
     v. Poor quality of fuel

vi. Poor design of furnace
vii. Overloading of furnace

- Slight white haze at the chimney top implies good combustion of fuel.
- Black/dark smoke from furnace chimney implies incomplete combustion of fuel.
- White smoke from furnace chimney indicates use of high percentage of excess air for combustion.
- Smoke from furnace chimney should come only during the lighting period of the furnace after repairs etc.
- Smoke sources include domestic heating, thermal plants, open fires, refuse incinerators, road, rail locomotives, ships, diesel engines and automobiles.

c. **Mists:**

- Fine liquid particles suspended in air are called mist.
- It relates to light dispersion of minute water droplets suspended in air.
- Its diameter is $< 2\ \mu m$.

d. **Fog:**

- It is visible suspension in which dispersed phase is liquid.
- Formed through condensation. It is dispersion of water/ice in the atmosphere near earth's surface.
- It reduces visibility to less than ½ km.
- Smoke & fog when mixed are called smog.

e. **Fumes**

- It is solid particles suspended in air produced by condensation from gaseous state.
- Formed due to action of volatilization from melted substances and is often accompanied by oxidation
- Its diameter is $< 1\ \mu m$

Main Sources of Air Pollution

Main sources of air pollution are listed below:

i) **Combustion of fossil fuels** (eg, oil, coal, natural gas) for domestic & industrial uses
ii) **Exhaust emission** from automobiles/vehicles
iii) **Dust, waste, gases & heat** from industries and thermal power plants
iv) **Natural pollutants** sources like volcanic eruptions, biological action in swamp areas and bacterial action in soil
v) **Radiation** pollution from a) ozone depletion in stratosphere, b) testing of nuclear bombs, c) nuclear processing plants and, d) radionuclides in the earth's crust

Globally, the major source of air pollution is the stack/chimney gases from thermal power plants, steel plants, petroleum refineries, chemical industries etc. Automobiles (like cars/trains/airplanes) which use fossil fuels is another important source of air pollution. Domestic use of fuels, evaporation of petroleum products, refuse incineration also contribute to air pollution

Classification of Main Sources of Air Pollution

Sources of air pollution are classified into two categories namely (A) natural sources, and (B) man-made sources

A. **Natural Sources:** They are listed below:

i. **Volcanic eruptions** emitting poisonous gases like CO, H_2S, SO_2 etc.

ii. **Pollen grains** of flowers

iii. **Vegetation decay**

iv. **Forest fire** and coal refuse fire

B. **Man-made sources** (Anthropogenic sources): they are listed below:

i. **Combustion of fossil fuels** (eg, coal, oil, natural gas) – They produce poisonous gases like CO, SO_2, NOx, CH_4, etc

ii. **Combustion of petrol/diesel in the automobiles**: They emit poisonous gases like CO, NOx & unburnt hydrocarbon besides lead particles, if leaded gasoline is used

iii. **Rapid industrialisation:** Industries like paper mill, sugar mills, metallurgical industries, leather industry, oil refineries, mining activities & host of chemical industries produce smoke, CO, CO_2, SO_2, H_2S, NO, NO_2 etc.

iv. **Agricultural activities:** The pesticides added to the soil or the spray done over crops are carried by winds which give a foul smell and effect the health of animals and human beings

v. **Deforestation:** Cutting of trees results in increase in CO_2 & decrease in O_2 percentage in air (because plants take up CO_2 for photosynthesis & give out oxygen). Increased CO_2 level in the atmosphere does not allow the heat radiation of the earth to go to outer space resulting in excessive heating of earth's atmosphere which is termed as global warming.

vi. **Nuclear radiation emission:** Testing & use of nuclear weapons emits out harmful radioactive radiations which affects the health of the people & may prove to be fatal also

vii. **Increase in global population:** Greater needs of enhanced population promotes use of unnatural methods, which disturbs the balance/equilibrium of the atmosphere

Five major air pollutants of concern are carbon monoxide, oxides of nitrogen (NOx), oxides of sulphur (SOx), hydrocarbons & particulates.

Some of the important air pollutants and their sources are given below in table 6.1

Table 6.1: Important air pollutants and their sources

S.No.	Pollutants	Major sources of air pollutants	Maximum permissible concentration (TLV) for exposure of 8 hours per day in ppm
1.	SO_2	Combustion of sulphur bearing fuels like coal & furnace oil, metallurgical plants, sulphuric acid manufacture, volcanic eruptions etc.	5
2.	H_2S	Cake ovens, petroleum refinery, biological decay of organic matter etc.	10

S.No.	Pollutants	Major sources of air pollutants	Maximum permissible concentration (TLV) for exposure of 8 hours per day in ppm
3.	CO	Incomplete combustion of carbonaceous fuels, automobile exhaust, iron & steel industry, petroleum refinery, solid waste disposal by pyrolysis & incineration, forest fire, volcanic eruptions	50
4.	NO, NO_2	Nitric acid manufacture, explosive industries, fertiliser plants, combustion of fuels like coal, fuel oil, natural gas etc., high temperature interaction of N_2 & NO_2 of atmosphere etc.	25 (for NO) & 5 (for NO_2)
5.	Cl_2	Chemical plants producing plastics, paper, dyes, chlorinated hydrocarbons, chloro chemicals etc.	1
6.	Hydrocarbons	Combustion of petrol/diesel/natural gas in automobiles, petroleum refineries polymer industry, solid waste disposal, anaerobic bacterial decomposition of organic matter etc.	500
7.	LPG	Leakage in cylinder	1000
8.	Particulates	Metallurgical smelting plant, incomplete combustion in industrial processes, fly ash, mining operations, dust from crushers & grinders, volcanic eruptions	-
9.	Cotton dust	Processing & handling of cotton	200 $\mu g/m^3$
10.	Chlorofluoro-carbons	Plants producing refrigerants & firefighting agents, their use as a solvent	1000
11.	Pesticides	Plants producing pesticides, application of pesticides & their evaporation from soil	-
12.	Ammonia (NH_3)	Nitrogenous fertiliser plants, ammonia works, part of cake oven gas, natural biological processes	25
13.	Benzene (C_6H_6)	Benzol distillation plant	5.75
14.	Naphthalene ($C_{10}H_8$)	Present in coke oven gas & tar	3.5
15.	$PbCl_2$ & $PbBr_2$	Addition of tetraethyl lead to petrol to improve its anti-knocking property	-

S.No.	Pollutants	Major sources of air pollutants	Maximum permissible concentration (TLV) for exposure of 8 hours per day in ppm
16.	CO_2	Combustion of fossil fuels, bacterial decay of plants/animals (unconsciousness prevails among men when exposed to atmospheric air containing more than 7-8% CO_2 in it). Usually CO_2 is not considered as a pollutant	5%
17.	Ground level ozone	–	0.1

Besides, the above mentioned air pollutants, certain metals like Be, Ba, Cd, Fe, Mn, Pb, Hg, Ni, Zn (emitted to atmosphere from their mining & metallurgy) and some non-metals like As, P, Se (emitted from combustion of fuels, fertiliser plants etc.) also pollute the air.

Threshold limit value (TLV) & the average levels found in urban atmosphere of certain toxic metals is tabulated in table 6.2.

Table 6.2: TLV of some metals found in atmospheric air

S.No.	Metals (in descending order of toxicity)	TLV, mg/m^3 of air (range denotes different toxic forms)	Average metal level found in urban atmosphere, mg/m^3 of air
1.	Berrilium (Be)	0.002	<0.00025
2.	Mercury (Hg)	0.01-0.05	0.002
3.	Cadmium (Cd)	0.10	0.0001
4.	Lead (Pb)	0.10-0.20	0.004
5.	Arsenic (As)	0.20-0.50	0.00004
6.	Vanadium (V)	0.50	0.0002
7.	Nickel (Ni)	0.007-1.0	0.00003
8.	Copper (Cu)	1.0	0.00009
9.	Iron (Fe)	1.0	0.0003
10.	Zinc (Zn)	1.0	0.0001

ILL Effects of Air Pollution

Adverse effect of air pollutants can be divided into following categories:

i. Effect on **human health**
ii. Effect on **vegetation & plants**
iii. Effect on **materials**

i. **Effect of air pollutant on human beings** – General health effects of air pollutants on human being are listed below:

- Eye, nose, throat & respiratory tract irritation
- Gases like H_2S, NH_3, RSH (mercaptans) cause odour nuisance even at low concentrations
- Nickel causes respiration problems
- Increase in mortality rate & morbidity rate
- Particulates, particularly pollen grains initiate asthmic attack
- High concentration of SO_2, particulate matter and photochemical smog aggravates chronic pulmonary diseases like bronchitis & asthma
- Mercury results in kidney, nerve and brain damage
- Carbon monoxide combines with haemoglobin in the blood & hence increase stress on those suffering from the cardiovascular & pulmonary disease
- Carcinogenic agents cause cancer
- Hydrogen fluoride causes bone disease (fluorosis) and mottling of teeth
- Respiratory disease is caused by dust particles eg, silicosis & asbestosis result from specific dust
- Pollen grains initiates asthmatic attacks
- Certain heavy metals like Pb may enter the body through lungs & cause poisoning and may lead to convulsion, comma and even death
- Cadmium particulates through cigarette smoking cause cardiovascular diseases, kidney & lever damage and even death
- Key health problems caused by air pollution are asthma, cancer, respiratory problems, heart disease, birth defects, tuberculosis, immune system damage & intellectual disorders

ii. **Effect of air pollutant on vegetation/plants** – Various forms of damages to the leaf of the plant due to the action of air pollutants is discussed below:

a. **Necrosis:** it is the killing or collapse of tissues

b. **Chlorosis:** It is the loss or reduction of green plant pigment, chlorophyll resulting in pale green or yellow leaf pattern. Chlorosis indicates a deficiency of some nutrient required by the plant. It is analogous to anaemia in animals

c. **Abscission:** Leaf abscission is dropping of leaves

d. **Epinasty:** Leaf epinasty is downward curvature of the leaf due to higher rate of growth on the upper surface

Effects of air pollution on plants/vegetation:-

- Sulphur dioxide bleaches the leaf surface and causes chlorosis
- Ozone causes necrosis and damage leaves
- Nitrogen dioxide causes premature leaf fall
- Peroxy acetyl nitrate (PAN) is responsible for suppressed growth, silvering of lower leaf surface, leaf abscission, leaf epinasty

Effects of various air pollutants on the plants is tabulated in table 6.3.

Injurious level of SO_3 exposure to trees have been tabulated in table 6.4.

Table 6.3: Effects of air pollutants on the plants

Pollutants	Injury threshold, ppm	Sustained exposure time	Effects
SO_2	0.03	8 hours	Interveinal chlorotic bleaching of leaves, necrosis in interveinal areas and skeltoni zed leaves, chlorosis, growth suppression, reduction in yield
O_3	0.03	4 hours	Flecks on upper surfaces, premature ageing & suppressed growth, collapse of leaf, necrosis and bleaching, tips of conifer needles become brown and necrotic
HF	0.0001	5 weeks	Necrosis of leaf tip, chlorosis, lower yield, leaf abscission, tiband margin bum, dwarfing
NO_2	-	-	Suppressed growth, leaf bleaching
Cl_2	0.01	2 hours	Bleaching between veins, tip and leaf abscission
C_2H_4	0.05	6 hours	Epinasty, leaf abscission, flower dropping and failure of flower to open, withering, leaf abnormalities
Peroxy acetyl nitrate (PAN)	0.01	6 hours	Glazing, silvering or bronzing of lower leaf surface, suppressed growth, young leaves more susceptible

Table 6.4: Injurious level of SO_3 to trees

Type of trees	Concentration of SO_3 (ppm) & time to produce injury
Pear	0.48, 6 hours
Apple	0.48, 6 hours
Mountain ash	0.54, 3 hours
Ponderosa pine	0.50, 7 hours

iii. **Effect of air pollutant on materials/metals** – five mechanisms which are responsible for the material damage in polluted atmosphere are described below:

a. **Abrasion**: solid particles travelling at high speed cause abrasive action. Also, large sharp edged particles embedded in fabrics can accelerate wear.

b. **Deposition and removal:** Solid/liquid particles deposited on a surface spoils its appearance. Removal of these particles by frequent cleaning/washing may cause some deterioration.

c. **Direct chemical attack:** Some air pollutants react directly or reversibly with materials to cause deterioration eg, bleaching of marble by SO_2, tarnishing of silver by H_2S & etching of a metallic surface by an acid mist.

d. **Indirect chemical attack:** Some materials absorb pollutants and get damaged, when the pollutants undergo chemical changes eg, SO_2 absorbed by leather is converted to H_2SO_4 causing its deterioration.

iv. **Effects of air pollution on Climate:**

- Increase of CO_2 in atmosphere causes global warming.
- Causes melting of ice and glaciers.

- Results in thinning of ozone layer.
- Allows penetration of more harmful ultraviolet (UV) rays.
- Causes formation of acid rain. Air pollutants mixing up with acid rain can cause high acidity in fresh water bodies (rivers, ponds, lakes etc.) resulting in the death of aquatic creatures including fish.

e. **Corrosion**: Atmospheric deterioration of ferrous metals by an electrochemical process i.e. corrosion is due to the action of air pollutants facilitated by the presence of moisture. Rate of attack of damaging pollutants are influenced by moisture, temperature, sunlight and air movement.

f. Because of their corrosiveness, particulates damage the exposed surface of materials. Presence of SO_2 and moisture accelerates corrosion of metallic surfaces. SO_2 affects fabrics, leather, paint, paper, marble and limestone. Ozone causes cracking of rubber. NOx causes fading of acetate, cotton and rayon fibres.

Damage inflicted on various materials due to air pollutants is tabulated below in table 6.5

Table 6.5: Air Pollution damage to various materials

S.No.	Materials	Principal air pollutants	Effects on materials	Other environmental factors affecting material damage
1.	Metals	SO_2, acid gases	Corrosion, spoilage of surface, loss of metal, tarnishing, soiling	Moisture, temperature
2.	Building materials	SO_2, acid gases	Discoloration, leaching corrosion	Freezing, moisture
3.	Paint	SO_2, H_2S, Particulates	Discoloration, softened finish strength	Fungus, moisture
4.	Textile	SO_2, acid gases	Spotting, deterioration, reduced tensile	Sunlight, fungus, moisture
5.	Dyes	NO_2, SO_2, oxidants	Fading of dyes	Moisture, sunlight
6.	Rubber	Oxidants, ozone	Cracking, weakening	Sunlight
7.	Leather	SO_2, acid gases	Disintegration, powdered surface, weakening	Physical wear
8.	Paper	SO_2, acid gases	Embrittlement	Sunlight
9.	Ceramics	Acid gases	Change in surface	Moisture

Sink mechanisms for air pollutants: Sink is a place or mechanism associated with the removal of pollutant. For example, pre-capitations remove particulates, soil absorbs carbon monoxide and oceans & vegetation absorb carbon dioxide.

There are basically three sink mechanisms for the removal of air pollutants, which are listed below:

Three Sink Mechanisms:

A. **Removal Process:**

a) **For removal of particles** from troposphere, the important processes are:

i. Wet removal by precipitation
ii. Dry removal by sedimentation
iii. Dry removal by impaction on vegetation

b. **For removal of gases** from the troposphere, important processes are:

i. Wet removal by precipitation
ii. Absorption or reaction at the earth's surface.
iii. Conversion into other gases or particulates by chemical reaction within atmosphere.
iv. Transport into the stratosphere

B. **Removal processes at the surface of the earth**: Gaseous pollutants & particles which cannot be affected by sedimentation (being small enough) can be transported to the earth surface by atmospheric motions and removed there by absorption or chemical reaction.

Efficiency of sinks depends upon the rate at which the substance is:

i. Supplied to the surface of the earth.
ii. Bound or transformed by the surface.

The nature of sink surface itself is also critical to the removal process of gaseous pollutants.

C. **Known sinks for common air pollutants**: Sink is a place or mechanism associated with the removal of an air pollutant. Sinks of selected gaseous air pollutant is tabulated in table 6.6.

Table 6.6: Sinks of selected gaseous air pollutants:

S.No.	Pollutants	Sinks for the pollutants
1.	SO_2	**Precipitation Scavenging:** Wash out, rain out. **Oxidation** in gas or liquid phase to sulphate. **Soil:** microbial degradation, physical & chemical reaction, absorption. **Vegetation**: sorption on surface, stomatal intake, oceans & lakes: absorption.
2.	H_2S	**Oxidation** to SO_2
3.	Ground level ozone (O_3)	**Chemical reaction** on vegetation, soil, snow and ocean surfaces.
4.	Carbon dioxide (CO_2)	**Vegetation:** photosynthesis, absorption. **Oceans:** absorption
5.	Carbon Monoxide (CO)	**Stratosphere:** reaction with hydroxyl (OH) radical and CH_4 **Soil:** Microbiological activity
6.	Nitrous oxide (N_2O)	**Soil:** Microbiological destruction **Stratosphere:** Photo dissociation **Ocean:** absorption
7.	Nitric Oxide, Nitrogen dioxide (NO, NO_2)	**Soil:** Chemical reaction **Vegetation**: Sorption, stomatal uptake **Chemical reaction** in gas & liquid phase (oxidation to nitrate)
8.	Hydrocarbons (HC)	**Chemical reaction** to particulates **Soil:** microbiological activity **Vegetation:** absorption, stomatal intake

S.No.	Pollutants	Sinks for the pollutants
9.	Non-methane hydrocarbons (NMHC)	Photochemical reaction with NO and O_3
10.	Ammonia (NH_3)	**Chemical reaction** to ammonium in gas & liquid phase. **Precipitation Scavenging:** wash out, rain out **Surface uptake**: Physical & chemical reaction, absorption
11.	Methane (CH_4)	Soil: Microbiological activity Vegetation: Chemical reaction, bacterial action. Troposphere & Stratosphere: Chemical reaction
12.	HCl, Cl_2	Precipitation
13.	CH_3Cl	Stratospheric reaction

- Utilisation of energy resources in industries causes environmental damage by polluting the atmosphere. Few examples of air pollution are SO_2, NOx & CO emission from boilers & furnaces, Chlorofluorocarbon (CFC) emission from refrigerants use etc. chemical & fertiliser industries release toxic gases while thermal power plants & cement plants release particulate matter in the atmosphere.
- Air pollutants mainly the product of combustion from fossil fuel use may not only create problems near to the sources but can also cause problems far away as they can travel long distances, chemically react in the atmosphere to produce secondary pollutants such as acid rain or ozone.
- In both developed & developing countries, major historic air pollution problem has been high level of smoke & SO_2 arising from the combustion of sulphur containing fossil fuels such as coal, oil & gas for domestic & industrial purposes.
- The major threat to clean air is now posed by traffic emissions. Motor vehicles emit CO, NOx, VOC's & particulates which is badly affecting the urban air quality. Besides, photochemical reactions resulting from the action of sunlight on NOx & VOC's from vehicles leads to formation of ozone which impacts rural areas often far from the original emission site. Acid rain is another long range pollutant influenced by vehicle NOx emissions.
- Domestic & industrial pollutant sources, together with their impact on air quality tend to be steady state or improving overtime.
- Traffic pollution problems are worsening worldwide. It is getting severe in the developing countries with increasing vehicle population, infra-structural limitation, poor engine/emission control technologies and limited provision for maintenance or vehicle regulation.
- Main air pollutants produced by industrial, domestic & traffic sources are SO_2, NOx, CO, O_3, C_6H_6, 1-3 butadiene, hydrocarbons (HC), particulate matter, toxic organic micro-pollutants, lead & heavy metals.

Measures to control Air pollution

- Controlling pollution at the source.
- Construction of tall chimneys
- Development of green belt around industrial areas,
- Use of better quality fuel and use of catalytic converters in vehicles
- Sprinkling water in mining areas to reduce the dust particles

- Siting of industries after proper Environmental Impact Assessment studies.
- Using low sulphur coal/fuels in industries.
- Removing sulphur from coal (by washing or with the help of bacterial desulphurisation).
- Removing NOx during the combustion process of fuels.
- Removing particulate from stack exhaust gases by employing electrostatic precipitators, bag-house filters, cyclone separators, wet scrubbers etc.
- Vehicular pollution can be checked by regular tune-up of engines, replacement of old, more polluting vehicles, installing catalytic converters, by engine modification to have fuel's complete combustion.

Natural Air Pollutants

Nature itself is a source of air pollution. Sources & effects of some natural air pollutants are described below:

Atmospheric pressure

- Atmospheric pressure is $1 kg/cm^2$ (760 mmHg), which is comfortable for man.
- 30 km above mean sea level (MSL), atmospheric pressure is only 10mmHg. It increases @ $1 kg/cm^2$ for each 10 meters depth under water below sea level.
- Atmospheric pressure, air density & temperature (in troposphere) decreases with altitude.
- At higher altitude, man suffers from headache, insomnia (loss of sleep), breathlessness, nausea, vomiting & impaired vision due to increase in respiration, haemoglobin concentration & cardiac output termed as "mountain sickness".
- At an altitude of 8 km, breathing equipment is required for surviving. Persons above 4km altitude experience irregular breathing, hallucination, mental confusion, stupor, seizures & coma. This condition termed as high altitude pulmonary odema does not respond to antibiotics even.
- On being exposed to high pressure (say 10 kg/cm^2) gases like N_2, CO_2 & O_2 are dissolved in the blood and tissues causing narcotic action leading to loss of mental functions/consciousness, convulsion & death. Persons working in diving wells and compressed air chambers suffer from this disease called "Caisson disease".

Humidity: It is the percentage of water vapor present in atmospheric air and is caused by evaporation of water from river/pond, sea. In indoors humidity increases due to evaporation of moisture from skin & lungs. Exhaled air contains 6% moisture. Relative humidity above 65% is uncomfortable/sticky, while less than 35% is also unpleasant and causes nasal mucasa, sore throat & cough. In coastal region humidities upto 80% is common. While in arid regions with no water sources, low humidity upto 5% is observed.

Water vapor present in atmosphere causes isnophelia, asthma and brings "electrical storm" besides enhancing the action of pollutants like SO_2, NO_2 & particulates on men, materials & vegetation.

Electricity is produced due to physio-chemical action as earth's crust & clouds are positively charged whereas the atmosphere in between is negatively charged. Electricity travels through water vapor laden atmosphere (highly conductive) and is observed as "electric storm" with temp reaching upto 10000°C at which surrounding air gets suddenly heated and expands. It is replaced by cooler air forming lightening & thunders. Thousands of electrical storms takes place every second. Depending upon temperature, we visualise only few electrical storms.

Temperature & Heat: Atmospheric temperature plays a vital role in controlling the concentration of air pollutants in troposphere in which temperature decreases with elevation.

- Average earth temperature is 16°C, maximum temperature. is 75°C (in Turkisthan) & minimum temperature is - 90°C (in Greenland). Mean global temperature. is increasing due to emission of green house gases (mainly CO_2) into the atmosphere.
- Air temperature. of 25-27°C & relative humidity of 40-60% is desirable for thermal comfort of human being.
- A man working in hot environment is subjected to **heat stress** resulting in increased sweat rate. Maximum allowable sweat rate for healthy human being is 4.5 litres of sweat excreted in four hours.
- Sweat rate for adults in 4 hours for comfort zone is 1-3 litres, just tolerable condition is 3-4.5 litres & intolerable condition is 4.5 litres. A sweat rate of 2.5 litres in 4 hours is optimal for a working man in hot environment
- When human being are exposed to hot/cold environment, then disorders like **heat stroke**, **heat cramps**, **heat syncope & cold stress** are observed
- **Heat stroke** results from failure of heat regulating mechanism and is characterised by dry/hot skin, high body temperature, delirium, convulsion & unconsciousness. Heat injured red blood cells release potassium (called hyperpotassemia) resulting in death. Body of patient subjected to heat stroke should be cooled to <103°F in ice water bath to avoid convulsion
- **Heat cramps** results from loss of chlorides & sodium from the blood of human beings who are doing heavy muscular work in high temperature & humidity. Cricket players playing in hot sunny weather are subjected to this disorder
- **Heat syncope** is common in soldiers standing in sun for parades in which man becomes pale, his blood pressure falls (due to lack of blood supply to the brain) & he collapses
- **Cold stress** results from exposure to freezing cold and is characterised by numbness, loss of sensation, muscular weakness, desire for sleep, coma & death. Hot fluid intake helps rewarming the body. Proper dress is required to avoid local cold injury as in immersion or trench foot
- **Heat stress** can be prevented by drinking cool water, cutting down the duration of exposure to hot environment, enhancing rest period between work in hot environments, wearing light/loose/light color clothes & controlling temp. & humidity (by proper ventilation & air conditioning) in the work environment

Sea salt particles: In coastal areas, salt particles are emitted from the sea (about 900 million tons every year) which are highly corrosive to the structures. Besides, these sea spray particulates have abrasive action on building materials (of temples, mosques, art treasures & historical monuments) thereby reducing their life span. Besides the abrasive action of sea breezes, coastal areas are sweaty & uncomfortable for the inhabitants

Aeroallergens

- Air borne substances like pollens, glues, spores, bacteria, viruses, fungi, some dusts algae, protozoa, terpenes emanated from plants, vegetation & animals which cause allergy are known as **aeroallergens**. Allergy is a misdirected activity of the immune system. It is the body's reaction to foreign substances, which otherwise do not harm the system.
- Air borne pollen & spores cause bronchial asthma. Turpene molecules combine to form aerosol that produces blue haze over forested areas.
- Aeroallergens also cause respiratory problems, dermatitis & other skin diseases.
- Ragweed pollens of 20 microns are deposited within 100 metres of the parent plant accelerating hay fever or asthma.

- Aeroallergens <10 micron diameter are more dangerous for human health as they cannot be filtered out by the hairs in the nose.
- Industrial processes contributing significant quantity of aeroallergens includes tobacco, jute, hair & feather processing units, cotton mills, wood working, floor milling etc. finely powdered industrial materials also may be allergic.

Dust, smoke and other particulates

- Natural dust storms emit about 200 million ton annually which can travel from one country to another.
- Dust from thar desert of Rajasthan can travel upto Delhi. Similarly, inter-continental dust storm from sahara desert of north Africa travels upto U.S.A.
- About 8 million tons of smoke, soot & other particulates are emitted annually from forest fire & volcanoes. Black smoke affects visibility and is considered as 'layman's index of air pollution'.

Carbon dioxide

- CO_2 is emitted into atmosphere due to respiration by human beings, animals & vegetation and also by erruptions of volcanoes & forest fires
- It being the green house gas, increase the ambient temperature & mean sea level (due to ice melting)

Hydrocarbons

- It is released into atmosphere from swamps, marshee & water bodies and anaerobic decomposition of organic matter. It is a green house gas having its deleterious effects. About 175 million tons of terepene is released annually
- Exploitation of natural sources for methane generation is increasing the methane concentration in atmosphere. Water logging in thousands of water reservoirs releases marshy gas (CH_4) into atmosphere
- Hydrocarbons released into atmosphere by natural sources is 1600 million tons/yr (including 1450 million tons/yr of CH_4) against only 90 million tons/yr released by artificial sources

Sulphur products

- Sulphur products like S, SO_2, SO_4 & H_2S are produced naturally by volcanoes eruptions, salt spray from sea (containing 120 million tons of SO_4/yr) & biological decay by sulphide bacteria (100 millions tons of H_2S/year)
- Natural air pollutants emit a total of 130 millions tons of sulphur/year which is almost equal to that released by artificial sources.

Ozone

- It is a photochemical oxidant produced in upper atmosphere (25-40 km from earth) by solar radiation small amount of ozone is produced by lightening & forest fires. Its background concentration in air is 40-80 $\mu g/m^3$
- It protects us from dangerous incoming ultraviolet rays (by absorbing it)

Noise

- Lightening & thunders produce noise of 110 decibel, which is nearer to threshold of pain

Economic Losses caused by Air Pollution: Apart from direct health related and environmental issues, air pollution brings with it economic losses as well which are listed below:

- Decreased productivity.
- Direct medical losses because of increased healthcare costs.
- Lost income from being absent from work.
- Travel time losses due to reduced visibility.
- Increased cost of cleaning.
- Losses due to damage to crops and plants.
- Losses from repair of damage to buildings.
- Absentism from schools/colleges/offices/factories.
- Missed work and lost income potential from premature deaths.

We can only imagine (cannot calculate) the economic losses incurred by developing countries such as China where air pollution levels are among the highest in the world.

Appendices

		Contents	Page No.
Appendix 6.1	:	Characteristics of Atmospheric Particles	771
Appendix 6.2	:	Sizes Ranges of Atmospheric Particulate Matter	771
Appendix 6.3	:	Types of Particulates in Suspended Matter	772
Appendix 6.4	:	Definition of Various Air Borne Particles Causing Air Pollution	772
Appendix 6.5	:	Particle Formation Mechanisms, Surface Area and Size	773
Appendix 6.6	:	Aerosol Formation Mechanism	774
Appendix 6.7	:	Aerodynamic Behavior of Aerosols	774
Appendix 6.8	:	Dry and Wet Deposition of Air Pollutants	775
Appendix 6.9	:	The Atmospheric Air-Pollution System	776
Appendix 6.10	:	Deposition of Particulate Matter (PM) of Various Diameters (in microns) in Various Organs of Human Respiratory System	777
Appendix 6.11	:	Natural and Man-Made (Anthropogenic) Sources of Air Pollution	778
Appendix 6.12	:	List of Major Stationary Sources of Air Pollution	780
Appendix 6.13	:	Major Sources of Air Pollution	781
Appendix 6.14	:	Sources of Air Pollutants Grouped by their Origin	782
Appendix 6.15	:	Examples of Various Air Contaminant/Pollutants Compounds	783
Appendix 6.16	:	Important Air Pollution Category of Sources & Emissions of Specific Air Pollutants	783
Appendix 6.17	:	Types of Air Pollution by Chemical Characteristics and Source of Pollutants	784
Appendix 6.18	:	The Common Air Pollutants Found in the Atmosphere	784
Appendix 6.19	:	Sources of Dust Emission in the Atmosphere	786
Appendix 6.20	:	Average Smoke Emission Rates from Different Sources	786
Appendix 6.21	:	Health Effects Associated with Air Pollution	787
Appendix 6.22	:	Maximum Permissible Concentration of Common Impurities in Polluted Air	787
Appendix 6.23	:	Allowable Limit of Some important Air Pollutants in Air	788
Appendix 6.24	:	Air Pollutants Emitted from a 200 MW Coal Fired Thermal Power Plant	788
Appendix 6.25	:	Total Cost of Air Pollution as a Sum of Control Costs and Damage Costs	789
Appendix 6.26	:	Common Air Pollutants, their Sources and Pathological Effects on Human Beings	789

Contents			Page No.

Appendix 6.1: Characteristics of Atmospheric Particles

S.No.	Parameters	Unit	Atmospheric particles		
			Aitken particles	Large particles	Giant particles
1.	Typical size	μm	<0.1	0.1–1	>1
2.	Principal nature/ sources	–	Combustion aerosols	Combustion products and photochemical aerosols	Natural & industrial dust
3.	Settling speed	m/sec	$< 8 \times 10^{-7}$	Intermediate	$>4 \times 10^{-5}$

1 mm= 1000 microns = 1000 micrometers (μm)

Appendix 6.2: Sizes Ranges of Atmospheric Particulate Matter

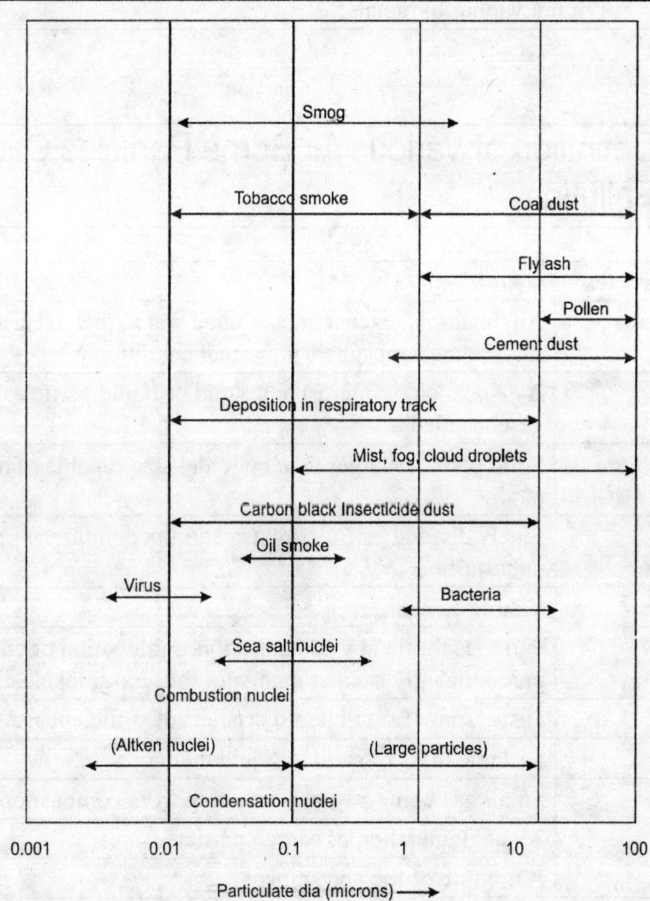

Appendix 6.3: Types of Particulates in Suspended Matter

S.No.	Term	Meaning	Examples
	Liquid particles		
1.	Mist	Aerosols consisting of liquid droplets.	Sulphuric acid mist.
2.	Fog	Aerosols consisting of water droplets.	
	Solid particles		
3.	Dust	Aerosols consisting of solid particles that are blown into the air, or are produced from larger particles by grinding them down.	Dust storm.
4.	Smoke	Aerosols consisting of solid particles or a mixture of solid and liquid particles produced by chemical reaction, such as by fires.	Cigarette smoke, smoke from burning of garbage, etc.
5.	Fumes	Generally means the same as smoke, but often used to indicate aerosols produced by condensation of hot vapour of metals.	Zinc/lead fumes.

Appendix 6.4: Definition of Various Air Borne Particles Causing Air Pollution

S.No.	Air borne particles	Definition
1.	Particulate matter	Any material, except uncombined water, that exists in the solid or liquid state in the atmosphere at normal condition.
2.	Aerosol	A dispersion of microscopic solid or liquid particles in the atmosphere or gaseous media.
3.	Dust	Solid particles larger than colloidal size capable of temporary suspension in air.
4.	Fly ash	Finely divided particles of ash entrained in flue gas particles may contain unburnt fuel.
5.	Fog	Visible aerosol.
6.	Fume	Particles formed by condensation, sublimation or chemical reaction predominantly smaller than 1 μ (Tabacco smoke, acid fume).
7.	Mist	Dispersion of small liquid droplets of sufficient size to fall from the air.
8.	Particle	Discrete mass of solid or liquid matter.
9.	Smoke	Small gas borne particles resulting from combustion.
10.	Soot	An agglomeration of carbon particles.
11.	Smog	A mixture of fog and smoke.

Appendix 6.5: Particle Formation Mechanisms, Surface Area and Size

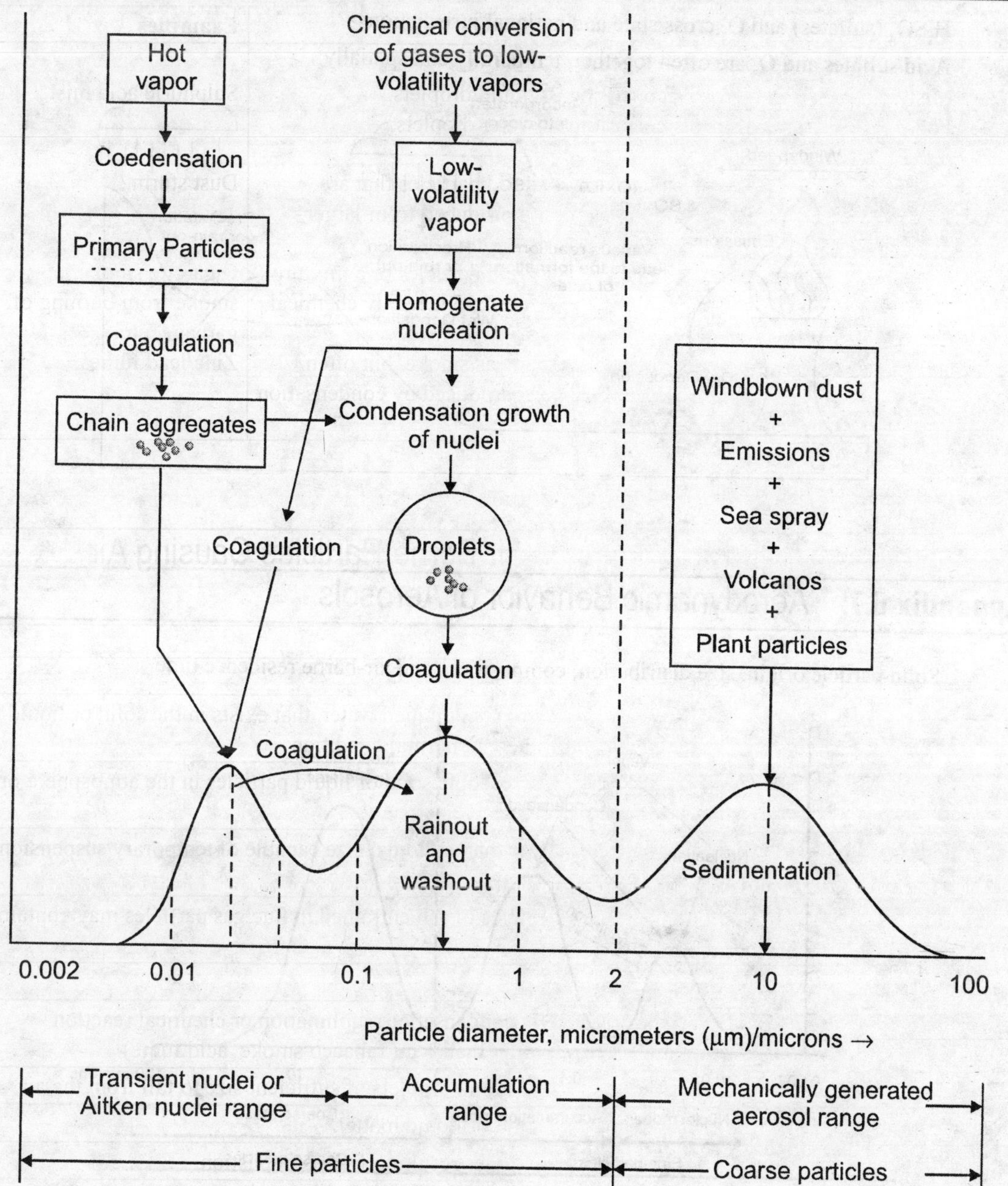

Appendix 6.6: Aerosol Formation Mechanism

- H_2SO_4 (sulfates) and O_3 cross state and national jurisdictions
- Acid sulfates and O_3 are often together, temporally and spatially

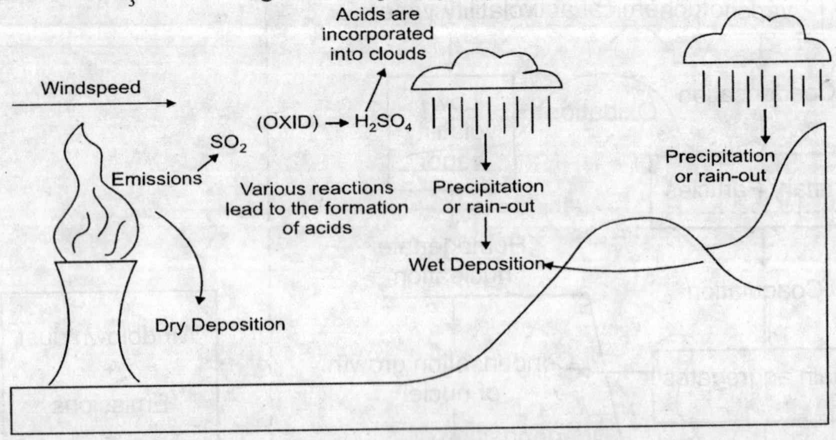

Appendix 6.7: Aerodynamic Behavior of Aerosols

- Solid particle origin, size distribution, composition, and air-borne residence time

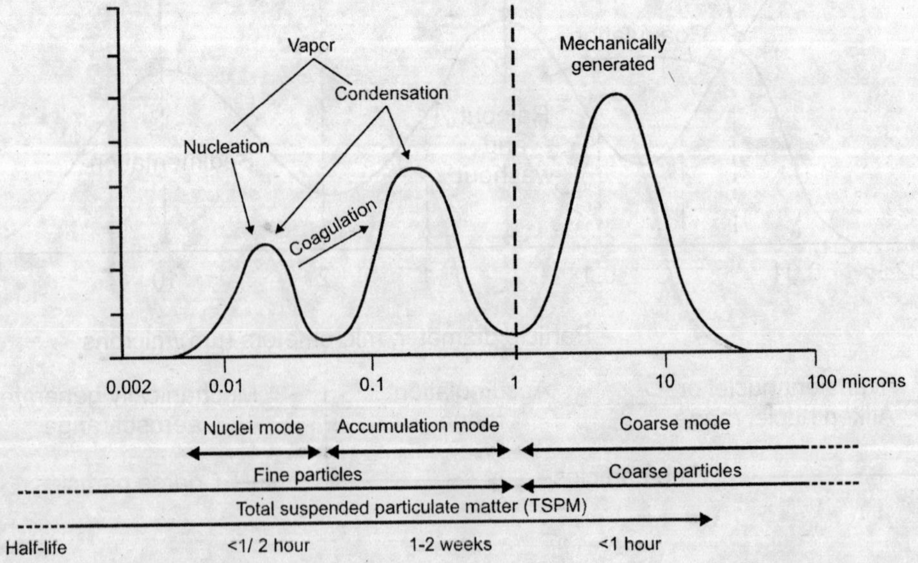

Appendix 6.8: Dry and Wet Deposition of Air Pollutants

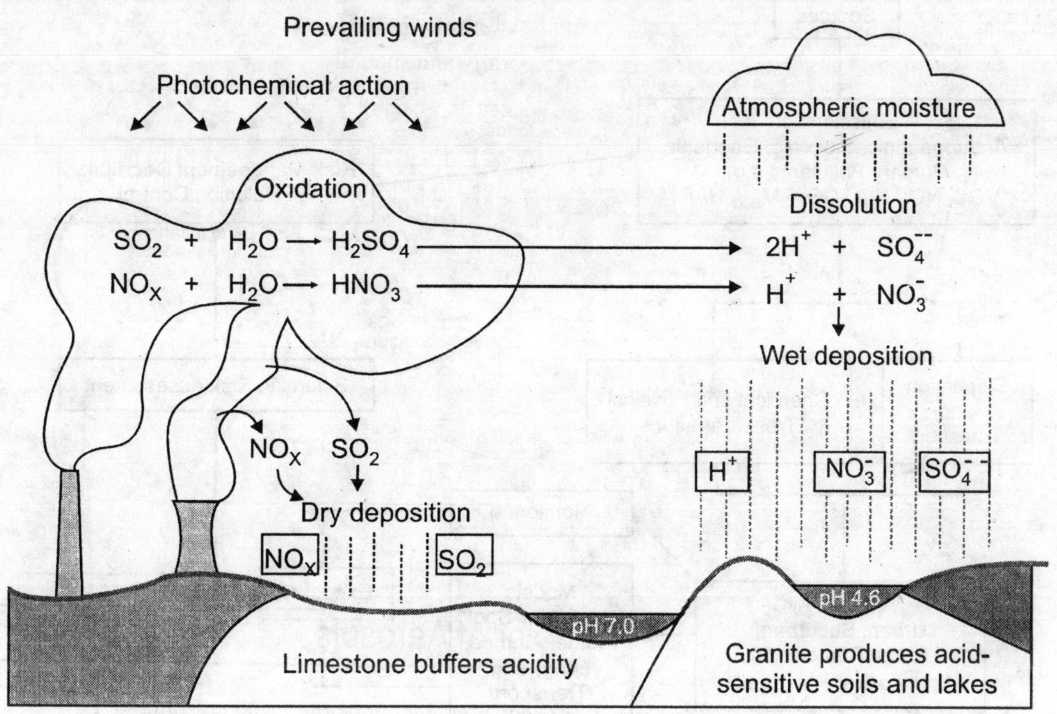

Appendix 6.9: The Atmospheric Air-Pollution System

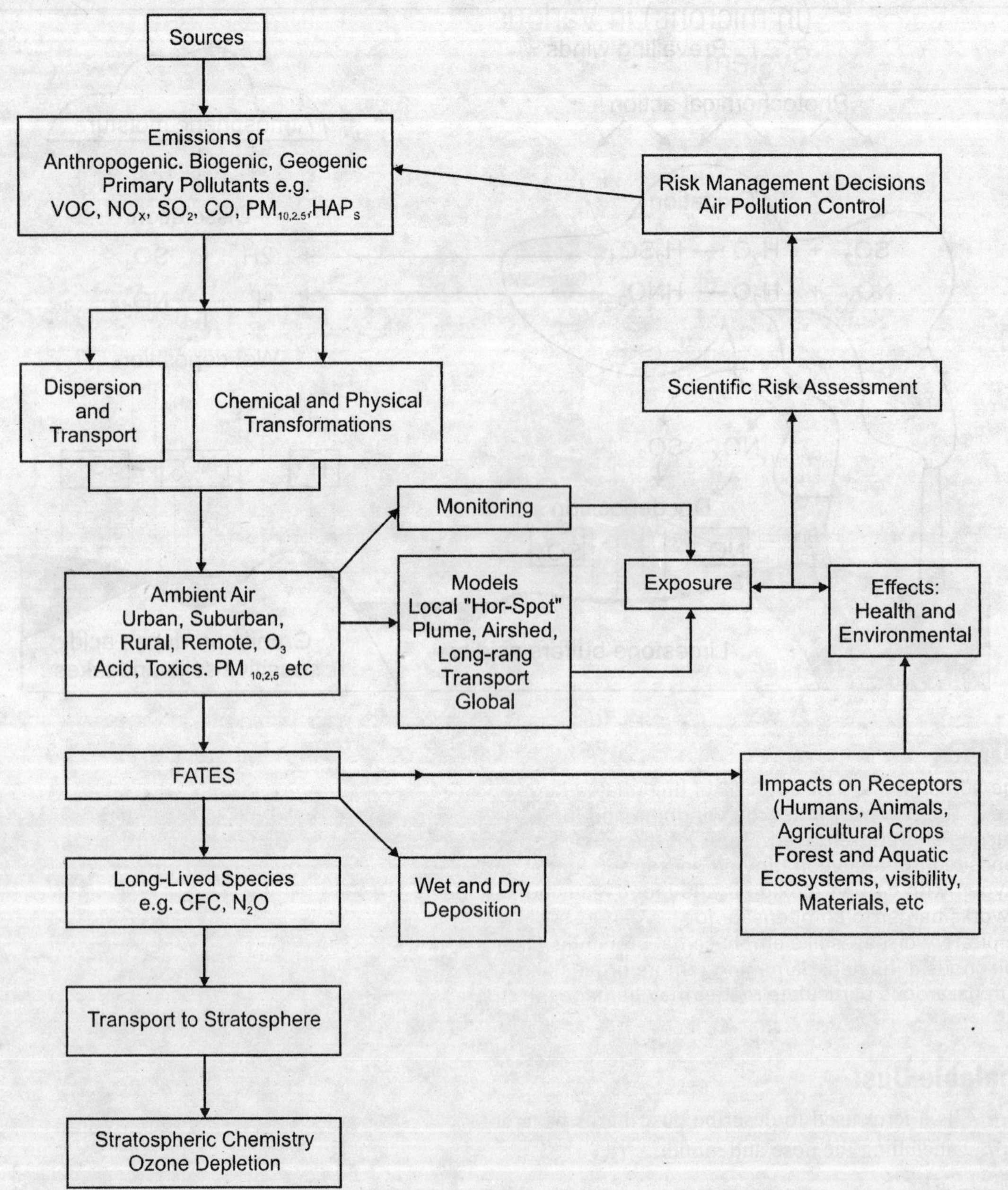

Appendix 6.10: Deposition of Particulate Matter (PM) of Various Diameters (in microns) in Various Organs of Human Respiratory System

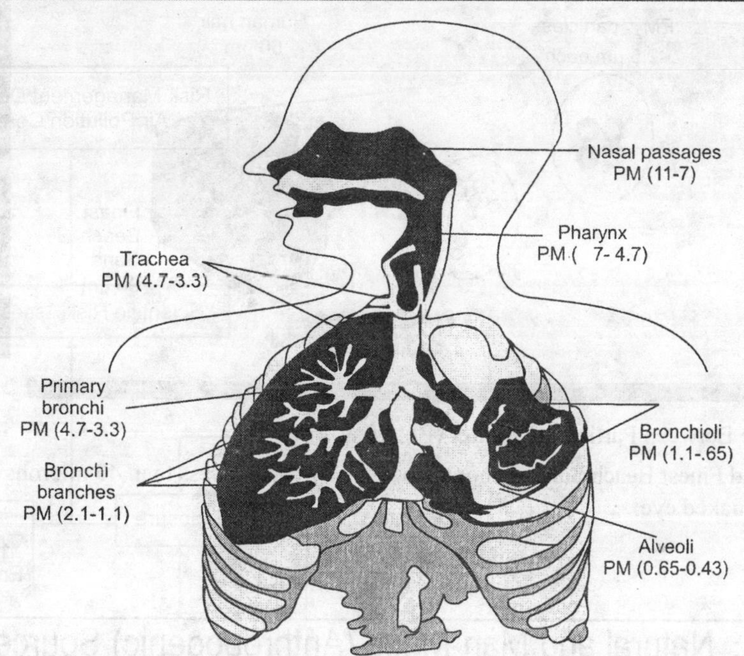

NOTES

The human nose filter out 99% of the inhaled large and medium sized particles (greater than 10 microns size). The rest may enter the windpipe and lungs where some inhalable particulates cling to protective mucous and are removed. Some of the smallest particles, called respirable particulates may tend to be deposited in the alveoli (tiny air sacs in the lungs). In the lungs particulates slow down the exchange of oxygen with carbon dioxide in the blood, causing shortness of breath. The heart gets strained because it works harder to componsatc for oxygcn loss. Usually, people most sensitive to these conditions have respiratory diseases like emphysema, bronchitis, asthma or heart problems. Particles themselves may be poisonous if inhaled, damaging remote organs like the kidneys or liver. Swallowed mucous that is laden with hazardous particulate matter may damage the stomach.

Inhalable Dust

- Is a term used to describe dust that is hazardous when deposited anywhere in the respiratory tree including the nose and mouth.

Thoraic dust

Is defined as those materials that are hazardous when deosited anywhere within the lung airways and the gas exchange region

Respirable dust

Respirable dust is defined as that fraction of the dust reaching alveolar region of the lungs.

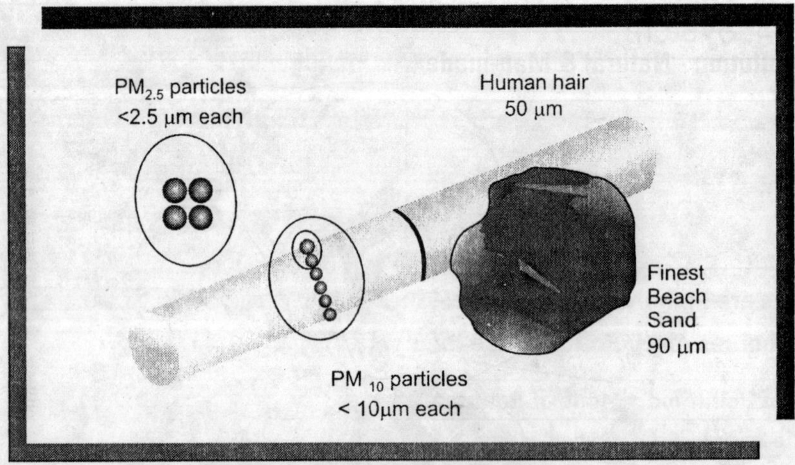

Size Difference Between Particulate Matter (PM 10 & PM 2.5),

Human Hair and Finest Beach Sand is shown above. Any particle less than 40 microns (μm) in diameter is not visible to the naked eye.

Appendix 6.11: Natural and Man-Made (Anthropogenic) Sources of Air Pollution

Primary Pollutants in Air Account for 90% of the Air Pollution. They are given below:

S.No.	Class of Pollutants	Specific pollutants generated
1.	Oxides of Carbon	Carbon-mono-oxide (CO)
2.	Oxides of Nitrogen	Nitric Oxide (NO) and Nitrogen-di-oxide (NO_2)
3.	Oxides of Sulfur	Sulfur-di-oxide (SO_2) and Sulfur-tri-oxide (SO_3)
4.	Hydrocarbons/VOCs	HC
5.	Particulate matter (dust)	Suspended particulate matter (< 500 μ) (SPM) and Respirable suspended particular matter (RSPM) (< 10.0 μ)

Air Pollution Generation activity wise

Activities generating air pollutants	Air pollutants generated (%)					Overall %
	CO	NO_X	SO_X	HC	Dust	
Transportation	64.0	48.0	5.0	51.0	4.2	42.0
Fuel combustion (stationary sources)	2.0	40.0	64.0	2.5	26.0	21.0

Industrial Processes	10.0	1.0	22.0	14.0	31.0	14.0
Solid Waste Disposal	8.0	3.0	5.0	9.0	3.8	5.6
Miscellaneous including domestic	16.0	8.0	4.0	23.5	35.0	17.4

Sources of Air Pollution - Natural & Man-made (Anthropogenic)

Natural Sources -	Volcanic Eruptions Forest Fires Dust Storms Gases of Decay Spores and Pollens

Natural pollutants are higher than man made – but they are distributed Globally

Anthropogenic Sources of Air Pollution are tabulated below :

S.No.	Sources of Pollution	Mode of Pollutants generation	Specific pollutants generated
1.	**Domestic Sources**	Space heating	CO, CO_2, NO_X, Soot, Smoke
		Cooking	Volatile organics including fats
		Cleaning	Solent vapors, dust, spray and propellants
		Gardening	Pesticides, fertilizers
		Painting	Solvent vapors
		Washing	Detergents,
2.	**Commercial Sources:**	Public service	Dry cleaning
		Industries	Petro-dispensing
		Printing	Solvent vapors
		Painting	solvent vapors
3.	**Agricultural sources**	Animal feed operations	Grain dust, Cellulosic dust, fibres
		Harvesting	Dust, Cellulosic fibres, grain dust
		Pest Control	Insecticides, Pesticides, Fungicides
4.	**Industrial Sources:**	Depends largely on the type of raw materials used and products manufactured. In addition process common to many industries are :-	
		Combustion (stationary sources)	CO, CO_2, NO_X, SO_2(fossil fuel), HC and dust
		Degreasing operations	Volatile organic chemicals (VOCs)
		Petrochemical/refineries	Volatile organic chemicals (VOCs)
		Other common pollutants are	Dust, carbon, smoke, H_2S, fibres (asbestos) chrome pigments
5.	**Transportation Sources**	Combustion (mobile sources)	CO, CO_2, NO_X, SO_2 (fossil fuel), HC and dust

S.No.	Sources of Pollution	Mode of Pollutants generation	Specific pollutants generated
6.	**Air conditioning and Refrigeration, Cold storages**	–	Chloroflouro carbons (CFC)

Appendix 6.12: List of Major Stationary Sources of Air Pollution

S.No.	Main stationary sources of air pollution
1.	Coal washeries
2.	Coke oven batteries
3.	Carbon black plants
4.	Chemical process plants
5.	Charcoal production facilities
6.	Fuel conversion plants
7.	Fossil fuel fired biolers
8.	Glass fibre processing units
9.	Hydrofluoric acid plants
10.	Iron & Steel plants
11.	Kraft paper pulp mills
12.	Lime plants
13.	Municipal incinerators
14.	Nitric acid plants
15.	Portland cement plant
16.	Primary zinc smelters
17	Primary aluminium reduction plant
18.	Primary copper smelters
19.	Petroleum refineries
20.	Phosphate rock processing plants
21.	Primary lead smelters
22.	Petroleum storage & transfer facilities
23.	Sulfuric acid plants
24.	Sintering plant
25.	Secondary metal production facilities
26.	Taconite ore processing facilities

Appendix 6.13: Major Sources of Air Pollution

S.No.	Sources of Air Pollution	Category of Pollution sources	Examples	Pollutants produced
1.	Combustion Process	• Fossil fuel combustion	Domestic ovens, Thermal power Plants, industrial operations	Oxides of sulphur (SO_x) oxides of nitrogen (NO_x), Particulate matter, carbon monoxide, smoke, flyash
		• Automobiles/ Aircrafts	Cars, buses, trucks, aeroplane	Particulates, carbon monoxide, smoke, organic hydrocarbon vapour, SO_x, NO_x, Odours
		• Refuse burning	Open burning of refuse	SO_x, NO_x, particulate
2.	Dust producing processes	Crushing grinding, screening, demolition milling	Road mix plants, construction work grain elevators, industrial operations	Mineral & Organic particulates
3.	Manufacturing Process	• Chemical Plants	Cement plants, petroleum refineries, pulps & paper plants, Fertiliser plants etc.	H_2S, SO_x, NO_x. odours, fluorides, particulates, organic vapour
		• Metallurgical Plants	Smelters, steel plants, non-ferrous industries etc.	SO_x, metal fumes, Pb, As, ZnF, other metals
4.	Agricultural activities	• Crop spraying, dusting etc	Weed and Pest Control	Organic phosphates, Chlorinated hydrocarbons, lead, arsenic etc.
		• Field burning	Forest fure, dead agricultural plant refuge burning etc.	Smoke, soot, flyash etc.
5.	Nuclear energy activities	• Nuclear pre preparation	Crushing, grinding, screening	Uranium and beryllium dust
		• Nuclear fissions	Nuclear reactors	Radioactive fission products, harmful radiation
		• Nuclear device testing	Atmospheric nuclear explosions	Radioactive fallouts

Appendix 6.14: Sources of Air Pollutants Grouped by their Origin

S.No.	Air Pollutants	Sources
Group I. Sources predominantly outdoor:		
1.	Lead (Pb)	Automobiles, industrial emission
2.	Ozone (O_3)	Photochemical reactions in the atmosphere
3.	Sulphur oxides (SO_x)	Fuel combustion, smelters
4.	Pollens	Trees and others plants
5.	Organic substances	Petrochemical solvents, natural sources, vaporization of unburnt fuels
Group II. Sources both indoor and outdoor:		
6.	Carbon dioxides (CO_2)	Metabolic activity, combustion
7.	Carbon monoxide and nitrogen oxides (CO_2 & NO_x)	Fuel-burning and tobacco smoke
8.	Organic substances	Volatilization, combustion, metabolic actions, pesticides
9.	Polyoyclic hydrocarbons	Fuel-burning, tobacco smoke
10.	Suspended particulate matter (SPM)	Resuspension, condensation of vapours, and combustion of products
11.	Water vapour (H_2O)	Biological activity, combustion, evaporation
Group III. Sources predominantly indoor:		
12.	Aerosols	Consumer products
13.	Allergens	House dusts
14.	Ammonia (NH_3)	Metabolic activity, cleaning products
15.	Asbestos and synthetic fibres	Fire-retardant, acoustics/ thermal, or electrical insulation
16.	Formaldehyde (HCHO)	Particleboard, insulation, furnishing, tobacco smoke
17.	Mercury (Hg)	Spills in dental-care facilities or laboratories, thermometer breakage
18.	Organic substances	Adhesives, cooking, cosmetics solvents
19	Radon (Ra)	Building materials (concrete, stone), water

Appendix 6.15: Examples of Various Air Contaminant/Pollutants Compounds

S.No.	Air contaminant/Pollutant compounds	Examples of various categories of air pollutant compounds
1.	Nitrogen compounds	NH_3, NO, NO_2
2.	Sulphur compounds	H_2S, SO_2, SO_3, RSH (mercaptans)
3.	Oxygen compounds	O_3
4.	Carbon compounds	CO, CO_2
5.	Halogen compounds	HF, HCI
6.	Organic compounds	HCH CH_4, HC (hydrocarbons)
7.	Radioactive Compounds	Radioactive gases

Appendix 6.16: Important Air Pollution Category of Sources & Emissions of Specific Air Pollutants

S.No.	Category of sources	Example of causative agents	Important air Pollutants produced
1.	Chemical Plant	Petroleum refineries, fertilizers, cement, paper mill, ceramic, clay products and glass manufacture	Hydrogen Sulfide (H_2S), Sulfur Oxides (SO_2, SO_3), fluorides (F)
2.	Crop Spraying	Pest and Weed control	Organochlorines, organophosphates, carbamates, lead (Pb), Arsenic (As)
3.	Fuel combustion	Domestic burning, thermal power plants, boilers	Carbon-mono-oxide (CO), Carbon dioxide (CO_2), Sulfur Oxides (SO_2, SO_3), Oxides of Nitrogen (NO_X), Unburnt Carbon/hydrocarbons
4.	Metallurgical plants	Aluminium, copper, steel plants	Metal fumes, (Lead, Zn), Flourides and particulates
5.	Nuclear device testing	Bomb explosions	Radio active fall out, Sr 90, Ca^{137}, C^{14} etc.
6.	Ore preparation	Crushing, grinding, screening	Uranium, Berryllium dust, other particulates, $Argon^{41}$, $Iodine^{131}$
7.	Spray painting, solvent extraction, ink, solvent cleaning	Furniture and appliances, dyeing, printing, chemical separations, dry cleaning	Hydro Carbons and other volatile organic carbon
8.	Transportation	Cars, trucks, buses, railways and aeroplanes	Carbon mono and di-oxides (CO, CO_2), Oxides of Nitrogen (NO_X), lead smoke, organic vapors (VOCs)
9.	Waste Recovery	Scrap metal yards, rendering plants	Smoke, soot, odors, VOC, metal, fumes

Appendix 6.17: Types of Air Pollution by Chemical Characteristics and Source of Pollutants

S.No.	Pollution Type	Composition of pollutants	Pollutant Sources
1.	Reducing	Sulfur dioxide, particulates.	Stationary combustion sources, such as fossil fuel power plants, industrial furnaces, home heating units.
2.	Photochemical	Hydrocarbons and nitric oxide emitted by sources such as internal combustion engine undergo automobiles, fossil fuel power Complex photochemical reactions in the plants and oil refineries. Presence of sunlight resulting in an atmosphere with significant concentrations of ozone, nitrogen dioxide, aldehydes, and organic nitrates.	Mobile sources of pollutants emission
3.	Point Sources of pollutants emission	Specific to source of emission, e.g. lead Specific industries; near a smelter.	Industrial or transportation accidents.

Appendix 6.18: The Common Air Pollutants Found in the Atmosphere

S.No.	Pollutants commonly found in atmosphere	Primary	Secondary
1.	Sulphur dioxide (SO_2)	✓	
2.	Particulate Matter (PM_{10} *, $PM_{2.5}$ **)	✓	✓
3.	Nitrogen oxides (NOx) (NOx = NO + NO_2)	✓	✓
4.	Volatile Organic Compounds (VOCs)	✓	✓
5.	Ozone(O_3)		✓
6.	Heavy metals, including chromium, nickel vanadium and lead compounds	✓	
7.	Carbon monoxide (CO)	✓	

*PM_{10}: Particulate matter with aerodynamic diameter < 10μm

**$PM_{2.5}$: Particulate matter with aerodynamic diameter <2.5μm

NOTES

Primary pollutants: SO, CO_2, CO, NO, NO_2, Particulate Matter (PM) and VOCs are primary pollutants - they are released directly into the atmosphere. Vehicle emissions and emissions from other combustion sources may be significant primary sources of fine faction ($PM_{2.5}$) particulate emissions. The combustion of the hydrocarbons - fuel oil, diesel and petrol -produces elemental carbon as a primary particulate. Windblown dust contributes to the coarse fraction of PM_{10} as a primary pollutant.

Secondary pollutants are formed in the atmosphere through chemical reactions and physical processes. For example, SO_2 and NO_2 react with ammonia or other alkaline species, atmospheric oxygen and water vapour to form sulphates (ammonium bisulphate and/or sulphuric acid) and nitrates (ammonium nitrate, peroxyacetylnitrate (PAN) and/or nitric acid). The nuclei that form when these substances condense may grow through the physical processes of deposition and agglomeration.

Types and sources of air pollutants (primary & secondary pollutants have been shown below diagramatically)

- **Primary air pollutants:** Materials that when released pose health risks in their unmodified form or those emitted directly from identifiable sources.
- **Secondary air pollutants:** Primary pollutants interact with one another, sunlight, or natural gases to produce new, harmful compounds.

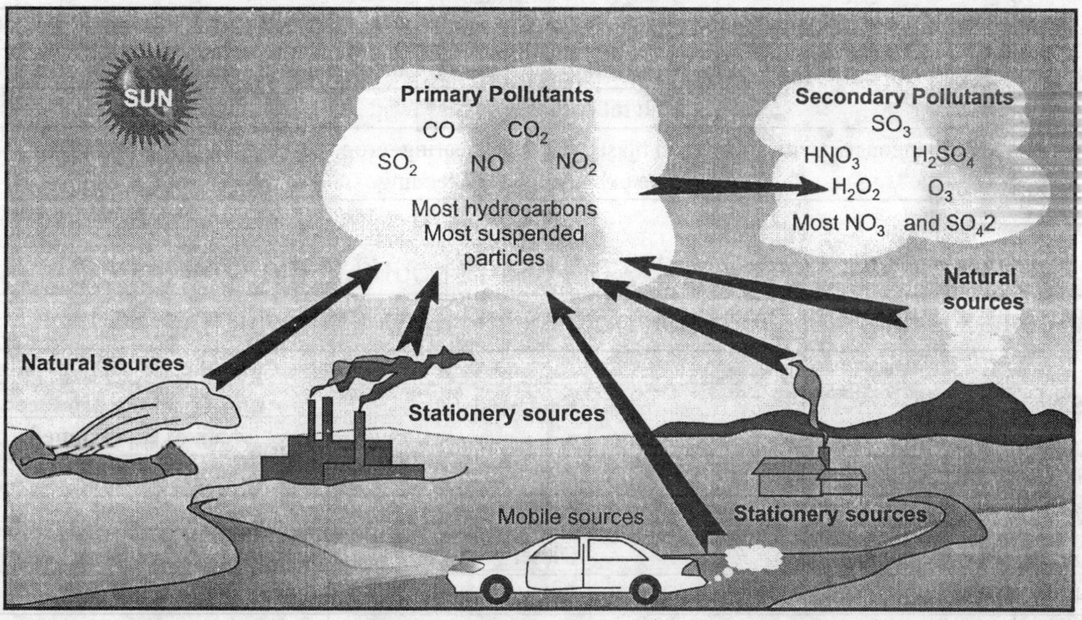

Appendix 6.19: Sources of Dust Emission in the Atmosphere

S.No.	Sources of dust generation	Activities responsible for dust generation
1.	Combustion	Fuel burning, open fires, incineration, forest fire, burning dumps
2.	Materials handling & processing	• Loading & unloading of sound, ores, coal, gravel, bulk chemicals, lime etc.
		• Crushing & grinding of cement, ores, chemical , gravel etc.
		• Mixing & packaging of feed, fertilizers, chemicals etc.
		• Food processing (milling, drying, grain handling)
		• Cutting & forming (saw mills, walls boards, plastics etc.)
		• Manufacturing & processing solids (e.g. carbon black, cement, chemicals etc.)
		• Metallurgical processes (smelters, blast furnaces, foundaries)
3.	Earth moving	• Mining operation (blasting, sorting, refuse disposal etc)
		• Construction activities (roads, dams, buildings, site clearance etc)
		• Agricultural operations (land preparation, soil tilling etc)
		• Natural (winds)
4.	Miscellaneous activities	Sand blasting, house clearing, crop spraying, rubber tyre abrasion, engine exhaust, poultry feeding

Appendix 6.20: Average Smoke Emission Rates from Different Sources

S.No.	Sources	Fuel used	tons of smoke produced/ ton of fuel burned
1.	Petrol engines	Petrol	2.1
2.	Industrial burners	Fuel oil	2.7
3.	Diesel engines	Diesel	3.9
4.	Domestic & municipal incineration	Rubbish	4.3–4.5
5.	Open fires	Trash	11
6.	Industrial installation	Coal	12
7.	Domestic grate	Coal	26

Appendix 6.21: Health Effects Associated with Air Pollution

- Loss of lung function
- Disability (absenteeism, increased need for medication, hospitalization)
- Increased mortality rate (respiratory and cardiovascular deaths)
- Symptoms of irritation (cough)

Chronic effects:

- Impaired lung growth
- Accelerated lung aging
- Damage to other organ systems
- Cancer

Appendix 6.22: Maximum Permissible Concentration of Common Impurities in Polluted Air

S.No.	Air pollutants/contaminants	Maximum permissible concentration (mg/m^3)	
		Average over 24 hrs	At any instant
1.	Dust (20% silica)	0.15	0.50
2.	Sulfur Dioxide (SO_2)	0.15	0.50
3.	Carbon monoxide (CO)	2.00	6.00
4.	Oxides of nitrogen (NO_X)	0.13	0.40
5.	Sulfuric acid (H_2SO_4)	0.10	0.30
6.	Chlorine (Cl_2)	0.02	0.06
7.	Hydrogen sulfide (H_2S)	0.005	0.015
8.	Lead (Pb)	0.0007	0.002
9.	Carbon disulfide (CS_2)	0.15	0.59
10.	Phenol(C_6H_5OH) (as an indicator of pollution by phenolic substance	0.10	0.30
11.	Arsenic (As)	0.003	0.01
12.	Fluorine (F) (gaseous and liquid inorganic compounds)	0.01	0.03
13.	Ammonia (NH_3)	0.10	0.30
14.	Soot	0.05	0.15
15.	Tarry substances	According to the composition	0.10
16.	Mercury (Hg)	0.0003	0.001
17.	Manganese (inorganic compound as MnO_2)	0.01	0.03

18.	Formaldehyde (HCHO)	0.015	0.05
19.	Acrolein	0.10	0.30
20.	Hydrogen-Chloride (HCl)	0.02	0.06

Appendix 6.23: Allowable Limit of Some important Air Pollutants in Air

S.No	Concentration in ambient air in microgram/m³ (μg/m³), permissible limit				
	Air Pollutant	**Time weighted average**	**Industrial area**	**Residential, rural and other areas**	**Sensitive area**
1.	Sulphur dioxide (SO_2)	Annual Avg.* 24 hours**	80 120	60 80	15 30
2.	Oxides of nitrogen as NO_2	Annual Avg. 24 hours	80 120	60 80	15 30
3.	Suspended particulate matter (SPM)	Annual Avg. 24 hours	360 500	140 200	70 100
4.	Respirable particulate matter (size less than 10 mm) (RSPM)	Annual Avg. 24 Hours	120 150	60 100	50 75
5.	Lead (Pb)	Annual Avg. 24 Hours	1.0 1.5	0.75 1.0	0.5 0.75
6.	Carbon monoxide(CO)	8 hours 1 hours	5.0 10.0	2.0 4.0	1.0 2.0

Appendix 6.24: Air Pollutants Emitted from a 200 MW Coal Fired Thermal Power Plant

S.No.	Air pollutants generated	Emission factor, kg/ton of coal	Emitted quantity, tons/day
1.	Oxides of sulphur (SO_X)-[0.5% S]	19 S	13.30
2.	Oxides of nitrogen (NO_X)	10.0	14.0
3.	Carbon monoxide (CO)	0.25	0.35
4.	Hydrocarbons (HC)	0.10	0.14
5.	Aldehydes (RCHO)	0.0025	0.0035
6.	Particulate matter (33% ash)	8 A	369.60
7.	Fly ash	2 A	92.40

A = % of ash in coal; S= % of sulphur in coal

Appendix 6.25: Total Cost of Air Pollution as a Sum of Control Costs and Damage Costs

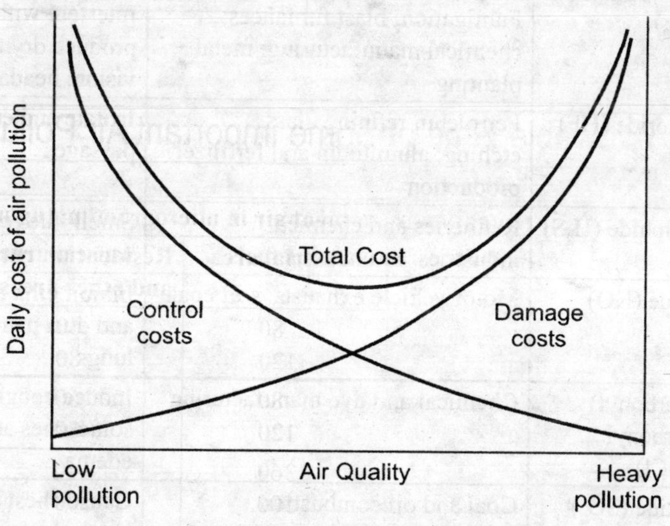

NOTES

a) **Damage costs:** The cost to the public of living in polluted air eg, tangible losses such as crop damage & deteriorated materials and intangible losses such as reduced visibility and eye & nasal irritation

b) **Control costs:** The cost incurred by polluters and the public in order to reduce the emissions eg, direct costs such as the price of the pollution control equipment which must be purchases and indirect cost such as induced unemployment as a result of plant shutdown or relocation

Appendix 6.26: Common Air Pollutants, their Sources and Pathological Effects on Human Beings

S.No.	Air Pollutants	Sources of pollutants	Pathological effect on human beings
1.	Aldehydes (RCHO)	Thermal decomposition of fats, oil or glycerol	Irritant nasal and respirator tract
2.	Ammonia (NH_3)	Chemical processes dye-making; explosives; lacquer; fertilizer	Inflame upper respiratory passages
3.	Arsenic (As)	Processes involving metals or acids containing arsenic soldering	Break down red cells in blood, damage kidneys, cause jaundice
4.	Carbon monoxide (CO)	Gasoline motor exhausts; burning of coal	Reduce oxygen-carrying capacity of blood

5.	Chlorine (Cl_2)	Reach cotton and flour, many others chemical processes	Attack entire respiratory tract and mucous membranes of eyes; cause pulmonary edema
6.	Hydrogen (H_2)	Fumigation; blast furnances; chemical manufacturing; metal planting	Interfere with nerve cells; produce dry throat, indistinct vision, headache
7.	Hydrogen fluoride (HF)	Petroleum refining glass etching; aluminium and fertilizer production	Irritate and corrode all body passages
8.	Hydrogen sulphide (H_2S)	Refineries and chemical industries; bituminous fuels	Smell like rotten eggs; cause nausea; irritate eyes and throat.
9.	Nitrogen oxide (NO)	Motor vehicle exhausts; soft coal.	Inhibit cilia action so that soot and dust penetrate far in to the lungs
10.	Phosgene (carbonyl) chlorine (Carbonyl chloride–COCl)	Chemical and dye manufacturing	Induce coughing, irritation, and sometimes and fatal pulmonary edema
11.	Sulphur dioxide (SO_2)	Coal and oil combustion	Cause chest construction, headache, vomiting and death from respiratory ailments
12.	Suspended particles (ash, soot, smoke)	Incinerators; almost any manufacturing plant	Cause emphysema, eye irritants and possibly cancer

Appendix 6.27: Harmful Effects of Some Air Pollutants on Human Health

S.No.	Air Pollutants	Harmful effects of pollutants
1.	Asbestos	Disease of lungs (Fibriosis, asbestosis) tumors, lung cancer
2.	Arsenic	Bronchitis, dermatitis, skin cancer
3.	Aldehydes	Irritation of eyes and respiratory track
4.	Beryllium	Systemic poison causes berilliosis, skin damage
5.	Chlorine	Irritation to eye, nose, throat. Bronchitis, pneumonitis
6.	Cadmium	Highly toxic, causes bronchilis, fibriosis of lungs hypertension, carcinogenic
7.	Lead	Systemic poison, causes anemia (reduction of haemoglobin) affects kidney function, damages brain.
8.	Mercury	Highly toxic, enzyme poison, nervous system disorder
9.	Nickel	Respiratory disorder
10.	Pesticides	Various insecticides, herbicides, fungicides, nematicides and repellants are toxic and produce headache, stomach disturbances and irritations.

11.	Radioactive	Somatic (on body cells) effects including lukemia (blood cancer), and other cancers. Genetic effects are mainly mutations in gem cells
12.	Selenium	Irritation to eyes, nose, respiratory tract and gastrointestinal tract, badly affects kidney, lungs and liver
13.	Vanadium	Affects gastrointestinal tract and respiratory tract. Cholesterol synthesis inhibition, carcinogenic

Appendix 6.28: Harmful Impacts on Human beings by Vehicular Emissions

S.No.	Pollutants from vehicle	Health effects of pollutants	Impacts on environment of the pollutants
1.	Carbon monoxide (CO)	Fatal at high doses. Attacks the nervous system.	Acts like CO_2, i.e. traps the earth's heat.
2.	Hydrocarbons (HC)	Cause eye irritation, coughing and drowsiness. High molecular	Produces the dangerous photochemical smog.

Appendix 6.29: Sources and Acute Health Effects of Indoor Air Pollutants as per OSHA

(Although asbestos and radon have been listed below, acute health effects are not associated with these contaminants. These have been included due to recent concerns about their health effects.)

1. **Acetic Acid (CH_3COOH):**

   Sources: X-ray development equipment, silicone caulking compounds.

   Acute health effects: Eye, respiratory and mucous membrane irritation.

2. **Carbon Dioxide (CO_2):**

   Sources: Unvented gas and kerosene appliances, improperly vented devices, processes or operations which produce combustion products, human respiration.

   Acute health effects: Difficulty concentrating, drowsiness, increased respiration rate.

3. **Carbon Monoxide (CO):**

   Sources: Tobacco smoke, fossil-fuel engine exhausts, improperly vented fossil-fuel appliances.

   Acute health effects: Dizziness, headache, nausea, cyanosis, cardiovascular effects, and death.

4. **Formaldehyde (HCHO):**

   Sources: Off-gassing from urea formaldehyde foam insulation, plywood, particle board, and paneling; carpeting and fabric; glues and adhesives; and combustion product including tobacco smoke.

Acute health effects: Hypersensitive or allergic reaction; skin rashes; eye, respiratory and mucous membrane irritation; odor annoyance.

5. **Nitrogen Oxides (NO_x):**

   Sources: Combustion products from gas furnaces and appliances; tobacco smoke, welding, and gas and diesel-engine exhausts.

   Acute health effects: Eye, respiratory and mucous membrane irritation.

6. **Ozone (O_3):**

   Sources: Copy machines, electrostatic air cleaners, electrical arcing, smog.

   Acute health effects: Eye, respiratory tract, mucous membrane irritation; aggravation of chronic respiratory diseases.

7. **Radon (Rn):**

   Sources: Ground beneath buildings, building materials, and groundwater.

   Acute health effects: No acute health effects are known but chronic exposure may lead to increased risk of lung cancer from alpha radiation.

8. **Volatile Organic Compounds (VOC's):** Volatile organic compounds include trichlorethylene, benzene, toluene, methyl ethyl ketone, alcohols, methacrylates, acrolein, polycyclic aromatic hydrocarbons, and pesticides.

   Sources: Paints, cleaning compounds, moth-balls, glues, photocopiers, "spirit" duplicators, signature machines, silicone caulking materials, insecticides, herbicides, combustion products, asphalt, gasoline vapors, tobacco smoke, dried out floor drains, cosmetics and other personal products.

   Acute health effects: Nausea; dizziness; eye, respiratory tract, and mucous membrane irritation; headache; fatigue.

9. **Miscellaneous Inorganic Gases:** Includes ammonia, hydrogen sulfide, sulfur dioxide.

   Sources: Microfilm equipment, window cleaners, acid drain cleaners, combustion products, tobacco smoke, blueprint equipment.

   Acute health effects: Eye, respiratory tract, mucous membrane irritation; aggravation of chronic respiratory diseases.

10. **Asbestos:**

    Sources: Insulation and other building materials such as floor tiles, dry wall compounds, reinforces plaster.

    Acute health effects: Asbestos is normally not as source of acute health effects. However, during renovation or maintenance operations, asbestos may be dislodged and become airborne. Evaluation of employee exposure to asbestos will normally be covered under the OSHA Asbestos standard.

11. **Synthetic Fibers:**

    Sources: fibrous glass and mineral wool.

    Acute health effects: Irritation to the eyes, skin and lungs; dermatitis.

12. **Environmental Tobacco Smoke (ETS):**

    Sources: Cigars, cigarettes, pipe tobacco.

    Acute health effects: Tobacco smoke can irritate the respiratory system and, in allergic or asthmatic persons, often results in eye and nasal irritation, coughing wheezing, sneezing, headache, and related sinus problems. People who wear contact lenses often complain of burning, itching, and tearing eyes when exposed to cigarette smoke. Tobacco smoke is a major contributor to indoor air quality problems. Tobacco smoke contains several hundred toxic substances including carbon monoxide nitrogen dioxide, hydrogen sulfide, formaldehyde, ammonia, benzene, benzo(a)pyrene, tars, and nicotine. Most indoor air particulates are due to tobacco smoke and are in the respirable range.

13. **Microoganisms and Other Biological Contaminants (Microbials):** Includes virus, fungi, mold, bacteria, nematodes, amoeba, pollen, dander, and miltes.

    Sources: Air handling system condensate, cooling towers, water damaged materials, high humidity indoor areas, damp organic material and porous wet surfaces, humidifiers, hot water systems, outdoor excavations, plants, animal excreta, animals and insects, food and food products.

    Acute health effects: Allergic reactions such as hypersensitivity diseases (hypersensitivity pneumonitis, humidifier fever, allergic rhinitis, etc.) and infections such infections such as legionellosis are seen. Symptoms include chills, fever, muscle ache, chest tightness, headache, cough, sore throat, diarrhea, and nausea.

NOTES

Modern office buildings are generally considered safe and healthful working environments. However energy conservation measures have minimized the infiltration of outside air and contributed to the buildup of indoor air contaminants. Often employee complaints result from items such as cigarette smoke, odors, low-level contaminants, poor air circulation, thermal gradients, humidity, job pressures, lighting, work-station, or noise.

Type of Building Problems. Inhabitants complaints can be due to two types of building problems: sick/tight building syndrome (SBS) and building related illnesses (BRI).

Sick building syndrome (SBS) is a condition associated with complaints of discomfort including headache; nausea; dizziness; dermatitis; eye, nose, throats & respiratory irritation; coughing difficulty in concentrating; sensitivity to odors; muscle pain and fatigue. The specific causes of the symptoms are often not known but sometimes are attributed to the effects of a combination of substances or individual susceptibility to low concentrations of contaminants. The symptoms are associates with periods of occupancy and often disappear after leaving the building/office.

Building-related illnesses (BRI) are those for which there is a clinically defined illness of known etiology and include infections such as legionellosis and allergic reactions such as hypersensitivity diseases

To counteract the above building problems, American society of Heating, Refrigerating and Air-Conditioning Engineers (ASHARAE) has recommended the following ventilation rates for buildings:

Type of building	Outdoor air requirement per person, ft^3/minute
1. Minimum value to be used during building design	5
2. Reception area for offices	15
3. General office space with moderate amount of smoking	20
4. Smoking lounges with local mechanical exhaust ventilation & no air circulation	60

Appendix 6.30: Indoor Gaseous and Particulate Pollutants: Sources and Health Effects

S.No.	Indoor gaseous pollutants	Sources	Health effects	Remarks
1.	Carbon Monoxide (CO)	CO may originate inside or outside buildings. **Inside:** CO is a component of tobacco smoke and is produced by combustion appliances such as kerosene heaters and gas stoves. **Outside:** Vehicle exhaust is the major source of CO, and may enter buildings when outdoor air intakes are located near busy streets, loading bays, adjacent or underground car parks.	Carbon Monoxide interferes with oxygen transport in the bloodstream, reducing delivery to various organs such as the heart and brain. It can cause coronary artery disease, peripheral vascular disease, chronic obstructive pulmonary disease and has other non-specific effects. Exposure to high concentrations of CO may irreversibly damage the central nervous system and cause death.	Little is known about exposure to the low levels generally found inside buildings. However, it has been associated with: • Decreased work capacity in adult males; • Heachaches, decreased alertness and flu-like symptoms in healthy adult; and • Nausea, breathlessness, dizziness and fatigue.
2.	Nitrogen Oxide (NO and NO_2)	The major sources of nitrogen oxides are tobacco smoke and combustion appliances such as gas stoves.	Mild effects of exposure include eye and upper respiratory tract irritation. NO_2 is a deep lung irritant, and in some cases can cause more severe effects such as pulmonary oedema (inflammation and swelling of the lung) and reduced lung function in asthamatics.	Levels are generally low in commercial office buildings.
3.	Sulphur Dioxide (SO_2)	SO_2 is released in the combustion of fuels containing sulphur, either from gas stoves or from outside vehicle exhaust and the environment generally.	Irritation of the upper respiratory tract and reduced lung function in asthmatics.	In combination with particulates in the air, there may be synergistic effects as the lung is already partially blocked. SO_2 has a very strong and unpleasant odour which may cause further irritation at higher concentrations.
4.	Polycyclic Aromatic Hydrocarbons (PAHs)	These are a large family of complex organic substances produced by incomplete combustion processes such as wood burning, cooking and, most commonly in the indoor office environment, tobacco smoking.	Cancer, cardiovascular disease and general sensory irritation.	Although individual PAHs have specific effects, they can be considered as a group since they generally occur in complex combinations.

S.No.	Indoor gaseous pollutants	Sources	Health effects	Remarks
5.	Volatile Organic Compounds (VOCs)	There are thousands of organic compounds that are volatile at room temperature. Consumer products and various building materials emit VOCs. This process is known as 'off-gassing'. **Consumer Products:** dry cleaned clothes, cosmetics, spray cans, tobacco smoke, deodorizers, felt markers, detergents, cleaning agents, carbonless copy paper, etc. **Building and Office Materials**: furnishings, carpets, printed material, paints, adhesives, particle board, waxes and polishes, sealants, foam, tape etc.	Some of the most commonly found VOCs are established and suspected mutagens and carcinogens. Many are mucous membrane irritants, causing eye and upper respiratory tract problems. Other related effects are headache, nausea and dizziness, as well as possible neurotoxic (nervous system) and hepatoxic (liver) effects.	More than 900 VOCs have been identified in indoor air. While always in higher concentration indoors than outdoors, they are also always well below the levels set as official occupational health standards. However, there is great potential for synergistic impact with other VOCs, other pollutants types and other environmental factors. There can be literally hundreds of VOCs in one building, and they may be highly localized depending on the equipment and activities of various tenants, and the adequacy of the ventilation system. The significance of these gases being present is not known, but there is no doubt that ventilation systems can reduce the concentrations of such pollutants.

<table>
<tr><th>S.No.</th><th>Indoor gaseous pollutants</th><th>Sources</th><th colspan="2">Health effects</th><th>Remarks</th></tr>
<tr><td rowspan="8">6.</td><td rowspan="8">Formaldehyde (HCOH)</td><td rowspan="8">Formaldehyde is the most notorious of the VOCs, also produced by the 'off-gassing' of many materials found in buildings such as:
• Plywood & particle board:
• Urea formaldehyde foam insulation (UFFI);
• Furniture;
• Carpets;
• Dry cleaned clothes/curtains;
• Resins;
• Kerosene;
• Adhesives;
• Carbonless copy-paper;
• Other paper products;
• Cosmetics; and
• Cleaning agents etc.
Formaldehyde is also a component of tobacco smoke.</td><td colspan="2">Asthmatics may have reactions to formaldehyde and it is a suspected human carcinogen. It has also been linked (inconclusively) to headache, fatigue, memory lapse, and difficulty in sleeping & concentrating. Acute and chronic effects are possible.</td><td rowspan="8">Levels are almost always higher indoors than outdoors. Concentration depends on :
• Type and age of materials present (new materials produce the gas at a high rate and slow down with age);
• Temperature and humidity (rates of off-gassing are greater when such parameters are high);
• The concentration of other gases (SO_2 lowers the emission rate); and
• The presence of particulates (for absorption of the gas).
Even at low concentrations, the effect of formaldehyde may be increased by other factors.</td></tr>
<tr><td>Possible Effects</td><td>Formaldehyde Concentrations (PPM)</td></tr>
<tr><td>Odour threshold</td><td>0.05-1.0</td></tr>
<tr><td>Neurophysiologic Effects</td><td>0.05-1.0</td></tr>
<tr><td>Eye Irritation</td><td>0.01-2.0</td></tr>
<tr><td>Upper air way Irritation</td><td>0.1-25</td></tr>
<tr><td>Lower airway irritation & pulmonary effects</td><td>5-30</td></tr>
<tr><td>Pulmonary oedema, inflammation, pneumonia

Death</td><td>50-100

>100</td></tr>
<tr><td>7.</td><td>Ozone (O_3)</td><td>Ozone is produced by electrical discharges during the operation of equipment such as photocopying machines, laser printers, ionisers and electrostatic air cleaners.</td><td colspan="2">Ozone is a very toxic gas, and has effects at very low concentrations. It causes eye and respiratory irritation, headache, dizziness, coughing, shortness of breath and severe fatigue.</td><td>Although the health effects are serious, ozone is a very reactive gas and decays quickly, with a half-life of minutes. Since it is only produced locally and is so unstable, it is generally not considered a major indoor pollutant problem.</td></tr>
</table>

S.No.	Indoor gaseous pollutants	Sources	Health effects	Remarks
8.	Radon(Rn)	Radon gas is produced by the breakdown of Uranium and radium in the Earth's crust. It diffuses out from the soil and rocks under buildings and from some masonary building materials. Also, it is soluble in water and can enter buildings through the water supply. Radon decays into 'daughter' products, which are isotopes of various other elements that undergo further decay: Radon-222→Polonium-218→Lead-214→Bismuth-214→Polonium-214→leead-210→bismuth-210→Polonium-210→lead-206 Radioactive 'alpha particles' are released during such decay stages.	Radioactive (alpha active) 'daughters' can attach to aerosols or airborne particles and be inhaled. Radiation penetraters the epithelial lining of the lung directly, damaging the genetic material of the basal cells. Prolonged exposure to Radon at some concentrations is known to cause lung cancer, but little is known about the health effects of low level 'background' exposures.	The concentration of Radon depends on geographical location. Two factors mitigate against Radon being a significant contributor to 'building sickness': • Even in the most highly contaminated buildings, there has been no corresponding increase in the frequency of reported 'sick building' complaints of symptoms; and • Multi-storey buildings usually have Radon concentrations much lower than worldwide average values for single storey dwellings.

S.No.	Indoor gaseous pollutants	Sources	Health effects	Remarks
9.	**Asbestos**:- • Anosite • Chysotile • Crocidolite • Other Forms	Asbestos is a term for a group of naturally occurring silicate rock fibres that have been used in over 3000 different products. Asbestos has mainly been used in products for fire and heat resistance, such as: • heat resistant textiles (cloth/ padding); • cement products (sheets and pipes); • special industrial chemical filters; • thermal insulation products (pipe and boiler insulation); • floor and ceiling tiles; and • rooting material. It can be mixed with many materials, and in buildings it has mainly been used in cement mixtures, sprayed onto ceilings and structural components for fire protection and insulation.	The inhalation of loose asbestos fibres that are 'respirable' (less than about 3 micrometres) poses serious health risks. *Asbestosis* is the progressive scarring of the lung tissue, which may develop further after exposure (pulmonary fibrosis). ***Cancer*** can be caused by the inhalation of fibres <3 $\propto$ m in diameter and longer than 8 $\propto$ m in length. The risk increases with increasing diameter and exposure levels. There are two types of cancer. 1. **Lung Cancer** – there is considerably more risk for smokers; and 2. **Mesothelioma** – cancer of the lining of the chest cavity (pleura) or the abdominal cavity (peritoneum). Chances are not increased by smoking, but correlate with exposure.	Asbestos is not dangerous unless fibres are airborne and are small enough to enter the lung. Exposure results from activities in buildings which are dust forming such as drilling, sawing, grinding, maintenance and renovation, ceiling work etc. the amount of fibre released depends on the cohesiveness of the material and the size of the disturbing force. Cancers have a 10-50 year latency period.

S.No.	Indoor gaseous pollutants	Sources	Health effects	Remarks
10.	**Other Fibres**: • Glass Fibre • Rock/Mineral Wool fibres	• **Glass Fibre** Fibre glass has commonly been used in the lining of ventilation duetwork. • **Rock/Mineral Wool** Rock wool is a fibre that has mainly been used in buildings as a fire proofing material. • **Other** Some other synthetic fibres are found in ceiling tiles.	Such fibres can be the cause of itching and other irritation to exposed skin areas when airborne.	In general, such synthetic fibres are too long and wide to be inhaled and cause the kind of damage associated with asbestos. Fiberglass can be introduced into the air stream due to weakeing from the buildup of moisture and microbial growth fibres may also become loosened with age. Rock Wool may become loosened by the same sorts of dust forming activities mentioned for asbestcs.
11.	Combustion Particles	In the indoor office environment, combustion particles are generated principally by tabacco smoking, but also by some combustion appliances such as gas stoves.	Tobacco smoke particles are small enough to be inhaled, and are therefore a source of respiratory irritation (particularly for asthmatics and particularly in the presence of CO_2). They can cause nasal and eye irritation, and these effects may be exacerbated by other chemicals in the smoke. Combustion particles can be a cause of cancer and may also be the cause of allergic reactions in some people.	
12.	Other Particulates	There are numerous sources of other particulate that may cause irritation, such as paper, clothing, and various biological sources.	Upper respiratory and eye irritation and possibly allergic responses in some people.	Biological sources are considered separately.

Appendix 6.31: Effects of Various Air Pollutants on Vegetation (Plants/ Trees)

S.No.	Pollutants	Sources of air pollutants	Effects on vegetation
1.	NO_2 (Nitrogen dioxide)	Combustion of fossil fuels (coal, oil & natural gas), automobile exhaust	Presence of 2.5 ppm for one hour or 1 ppm for 24-48 hours or 0.25 ppm for 8 months duration affects middle aged mesophyll cells of citrus, cotton & bean plants, tomatoes, tobacco seedlings, pine and potatoes, suppressed growth observed in many plants, irregular, white or brown collapsed lesion on intercostal tissue and near leaf margin
2.	Ground level O_3 (Ozone)	Photo chemical reaction of hydrocarbon & NO_x form fossil fuel combustion, incineration of refuse and evaporation from petroleum products and organic solvents	Fleck, stipple, bleaching, bleached spotting, pigmentation, growth suppression & early abscission are the effects observed on all ages of tobacco leaves, tomatoes, grapes, beans, pine, pumpkins & potatoes
3.	NH_3 & SO_2 (Ammonia & Sulfur dioxide)	Thermal power plants, petroleum refineries, nitrogenous fertilizer plants	0.3 ppm of SO_2 in air for 8 hours affects middle aged mesophyll cells & young spongy cells of leafy Vegetables (eg. Spinach), apples, pumpkins etc. Bleached spots, bleaching areas between veins, bleached margins, chlorosis, growth suppression, early abscission, yield reduction & tissue collapse also occur.
4.	HF & SiF_4 (Hydrogen Fluoride & Silicon tetrafluoride)	Phosphate rock processing, aluminium industry, ceramic works & fibre glass manufacturing	Presence of 0.001 ppm of HF in air for 5 weeks causes epidermis & mesophyll of grapes, large seed fruits, pines and fluorosis in animals.
5.	Cl_2 (Chlorine)	Leakage in chlorine storage tanks, hydrochloric acid mists	Presence of 0.1 ppm for at least 2 hours affects epidermis 2 mesophyll of plants by bleaching between veins, tip and margin burn and leaf abscission.
6.	HCl (Hydrochloric acid)	Alkali & glass manufacturing units	Browning of tips of gasses at 50 ppm, interveinal bronzing of tomatoes leaves at 5 ppm and necrosis in 3 days in presence of humidity, Leaf burning of broad leaf plants, bleaching & necrosis occurs.

S.No.	Pollutants	Sources of air pollutants	Effects on vegetation
7.	C_2H_4 (Ethylene)	Incomplete combustion of coal, automobile exhaust	Flowers of all ages are affected; chlorisis, epinasty leaf abnormalities, flower dropping, failure of flower to open properly, necrosis & chlorosis of plant will happen; indirect attack on plant hormones and growth regulators resisting in morphogenetic and physiological changes in tissue.
8.	Hg (Mercury)	Combustion of coal & oil, processing of mercury containing ores.	Greenhouse crops and floral parts of all vegetation are affected; growth reduction & abscission occur in most of the plants.
9.	Se (Selenium)	Incineration of wastes and combustion of industrial & residential fuels	Growth reduction occurs mostly in wheat, corn, rice & barley.
10.	Aldehydes (RCHO)	Photochemical reactions	Presence of 250 ppm of aldehydes in air for 2 hours will affect the upper portions of alfalfa to necrosis.
11.	PAN (Peroxy acetyl nitrate)	Photochemical reactions of hydrocarbons & NO_X from fuel combustion incineration of refuse.	Presence of 0.01 ppm of PAN in air for > 6 hours affects the young spongy cells of plants; glazing, silvering or broanzing on lower surfaces of leaves occurs.
12.	Particulates	Thermal power plants, cement plants, Blasting, Crushing and processing industries.	Plants quality affected; Reduces vigor & phardness; interferes with photosynthesis due to plugging of leaf stomata & blocking of light.
13.	Pesticides & herbicides	Agricultural activities	Defoliation, dwarfing, curling, twisting, growth reduction & killing of plants may occur; taste & flavour of plants and cotton are affected if adequate precautions are not taken.

Appendix 6.32: Air Pollution Damage to Trees

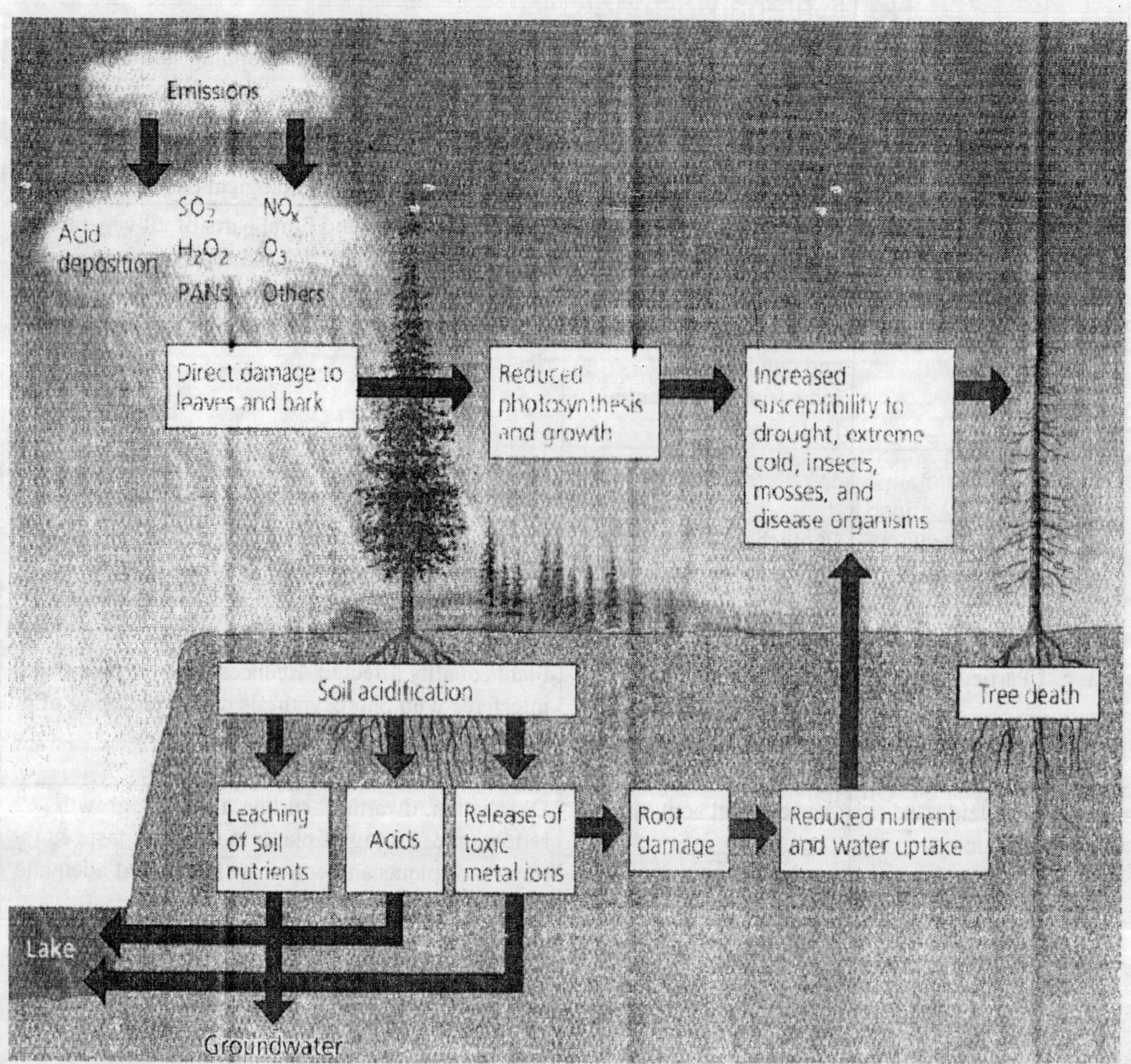

Appendix 6.33: List of Regulated Air Pollutants (Criteria Pollutants and Non-Criteria Pollutants) in USA

S.No.	Criteria pollutants	Non-criteria pollutants
1.	Carbon monoxide (CO)	Asbestos
2.	Nitrogen oxides (NO_x)	Beryllium
3.	Sulfur dioxide (SO_2)	Fluorides
4.	Particulate matter	Hydrogen sulphides
5.	Ozone (O_3) [regulate volatile organic compounds (VOC's)]	Mercury
		Reduced sulphur compounds sulphuric acid mist
6.	Lead	Total reduced sulphur
		Vinyl chloride

NOTES

a) The pollutants which are used to define the quality of air are termed as "criteria pollutants" (in U.S.A).
b) Additional pollutants called "non-criteria pollutants" are regulated under the clean air act. However, "clean air " is defined in terms of only six criteria pollutants listed above.
c) National Ambient Air Quality Standards (NAAQS) for 6 nos criteria pollutants (which is a part of clean air act) along with standard for non- methane hydrocarbon defines the desired or maximum permissible levels of those pollutants in air throughout U.S.A.

Appendix 6.34: Sources, Health and Welfare Effects for Six Criteria Pollutants

S.No.	Pollutant	Description	Sources	Health Effects	Welfare Effects
1.	Carbon Monoxide (CO)	Colourless, odourless gas	Motor vehicle exhaust, indoor sources include kerosene or wood burning stoves.	Headaches, reduced mental alertness heart attacks, cardiovascular diseases, impaired fetal development, death.	Contribute to the formation of smog.

S.No.	Pollutant	Description	Sources	Health Effects	Welfare Effects
2.	Sulphur Dioxide (SO_2)	Colourless gas that dissolves in water vapour to form acid, and interact with other gases and particles in the air.	Coal-fired power plants, petroleum refineries, manufacture of sulphuric acid and smelting of ores containing sulphur.	Eye irritation, wheezing, chest tightness, shortness of breath, lung damages.	Contribution to the formation of acid rain, visibility impairment plant and water damage, aesthetic damage.
3.	Nitrogen Dioxide (NO_2)	Reddish brown, highly reactive gas.	Motor vehicles, electric utilities, and other industrial, commercial, and residential sources that burn fuels.	Susceptibility to respiratory infections, irritation of the lung and respiratory symptoms (e.g., cough, chest pain, difficulty breathing).	Contribute to the formation of smog, acid rain, water quality deterioration, global warming, and visibility impairment.
4.	Ozone (O_3)	Gaseous pollutant when it is formed in the troposphere	Vehicle exhaust and certain other fumes. Formed from other air pollutants in the presence of sunlight	Eye and throat irritation, coughing, respiratory tract problems, asthma, lung damage	Plant and ecosystem damage.
5.	Lead (Pb)	Metallic element	Metal refineries, lead smelters, battery manufactures, iron and steel producer.	Anaemia, high blood pressure, brain and kidney damage, neurological disorders, cancer, lowered IQ.	Affects animals and plants, affects aquatic ecosystems.
6.	Particulate Matter (PM) PM_{10} $PM_{2.5}$ RSPM	Very small particles of soot, dust, or other matter, including tiny droplets of liquids.	Diesel. Engines, Industries, windblown dust, wood stoves.	Eye irritation, asthma, bronchitis, lung damage, cancer, heavy metal poisoning cardiovascular effects.	Visibility impairment atmospheric deposition, aesthetic damage.

Appendix 6.35: Air Pollution Effects on Various Materials (Material Effect and Air Pollutant Relationship)

S.No.	Air pollutants	Materials Affected	Effect on Materials
1.	SO_X (Sulfur Oxides)	• Metals, Building materials, • Paper & leather, • Textiles	Corrosion chemical reaction, embrittlement weakening, disintegration
2.	NO_X (Oxides of nitrogen)	• Building materials, • Dyes	Chemical reaction, fading, discoloration
3.	H_2S (hydrogen sulfide)	• Metals, • Lead base points	Tarnish Darkening
4.	HF (Hydrofluoric acid)	• Metals • Glass	Corrosion etching
5.	O_3 (Ozone)	• Rubber • Dye	Embrittlement Fading, discoloration
6.	CO_2 (Carbon dioxide)	• Building stone	Formation of soluble carbonates
7.	Particulates	• Metals • Building materials • Painted surfaces • Textiles	Abrasion, chemical reaction Soiling Abrasion Chemical reaction

Appendix 6.36: Examples of Common Conditions to which Air pollutants Exposure May Contribute

S.No.	Disease or condition	How air pollution may affect it	Associated factors
1.	Eye irritation	Specific effect of photochemical oxidants, Possibly aldehydes or peroxyacetyl nitrates; particulate matter (fly ash) as a foreign body	Susceptibility differs
2.	Acute bronchitis	Direct irrigative effects of SO_2, soot and Petrochemical pollution	Cigarette smoking may have a more than Additive interaction
3.	Chronic bronchitis	Aggravation (increase in Frequency or Severity) of cough or sputum Associated with any sort of pollution	Cigarette smoking, occupation
4.	Asthma	Aggravation from respiratory irritation, Possibly on reflex basis	Usually pre-existing respiratory allergy or airway hyperactivity
5.	Headache	Carbon monoxide sufficient To lead to more than 10% carboxyhaemoglobin	Smoking may also increase carboxyhaemoglobin but not Enough to lead to headache
6.	Lead toxicity	Add to body burden	Close proximity to lead source; Exposure at home

Appendix 6.37: List of Major known Sinks for Common Air pollutants

S.No.	Air Pollutants	Major known sinks for the removal of contaminants
1.	SO_2 (Sulfur dioxide)	Scavenging, chemical reactions, soil and surface water absorption,
2.	H_2S(Hydrogen Sulfide)	Oxidation to SO_2
3.	N_2O(Nitrous oxide)	Photodissocation in stratosphere, surface water and soil absorption
4.	NO(Nitric Oxide)	Oxidation to NO_2
5.	NO_2(Nitrogen dioxide)	Photochemical reactions, oxidation to nitrate, scavenging
6.	NH_3(Ammonia)	Reaction with SO_2, oxidation to nitrate, scavenging
7.	CO(Carbon monoxide)	Soil absorption, chemical oxidation
8.	O_3(Ozone)	Photochemical reactions, absorption by surfaces (soil and vegetation)
9.	Non reactive Hydrocarbons (HC)	Biological action
10.	Reactive Hydrocarbons (HC)	Photochemical oxidation
11.	CO_2 (Carbon dioxide)	Biogenic
12.	HCl, Cl_2 (Hydrochloric acid, chlorine)	precipitation

Appendix 6.38: Particle Deposition in the Lung – Deposition Mechanism and Sites of Deposition

Particle size	Respiratory rate	Deposition mechanism	Deposition site
Coarse	High	Impaction	Large airways
Coarse	Low	Sedimentation	Large and smaller airway
Fine	High	Diffusion	Large and smaller airways
Fine	Low	Diffusion	Alveoli

Appendix 6.39: Definition of Various Exposure Limits of Toxic Air Pollutants/Chemicals

S.No.	Exposure limits	Description/ definition
1	TWA (Time weighted average) - OSHA	Time weighted average for a normal 8 hours work-day and 40 hours work-week unless otherwise indicated
2	TLV (Threshold limit value) - NIOSH	This threshold limit value is the time weighted average adopted by the American conference of Governmental Industrial Hygienists (ACGIH) for a normal 8 hours work-day and 40 hours work-week.
3	Ceil (ie, ceiling)	This is the concentration of toxic pollutants/chemicals that should not be exceeded ever instantaneously
4	Skin	This indicates the possibility of absorption of toxic pollutants/ chemical is significant amount through the skin, mucous members and eyes
5	STEL (short term exposure limit) of ACGIH	It denotes the maximum concentration of toxic pollutants/chemicals to which workers can be exposed for a period of upto 15 minutes continuously, provided that no more than 4 excursions per day are permitted, with at least 60 minutes between exposure period and provided that the daily TLV/TWA is not exceeds.
6	IDLH (Immediately dangerous to life or health)	It is the concentration of pollutants immediately dangerous to life or health (IDLH) from which a worker could escape without any escape-impairing systems or any irreversible health effects (NIOSH/OSHA standards completion programme)
7	MAC, USSR	It is the maximum allowable concentration (MAC) not to be exceeded.
8	RAL, USSR	It is the temporary safe reference action level (TSRAL)

NOTES

Above exposure limits include the United states Occupational Safety and Health Administration (OSHA) values and when they differ or are the only ones in existence, those of the National Institute of Occupational Safety and Health (NIOSH) and the American Conference of Govermental Industrial Hygienists (ACGIH), as well as the MAC values (Maximum allowable Concentration fixed by ministry of public health in the USSR. These values of explosive limits are subject to periodical adjustment.

Appendix 6.40: Mechanism of Transboundary Air Pollution

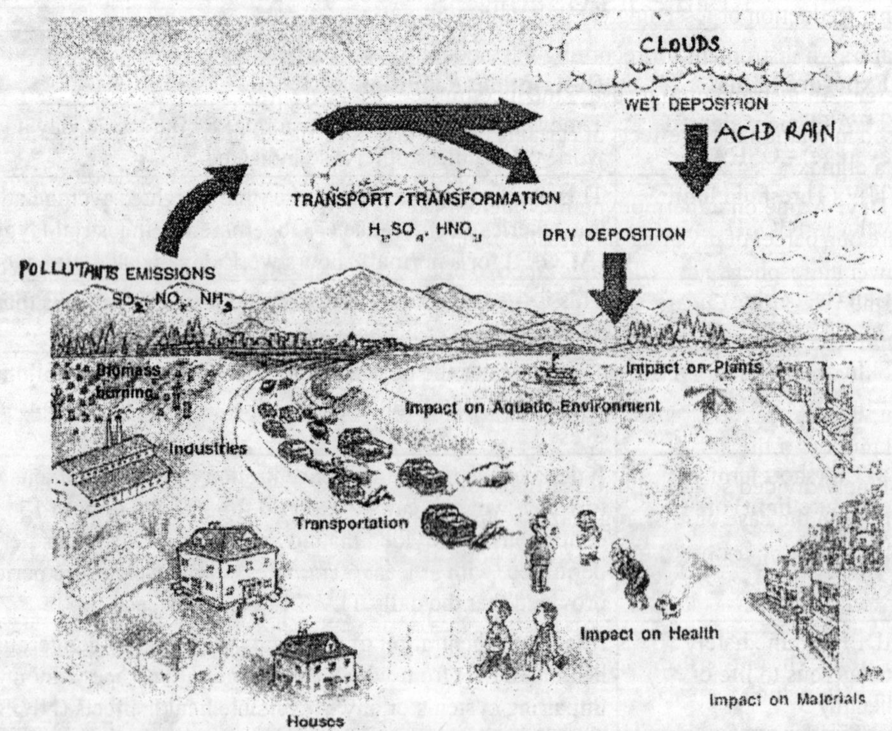

Appendix 6.41: Aerosols Production Rate from Combustion of Various Fuels

S.No.	Sources of aerosol production	Aerosol production rate
1.	Petrol engine exhaust: • During idling • During cruising • During acceleration/decelaration	 0.86 kg/m^3 of petrol 1.09 kg/m^3 of petrol 2.34 kg/m^3 of petrol
2.	Disel bus exhaust: • During cruising	 9.12 kg/m^3 of diesel
3.	Boilers: • Fuel oil fired boiler • Natural gas fired boiler	 2.45 kg/ton of fuel oil 0.18 kg/1000 m^3 of natural gas
4.	Waste incinerator: • Domestic incinerator • Municipal incinerator	 9.75 kg/ton of rubbish 3.85 kg/ton of rubbish

Appendix 6.42: Glossary of Terms Related to Air Pollution

Abatement: Reduction or lessening (of pollution) or doing away with a nuisance.

Actinograph: An instrument for recording the incoming solar radiation.

Actinometer: An instrument to measure the incoming solar radiation.

Acute: Used in relation to the effects of certain concentrations of air pollutants to describe a response coming rapidly to a climax.

Adiabat, Dry: A line on a thermodynamic diagram (for example Tephigram) depicting the variation of the temperature of a parcel of dry air as its pressure changes. Since in atmosphere pressure decreases with height, often in lower atmosphere a dry adiabat is taken to be a line on a height-temperature graph such that the slope is 10 °C/Km.

Adsorption: The phenomenon in which molecules of a substance (the adsorbate) are taken up and held on the surface of another material (the adsorbent).

Advection: The process of transfer of an atmospheric property (for example, heat, moisture, pollution) by horizontal motion of the air.

Aerosol: Finely divided solid or liquid particles of microscopic size suspended and dispersed in a gas or in atmosphere as in the case of mist, fog, haze or smoke.

Afterburner: A device fitted after the main point of fuel combustion in a stationary plant or a vehicle to produce more nearly complete combustion of the exhaust gases. It may be a burner fitted to incinerate polluted process gas.

Agglomeration: The clustering or adhering together of a number of small particles to form a body or structure (an agglomerate) which then acts as a larger single particle.

Aggregate: Similar to agglomerate, but having weaker bonding.

Air Quality: The term usually refers to the concentration in air of one or more pollutants. For many pollutants the air quality is expressed as a concentration over a certain period of time, for example, $\mu g/m^3$ over 24 hours.

Aitken Nuclei: Particles ranging in size from 0.005 µm to 0.1 µm with or without electrical charge.

Albedo: The ratio of the amount of electromagnetic radiation reflected by a body or a surface to the amount incident upon it.

Ambient: It is used to describe physical properties of air (temperature, humidity, pressure, etc) or air pollution concentration in the open air, as against, at the point of emission or indoors.

Anemometer: An instrument for measuring wind/air speed.

Anoxia: Deficiency of oxygen in the blood or thereby in body tissues.

Anticyclone: The condition of atmospheric pressure distribution in which pressure increases towards the centre. This is usually associated with fine, calm weather and with fog in winter.

Antigens: Foreign substances of different chemical types which when introduced into the body, cause the production of specific antibodies capable of reacting with them.

Arrestment: Arrestment of a pollutant from a gas stream is the process of its trapping or removal by an arrester or cleaning device.

Atmosphere: The atmosphere is the envelope of air surrounding the earth.

Atmosphere, Neutral: A neutral atmosphere is one in which lapse rate is dry adiabatic, that is, there is no change in potential temperature with height

Atmosphere, Stable: State of the atmosphere in which vertical air movements are restricted. This happens when there is temperature inversion.

Atmosphere, Unstable: An unstable atmosphere is one in which vertical movements tend to grow thereby substantially increasing turbulence and aiding the dispersion of pollution. Atmosphere is unstable when lapse rate is higher than adiabatic lapse rate.

Atomization: The process by which a solid or liquid is reduced, not to atoms but, to very small particles or droplets as in a fine spray. Good atomization of fuel oil allowing increased fuel/air mixing is necessary to obtain satisfactory combustion. Atomized pollutants may persist remaining suspended in air because of their small particle size.

Background: The levels of pollution (in terms of concentration) present already in the atmosphere exclusive of that due to the pollution source under consideration. In general the term is used to imply the level of any entity present exclusive of the specified source for example, background noise, background illumination.

Sometimes the term is used to mean the concentration of the substance some distance from the particular source and therefore largely uninfluenced by it.

Baghouse/Bag Filter: Fabric filter used for dust removal usually made into bag of tubular shape. The entire structure housing the bags is called a baghouse.

Balloon, Pilot: Balloon filled with a light gas (for example hydrogen or helium) released to determine the wind at various heights in the atmosphere. The balloon is tracked by theodolites and successive locations of the balloon are computed, from which winds can be estimated. This is done on routine basis at several stations by national weather services. Radar is also used for this purpose.

Beaufort Scale: A scale for estimating and reporting wind force ranging from zero for calm to 12 for a hurricane. The latter corresponds to a wind speed range of 133 to 147 km/h. Wind speeds above this and up to 218 km/h are also classed under hurricane having Beaufort scale up to 17 depending upon wind speed.

Boundary Layer - Where a fluid flows past a physical boundary, the boundary layer is that part of the fluid extending from the boundary into the bulk of the fluid in which the motion of the fluid is significantly affected by the frictional drag at the boundary.

The earth's atmospheric boundary layer is typically the lowest 1000 m or so of the atmosphere. This is also called planetary boundary layer or friction layer.

Breathing Zone - That location in the atmosphere at which persons breathe. Generally, a layer 0–1.5 m above ground level is considered as breathing zone.

Breeze, Sea - A local wind blowing on shore when (and because) the land surface is warmer than the sea surface, for example, on a sunny day. The situation may be reversed at night, producing an off-shore land breeze.

Bronchitis: Basically, bronchitis is inflammation of bronchi, the two branches of the air pipe leading to the lungs. It is a reaction of the mucosal lining to an irritant. The word is less exactly, but more widely used to describe a variety of lung disablements which can be associated with air pollution and smoking.

Bubbler: An apparatus charged with water or a reagent selected to absorb specific pollutants when a sample of air is made to bubble through it.

Buoyancy: The tendency of material immersed in a fluid to float up. In air pollution it is frequently applied to the tendency or otherwise of the hot gases emitted from a chimney to rise in view of their lower density relative to that of the surrounding air. In the atmosphere it implies the upward force experienced by parcels of air which are heated and less dense as compared with the surrounding air.

Calm: The air is described as calm in meteorological terms if the wind speed is less than 1 knot. (1 knot = 1.8 km per hour = 0.5 m per second approximately.)

Sometimes the term is also used when the wind speed is below the measurement threshold of the anemometer.

Cancer: A malignant growth, a tumour.

Carburation: The mixing of air with a volatile fuel to provide a combustible mixture for use in the internal combustion engine. The device in which this occurs is the carburator.

Carcinogen (Carcinogenic Substance): A substance which causes carcinoma or cancer.

Carcinoma: Malignant tumor of epithelium, regardless of its derivation.

Channelling Effect: Change of normal flow of air when it passes through peculiar constructed topographic features such as valleys or streets and lawns flanked by buildings. Often channelling leads to increase in local wind velocity. Channelling is sometimes also called canyon effect.

Chimney (Stack) Height, Effective: The effective height of a chimney is the sum of its actual physical height and the plume rise. The latter is the rise of the effluent due to buoyaucy and efflux velocity.

Chronic: Referring to injurious effects of air pollution, particularly effects on health. It implies lasting or long term morbid effects as against acute effects.

COH: Coefficient of haze. This is a unit for measuring smoke stains which depends on the reduction of light transmission through the filter after it has collected the smoke sample. Its object is to measure haze over a distance of about 300 metres.

Collection Efficiency (of a Device): Ratio of the quantity of a substance retained by a device to that passed through it. In any air pollution control equipment, if M_1 is the mass of incoming dust, M_e the mass of collected dust and M_o the mass of outgoing dust, then collection efficiency is:

$$\frac{M_c}{M_1} = \frac{M_1 - M_o}{M_1} = 1 - \frac{M_o}{M_1}$$

Collector: A device for removing and retaining contaminants from air or other gases. Usually, this term is applied to cleaning devices in exhaust systems.

Combustion, Catalytic: Combustion in the presence of a catalyst, to increase the rate of combustion, or affect combustion at a lower temperature.

Concentration, Ground Level: The concentration (amount per unit volume) of a pollutant in air near ground. For physiological purposes this may refer to between 1 to 2 m above the ground, that is, at breathing level. For computation, however, zero height is considered. In most practical cases, there is no difference between the two.

Concentration, Mass: The amount of a substance in unit volume of the medium, for example air, flue gas. Generally ambient air concentrations are expressed in mg/m^3 or µg/m^3. In case of stack emissions, normal m^3 (Nm^3) is used instead of m^3.

Concentration, Odour: The number of volume units which a unit volume of sample will occupy when diluted to the odour threshold.

Concentration, Particle: Concentration expressed in terms of number of particles per unit volume of the medium, for example, air, flue gas. Also called 'number concentration'.

Concentration, Time Integrated (TIC): The concentration at a point integrated with respect to time over a given period. If concentration is plotted against time, TIC is the area under this curve. It has units of µg s/m^3 (or Ci s/m^3 for radioactive pollutants). TIC divided by the time period over which integration is done is therefore equal to average concentration over that period.

Coning: A type of plume behaviour in which both vertical and lateral diffusion occurs. The plume appears like a cone. This takes place in a near neutral atmosphere when the wind velocity is moderately high or high.

Contaminant/Contamination: Synonyms for indicating the presence of a foreign component or pollutant in air, gas, liquid or on surfaces.

Control, Air Pollution: The process of effecting reduction in the concentrations and extent of contaminants in atmosphere within permissible limits, by whatever means, such that they do not become injurious to human, plant or animal life, or property.

Cyclone: a. A mechanical device for the removal of particles from gas streams by centrifugal force imparted by a whirling motion of the gas inside the device. b. A natural wind system rotating around a centre of low atmospheric pressure. A cyclonic system (in contrast to anticyclonic system) has a low pressure centre around which strong winds flow anticlockwise in northern hemisphere and clockwise in southern hemisphere

Decay: In air pollution, the decay of an airborne pollutant usually implies the rate at which it is lost from the atmosphere, for example, by absorption or precipitation, including its loss by transformation into some other chemical. For radioactive pollutants it implies reduction by radioactive transformation.

Deposit Gauge: An instrument designed to assess the rate of deposition of material from the air. In most instances, the gauge receives material through a horizontal opening and the results can be expressed in terms of amount deposited on a given area in a given time.

Deposition Velocity: A parameter describing the rate at which a pollutant from air deposits on a surface. It is the ratio of the rate of deposition per unit area of the surface to the concentration near the surface. Though having the dimensions of velocity, it is only a parameter and not a real velocity.

Desorption: The removal of absorbed or adsorbed material.

Desulphurisation: Literally the removal of sulphur. To reduce sulphur dioxide emissions it is possible to use 'fuel desulphurisation' so as to reduce the sulphur present in the fuel when it is burnt or 'flue gas desulphurisation' in which the sulphur dioxde is removed from the gases formed by combustion.

Dew Point: The temperature to which the sample of air has to be cooled to make it saturated with respect to water at its existing pressure and humidity mixing ratio. It is a measure of moisture content of air. The higher the moisture content, the higher is the dew point.

Diameter, Mass Median: The mass of a sample of particulate' matter in terms of the diameter above and below which the masses of all large and smaller particles are equal.

Diffusion: a. On the scale of molecules, diffusion is a mixing of substances caused by molecular motion (Molecular Diffusion). b. In atmosphere it implies the spread of pollutants by turbulence or eddy motion (also called Atmospheric Diffusion).

Dispersion: The method by which a pollutant spreads from its point of emission and becomes diluted in the atmosphere. It includes transport by winds and simultaneous spread by turbulence (atmospheric diffusion).

Disposal: Getting rid of, by suitable process and treatment, for example, disposal of the gases to atmosphere and disposal of gaseous wastes after treatment for desulphurisation.

Diurnal: An adjective describing the variation of some parameter during a typical day or over 24 hours.

Dose: The amount taken in or received. Time integrated exposure to pollutants, for example, dose of exposure for 1 h to an average sulphur dioxide concentration of 1 000 μg/m^3. In radiology, it refers to the amount of energy absorbed by a unit mass of body tissue from radiation such as *gamma* or X-rays.

Down Wash (Down Draught): A downward movement of smoke or other emissions on the leeward side of stack (chimney) or other obstacles such as buildings, hills, etc, due to aerodynamic characteristics of flow past obstacles (bluff bodies). The downward motion of air itself may be called downdraught whereas downwash applies to the pollutant. Downwash causes reduction in the effective height of emission.

Driving Cycle: In relation to the testing of motor vehicles, particularly to estimate their pollution emission characteristics, they may be made to perform standard driving cycles which are fixed patterns of steady speeds, accelerations and decelerations typical to some particular form of driving, for example, urban use.

Droplet: A small liquid particle of such size and density that it falls under still conditions but may remain suspended under turbulent conditions.

Dust: Small particles in the size range 1 to 200 micro metre(um) in diameter. Particles above 75 μm size would come into the category of grit. Below about l μm the particles, if air borne, would tend to remain in suspension rather than settle due to gravity.

Dust Fall: The rate at which dust is deposited from the air on a horizontal surface.

Dust Loading: Mass of dust per normal unit volume in a gas or air stream. When applied to equipment, it is expressed as mass per unit time.

Ecology: The relation or interaction of living systems with each other and their environment.

The term is also used to denote the branch of science dealing with the above.

Eddy: A fluctuation from the mean motion of a fluid; also used for parcels of air having circular whirlpool type of motion in the atmosphere. Eddies are the result of turbulent motion of the atmosphere.

Effect, Cumulative: Increasing or building up of the effect with successive or continuing exposures.

Efficiency, Fractional: The emission collection efficiency of a specific size fraction of a contaminant. This term is commonly applied to the performance of air cleaning equipment intended for particulate matter in various sizes and ranges.

Effluent: Literally anything which flows out or is discharged. Usually applied to the discharge of a waste material into a water body or the atmosphere.

Efflux Velocity: The speed with which' something his discharged from an orifice, for example gases from the top of a chimney. Also called discharge velocity.

Ejector: A device which uses a fluid under pressure, such as steam, air or water to move another fluid by developing suction. Suction is developed by discharging the fluid under pressure through a venturi.

Electrostatic Precipitator: A device for removing particles from a gas by making them electrically charged and then making them to adhere to a plate or plates by application of electrostatic field.

Elutriation: A process for separating tie particles into sized fractions by their different rate of settling down under gravity in a stream of fluid.

Emission: Discharge of effluent into the atmosphere. The term is often used to express an emission rate, for example mass per unit time or an emission concentration, in terms of volume or mass of material per unit volume of total emitted gas.

Emission Standard: The amount of pollutants permitted to be discharged from a pollution source. Emission standards are commonly described in one or more of the following ways: a. Mass of pollutants over a certain time, for example kg/h or tonnes/day; b. Mass of pollutants per unit mass of material processed, for example, 2.5 kg per tonne (approximately 0.25 percent); c. Volume of pollutants (if gaseous) per unit volume of; discharged gas, for example, parts per million.

Emphysema: Pulmonary or lung emphysema is a distension or swelling of the lung air spaces due to destruction of the alveoli which diminishes the area for exchanging oxygen and carbon dioxide between the air and the blood.

Engine, Internal Combustion: An engine in which- a fuel/air mixture is burnt within a chamber and the resulting .pressure of hot gas is converted to mechanical energy. In contrast, external combustion as in a boiler furnace involves the transmission of energy by a separate agent, for example, steam cycle.

Environment: Literally surrounding; may apply to the indoor or-working conditions but commonly refers to man's total surroundings, natural and manmade. Parts of the environment are often specified. For example biological environment, water environment, air environment, marine environment, etc.

Epidemiology: The scientific study of the frequency and distribution of a disease within a population and often extended to include the relationship of the disease with external factors such as air pollution.

Equivalent Diameter: In relation to airborne particles, it is used to describe the way in which a particle of generally irregular shape behaves as if it were a sphere of a certain diameter (arid often of a standard density).

The settling rate equivalent diameter of a dust particle is defined as a diameter of a sphere with unit density which has the same settling rate as the particle in question in air at 20 °C and 1.103 bar (760 mm mercury).

Exhaust: Vent or aperture from which gases are discharged out.

Exposure: A measure or description of the extent to which an object or organism has been subjected or exposed to a particular pollutant.

Extrapolation: Estimating the value of something beyond the range in which it has been measured, for example, estimating pollution levels in the near future from a number of values over past years may be called extrapolation of pollution levels in time. Estimation of winds at higher levels from those measured at two or more levels below may be called extrapolation in height.

Fanning: A type of plume behaviour under extreme inversion conditions where mostly lateral diffusion takes place with very little vertical mixing.

Fallout: A radioactive pollutant in the air caused after the explosion of a nuclear device; its degree of contamination depending on several factors, such as distance, wind and power of the device.

Sometimes also used to mean heavy grit, dust and ash depositing near the source (chimney).

Fluidised Bed: A term to describe a bed of solid particles through which air or any other gas is made to flow upwards in such a fashion as to support the particles and to make them act, in many ways, like a liquid. Furnaces have been devised in which oil or powdered or granulated coal can be introduced into such a (heated) bed where it can burn evenly and produce an even temperature throughout the bed.

Fluorine (F)/Fluorides: In air pollution the term fluorine is often used to cover both the element of that name and its compounds, mostly fluorides. Fluorine and its soluble compounds can be toxic. They are pollutants particularly associated with brick and aluminium production.

Fly Ash: The finely divided particles of ash readily entrained in flue gases arising from the combustion of fossil fuels (mainly coal). The particles of ash may contain partially burnt fuel.

Fog: Atmospheric obscuration caused by liquid water droplets. In meteorological terms fog implies reduction of visibility to less than 1 km.

Fog, Radiation: With a clear sky and light wind there can be substantial cooling of the air during the evening and night due to earth losing heat by radiation. Where the air temperature falls below the dew point, a fog can be formed. Fog formed by this process is called radiation fog.

Fossil Fuels: A general term covering coal, mineral oil, natural gas, etc, which are fuels dervied from organic deposits in ancient geological periods. They consist mainly of carbon and hydrocarbons. By comparison wood and nuclear fuel are not fossil fuels.

Freeze-Out Trap: A device to retain specific pollutants by cooling to low temperatures.

Freons: A range of hydrocarbons with hydrogen replaced partly or totally by fluorine or chlorine. These wholly manmade compounds are used as refrigerants, in fire extinguishers and as aerosol propellants.

Frequency Distribution: A graphical or mathematical method of showing the frequency of occurrence of values of some parameter against the value of that parameter.

Front, Cold: The sharp boundary between two extensive air masses where cold air is replacing warm air.

Front, Warm: The sharp boundary between two extensive air masses where warm air is replacing the cold air.

Fume: Very fine solid or liquid particles, 0.001 to 1 μm in diameter, arising from chemical reactions or condensation of vapour (often metal vapour).

The word fumes is commonly used to imply unpleasant and smelling airborne effluents.

Fumigation: When atmospheric conditions are such that there is an unstable layer of atmosphere above ground, topped by a stable inversion layer, the pollutants in the unstable layer often experience a thorough vertical mixing raising the ground level concentration. This is called fumigation. It may occur early in the morning when night time inversion breaks up after sunrise or may exist near the coast when sea breeze brings cold air over the warm air on land.

Green House Effect: Retention of heat by the earth and the atmosphere owing to the carbon dioxide in air being 'transparent' to incoming shortwave solar radiation but 'opaque' to long wave radiation back from the earth. This term comes from analogy of green houses where glass and water vapour inside cause shortwave sun's radiation to enter but does not allow long wave radiation to pass out.

Grit: Particles of solid ranging from 75 to 200 μm in diameter in the air or flue gases which fall out or deposit under their own weight.

Guideline, Air. Quality: Air quality guideline indicates the level and duration of air pollution by a pollutant above which effects on materials, vegetation or health are produced.

Haze: Fine aerosol dispersed through a portion of the atmosphere. The particles are so small that they cannot be felt or individually seen with the naked eye, but they diminish the horizontal visibility and give the atmosphere a characteristic opalascent appearance which subdues all other colours.

Blue haze is found due to the combustion of sulphur or its compounds or some other organic compounds.

Homeostasis: The maintenance of internal constancy and independence of the environment within a living substance. Thus, despite changes in ambient temperature, body maintains a nearly constant temperature. Similarly, in spite of deteriorated air quality, the body maintains its functions through proper compensation. There is, however, a limit beyond which the compensation is not possible leading to permanent damage.

Humidity/Relative Humidity - The water vapour content of atmosphere is its humidity. Relative humidity is the ratio (expressed as percentage) of the quantity of water vapour in the atmosphere (per unit volume) to that of the quantity at that temperature if it was saturated with water vapour.

Immission: Strictly this means intake by a receptor of pollution. Since this is often related to the concentration of the pollutant in the air there has been a tendency to speak of the immission concentration or even immission as the same as ground level concentration.

Impaction: A forcible contact of particles of matter on any surface. This term is often used synonymously with 'Impingement'. The impacting particles adhere to the surface; a property used in air sampling.

Impactor: A sampling device which employs the principle of impaction (impingement). 'Cascade Impactor' is a specific instrument which employs several impactions in series to collect successively smaller sizes of particles.

Impingement: The act of bringing matter forcibly in contact with any surface. In air sampling, it refers to a process for the collection of particulate matter in which the gas being sampled is directed forcibly against a surface.

Impingement, Dry: The process of impingement so carried out that particulate matter carried in the gas Stream is retained upon the surface against which the stream is directed. The collecting surface may be treated with a film of adhesive.

Impingement, Wet - The process of impingement carried out within a body of liquid, the latter serving to retain the particulate matter.

Impinger: A sampling instrument employing impingement for the collection of particulate matter and gases.

Incineration: A waste disposal process by means of which solid, liquid and gaseous combustible wastes are converted, through controlled combustion, to a residue, which contains virtually no combustible matter, and to gases.

Ingestion: The taking is of a substance to the stomach.

Insolation: Incoming solar radiation falling on a unit horizontal surface (of the earth). This may consist of the radiation directly received from the sun plus that received after scattering by the atmosphere and clouds.

Interference: Effects on the determination of a pollutant, due to the presence of other pollutants or impurities.

Interpolate: To estimate the value of an entity at points between actual, measured or known values, for example estimating pollution levels at points between sites at which pollution measurements are actually made. Interpolations can also fee done in time, that is, in between two instants.

Inversion: In meteorology an inversion exists when the air temperature increases with height. Under such conditions, vertical movements are damped and vertical spread of the pollutants is attenuated. Inversion may occur in isolated layers also. During inversion, atmosphere is said to be stable.

Isokinetic: Isokinetic sampling implies sampling In such a way that the velocity of the gas entering the sampler is the same as that of the gas being sampled. The sample can thus be expected to have the same particulate content as the gas being sampled.

Isopleth: A line drawn on a map linking places with the same value of some parameter, for example, height, annual rainfall, pollutant concentration.

Isothermal: Literally, at constant temperature. An isothermal layer in the atmosphere is one in which there is no temperature variation with height within the layer. An isothermal process is that in which temperature does not change.

Knock: The situation in a spark ignition internal combustion engine where part of the air/fuel mixture detonates on compression before the flame front has reached it. The tendency of a fuel to knock in a petrol engine can be reduced by the blending of suitable compounds in the fuel or by antiknock additives.

Lachrymator: A substance which, on contact with eyes, causes irritation and flow of tears.

Lapse Rate: The decrease of an atmospheric variable with height, the variable being temperature unless otherwise specified.

Lapse Rate, Dry, Adiabatic: Rate of decrease of temperature at the rate of 10 °C/km. It is the rate at which parcel of unsaturated air will cool if lifted adiabatically. Also called neutral lapse rate.

Lapse Rate, Super Adiabatic: Atmospheric lapse rate in which the decrease in air temperature with elevation is greater than the normal adiabatic lapse rate. Under this condition, vertical motions are accelerated, and the atmospere is said to be unstable.

Layer, Mixing: A layer bounded between two surfaces in which pollutants can disperse and mix well in the vertical direction. Usually, the lower surface is ground and the upper surface may be the lower boundary of a layer having temperature inversion.

Lead Dioxide (Peroxide) Candle: A device in which sulphur dioxide in the air reacts with a film of lead dioxide to produce lead sulphate which is subsequently measured. The device thus gives an indication of sulphur dioxide in the air. Its performance is influenced by condition such as wind and humidity.

Lean Mixture: The air/fuel mixture in an internal combustion engine is described as lean if it contains a low proportion of fuel while a mixture with a high proportion of fuel is described as rich.

Level, Adverse: Referring to level of concentration of a pollutant in the atmosphere at which remedial action is necessary.

Lidar: Laser radar used to study the location and concentration of pollutants, both particulate and gaseous by the reflection or absorption of a laser beam.

Lofting: A type of plume behaviour characterised by vertical mixing above a specific level, caused by an unstable condition above and a stable (inversion) condition below the level.

Looping: A type of plume which has a wavy character. It occurs in a highly unstable atmosphere because of rapid mixing.

Lung Function: Literally the function of a lung, that is, the transfer of oxygen from the air to the blood and the disposal to the air of carbon dioxide from the blood. More specifically it is a measure of efficiency of lung performance.

MAC: Maximum Allowable Concentration. A term similar to the TLV.

Mean, Arithmetic: Sum of the values divided by the number of observations.

Microclimate: The distinctive pattern of meteorological variables within the lowest layers of atmosphere or (sometimes) within a limited area, for example, a town.

Micrometeorology: Study of physical phenomena in the very lowest layers of the atmosphere (of the order of 100 metres) or occasionally, over a restricted area, generally of a few kilometres in extent.

Mist: A suspension or dispersion of liquid droplets in a gas. Usually applied to water droplets in air where visibility has a value of 1 km or more. Natural mists may be aggravated by air pollutants.

Mist Eliminator: A device used for removing mist. It generally consists of wire mesh for mist removal.

Mixing Height: The height in the atmosphere within which emitted pollutants are dispersed and mixed vertically, the vertical mixing being attenuated by presence of inversion layers above that height.

Model, Mathematical: An air pollution mathematical model is a numerical simulation of the emission, dispersion and chemical and other processes controlling ambient pollutant concentrations.

Model, Physical: A physical simulation, like in a wind tunnel, of the emission and dispersion processes controlling ambient pollutant concentrations in a model, used as an aid to understanding the process.

Monitoring: Generally this is used interchangeably with surveillance to mean measuring concentration levels continuously or regularly over a period of time. More specifically it may imply checking that the level of an air pollutant is within prescribed limits. Stack monitoring implies the monitoring of the concentrations in the stack effluents from which total emission rates can be estimated.

Necrosis: Localized death of a cell or ceils in the living body tissues.

Nephelometer: An instrument for measuring the amount of light scattered or absorbed by a suspension of particles (aerosols). It is used to determine suspended particle concentrations or particle size.

NOx (Nitrogen Oxides): Refers to all the oxides of nitrogen, but usually to nitric oxide (NO) and nitrogen dioxide (NO_2) in matters of air pollution. Nitrous oxide (N_2O) the anaesthetic laughing gas is a third reasonably stable oxide of nitrogen.

Noxious: Harmful.

Obscuration: Concealing from sight, In meteorology it relates to restriction of visibility.

Octane Rating: An indicator of the tendency of a motor gasoline to knock in a spark ignition internal combustion engine, higher octane ratings being required for engines of higher compression ratios. The octane number is the percentage by volume of the hydrocarbon iso-octane in a mixture with normal heptane which has the same knocking characteristics as the fuel under test. The higher the octane number, the lower the tendency to knock.

Opacity Rating: A measurement of the opacity of emissions, defined as the apparent obscuration of an observer's vision to a degree equal to the apparent obscuration of smoke of a given rating on the Ringelmann Chart.

Organoleptic: Affecting an organ. Thus organoleptic methods of estimating odour depend upon its perception by the human nose.

Oxidants: Material which have an oxidising action on other materials, that is, generally adding oxygen to them chemically or extracting hydrogen.

Typical atmospheric oxidants are ozone and nitrogen dioxide whose action on vegetable material can be damaging.

PAN: Peroxyacetyl nitrate. This is an eye irritant compound associated with photochemical smog.

Particle: A small discrete mass of solid or liquid matter.

Particulates: Particulate matter in the air usually refers to small solid particles of material found in the atmosphere in addition to gases. Less frequently, it is applied to small liquid aerosol particles.

Peak, Instantaneous: It refers to a short term high value, for example of a pollutant concentration.

Photochemical Smog: The result of reactions on the atmosphere between nitrogen oxides, organic compounds and oxidants under the influence of sunlight, leading to the formation of oxidizing compounds or possibly causing poor visibility, eye irritation or damage to material and vegetation if sufficiently concentrated.

Phytotoxic: Poisonous to plants.

Plume: The flow or path of effluent from a specific outlet such as a chimney stack or vent. While for smoke, the plume is visible, for many pollutants it may not be so.

Pneumoconiosis (or Pneumonoconiosis): A term applied to various diseases of the lung caused by the inhalation of dust particles in such occupations as coal mining, quarrying and asbestos working; an excess of fibrous tissue forms in the lungs round the dust particles.

Pollen: The small grains from the anthers of flowers which contain the male reproductive nuclei. Transported in the air certain of these grains in the air can induce hay fever or asthma in some people by action in the nose. They are thus natural air pollutants.

Pollutant (Primary & Secondary): A pollutant is something which pollutes the air. A primary air pollutant is one which is put into the air in that form. A secondary pollutant is not emitted as such, but forms in the air from primary pollutants.

Pollution, Air: The presence in ambient atmospheres of substances, generally resulting from the activity of man, in sufficient concentration, present for a sufficient time and under circumstances such as to interfere with comfort, health or welfare of persons or with reasonable use or enjoyment of property.

ppb: A billion in European usage is a million times a million (10^{12}) so ppb is parts per billion that is one part in 10^{12}.

ppm: Parts per million. In pollution of the air by a gaseous pollutant, it refers to the number of parts by volume of the gaseous pollutant in one million parts of the total air.

The quantity ppm is related to µg/m³ by the relation.

$$\text{Concentration in } \mu g/m^3 = 44.64\ M\left(\frac{273}{273+T}\right)\left(\frac{P_2}{P_1}\right)(\text{ppm})$$

where

M = molecular weight of the pollutant,
T = temperature in degrees centigrade,
P_1 = Pressure at NTP, and
Ps = Air pressure

It is recommended that instead of ppb or ppm, µg/m³ should be used.

Precipitation

In meteorological parlance, moisture falling on the earth's surface as rain, hail or snow.

Precipitation, Thermal: A process consisting of the separation of particulate matter from air and other gases under the influence of a sufficiently steep thermal gradient In the thermal precipitator (a sampling instrument), the air or gas is drawn slowly through a narrow chamber across which (»tends a heated wire, the particulate matter being deposited upon the adjacent collecting surface.

Precursor: Something which goes before. Commonly applied to substances in the air which subsequently react or are transformed to produce new air pollutant, for example hydrocarbons are precursors of ozone and other compounds associated with photochemical smog.

Probe: A tube used for sampling or for measuring pressures at a distance from the actual collection or measuring apparatus. It is commonly used for reaching inside stacks and ducts.

Pulmonary: Related to lungs

Radiosonde: A balloon borne instrument which transmits data on temperature, pressure and humidity as it rises and moves with the wind.

Rain, Orographic: Rain caused by the raising and cooling of moist air as it passes over mountains.

Rainout: The mechanism whereby small particles in the air are removed by raindrops as they form in the cloud.

Recycling: Recovery and reuse of the material generally to make the same sort of products. In some cases reuse for another purpose is included in the term recycling in that it assists material conservation. Exhaust gases from motor vehicles may be recycled in the air input to reduce the overall production of air pollutants.

Reflectance: The reflectance of a surface is a measure of the extent to which it reflects light or other radiation of specific wavelength. When the radiation is not monochromatic, the term Albedo is used.

Remote Sensing: Detection or measurement of a property of a medium (like atmosphere) such as temperature, concentration of pollutants, etc from a distance. For example, using Radar, rain can be detected or using Laser beam, concentration of pollutants can be measured along the path of the beam by analysing the scattered radiation from the beam. This is in contrast to *in situ* sampling or measurement where the sensor measures the property in the immediate vicinity of the sensor.

Residence Time: Average time for which any pollutant or particulate matter remains in the atmosphere before settling on earth's surface. It is equal to the time requited for the removal of half the quantity of matter divided by 0.693 or the time required for the contents to reach 1/e or 36.8 percent of original value, where e is the base of $\log_e$.

Respirator: A device used by a person for breathing filtered air, or artificial air from a cylinder while working in an area having air contamination.

Retention Factor: Fraction of a contaminant breathed that is retained in the body of the person.

Ringelmann Chart (Scale): A chart for assessing the darkness of a plume of smoke by visual comparison with a numbered set or grade differing from each other in the width and spacing of black lines printed on a white background. Thus Ringelmann 1 is equivalent to 20 percent black. The percentage of blackness represented by Ringelmann numbers is as follows:

Ringelmann number	Percent of Blackness
1	20
2	40
3	60
4	80
5	100

'Dark smoke' is smoke which is as dark or darker than Ringelmann 2 and 'Black smoke' is as dark or darker than Ringelmann 4.

Sample, Cumulative: A sample obtained over a period of time, the collected atmosphere being retained in a single vessel, or with a separated component accumulating into a single whole such as dust sampling in which all the dust separated from air is accumulated in one mass of fluid; acid gas is absorbed in an alkaline solution; and air is collected in a plastic bag or gasometer. Such a sample does not reflect variations in concentration during the period of sampling.

Sample, Grab: A sample taken or grabbed at a single point in a short time rather than accumulated over a period. This is representative of only the conditions at the particular time of sampling. This is also known as *Snap sample.*

Sampling: Taking or withdrawing a fraction of a substance usually for study as representative of the whole.

Sampling, Continuous: Sampling without interruptions throughout an operation or for a predetermined time, that is, uninterrupted sampling of air, usually at a fixed rate. Where the sample can be analysed continuously, the stream of gas may be passed through a measuring instrument continuously. Otherwise, the sample is collected in an uninterrupted fashion for a given period and the entire sample is finally analysed to give the mean composition of the air over whole period.

Sampling, Cumulative: A process or system in which the sample is accumulated either by sampling continuously or for finite periods at regular intervals to give a single sample whose composition is regarded as representative of the whole period of its accumulation.

Sampling, ***In-situ*****:** Sampling of a property of a medium (for example atmosphere) by using sensors or detectors at the actual place where sampling is desired.

Sampling, Instantaneous: Obtaining a sample of an atmosphere in a very short period of time, such that this sampling time is insignificant in comparison with the duration of the operation or the period being studied. It implies sampling at one point in time. (Instantaneous sampling is possible with electrical and optical devices.)

Sampling, Intermittent: Sampling successively for limited periods of time throughout an operation or for a predetermined period of time. The durations of sampling periods and of the intervals in between are not necessarily regular and are not specified. In contrast to continuous sampling, intermittent sampling means that sampling occurs at intervals for only a fraction of the total time.

Sampling, Random: Random sampling is a sampling scheme without a predetermined pattern. Proper random sampling involves the express selection of sampling units or times so that there is no bias towards selection of particular units or times and the sample is likely, as far as possible, to be representative of the whole.

Sampling, Running: Withdrawal of a portion of the atmosphere over a period of time with continuous analysis or with separation of the desired material continuously and in a linear form such as withdrawal of the atmosphere accompanied by absorption of a component in a flowing stream of absorbent or by filtration in a moving strip of paper. Such a sample may be obtained with a considerable concentration of the contaminant but it still indicates fluctuations in concentration which occur during the period of sampling.

Sampling, Source: Sampling of air containing contaminants at the place of origin of the contaminant.

Sampling Devices, Inertial: Devices, such as depository trays, dust fall gauges and collectors, to determine the total amount of material settling in them in a given time.

Scrubbing/Scrubber: An absorption operation in which gaseous or fine particulate pollutants are removed from a stream of air or gas by contact with a liquid spray or a bath or wetted packing/wet surfaces in a tower. The apparatus/equipment used is known as scrubber.

Settling Velocity, Terminal: The maximum velocity reached by a falling particle when the gravitational force on it (its weight) is balanced by the viscous drag of the fluid through which it is falling.

Size Distribution, Particle: A method of expressing the make up of a population of particles (in any dust, aerosols, etc) in terms of either the mass or number of particles between certain size limits, for example equivalent diameter of 0 to 1 μm, 1 to 5 μm, 5 to 10 μm.

Smelter: A place where fusion of an ore or concentrate is carried out with suitable fluxes, to produce a melt consisting of two layers; on top a slag of the flux, and below, molten impure matter.

Smog: Term derived from the terms 'smoke' and 'fog'. Extensive atmospheric pollution by aerosols, arising partly through natural processes and partly from human activities.

Smoke: a. Small gas borne liquid and/or solid particles formed by incomplete combustion of fuel and in the size range 0 001 μm to 10 μm. b.In some cases soot, ash, grit and gritty particles emitted in smoke are also included. c.Suspended matter collected on filter paper.

Smoke Scale, Bacharach: A scale of 10 shades from white to black for the assessment of smoke in flue gases. A smoke stain obtained by sampling flue gases through a filter paper in a prescribed manner is graded according to the scale.

Smuts: Small aggregates of soot, unburnt solids or fly ash which may be emitted from chimneys and then be deposited in the vicinity. In certain circumstances, the production of sulphur trioxide during combustion may give rise to add smuts which are corrosive.

Soot: Aggregates of carbonaceous particles impregnated with tar formed by the incomplete combustion of carbonaceous material, particularly bituminous coal. They tend to adhere to the inside of the chimney, etc.

Sorbent: A liquid or solid medium in or upon which materials are retained by absorption or adsorption.

Sorption: A process consisting of either absorption or adsorption.

Source (of Pollution): A point or place from which effluents are discharged into air.

Spectrochemical Method: The method using optical absorption or emission properties of a substance. The light absorbed by or emitted from a substance when made to do so is measured using a spectrometer. The wavelength and quality of light absorbed or emitted is used to detect the substance and measure its quality in a sample.

Spectrometry, NDIR: 'Non Dispersive Infra Red' spectrometry. While in conventional spectrometry, a light beam is first dispersed using a prism or grating and a (effectively) monochromatic beam is selected, by changing a suitable angle, in the NDIR spectrometry a beam in infrared region of the spectrum is made monochromatic using an interference filter. Absorption of this beam by a sample gas is measured to determine the quality of gas in the sample.

Spectrophotometry: The measurement of the concentration of a substance in solution by the amount of light emitted or absorbed of a colour (wavelength) characteristic of the substance being measured.

Stable (Compound): A stable compound is one which is not easily decomposed or does not react readily.

Stack Solids: The solid content of the gas stream going up a stack or chimney.

Stain, Smoke: This term is applied to the darkness of a filter through which air has been sampled. The filter retains smoke particles and the stain can be expressed as a concentration of equivalent standard smoke.

Standard Level: In respect of air quality it may imply a commonly agreed level above which effects of some type or other are believed to occur. It also represents an air quality or emission level which should not be exceeded at all or for specified periods. It may also refer to the maximum permitted level of air contaminant as established by a legal authority.

Standard Smoke: For certain smokes the relationship has been measured between the darkness of stain produced by sampling on a filter and separate gravimetric measurements of the smoke concentrations. This allows filter stains to be expressed in terms of a concentration of equivalent standard smoke.

Survey (Surveillance): An air pollution survey is an examination of the condition of the air usually at a number of points and/or over a length of time. Surveillance, usually from safety consideration, implies surveying over time. Monitoring is often used in the same sense.

Survey, Damage: Assessment of damage caused by the pollutant in a specified area.

Suspended Particulates: Particles which are too fine to have an appreciable falling velocity and therefore tend to stay suspended in atmosphere for a considerable period.

Synergistic: Referring to additive or mutually supporting or aggravating effect of toxic substances on living things.

Synoptic: Affording an overall view. In meteorology a synoptic chart displays the state of atmosphere over a large area at one time. By extension, synoptic is used to refer to features, for example cyclones or anticyclones, which appear in such charts.

Telemetering: The transmission of signals from a measuring instrument by cables or by wireless to a distant point where the results can be displayed or recorded. Telemetering enables the levels of pollution at a number of locations to be displayed at a single point.

Threshold, Odour: The lowest concentration of an odour bearing gas at which only half of a panel of sniffers can detect its presence.

TLV: Threshold Limit Value. This is the concentration of any of a number of pollutants recommended as a maximum to which healthy adult workers may be exposed 8 hours a day for five continuous days a week without adverse effects.

Topography: The layout of an area in terms of height, the natural features and/or buildings, etc.

Tower, Flare: An outlet, generally elevated, to exhaust inflammable gases after igniting them. Also called 'flare stack', or 'flare'.

Toxicity: A measure of poisonous or damaging effects of a substance on a living organism.

Tracer Technique: A method of studying pollutant behaviour in the air by the injection or release of a specifically identifiable substance. The subsequent distribution of the tracer is taken to be similar to that of the pollutant from the same starting point.

Troposphere: The lowest part of the earth's atmosphere, (about 9 to 20 km deep depending upon latitude), in which the temperature generally decreases with height.

Turbidity: In meteorology it refers to any condition of the atmosphere which reduces its transparency to light; a measure of reduction of transparency normally excluding cloud.

Turbulence: A motion of the wind or air flow in a highly irregular manner both in speed and direction. It is the main mechanism of diffusion of pollutants in the air

Urban Heat Island: A zone of excess warmth within a city over the suburban and rural environs. The warmer air in this region rises upwards and then spreads radially outwards. The colder air from the surrounding countryside at the surface flows inwards towards the heat island to replace the rising and outspreading warm air. This occurs when the heat islands are intense. This circulation is known as Urban Heat Island Effect.

Unstable Compound: An unstable chemical compound is one which is easily decomposed or which reacts readily.

Vapour: A gas is said to be vapour if, at the temperature concerned, it can be liquefied by pressure alone.

Vapour Pressure: The pressure exerted by a vapour, either alone or mixed with other gases. The saturated vapour pressure of a substance at any temperature is pressure of the vapour of the substance in equilibrium with its liquid form.

Ventilation: a. The process of replacing contaminated or stale air by fresh air. b. The movement and circulation of outdoor air. c. The volume of air entering or leaving the lungs in one respiratory cycle.

Ventilation Coefficient: It is defined as a product of mixing height and average wind speed.

Venturi: A tube which is constricted and then opens out again and can be used to meter the flow of gas or to assist the scrubbing of a gas by a liquid.

Visibility: A measure of cleanness of the air expressing how far objects can be clearly seen under specified conditions. Light absorption and scattering by aerosols are the main causes of reduced visibility.

Washout: The removal of gases, and sometimes particles, from the atmosphere by their solution or attachment to rain drops as they fall.

Wind Rose: A diagram showing the frequency and strength of winds of various directions over a period in a particular place.

Wind Shear: The change in speed and direction of the wind, usually with height.

Year: For reporting analyses of outdoor air on a yearly rate, twelve 30-day months are to be used.

Zoning: A method of abating air pollution at a location by a systematic planning of sites for dwellings and for factories/industries which discharge effluents taking into account the micrometeorological conditions in and around the location.

Appendix 6.43: Effects of Specific Air Pollutants on Human, Animals, Vegetation and Materials

A. Effect On Human

Hydrogen sulfide and Mercaptans

Exposure hydrogen sulfide for short periods can result in result in fatigue of the sense of smell. In fact merceptanes are often added to natural gas, LPG and manufactured gas supplies so that leakages of gas will be detected.

Lead

The main source of lead in urban atmosphere is the automobile using leaded petrol. It creates urban concentrations of lead of about 1–3 g/m^3. The inorganic lead causes gastro intestinal damage, liver & kidney damage, abnormalities in fertility & pregnancy and effects mental development of children. However, in India leaded petrol is not being used in automobiles.

Hydrocarbon Vapours

The effect of formaldehyde is primarily irritating. It is a major contributor to the eye and respiratory irritation caused by photo chemical smog.

The polycyclic organic compound 3,4 benzopyrene is formed due to incomplete combustion of hydrocarbons. They also present in exhaust discharge from IC Engines. Some aliphatic hydrocarbons are also carcinogenic (cancerous).

Insecticides

They are not only harmful for insects but also for man, eg, DDT (Dichloro Diphenyl Trichloro ethane). They can effect the central nervous systems and may attack other vital organs.

B. Effect on Animal

Fluorine

Cattle and sheep are most susceptible to fluorine toxicity. Horse appear to be resistant to fluorine poisoning. Poultry are most resistant to fluorine.

Symptoms of acute fluorine poisoning: Lack of appetite, rapid loss in weight, lameness, periodic diarrhoea, muscular weakness and death.

Symptoms of chronic fluorine poisoning: Teeth in the process of formation are easily affected. Hence tooth symptoms are a sensitive and unique criterion of chronic fluorosis. Bone lesions may develop at any age. A bony overgrowth may be observed on the leg bones, few bones and ribs. This may lead to lameness and stiffness of points. Lack of appetite, lowered fertility, reduced milk production and growth retardation are other effects.

Arsenic

Arsenic occurs as an impurity in many ores and in coal. Besides industrial sources, the arsenic is present in some of the dust and sprays on plant that can lead to poisoning of cattle. Very deep bore well water also contains arsenic.

Symptoms of acute arsenic poisoning: Sever salivation, thirst, vomiting, uneasiness, feeble and irregular pulse and respiration. There is diarrhoea and the faces have a garlic odor and are some times bloody. The ears become cold, the body trembles and develops abnormal temp. and convulsion.

Symptoms of chronic arsenic poisoning: Depressing effect on the central nervous system. The animal becomes dull and exhibits a lack of appetite with a resulting weight loss. Chronic poisoning can result in eventual paralysis and death.

Arsenic Tolerance Limits: Sheep have been poisoned by as little as 0.25–0.50 gm of arsenic daily, where as cattle and horses may tolerate 1.3–1.9 gm daily. It has of been reported that in experiments on guinea pigs that 10 mg of arsenic per kg of body weight per day has produced chronic arsenic poisoning.

Lead

Lead contamination of the atmosphere takes place on account of various industrial sources such as smelters, coke ovens and other coal combustion processes. Lead is also used in dust and sprays containing lead arsenate.

Symptoms of acute lead poisoning: Prostration, staggering and inability to rise, the pulse is fast but weak stiffening of legs and convulsions; loss of appetite, paralysis of digestive tract and diarrhoea. Other nervous symptoms in cattle are grinding of the teeth and rapid chewing of the cud. In case of horses, it may result in compete loss of appetite, nervous depression, lethargy and death.

Symptoms of Chronic lead poisoning: Paralysis of the larynx (breathing pipe) and difficulty in breathing are the main symptoms. Convulsions may occur due to paralysis of the throat.

Lead Tolerance Limits: Lead is cumulative poison. Inadequate calcium in diet increases the retention of lead in the animal body. Low calcium diets may result in lead storage of as such as five times as compared to that found in animals that received sufficient amount of calcium.

C. Effect on Vegetation

The primary factor which controls gas absorption by the leaves is the degree of opening of the stomata. Many plants open their stomata during day time and close it during night time. But some plants (eg. potato) do not close their stomata in the nights (also).

Damage to leaves are of following forms:

- Necrosis — Necrosis is the killing or collapse of tissue.
- Chlorosis — Chlorosis is the loss or reduction of the green plant pigment chlorophyll. It is similar to anaemia in animals.
- Abscission — Leaf abscission is dropping of leaves.
- Epinasty — Leaf epinasty is downward curvature of the leaf due to higher rate of growth on the upper surface.

Ozone

Ozone is photo-toxic in exposure of a few hour at about 0.2 ppm. The lesions are generally confined to the upper surface of the leaves.

Chlorine:

Chlorine is more toxic to vegetation than SO_2 by a factor of two or three. Lesions are generally marginal and intermittent.

Hydrogen chloride

Hydrogen chloride is less toxic than SO_2 towards vegetation. Marginalisation followed by necrosis at low concentrations. At higher concentrations, lesions are produced. The threshold concentrations is about 10 ppm for a few hours exposure.

Nitric Oxides (NO_x)

They cause brown margins and brownish-black spots on the leaves. Concentration of about 25 ppm will cause these effects.

Ammonia: It is gas of intermediate toxicity. It has the same toxicity characteristics as that of hydrogen chloride.

Hydrogen Cyanide: This is used to fumigate green houses and trees in orchards for pest control. This damages the vegetation.

Herbicides/Pesticides: Injury to sensitive vegetation may be caused in the field through careless or uncontrolled use of herbicides and pesticides. Herbicides are applied to the weedy grain fields.

D. Effect on Materials

Sulphur Oxides (SO_x)

Threshold Limit value (TLV)L: 5ppm

Primary air standards: 80 g/m³ (annual) and 365 g/m3 (24hrs)

Sulphur Oxide through acid rain or otherwise, causes a variety of damages to materials. It is responsible for the corrosion tarnishing of metals, surface erosion and decolourisation of building stones and paints; soiling of textiles and powdered surface of leather materials. It also reduces the tensile strength of textiles, causes embrittlement in paper, and weakens leather products.

Nitrogen Oxides (NO_x)

TLV: for NO: 25 ppm, for NO_2: 5 ppm

Primary air standards: 100 g/m³ (Annual)

Nitrogen oxides, through acid rain or otherwise, reduce tensile strength of textile materials. It is also responsible for soiling in textiles, fading and color change of textile dyes.

CHAPTER 7

Effects of Environmental Pollution on Plants/ Trees/Vegetation

Plants/trees/vegetation are very important for man's survival and the effects of various environment pollutants on them is discussed below:

- **Effects of various primary air pollutants on plants/trees/vegetation.** This includes the effects of major gaseous pollutants SO_2, NO_2, Fluorides, Cl_2, HCl, NH_3, H_2S, CO, Br_2, I_2, Hg vapour etc.
- **Effects of particulates pollutants on plants/trees/vegetation.** This includes the effects of cement kiln dust, lime & gypsum, soot, magnesium oxide, boron, chlorides of Na, K & Ca, Na_2SO_4 etc.
- **Effects of secondary air pollutants on plants/trees/vegetation.** This includes the effects of photo-oxidants (e.g. PAN, Ozone, Photochemical smog)
- **Effects of water pollution on plants/trees/vegetation.** This includes the effects of acid deposition, organic matter deposition, detergent deposition, industrial deposition, industrial effluents deposition, oil deposition, waste heat etc.
- **Effects of soil/land pollution on plants/trees/vegetation.** This includes the effects of soil acidification, pesticides, herbicides, sewage & ash pollution, chemical fertilisers, industrial effluents etc.

Effects of Various Primary Air Pollutants on Plants/Trees/Vegetation

Major primary air pollutants gases are sulphur dioxide, oxides of nitrogen particularly NO_2, HF, HCl, chlorine, ammonia, ethylene and other organic substances. Particulate air pollutants are soot, dust, fine particles of cement and various other substances. Various fertilizers, pesticides and insecticides used are also important air pollutants. The common sources of the pollutants, factors affecting the effect of pollutant and the injury symptoms produced in plants are discussed below.

Major Gaseous Pollutants & Their Effect on Plants

The gaseous pollutants are emitted in large amounts into the atmosphere due to a variety of anthropogenic activities and form most important atmospheric pollutants that injuriously affect the plants. Important such pollutants are discussed below.

Effects of Sulphur dioxide (SO_2) on Plants

It is the most important and common air pollutant produced in huge amounts in combustion of coal and other fuels in industrial & domestic use. It is also produced during smelting of sulphide ores. SO_2 effects increase in high humidity, windy conditions, in the early morning, in the deficiency of K & Cl_2 and excess of sulphur in the soil. It interacts with ozone, NO_2 and HF. The nature of interaction depends on the relative proportion of gases. The impact of SO_2 decreases in low soil moisture, low temperature, deficiency of nitrogen, sulphur & phosphorus and sometimes in excess of nitrogen also.

In angiosperms, young leaves and in conifers, needles are most sensitive to SO_2 pollution. In general, seedlings are more sensitive than older plants. The effect of the gas usually decreases with age of the plant and lesser morphological & physiological symptoms appear in older plants.

Plants Injury symptoms: The SO_2 is a strong reducing agent. In low concentration, it is oxidized and used in protein synthesis of the plant. However, in high concentration, it causes swelling of thylakoids and interferes with electron transport chain. In SO_2 pollution, plants show initial reduction of photosynthesis and increased respiration. The gas reduces stomatal opening and thus causes general water stress in plants. SO_2 replaces oxygen in cellular materials and changes their nature. It affects structural proteins in the cell membrane and thus changes the membrane permeability. High concentration of the gas causes accumulation of sulphydril and decrease of sulphides in plants. SO_2 interferes with amino acid metabolism and reduces the synthesis of proteins and enzymes. It stimulates the oxidation of PGA and increases the pentose phosphate cycle activity. It reduces the level of keto acids, sucrose and glutamate in plants and increases the level of glucose, fructose & glycolate. It inactivates many enzymes either by breaking their S-S bonds or by changing their stereo structure. In lichens, the gas induces photo-oxidation in the phycobiant part.

Most common visible symptom of SO_2 injury is water-soaked appearance of leaves which later become necrotic changing into brown spots. Colour and shape of necrotic spots is characteristic in different species and NO_2 concentrations. In some species, characteristic intraveinal chlorosis is caused. In general, SO_2 pollution results in abscission of older leaves and tip necrosis in flower and sepals.

Effects of Nitrogen dioxide (NO_2) on Plants

Major sources of this gas are nitrate fertilizer factories. The NO gas produced during combustion of fossil fuels and other oxides of nitrogen viz. N_2O_3, N_2O_4 and N_2O_5 are all converted gradually to NO_2 in the atmosphere.

The impact of the gas on plants increases with humidity, low light intensity and deficiency of nitrogen & iron in the soil. The effect of gas decreases in the conditions of drought. Sensitivity of plants to this gas changes by a factor of six during day and night. NO_2 interacts with SO_2, O_3 and HF and the nature of interaction varies with relative proportion of gases.

NO_2 mostly affects the leaves and seedlings. Its effects decrease with increasing age of the plant and tissue. Conifers are found to be more sensitive to this gas during spring and summer than in winters. Older needles are more sensitive to the gas than young ones.

Plants Injury symptoms: The NO_2 gas causes formation of crystalloid structures in the stroma of chloroplasts and swelling of thylakoid membrane. As a result, the photosynthetic activity of the plant is reduced.

Most common visible injury symptoms are chlorosis in angiospermic leaves and tip burn in conifer needles. In angiosperms, most of the species produce water-soaked intraveinal areas that later become necrotic. Tip burn is common symptom in bracts, sepals and awns.

Effects of Fluorides on Plants

Many particulate and gaseous fluorides are produced when ores containing fluorine are processed and used in industries. Common gaseous fluoride pollutants are HF, SiF_6, CF_4 and F_2. Particulate fluoride pollutants include Ca_3AlF_6 (Cryolite), CaF_2, NH_3F, AlF_6, CaSiF, NaF and Na_2SiF_6. Aerosols are often formed from NaF, NaAlF6 and AlF_6. Chief sources of fluoride pollutants are brickworks, aluminium factories, glassworks, steelworks, ceramic factories, phosphate fertilizer plants and uranium smelters. Some fluorine pollution also occurs during combustion of coal. Most injurious fluoride pollutant is gaseous hydrogen fluoride. (HF).

Fluorides in general, are accumulated in the plant tissues over long times. They are first accumulated in the leaves and then are translocated towards tips and margins of the leaves. The injury symptoms are produced only after a critical level of fluoride is attained. Due to such accumulation over long times, fluorides generally and HF particularly can induce injury at very low atmospheric concentrations. Critical concentration for-fluoride injury is 0.1 ppm for several days. Toxicity of particulate fluorides depends upon the particle size, their solubility and humidity of the atmosphere.

HF gas is much lighter than air and so can cause damage in plants even at a distance of 30 km from the source. It is a hygroscopic gas and forms acidic cloud near the source. Generally, the impact of HF pollution increases with humidity and excess of phosphorous(P) in soil, while it decreases in low temperature, drought and deficiency of N & Ca in the soil. In some species, impact of HF has been reported to decrease with excess of N and Ca in the soil.

In most of the species, recovery from moderate fluoride injury can occur within few days if exposure to pollutant stops. However, some highly sensitive species e.g. pine and spruce can never recover fully. HF generally affects immature leaves in angiosperms and needles in conifers.

Plants Injury symptoms: Fluorides combine with metal components of proteins or inhibit them otherwise and thus interfere with the activity of many enzymes. As a result the cell wall composition, photosynthesis, respiration, carbohydrate synthesis, synthesis of nucleic acids and nucleotides and energy balance of the cell are affected. In the leaves subjected to HF exposure, endoplasmic reticulum is reduced, ribosomes are detached from ER, number of ribosomes is reduced and mitochondria become swollen. Chlorophyll synthesis and cellulose synthesis are inhibited. Activities of UDP-glucose-fructose transglucosylase, phosphoglucomutase, enolase and polyphenol oxidase are reduced. On the other hand activities of catalase, peroxidase, pyruvate kinase, PEP-carboxylase, glucose 6-phosphate dehydrogenase, cytochrome oxidase and pentose phosphate pathway are stimulated.

In conifer needles common visible injury symptoms are chlorosis later turning into red/brown discolouration, tip burn later turning into necrosis of whole needle, formation of sharply defined red/purple bands between healthy and injured tissue. Similar symptoms are common in angiospermic leaves also. In addition, the angiospermic leaves in many species also show zonation of necrotic areas, leaf cupping, curling of leaf edges and ragged leaf margins. In sepals, petals, bracts and awns, water-soaked margins and later tip and marginal necrosis are observed.

Effects of Chlorine (Cl_2) Gas on Plants

Many industrial activities are the sources of chlorine pollution. Although chlorine concentrations change very rapidly in the atmosphere due to atmospheric chemistry and light rain can remove all the chlorine from the air in a very short time, chlorine injury can occur to plants near the source of pollution.

The impact of chlorine pollution increases in bright sunlight and decreases in drought and low temperature. Older plants are more sensitive to chlorine than seedlings. The age of tissue has little effect on the sensitivity and older as well as young tissues are almost equally affected by chlorine pollution.

Plants Injury symptoms: Chlorine injury symptoms can appear from 18 hours to 8 days after exposure. In plant species, recovery from chlorine injury can occur 3 to 4 days after exposure is stopped. Chlorine injury symptoms are quite variable in different species. Most common visible symptoms in conifers are chlorosis, tip burn and necrosis in needles. In angiosperm leaves, marginal or intraveinal necrosis, water-soaked appearance, leaf cupping and abscission are common.

Effects of Hydrogen Chloride (HCl) on Plants

HCl gas is released in large quantities in combustion of PVC and all chlorinated hydrocarbon material in large fires or incinerators. The HCl gas is very hygroscopic and quickly changes to hydrochloric acid by reacting with atmospheric moisture and forms aerosol droplets.

The HCl injury can be caused to plants even at a distance of 800 meter from the source. Like fluorides, the chloride from HCl is accumulated in the leaves and translocated towards their margins and tips. Symptoms of HCl injury appear after a critical concentration is reached, usually between 24 and 72 hours after the exposure.

Impact of HCl pollution decreases with increase in humidity, deficiency of Mg and excess of Ca. Mature plants are more sensitive to HCl than seedlings. Similarly, young fully expanded leaves are more sensitive than immature unexpanded leaves.

Plants Injury symptoms: Most common visible injury symptoms in conifer needles are red or brown discolouration and tip burn. In angiosperm leaves, common symptoms are intraveinal water-soaked streaks, yellow or brown necrosis, tip necrosis, bleached areas around the necrosis and shot-holing. Tip burn, necrotic stipple and discolouration in sepals and petals are also observed.

Effects of Ammonia (NH_3) on Plants

Continuous releases of ammonia from the sources are rarely high enough to cause acute injury, but occasional high release or spillage may cause ammonia pollution. High concentrations of ammonia are sometimes found around intensive farm units e.g. chicken batteries. Extent of injury reduces rapidly with increase in distance from the source. Under certain conditions the ammonia may remain as a cloud above ground level causing more injury to trees than to the ground flora. Injury symptoms may take upto 9 days to develop. In most plant species, recovery may occur in about 2 weeks after exposure is stopped.

Impact of ammonia on plants generally increases with humidity and decreases with drought. Effect of darkness on ammonia sensitivity is highly variable among species. Some species are more sensitive to low concentrations of ammonia than to its high concentration. Age of tissue has little effect on sensitivity and both young and old tissues are equally sensitive to ammonia.

Plants Injury symptoms: Most common visible symptoms in conifers are black discolouration, usually sharply bordered tip burn and abscission of needles. In angiosperm leaves, common symptoms are water-soaked appearance later turning black, intercostal necrosis, slight marginal and upper surface injury, glazing bronzing of upper surface, desiccation and abscission.

Effects of Organic gases (e.g., Ethylene) on Plants

Among organic gaseous pollutants, ethylene is most common. Other organic gases are propylene, butylene and acetylene. Ethylene is continuously emitted from many sources involving combustion or processing of petroleum or its products or burning of organic materials e.g. straw burning. Other organic gases are also produced in various chemical industrial processes.

Ethylene is a natural plant growth substance so the injury effects produced by it on plants are very similar to growth abnormality symptoms. Other organic gases also produce symptoms similar to those of ethylene pollution. However, the sensitivities of species to different gases are variable.

Ethylene injury symptoms develop in plants only in exposure to high concentrations and take several days to develop. After exposure to the gas is stopped, level of recovery is variable in different species. Generally, younger plant parts recover but older parts do not. Much 'acute' damage to plants is caused on the fringes of polluted area or by a steady leakage of gas in low concentration.

Impact of ethylene on the plants increases with high temperature. The gas interacts with SO_2 and CO_2 in atmosphere and the interaction is antagonistic i.e. high levels of these latter gases inhibit the development of ethylene injury.

Plant Injury symptoms: In injuriously high concentrations of ethylene, growth of plants is stopped. In low concentrations, growth abnormalities appear. In conifers, yellow tips in needles and abscission of branches and cones is common. In angiosperms, common symptoms are epinasty or hyponasty, loss of bark, abscission of leaves and flowers, premature flower opening and fruit ripening.

Effects of Minor Gaseous Pollutants on Plants

Many other air pollutants which are highly injurious to animals and human beings also cause damage to plants. However, plants are affected by these gases at quite higher concentrations than the animals. Common such gaseous pollutants are CO, H_2S, Br_2, I_2 and Hg-vapour.

Effects of Hydrogen sulphide (H_2S) on plants

Plants show wilting on exposure to this gas, but the symptoms develop after about 48 hours. No injury occurs below the exposure of 40 ppm for 4 hours.

Effects of Carbon Monoxide (CO) on Plants

Like ethylene, this gas produces epinasty, chlorosis and abscission. However, concentration of over 1000 times that of ethylene is needed to produce same degree of damage. No injury to plants occurs below exposure of 100 ppm for 1 week.

Effects of Bromine (Br_2) and Iodine (I_2) on Plants

Studies show these gases are highly toxic to plants. Br_2 and I_2 are readily absorbed and accumulated by plants producing visible injury symptoms similar to those of SO_2. Injury occurs at exposure of 0.1 ppm for 18 hours.

Common injury symptoms of bromine in angiosperms are necrosis of leaf margins, leaf tips & tendrils; brown discolouration and black spots later spreading to entire leaf. In conifers, yellow/white needle tips or red/brown discolouration later becoming grey/brown are common symptoms.

Effects of Mercury vapour (Hg) on Plants

Unlike other pollutants, flowers are more sensitive to Hg than leaves-Injury symptoms usually appear within 24 hours of Hg exposure, but often go on increasing upto 5 days.

Common injury symptoms due to Hg-vapour pollution are abscission of oldest leaves, interveinal necrosis, chlorosis around veins, flower abscission, loss of petal colour, buds remaining closed and later becoming necrotic, blackening of stamens, pistils and peduncles.

Effects of Particulate Pollutants on Plants/Trees/Vegetation

Different types of solid particulate materials are also important air pollutants. Each of these affects the plants in characteristic manner. Effects of some common particulate air pollutants have been discussed below.

Effects of Cement-kiln Dust on Plants

Cement factories are the chief source of cement dust pollution. The composition of such dust varies with the source. Main component of cement dust is CaO and varying amounts of K_2O, Na_2O, KCl and traces of Al, Fe, Mn, Mg, S & silica. Dust with more than 24% CaO is more injurious to plants. Fine particles cause more damage than larger particles. Cement-kiln dust is alkaline in nature and dissolves in atmospheric moisture forming a solution of pH 10-12.

In generals, plants having hairy surface of leaves trap more dust and are, therefore, damaged more than the plants with shiny leaf surface The cement dust forms crusts on the surface of leaves, twigs & flowers. This inhibits gaseous exchange from the surfaces of plant parts. Such crust on the leaves also inhibits light penetration and consequently reduces photosynthesis. Such crusts are especially thicker in conditions of dew, mist or light rains. In dry conditions, dust blowing with wind is highly abrasive and damages the cuticle of leaves. Cuticle is also damaged due to alkalinity of cement dust. Due to damaged cuticle, plants become more susceptible to infection by pathogens.

Effects of Lime and Gypsum Dust on Plants

Lime and gypsum processing industries and mining deposits are chief sources from where fine particles of these substances are blown away to great distances. Deposited on the soil from the air, these change the pH of the soil and thus affect the nutrient availability to plants. Such deposition usually causes appearance of various nutrient deficiency symptoms in the plants. Lime and gypsum are less adhering as compared to cement-kiln dust. However, these are also trapped and deposited on the surface of plant parts particularly the leaves with hairy surfaces and produce injury symptoms similar to cement dust. Lime and gypsum particles blowing with wind are also highly abrasive for cuticle.

Effects of Soot on Plants

Burning of fossil fuels, organic matter or natural forest fires produce huge quantities of fine carbon particles which form the soot pollution. Soot can be dispersed over a quite wide area and transported to great distances by blowing winds.

Soot deposited on the surface of leaves may be washed away by rains, so its damage may be reduced. However, in bright sunlight and high temperature, the damage is increased.

Soot deposited on the surface of leaves inhibits light penetration, increased surface temperature due to absorption of heat and clogging of stomata. The result of these is reduced gaseous exchange, reduced photosynthesis and general weakening of the plant growth. Necrotic spots also develop in many species due to soot deposition.

Effects of Magnesium Oxide Dust on Plants

Magnesium roasters are the chief sources of such pollution. The magnesium oxide dust may be carried by winds and deposited even at a distance of 5 km from the source.

In the atmosphere, magnesium sulphate ($MgSO_4$) combines with carbon dioxide and water to form ($MgCO_3$). Both these compounds are alkaline and slightly soluble in water. Deposited on the soil these compounds can soon increase the soil pH to levels injurious to plants. Deposition of these substances on the soil prevents germination of seedlings. The seedlings that are able to emerge usually have yellow/brown tips of leaves and their roots are stunted. In areas of heavy pollution, composition of the vegetation changes completely.

Effects of Boron Dust on Plants

Boric acid and borax are common raw materials in many industries. Oven and refrigerator manufacturing industries are chief sources of boron pollution. Severe injury to plants is observed even at a distance of 200 meters from the source and mild injury may be observed upto 500 meters in all the directions from the source.

Impact of boron pollution is more severe on older leaves than on younger leaves. Boron is also accumulated in the leaves and produces injury symptoms quite similar to fluoride pollution.

Effects of Dust of Chlorides of Sodium, Potassium & Calcium on Plants

Sodium and calcium chlorides are commonly used in cold countries on the roads during winters to melt ice and snow. Potash industry produces aerial emission of KCl and NaCl in ratio of 3:1. All such chlorides are carried away by winds and deposited on the soil and plants. Injury symptoms produced by these chlorides in plants are very similar to those produced by SO_2 and fluoride pollution.

Effects of Sodium Sulphate [Na_2SO_4] Dust on Plants

Sodium sulphate dust can cause necrosis of leaves. The damage increases in moist condition.

Effects of Secondary Air Pollutants on Plants/Tree/Vegetation

Many of the primary pollutants under specific environmental conditions may interact with each other and produce secondary environmental pollutants or certain complex environmental conditions that are injurious to plants. Such secondary pollutants and pollution conditions are discussed below.

Effects of Photo-Oxidants on Plants

In presence of strong sunlight and in hot weather a series of complex chemical reactions involving nitrogen oxides and hydrocarbons may produce certain photo-oxidant chemicals. These chemicals do not have any specific anthropogenic source but are formed over wide areas in which suitable environmental conditions are prevailing. Two such photo-oxidants that can reach ambient concentrations toxic to plants are PAN (Peroxyacetylnitrate) and ozone.

Effects of PAN (Peroxyacetylnitrate-$CH_3CO.O_2.NO_2$) on Plants

Impact of this secondary pollutant is not affected by humidity. However, the impact decreases with lowering of temperature and increasing drought conditions. The impact also increases in the morning and in bright sunlight. Young plants and young rapidly expanding leaves are more sensitive to this pollutant. PAN interacts with SO_2 and O_3 in complex manner producing variable impact conditions.

The common visible symptoms of exposure to PAN are chlorosis and necrosis in leaves. It also interferes with photosynthesis, respiration and absorption & synthesis of carbohydrates & proteins. It inhibits photorespiration, NADP reduction, carbon dioxide fixation, cellulose synthesis and the enzymes associated with photosynthesis & respiration.

Effects of Ozone (O_3) on Plants

The impact of ozone on plants increases with humidity and decreases with drought, darkness, low temperature, high soil salinity, deficiency of soil phosphorus & excess of soil sulphur. Middle aged leaves and young plants are more sensitive to ozone. This pollutant interacts with SO_2, NO_2, PAN and heavy metals in complex manner.

Common symptoms of ozone pollution are yellowing, flecking and blotching in leaves, premature senescence and early maturity. It interferes with pollen formation, pollination, pollen germination and growth of pollen tubes. Increase in the level of RNA, starch, polysaccharides and number of polysomes is observed in ozone pollution. Ozone stimulates respiration, inhibits oxidative phosphorylation and changes membrane permeability. In some species, it inhibits the synthesis of glucon & cellulose and reduces the level of reducing sugars & ascorbic acid, while in other species the effect is opposite to it.

Effects of Secondary Pollution Conditions on Plants

Certain primary inorganic and organic pollutants in the atmosphere under certain specific environmental conditions, undergo a variety of complex photochemical and other chemical reactions. These reactions produce certain-specific secondary atmospheric pollution conditions that also adversely affect plants. Important such secondary atmospheric pollution conditions are acid rains, photochemical smog, ozone depletion and greenhouse problem.

Effects of Water Pollution on Plants/Trees/Vegetation

The water bodies of the earth are being continuously polluted by a variety of sources. The pollution is occurring in all types of water bodies; both freshwater bodies like ponds, lakes & rivers as well as marine bodies like coastal & deep-water seas. Major causes of water pollution are deposition of add, organic sewage, detergents, agricultural chemicals, industrial effluents, silt, oil and heat into the water bodies.

Effects of Acid Deposition on Plants

Various acid gases, aerosols and other acidic substances released into the atmosphere from the industrial or domestic sources of combustion of fossil fuels eventually come down to the ground. These substances are deposited directly on the water bodies. In addition, these substances also reach the water bodies along with run-off rainwater from the polluted soil. Deposition of acidic substances causes acidification of water by lowering its pH below 6.0. The sulphates, nitrates and chlorides make water bodies like lakes, rivers & ponds acidic.

Nutrient deficiency in aquatic ecosystem: The decomposing bacteria and fungi decrease in acidified water. This reduces the rate of decomposition of organic matter and, therefore, the nutrient cycling in the aquatic ecosystem. Thus, low pH causes nutrient deficiency and consequent general reduction in abundance of aquatic plants in the affected water body.

Decrease of species diversity: Critical pH for most of the aquatic species is 6.0. The number and variety of aquatic species in the water body generally decreases below this pH.

Change in species composition: The number and abundance of acid tolerant species increases while that of sensitive species decreases. In the initial phase of water acidification, filamentous algae grow very fast and form thick mats. However, most of the diatoms and green algae disappear below the pH 5.8. Diatoms and small siliceous phytoplankton populations are highly sensitive to pH changes and species composition of their communities shows highly specific changes with pH change of the water body. Among green algae, *Cladophora* is highly acid tolerant species and becomes abundant in highly acidic freshwater bodies. *Euglena* & some other unicellular algae are found up to pH of 1.6 while *Chlamydomonas acidophila* is found in water up to pH of 1.0.

Macrophytes are generally absent in extremely acidic water. Potamogeton pectinalis is only aquatic macrophyte found in heavily acdified water. At pH lower than 4.0, angiosperm species using dissolved carbon dioxide e.g. *Juncus bulbosus, Juncus effusus, Sparganium emersurn, Gyceria fluitans, Eleocharis acicularis, Typha latifolia* and bryophytes like *Polytrichum, Anisothecium, Fontanalis, Catharina* become the only survivors. The roots of macrophytes are generally affected adversely in acidic water and result in poor plant growth. Plants with deep roots and rhizomes are less affected while plants with short root systems are severely affected. Yellowing of plants is common in polluted water.

Effects of Organic Matter Deposition on Plants

Large amounts of dead and decaying animal and plant material, fecal material and other organic material is deposited directly from sewage discharges or is washed along with rainwater into the water bodies. Most important consequences of such organic matter deposition are as following.

Increase in decomposer microbes: Increased addition of organic matter into the water body results in rapid multiplication and increase in decomposer aerobic and anaerobic bacteria. The rapid decomposition of organic matter by these increases nutrient availability in the water.

Eutrophication: Addition of organic matter and its rapid decomposition resulting in increased nutrient supply causes much nutrient enrichment (eutrophication) of water body. In such a condition, planktonic green and blue-green algae grow very rapidly causing water blooms. In addition to these many types of hydrophytes like Salvinia, Azolla, Eicchhornia etc. also become abundant. All this rapid growth of planktonic and free-floating hydrophytes reduces light penetration into deeper layers of water body and submerged flora gradually declines. Abundant flora after death further increases supply of organic matter in the water body.

Oxygen depletion: Rapid decomposition of organic matter by aerobic bacteria during eutrophication phase consumes much water-dissolved oxygen. On the other hand, gradual decrease of submerged aquatic flora results in reduced oxygenation of water. Both these phenomena together result in increase in biological oxygen demand **(B.O.D.) of the water-** Biological oxygen demand (B.O.D.) of water is defined as the amount of oxygen needed by a unit volume (usually one litre) of water sample to completely decompose the organic matter present in it by microbial activity, measured at 20ºC and tested at least five days after sampling. The B.O.D. value of fresh, unpolluted water is usually below 1 ppm while in organic matter polluted water, it may be more than 400 mg/litre (ppm).

Effects of Detergent Deposition on Plants

Various detergents from domestic or industrial use directly released or washed down into the water bodies cause serious effects on plants.

Most of the domestic and industrial detergents contain high (up to 40%) phosphate content. Addition of such detergents into water results in phosphate-enrichment of water.

Most of the detergents that are toxic, enter the plants through roots or surface absorption. **Common effects of detergents** on plants are as follows:-

- **Retardation of plant growth**, root elongation, carbon dioxide fixation, photosynthesis, cation uptake, pollen germination and growth of pollen tubes.
- **Destruction of the chlorophylls** and cell membranes.
- **Alteration of the absorption maxima** of chlorophylls.
- **Binding of membrane lipids** and proteins.
- **Denaturation of proteins** and thus causing enzyme inhibition in various metabolic processes.
- **Non-degradable alkyl benzene sulphonates and phosphate-rich detergents interfere with gaseous exchange** even in very low concentrations.
- **Cation-active compounds hinder algal growth** between 0.1 and 10.0 ppm while non-ionic compounds hinder algal growth between <1.0 and around 10,000 ppm concentrations depending upon the species and the compound.
- Macrophytes are most sensitive to damage by anionic surfactants.

Effects of Agricultural Chemical Deposition on Plants

Many chemical fertilizers, pesticides, insecticides, herbicides etc. are applied to crops far in excess. These excess chemicals are washed away with rainwater, first into the soil then finally into the water bodies.

Chemical fertilizers entering the water bodies result in eutrophication by enriching the water with major plant nutrients. Many of these fertilizers are acidic in nature e.g. ammonium sulphate. These cause acidification of water.

Pesticides, herbicides and insecticides also cause pH changes in the water bodies. The effects of these plants on aquatic plants are similar to those of their overdose in foliar application. The herbicides act directly on aquatic flora but insecticides act indirectly by allowing algal blooms to develop in the water body. Different substances have different patterns of their toxic action, decomposition pathways and environmental persistance. Most common effect of these substances is reduced photosynthesis. Some may uncouple oxidative phosphorylation or inhibit nitrate reductase enzyme. The uptake and bioaccumulation of these substances in aquatic plants is great due to their low solubility in water.

Effects of Industrial Effluent Deposition on Plants

Various inorganic and organic waste products from industries and mining activities are directly deposited into the water bodies. A large amount of these substances deposited on the soil, comes into water bodies indirectly along with surface run-off. Fly-ash, various organic/liquid effluents and heavy metals e.g. Hg, Cu, Cd, Pb, Zn, Ni, Ti, Se etc. are important industrial pollutants of water.

Fly ash forms thick, floating covering over the water surface. This reduces the penetration of light into deeper layers of water body. Fly-ash increases the alkalinity of water and thus, causes reduced uptake of essential bases. All these phenomena cause death of aquatic plants.

Organic/liquid effluents disturb the pH of water and depending upon their chemical composition, cause specific toxicity effects on the aquatic plants. Change in the floristic composition of the water body is most obvious and direct effect of pollution by such effluents. Heavy metals usually occur together and with many other types of pollutants so the effects of single metal are usually difficult to Interpret. There may be synergistic, additive or antagonistic interactions between metals regarding their effects on plants. Impact of metals is reduced in hard, well-buffered freshwater systems. For example, Cd-uptake by Nitella and Elodea is less in hard water, Zn-toxicity is less with high Ca for Stigeodonium and Hormidium while is less with high pH for Hormidium. Bioaccumulation of metals is more in mosses than in angiosperms and is usually more in lower plant parts. Chelators decrease while methylated forms increase the metal toxicity to aquatic plants. Reduced oxygen and low temperature also increase metal toxicity. While bryophytes appear to be highly resistant to heavy metal toxicity, in all classes of algae, strains vary in tolerance to metals. Photosynthesis and growth in most of the algae is inhibited at 1-2 ppm of Cu^{++}. Cholrella is more sensitive to Cu^{++} than Scendensmus. Cholrella is retarded more by Ag than by Cd, Hg or Ni while cell division in the genus is reduced more by Cd than by Cu or Hg at 0.32 ppm. Lemanea is quite resistant to Zn and Pb.

Effects of Silt Deposition on plants

The top soil removed due to erosion is carried with rainwater or flood water and deposited into the water bodies causing silt deposition in them.

The deposition of silt increases the turbidity of water and reduces light-penetration deep into the water, causing decline in submerged flora.

Silt deposition, in general, inhibits growth of aquatic plants. Phytoplankton is particularly affected by silt deposition due to reduction in surface exchange of gases and nutrients.

Species tolerant to turbidity (e.g. Ceratophyllum demersum, Lemna minor agg., Nuphar lutea, Polygonum amphibium, Sagittaria sagittifolia, Scirpus lacustris) becomes highest followed by species of intermediate tolerance (e.g. Callitriche spp., Myriophyllum spicatum, Potamogeton natans, P. pectinatus, Sparganium emersum, S. erectum) while least tolerant species (e.g. Elodea canadensis, Potamogeton perfoliatus, Ranunculus spp., Mosses) are much reduced.

Effects of Oil Depositions on Plants

Washing of oil tankers and storage containers in many rivers, large lakes and particularly near sea coasts causes deposition of oil slicks on water. Oil spills in water are also common during normal transport operations or during accidents involving oil tankers.

- Oil pollution of water body prevents oxygenation of water.
- Oil depletes oxygen of the water body by consuming dissolved oxygen in oil degradation.
- Oil inhibits planktonic growth and photosynthesis in aquatic macrophytes.
- Oil may even cause destruction of aquatic flora if it catches fire.

Effects of Waste Heat on Plants

Many industries, particularly thermal power plants, take water from rivers, lakes or sea to cool the heat-producing boilers and equipment. The heated water is then returned to the water body. The deposition discharge of waste heat into the water body has many consequences for the plants in it:-

- The solubility of oxygen in water is reduced at higher temperature. Thus, the reduced oxygenation of water adversely affects the aquatic flora.
- Reduced oxygenation and high temperature of water causes reduction in the activity of aerobic decomposers. The reduced decomposers result in decreased organic matter decomposition and consequently, reduced nutrient availability in the water body.
- In high temperature, aquatic plants show increased respiration, reduced photosynthesis and general inhibition of enzyme activity with increasing temperature.
- The aquatic flora and primary productivity of the aquatic ecosystem declines with increasing temperature. Green algae are mostly replaced by blue-green algae, which have comparatively less primary productivity.

With increase in temperature, species diversity of the water body declines and heat-tolerant species gradually become dominant.

Effects of Soil Pollution on Plants

Most of the atmospheric pollutants finally come down and settle on the ground surface, causing soil pollution. Pollutants also come back to soil when polluted water is used for irrigation. Many pollutants are directly deposited on the soil as sewage, industrial effluents and chemical fertilizers. The soil pollution by various pollutants has far reaching effects on plants and vegetation.

Effects of Soil Acidification on Plants

The oxides of sulphur & nitrogen, chlorides, fluorides, ammonium etc. emitted into the atmosphere in combustion from various industries come down as dry or wet deposition (acid rain) onto the soil and lower the soil pH. Increased acidity of soil results in following effects:-

- **Exchangeable bases** become deficient in the soil, plant growth is reduced due to nutrient deficiency.
- **The roots**, particularly the root hairs, are damaged resulting in reduction in nutrient uptake by plants.
- **Increased acidity** mobilizes heavy metals like Al, Cd, Zn, Hg, Mn, Fe etc. These spread rapidly in the soil along with soil water and reach concentrations toxic to plants. Consequently, plants show

species specific metal toxicity symptoms. Al-toxicity generally damages root hairs and reduces nutrient uptake while Fe-toxicity has general adverse effect on plant growth.

- **In some soils**, acidification increases weathering of silicate minerals destroying the mineral structure of the soil. This leads to poor growth of vegetation in general.
- **In some marginal soils** and grasslands, acidification increases the supply of plant nutrients like sulphur and nitrogen. The vegetation is thus, benefited by soil acidification and plants may show better growth.

Effects of Pesticides on Plants

Various pesticides, insecticides, fungicides etc are used in agriculture as foliar spray or are applied to soil far in excess to the requirement causing soil pollution. These substances pollute the soil depending upon their volatility, biodegradability, persistence, leaching, chemical reactivity and adsorption on the soil particles. Many of these substances form cations, are adsorbed on silicate clay micelle or humus molecules on the pH-dependent exchangeable charge sites and are later absorbed by the plants.

Absorbed pesticides substances produce characteristic species-specific toxicity symptoms in plants just like their aerial overdose.

Fungicides reduce abundance of soil fungi and actinomycetes and interfere with decomposition of soil organic matter adversely affecting the nutrient cycling.

Pesticides increase the abundance of some bacterial species, particularly of ammonifying bacteria while reduce the abundance of some susceptible bacteria.

Insecticides reduce the abundance of predator soil microbes and consequently increase the abundance of their prey species.

In general, species composition of the soil microflora and fauna is changed by pesticide substances.

The inorganic pesticides contain arsenic and sulphur. These substances cause toxicity symptoms like yellowing, necrosis, shot holes, premature defoliation etc. in plants.

Effects of Herbicides on Plants

Various herbicides are used for weed control in agricultural practice. General effects of some common herbicide substances on the plants are as follows. The symptoms in response to a particular herbicide may be characteristic but their development depends upon the dose to which plants are exposed, rate of growth of plant, weather conditions and the plant species.

Acetanilides (e.g. Prochlor, Metachlor) cause stunting of plants and roots. In brassicas, these cause yellow, red and blue colouration of leaves with pronounced stunting.

Amides and carbamates (E.g. Diphenamide. Propyzamide. Chlorpropham and Asulam) cause stunting without chlorosis or leaf-scorch in mild doses. Crops generally become greener in colour. Roots show thickening and stunting. In the cereals, coleptile becomes stunted and swollen.

Benzonitriles (e.g. Dichlorobenil) cause stunting of plants and thickening of roots. In shrubs and bushes, bark at the ground level develops necrosis. In some cases, marginal yellowing occurs in the leaves.

Aliphatic acids cause stunted growth of shoot, loss of leaf wax and mild irregular chlorosis.

Growth regulators (e.g. 2, 4 D, MCPA, Chlopyralid, and Mecoprop) cause hormone-type distortion in leaves and fruits. In the brassicas, stem splitting occurs. In the cereals, ears become distorted, becoming blind or with shriveled grains.

Ureas and Uracils (e.g. Diuron, Linuron, Bromocil, Lencil) cause inhibition of photosynthesis leading to yellowing along the veins that later extends across whole leaf. In the cereals, chlorosis starts in the middle of leaf but extends quickly to the tip.

Sulphonyl ureas (e.g. Chlorsulfuron) cause inhibition of growth in susceptible plants and these ultimately die due to root growth being stopped.

Triazinones (e.g. Metribuzin) cause inhibition of photosynthesis that leads to yellowing and tip-scorch in the leaves.

Triazines (e.g. Atrazine, Simazine) cause chlorosis and necrosis spreading towards tips of leaves and inhibit photosynthesis. This is more severe in older leaves.

Thiocarbamates and ethofumesates cause chlorosis, stunting and leaf-curling. Younger leaves become stuck to older leaves. In the cereals, this condition often leads to characteristic pink colouration at the base of the stem.

Effects of Sewage and Ash Pollution on Plants

Sewage matter is commonly used as fertilizer or deposited as waste on the soil. Effects of such pollution are mostly common to all plants. The organic matter in sewage decomposes and produces nitrogenous substances that become excessive in the soil and harm the vegetation.

Decomposition of sewage may also release various toxic heavy metals that cause characteristic heavy metal toxicity symptoms'fn plants. Detergent substances may also be released from sewage causing characteristic injury to plants.

Ash produced mainly from combustion of coal in thermal and industrial plants used for landfilling or deposited on soil makes the soil unfit for vegetation. It may release many toxic substances in the soil causing characteristic plant injuries.

Effects of Chemical Fertilizers on Plants

Chemical fertilizers are generally used far in excess of the requirements of the crop. The unutilized fertilizers cause soil pollution.

Toxic concentrations of nitrogen fertilizers cause characteristic symptoms of nitrite or nitrate toxicity in plants, particularly in the leaves. Nitrogenous fertilizers generally cause deficiency of potassium, increased carbohydrate storage and reduced proteins, alteration in amino acid balance and consequently change in the quality of proteins.

Ammonium fertilizers produce ammonia around the roots that may escape the soil and cause ammonia injury to plants.

Ammonium nitrate produce acids in the soil and increase soil acidity.

Nitrate and nitrite bacteria are reduced while ammonifying bacteria are increased in the soil disturbing the nitrogen cycle.

Excessive potash in the soil decreases ascorbic add and carotene in the plants.

Superphosphates cause deficiency of Cu and Zn in plants by interfering with their uptake.

Excessive lime prevents the release of Co, Ni, Mn and Zn from the soil and their uptake by plants is reduced causing their deficiency symptoms.

Excessive deposition of various substances released from chemical fertilizers into the soil generally causes their over-absorption by plants. These over-absorbed substances become accumulated in plant parts (bioaccumulation) e.g. nitrogen and sulphur are deposited in the leaves.

Effects of Industrial Effluents on Plants

Various inorganic and organic substances are present in the industrial effluents. These substances mostly remain tied up in the soil and are not readily available to plants. However, they affect various soil characteristics.

Effluents affect the mineral structure, soil pH, exchangeable base status etc. of the soil and thus indirectly affect the plants.

The pH of the soil is disturbed making soil either acidic or alkaline. Various inorganic and organic chemicals are accumulated in the soil up to levels toxic to plants.

In highly polluted soils, plants absorb and accumulate toxic substances (bioaccumulation). These substances may or may not produce direct injury symptoms in plants but are passed on to higher trophic levels (biomagnification).

Effects of Radioactive Pollutants on Plants/Trees/Vegetation

A variety of radioactive waste materials like Strontium-90, Cesium-137, Iodine-131, Plutonium, Uranium, Americum, Curium, Neptunium etc. are added to soil from nuclear activities. These ' substances may be washed into water or may be directly added to water that is used as coolant in nuclear power plants.

Uptake of radioactive substances by plants depends upon pH and organic matter status of the soil.

Various radioactive materials show different solubility and absorption by plants. Plant uptake is generally highest for Neptunium, lowest for Plutonium and intermediate for Americum and Curium.

Strontium-90 behaves as Ca in the soil and is absorbed by plants like it.

Cesium-137 behaves like K but its uptake by plants is very limited. Absorbed radioactive substances are generally accumulated in the vegetative parts of the plants, particularly the leaves. Their accumulation in crop fruits and seeds is very low.

Accumulation of absorbed radioactive substances in the plants may be up to 100 times their concentration in the soil of water.

Radiation from radioactive substances may also adversely affect the plants. Pines are eliminated on exposure to 20-30 roentegen/day while most plants die at 200 roentgen/day. Only lichens and mosses are highly resistant to radiation.

Radiation also damages chromosomes, it increases the frequency of chromosomal aberrations and causes genetic mutations. Such genetic changes may adversely affect plant metabolism or change their characteristics in subsequent generations.

Appendices

Contents			Page No.

Appendix 7.1: Characteristics of symptoms of visible damage to plant leaves by various air pollutants gases.

S.N	Symptoms of Damage / Air Pollutant gases	Yellowing to browning of leaf lip and leaf margin	Spots between leaf veins	Small spots on leaf surface	Shininess of abaxial leaf surface change in color silver gray - bronzing
1.	Hydrogen Fluoride	++	+		
2.	Chlorine	++	+	+	
3.	Ozone		+	++	
4.	PAN		+		++
5.	Sulfur Dioxide		++	+	
6.	Sulfuric Acid Mist	+	+	++	++ observed frequently
7.	Nitrogen Dioxide		++	+	+ observed sometimes

Appendix 7.2: Concentration Threshold Values for Various Gaseous Pollutants

S.No.	Gas	Threshold value concentration of NO_2, ppm	Exposure period, days	Type of Damage to plant	Plant name which is affected
1.	NO_2	0.068	56~140	Smaller leaf area and decrease of plant dry weight	Kentucky blue grass
		0.15~0.26	10~22	Reduced leaf area and decrease of plant dry weight	Tomato
2.	O_3	Threshold value concentration of O_3, ppm	Exposure time	Rate of inhibition	Plant
		0.04	During entire growth period	Approximately 15% reduction in weight of rough rice	Paddy rice
		0.05	During entire growth period	0.05 ppm day/0.025 ppm night, 23% less in yield	Paddy rice
		0.05	8hrs/day × 22 days	Growth inhibition (growth in middle)	Tamoto

S.No.	Gas	Threshold value concentration of NO_2, ppm	Exposure period, days	Type of Damage to plant	Plant name which is affected
		0.05	8hrs/day × 26 days	Growth inhibition (leaf and root), thinning of leaf	Cucumber
3.	PAN	Threshold value concentration of PAN, ppm	Exposure time, months	Damage	Plant
		0.03	16	Root growth, defoliation	Navel orange
4.	HF	Threshold value concentration of HF, ppm	Exposure time	Damage	Plant
		2~3	13~26 months	Reduction of trunk diameter, tree height, width of branch, trunk volume and leaf surface; decrease of amount of fruit and particularly, 25%~35% reduction in leaf surface	Navel orange
		2.5	Until harvest	Reduced amount of fruit and seeds	4 types of kidney bean
		0.7	43 days	25% reduction in wet weight of bean	Kidney bean

NOTES

PAN: Peroxy acetyl nitrate
ppb: Parts per billion (10^9)
ppm: Parts per million (10^6)

Appendix 7.3: Sources and Effects of Various Air Pollutants on Vegetation

S.No.	Air pollutants	Sources	Effects on vegetation
1.	Aldehydes	Photochemical reactions	The upper portions of Alfalfa etc. will be affected , Necrosis, if 250 ppm of aldehydes is present for 2 hrs duration.
2.	Ozone (O_3)	Photochemical reaction of hydrocarbon and nitrogen oxides from fuel combustion, refuse burning, and evaporation from petroleum products.	All ages of tobacco leaves, beans, grapes, pine, pumpkins and potato are affected. Fleck, stipple, bleaching, bleached spotting, pigmentation, growth suppression, and early abscission are the effects.

S.No.	Air pollutants	Sources	Effects on vegetation
3.	Peroxy Acetyl Nitrate (PAN)	The sources of PAN are the same as ozone	Young spongy cells of plants are affected if 0.01 ppm of PAN is present in the ambient air for more than 6 hrs.
4.	Nitrogen dioxide (NO_2)	High temperature combustion of coal, oil, gas, and gasoline in power plants and internal combustion engines.	Irregular, white or brown collapsed lesion on intercostals tissue and near leaf margin. Suppressed growth is observed in many plants.
5.	Ammonia(NH_3) and Sulfur dioxide(SO_2)	Thermal power plants, oil and petroleum refineries.	Bleached spots, bleached areas between veins, bleached margins, chlorosis, growth suppression, early abscission, and reduction in yield and tissue collapse occur.
6.	Chlorine (Cl_2)	Leaks in chlorine storage tanks, hydrochloric acid mists.	If 0.10 ppm is present for at least 2 hrs, the epidermis and mesophyll of plants will be affected.
7.	Hydrogen fluoride(HF), Silicon tetrafluorideSiF_4	Phosphate rock processing, aluminium industry, and ceramic works and fiberglass manufacturing.	Epidermis and mesophyll of grapes, large seed fruits, pines and fluorosis in animals occur if 0.001 ppm of HF is present for 5 weeks.
8.	Pesticides & Herbicides	Agricultural operations	Defoliation, dwarfing, curling, twisting, growth reduction and killing of plants may occur.
9.	Particulates	Cement industries, thermal power plants, blasting, crushing and processing industries	Affects quality of plants, reduces vigor & hardness and interferences with photosynthesis due to plugging leaf stomata and blocking of light.
10.	Mercury (Hg)	Processing of mercury containing ores, burning of coal and oil.	Greenhouse crops, and floral parts of all vegetations are affected; abscission and growth reduction occur in most of the plants.

Appendix 7.4: Algae: As Water Pollutant, as Pollution Indicator and in Pollution Control

Most of the algae are aquatic and are found in fresh water or marine habitat. Therefore, they are related with different aspects of fresh-water or marine pollution.

Algae as Water Pollutants

During the favourable season of algal growth, water bodies like lakes, ponds and rivers show so much growth of various species of blue-green algae, green algae and golden brown algae that the water becomes turbid, cloudy and yellowish/greenish in colour. Filamentous planktonic algae form thick floating mats on the water surface. Such excessive algal growth is called water bloom. This bloom cuts off the light to deeper layers of water body and thus inhibits decomposition of organic matter in that water body. The algae further add a large amount of organic matter after death and decay to the water body which is not decomposed quickly due to prevailing conditions in that water body. This causes serious water pollution.

Many algae like *Microcystis, Aphanizomenon* produce toxic substances that are harmful to fishes and aquatic animals. These toxins also harm the land animals drinking this polluted water.

In general, water blooms make water oily, unpleasant in smell, fishy in taste and unfit for drinking.

Many species of blue-green and green algae that form excessive growths also choke the water tanks, pipe lines and other associated installations causing undesirable problems.

Algae as Pollution Indicators

Analysis of the composition and growth pattern of the algal flora in a water body can be used to identify the type and level of water pollution. Such studies of algal flora have been used in identification of following types of water pollution problems:

- **Water acidity:** Increase in the acidity of water initially causes general increase in filamentous algae. However, high levels of water acidity due to pollution by acid forming chemicals or acid rains results in decrease in planktonic algae in the water body. Most algae and diatoms disappear completely in water below the pH 5.8.

  Diatoms are highly sensitive to pH and different species of diatoms are found at different pH values of water body. Thus changes in the species composition of diatoms very accurately indicates the pH level of the water body.
- **Sewage, organic matter and chemical fertilizers:** Increase in the organic matter or chemical fertilizers that are washed off into the water body results in increased nutrient supply for algal growth. Such a condition of increased nutrient supply is termed **eutrophication** and results in water blooms of various types of algae. The water blooms of planktonic algae like *Microcystis, Scendensmus, Hydrodictyon* and *Chlorella* indicate pollution of water body due to excessive addition of organic matter, nitrates or phosphates.
- **Heavy metals:** Some algae like *Cladophora* and *Stigeoclonium* absorb and accumulate many heavy metals from the water. Thus the excessive growth of these algae in the water indicates pollution due to heavy metals.
- **Oil pollution:** Excessive growth of algae like *Duniella tertiolacta, Skeletonema costatum. Cricosphaera carterae, Amphidium carterae*, *Cyclotella cryptica* and *Pavlova lutheri* indicate oil pollution of water bodies.

Algae in Pollution Control

Algae like Chlorella, Chlamydomonas, Scendensmus and *Spirullina* are grown in sewage treatment plants along with suitable bacteria (algal-bacterial systems). Organic sewage degraded by bacteria is used up by algae in their photosynthesis and growth. The abundant algal growth in the treatment plant is periodically removed and used as animal feed or source of protein. The polluted water is thus cleaned by the combined action of bacteria and algae.

For the treatment of water containing metals as pollutants, algae like *Chlorella* are cultured in the polluted water. These algae absorb the metals from the water. The algal growth is periodically removed and destroyed.

Note: Toxic Algae Factsheet

- Algae are small, microscopic plants, which lack specialized structures like leaves, stems, and roots.
- Only certain species of blue-green algae are capable of producing toxins, and even these species are harmless most of the time.
- Blue-green algae are the most primitive form of algae, similar in structure to bacteria. Algae are microscopic organisms, which are common in surface waters and wet areas throughout Montana.
- Potentially toxic blue-green algae occur in any standing bodies of water exposed to the sun (e.g., lakes, reservoirs, stockponds, and roadside ditches). They normally are not found in rivers, streams, springs, irrigation canals, or wells.
- Potentially toxic blue-green algae prefer warm, clear waters that are rich in plant nutrients, especially phosphorus.
- An algae blood is a rapid and massive build-up of algae cells that imparts a green color to the water. Sometimes the algae can be further concentrated along the shore by wind and wave action.
- Blue-green algae blooms can occur anytime from May through October, but they normally occur during the hottest part of the summer – June, July, and August.
- A blood of potentially toxic blue-green algae appears as "pea soup," "grass clipping," or "green latex paint." The algae usually are suspended in the water column or aggregated into floating mats; they do not grow from the bottom as do mosses or "water weeds."
- There is no way to tell for sure if the water is toxic unless some of it is actually injected into an experimental laboratory animal. Harmless strains of potential toxin-producing algae look the same as deadly strains under a microscope. A toxin will impart no distinguishing odor, test, or color to the water in which it is dissolved.
- Conventional treatment and disinfection afforded most public drinking water supplies are not effective in removing or deactivating blue-green algae toxins. Water that is free of blue-green algae may not be free of the toxin. Boiling is similarly ineffective.
- Essentially, all warm-blooded animals are susceptible to blue-green toxins, including people, waterfowl, furbearers, game and non-game animals, livestock, poultry, and household pets.
- Fish kills associated with algae blooms are most likely due to depletion of dissolved oxygen rather than that toxic blue greens, although fish kills resulting directly from algae toxins have been reported.
- Although fish taken from waters infested with toxic algae are not likely to cause secondary poisoning if eaten in moderation, they may taste "weedy". It is advisable to shun fish taken from such waters, particularly if they appear sickly or sluggish.
- There are two basic kinds of algae toxins: (1) the milder peptide type is rarely fatal but may produce liver damage and general long term debility; and (2) the more potent alkaloid type is usually fatal within a short time.
- The peptide toxin may produce diarrhea, nausea, cramps or general lethargy and unthriftiness in animals. The alkaloid toxin may cause staggering, muscle spasms, labored breathing, or convulsions. A very rigid neck is characteristic at death; death is caused by respiratory arrest. In people who unwittingly ingest the toxin, sensations of numbness, dizziness, tingling, and fainting may be manifest.
- There is no antidote known to be effective at counteracting the effects of the toxin once it has been ingested.
- The time to control a toxic algae bloom is before the bloom develops. Assuring that fertilizers, animal wastes and other sources of nutrients do not reach the water is the best preventative.

Appendix 7.5: Diagrammatic Representation of the Effects of Sulphur Dioxide (SO_2) on Plants

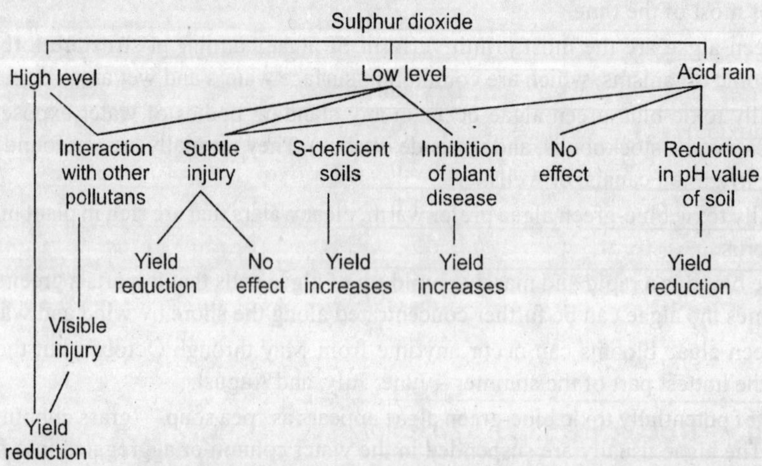

Appendix 7.6: Absorption Rate of SO_2 By Seedlings Fumigated at 1.0 ppm for 1 hour in a Controlled Chamber (27± 1ºC, 51 ± 7 R.H., 1300 fc)

Species	SO_2 uptake	
	mg SO_2 dm^{-2} hr^{-1}	mg SO_2 g^{-1} hr^{-1}
Red maple	0.088	0.260
White birch	0.086	0.268
Sweetgum	0.074	0.267
Firethorn	0.072	0.213
Privet	0.068	0.134
Rhododendron	0.056	0.079
White ash	0.046	0.118
Azalea	0.044	0.078

Appendix 7.7: Guesstimated Gaseous Pollutant Flux Rates for Dry Soil and Vegetative Surface

Pollutant	Flux rate, μg m^{-2} hr^{-1}	
	Dry soil surface	Vegetative surface
CO (carbon monoxide)	1.9×10^4	2.6×10^3
NO_x (Nitrogen oxides)	2.0×10^2	2.3×10^3
O_3 (Ozone)	1.0×10^3	6.2×10^4
PAN (peroxyacetyl nitrate)	–	1.2×10^3
SO_2 (sulphur dioxide)	7.7×10^6	4.1×10^4

Appendix 7.8: Tree Species for Wasteland Development

1. **Common Name:** Babul

   Soil requirements: Grows in arid, low rainfall areas having long dry seasons. Thrives on a wide range of soils from heavy clayey to rocky soils.

   Uses: It is a high calorific value food. It is useful for fuel, to make farm equipment, and in timber and pulp production. The foliage and pods are good food for livestock.

2. **Common Name:** White Babul

   Soil requirements: Growth in high pH soils and limestone and rocky soils but cannot tolerate high acidic or water-logged soils.

   Uses: Salvador type is the fast growing one. The wood is good for fuel, pole, timber and furniture. Foliage is used as fodder for livestock. It controls soil erosion and is useful in reforesting degraded pastures.

3. **Common Name:** Mesquite

   Soil requirements: Grows in all kinds of soils.

   Uses: Useful as fuel, light timber and in fencing; pods are source of food for livestock. Foliage is not browsed.

4. **Common Name:** Casuarina

   Soil requirements: Adapted to a wide range of soils including calcareous, saline and seashore areas.

   Uses: Good for fuel, poles, timber and as raw material for pulp. Foliage is not browsed by livestock. It controls soil erosion, acts as a wind breaker and can withstand a polluted atmosphere.

5. **Common Name:** Siris

   Soil requirements: It requires well drained soil, and the trees grow well near sea shores and on dry, alkaline soils.

Uses: It is useful for fuel, furniture and timber. The bark is used in tanning. It has high medicinal properties. The foliage is browsed on by animals.

6. **Common Name:** Jack Fruit

   Soil requirements: Requires deep soils, but adapted to alkaline, acidic, laterite, sandy and clayey soils.

   Uses: Useful for furniture, foliage browsed by livestock, fruits are edible.

7. **Common Name:** Persian Lilac or China Berry

   Soil requirements: Grows in saline, alkaline and slightly acidic soils.

   Good for fuel, timber and paper production. Foliage is browsed by livestock and it is good for agro-forestry.

8. **Common Name:** Rain Tree

   Soil requirements: Grows on poor soils, alkaline conditions. Not in water-logged areas.

   Uses: Good as fuel wood and for furniture making. Foliage is browsed. Pods make nutritious food for livestock.

9. **Common Name:** Portia Tree

   Soil requirements: Grows in sandy, laterite, alluvial black clay, marshy and acidic soils and on land flooded by salt water

   Uses: The trunk is used in making bullock carts, central bars, match sticks, etc. Foliage is a good fodder.

10. **Common Name:** Mexican Lilac

    Soil requirements: Grows well in all kind of soils like dry, moist, eroded, sandy and limestone soils.

    Uses: Good for fuel and to make furniture and tools. Foliage is browsed by animals. This is good for live fencing and for control of soil erosion.

11. **Common Name:** Bamboo

    Soil requirements: Grows in deep, well drained, sandy loams, but can be grown in clayey, acidic and alkaline soils too. Does not grow well on shallow and gravelly soils.

    Uses: Used for constructing houses, basket making and for making quality paper. Foliage is a fodder for livestock.

12. **Common Name:** Arjun

    Soil requirements: Grows in humid, tropical conditions, cannot withstand drought, but grows well in a variety of soils like saline, alkaline, gravelly, etc.

    Uses: Leaves are a good fodder for livestock. The wood is good for fuel and to make furniture.

13. **Common Name:** Teak

    Soil requirements: Grows on a variety of soils, laterite, gravelly, acidic and alkaline.

    Uses: Wood is used for timber and is highly durable.

14. **Common Name:** Drumstick Tree

    Soil requirements: Grows in all soil conditions but not in water-logged soils

    Uses: Eaten as a vegetable, good species for agro-forestry and for control of soil erosion.

15. **Common Name:** Neem

Soil requirements: It is suitable for areas with long, dry seasons. It grows on a wide range of soils including dry, clayey, shallow, rocky and sandy soils. It does not grow in water-logged, saline or deep, dry areas.

Uses: Resistant to pests, good for fuel and timber and to make furniture. The oil is used for spa. The cake is used for manure and to control soil-borne pests.

16. **Common Name:** Mango

    Soil requirements: Grows on a variety of soils but prefers deep, well drained soils.

    Uses: Good for fuel, timber and medium quality furniture. Foliage is browsed by livestock. Fruits provide a substantial income.

17. **Common Name:** Indian Jujube

    Soil requirements: Grows in a variety of soils from sandy, gravelly to clayey, saline, alkaline and dry soils.

    Uses: Wood is good for fuel. The wood is also used for making agricultural implements. Fruits are edible and the leaves are browsed by livestock.

18. **Common Name:** Tamarind

    Soil requirements: Grows in a variety of soils from gravelly, sandy and deep clayey to acidic and alkaline soils.

    Uses: Wood is heavy and is excellent timber. The fruits are dibble. The pulp is used in cooking. Foliage is browsed by livestock.

19. **Common Name:** Gooseberry

    Soil requirements: Grows in all kins of soil.

    Uses: Good fuel, foliage is browsed by livestock. Fruits are rich in Vitamin C and are edible.

20. **Common Name:** Jamun

    Soil requirements: Grows in sandy, limestone and alkaline soils. A full-grown tree can withstand drought and flood conditions.

    Uses: Wood is used for fuel. The bark contains tannin. Fruits are edible and the foliage is browsed by livestock.

21. **Common Name:** Pungam

    Soil requirements: Grows in a variety of soils and can even tolerate acidic soils.

    Uses: Used as firewood. Foliage is browsed by livestock and it has medicinal properties.

Appendix 7.9: Decline Scale of Trees in Japanese Cedar and Zelkova

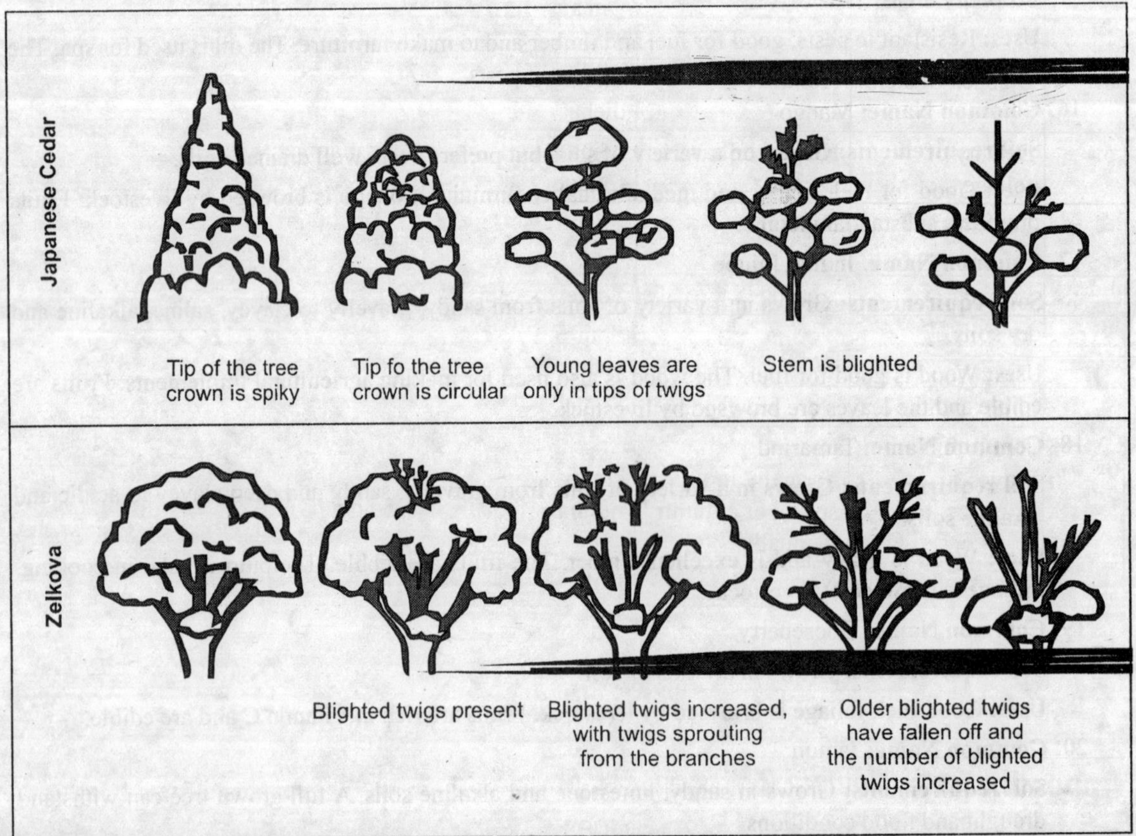

Appendix 7.10: Evaluation standards for tree vitality index

S.No.	Measurement parameters	Trees Evaluation standards			
		1. Good, normal	2. Affected, but nearly normal	3. Deterioration is considerably advanced	4. Striking deterioration
1.	Tree vitality	No damage	Slightly damaged	Evidently damaged	Severely damaged
2.	Form of tree	Normal	Some branches lacking	Evidently wanting	Severely wanting
3.	Branch growth	Normal	Somewhat reduced	Short and slender	Extremely short and slender

S.No.	Measurement parameters	Trees Evaluation standards			
		1. Good, normal	2. Affected, but nearly normal	3. Deterioration is considerably advanced	4. Striking deterioration
4.	Dieback of stem	None	Slightly diebacked	Evidently diebacked	Completely dead
5.	Density of foliage	Healthy foliage	Unbalanced foliage	Sparse foliage	Very sparse with dead leaves
6.	Leaf shape	Normal	Slightly deformed	Evidently deformed	Completely deformed
7.	Leaf size	Normal	Somewhat small	Evidently small	Strikingly small
8.	Leaf color	Normal	Somewhat abnormal	Abnormal	Strikingly abnormal
9.	Injury of leaves	None	Slight	Considerable	Striking
10.	Red (yellow) leaves	Leaves shed normally in spring or autumn Annually	Slightly early compared with normal Annually	Leaves shed at odd times Biannually	Leaves shed at odd times Three or more times a year
11	Leaf shedding	Normal	Color slightly off	Leaves partially red and yellow, but color is bad	No red or yellow colors; leaves shed in dirty state.

Appendix 7.11: Nutrient (Nitrogen – N, phosphorous – P and Potassium – K) Contents of Some Types of Organic Mannures

Mannure type	Nutrient content, %			Organic matter, %
	N	P	K	
Farm yard mannure	0.62	0.13	0.49	24
Pig slurry	0.21	0.10	0.18	5
Poultry mannure	2.30	0.90	0.65	65
Sewage sludge	1.82	0.43	0.46	39
Mushroom compost	2.80	0.20	0.80	65
Domestic refuse	0.50	0.20	0.30	85
Straw	0.48	1.62	0.85	95

References/Bibliography/Further Reading

1. M.N. Rao & H.V.N Rao, "Air pollution", Tata McGrow Hill, New Delhi.
2. Arther C.Stern "Air pollution", Volume I , Academic press New York.
3. C.S Rao "Environmental Pollution Control Engineering", Wiley Eastern Ltd. New Delhi.
4. "Industrial Safety and pollution Control Handbook", Associate (DATA) Publishers Pvt Ltd, Secunderabad
5. Dr. D.Venkateswaralu (editor), "CHEMTECH-volume I", CEEDC I.I.T. Madras
6. S.C. Bhatia, "Environmental pollution and control in chemical process industries", Khanna Publishers, Delhi.
7. KVSG Muralikrishna, "Air Pollution and Control", Kaushal & Co. Kakinada
8. Kirk Othmer (editor), "Encyclopedia of chemical Technology", John Wiley and Sons, New york.
9. Perry & Chilton, "Chemical Engineers Handbook", Tata McGrow Hill, New york.
10. B.K. Bhaskara Rao, "A Text on Petrochemicals", Khanna Publisher, New Delhi
11. G.R. chatwal, M.C Mehra etal, "Encyclopedia of environmental pollution and its control", Anmol Publication Daryaganj, Delhi
12. P.R. Trivedi & Gurdeep, "Encyclopedia of environmental sciences", Akashdeep publishing House, New Delhi
13. N. Irving Sax (Editor), "Industrial Pollution", Van Nostrand Reinhold Company, New York
14. McGrow Hill Encyclopedia of science & Technology, McGrow Hill Book co. New York
15. Robert A. Meyers (Editor), Encyclopedia of Physical science and Technology, Academic Press Inc. Orlando.
16. S.P. Sharma and chander Mohan fuels and combustion Tata McgGrow Hill publishing company Ltd. New Delhi
17. P.L. Diwakar Rao, "Pollution Control Handbook 1986", Utility Publications Ltd. Secunderabad
18. R.K. Saraf (Editor), "Proceedings of symposium on Environmental Pollution", Central Public Health Research Institute, Madras
19. "Industrial Air Pollution, Problems and Control", National Seminar Proceedings, April 1981, National Productivity Council, Lodi Road, New Delhi.
20. Ghoshal, Sanyal, Dutta, "Introduction to chemical Engineering", Tata McGrow Hill publishing Co. Ltd. New Delhi.
21. John. H. Seinfeld, "Air Pollution- physical and chemical fundamentals", McGrow Hill Book Co. New York, 1975
22. Attitio Bisio and sharoon boots)Editors), "Encyclopedia of Energy Technology and the Environment" Volume I, John Wiley & Sons Inc. New York
23. S.M. Khopkar, "Environmental Pollution Analysis", Wiley Eastern Limited, New Delhi.
24. Anand Gopal Mukherjee, "Environmental Pollution and health hazards causes and control", Galgotia Publication Pvt Ltd, New Delhi.
25. Keith Tearle (editor in chief), "Industrial Pollution Control, the practical implications", Business Books Limited, London 197.

26. S.V.S. Rana, "Essentials of Ecology and Environmental Science", Prentice Hall of India Private Ltd; New Delhi.
27. Sybil P. Parkar (edito in chief), "McGrow Hill Encyclopedia of Science and Technology", McGrow Hill Book Co. New York.
28. Roberts A.Meyer (editor), "Encyclopedia of physical science and Technology", Academic press Inc, Orlando, Florida
29. "Air Pollution Control", CPHERI – short course manual, Nagpur, 1973.
30. "Air quality monitoring", A course manual, NEERI, Nagpur, 1978
31. Faith, W.L, "Air pollution Control", John Wiley and sons, Inc., Newyork
32. Lund, H.F, "Industrial Pollution Control Handbook", McGrow Hill Book Co., New York, 1971
33. Stern A.C (editor), "Air pollution (volume 1 to 5)", 2nd edition, Academic Press, New York
34. Strauss, W. (editor)," Air Pollution", Part II & III, Wiley Interscience, Newyork.
35. Yennamwar, P.K, et at, "A course manual on Air pollution control, "NEERI, Nagpur, 1979
36. Hesketh, E. Howard, "Air pollution Control", ANN ARBOR SCIENCE, publishers INC, Michigan, U.S.A
37. N. Manivasakam, "Environmental pollution", National Book Trust, India
38. Rass, R.D. "Air Pollution and Industry" Van Nostrand Reinhold Company, New York
39. W. Summer, "odour pollution of air causes and control", Leonard Hill Books, London
40. S.K. Garg, "Sewage disposal and air pollution engineering", Khanna Publishers, Delhi
41. A.K. Dey, "Environmental chemistry", Wiley Eastern Limited, New Delhi
42. Calvin R. Brunner, P.E. "Hazardous air emission from incineration", chapmenond Hall, New york
43. Bethea R.M, "Air Pollution Control Technology", Van Nostrand Reinhold Co. New York, 1978
44. V.P. Kudesia, "Air Pollution, "Pragati Prakashan, Merrut, India
45. Mahajan, S.P, "Pollution Control in process Industries, "Tata McGrow Hill Publishing Co. Ltd, New Delhi
46. Perkins, H.C, "Air Pollution, "McGrow Hill kogakusha Limited," Tokyo, 1974
47. G.N. Pandy & G.C. Karney, "Environmental Engineering," Tata McGrow Hill Publishing Co. Ltd.
48. Conway & Richard D. Ross, "Handbook of Industrial waste disposal, "Van Nostrand Reinhold, New york .
49. Conway & Richard D. Ross, "Handbook of Industrial waste disposal, "Van Nostrand Reinhold, New york
50. Calvin, R. Brunner, "Hazardous air emissions from incineration", P.E. Chapman Hall, New York.
51. Sax. N.I, "Industrial Pollution, "Van Nostrand Reinhold Ltd, New York
52. Dix. H.M, "Environmental Pollution, "John Wiley & Sons, New York
53. Masters, G.M, "Introduction to environmental science and technology, John Wiley & sons, New York.
54. Rosen, M.J, "Industrial noise Pollution", John Wiley & sons, New York
55. Gupta, O.P, "Elements of fuels, furnaces & refractories", Khanna publishers, Delhi
56. James. A. Kent(Editor),Reigel Handbook of Industrial Chemistry , Van Nostrand Reinhold Company , Newyork.

Subject Index